Proceedings
of the
International
Clay Conference

Denver, 1985

Proceedings of the
International Clay Conference
1985

Denver, Colorado, July 28 to August 2, 1985

Organized by
The Clay Minerals Society and
the United States Geological Survey
under the auspices of
Association Internationale pour l'Etude des Argiles

Editors

Leonard G. Schultz
U.S. Geological Survey
Federal Center
Denver, Colorado 80225

H. van Olphen
Dalweg 114
6865 CW Doorwerth
The Netherlands

Frederick A. Mumpton
Department of the Earth Sciences
SUNY-College at Brockport
Brockport, New York 14420

Published by
The Clay Minerals Society
P.O. Box 2295
Bloomington, Indiana 47402, U.S.A.

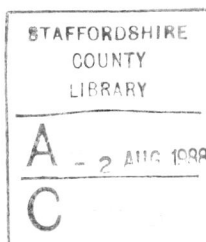

ISBN No. 0-935868-29-1

For information, write The Clay Minerals Society, P.O. Box 2295, Bloomington, Indiana 47402, U.S.A.

Citations of articles contained in this volume are properly referenced as follows:

Author (1987) Title: in *Proc. Int. Clay Conf., Denver, 1985,* L. G. Schultz, H. van Olphen, and F. A. Mumpton, eds., The Clay Minerals Society, Bloomington, Indiana, 000–000.

CONFERENCE ORGANIZING COMMITTEE

John B. Hayes, General Chairman
Paul D. Blackmon, Treasurer
Haydn H. Murray, Fund Raising
Richard M. Pollastro, Poster Sessions, Exhibits
David R. Pevear, Workshop
Leonard G. Schultz, Technical Program Chairman
   Marc W. Bodine, Jr., Technical Program
   Bruce Bohor, Technical Program
   Dennis D. Eberl, Technical Program
   Richard M. Pollastro, Technical Program
   Gene Whitney, Technical Program
Marc W. Bodine, Jr., Facilities
Harry C. Starkey, Facilities
Harry A. Tourtelot, Field Trips
Caroline A. Watkins, Social and Guest Programs

PREFACE

The 8th Conference of the Association Internationale pour l'Etude des Argiles (AIPEA) was held at the Marriott Hotel, Denver, Colorado, from July 28 to August 2, 1985. The Conference was organized by The Clay Minerals Society (CMS) and the United States Geological Survey (USGS) and attended by 421 professional members from 37 countries. Welcoming addresses were given by John B. Hayes, Chairman of the Organizing Committee; Wayne M. Bundy, President of CMS; Harry A. Tourtelot, USGS; and Lisa Heller-Kallai, President of AIPEA. As part of the technical program, three plenary lectures were given: (1) "Internal Surface of Clays and Constrained Chemical Reactions" (the CMS 1985 George W. Brindley Lecture) by J. J. Fripiat; (2) "The Clay-Water Interface" by Philip F. Low; and (3) "Mixed-Layer Minerals: Diffraction Methods of their Study and Structural Peculiarities" by Victor A. Drits.

About 180 oral and 75 poster papers were arranged into 13 technical categories with the help of the following session chairmen: S. W. Bailey, F. C. Loughnan, W. D. Keller, D. A. Spears, R. C. Reynolds, V. A. Drits, Jan Środoń, P. Klar, D. R. Pevear, F. M. Lippmann, R. F. Giese, Jr., M. J. Wilson, J. B. Dixon, Udo Schwertmann, and H. Kodama. Selection of papers for the Proceedings volume from manuscripts submitted was based on the recommendations of the session chairmen and two referees for each paper and on the judgement of the editors.

A CMS workshop on Quantitative Mineral Analysis, organized by David R. Pevear, was held on the first day of the Conference, and three field trips were made before and after the Conference, as follows: (1) Bentonite, Coal, and Uranium—Black Hills, South Dakota, led by J. W. Hosterman, I. E. Odom, and S. H. Patterson; (2) Clays and Clay Minerals—Western Colorado and Eastern and Central Utah, led by R. B. Hall; and (3) Clay Minerals and Geology, Montana Disturbed Belt, in and around Glacier National Park—Montana, led by C. G. Whitney and S. Altaner. A fourth field guide, Clays in the Petroleum Industry and Clay Minerals and Zeolites—California and Nevada, was prepared by A. J. Gude, 3rd and M. G. Reed, but the trip was not held. The Workshop Notes and the four Field Guides are available from the Central Office of The Clay Minerals Society, P.O. Box 2295, Bloomington, Indiana 47402. A fifth field trip, Clays in the Foothills Area West of Denver, Colorado, led by Bruce Bohor, had no field guide.

The papers included in this proceedings volume are representative of the research on clays being conducted in all parts of the world. Many of the subjects treated are controversial, and although some ideas expressed may not necessarily represent the views of the editors, the referees, or the publisher, they deserve to be brought to the attention of the international clay community. Our thanks are extended to the authors, many of whom had to prepare a manuscript in a language foreign to them; to the session chairmen, who helped organize the technical sessions and select referees for the papers; and to the technical referees themselves, who critically reviewed the submitted manuscripts. The Conference organizers gratefully acknowledge the financial assistance of The Clay Minerals Society, the United States Geological Survey, and the many industrial organizations listed below:

American Colloid Company
Anglo-American Clays Corporation
Aquafine Corporation
Atlantic Richfield Company
Chevron Oil Field Research Company
Conoco, Inc.
Cyprus Industrial Minerals Company
Engelhard Minerals & Chemicals Corporation
Eriez Manufacturing Company
Exxon Production Research Company
Floridin Company
Georgia Kaolin Company, Inc.

International Minerals & Chemical Corporation
International Trading Company
Wharton Jackson
Kentucky-Tennessee Clay Company
Lowe's Incorporated
A. J. Lynch & Company
Mobil Oil Corporation
Nord Kaolin Company
Oil-Dri Production Company
Old Hickory Clay Company
Shell Oil Company
Siemens-Allis, Inc.

W. R. Grace & Company
Halliburton Services
Harrison & Crosfield Pacific, Inc.
E. T. Horn Company
J. M. Huber Corporation

Southeastern Clay Company
H. C. Spinks Clay Company, Inc.
Thiele Kaolin Company
United Catalysts, Inc.
R. T. Vanderbilt Company, Inc.

U.S. Geological Survey   L. G. SCHULTZ

AIPEA   H. VAN OLPHEN

The Clay Minerals Society   F. A. MUMPTON

TECHNICAL REFEREES

The editors are deeply indebted to the following individuals who gave of their time and talent as technical referees of manuscripts submitted for publication in this volume.

April, R. H.
Atkins, M. P.
Bailey, S. W.
Bain, D. C.
Bain, J.
Banin, Amos
Bargeman, D
Barnhisel, R. I.
Bennett, R. H.
Bodine, M. W.
Bohor, B. F.
Bolt, G. H.
Bonne, A. A.
Borchardt, B. L.
Buie, B. F.
Bundy, W. F.
Calvert, C. S.
Childs, C. W.
Coey, Michael
Colten-Bradley, Virginia
Curtis, C.
van Damme, Henri
DeConinck, Frans
Delmon, A.
Dixon, J. B.
Drits, V. A.
Eberl, D. D.
Eckhardt, F. J.
Edil, T.
Eggleton, R. A.
Farmer, V. C.
Foscolos, A. E.
Fripiat, J. J.
Galan, Emilio
Giese, R. F., Jr.
Gilkes, R. J.
Goodman, B. A.
Grant, W. H.
Grim, R. E.
Guggenheim, S.
Güven, Necip
Hall, R. B.
Harrison, Wendy
Herbillon, A. J.
Hevlin, F. G.

Hossner, L. R.
Huang, P. M.
Hughes, R. E.
Hurst, Andrew
Jackson, M. L.
Jepson, W. B.
Johnson, J. W.
Jones, B. F.
Joswig, W. J.
Karathanasis, A.
Knudson, M. I.
Kodama, H.
Köhler, E. E.
Konta, Jiri
Koster, H. M.
Kühnel, R. A.
Lagaly, G.
Lemaître, J. L.
Lipsicas, Max
Locat, J.
Løken, Tor
Loughnan, F. C.
Lynn, Warren
Mackinnon, I. D.
Maksimovic, Zoran
Malla, P. M.
Martin, R. T.
Matijevic, E.
McAtee, J. L., Jr.
McCormick, Charles
Mitchell, Dennis
Moody, W. E.
Moore, Charles
Mortland, M. M.
Mosotti, Victor
Mulla, D.
Muller-Vonmoos, Max
Murad, E.
Murray, H. H.
Murray, R. S.
Nettleton, W. D.
O'Brien, N. R.
Occelli, M. L.
Odom, I. E.
Olsen, H. W.

Ottewill, R. H.
Paquet, Hélène
Patterson, S. H.
Pearson, M. J.
Penner, E.
Pinnavaia, T. J.
Prost, R.
Quigley, R. M.
Rand, B.
Raythatha, R. H.
Reynolds, R. C.
Roberson, H.
Robert, M.
Rosenqvist, I. Th.
Roth, C. B.
Russell, J. D.
Schmidt, Volkmar
Schokking, F.
Schüller, K.-H.
Schulze, D. G.
Schwertmann, Udo
Serratosa, J. M.
Shadfan, H.
Siffert, Bernard
Singer, A.
Slaughter, M.
Solin, S. A.
de Souza Santos, Helena
Spears, D. A.
Środoń, Jan
Stoessell, R. K.
Stucki, J. W.
Suquet, Helena
Tadros, T. F.
Tardy, Yves
Taylor, R. K.
Tchoubar, C.
Thompson, G. R.
Weaver, C. E.
Weaver, R. M.
White, J. L.
Whitney, Gene
Wilson, M. J.
Yamanaka, Shoji
Zelazny, L. W.

TABLE OF CONTENTS

STRUCTURE
AND
CRYSTAL CHEMISTRY

Proceedings of the International Clay Conference, Denver, 1985, L. G. Schultz, H. van Olphen, and F. A. Mumpton, eds.,
The Clay Minerals Society, Bloomington, Indiana, 3–8 (1987).

STRUCTURAL STUDIES OF CLAY MINERALS
DURING 1981–1985

S. W. BAILEY

Department of Geology and Geophysics, University of Wisconsin
Madison, Wisconsin 53706

Abstract—Nearly 40 papers involving structural refinements of phyllosilicate minerals, including those of 30 micas, have appeared during 1981–1985, plus a number of other papers involving structural information obtained from infrared, nuclear magnetic resonance, Mössbauer, and Raman spectra. Consequently, our knowledge of the geometry, energetics, crystal chemistry, and long-range cation order-disorder relationships in all the clay mineral groups has advanced enormously during this period. In addition to the explosion of information in the mica group, the other very important advance has been the precise location of the H^+ protons of the OH groups for examples of all the clay mineral groups, except smectite and sepiolite-palygorskite, either by detailed X-ray or neutron diffraction study. Information is just starting to emerge on local or short-range cation ordering in clay minerals.

Key Words—Crystal structure, Hydrogen positions, Mica, Neutron diffraction, Order-disorder, X-ray diffraction.

INTRODUCTION

This paper reviews structural studies of clay minerals and phyllosilicates that have appeared in the literature since the 1981 International Clay Conference in Italy. Nearly 40 papers involving structural refinements fall in this category, plus a number of others involving structural information obtained from infrared (IR), nuclear magnetic resonance (NMR), Mössbauer, and Raman spectra. It is not possible to review all of these results in the space allotted for this paper, so that considerable selectivity has been necessary. The results are considered below according to mineral groups.

KAOLIN-GROUP MINERALS

Much of the previous confusion regarding the nature of the interlayer hydrogen bonding in dickite and kaolinite has been removed by location of the H^+ protons of the OH groups. By using Rietveld profile refinement of neutron diffraction data of powdered material from Australia, Adams and Hewat (1981) located the H^+ protons in dickite for the first time from electron density and difference-electron density maps. Rozhdestvenskaya *et al.* (1982) and Sen Gupta *et al.* (1984) also located the proton positions on difference electron density maps during X-ray refinements using single crystals from the Kulanteubin porcelain clay deposit, U.S.S.R., and St. Clair, Pennsylvania, respectively. Despite differences in detail, all three studies agree that the OH vector of the inner hydroxyl points away from the octahedral sheet and is inclined 14°–20° to the (001) plane, similar to the situation in most of the dioctahedral micas. The OH vectors of the three surface hydroxyls form bent hydrogen bonds to the adjacent basal

oxygens. The involvement of all three surface hydroxyls in hydrogen bonding in dickite confirms earlier conclusions from IR patterns. Adams and Hewat (1981) also suggested that the inner-hydroxyl OH(1) may be involved in a very weak hydrogen bond to basal oxygen O(3). Refinement of the non-H^+ atoms in the first two studies led to atomic positions similar to those of the earlier X-ray refinement by Newnham (1961).

Adams (1983) used Rietveld profile refinement of neutron diffraction data for kaolinite from St. Austell, Cornwall, to show OH vectors similar to those in dickite, namely with the inner-OH vector pointing away from the octahedral sheet at an angle of 34° and all three surface hydroxyls involved in bent hydrogen bonds to basal oxygens. The possibility of a very weak hydrogen bond between the inner hydroxyl and a basal oxygen was suggested here also. Suitch and Young (1983) also used Rietveld profile refinement of neutron diffraction data of kaolinite from Keokuk, Iowa, to show a similar geometry for the surface hydroxyls and their hydrogen bonds. The involvement of all three hydroxyls in the interlayer bonding and the orientation of all their OH vectors quasinormal to (001) conflicts with some earlier interpretations of IR data. Resolution of the conflict appears to require that one of the three interlayer bonds be of different energy due to some structural difference rather than to being oriented parallel to (001).

The Suitch and Young refinement of kaolinite differs from that of Adams in using symmetry $P1$ instead of $C1$. The former investigators found the OH vectors of the inner hydroxyl to be significantly different in the two half cells of the same layer, pointing into the oc-

3

tahedral sheet by 23°–31° in the first half cell and away from the sheet by 10°–11° in the second half cell. The remainder of the structure is related by C-centering within experimental error. A tetrahedral rotation angle of $\alpha = 7.0°$ was found, so that the basic 1:1 layer of kaolinite is similar to that of dickite ($\alpha = 6.7°$). Previously, Brindley and Nakahira (1958) had assumed a value of $\alpha = 10°$ in their model of the kaolinite structure, and Zvyagin (1967) had experimentally found a value of $\alpha = 11.4°$ by oblique-texture electron diffraction (as calculated by the present writer). The Adams (1983) refinement of kaolinite, on the contrary, found no evidence from difference-electron density maps for splitting of the inner-hydroxyl peak to support the Suitch and Young (1983) model of two nonequivalent protons. Brindley *et al.* (1986) also pointed out that the presence of two differently oriented inner-hydroxyl vectors should produce a broader band than is observed for the inner-OH stretching frequency near 3620 cm^{-1} in IR patterns of kaolinite.

Several teams of French researchers have carried further the pioneering studies by Plançon and Tchoubar (1977) on structural defects in kaolinite, especially in relation to the effects on physical properties and commercial utilization of kaolins. Cases *et al.* (1982) classified different kaolinites according to the density of defects and showed that the presence of Fe is intimately related to defect density, particle size, and physical properties. Tchoubar *et al.* (1982) analyzed the profiles of several kaolinite X-ray powder patterns quantitatively in terms of displacement of the vacant octahedral site, random layer shifts, and domain sizes of coherent diffraction. The size of coherent domains is limited by lenticular voids seen to be present by lattice imaging. Mestdagh *et al.* (1982) concluded that the amount of substitution of Fe^{3+} for octahedral Al is directly related to the proportion of defects due to vacancy displacements. They interpreted the OH-stretching vibration bands of the IR spectra to indicate the formation of domains in which the OH configuration is similar to that in dickite with the vacant site alternating regularly between the B and C positions. Cruz-Cumplido *et al.* (1982) concluded that the OH-vector orientation appears to be a function of the distribution of octahedral vacancies between adjacent layers. The IR spectral characteristics indicative of crystallinity and the interlameller cohesion energy are intimately related to OH-vector orientation. Brindley *et al.* (1986) applied several techniques to study a series of samples of kaolinites and dickites showing a range of stacking order-disorder characteristics. They found a sequential change in the IR patterns of the half-width of the inner-hydroxyl stretching band and in the frequencies and intensity ratios of all of the hydroxyl stretching bands as a function of stacking order/disorder over the series. Four bands were found in all specimens, and the band

frequencies were believed to indicate that the hydrogen bonding system is weaker in dickite than in kaolinite.

TRIOCTAHEDRAL 1:1 LAYER SILICATES

Lizardite

The structure of lizardite-$1T$ containing 4.5% Al$_2$O$_3$ was refined by Mellini (1982) using X-ray diffraction of a single crystal. The silicate rings are nearly hexagonal ($\alpha = -1.7°$), and the basal oxygens rotate away from the nearest octahehdral Mg and toward the OH groups of the adjacent layers. The H$^+$ protons were located, and those of the surface OH groups were found to participate in three bent interlayer-hydrogen bonds. The OH vector of the inner-hydroxyl points perpendicularly away from the octahedral sheet. Octahedral Mg atoms are closer to the OH plane than to the apical oxygen plane, as in all 1:1 layer minerals, but the octahedral sheet is not buckled as in the model determined by Krstanović (1968).

Amesite

Anderson and Bailey (1981) refined the structure of a crystal of amesite-$2H_2$ from Saranovskoye, U.S.S.R., and found the tetrahedral Si,Al and octahedral Mg,Al to be highly ordered in a pattern that reduces the symmetry from the ideal hexagonal $P6_3$ space group to triclinic subgroup $P1$. The ordering pattern differs from that of amesite-$2H_2$ from Antarctica in that the ordered tetrahedral and octahedral Al form spirals around the c-axis. The likelihood of still other ordering patterns in amesite was also assessed. The basal oxygens are rotated by $+14.7°$ in a direction toward both the Mg cations of the same layer and the OH atoms of the adjacent layer below. The interlayer hydrogen bonds are slightly bent, and opposing layer surfaces are corrugated as a result of cation ordering and are keyed together for better steric fit.

Cronstedtite

Geiger *et al.* (1983) refined the structure of a crystal of cronstedtite-$2H_2$ from Pribram, Czechoslovakia. As in amesite-$2H_2$, the ideal hexagonal $P6_3$ symmetry is reduced to triclinic $P1$. From electron density maps, tetrahedral Si and Fe^{3+} were shown to be ordered and octahedral Fe^{2+}, Fe^{3+}, and Mg to be disordered. The presence of anti-phase domains, in which the tetrahedral ordering pattern is reversed and the tetrahedral sheet is distorted differently than in adjacent domains, made detailed refinement and interpretation of the structure difficult.

Greenalite

Guggenheim *et al.* (1982) showed that the structure of greenalite is not simply that of the iron analogue of lizardite, as previously postulated. A trigonal and a monoclinic phase were found in all samples studied of greenalite and of its Mn-analogue caryopilite, but the trigonal phase was dominant at the Fe end and the monoclinic phase dominant at the Mn end. Satellites present on electron diffraction photos indicated an incommensurate modulation of the structure that ranged from 23 Å = $2.4b_0$ in greenalite to 17 Å = $1.7b_0$ in caryopilite. High-resolution transmission electron micrographs (HRTEM) suggested saucer-shaped island domains, four tetrahedral rings in diameter in greenalite and three tetrahedral rings in diameter in caryopilite. One extra tetrahedron is inserted into the tetrahedral sheet after every eight tetrahedra to facilitate articulation with the large Fe,Mn-rich octahedral sheet, thereby creating antiphase relations between adjacent islands. Four- and three-member rings joining adjacent islands may be inverted, and rows of octahedral cations around the island edges are believed to be vacant.

TALC-PYROPHYLLITE GROUP

Talc

The structure of talc-1Tc from Zillertal in the Swiss Alps was refined by Perdikatsis and Burzlaff (1981). The tetrahedral rotation angle is $\alpha = 3.6°$. The potential misfit of the lateral dimensions of the Si-rich tetrahedral sheet and of the Mg-rich octahedral sheet is minimized primarily by a tetrahedral sheet that is considerably thinner (2.18 Å) than for any known mica (range 2.20–2.39 Å) and thereby has larger lateral dimensions. The octahedral sheet thickness of 2.17 Å is not unusual (Bailey, 1984a).

Pyrophyllite

Lee and Guggenheim (1981) refined pyrophyllite-1Tc from Ibitiara, Brazil, in space group $C\bar{1}$. It is interesting to note that the tetrahedral sheet is as thin as that in talc-1Tc. The octahedral sheet is thinner (2.08 Å) and contains smaller cations so that a larger tetrahedral rotation angle of $\alpha = 10.2°$ is required to match the lateral dimensions of the two sheets.

Minnesotaite

Guggenheim and Bailey (1982) determined that minnesotaite is not the Fe^{2+}-analogue of talc, but has a more complex structure containing superlattices along all three axes. Regular modulations of the Si-rich tetrahedral sheet and Fe^{2+}-rich octahedral sheet to alleviate their lateral misfit were postulated, but not specifically determined. Since the publication, HRTEM study has allowed elucidation of the structural modulations and of their variation with Fe,Mg content (S. Guggenheim, Department of Geology, University of Illinois at Chicago, personal communication, 1984).

MICAS

The greatest effort regarding structural analyses of phyllosilicates in the period 1981–1985 was with re-

lation to micas in that at least 30 mica refinements were published. In addition, review papers were published by Toraya (1981), Bailey (1984b), and Weiss et al. (1985), and a comprehensive book on micas appeared as Volume 13 of the Mineralogical Society of America series Reviews in Mineralogy (Bailey, 1984c). It is not possible to consider each structural refinement here, but the structural details of most of the refinements have been tabulated previously in Chapters 2 and 3 of the mica book noted above.

Bailey (1984b) showed that octahedral cation ordering is common in micas and is to expected if there is sufficient substitution of cations of different size or charge. The ordering pattern is always that the trans M(1) site is larger than the cis M(2) sites. Long-range ordering of tetrahedral cations is rare, but is favored by three situations: (1) large tetrahedral substitutions in which the tetrahedral Si:R^{3+} ratio approaches 1:1 and the local strain due to the substitutions is correspondingly large (examples: margarite, bityite, ephesite, anandite); (2) the 3T stacking sequence of layers (examples: 3T phengitic muscovite, paragonite, lepidolite, protolithionite); and (3) phengitic compositions involving different stacking sequences (1M, 2M_1, 2M_2, 3T), although some of the refinements are of lower reliability than desired for proof of this latter trend.

Cation ordering may take place in a long-range pattern that reduces the symmetry to a subgroup of that of the ideal space group, and the lower symmetry is often difficult to detect. It may also lead to ordering in small, local domains that are out-of-step with one another and undetectable by conventional X-ray refinement procedures. In addition to the earlier work of Gatineau and Méring (1966) on diffuse X-ray scattering by a lepidolite, additional evidence for such short-range order in micas was provided by Abbott (1985) by electron diffraction of a biotite and Sanz and Stone (1983) by NMR study of phlogopite and biotite.

VERMICULITE

Lagaly (1982) showed by alkylammonium ion exchange that some high-charge vermiculites, such as that from Llano County, Texas, have homogeneous charge distributions, whereas others are highly heterogeneous, presumably as a result of differences in the parent materials and/or reaction sequences for transformed materials. Slade et al. (1985) were able to obtain high-resolution electron density maps of the interlayer structures of regular 2-layer hydrates of Na- and Ca-vermiculites prepared from the Llano vermiculite. In Na-vermiculite the Na ions are at the mid-plane of the interlayer region on sites m_1 and m_2 and are octahedrally coordinated by six water molecules. The Na ions lie between triads of basal oxygens of the two adjoining layers. In Ca-vermiculite the Ca ions at the mid-plane are located not only between oxygen triads above and below, as in the Na-form, but also at site m_3 between ditrigonal rings where they have eight water nearest neighbors. In these regular Na- and Ca-hydrates, adjacent 2:1 layers have the mica orientation with the ditrigonal rings opposite one another. In Mg-vermiculite, where adjacent layers are displaced by $b/3$ relative to each other, a 2-layer stacking sequence with a large number of $\pm b/3$ stacking faults always seems to be realized. Thompson (1984) determined by NMR that the tetrahedral Si,Al cations are only locally ordered in the Llano material, rather than long-range ordered.

Slade and Stone (1983, 1984) showed that adjacent 2:1 layers adopt the mica orientation and a regular stacking sequence in a vermiculite-aniline intercalate of the Na-saturated Llano material and were able to determine the interlayer aniline configuration. Slade and Raupach (1982) were only able to determine a partial interlayer configuration for benzidine in a vermiculite-benzidine intercalate prepared from Na-saturated vermiculite of medium charge from Young River, Western Australia.

SMECTITE

In synthetic saponites with two water interlayers, Suquet et al. (1981a, 1981b) found that the b parameter is a result of interactions involving the compositions and lateral dimensions of both the tetrahedral and octahedral sheets. For specimens containing one water interlayer, the detailed nature of the interlayer space also has an effect that varies with the size and location of the interlayer cation. For dehydrated specimens, the interaction between the layers and the interlayer cations is even larger and can even affect the amount of tetrahedral rotation. The b dimension and $d(001)$ increase linearly with increase in tetrahedral Al-for-Si substitution, but the magnitudes of their increases depend on the water content of the interlayer. A linear relationship also exists between these parameters and the interlayer cation radius. Desprairies (1983) and Brigatti (1983) studied the effect of octahedral Fe on the b dimension of smectite, and Brigatti and Poppi (1981) applied statistical analysis to the compositions of a large number of dioctahedral smectites to distinguish the main solid solution ranges and the possible miscibility gaps.

Tsipursky and Drits (1984) used oblique-texture electron diffraction to study the octahedral cation distribution in dioctahedral smectites that had been K-saturated and subjected to 70–100 wetting-drying cycles to improve the degree of stacking order. Adjacent layers adopt the mica orientation with a $1M$ type unit cell. By comparing the observed intensities with those calculated from models having different distributions of cations and vacancies they found a wide variety of octahedral ordering patterns. There is an imperfect trend

that centrosymmetric layers [with the vacancy on the mirror plane at *trans* site M(1)] are characteristic of smectites in which the layer charge is localized in the tetrahedral sheet, as for the Black Jack Mine beidellite and most nontronites. This finding confirms the results of a study of the Garfield, Washington, nontronite by Besson et al. (1983) who used a variety of experimental techniques. Because M(1) on the mirror plane is larger than the M(2) sites, the intralayer shift is greater than $a/3$ in such smectites and the observed β angle is greater than ideal $[\beta = \arccos(-a/3c)]$. For montmorillonites having only an octahedral source of charge, the layer tends to be noncentrosymmetric because the *trans* M(1) site is occupied by a cation and the vacancy is in one of the *cis* M(2) sites, as in Wyoming bentonite. M(1) is now smaller than M(2) and β smaller than ideal. In some smectites the cation distribution is random and β is close to ideal. The trends noted above are imperfect because smectites from the same deposit with very close chemical compositions may have totally different octahedral cation distributions.

Lipsicas et al. (1984) determined by NMR that Loewenstein's rule of Al-avoidance is obeyed in a series of synthetic trioctahedral smectites and that some local ordering of tetrahedral Si,Al is present.

CHLORITE

Bish and Giese (1981) used the H^+ proton positions, as determined previously by X-ray diffraction for two chromian chlorites of the IIb-4 structural type and for another IIb-4 chlorite studied by neutron diffraction, to calculate the interlayer bond energy (ILBE) as a function of layer charge, the site of the charge, and the selective replacement of OH by F. Although the calculations indicate that chlorite could exist with neutral layers and still have an ILBE comparable to that of mica, its stability is dramatically increased by charged layers and by location of a trivalent cation in interlayer site M(4) on the symmetry plane of the layer (where it had been located experimentally). Because of this large effect, small changes in the chemistry of the interlayer sheet should lead to large changes in physical and chemical properties. The long hydrogen bonds between the layer and interlayer are strong and by themselves sufficient to create a stable structure. In contrast to trioctahedral micas, where repulsion exists between interlayer cation and the H^+ proton of the OH, substitution of F for OH in the chlorite structure does not lead to increased stability.

Ďurovič (1981) and Dornberger-Schiff et al. (1982) applied order-disorder (OD) theory in a systematic geometrical analysis and description of polytypes of layer silicates. Ďurovič et al. (1983) and Weiss and Ďurovič (1983) then applied these concepts specifically to chlorite polytypes and derived criteria for their identification by X-ray diffraction. Chlorite polytypism can be determined by HRTEM imaging when viewed in two projections, or in one projection when combined with X-ray diffraction data (Spinnler *et al.,* 1984). Small scale, semirandom stacking is present even in chlorites that are macroscopically well ordered according to X-ray diffraction study.

Bayliss (1983) showed that the Mn-chlorite pennantite occurs in both the II*b* and I*a* structural forms. Material previously called grovesite gives the I*a* pattern. Sudoite from Ottré, Belgium, a di,trioctahedral chlorite, was shown by Lin and Bailey (1985) to have a 2-layer II*b* structure of the *s* type with L-shifts along a_1 in each 2:1 layer and imperfect alternation of adjacent layers by $a_2/3$ and $a_3/3$ displacements.

REGULAR INTERSTRATIFICATIONS

Two new species names have been proposed during this period for 1:1 regular interstratifications—kulkeite for chlorite/talc (Schreyer *et al.,* 1982) and hydrobiotite for biotite/vermiculite (Brindley *et al.,* 1983). The AI-PEA Nomenclature Committee (Bailey, 1982; Bailey *et al.,* 1984) recommended rules regarding the usage of species names for regular interstratifications, including a quantitative criterion for defining sufficient regularity to merit a species name, and applied their criteria to published descriptions of aliettite, corrensite, kulkeite, rectorite, tarasovite, and tosudite. Brigatti and Poppi (1984) have extensively reviewed the crystal chemistry of corrensite specimens from many occurrences.

ACKNOWLEDGMENTS

This study has been supported, in part, by NSF grant EAR-8106124 and, in part, by Petroleum Research Fund grant 15932-AC2-C, administered by the American Chemical Society.

REFERENCES

Abbott, R. N. (1985) Al-Si ordering in 1*M* trioctahedral micas: *Can. Mineral.* **22,** 659–667.

Adams, J. M. (1983) Hydrogen atom positions in kaolinite by neutron profile refinement: *Clays & Clay Minerals* **31,** 352–356.

Adams, J. M. and Hewat, A. W. (1981) Hydrogen atom positions in dickite: *Clays & Clay Minerals* **29,** 316–319.

Anderson, C. S. and Bailey, S. W. (1981) A new cation ordering pattern in amesite-2*H₂*: *Amer. Mineral.* **66,** 185–195.

Bailey, S. W. (1982) Nomenclature for regular interstratifications: *Clay Miner.* **17,** 243–248.

Bailey, S. W. (1984a) Structure of layer silicates. Chap. 1, in *Crystal Structures of Clay Minerals and Their X-ray Identification,* G. W. Brindley and G. Brown, eds., Miner. Soc., London, Mono. 5, 1–123.

Bailey, S. W. (1984b) Review of cation ordering in micas: *Clays & Clays Minerals* **32,** 81–92.

Bailey, S. W., ed. (1984c) *Micas:* Reviews in Mineralogy 13, Mineral. Soc. Amer., Washington, D.C., 584 pp.

Bailey, S. W., Brindley, G. W., Fanning, D. S., Kodama, H.,

and Martin, R. T. (1984) Report of The Clay Minerals Society Nomenclature Committee for 1982 and 1983: *Clays & Clay Minerals* **32,** 239.

Bayliss, P. (1983) The polytypes of pennantite: *Can. Mineral.* **21,** 545–547.

Besson, G., Bookin, A. S., Dainyak, L. G., Rautureau, M., Tsipursky, S. I., Tchoubar, C., and Drits, V. A. (1983) Use of diffraction and Mössbauer methods for the structural and crystallochemical characterization of nontronites: *J. Appl. Crystallogr.* **16,** 374–383.

Bish, D. L. and Giese, R. F., Jr. (1981) Interlayer bonding in IIb chlorite: *Amer. Mineral.* **66,** 1216–1220.

Brigatti, M. F. (1983) Relationships between composition and structure in Fe-rich smectites: *Clay Miner.* **18,** 177–186.

Brigatti, M. F. and Poppi, L. (1981) A mathematical model to distinguish the members of the dioctahedral smectite series: *Clay Miner.* **16,** 81–89.

Brigatti, M. F. and Poppi, L. (1984) Crystal chemistry of corrensite: a review: *Clays & Clay Minerals* **32,** 391–399.

Brindley, G. W., Kao, C.-C., Harrison, J. L., Lipsicas, M., and Raythatha, R. (1986) On the relation between order-disorder and other clay characteristics in kaolinite and dickites: *Clays & Clay Minerals* **34,** 239–249.

Brindley, G. W. and Nakahira, M. (1958) Further consideration of the crystal structure of kaolinite: *Mineral. Mag.* **31,** 781–786.

Brindley, G. W., Zalba, P. E., and Bethke, C. M. (1983) Hydrobiotite, a regular 1:1 interstratification of biotite and vermiculite layers: *Amer. Mineral.* **68,** 420–425.

Cases, J.-M., Lietard, O., Yvon, J., and Delon, J.-F. (1982) Étude des propriétés cristallochimiques, morphologiques, superficielles de kaolinites désordonnées: *Bull. Mineral.* **105,** 439–455.

Cruz-Cumplido, M., Sow, C., and Fripiat, J. J. (1982) Spectre infrarouge des hydroxyles, cristallinité et énergie de cohésion des kaolins: *Bull. Mineral.* **105,** 493–498.

Desprairies, A. (1983) Relation entre le paramètre *b* des smectites et leur contenu en fer et magnésium. Application à l'étude des sediments: *Clay Miner.* **18,** 165–175.

Ďurovič, S. (1981) OD-Charakter, Polytypie und Identifikation von Schichtsilikaten: *Fortschr. Mineral.* **59,** 191–226.

Ďurovič, S., Dornberger-Schiff, K., and Weiss, Z. (1983) Chlorite polytypism. I. OD interpretation and polytype symbolism of chlorite structures: *Acta Crystallogr.* **B39,** 547–552.

Dornberger-Schiff, K., Ďurovič, S., and Zvyagin, B. B. (1982) Proposal for general principles for the construction of fully descriptive polytype symbols: *Crystal Res. Tech.* **17,** 1449–1457.

Gatineau, L. and Méring, J. (1966) Relations ordre-désordre dans les substitutions isomorphiques des micas: *Bull. Grpe. fr. Argiles* **18,** 67–74.

Geiger, C. A., Henry, D. L., Bailey, S. W., and Maj, J. J. (1983) Crystal structure of cronstedtite-2*H₂*: *Clays & Clay Minerals* **31,** 97–108.

Guggenheim, S. and Bailey, S. W. (1982) The superlattice of minnesotaite: *Can. Mineral.* **20,** 579–584.

Guggenheim, S., Bailey, S. W., Eggleton, R. A., and Wilkes, P. (1982) Structural aspects of greenalite and related minerals: *Can. Mineral.* **20,** 1–18.

Krstanović, I. (1968) Crystal structure of single-layer lizardite: *Z. Kristallogr.* **126,** 163–169.

Lagaly, G. (1982) Layer charge heterogeneity in vermiculites: *Clays & Clay Minerals* **30,** 215–222.

Lee, J. H. and Guggenheim, S. (1981) Single crystal X-ray refinement of pyrophyllite-1*Tc*: *Amer. Mineral.* **66,** 350–357.

Lin, C.-Y. and Bailey, S. W. (1985) Structural data on su-doite: *Clays & Clay Minerals* 33, 410–414.

Lipsicas, M., Raythatha, R. H., Pinnavaia, T. J., Johnson, I. D., Giese, R. F., Jr., Costanzo, P. M., and Robert, J.-L. (1984) Silicon and aluminium site distributions in 2:1 layered silicate clays: *Nature* 309, 604–607.

Mellini, M. (1982) The crystal structure of lizardite 1*T*: hydrogen bonds and polytypism: *Amer. Mineral.* 167, 587–598.

Mestdagh, M. M., Herbillon, A. J., Rodrique, L., and Rouxhet, P. G. (1982) Évaluation du rôle du fer structural sur la cristallinité des kaolinites: *Bull. Mineral.* 105, 457–466.

Newnham, R. E. (1961) A refinement of the dickite structure and some remarks on polymorphism in kaolin minerals: *Mineral. Mag.* 32, 683–704.

Perdikatsis, B. and Burzlaff, H. (1981) Strukturverfeinerung am Talk $Mg_3[(OH)_2Si_4O_{10}]$: *Z. Kristallogr.* 156, 177–186.

Plançon, A. and Tchoubar, C. (1977) Determination of structural defects in phyllosilicates by X-ray powder diffraction—II. Nature and proportion of defects in natural kaolinites: *Clays & Clay Minerals* 25, 436–450.

Rozhdestvenskaya, I. V., Bookin, A. S., Drits, V. A., and Finko, V. I. (1982) The proton position and the structural peculiarities of dickite according to the data of X-ray structural analysis: *Mineral. Zhurn.* 4, 52–58 (in Russian).

Sanz, J. and Stone, W. E. E. (1983) NMR applied to minerals: IV. Local order in the octahedral sheet of micas; Fe-F avoidance: *Clay Miner.* 18, 187–192.

Schreyer, W., Medenbach, O., Abraham, K., Gebert, W., and Müller, W. F. (1982) Kulkeite, a new metamorphic phyllosilicate mineral: ordered 1:1 chlorite/talc mixed-layer: *Contrib. Mineral. Petrol.* 80, 103–109.

Sen Gupta, P. K., Schlemper, E. O., Johns, W. D., and Ross, F. (1984) Hydrogen positions in dickite: *Clays & Clay Minerals* 32, 483–485.

Slade, P. G. and Raupach, M. (1982) Structural model for benzidine-vermiculite: *Clays & Clay Minerals* 30, 297–305.

Slade, P. G. and Stone, P. A. (1983) Structure of a vermiculite-aniline intercalate: *Clays & Clay Minerals* 31, 200–206.

Slade, P. G. and Stone, P. A. (1984) Three-dimensional order and the structure of aniline-vermiculite: *Clays & Clay Minerals* 32, 223–226.

Slade, P. G, Stone, P. A., and Radoslovich, E. W. (1985) The interlayer structures of the two-layer hydrates of Na- and Ca-vermiculites: *Clays & Clay Minerals* 33, 51–61.

Spinnler, G. E., Self, P. G., Iijima, S., and Buseck, P. R. (1984) Stacking disorder in clinochlore chlorite: *Amer. Mineral.* 69, 252–263.

Suitch, P. R. and Young, R. A. (1983) Atom positions in highly ordered kaolinite: *Clays & Clay Minerals* 31, 357–366.

Suquet, H., Malard, C., Copin, E., and Pezerat, H. (1981a) Variation du paramètre *b* et de la distance basale d_{001} dans une série de saponites à charge croissante. I. Etats hydrates: *Clay Miner.* 16, 53–67.

Suquet, H., Malard, C., Copin, E., and Pezerat, H. (1981b) Variation du paramètre *b* et de la distance basale d_{001} dans une série de saponites à charge croissante: II. Etats "zero couche": *Clay Miner.* 16, 181–193.

Tchoubar, C., Plançon, A., Ben Brahim, J., Clinard, C., and Sow, C. (1982) Caractéristiques structurales des kaolinite désordonnées: *Bull. Mineral.* 105, 477–491.

Thompson, J. G. (1984) ^{29}Si and ^{27}Al nuclear magnetic resonance spectroscopy of 2:1 clay minerals: *Clay Miner.* 19, 229–236.

Toraya, H. (1981) Distortions of octahedra and octahedral sheets in 1*M* micas and the relation to their stability: *Z. Kristallogr.* 157, 173–190.

Tsipursky, S. I. and Drits, V. A. (1984) The distribution of octahedral cations in the 2:1 layers of dioctahedral smectites studied by oblique-texture electron diffraction: *Clay Miner.* 19, 177–193.

Weiss, Z. and Ďurovič, S. (1983) Chlorite polytypism. II. Classification and X-ray identification of trioctahedral polytypes: *Acta Crystallogr.* B39, 552–557.

Weiss, Z., Reider, M., Chmielová, M., and Krajicek, J. (1985) Geometry of the octahedral coordination in micas: a review of refined structures: *Amer. Mineral.* 70, 747–757.

Zvyagin, B. B. (1967) *Electron Diffraction Analysis of Clay Mineral Structures:* Plenum Press, New York, 364 pp.

Proceedings of the International Clay Conference, Denver, 1985, L. G. Schultz, H. van Olphen, and F. A. Mumpton, eds.,
The Clay Minerals Society, Bloomington, Indiana, 9–16 (1987).

EXTENDED X-RAY ABSORPTION FINE STRUCTURE STUDY
OF COBALT-EXCHANGED SEPIOLITE

YOSHIAKI FUKUSHIMA AND TOKUHIKO OKAMOTO

Toyota Central Research and Development Laboratories, Inc., Nagakute-cho
Aichi-gun, Aichi-ken 480-11, Japan

Abstract—Natural sepiolite and synthetic loughlinite, prepared from sepiolite by treatment with an aqueous solution of NaOH, were immersed in a 1.0 N aqueous solution of $Co(NO_3)_2$ for about 30 hr at 20°C. Local structure around the Co ions was studied by X-ray powder diffraction and two types of X-ray absorption experiments—extended X-ray absorption fine structure (EXAFS) and X-ray absorption near-edge structure (XANES). The results can be explained well by a model in which Co ions substituted for Mg or Na ions in octahedral sites in the sepiolite or the synthetic loughlinite.

The d(110) values and Mg, Na, and Co contents of the sepiolite and synthetic loughlinite treated with aqueous NaOH and $Co(NO_3)_2$ solutions support this substitution. In synthetic loughlinite, about 15% of octahedral sites were substituted by Co after a 30-hr treatment, whereas only about 4% of the sites in the natural sepiolite were substituted.

Key Words—Cobalt, Extended X-ray absorption fine structure, Loughlinite, Sepiolite, X-ray absorption near-edge structure, X-ray powder diffraction.

INTRODUCTION

Although ionic substitution and ion exchange are important properties of clay minerals, the local structure around the substituted or exchanged ions has been difficult to determine because of either the low concentration of these ions or the difficulty in obtaining large enough crystals for X-ray or neutron diffraction investigation. X-ray absorption experiments, such as extended X-ray absorption fine structure (EXAFS) analysis or X-ray absorption near-edge structure (XANES) analysis, should be useful for studying the local structures in clay minerals, but few such studies have been made.

Although EXAFS and XANES have advantages, they also have limitations, such as uncertainties of phase shifts, backscattering amplitudes, absorption thresholds and backgrounds (Pettifer and Cox, 1983), the influence of sample thickness (Heald, 1983), and the effect of multiple scattering (Motta *et al.*, 1983).

To evaluate their usefulness in the study of the clay minerals, natural sepiolite and synthetic loughlinite prepared from it were treated with aqueous $Co(NO_3)_2$ solution and examined by EXAFS and XANES at the Co K-absorption edge. The results were compared with EXAFS and XANES spectra for various cobalt compounds.

THEORY OF EXAFS AND XANES

The X-ray absorption coefficient, μ, is defined as

$$\mu \cdot t = \exp(I_0/I), \qquad (1)$$

where I_0 and I are intensities of incident and transmitted X-rays and t is sample thickness. The value of μ increases discontinuously at the absorption edge, and μ is modulated as a function of X-ray photon energy to oscillate beyond the absorption edge. This phenomenon is due to an interference effect involving scattering of outgoing photoelectrons from neighboring atoms. The oscillation 8–40 eV above the absorption edge (XANES) has been identified as the multiple scattering resonance of the photoelectrons (Bianconi, 1983). XANES gives qualitative information of stereochemical geometries around the excited atom (Bianconi, 1983). On the other hand, EXAFS, that is, the oscillation within an energy range of about 30–1000 eV above the edge, is due to scattering of the photoelectron by only one neighbor atom in a single scattering process (Stern, 1974) and gives information on the local radial distribution of atoms around the excited atom. EXAFS spectra, $\chi(k)$ defined as

$$\chi(k) = (\mu - \mu_0)/\mu_0 \qquad (2)$$

are described by

$$\chi(k) = \sum_j (C_j/kR_j^2)\exp(-2k^2\sigma_j^2)$$
$$\cdot F_j(k)D_j\sin[2kR_j + \phi_j(k)], \qquad (3)$$

where μ is the K- or L-shell mass absorption coefficient, μ_0 is the smoothed background of the absorption coefficient, C_j is the coordination number of the j-th atom, σ_j is the Debye-Waller factor, $F_j(k)$ is the backscattering amplitude, D_j is the scaling factor, R_j is neighboring distance, $\phi_j(k)$ is the total phase shift experienced by the photoelectron, and k is photoelectron wave number defined as

$$k = \sqrt{2m(E - E_0)/\hbar^2} \qquad (4)$$

where E is photon energy, E_0 is the energy threshold of the absorption edge, m is the mass of the electron, and $\hbar$ is Planck's constant divided by 2π. Inasmuch as the energy of the absorption edge depends on the atomic number, it is possible to analyze the local structure round an interesting atom species by EXAFS and XANES analysis.

Although syncrotron radiation (SR) is an ideal X-ray source for EXAFS and XANES because of its high intensity and continuous spectrum, EXAFS using a conventional X-ray generator also can be used (Knapp and Georgopoules, 1980). An EXAFS spectrometer using such a conventional generator is called laboratory or in-house EXAFS equipment. In the present work, both SR and laboratory equipment are used, and the results for the same sample of $Co(NO_3)_2 \cdot 6 H_2O$ are compared.

EXPERIMENTAL

Sample preparation

Natural sepiolite from Spain, which was commercially available from TOLSA Co. Ltd., was used in this study. The granular sample was crushed to <100 μm with an agate mortar. The powdered sample was transformed into synthetic loughlinite by the method of Imai et al. (1969), whereby 50 g of the powdered sepiolite was immersed in 200 ml of 8.0 N aqueous NaOH solution for 20 hr at 60°C, washed with deionized water until the pH of the filtrate reached 7, and dried at 60°C. Thirty grams of the powdered sepiolite and synthetic loughlinite were immersed in 1 liter of aqueous 1.0 N $Co(NO_3)_2$ solution for 30 hr at 20°C, washed with deionized water until NO_3^- radicals were not detected by infrared spectroscopy, and dried at 60°C.

Chemical analyses were made by inductively coupled plasma (ICP) and atomic absorption (AA) analyses. Concentration of Mg^{2+} and Na^+ in the filtrates from the aqueous $Co(NO_3)_2$ solution, taken after various periods of immersion, were also analyzed by AA. The samples before and after the treatment with $Co(NO_3)_2$ were examined by X-ray powder diffraction (XRD) using Fe-filtered Co radiation at 30 kV and 20 mA, and divergence, receiving, and scattering slit widths of 0.5°, 0.15 mm, and 0.5°, respectively.

Measurement of X-ray absorption spectra

X-ray absorption spectra for the Co-treated sepiolite, the synthetic loughlinite, and $Co(NO_3)_2 \cdot 6 H_2O$ were measured using the synchrotron at the National Laboratory for High Energy Physics, Tsukuba, Japan. The SR X-ray beam was monochromatized by a channel-cut double crystal of Si(311), and the intensity of incident X-ray (I_0) and transmitted X-ray (I) were measured by ion-chambers. The energy resolution of the spectrometer was about 1 eV at 8000 eV (Oyanagi et al., 1984). Details for the spectrometer were reported by Oyanagi et al. (1984).

The EXAFS and XANES spectra for Co-metal foil (t = 8 μm), CoO, Co_3O_4, $Co(OH)_2$, and $Co(NO_3)_2 \cdot 6 H_2O$ and XANES spectra for Co-treated sepiolite and loughlinite were measured on our laboratory EXAFS spectrometer equipped with a conventional rotating Mo anode as an X-ray source. Tube voltage and current were 17 kV and 60 mA. X-rays were monochromatized by a Johansson-cut curved Si(220) crystal whose energy resolution was about 10 eV at 9000 eV. Intensities of incident and transmitted X-rays (I_0 and I) were measured by an ion chamber and a solid state detector. Toji et al. (1983) reported the details of similar EXAFS equipment.

Powdered samples were held in a rectangular sample holder made of thick paper (t $\approx$ 0.8 mm) held together by Scotch tape. Powders of reference samples were mixed with the finely powdered Si_3N_4 to achieve an optimum absorption coefficient for EXAFS measurements (Lytle et al., 1975).

Data processing of absorption spectra

EXAFS spectra, expressed as $\chi(k)$, were extracted from the observed absorption coefficient in the manner described by Lytle et al. (1975). The $\chi(k)$ values weighted with k^3 were Fourier-transformed to r-space to obtain $|\phi_3(r)|$ values. Fourier windows in k-space of 3.3 to 15.0 $Å^{-1}$ or 3.3 to 12.0 $Å^{-1}$ were used for the data measured by SR or by the laboratory EXAFS equipment, respectively.

Values of $\phi_3(r)$ for Co-treated samples were inverse-Fourier-transformed to k-space using a Fourier window of 1.0 to 2.1 Å for the 1st neighbor shell and 2.1 to 3.2 Å for the 2nd neighbor shell of Co atoms. Parameters, such as neighboring distance (R_i), coordination number (C_i), Debye-Waller factor (σ_i), and a shift of threshold energy (ΔE_0), were calculated by the least squares method using the inverse-Fourier-transformed spectra, the theoretical Eq. (2) proposed by Lytle et al. (1975), and the results of ab initio calculation for phase shifts and backscattering amplitudes reported by Teo and Lee (1979).

RESULTS

X-ray powder diffraction

The XRD results are shown in Figure 1 and Table 1. The pattern for synthetic loughlinite agrees well with data for natural and synthetic loughlinite reported by Fahey et al. (1960) and Imai et al. (1969). Treatment of sepiolite with aqueous NaOH solution increased the value of d(110) from 12.2 to 12.6 Å. The d(110) value, however, decreased to 12.3 Å after an aqueous $Co(NO_3)_2$ treatment. The XRD patterns near $2\theta = 22°-24°$, 34°, and in the small-angle region ($2\theta < 5°$) were also changed

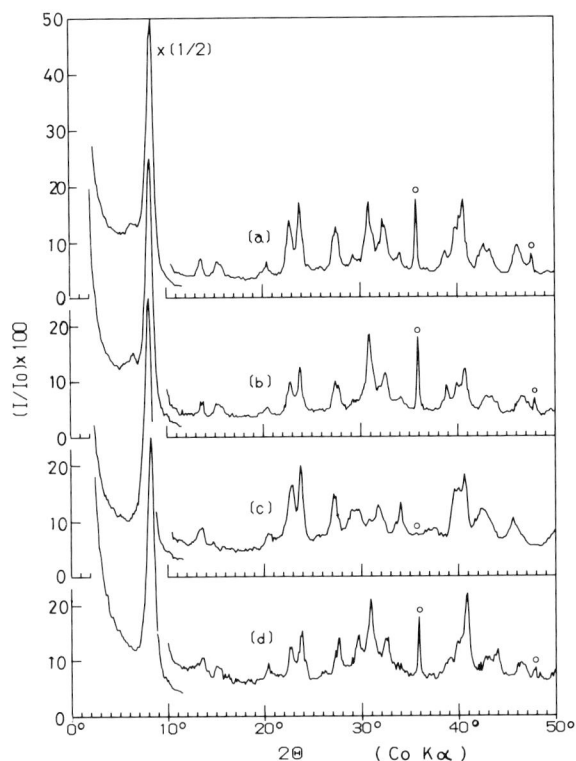

Figure 1. X-ray powder diffraction patterns for (a) natural sepiolite, (b) Co-treated sepiolite, (c) synthetic loughlinite and (d) Co-treated synthetic loughlinite. Dolomite (or calcian dolomite) reflections indicated by O.

by the treatments with aqueous NaOH and Co(NO$_3$)$_2$ solutions.

Chemical analysis

Chemical compositions of the Co-treated samples and natural sepiolite are listed in Table 2. The Mg content in the sepiolite and the synthetic loughlinite and Na content in the synthetic loughlinite decreased

Table 1. X-ray powder diffraction data for natural sepiolite, artificial loughlinite and the samples treated with aqueous Co(NO$_3$)$_2$ solution for 30 hr.

Natural sepiolite		Synthetic loughlinite		Co-treated synthetic loughlinite		Co-treated sepiolite	
d (Å)	I	d (Å)	I	d (Å)	I	d (Å)	I
16.1	8	—	—	—	—	15.8	12
12.2	100	12.6	100	12.3	100	12.4	100
7.56	4	7.63	4	7.62	4	7.59	3
6.71	3	—	—	6.77	3	6.72	3
5.05	3	5.02	3	5.06	4	5.06	2
4.53	10	4.53	5	4.63	8	4.54	7
4.32	14	4.32	18	4.41	11	4.34	10
—	—	3.79	10	—	—	—	—
3.75	8	—	—	3.75	9	3.74	6
3.53	3	3.51	7	3.54	10	3.55	3
3.35	13	—	—	3.36	18	3.37	17
—	—	3.26	7	—	—	3.25	9
3.20	9	—	—	3.20	8	—	—
3.07	3	3.05	8	—	—	3.10	3
2.90[1]	15	2.91[1]	1	2.91[1]	13	2.91[1]	16
—	—	2.79	6	—	—	—	—
2.68	4	2.63	10	2.68	5	2.69	6
2.61	9	—	—	2.62	9	2.63	6
2.60	11	2.60	10	2.58	9	2.60	6
2.56	15	2.56	13	2.56	19	2.57	10
2.44	6	2.46	6	2.45	6	2.45	4
—	—	—	—	2.40	7	2.42	4
2.27	6	2.28	5	2.28	5	2.27	4

[1] Reflection from dolomite (or calcian dolomite).

and the Co content increased by treatment with aqueous Co(NO$_3$)$_2$ solutions. The amounts of Mg^{2+} and Na$^+$ entering the aqueous Co(NO$_3$)$_2$ solutions, in which the powders of sepiolite or loughlinite were immersed, are shown in Figure 2 as a function of the immersion period. Ca^{2+} ions supposedly from impurity dolomite or calcian dolomite were also detected. Although Figure 2 shows more Mg^{2+} ions were dissolved from the synthetic loughlinite than from the sepiolite, dissolution of Mg^{2+} from dolomite should be taken into account.

Table 2. Chemical compositions and atom ratio of cations in natural sepiolite, synthetic loughlinite, and the samples treated with aqueous Co(NO$_3$)$_2$ solution for 30 hr.

	Natural sepiolite		Synthetic loughlinite		Co-treated sepiolite		Co-treated synthetic loughlinite	
	(Wt. %)	[Atom ratio]	(Wt. %)	[Atom ratio]	(Wt. %)	[Atom ratio]	(Wt. %)	[Atom ratio]
SiO$_2$ [Si]	52.5	[11.6]	47.0	[11.0]	52.1	[11.4]	50.6	[11.2]
MgO [Mg]	22.8	[7.5]	20.3	[7.1]	20.4	[6.7]	20.0	[6.6]
Al$_2$O$_3$ [Al]	1.7	[0.4]	2.3	[0.6]	3.2	[0.8]	2.2	[0.6]
Fe$_2$O$_3$ [Fe]	0.8	[0.1]	0.6	[0.1]	0.9	[0.1]	0.9	[0.2]
CaO [Ca]	0.8	[0.2]	3.1	[0.8]	2.8	[0.7]	1.9	[1.2]
Na$_2$O [Na]	0.3	[0.1]	4.7	[2.1]	0.3	[0.1]	0.8	[0.3]
K$_2$O [K]	0.4	[0.1]	0.5	[0.1]	0.6	[0.2]	0.5	[0.1]
CoO [Co]	0.0	[0.0]	0.0	[0.0]	1.9	[0.3]	6.7	[1.2]
TiO$_2$ [Ti]	0.0	[0.0]	0.1	[0.0]	0.1	[0.0]	0.1	[0.0]
L.O.I.	21.0		21.5		15.8		16.8	
Total	100.3		100.1		98.1		100.5	

Atom ratios were normalized by ideal total charge (+64) of cations in one formula unit of sepiolite.

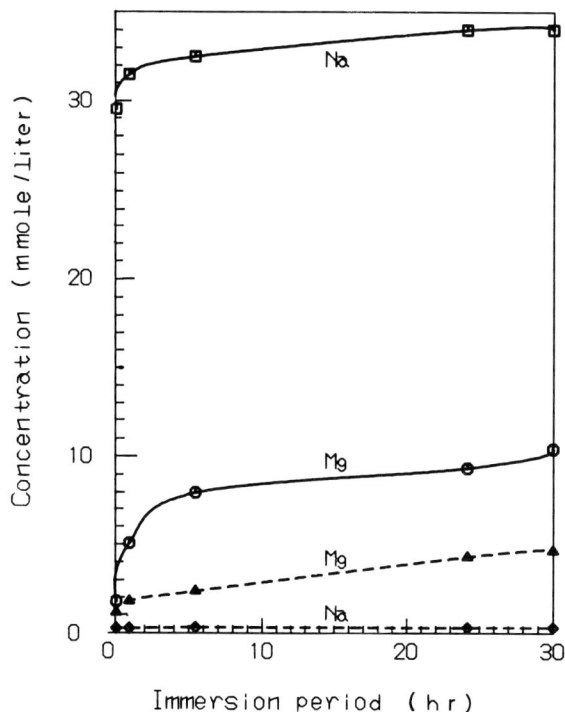

Figure 2. Concentration of Mg^{2+} and Na^+ in aqueous $Co(NO_3)_2$ solutions in which the synthetic loughlinite (solid line) or sepiolite (broken line) was immersed.

EXAFS and XANES spectra

The EXAFS spectra, $\chi(k)$, of the Co-treated samples and the reference samples are shown in Figures 3 and 4. $\chi(k)$ spectra for the Co-treated samples and $Co(OH)_2$ resemble each other.

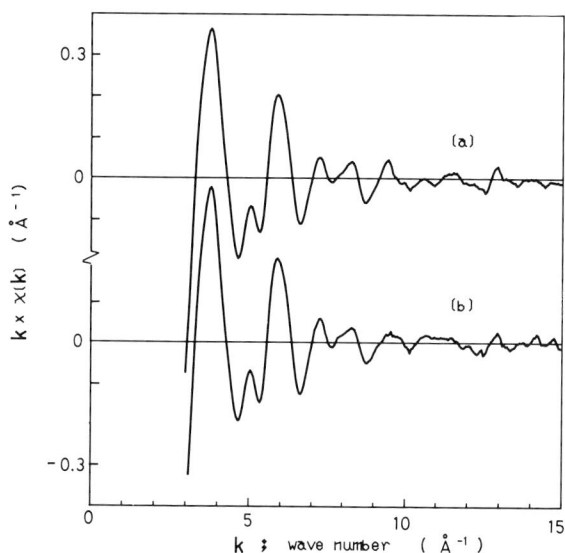

Figure 3. Extended X-ray absorption fine structure spectra ($\chi(k)$) for (a) Co-treated loughlinite and (b) Co-treated sepiolite obtained by using synchroton radiation.

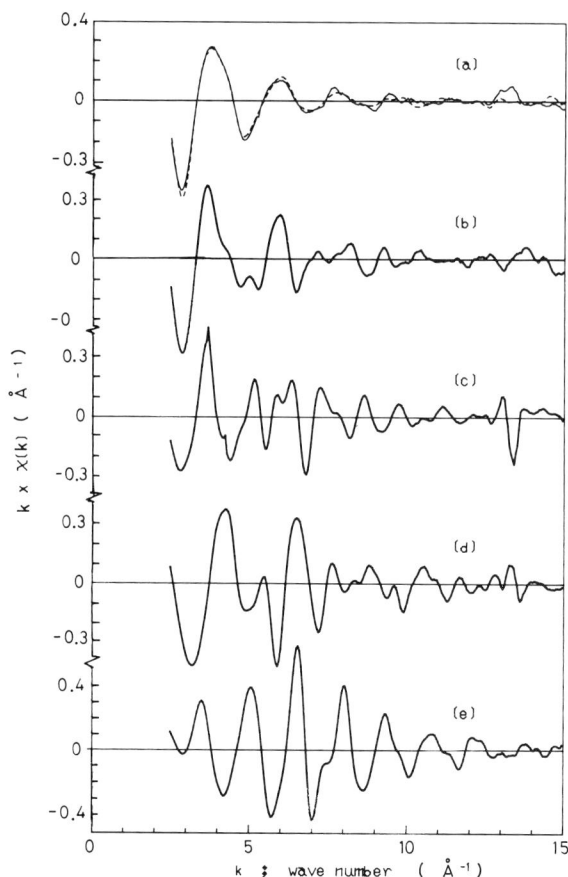

Figure 4. Extended X-ray absorption fine structure (EXAFS) spectra ($\chi(k)$) for (a) $Co(NO_3)_2 \cdot 6 H_2O$, (b) $Co(OH)_2$, (c) CoO, (d) Co_3O_4, and (e) Co metal. Solid lines are results obtained by using laboratory EXAFS equipment and broken line are those obtained by using synchroton radiation.

Amplitudes of Fourier transforms of $k^3\chi(k)$ spectra ($|\phi_3(r)|$) are shown in Figures 5 and 6. Although the $|\phi_3(r)|$ values for Co-treated samples also resemble each other, the detailed features, such as the shapes of first and second peaks, are different from each other. Although the peak positions in $|\phi_3(r)|$ values for Co-treated samples are close to those for $Co(NO_3)_2 \cdot 6 H_2O$ and $Co(OH)_2$, the precise positions and amplitudes are different.

The results for $Co(NO_3)_2 \cdot 6 H_2O$ obtained by using the SR and the laboratory equipment are compared in Figures 4a and 6a. Although the difference in the profiles of $|\phi_3(r)|$ that is supposedly due to the difference of Fourier windows in k-space was observed, $\chi(k)$ spectra and peak position and peak area of $|\phi_3(r)|$ obtained by using laboratory EXAFS equipment agree well with those obtained by SR.

XANES spectra obtained using the SR and laboratory equipment are compared in Figure 7. Because the energy resolution of the laboratory equipment was insufficient for XANES measurements, the detailed pro-

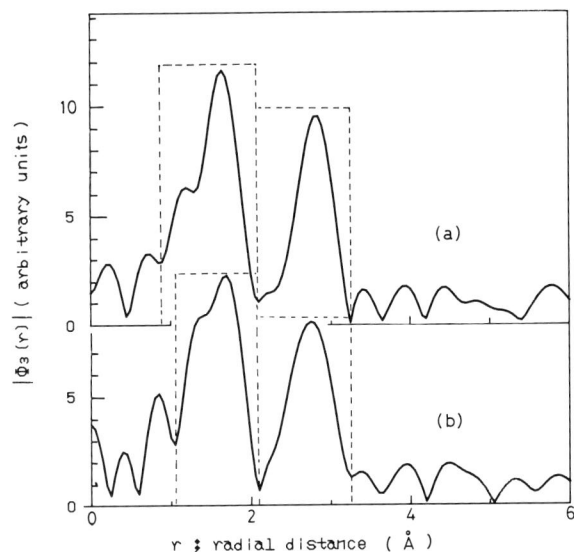

Figure 5. Amplitudes of Fourier transforms of $k^3 \cdot \chi(k)$ spectra for (a) Co-treated synthetic loughlinite and (b) Co-treated sepiolite. Broken lines show the Fourier window used in back-Fourier-transformation.

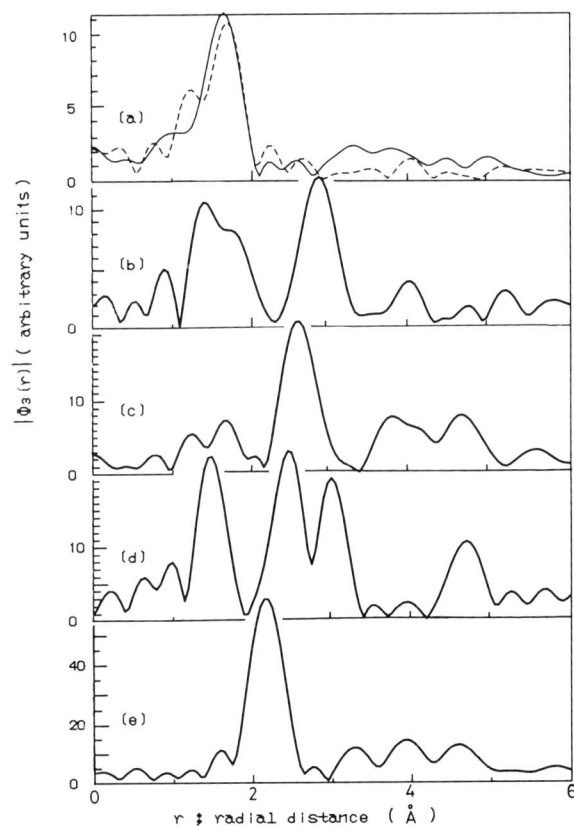

Figure 6. Amplitudes of Fourier transforms of $k^3 \cdot \chi(k)$ spectra for (a) $Co(NO_3)_2 \cdot 6 H_2O$, (b) $Co(OH)_2$, (c) CoO, (d) Co_3O_4, and (e) Co metal, where solid lines are results obtained by using laboratory extended X-ray absorption fine structure equipment and broken line is those obtained by using synchroton radiation.

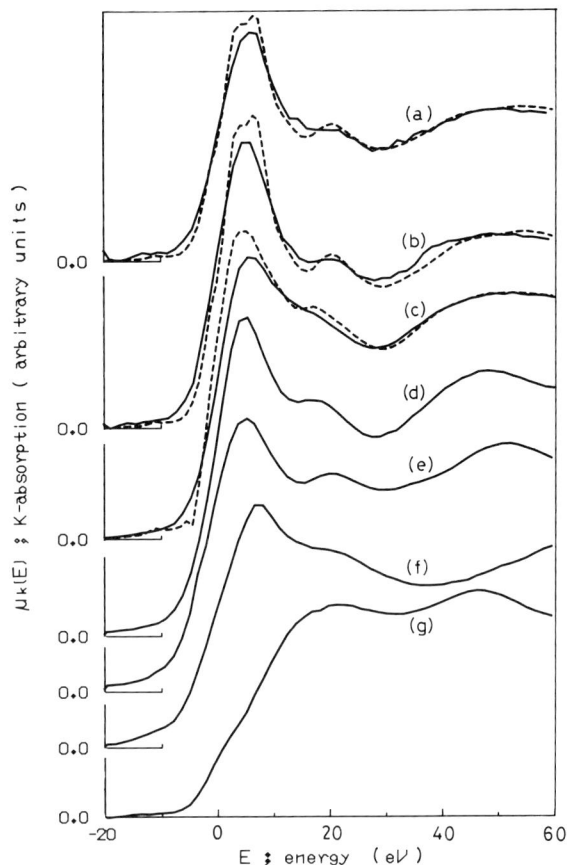

Figure 7. X-ray absorption near-edge structure spectra for (a) Co-treated synthetic loughlinite, (b) Co-treated sepiolite, (c) $Co(NO_3)_2 \cdot 6 H_2O$, (d) $Co(OH)_2$, (e) CoO, (f) Co_3O_4, and (g) Co metal. Solid lines are results using laboratory extended X-ray absorption fine structure equipment and broken lines are those by synchroton radiation.

files of XANES were not observed in the spectra obtained using the laboratory equipment. Outlines of XANES, however, such as peak positions and peak shapes, obtained by the laboratory equipment agree well with the results obtained by the SR. The XANES spectra for Co-treated samples resemble those for CoO and $Co(OH)_2$.

DISCUSSION

Compositional and structural change due to $Co(NO_3)_2$ treatment

The chemical data in Table 2 for sepiolite and loughlinite contain no corrections for dolomite, because the data for the untreated and treated minerals were not entirely consistent (note the increase in CaO of the sepiolite with treatment), nor were they entirely consistent with XRD data (note the small dolomite peak for loughlinite in Figure 1c, in contrast to its relatively large CaO content). However, if an ideal composition of dolomite, $CaMg(CO_3)_2$, is assumed and all CaO is

Table 3. Neighboring distance (R_i), Debye-Waller factor (σ_i), and coordination number (C_i) around Co atoms in Co-treated sepiolite and synthetic loughlinite.

Shell	Central atom	Scatter atom	R_i (Å)	σ_i (Å)	D_iC_i	C_i	ΔE_0[1] (eV)
Co-treated sepiolite							
1	Co	O	2.15	0.060	2.00	5.7	−5.0
1	Co	O	1.98	0.060	0.42	1.2	−5.0
2	Co	Co	3.17	0.070	1.80	1.8	−5.5
2	Co	Mg or Si	2.82	0.070	0.32		−5.5
Co-treated synthetic loughlinite							
1	Co	O	2.15	0.045	1.83	5.2	−5.0
1	Co	O	1.97	0.045	0.42	1.1	−5.0
2	Co	Co	3.16	0.045	0.52	0.9	−6.0
2	Co	Mg or Si	3.45	0.045	0.74		−6.0
Brucite; calculated using data of Wycoff (1982)							
1	Mg	O	2.17			6.0	
2	Mg	Mg	3.15			6.0	

[1] ΔE_0 = shift of threshold energy obtained by least square curve fitting and calculated values in brucite. D_i is scaling factor.

assigned to dolomite, then dolomite corrections decrease the raio of Mg/(Si + Al) for sepiolite from 0.625 to 0.615 and (Mg + Na/2)/(Si + Al) for the synthetic loughlinite from 0.703 to 0.685. These ratios are all close to the ideal ratio of 0.667 for sepiolite, Mg_8-$Si_{12}O_{30}(OH)_4(OH_2)_4$ (Caillère and Hénin, 1961). For the Co-treated samples, adding the atomic raio of Co to that of Mg and Na gives uncorrected and corrected

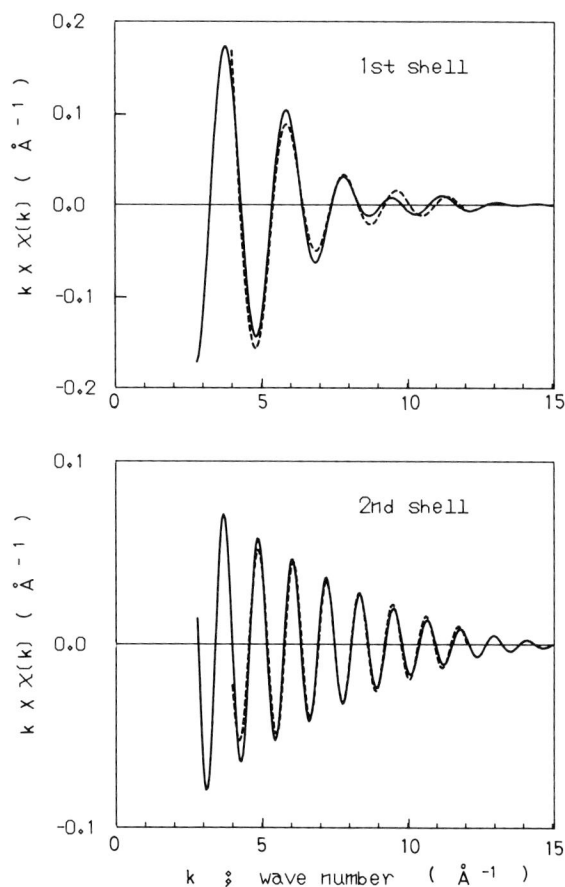

Figure 8. Fourier-filtered $\chi(k)$ spectra for first shell and second shell of Co-treated sepiolite (solid lines) and those calculated by using the best fit parameters (broken lines).

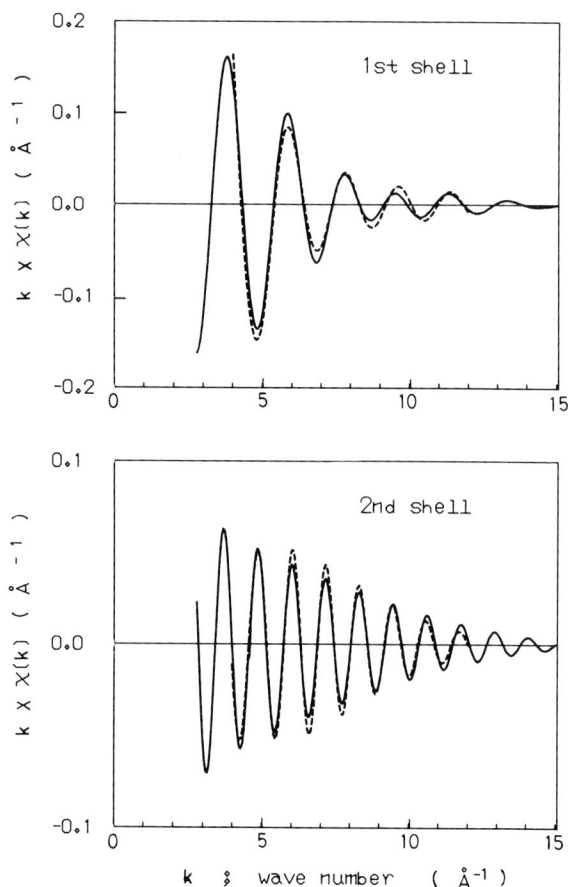

Figure 9. Fourier-filtered $\chi(k)$ spectra for first shell and second shell of Co-treated synthetic loughlinite (solid lines) and those calculated by using the best fit parameters (broken lines).

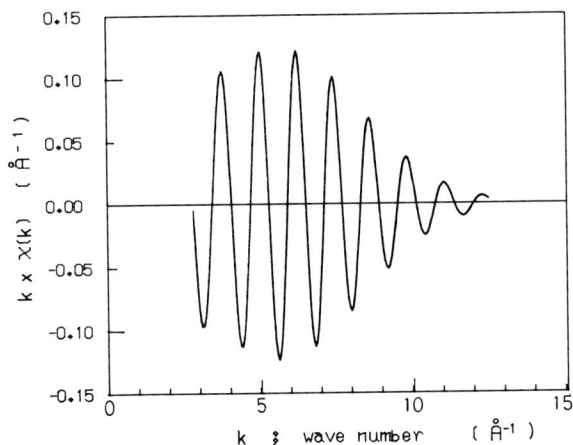

Figure 10. Fourier-filtered χ(k) spectrum for second shell of CoO.

values of 0.574 and 0.520 for sepiolite and 0.674 and 0.577 for loughlinite—lower than the ratios before Co treatment, but still in reasonably good agreement with the ideal mineral. These results suggested that about 15% of octahedral cations in the synthetic loughlinite and about 4% of those in sepiolite were substituted by Co^{2+} during the $Co(NO_3)_2$-treatment. The chemical analysis of the filtrates shown in Figure 2 and the lattice parameters obtained from XRD shown in Figure 1 and Table 1 also suggest Co substitution.

Local structure around the Co ion

The XANES and EXAFS spectra and their Fourier transforms for Co-treated sepiolite and synthetic loughlinite all resemble each other, which suggests that the local structure around the Co ions in the samples are similar. The peak positions of Fourier transforms for $Co(NO_3)_2$-treated samples resemble those for $Co(NO_3)_2 \cdot 6 H_2O$ and $Co(OH)_2$. This similarity suggests that the distances from the Co ion to the first and the second neighbor ions are about 2.1 and 3.2 Å, respectively, as calculated from the structure models for $Co(NO_3)_2 \cdot 6 H_2O$ (Prelesnik *et al.,* 1973) and $Co(OH)_2$ (Wycoff, 1982). As Bianconi (1983) suggested, EXAFS gives information about local structures in terms of atomic radial distribution, whereas XANES gives information on stereochemical geometries. That the XANES spectra for Co-treated samples resemble those for CoO and $Co(OH)_2$ is an indication of the octahedral local geometry round the Co ion in the samples.

To refine these results, $|\phi_3(r)|$ values for the Co-treated sample were subjected to inverse-Fourier-transformation into k-space, using the windows shown in Figure 5. The best-fit parameters obtained by the least-squares curve fitting are listed in Table 3. The EXAFS spectra calculated by using the best-fit parameters shown in Table 3 are compared with the inverse-Fourier-

transformed spectra for the first and second shells of Co-treated sepiolite and synthetic loughlinite in Figures 8 and 9, respectively. Coordination numbers of the first shell and the Co–Co distance in the second shell were calculated using the scaling factor $D_{Co-O} = 0.35$ and $D_{Co-Co} = 0.58$, which were estimated from the EXAFS and structure models for $Co(NO_3)_2 \cdot 6 H_2O$ and CoO, respectively.

The distinction between Mg and Si is difficult in X-ray absorption studies; hence, parameters were obtained only for Co–(Mg or Si). In the curve-fitting calculation, backscattering amplitude and backscattering phase shift for Mg were used. The coordination number for Co–(Mg or Si) could not be calculated because of the lack of suitable references to estimate the scaling factor. The neighbor distances and coordination number around Mg atoms in brucite are also shown in Table 3. The parameters for Co atoms obtained by EXAFS are similar to those for brucite. This fact also indicates the octahedral coordination of Co atoms. Although Co atoms in CoO or $Co(OH)_2$ are also in an octahedron of oxygen atoms, the second neighbor atoms of Co in these compounds are only Co atoms. Because the backscattering amplitude for Co has a maximum value at about $k = 6$ Å^{-1} (Teo and Lee, 1979), EXAFS spectrum should have a maximum as shown in Figure 10. The back-Fourier-transformed spectra of second neighbor shell for Co-treated samples are different from the spectrum shown in Figure 10, which indicates that second neighbor atoms of Co in these samples are not only Co but also Mg or Si. These facts suggest substitution of Co in octahedral sites in the sepiolite and synthetic loughlinite. The results of curve fitting for the first shell suggest two different but close neighboring distances in the first shell for both Co-treated sepiolite and synthetic loughlinite. A more precise analysis, however, will be necessary to confirm this result.

ACKNOWLEDGMENTS

The authors are grateful to Dr. Nomura of the National Laboratory for High Energy Physics, and Mrs. Suda, Mr. Mizuta, Mr. Mori, and Mr. Inagaki of Toyota R & D Center for their assistance in X-ray absorption measurement, sample preparation, and chemical analyses. Discussions with Drs. Kamigaito and Morimoto of Toyota R & D Center were also very helpful.

REFERENCES

Bianconi, A. (1983) XANES spectroscopy for local structures in complex systems: in *Proc. Int. Conf. EXAFS and Near Edge Structure, Frascati, Italy, 1982*, A. Bianconi, L. Incoccia, and S. Stipcich, eds., Springer-Verlag, Berlin, 118–129.

Caillère, S. and Hénin, S. (1961) Sepiolite: in *The X-ray Identification and Crystal Structures of Clay Minerals*, 2nd ed., G. Brown, ed., Mineralogical Society, London, 325–342.

Fahey, J. J., Ross, M., and Axelrod, J. M. (1960) Loughlinite, a new hydrous sodium magnesium silicate: *Amer. Miner.* **45,** 270–281.

Heald, S. M. (1983) Thickness effects in X-ray absorption measurements: in *Proc. Int. Conf. EXAFS and Near Edge Structure, Frascati, Italy, 1982,* A. Bianconi, L. Incoccia, and S. Stipcich, eds., Springer-Verlag, Berlin, 98–99.

Imai, N., Otsuka, R., and Nakamura, T. (1969) Artificial transformation of natural sepiolite into loughlinite: *Bull. Sci. Engineer Res. Lab. Waseda Univ.* **46,** 49–59.

Knapp, G. S. and Georgopoules, P. (1980) General consideration for a laboratory EXAFS facility: in *Laboratory EXAFS facilities—1980,* Amer. Inst. Physics Conf. Proc. **64,** E. A. Stern, ed., 2–20.

Lytle, F. W., Sayers, D. E., and Stern, E. A. (1975) Extended X-ray absorption fine structure technique. II. Experimental practice and selected results: *Phys. Rev. B* **11,** 4825–4835.

Motta, N., DeCrescenzi, M., and Balzarotti, A. (1983) Multiple scattering effects in the EXAFS of Fe and TiFe: in *Proc. Int. Conf. EXAFS and Near Edge Structure, Frascati, Italy, 1982,* A. Bianconi, L. Incoccia, and S. Stipcich, eds., Springer-Verlag, Berlin, 103–106.

Oyanagi, H., Matsushita, T., Ito, M., and Kuroda, H. (1984) An EXAFS spectrometer on beam line 10B at the Photon Factor, National Laboratories for High Energy Physics: *KEK Reports* **83-30,** Tsukuba, Japan, 1–27.

Pettifer, R. F. and Cox, A. D. (1983) The reliability of *ab initio* calculations in extracting structural information from EXAFS: in *Proc. Int. Conf. EXAFS and Near Edge Structure, Frascati, Italy, 1982,* A. Bianconi, L. Incoccia, and S. Stipcich, eds., Springer-Verlag, Berlin, 66–72.

Prelesnik, P. V., Cabela, F., Ribar, B., and Krstanovic, I. (1973) Hexaquacobalt(II) nitrate, $Co(OH_2)_6(NO_3)_2$: *Cryst. Struct. Commun.,* **12,** 581–583.

Stern, E. A. (1974) Theory of the extended X-ray absorption fine structure: *Phys. Rev. B* **10,** 3027–3037.

Teo, B. and Lee, P. A. (1979) *Ab initio* calculation of amplitude and phase functions for extructed X-ray absorption fine structure spectroscopy: *J. Amer. Chem. Soc.* **101,** 2815–2822.

Toji, K., Udagawa, Y., Kawasaki, T., and Nakamura, K. (1983) Laboratory EXAFS spectrometer with a bent crystal, a solid-state detector and a fast detection system: *Rev. Sci. Instrum.* **54,** 1482–1487.

Wycoff, R. W. G. (1982) *Crystal Structures. Vol. 1:* 2nd ed., Krieger Publishing, Malobar, 266 pp.

Proceedings of the International Clay Conference, Denver, 1985, L. G. Schultz, H. van Olphen, and F. A. Mumpton, eds.,
The Clay Minerals Society, Bloomington, Indiana, 17–23 (1987).

INFRARED STUDY OF STRUCTURAL OH IN KAOLINITE, DICKITE, AND NACRITE AT 300 TO 5 K

R. Prost, A. Damême, E. Huard, and J. Driard

Station de Science du Sol, Institut National de la Recherche Agronomique
Route de Saint Cyr, 78000 Versailles, France

Abstract—To obtain a better understanding of the structure of kaolin-group minerals, infrared spectra of structural OH in 1:1 dioctahedral clays were recorded as a function of temperature between 300 and 5 K. The low- and high-frequency components of the absorption bands of the structural OH groups shifted in opposite directions as the temperature changed and gave better resolution of the spectra. This behavior with temperature was used as a basis for assigning the stretching and bending modes of structural OH in the studied clays. Dichroic components, which are the result of coupling phenomena, were found at 3692, 3650, and 3644 cm^{-1} for kaolinite, dickite, and nacrite, respectively, and were attributed to vibrations of inner-surface OH groups which are, even in kaolinite, all nearly perpendicular to the sheet. The frequency and behavior with temperature of the corresponding decoupled OD groups indicate that inner-surface hydroxyls are involved in stronger hydrogen bonds in dickite than in kaolinite.

Key Words—Dickite, Hydroxyl, Infrared spectroscopy, Kaolinite, Nacrite.

INTRODUCTION

It is now well established that the principal physicochemical properties of kaolin materials are related to their crystallinity. Thus, in addition to crystal structure determinations which provide data on the arrangements of atoms in the structure, it is equally and perhaps more informative to complement this approach by an infrared (IR) study of the structural OH groups, which makes it possible to probe the inner structure of the clay. Publications by Serratosa *et al.* (1963), Farmer and Russell (1964), Ledoux and White (1965), Rouxhet *et al.* (1977), and Cruz-Cumplido *et al.* (1982) have established that the low-frequency stretching band in IR patterns of kaolin-group minerals is related to internal OH groups, or inner-hydroxyls according to the nomenclature proposed by Ledoux and White (1965). This band has no dichroic character and is not the result of a coupling phenomenon between structural OHs. The high-frequency stretching bands are generally attributed to external OH groups which correspond, according to Ledoux and White (1965), to outer- and inner-surface hydroxyls. Some of these bands are dichroic, and progressive deuteration experiments show that some of them are the result of coupling (Rouxhet *et al.*, 1977). Prost (1984) showed that a decrease of temperature from 300 to 150 K led to a better resolution of the spectra and to the appearance of more components, which he used to study the structural OH in kaolinite, dickite, and nacrite. The present study reports the change of the stretching and the bending bands between 300 and 5 K. The higher resolution IR spectra of the structural OH groups is then used to analyze similarities and differences between these minerals.

MATERIALS AND METHODS

Three kaolin-group samples were used: kaolinite from Keokuk, Iowa, from W. D. Keller; dickite from Wisconsin, from C. B. Roth; and nacrite from Germany, from H. Paquet. All samples were well crystallized and gave IR stretching and bending bands typical for their structural OH groups.

IR spectra recorded on a Perkin Elmer 580 spectrometer equipped with a data station were obtained using a helium cryostat with KRS5 windows built by the Oxford Instrument Company. This arrangement made it possible to control the temperature of the sample between 300 and 5 K to a precision of ±1 K. The sample holder could be tilted around an axis perpendicular to the electric field of the polarized light of the IR beam. Samples were studied in KBr discs or as films deposited on CaF$_2$ windows. Partially deuterated samples, used to study the coupling process of the structural OH groups, were obtained by putting clay in sealed glass tubes with appropriate mixtures of D$_2$O and H$_2$O and heating the mixtures at 310°C for two days. Under these conditions, there is no selectivity in the deuteration process (White *et al.*, 1970).

RESULTS

Frequency shift with temperature

Figures 1a–1c show the stretching and the bending IR spectra of the structural OH groups of kaolinite, dickite, and nacrite as a function of temperature.

For the Keokuk kaolinite (Figure 1a), the decrease of temperature from 300 to 5 K led to a better resolution of the stretching and the bending modes. The low-frequency component shifted from 3615 to 3612

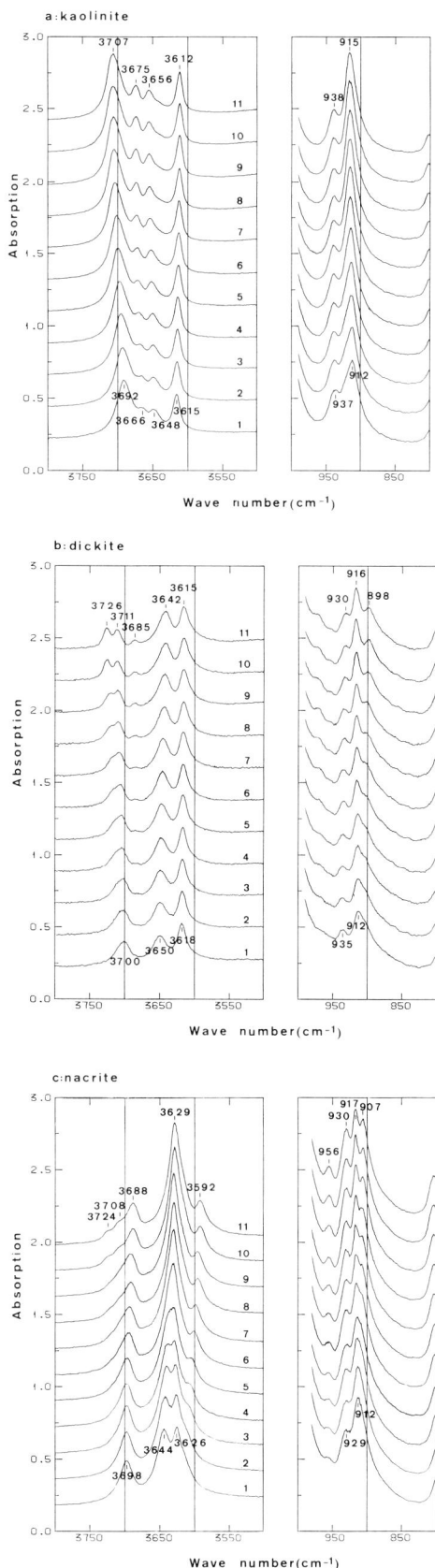

cm^{-1}, whereas the high frequency components shifted from 3648, 3666, and 3692 cm^{-1} to 3656, 3675, and 3707 cm^{-1}, respectively. The shift was larger (15 cm^{-1}) for the highest frequency component. In the bending range only a small shift from 912 to 915 cm^{-1} was observed. The component at 937–938 cm^{-1} remained essentially at the same frequency; however, the intensity of the band at 912–915 cm^{-1} was greater at the lower temperature.

For the Wisconsin dickite (Figure 1b), three components at 3618, 3650, and 3700 cm^{-1} were noted at 300 K, whereas five were visible in the spectrum obtained at 5 K at 3615, 3642, 3685, 3711, and 3726 cm^{-1}. The low-frequency component shifted slightly from 3618 to 3615 cm^{-1}. The 3650-cm^{-1} component shifted only a few cm^{-1} (8 cm^{-1}) to lower frequencies. The 3685-cm^{-1} component remained at the same frequency, but new components appeared at 3711 and 3726 cm^{-1} at the lower temperature, corresponding to a shift of about 11 and 26 cm^{-1}, respectively. In the bending range the 935-cm^{-1} component shifted about 5 cm^{-1} to 930 cm^{-1}, whereas the 912-cm^{-1} band split into two components at 916 and 898 cm^{-1}.

For nacrite (Figure 1c), the decrease in temperature from 300 to 5 K led to the resolution of new components. At the lower temperature, a new band appeared on the low-frequency side of the 3626-cm^{-1} band, which shifted to 3592 cm^{-1}. Concomitantly the doublet at 3644–3626 cm^{-1} merged into a single strong component at 3629 cm^{-1}. The high-frequency band at 3698 cm^{-1} split into three components at 3688, 3708, and 3724 cm^{-1}. In the bending range, the 929–930-cm^{-1} band did not shift significantly with temperature, but the 912-cm^{-1} band split into two components at 917 and 907 cm^{-1}.

Dichroic character of the bands

The dichroic character of the bands (the increase of their intensity as a function of the incidence angle of the IR beam on the sample) was studied by tilting the clay film deposited on a CaF$_2$ window in the polarized beam of the spectrometer. The tilting angles with respect to the normal of the plane of the deposit were successively equal to $-40°$, $-30°$, $-20°$, $-10°$, $0°$, $+10°$, $+20°$, $+30°$, and $+40°$. The experiments were performed at different temperatures. Figures 2a–2c show the results obtained for kaolinite, dickite, and nacrite at 5 K. The small difference observed between band frequencies of spectra obtained in KBr and with those of the film deposited on CaF$_2$ is related to the change

←

Figure 1.   Infrared spectra for (a) kaolinite, (b) dickite, and (c) nacrite in KBr vs. temperature. Numbers from 1 to 11 correspond to decreasing temperatures from 300 to 5 K by steps of 30 K.

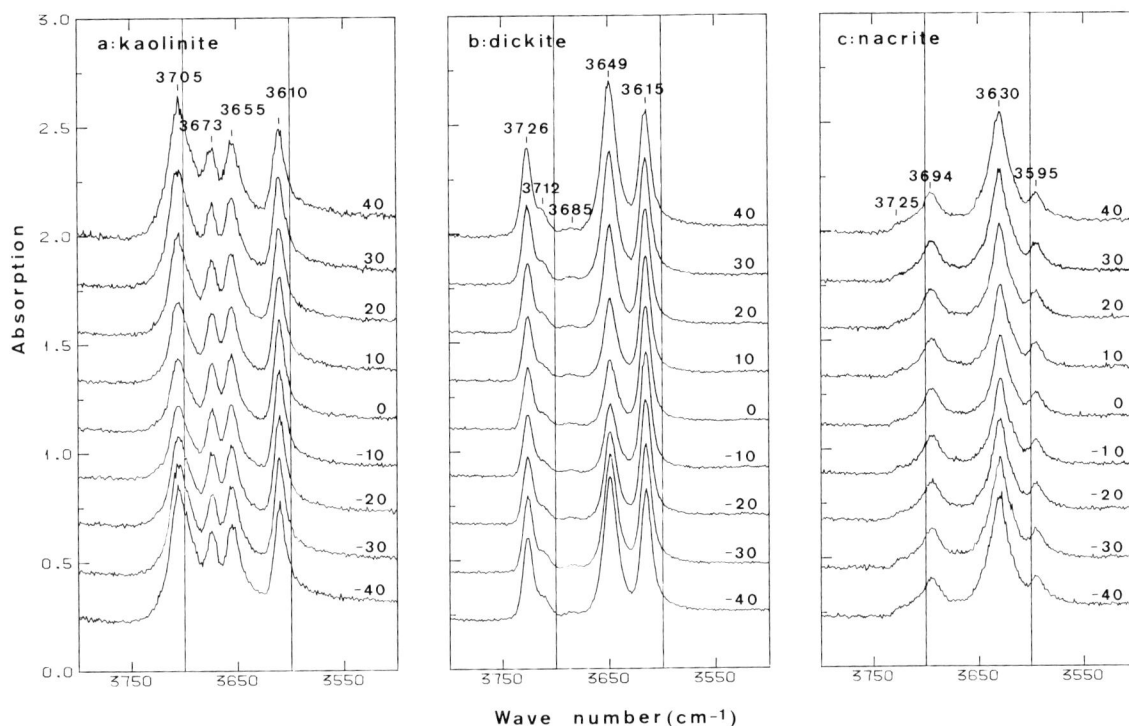

Figure 2. Infrared spectra at 5 K for (a) kaolinite, (b) dickite, and (c) nacrite deposited on CaF_2 window. Numbers indicate different incidence angles of the beam.

of the refractive index of the matrix in which the sample is mixed (Prost, 1973). To know the dichroic character of each band more precisely, spectra were deconvoluted into Lorentzian components. Fitting was better for spectra obtained from KBr discs than from films. The Christiansen effect probably altered the profile of absorption bands when the spectra were from films more than those obtained using KBr (Prost, 1973).

Curves plotted on Figures 3a–3c give deconvolution examples for the spectra of kaolinite, dickite, and nacrite obtained from films. The deconvoluted spectra of kaolinite and dickite correspond to spectra obtained at 5 K and that of nacrite, to spectra obtained at 210 K. The temperature of 210 K was chosen for nacrite because the resolution was much better for the low-frequency components of the spectrum. The data clearly show that the 3705- and 3649-cm^{-1} components of kaolinite and dickite cooled to 5 K were dichroic and that the 3643-cm^{-1} component of nacrite cooled to 210 K was dichroic. Therefore, for each sample studied only one component has a strong dichroic character, showing that the corresponding normal vibrations are oriented perpendicularly to the sheet.

The coupling effect

Deuteration experiments were performed to identify components that arose as a result of coupling. Indeed, when the rate of deuteration was low enough to have

a slight probability of finding two structural OD groups near each other, the OD groups vibrated as isolated functional groups. In this situation, each band visible on the spectrum corresponded to a particular kind of structural OH group in the mineral. When the deuteration rate increased, coupling between structural OD groups appeared, and the single band was replaced by several bands at different frequencies (Prost, 1975).

The rate of deuteration was estimated from the ratio $2I_{OD}(I_{OH} + 2I_{OD})^{-1}$, where I_{OH} and I_{OD} were the integrated intensity of OH and OD groups (Rouxhet et al., 1977). Under the experimental conditions, the rate of deuteration was 0.045 for kaolinite and dickite in contact with a 99.8% D_2O solution.

The low deuteration rate was probably due to the fact that only a small fraction of the volume of the mineral was accessible to deuterium atoms. It was assumed that at the experimental temperature and pressure the same fraction of the mineral was always accessible. Thus, the progressive deuteration took place in this fraction of clay, which is assumed to be representative of the material. The low rate of deuteration obtained with this method was preferred to the higher rates reached by intercalation with dimethyl sulfoxide or hydrazine, because the intercalation process can induce faults in the stacking of the sheets (Barrios et al., 1977).

Results obtained for kaolinite and dickite are shown

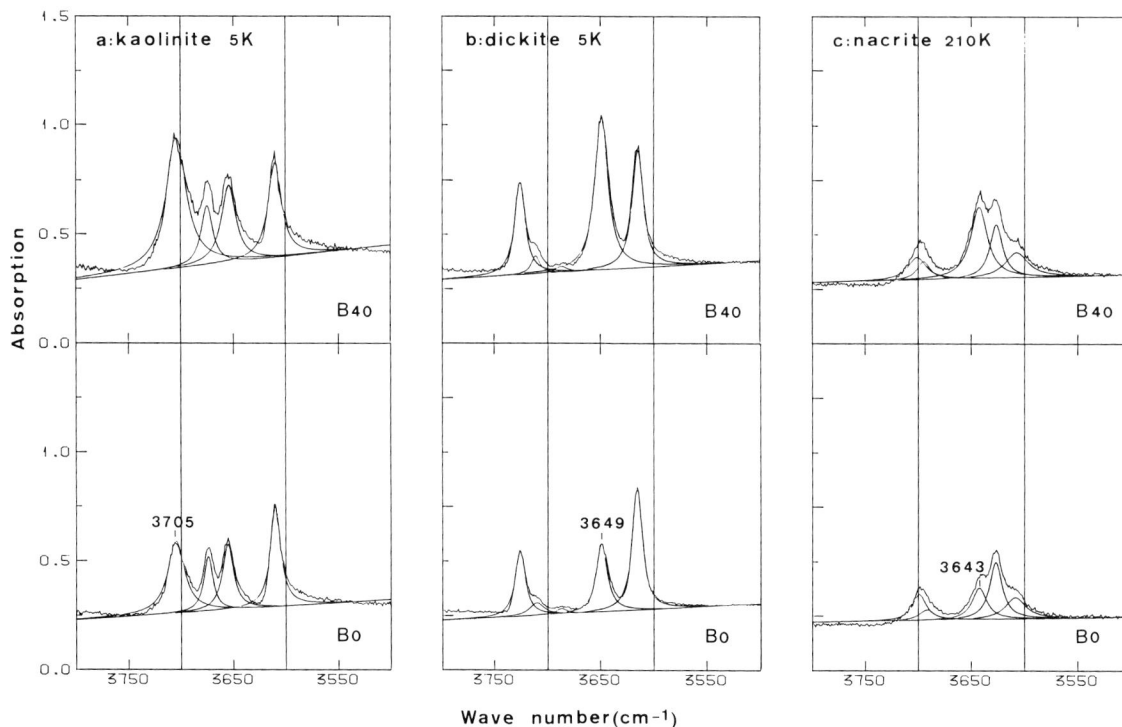

Figure 3. Deconvoluted infrared spectra of (a) kaolinite and (b) dickite films at 5 K and (c) nacrite film at 210 K. (B_0) and (B_{40}) spectra correspond to an incidence angle of the beam on the clay film of 0° and 40°, respectively.

in Figures 4a and 4b. The progressive deuteration experiment was not performed with nacrite because of a lack of material. Spectra obtained in KBr for kaolinite (Figure 4a) showed the presence of two components which remained at 2665 and 2700 cm⁻¹ as the rate of deuteration increased. The 2714-cm⁻¹ band shifted to higher frequencies, and three components at 2712, 2724, and 2735 cm⁻¹ were noted for the highest degree of deuteration. The isotopic ratio determined by dividing the OH and OD frequencies of the low frequency component was 3612:2665 = 1.355. Therefore, the corresponding frequencies in the OH range determined for material having this isotopic ratio were 3659, 3675, 3691, and 3706 cm⁻¹. Except for the 3691-cm⁻¹ component, which corresponds to the 2724-cm⁻¹ band, all these bands were visible in the spectrum of kaolinite at 5 K in KBr (Figure 1a). Spectra obtained on films (Figure 4a) show that these high-frequency components had a dichroic character. For dickite (Figure 4b), the progressive deuteration experiment showed that the 2668, 2720, 2735, and 2750 cm⁻¹ components kept the same frequency as the rate of deuteration increased. Calculation of the corresponding OH-group frequencies gave the following values: 3615, 3686, 3706, and 3726 cm⁻¹, which are close to those seen in Figure 1b (3615, 3685, 3711, and 3726 cm⁻¹). In addition, the increase and the shift of the 2688-cm⁻¹ (3642-cm⁻¹) component was obvious. The shift was in the same

direction as that for kaolinite. Spectra of Figure 4b obtained for films indicate that this component had a strong dichroic character. Thus, for kaolinite and dickite the component which had a strong dichroic character was also strongly affected by progressive deuteration, showing that the components were due to a coupling phenomenon.

DISCUSSION

The lowering of temperature affected both the intensity and the frequency of absorption bands. The intensity change was apparently due to small displacements of the atoms inducing modifications in the interatomic distances and the angles between bonds. This produced a change of the dipole moment derivative $d\mu/dQ$, where μ is the dipole moment and Q is the normal coordinate. As reported by Freund (1974), the integrated intensity generally decreased as the temperature increased, a phenomenon more pronounced in the long-wavelength range than in the short-wavelength range. Considering the spectra obtained in this work, a greater increase of the integrated intensity was observed as the temperature decreased for the bending than for the stretching modes of kaolinite.

The frequency shifts presented here at liquid helium temperature (~5 K) are an extension of the results reported by Freund (1974) and Prost (1984). No discontinuity appeared which could be interpreted as a

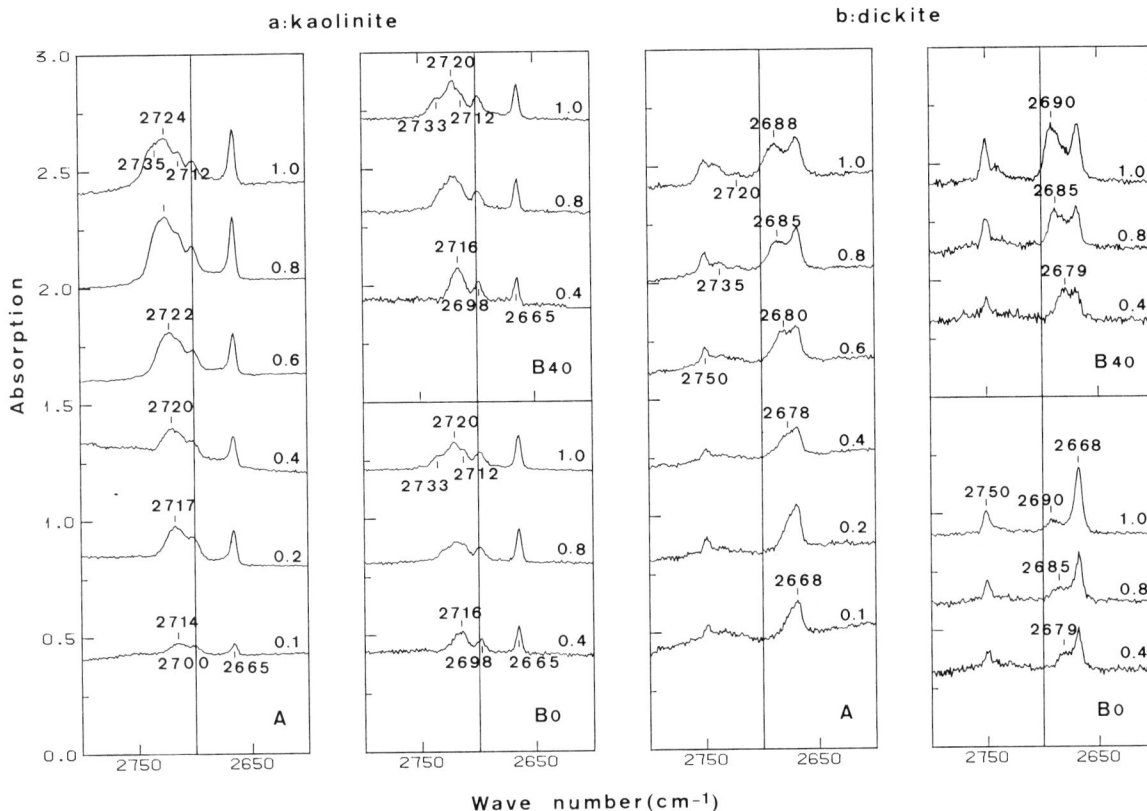

Figure 4. Infrared spectra at 5 K of (a) kaolinite and (b) dickite for different rates of deuteration. Numbers indicate the ratio of $D_2O/(H_2O + D_2O)$ in the solution used for the deuteration experiment. A = spectra of clay in KBr, B_0 and B_{40} = spectra obtained at an incidence angle of the beam on the clay film of 0° and 40°, respectively.

phase transition. Therefore, spectra obtained at the temperature of the spectrometer can be considered as an instantaneous measurement in the temperature scale. Data obtained at the temperature which gives the best resolution can be used to study the structural OH groups of these three minerals.

Internal OH groups of the minerals studied are less affected by temperature than external OH groups, possibly because changes in the interatomic distances within the structure are less important than changes between adjacent sheets. Because the frequency shifts observed with temperature for the stretching bands of kaolinite, dickite, and nacrite (at 3615, 3618, and 3626 cm^{-1}, respectively) are small, these bands may be associated with the 937-, 935-, and 929-cm^{-1} bending components which shift only a few cm^{-1} (Figure 1). Under these conditions, the bands are probably due to internal OH groups of the 1:1 dioctahedral minerals studied. This analysis could be continued using the general rule of opposite frequency shifts for the stretching and the bending modes. Thus, the shift from 3700 to 3726 cm^{-1} of the stretching mode of dickite (Figure 1b), observed as the temperature decreased, suggests that the corresponding bending mode was near 912–

898 cm^{-1}. The shift from 3650 to 3642 cm^{-1} indicates that the corresponding band in the bending range was the 912–916-cm^{-1} component. For nacrite (Figure 1c), the 3644–3629-cm^{-1} band was associated with the 912–917-cm^{-1} component and the high-frequency bands (3688, 3708, and 3724 cm^{-1}) with the 912–907-cm^{-1} component. Similarly, the high-frequency stretching mode of kaolinite (3692–3707 cm^{-1}) (Figure 1a) was associated with the low-frequency bending mode (912–915 cm^{-1}).

The following assignments are therefore suggested. Internal OH groups of kaolinite, dickite, and nacrite give 3615-, 3618-, and 3626-cm^{-1} bands for the stretching modes and 937-, 935-, and 929-cm^{-1} bands for the bending modes, respectively. External OH groups give two sets of absorption bands in the stretching range. One has a strong dichroic character, and corresponds for kaolinite, dickite, and nacrite to the 3692-, 3650-, and 3644-cm^{-1} components, respectively, with which the 912–915-, 912–916-, and 912–917-cm^{-1} bands are associated in the bending range. The second set of absorption bands corresponding to external OH groups are strong and well resolved for dickite and nacrite and give the 3700- and 3698-cm^{-1} bands in the stretching range and the 912–898- and 912–907-cm^{-1} components in the bending range, respectively.

The dichroic bands at 3692–3705, 3650–3649, and 3644–3630 cm^{-1} of the structural OH-stretching modes of kaolinite, dickite, and nacrite (Figure 2) probably correspond, according to the nomenclature of Ledoux and White (1965), to inner-surface OH groups. For kaolinite and dickite these bands are the result of a coupling phenomenon between inner-surface OH groups (Figures 4a and 4b). Rouxhet et al. (1977) show that the 3695- and 3670-cm^{-1} components of kaolinite correspond to the coupling of two inner-surface OH groups which are nearly perpendicular to the sheet. The symmetric mode gives the 3695-cm^{-1} component which is dichroic, and the asymmetric modes give the 3670-cm^{-1} component which is not dichroic.

Low-temperature spectra of partially deuterated samples obtained in KBr suggest that the 2724 (3691)-cm^{-1} (Figure 4a) component could be a decoupled vibration of inner-surface OH groups. The 3691-cm^{-1} band is not visible in the undeuterated spectrum of kaolinite (Figure 1a). The coupled vibrations could correspond to the 2712 (3675)- and 2735 (3706)-cm^{-1} bands which are symmetric with respect to the 2724 (3691)-cm^{-1} component. The dichroic character of the 2724 (2720)- and the 2735-cm^{-1} components (Figure 4a, B$_{40}$) confirms the analysis by Rouxhet et al. (1977), who showed that the 2700 (3659)-cm^{-1} band is not involved in this coupling phenomenon. This analysis is somewhat questionable if the lower rates of deuteration are considered. Indeed, a continuous shift from 2724 to 2714 cm^{-1} was observed as the rate of deuteration decreased. If the 2724-cm^{-1} component is considered as the decoupled vibration of inner-surface OH groups, an absorption band at 2724 cm^{-1} should exist for the lowest rate of deuteration; however, this band was not observed (Figure 4a, A).

As noted by Farmer and Russell (1964), however, the coupling phenomenon could be not only between two inner-surface OH groups, as Rouxhet et al. (1977) suggested, but between three inner-surface OH groups nearly perpendicular to the sheet. Here, as the rate of deuterium atoms increases, coupling would occur in two steps: first, between two inner-surface OH groups, then between three. In other respects, depending on the structure, the coupling phenomenon could involve more than three inner-surface OH groups at high deuterium content. This possibility, that more than two inner-surface OH groups are involved in the coupling phenomenon, is confirmed by recent data that show inner-surface OH groups of kaolinite to be perpendicular to the sheets (Suitch and Young, 1983; Adams, 1983). The same reasoning is also valid for dickite where the 2688-cm^{-1} component (Figure 4b) shifted continuously to lower frequencies as the deuteration rate decreased. It is now well established that inner-surface OH groups of dickite are perpendicular to the

sheet (Adams and Hewat, 1981; Bookin et al., 1982; Sen Gupta et al., 1984). Thus, the same behavior observed with kaolinite and dickite for increasing rates of deuteration suggests that the inner-surface OH groups are perpendicular to the sheets in both samples.

The frequencies of the 3692–3707-, 3650–3642-, and 3644–3629-cm^{-1} bands (Figure 1) (which have a dichroic character and correspond to inner-surface OH groups) are different for each sample, and, as the temperature decreased, they shifted to higher frequencies for kaolinite and to lower frequencies for dickite and nacrite. The 2714-cm^{-1} decoupled band of kaolinite (Figure 4a), which corresponds to isolated inner-surface OD groups, shifted to 2712 cm^{-1} when the spectra were recorded at 300 K. In addition, the decoupled OD band of dickite, which was not resolved at 5 K (Figure 4b), gave a shoulder near 2680 cm^{-1} at 300 K, thereby suggesting that the temperature change induced a stronger frequency shift for decoupled OD groups of dickite than for kaolinite. The absolute values and the shift of decoupled OD frequencies with temperature are possibly related to the strength of the hydrogen bonds in which the corresponding OH groups are involved. Indeed, where OH groups are involved in hydrogen bonds, a decrease of temperature induced a shift of the stretching modes to lower frequencies. Thus, inner-surface OH groups of kaolinite are probably involved in weaker hydrogen bonds than hydrogen bonds in dickite and nacrite. This point is probably related to the general behavior of these minerals with respect to the intercalation process (Cruz-Cumplido et al., 1982). Intercalation of hydrazine is, for example, easier in well-crystallized kaolinite than in poorly crystallized materials or in dickite.

CONCLUSION

These results show that temperature is an important parameter in the study of structural OH groups of 1:1 dioctahedral minerals. By appropriate selection of temperature, IR spectra with optimum resolution have permitted: (1) a better assignment of the stretching and the bending modes of structural OH groups of kaolinite, dickite, and nacrite; (2) more precise determination of the dichroic character of each component; and (3) identification of those bands which are the result of coupling of vibrations. The dichroic component which exists in the stretching modes of structural OH groups of kaolinite, dickite, and nacrite does not correspond to a functional group, but is the result of a coupling phenomenon between inner-surface OH groups which are nearly perpendicular to the sheets. The frequency and behavior of that dichroic band with changes in temperature are different for kaolinite, dickite, and nacrite. A better understanding of the origin of that band could be the next step towards understanding the bonding between sheets, and therefore the fundamental concept of crystallinity.

REFERENCES

Adams, J. M. and Hewat, A. W. (1981) Hydrogen atom positions in dickite: *Clays & Clay Minerals* **29**, 316–319.

Adams, J. M. (1983) Hydrogen atom positions in kaolinite by neutron profile refinement: *Clays & Clay Minerals* **31**, 352–356.

Barrios, J., Plançon, A., Cruz, M. I., and Tchoubar, C. (1977) Qualitative and quantitative study of stacking faults in a hydrazine treated kaolinite-relationship with the infrared spectra: *Clays & Clay Minerals* **25**, 422–429.

Bookin, A. S., Drits, V. A., Rozhdestvenskaya, I. V., Semenova, T. F., and Tsipursky, S. I. (1982) Comparison of orientations of OH-bonds in layer silicates by diffraction methods and electrostatic calculations: *Clays & Clay Minerals* **30**, 409–414.

Cruz-Cumplido, M., Sow, C., and Fripiat, J. J. (1982) Spectre infrarouge des hydroxyles, cristallinité et énergie de cohésion des kaolins: *Bull. Minéral.* **5**, 493–498.

Farmer, V. C. and Russell, J. D. (1964) The infra-red spectra of layer silicates: *Spectrochim. Acta* **20**, 1149–1173.

Freund, F. (1974) Ceramics and thermal transformations of minerals: in *The Infrared Spectra of Minerals*, V. C. Farmer, ed., Mineralogical Society, London, 465–482.

Ledoux, R. L. and White, J. L. (1965) Infrared studies of the hydroxyl groups in intercalated kaolinite complexes: in *Clays and Clay Minerals Proc. 13th Natl. Conf., Madison, Wisconsin, 1964*, W. F. Bradley and S. W. Bailey, eds., Pergamon Press, New York, 289–315.

Prost, R. (1973) The influence of the Christiansen effect on I.R. spectra of powders: *Clays & Clay Minerals* **21**, 363–368.

Prost, R. (1975) Etude de l'hydratation des argiles: interactions eau-minéral et mécanisme de la rétention de l'eau: *Thèse de doctorat d'état ès sciences physiques*, Univ. Paris, 135 pp.

Prost, R. (1984) Etude par spectroscopie infrarouge à basse température des groupes OH de structure de la kaolinite, de la dickite et de la nacrite: *Agronomie* **4**, 403–406.

Rouxhet, P. G., Samudacheata, N., Jacobs, H., and Anton, O. (1977) Attribution of the OH stretching bands of kaolinite: *Clay Miner.* **12**, 171–179.

Sen Gupta, P. K., Schlemper, E. O., Johns, W. D., and Ross, F. (1984) Hydrogen positions in dickite: *Clays & Clay Minerals* **32**, 483–485.

Serratosa, J. M., Hidalgo, A., and Vinas, J. M. (1963) Infrared study of the OH groups in kaolin minerals: in *Proc. Int. Clay Conf., Stockholm, 1963, Vol. 1*, I. T. Rosenqvist and P. Graff-Petersen, eds., Pergamon Press. New York, 17–26.

Suitch, P. R. and Young, R. A. (1983) Atom positions in highly ordered kaolinite: *Clays & Clay Minerals* **31**, 357–366.

White, J. L., Laycock, A., and Cruz, M. (1970) Infrared studies of proton delocalization in kaolinite: *Bull. Groupe Franç. Argiles* **22**, 157–165.

Proceedings of the International Clay Conference, Denver, 1985, L. G. Schultz, H. van Olphen, and F. A. Mumpton, eds.,
The Clay Minerals Society, Bloomington, Indiana, 24–30 (1987).

MONTE CARLO SIMULATION AND CALCULATION OF ELECTROSTATIC ENERGIES IN THE ANALYSIS OF Si-Al DISTRIBUTION IN MICAS[1]

CARLOS P. HERRERO

Instituto de Físico-Química Mineral, C.S.I.C., Serrano, 115 dpdo.
Madrid 28006, Spain

Abstract—The Si-Al distribution in the tetrahedral sheet of phyllosilicates having Si:Al ratios near 3:1 has been analyzed by a combination of magic angle spinning-nuclear magnetic resonance spectroscopy, Monte Carlo simulation, and calculations of electrostatic energies. The results of this analysis indicate that Si-Al distribution is mainly controlled by electrostatic requirements of homogeneous dispersion of Al and local balance of interlayer charges. Long- and short-range ordered Si-Al distribution models, complying with the above requirements, differ little in energy, and no evidence for long-range order has been found. In muscovite-$2M_1$, phlogopite-$1M$, and Llano vermiculite, meta and para dispositions for Al in hexagonal rings of tetrahedral cations occur in a ratio meta:para of about 2:1.

Key Words—Electrostatic energy, Mica, Monte Carlo simulation, Nuclear magnetic resonance, Ordering, Si-Al distribution, Vermiculite.

INTRODUCTION

The Si-Al distribution in the tetrahedral sheet of phyllosilicates is an important crystallochemical factor that controls the thermodynamic stability (Saxena, 1973) and the physico-chemical characteristics of these minerals (Méring and Pedro, 1969; Serratosa *et al.,* 1984). This distribution is not easily determined by X-ray diffraction methods because of the similarity of the scattering factors of Si^{4+} and Al^{3+}, and the Si-Al ordering patterns that have been proposed are essentially based on the differences observed in the mean T–O (T = Si or Al) bond lengths between nonequivalent tetrahedra. A statistical analysis of the crystallographic data (Bailey, 1984) indicates that only in a few samples are tetrahedral size differences significant; the most reliably determined ordered structures are those corresponding to margarite-$2M_1$ (Guggenheim and Bailey, 1978) and intermediate margarite-bityite-$2M_1$ (Lin and Guggenheim, 1983), both having a Si:Al ratio near 1:1. For a Si:Al ratio of 3:1, some ordered examples have been proposed, but the evidence is not unequivocal.

The use of Mössbauer, infrared, nuclear magnetic resonance (NMR), and Raman spectroscopic methods has permitted the study of short-range ordering of cations and vacancies in phyllosilicates. In particular, different environments of ^{27}Al and ^{29}Si have been recently distinguished by magic angle spinning (MAS) NMR spectroscopy (Sanz and Serratosa, 1984; Lipsicas *et al.,*

1984). From the intensities of the components in the ^{29}Si spectra, Herrero *et al.* (1985a) checked the validity of different distribution models. Although NMR information has a local character, it may also be used to test long-range order schemes, such as those suggested by X-ray diffraction methods.

Other attempts to investigate cation order-disorder are based on the calculation of electrostatic energies. Giese (1984) recently reported the Si-Al distribution for muscovite and compared the energies of many long-range order schemes. He concluded that, according to his calculations, muscovite $2M_1$ should be long-range disordered.

The present contribution reports tetrahedral cation order in micas by the use of simulation methods and calculation of electrostatic energies. Based on ^{29}Si NMR spectra, different Si-Al distribution models are considered for micas having Si:Al ratio near 3:1. The application of Monte Carlo simulations has permitted the determination of Si-Al distribution in such minerals, and the extension of the electrostatic energy calculations to distributions in which restrictions of short-range order character are imposed (Loewenstein's (1954) rule and local balance of charge). Finally, electrostatic energies have been calculated taking into account the interaction between contiguous sheets across the interlayer space.

MATERIALS AND METHODS

Two natural micas ($1M$-phlogopite from Greenville, Quebec and $2M_1$-muscovite from Miyori, Tochigi, Japan) and a vermiculite (from Llano County, Texas) having Si/Al ratios near 3 and very low iron contents

---

[1] Recipient of W. F. Bradley award, 1985 International Clay Conference, for paper with best technical content by author of less than 35 years of age.

were selected for this study. The structural compositions of these samples were given by Sanz and Serratosa (1984). The Si-Al distribution in a synthetic sample analyzed by Lipsicas *et al.* (1984) was also studied. This sample has an Al atomic tetrahedral fraction of 0.27.

High-resolution ^{29}Si NMR spectra of the powdered samples were recorded at 59.6 MHz by spinning the sample at the magic angle in a Brucker CXP 300hp spectrometer. The obtained spectra were presented by Herrero *et al.* (1985a), where a more detailed description of the experimental procedure was given.

Decompostions of the NMR spectra were carried out assuming a gaussian profile for each component. The total intensity of the spectra was normalized to 100, and the intensity of each component was obtained by integration over the corresponding curve. Models of the Si-Al distribution were generated in an HP 9845 B calculator in which a random-number generator is available.

STATISTICAL ANALYSIS OF THE NMR RESULTS

^{29}Si MAS-NMR spectra of phyllosilicates are composed of different absorption lines corresponding to different tetrahedral environments of Si in the structure. In principle, four Si environments are possible; i.e., Si surrounded by 3 Al, 2 Al and 1 Si, 1 Al and 2 Si, or 3 Si.

For natural samples of phlogopite, muscovite, and vermiculite, three lines were observed corresponding to the associations Si(2Al1Si), Si(1Al2Si), and Si(3Si) (Sanz and Serratosa, 1984). The synthetic phyllosilicate sample having a tetrahedral Al atomic fraction of 0.27 gave a spectrum similar to that of the natural specimens (Lipsicas *et al.,* 1984). Intensities of the lines in the spectra of the four samples are given in Table 2. Relative intensities of the different lines are a consequence of Si-Al distribution; hence, information about this distribution may be obtained from an analysis of those intensities for each sample.

Information about first tetrahedral neighbors was obtained by comparing the values of the Al fractional contents given by the mineralogical formulae with those calculated from the spectra assuming different distribution models. In fact, expressions relating the Al content to the line intensities can be easily deduced. Thus, if the Si-Al distribution was completely random (Herrero *et al.,* 1985b):

$$\frac{1 - x}{x} = \frac{3I_0 + 2I_1 + I_2}{3I_3 + 2I_2 + I_1},\qquad(1)$$

where x and $1 - x$ are, respectively, the Al and Si fractional contents, and I_i (where i = 0, 1, 2, or 3) is the intensity of the line produced by a Si environment that contains i Al ions (Table 1, first column).

Based on electrostatic considerations, Loewenstein

Table 1. Fractional aluminum contents (x) and probabilities P_2 for the different models of Si-Al distribution in the tetrahedral sheet of phyllosilicates.[1]

Samples	Fractional Al content (x)			Probabilities P_2		
	r.	L.r.	s.f.	L.r.	h.d.c.	NMR
Muscovite[2]	0.28	0.22	0.21	0.28	0.16	0.14
Phlogopite[2]	0.36	0.27	0.29	0.37	0.25	0.21
Vermiculite[2]	0.38	0.28	0.28	0.39	0.28	0.27
Synthetic[3]	0.35	0.26	0.27	0.35	0.24	0.23

[1] Values of x obtained from the structural formulae and of P_2 calculated from experimental ^{29}Si nuclear magnetic resonance (NMR) spectra are given for comparison. r. = random distribution; L.r. = Loewenstein's restriction; s.f. = structural formula; h.d.c. = homogeneous dispersion of charge.
[2] Herrero *et al.* (1985).
[3] Lipsicas *et al.* (1984).

(1954) postulated the principle of avoidance of Al–O–Al linkages in the tetrahedral networks of aluminosilicates. If this rule is introduced in the Si-Al distribution in phyllosilicates, the Si:Al ratio is given by (Sanz and Serratosa, 1984):

$$\frac{1 - x}{x} = \frac{\sum_{i=0}^{3} I_i}{\frac{1}{3}\sum_{i=0}^{3} iI_i}.\qquad(2)$$

Values of x deduced from Eq. (2) (Table 1, second column) agree reasonably well with those of the structural formulae (Table 1, third column), confirming the avoidance of Al in neighboring tetrahedra.

For a distribution in which the only restriction is Loewenstein's rule, values of I_0, I_1, I_2, and I_3 are respectively proportional to

$$s^3,\ 3s^2a,\ 3sa^2,\ \text{and}\ a^3,\qquad(3)$$

where a = x/y and s = 1 − a, where y = 1 − x (Si tetrahedral fraction) (see Herrero *et al.,* 1985a). Comparison between the values deduced from Eq. (3) and those calculated from the spectra indicates that Loewenstein's rule is not sufficient to explain the NMR experimental data (Table 2).

Additional information about Si-Al distribution was obtained from the NMR spectra by calculation of the probability P_2 that the second neighbor of an Al be also an Al. Assuming that Loewenstein's rule is verified, P_2 values are given by the expression:

$$P_2 = \frac{I_2 + 3I_3}{I_1 + 2I_2 + 3I_3},\qquad(4)$$

where $I_2 + 3I_3$ is proportional to the number of Al-Si-Al groups and $I_1 + 2I_2 + 3I_3$ gives the normalization over Al-Si-Al and Al-Si-Si triads. For the intensities given by Eq. (3), P_2 takes the particular value:

$$P_2 = a.\qquad(5)$$

Values of the P_2 parameter obtained from NMR data

Table 2. Experimental and calculated relative intensities of the possible environments of Si in the tetrahedral sheet of phyllosilicates for different Si-Al distribution models.[1]

Distribution model	Muscovite[2]					Phlogopite[2]				
	3Al	1Si2Al	2Si1Al	3Si	S	3Al	1Si2Al	2Si1Al	3Si	S
Loewenstein's restriction	2.2	16.7	43.5	37.5	33.3	4.7	24.9	44.2	26.2	35.8
Homogeneous dispersion of charge	0.3	12.1	58.2	29.4	4.0	0.7	24.8	56.5	18.0	11.2
Experimental	—	11.8	60.2	28.0		—	23.1	62.1	14.8	

Distribution model	Vermiculite[2]					Synthetic sample[3]				
	3Al	1Si2Al	2Si1Al	3Si	S	3Al	1Si2Al	2Si1Al	3Si	S
Loewenstein's restriction	5.6	27.2	43.7	23.5	27.2	4.3	24.0	44.4	27.3	25.2
Homogeneous dispersion of charge	0.8	29.5	53.1	16.6	3.8	0.7	23.6	56.2	19.5	2.4
Experimental	—	30.5	54.0	15.5		—	24	57	19	

[1] Values of the parameter S for each model have been included.
[2] See footnote 2, Table 1.
[3] See footnote 3, Table 1.

through Eq. (4) and those calculated from Eq. (5), which assumes avoidance of Al–O–Al linkages, are given respectively in the fourth and sixth columns of Table 1. The values deduced from the experimental data were always significantly smaller than those calculated if the only restriction was Loewenstein's rule, meaning that Al ions are, on average, more separated than that required by the avoidance of Al in neighboring tetrahedra. In other words, in addition to Loewenstein's rule, other factors governed Si-Al distribution in the considered phyllosilicates.

SIMULATION OF Si-Al DISTRIBUTION BY MONTE CARLO METHODS

As suggested in the statistical analysis of the NMR data, actual Si-Al distributions in the analyzed samples require a greater separation between Al ions than that imposed by Loewenstein's rule. This requirement suggests that homogeneous dispersion of Al cations through the sheet may be an important characteristic of the distribution, which, in principle, agrees with the dispersion of charge deficits required by electrostatic considerations.

For phyllosilicates having Si:Al ratio equal to 3:1, the average number of Al per hexagonal ring is 1.5. Consequently, an effective compensation of interlayer ions should be obtained if the two hexagonal rings of contiguous layers contain either one and two Al ions or zero and three Al ions around each interlayer cation. The last disposition is unfavorable for electrostatic reasons, whereas the disposition having one and two Al ions not only satisfies the principle of local balance of charges, but favors a more homogeneous charge distribution on each individual sheet.

Si-Al distribution models complying with the above conditions, i.e., avoidance of Al–O–Al linkages and hexagonal rings containing 1 or 2 Al ions, have been simulated in a computer by means of a Monte Carlo procedure. A tetrahedral sheet composed of 10,000

tetrahedra (5000 hexagonal rings) has been considered to keep the standard deviations below 1% for each calculated value. Periodic boundary conditions were introduced. Al ions, in the number required by the chemical composition of the tetrahedral sheet of each sample, were distributed in a random sequential way. When the total number of Al ions was exhausted, the remaining vacant positions were filled with Si. The constraint of Loewenstein's rule was introduced by filling with Si the three tetrahedra around each Al site. The second restriction (avoidance of hexagonal rings with zero and three Al) was taken into account by introducing at least one Al ion in each hexagonal ring and considering, for a given outcome, the three hexagonal rings sharing the tetrahedron: if one of these rings already contained two Al ions, the tetrahedral site was filled with Si, otherwise the position was filled with Al.

Probabilities of the different tetrahedral environments of Si calculated from the simulated models are given in Table 2. Reliability of the different proposed models is deduced directly by comparing the experimental I_i^e and calculated I_i^c intensities. A quantitative estimation of the fitting of the spectra is given by the S values calculated from the expression (Herrero *et al.*, 1985b):

$$S = \sum_{i=0}^{3} |I_i^e - I_i^c|, \qquad (6)$$

where intensities have been normalized to 100.

Values of S for the different Si-Al distribution models are given in Table 2. As expected, S values calculated if only Loewenstein's rule was considered were too high. If the restriction of homogeneous dispersion of charges was introduced in the distribution, the S values of the four considered samples decreased considerably. Moreover, P_2 values calculated using Eq. (4) and the intensities obtained in the simulation (Table 1, fifth column) are much closer to those calculated from NMR results than the values obtained from Eq. (5), for which the only restriction was Loewenstein's rule. These results indicate that the two conditions listed above were the main factors governing the Si-Al distribution in micas. The resulting distributions were long-range disordered and, for hexagonal rings containing two Al ions, the occurrence of meta and para dispositions of Al (solid circles in Figure 1) is given by the ratio meta/para ~2. For the tetrahedral sheet having a Si₃Al configuration, long-range order models may be constructed in which Al ions are distributed either in meta or in para dispositions. The calculated NMR intensities for both models (Herrero et al., 1985a) departed considerably from the experimental ones, and the agreement was improved only if the meta:para ratio was near 2:1.

Therefore, as far as NMR data and simulation methods are concerned, one may conclude that for the natural and synthetic phyllosilicates analyzed in this work, Si-Al distribution is mainly controlled by the requirements of a homogeneous dispersion of Al in order to decrease the electrostatic energy of the crystalline lattice and a need to balance locally the charge of the interlayer cations.

ELECTROSTATIC ENERGY CALCULATIONS

The above analysis suggests that the Si-Al distribution model that fits the experimental NMR spectra of micas contains Al ions in meta and para dispositions according to their relative statistical multiplicity in the hexagonal rings of the structure (meta:para = 2:1). This result seems to be in apparent contradiction to what should be expected if simple electrostatic considerations are applied to an isolated tetrahedral sheet. For a single sheet, after elimination of Al ions in ortho positions of the hexagonal rings (Loewenstein's rule), the next step should be to minimize the number of Al ions in meta disposition (Klinowski et al., 1982; Vega, 1983). If the whole structure of micas is considered, however, the hexagonal rings of adjacent layers superimpose around the interlayer cations, and the distance between tetrahedral cations across the interlayer space is about 1.5 times the distance T–T between first neighbors of the same sheet, and is shorter than the separation between second neighbor cations (~1.7 × T–T). Thus, tetrahedral sheets cannot be treated as isolated entities, and the possibility of correlations between sheets of adjacent layers must be considered.

The most stable configurations for two superimposed isolated hexagons containing one and two charge deficits, respectively, are those represented in Figure 1. Calculation of the electrostatic energy as a function of the distance between hexagons shows that the most favorable disposition is *m* for short distances and *p* for large ones. This fact suggests that the separation between tetrahedral sheets may be an important factor if the relative stabilities of different Si-Al distribution models must be compared.

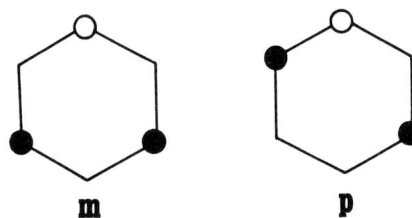

Figure 1. Electrostatistically most favorable configurations for three Al compensating locally the charge of the interlayer cation. Solid and open circles represent Al in adjacent sheets; m and p stand for meta and para dispositions of Al ions.

Method of calculation

The electrostatic calculations were carried out for the tetrahedral composition Si₃Al on the assumptions that ions in the structure were completely ionized with point charges and that the interactions between them were exclusively of the Coulomb type. The repulsion and van der Waals short-range interactions were ignored in these calculations because their contributions to the lattice energy do not vary appreciably for the different models considered for the cation distribution (Herrero et al., 1986; Herrero, 1986).

Different short- and long-range order distribution schemes were analyzed to determine the electrostatically most stable models and the energy variations associated with the different types of restrictions imposed in the Si-Al distribution were calculated. For the long-range order models, the electrostatic energy was calculated by the method of Ewald, using the expressions given by Kittel (1953); the convergence parameter η was chosen according to the criterion of Bonnin and Legrand (1975), as indicated by Herrero (1986).

For the short-range order distributions, there is no periodicity in tetrahedral occupancy and no unit cell can be defined that is reproduced throughout space; thus, a different approach had to be followed. In that approach, the different types of short-range order distributions were described in terms of the probabilities of occupancy by Si or Al of the tetrahedral positions surrounding a so-called reference site (Figure 2). Thus, P_i will be the probability of finding Al in the ith tetrahedral site if the reference position is occupied by Al. These probabilities were calculated from the computer simulated Si-Al distributions for the different considered models. In this statistical approach, it is convenient to refer the calculated energies to that of the random Si-Al distribution that will be taken as the zero level for the values presented below.

Thus, for a short-range order distribution, the average variation of electrostatic energy per tetrahedral cation, with respect to the random distribution, is given by the expression (see Herrero et al., 1986):

$$\Delta E_T = \frac{x}{2} \sum_{i=1}^{\infty} \frac{P_i - x}{d_i}, \qquad (7)$$

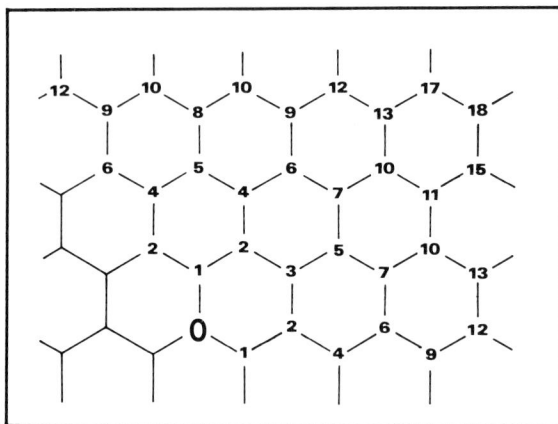

Figure 2. Values of n for the different tetrahedral neighbors of a tetrahedral site (O). For simplicity, only a part of the positions are given.

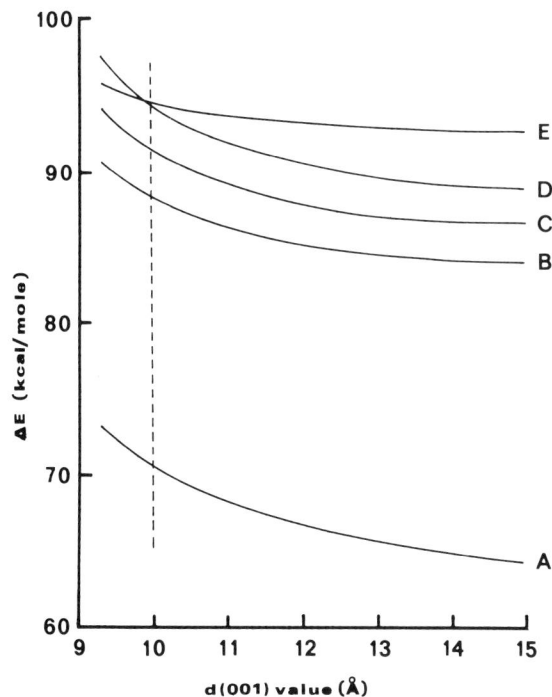

Figure 3. Dependence of electrostatic energy on basal spacing for different Si-Al distribution models described in the text. Stabilization energies are given in kcal/mole. Vertical line corresponds to the basal spacing of Franklin, New Jersey, phlogopite: d(001) = 9.998 Å (Hazen and Burnham, 1973).

in which the sum is extended to the lattice tetrahedral positions; x is, as noted above, the Al fractional content, and d_i is the distance from the ith tetrahedral site to the reference. For the random Si-Al distribution, as expected, $\Delta E_T = 0$, because $P_i = x$ for all values of i. If different restrictions were imposed on the cation distribution, P_i values differed from the average value x, and different ΔE_T were obtained.

Electrostatic energy calculations were made according to the methods described above for different Si-Al distribution models and for a phlogopite-$1M$ from Franklin, New Jersey, analyzed by X-ray diffraction methods by Hazen and Burnham (1973). For the calculation, the interlayer cations were assumed to have $1+$ charge and to occupy all the available interlayer positions, and OH groups in the structure were considered as point $1-$ charges to avoid the problem of uncertainty in their orientation (Giese, 1984).

The influence on the electrostatic energy of the interaction between tetrahedral cations of adjacent sheets is realized if this energy is calculated as a function of interlayer separation. For this purpose, the phlogopite basal spacing was varied in the calculations by separating the layers perpendicularly to the ab plane, starting from a spacing of 9.3 Å, for which basal oxygens should be in contact. Interlayer cations were maintained in a plane equidistant from the layers. Variations of the electrostatic energy with respect to the random distribution were calculated as a function of the basal spacing for the different distribution schemes. Energies were referred to an $a \times b$ cell containing eight tetrahedral positions.

Results of the electrostatic calculations

In the absence of a correlation between tetrahedral sheets, the electrostatic energy with respect to the random distribution did not change with the basal spacing.

If the only restriction in cation distribution was Loewenstein's rule, a stabilization energy of 58.6 kcal/mole was obtained. If the additional condition that hexagonal rings contain one or two Al ions was imposed, the stabilization energy was 80.8 kcal/mole. Thus, the elimination of Al–O–Al linkages stabilized the structure notably, but an important additional decrease (22.2 kcal/mole) in the electrostatic energy was obtained if the condition of homogeneous dispersion of charges over the sheets was considered. This result confirms the conclusion obtained from NMR data in the sense that interactions between tetrahedral cations goes further than the first neighbors.

If correlations between sheets were introduced in the Si-Al distribution, the electrostatic energy varied with layer separation. The considered correlations were chosen in order to obtain additional lattice stabilization. The first restriction introduced in the occupancy of tetrahedral sites in neighboring sheets was the avoidance of Al ions in superimposed tetrahedra. If this restriction was added to Loewenstein's rule, a stabilization energy of 70.5 kcal/mole was obtained for the real basal spacing of phlogopite. This value represents a stabilization of 11.9 kcal/mole with respect to the case in which only the Loewenstein's rule was considered. Curve A of Figure 3 shows the variation of this stabilization energy as a function of the basal spacing.

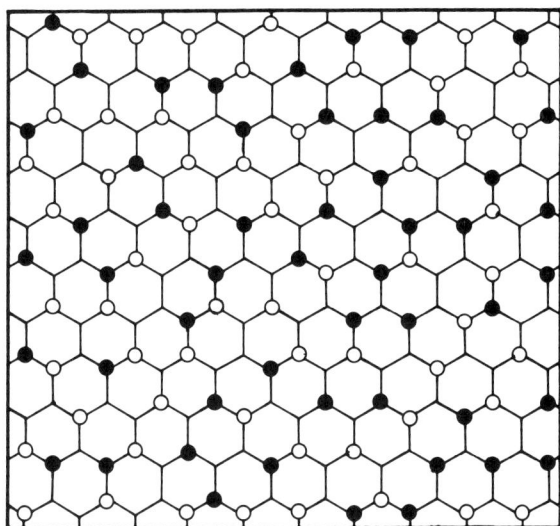

Figure 4. Simulated Si-Al distribution for two adjacent tetrahedral sheets that comply with Loewenstein's rule, avoidance of Al in superimposed tetrahedra, and local charge balance. Solid and open circles represent Al in different sheets.

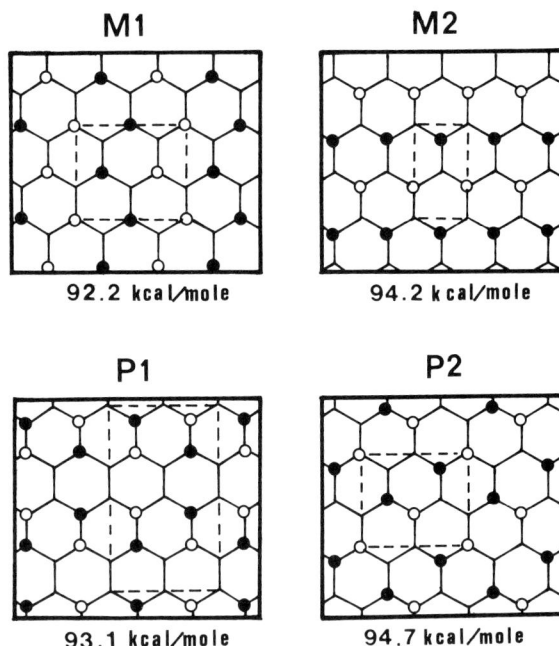

M1   **M2**

92.2 kcal/mole   94.2 kcal/mole

P1   **P2**

93.1 kcal/mole   94.7 kcal/mole

Figure 5. Electrostatically most stable, long-range order models for two superimposed tetrahedral sheets. The unit cell used for the electrostatic calculations is shown for each model. Values of stabilization energy, normalized to an $a \times b$ cell, are given.

A more important energetic stabilization was obtained if the condition of local charge balance was added to the restrictions of the previous model. This new condition required that superimposed hexagonal rings across the interlayer space contain one and two Al ions. If neighboring sheets were simulated according to the three specified conditions (Loewenstein's rule, avoidance of Al in superimposed tetrahedra, and local balance of charge), the calculated energy was 88.4 kcal/mole for $d(001) = 9.998$ Å, that is, 17.9 kcal/mole more stable than the previous case (curve A) in which local charge balance was not considered. The variation of ΔE with interlayer separation is represented by curve B of Figure 3. Two adjacent tetrahedral sheets simulated with the three above conditions are pictured in Figure 4.

To obtain a higher lattice stabilization, neighboring adjacent sheets containing hexagonal rings with Al exclusively in either of the dispositions represented in Figure 1 may be attempted. Due to the randomness of the method used to fill the tetrahedral positions, however, such a configuration can only be achieved by about 95% of the hexagonal rings. For this Si-Al distribution model, the variation of electrostatic energy with distance is given by curve C of Figure 3. The real basal spacing of phlogopite yields a stabilization of 3.0 kcal/mole with respect to the previous model (curve B).

All the short-range order distributions considered above contain Al in meta and para dispositions in the statistical ratio 2:1. For long-range order models, different schemes may be constructed on the basis of hexagonal rings containing Al either in meta or in para dispositions. Figure 5 represents the most stable long-range order patterns found for two neighboring tetrahedral sheets. Two of them (M1 and M2) have Al in meta disposition, whereas Al is exclusively in para positions in the others. Calculation of ΔE as a function of interlayer separation indicated that, for all distances, M2 and P2 were slightly more stable than M1 and P1, respectively. Considering only the most stable models, M2 is more stable than P2 for small distances, and P2 is more stable than M2 for large separations (curves D and E, Figure 3). For $d(001) = 9.998$ Å, the stabilization energies were almost equal (94.2 kcal/mole for M2 and 94.7 kcal/mole for P2). Differences between energies for these long-range order schemes and the short-range order model C (Figure 3) were about 3 kcal/mole.

Although the highest energy of stabilization corresponds to the long-range order models, the small difference between their energy values and that corresponding to the model C suggests that the drop in configurational entropy of the cation distribution associated with an increase in the cation ordering may modify the relative stability of these models from the point of view of the lattice free energy. When configurational entropy is calculated by the Kikuchi method (see Herrero, 1986), one finds, in fact, that the long-range order models are not favored over the short-range ones. The details of these entropy and free energy calculations will be the subject of a forthcoming paper.

CONCLUSIONS

The combination of NMR spectroscopy, Monte Carlo simulations, and lattice electrostatic energy calculation has proven to be an efficient way to obtain insight into the tetrahedral cation ordering in phyllosilicates. For the natural and synthetic samples analyzed here, having fractional tetrahedral Al contents near 0.25, the Si-Al distribution is mainly governed by the criteria of local balance of charge of interlayer cations and of homogeneous dispersion of Al over the sheets. Loewenstein's rule is included as a first step in that dispersion of charge.

Calculations of the electrostatic energy show that long-range order models having Al ions either in para or in meta dispositions are more stable than the long-range disordered schemes. The fact that long-range ordering does not exist in the studied mica samples, as shown from NMR spectroscopic data, may be attributed to the similarity of the electrostatic energies of models M2 and P2 and to the small difference between the calculated energies for the long- and short-range order models. This justifies also that meta and para dispositions of Al ions were found in a ratio meta : para = 2:1, in agreement with the relative statistical multiplicities of both cation dispositions in the hexagonal rings of the structure.

ACKNOWLEDGMENTS

The author is indebted to J. M. Serratosa, J. Sanz, and W. E. E. Stone for critical reviews of the manuscript.

REFERENCES

Bailey, S. W. (1984) Review of cation ordering in micas: *Clays & Clay Minerals* **32**, 81–92.

Bonnin, D. and Legrand, A. P. (1975) Calculation of the electric potential in heulandite-type zeolite: *Chem. Phys. Lett.* **30**, 296–299.

Giese, R. F., Jr. (1984) Electrostatic energy models of micas: in *Micas, Reviews in Mineralogy* **13**, S. W. Bailey, ed., Miner. Soc. Amer., Washington, D.C., 105–144.

Guggenheim, S. and Bailey, S. W. (1978) Refinement of a margarite structure in subgroup symmetry: correction, further refinement and comments: *Amer. Mineral.* **63**, 186–187.

Hazen, R. M. and Burnham, C. W. (1973) The crystal structures of one-layer phlogopite and annite: *Amer. Mineral.* **58**, 889–900.

Herrero, C. P. (1986) Orden local de la distribución catiónica en redes tetraédricas bidimensionales: Ph.D. thesis, Univ. Autónoma de Madrid, Madrid, Spain, 149 pp.

Herrero, C. P., Sanz, J., and Serratosa, J. M. (1985a) Si,Al distribution in micas: analysis by high-resolution ^{29}Si NMR spectroscopy: *J. Phys. C: Solid State Phys.* **18**, 13–22.

Herrero, C. P., Sanz, J., and Serratosa, J. M. (1985b) Tetrahedral cation ordering in layer silicates by ^{29}Si NMR spectroscopy: *Solid State Comm.* **53**, 151–154.

Herrero, C. P., Sanz, J., and Serratosa, J. M. (1986) Electrostatic energy of micas as a function of Si-Al tetrahedral ordering: *J. Phys. C: Solid State Phys.* **19**, 4169–4181.

Kittel, C. (1953) *Introduction to Solid State Physics:* 2nd ed., Wiley, New York, 396 pp.

Klinowski, J., Ramdas, S., Thomas, J. M., Fyfe, C. A., and Hartman, J. S. (1982) A re-examination of Si-Al ordering in zeolites NaX and NaY: *J. Chem. Soc., Faraday Trans. 2* **78**, 1025–1050.

Lin, J. C. and Guggenheim, S. (1983) The crystal structure of a Li,Be-rich brittle mica: a dioctahedral-trioctahedral intermediate: *Amer. Mineral.* **68**, 130–142.

Lipsicas, M., Raythatha, R. H., Pinnavaia, T. J., Johnson, I. D., Giese, R. F., Jr., Costanzo, P. M., and Robert, J. L. (1984) Silicon and aluminium site distributions in 2:1 layered silicate clays: *Nature* **309**, 604–607.

Loewenstein, W. (1954) The distribution of aluminum in the tetrahedra of silicates and aluminates: *Amer. Mineral.* **39**, 92–96.

Méring, J. and Pedro, G. (1969) Discussion à propos des critères de classification des phyllosilicates 2/1: *Bull. Groupe Fr. Argiles* **21**, 1–30.

Sanz, J. and Serratosa, J. M. (1984) ^{29}Si and ^{27}Al high-resolution MAS-NMR spectra of phyllosilicates: *J. Amer. Chem. Soc.* **106**, 4790–4793.

Saxena, S. K. (1973) *Thermodynamics of Rock-Forming Crystalline Solutions:* Springer Verlag, New York, 200 pp.

Serratosa, J. M., Rausell-Colom, J. A., and Sanz, J. (1984) Charge density and its distribution in phyllosilicates: effect on the arrangement and reactivity of adsorbed species: *J. Mol. Catal.* **27**, 225–234.

Vega, A. J. (1983) A statistical approach to the interpretation of silicon-29 NMR of zeolites: *Amer. Chem. Soc. Symp. Ser.* **218**, 217–230.

MINERALOGY

Proceedings of the International Clay Conference, Denver, 1985, L. G. Schultz, H. van Olphen, and F. A. Mumpton, eds.,
The Clay Minerals Society, Bloomington, Indiana, 33–45 (1987).

MIXED-LAYER MINERALS: DIFFRACTION METHODS AND STRUCTURAL FEATURES

V. A. Drits

Geological Institute, Academy of Sciences, Moscow, U.S.S.R.

Abstract—The study of structural features of mixed-layer minerals by X-ray powder diffraction (XRD) is based on the analysis of the intensity distribution of basal reflections using direct and indirect Fourier-transform methods. The direct Diakonov method, in contrast to that of MacEwan, provides information on a mixed-layer mineral structure based only on d-values and intensities of basal reflections. The indirect methods are difficult because a great number of independent parameters are needed to model XRD patterns. An *a priori* precise determination of these parameters is not always possible. A combination of direct and indirect methods may prove fruitful for studying "monomineralic" interstratified clays.

The degree of homogeneity is an important characteristic of mixed-layer samples consisting of thin particles (e.g., containing 5–15 layers). Finely dispersed mixed-layer samples can be classified as homogeneous, quasihomogeneous, and heterogeneous. All particles in homogeneous mixed-layer samples have the same composition and differ from one another only in the pattern of alternation of layer types. Mixed-layer samples whose thin individual particles differ both in composition and in the pattern of alternation of layer types are called quasihomogeneous if the structural features of the samples can be described by a small number of independent probability parameters for each given short-range order factor, R. These parameters are the total number of layers, N, the contents of different layer types, and the pattern of their alternation. If these factors cannot be described by a small number of independent probability parameters, the mixed-layer samples are termed heterogeneous. XRD effects calculated for quasihomogeneous and homogeneous samples and for mixtures of quasihomogeneous samples are generally not sensitive to wide variations in the degrees of homogeneity.

Selected-area electron diffraction is an independent tool for structural studies of poorly crystalline minerals, including mixed-layer phases, and has been used to reveal ordered and random mixed-layer phases among manganese minerals in Fe-Mn nodules.

Analyses of the structural features of interstratified illite/smectite (I/S) from diagenetic and hydrothermal environments suggest a two-stage mechanism for the transformation of smectite into illite. In the first stage, smectite is believed to be transformed into I/S having predominantly smectite layers by a solid-phase mechanism. At higher temperatures, I/S phases having predominantly illite layers and R > 1 may be formed by dissolution-precipitation. Identical azimuthal orientations of 2:1 layers in the crystal favor the irreversible fixation of K during the smectite-into-illite transformation. Illitization of smectite is accompanied not only by the increase in the layer charge brought about by the Al-for-Si substitution in tetrahedra, but also by the redistribution of cations over *trans*- and *cis*-octahedra in the 2:1 layers.

Key Words—Basal reflections, Illite, Interstratification, Ordering, Selected area electron diffraction, Smectite, X-ray powder diffraction.

INTRODUCTION

Mixed-layer minerals are abundant in various geological environments, including weathering crusts, soils, modern sediments, diagenetically altered rocks, and regions of hydrothermal activity. The structural and composition of these minerals reflect the dynamics of the change in physicochemical conditions of various geological processes. The successful use of mixed-layer minerals to indicate geological environments (e.g., different stages of burial diagenesis) requires that their structural features be determined as thoroughly as possible.

The solution of this problem prompted the development of special methods based on X-ray powder diffraction (XRD) (Méring, 1949; MacEwan, 1956, 1958; Kakinoki and Komura, 1954a, 1954b, 1965;

Diakonov, 1961, 1962, 1983; Reynolds, 1967, 1980; Reynolds and Hower, 1970; Drits and Sakharov, 1976; Plançon, 1981; Plançon *et al.,* 1983, 1984; Sakharov *et al.,* 1982a, 1982b, 1983). The number of papers that treat various aspects of mixed-layer minerals is now so great that listing them here is not practical. A survey of XRD methods and the results of the study of various mixed-layer minerals were published recently by Reynolds (1980) and Diakonov (1983).

The following problems will be reviewed in this paper: (1) XRD methods for the study of mixed-layer minerals; (2) results demonstrating the possibilities and limitations of different methods; and (3) structural aspects of the formation of some mixed-layer minerals. Major attention will be paid to results obtained in the Soviet Union.

X-RAY POWDER DIFFRACTION METHODS
FOR MIXED-LAYER MINERALS

Direct methods

Information on the types of layers in mixed-layer minerals, their proportions, and the pattern of interstratification can be obtained directly from the analysis of the positions and the intensities of basal reflections on XRD patterns. The method of MacEwan (1956) is the best known. Its main drawback is that the formula determining the function for the distribution of interlayer distances, w(z), was deduced for idealized mixed-layer structures consisting of identical layers separated by vacant interlayers of different thicknesses. In addition, the divergence of the integral in the formula involved and the need to know exact values for structural layer factors can lead to considerable difficulties, as can the choice of base line for the analysis of the w(z) function.

Diakonov (1961, 1962, 1974, 1977, 1978) proposed a simpler and more reliable method for the study of mixed-layer minerals using the formula:

$$\mathcal{P}(z) = \sum_{k=1}^{n} J_k \cos 2\pi \frac{z}{d_k}, \qquad (1)$$

where J_k is the integral intensity of the reflection having a spacing of d_k corrected for the Lorentz factor, and z is the distance in the crystal. $\mathcal{P}(z)$ is the generalized one-dimensional Patterson function.

Assuming that $\rho(z)$ is electron density distribution projected on the normal to the layers and z is the coordinate along this direction, $\mathcal{P}(z)$ is defined mathematically as the autocorrelation function, or as the convolution of $\rho(z)$ and $\rho(-z)$, that is,

$$\mathcal{P}(z) = \int \rho(z')\rho(z' + z) \, dz'.$$

Integration is taken over the volume of the crystal. The basic peaks in $\mathcal{P}(z)$ correspond to zero peaks in the Patterson function. For crystals consisting of identical layers, these peaks appear for those superpositions of $\rho(z')$ and $\rho(z' + z)$, where z is equal to a whole number of the repeat distances along the normal to layers. For mixed-layer crystals, $\mathcal{P}(z)$ will contain basic peaks if z is equal to the distance between identical layers. For example, for interstratified illite/smectite (I/S) formed by layers A (d = 10 Å) and B (d = 18 Å), $\mathcal{P}(z)$ contains basic peaks at z values of 10 and 18 Å, as well as at $z = md_A + nd_B$, corresponding to combinations of mA layers and nB layers (m and n being whole numbers).

The interpretation of $\mathcal{P}(z)$ is difficult due to the presence of the additional peaks that correspond to ordinary peaks of the Patterson function and depend on the arrangement of atoms within layers. Diakonov (1977) proposed a method for the quantitative treatment of $\mathcal{P}(z)$ that diminishes the influence of the additional peaks. He also deduced formulae for those materials in which the basal reflections on the XRD patterns are broad and diffuse or form a rational series and have an asymmetrical form (Diakonov, 1978).

Mathematical analysis of $\mathcal{P}(z)$ for mixed-layer crystals gives layer contents W_A and W_B and layer thicknesses d_A and d_B that are accurate to ± 0.03 and ± 0.03 Å, respectively (Diakonov, 1974, 1978), if all basal reflections having d values ≥ 2 Å are involved in the calculation. The accuracy in the determination of the parameters in question may vary depending on the accuracy in the measurements of J_k and d_k and the complexity of $\mathcal{P}(z)$. The method, however, permits evaluation of accuracy in determination of W_A, W_B, d_A, and d_B for each individual $\mathcal{P}(z)$.

The Diakonov method has proven to be an effective tool in the study of various mixed-layer minerals, as has been demonstrated by its use in the Soviet Union (see, e.g., Diakonov, 1983; Gradusov, 1976; Drits and Sakharov, 1976). For example, in his comprehensive study of hydrobiotites, Diakonov (1981) was able to determine the finest features of their structures, as follows: (1) The proportion of vermiculite layers (W_B) ranged from 0 to 50%. Hydrobiotites with a greater proportion of vermiculite layers were not found. (2) Two groups of hydrobiotites appeared to be the most abundant, having W_B values of 0–15% and 30–50%. (3) Regardless of W_B, pairs of adjacent vermiculite layers were not found. (4) If W_B was small, vermiculite layers were separated by large blocks of mica-like layers. Two B layers were commonly separated by 8–10 A layers (A = mica-like layer), implying that even in the earliest stages of leaching, the newly formed vermiculite interlayers formed as far from each other as possible. (5) At W_B values of 30–50%, subsequences BAAB and BAB prevailed in the structure. At $W_B \approx$ 33%, an almost ordered structure commonly formed in which the layer types alternated according to the pattern AABAAB.... If $W_B \approx 50\%$, structures formed that had an almost ordered alteration of layer types according to the pattern ABAB.... (6) Simple criteria were established that permitted the determination of the fine structural features of homoionic Mg-, Na-, and Ca-hydrobiotites, as well as hydrobiotites having complex compositions of exchangeable cations.

Indirect methods

General outline. Despite the obvious advantages, the direct method has certain limitations. For example, it is difficult to use if an XRD pattern contains an insufficient number of basal reflections, e.g., if the smallest d-value is greater than 2 Å (Diakonov, 1974, 1978). Problems exist in the interpretation of $\mathcal{P}z(z)$ due to the presence of additional peaks. Certain difficulties also arise if layers of different types are segregated in a mixed-layer structure. Finally, the method cannot be used if a sample is polymineralic such that basal reflections of different phases overlap on the XRD pattern.

In the last two decades, indirect methods have been developed that permit the calculation of XRD curves from structural models; the calculated profiles are then compared with those obtained experimentally (Reynolds, 1967, 1980; Reynolds and Hower, 1970; Drits and Sakharov, 1976; Plançon, 1981; Sakharov et al., 1982a, 1982b, 1983). The advantage of these methods is that they permit a systematic and purposeful study of diffraction effects from all structural models possible. Thus, the influence on XRD patterns of each parameter that characterizes a particular structural feature of a mixed-layer sample can be demonstrated (e.g., the contents and thicknesses of different layer types, different types of interstratification, thicknesses and distribution of coherent domains, and degrees of orientation of particles). Analysis of these data permits elaboration of XRD criteria that may be used to determine the main structural features of real mixed-layer minerals. This approach is especially useful if the direct method cannot be used. One must bear in mind, however, that the successful application of indirect methods requires that the initial structural models are as close to the real structure of the mineral as possible. This requirement is not easily met, because numerous parameters are needed to characterize an interstratified model. For example, Środoń (1980) showed that in I/S the thickness of glycolated smectite layers varies within certain limits depending on the nature of the exchangeable cations, the layer charge, and other factors. Diakonov and Volosnykh (1979) described a similar effect for I/S saturated by glycerol. The thickness of K-mica-like layers was also found to depend on their composition, varying from 10 Å for illite to 9.85 Å for leucophyllite (Sokolova et al., 1976, 1978).

Certain difficulties also arise due to an ambiguity in the determination of the mean thickness of coherent domains[1] and distribution of thicknesses. For example, increase of the average total number of layers in the crystals (N) from 10 to 20 changes the positions of basal reflections. As a result, mixed-layer models having different structural features may yield XRD curves having identical d-values if the thicknesses of mica-like and smectite-like layers and N values are manipulated. Bearing in mind all of these difficulties, the indirect approach can be used to determine a set of probability parameters that describe adequately the distribution of layer types in the crystals of a mixed-layer mineral.

Quasihomogeneous mixed-layer models. Several structural models may be envisaged. The existing indirect methods are based on a model which implies that the probability of the occurrence of a layer of a given type depends on the nearest neighbors only. To describe the sequence of layer types in terms of this model, it is convenient to introduce the nearest-neighbor ordering factor R (Jagodzinski, 1949). R signifies the number of the preceding layers that affect the probability of occurrence of the final layer. The use of this model leads to the assumption that a powder of a mixed-layer sample is a physical assemblage of statistically weighted crystals having all the possible compositions (Drits and Sakharov, 1976; Drits et al., 1984; Reynolds, 1980).

Histograms showing the change of the degree of heterogeneity of the composition of coherent domains in mixed-layer samples as a function of the total number of layers N, the contents of layers A and B, and the pattern in the sequence of layer types are given in Figure 1.

If N = 5, R = 0, and W_A = 60% (Figure 1a), a mixed-layer sample will contain crystals having all the possible compositions; of all the crystals, those containing only A layers will comprise 8%, those containing only B layers 1%, those containing A^3B^2 domains 35%, etc. (A^nB^m signifies a crystal containing n·A layers and m·B layers regardless of the sequence in their alternation). An increase of the portion of A layers to as much as 80% with R = 0 (Figure 1b) increases the content of crystals containing A layers only to 32%, and the degree of the heterogeneity of the sample decreases due to the decrease of the abundancies of A^2B^3 and AB^4 crystals. If limitations are imposed on the sequence of layer types by excluding BB pairs, the degree of homogeneity of mixed-layer sample having N = 5, W_A = 60%, W(BB) = 0, and R = 1 (Figure 1c) increases considerably, as the portion of crystals A^3B^2 is 65%.

If, for R = 0 and W_A = 60%, the total number of layers in the crystal is increased to N = 10 (Figure 1d), the degree of heterogeneity decreases only slightly as compared with N = 5. Here, the portion of crystals having the composition A^6B^4 is only 25%, and the remaining 75% of the crystals have other compositions. If N = 10, W_A = 60%, W(BB) = 0, and R = 1 (Figure 1e), crystals having $N_B > N_A$ should be excluded. As a result, the homogeneity of the sample increases, although the portion of domains having the composition A^6B^4 remains less than 50%. An increase of W_A to 80% and R to 2 (Figure 1f) leads to a further decrease of the degree of heterogeneity due to loss of A^5B^5 crystals. This decrease in heterogeneity, however, is accompanied by the increase (from 0 to 4%) of the portion of crystals consisting of A layers only. If W_A = 90% and R = 3 (Figure 1g) 27% of the crystals have A layers only.

Thus, in terms of the model discussed, the heterogeneity of a powder of a mixed-layer sample results both from the heterogeneity of the composition of coherent domains (A:B ratio) and from the different dis-

---

[1] The terms "coherent domain" and "crystal" are used here synonymously; the term "particle" is used here for an assemblage of coherent domains.

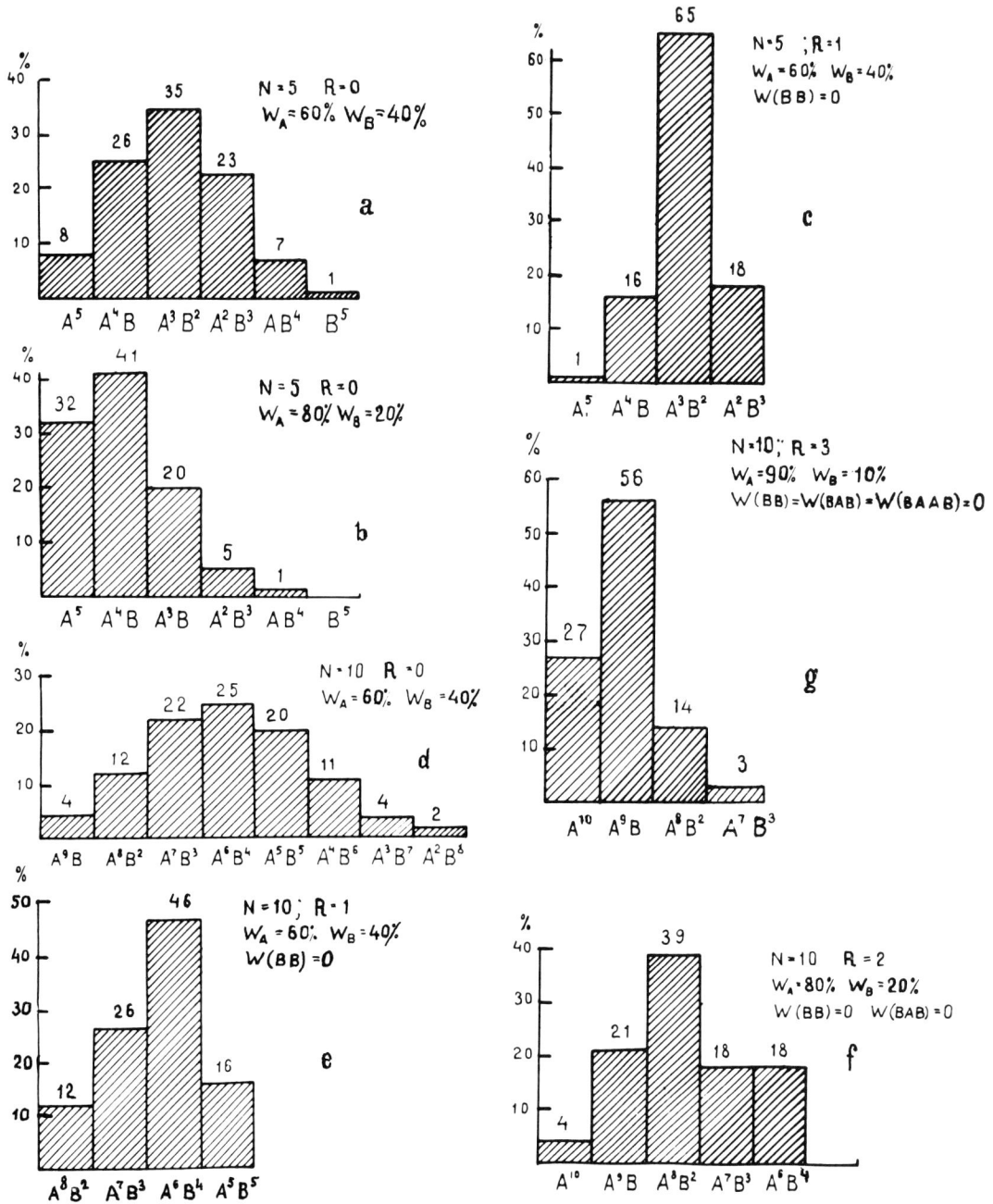

Figure 1.   Distribution of the composition of coherent domains ($A^m B^n$) in quasihomogeneous mixed-layer samples as a function of the total number (N) of layers in the crystals, the portions of the different types of layers W_A and W_B, and the pattern of their alternation. To characterize the latter, independent probability parameters are given for each R, e.g., $W(BB) = 0$ for $R = 1$, $W(BB) = W(BAB) = W(BAAB) = 0$ for $R = 3$. Ordinate axes = percentage of layers.

tributions of layer types in crystals of a given composition (e.g., $A^2 B^1$ crystal can be AAB or ABA).

In addition, the heterogeneity of mixed-layer samples depends on the range within which thicknesses of coherent domains vary and on the distribution of thicknesses. Because the structure and composition of as-

semblages of mixed-layer crystals may be described unambiguously by a small number of independent parameters (MacEwan, 1958; Drits and Sakharov, 1976), such models are herein termed quasihomogeneous.

In terms of the quasihomogeneous model, an XRD pattern of a powder of a mixed-layer sample is a statistically weighted sum of XRD patterns of coherent domains having different compositions and different distributions of layer types. Calculations show that, despite the wide variation in composition of the domains, integral XRD curves, even from random mixed-layer samples, contain fairly intense and sharp peaks. Thus, XRD reflects primarily the averaged structural characteristics of the mixed-layer samples and has a relatively low sensitivity to the specific forms of local structural disorder in the c direction. Interstratification as such, however, has a relatively small influence on the positions and the intensities of nonbasal reflections. For example, calculations show that an XRD curve of interstratified phlogopite/vermiculite having random alternation of equal portions of 10- and 14-Å layers contains not only basal and $hk0$ reflections, but also other intense nonbasal reflections (Sakharov et al., 1983).

Finally, the degree of heterogeneity in the composition of coherent domains will be the same as that of the particles of a mixed-layer sample only if their thicknesses are the same. According to the XRD data, the total number of layers in coherent domains in disperse mixed-layer minerals usually does not exceed 10–15. These data, however, do not permit the relation between the thicknesses of coherent domains and those of particles to be estimated. Therefore, if the total number of layers in the particles of a mixed-layer sample is much greater than that in the coherent domains, the degree of heterogeneity in the composition of the particles may be much lower than that of the coherent domains.

Physical mixtures of quasihomogeneous mixed-layer samples. The degree of heterogeneity of mixed-layer minerals obviously reflects the degree of heterogeneity in the microenvironment at the time of their formation or at the time of their deposition. For example, mixing of mixed-layer particles may take place during their transportation and sedimentation. Thus, it is worthwhile to analyze the diffraction effects that would be observed from a physical mixture of quasihomogeneous mixed-layer samples.[2] It is first necessary to know whether a physical mixture of two quasihomogeneous mixed-layer samples can always be distinguished from a quasihomogeneous sample.

Figure 2c is a histogram showing the distribution of the compositions in the crystals in a physical mixture of equal portions of two randomly interstratified samples, Figures 2a and b, having the same N = 10 and R = 0, but different $W_A:W_B$ ratios; sample I = 0.6:0.4

[2] Physical mixtures of minerals consisting of particles having ordered structures and homogeneous compositions, such as smectite, illite, rectorite (ABAB), tarasovite (AABAAB) will not be considered here, as they can be unambiguously identified using XRD patterns.

and sample II = 0.8:0.2. The bottom histogram (Figure 2d) represents a quasihomogeneous assemblage of crystals having $W_A:W_B$ = 0.7:0.3 and R = 0. Because the bottom two histograms are quite similar, an XRD curve having $W_A:W_B$ = 0.7:0.3, R = 0, and N = 10 probably cannot be distinguished from one which is a mean-weighted sum of the curves calculated for R = 0, N = 10, and $W_A:W_B$ = 0.6:0.4 and 0.8:0.2, respectively.

To test this conclusion, calculations of XRD curves were made for the four types of I/S (Figure 2) having 10-Å illite-like layers, and 17.8-Å, glycolated smectite-like layers. Curves c and d coincide, confirming the conclusion from the histograms. A similar result was obtained for the curves calculated for a quasihomogeneous sample having W_A = 30% and R = 0 and a mixture of equal portions of two I/S samples having W_A = 40% and 20% and R = 0. The same comparison carried out for I/S samples saturated by ethylene glycol yielded similar results.

Curves calculated for samples like those in Figure 2 but having the additional limitation imposed on layer stacking of R = 1 ($W_A > W_B$, P_{BB} = 0) are shown in Figure 3. Comparing curves c and d, the peaks also coincide for all the scattering angles (θ); only the intensity of the 13.9-Å peak differs. This difference, however, can hardly be distinguished in practice due to a number of factors affecting the intensity distribution in the low-angle region which are difficult to take into account (e.g., the change of the degree of orientation of particles with sample thickness and different modes in the distribution of thicknesses of coherent domains). Moreover, it should be taken into account that the comparison is being carried out under ideal conditions, i.e., the degree of heterogeneity of the samples is known exactly; only two quasihomogeneous samples taken in equal portions were mixed; and parameters characterizing structural peculiarities of mixed-layer samples were fixed.

Figure 4a shows a curve calculated for a mixture of two quasihomogeneous samples having the same proportion of A layers (W_A = 70%), but differing in the sequence of layer types. One part of the mixture has R = 0; the other has R = 1 and P_{BB} = 0. Figure 4b corresponds to a quasihomogeneous sample having R = 1, W_A = 70%, P_{BB} = 0.15 (50% of maximum possible ordering). The curves again coincide in the whole range of θ except for a minor difference in the higher order peaks and in the shape of the first low-angle reflection.

Calculations show that a similar diffraction effect appears if the curve simulated for the mixture of two quasihomogeneous I/S samples ($W_B = P_{BB}$ = 0.4, R = 0 for the first sample and W_B = 0.4, P_{BB} = 0, R = 1 for the second) is compared with that for a quasihomogeneous I/S having W_B = 0.4, P_{BB} = 0.2, R = 1. A small difference is observed here only in the region of the low angles of theta, regardless of whether the I/S is saturated with glycerol or ethylene glycol.

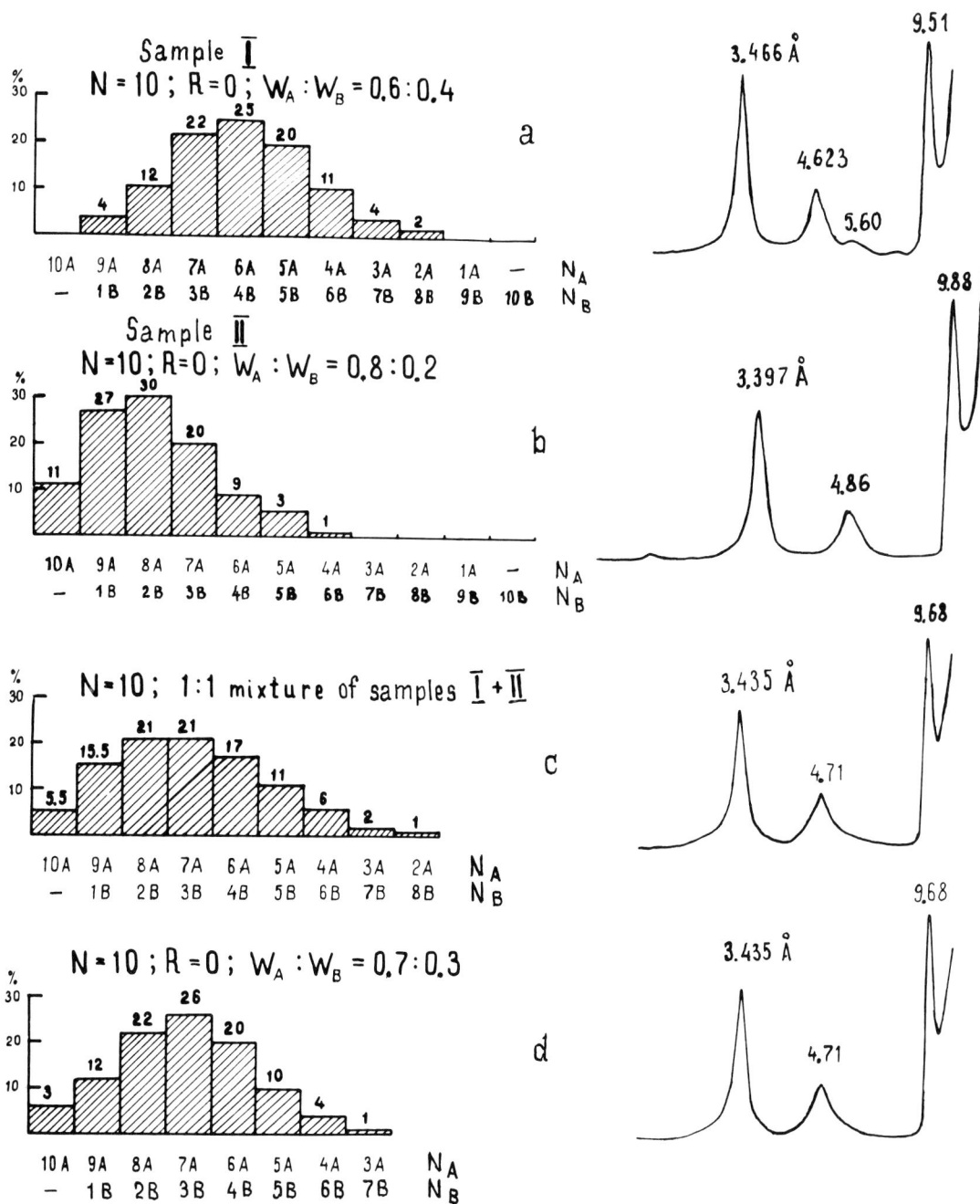

Figure 2. Comparison of the histograms and X-ray powder diffraction curves of the quasihomogeneous mixed-layer samples I/S and a mixture of sample I (N = 10; R = 0; $W_A:W_B$ = 0.6:0.4) and sample II (N = 10; R = 0; $W_A:W_B$ = 0.8:0.2). The histograms show the distribution of concentrations for coherent domains having N_A layers A and N_B layers B for the constant total number of layers in the crystal, $N = N_A + N_B$. Values for N and independent probability parameters are shown above each histogram. Smectite layers in I/S were saturated with glycerol; d_A = 10.0 and d_B = 17.8 Å. For quasihomogeneous I/S, domains of different thicknesses having N = 7–13 layers were assumed to be equally distributed. Ordinate axes = percentage of layers.

If equal quantities of two mixed-layer samples are mixed, each with N = 10 and R = 0, one containing 60% and the other 20% of A layers, the histogram for the mixture should be markedly different from that for a quasihomogeneous sample having the averaged parameters of W_A = 40%, N = 10, and R = 0. The histogram for the mixture should indeed contain two maxima. One of these corresponds to the composition

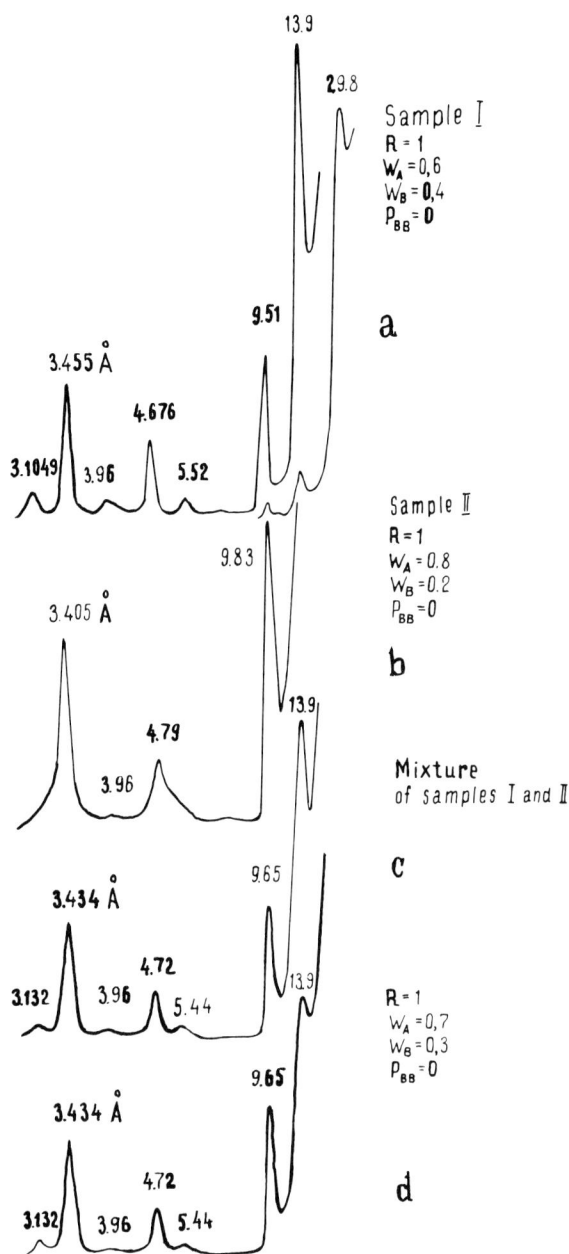

Figure 3. Comparison of the X-ray powder diffraction curves calculated for the quasihomogeneous ordered (R = 1) mixed-layer illite/smectite (I/S) and for the mixture of the sample I (N = 10; R = 0; W_A:W_B = 0.6:0.4) and sample II (N = 10; R = 0; W_A:W_B = 0.8:0.2). Smectite layers in I/S were saturated by glycerol; d_A = 10.0 and d_B = 17.8 Å. For quasihomogeneous I/S, domains of different thicknesses with N from 7 to 13 layers were assumed equally distributed.

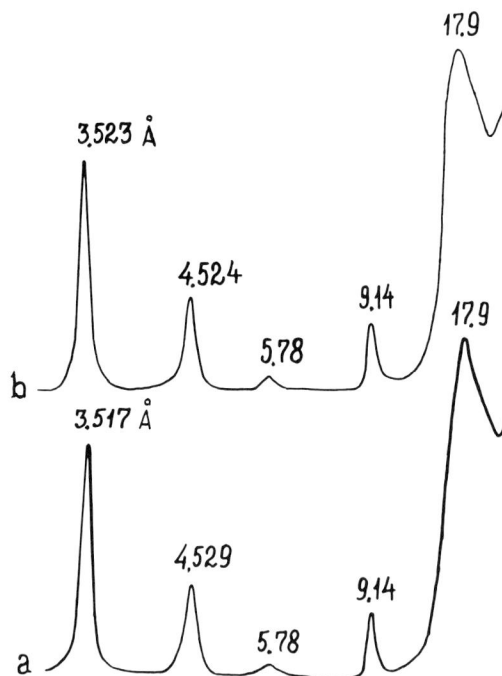

Figure 4. Comparison of X-ray powder diffraction curves calculated for illite/smectite (I/S) samples saturated by glycerol (d_A = 10 Å and d_B = 17.8 Å). The upper curve corresponds to quasihomogeneous I/S, having W_A = 70%, P_{BB} = 0.15, and R = 1. The lower curve corresponds to a 1:1 mixture of quasihomogeneous I/S having W_A = 70%, R = 0, and W_A = 70%, P_{BB} = 0, R = 1. The quasihomogeneous samples were supposed to have equally distributed coherent domains with thicknesses ranging from 7 to 13 layers.

6. On the other hand, a quasihomogeneous I/S having W_A = 40% may be described by a histogram with two maxima if different layer types alternate with a certain degree of segregation. Here the value of R = 1 should be used to describe the pattern in the interlayering.

Figure 5 shows two XRD curves for the glycerol-saturated I/S described above. The lower curve (a) corresponds to a mixture of equal quantities of quasihomogeneous I/S having R = 0, W_A = 60 and 20%, respectively. The upper curve (b) corresponds to a quasihomogeneous sample having R = 1, W_A = 40%, P_{AA} = 0.5, indicating a tendency toward segregation. The intensity distribution and positions of the basal reflections are similar in both curves. Thus, in certain cases, the degree of heterogeneity for a physical mixture of quasihomogeneous samples may be similar to that for a quasihomogeneous sample. It is evident that in these cases it is difficult to distinguish between a mixture of samples and a quasihomogeneous sample having the same average W_A using XRD patterns.

Interpretation of XRD patterns from physical mixtures of quasihomogeneous samples is less ambiguous if these patterns contain "split" basal reflections cor-

of the crystals having N_A = 2 and N_B = 8; the other corresponds to N_A = 6 and N_B = 4. The histogram for the quasihomogeneous sample having W_A = 40% and R = 0 contains a single maximum corresponding to the composition of the crystals with N_A = 4 and N_B =

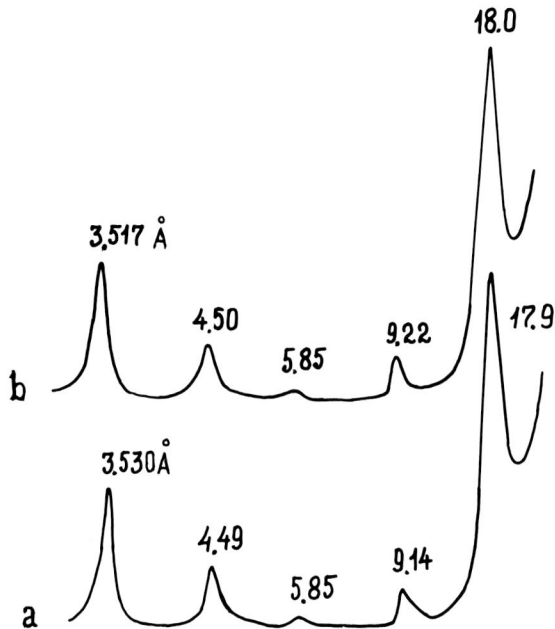

Figure 5. Comparison of X-ray powder diffraction curves calculated for illite/smectite (I/S) saturated by glycerol (d_A = 10 Å and d_B = 17.8 Å). The upper curve corresponds to a quasihomogeneous I/S having W_A = 40%, P_{AA} = 0.5, and R = 1. The lower curve corresponds to the physical mixture of random quasihomogeneous I/S samples having W_A = 60 and 20%, respectively, and R = 0. For quasihomogeneous I/S, domains of different thicknesses having N from 5 to 15 layers were assumed equally distributed.

Figure 6. Comparison of X-ray powder diffraction curves calculated for mixtures of equal quantities of smectite and illite/smectite (I/S) having W_A = 45%, P_{AA} = 0.15, and R = 1 that differ in the range for coherent domain thicknesses marked above each curve. Domain thicknesses were assumed to be equally distributed. Smectite layers in I/S were saturated by ethylene glycol; d_A = 10 Å, d_B = 16.86 Å.

responding to the components of a mixture as exists if quasihomogeneous samples composing the mixture differ essentially in proportions of different layer types and/or the pattern in the interlayering and in the intensities of the reflections being "mixed". For I/S, such patterns were observed from mixtures of quasihomogeneous samples having, for example, the following parameters: W_A = 40%, R = 0, and W_A = 40%, P_{AA} = 0, R = 1; W_A = 20%, R = 0, and W_A = 80%, R = 0; W_A = 60%, P_{BB} = 0, R = 1, and W_A = 20%, P_{AA} = 0, R = 1, etc. Interpretation of diffraction patterns may be complicated and ambiguous even for physical mixtures of samples having essentially different degrees of heterogeneity.

Figure 6 shows curves calculated for a physical mixture of pure smectite and a quasihomogeneous I/S sample with W_A = 45% and W_B = 55%, P_{AA} = 0.15, R = 1. Smectite and I/S were mixed in equal quantities, but it was assumed in Figure 6a that the total number of layers in coherent domains ranged from 4 to 8, whereas in Figure 6b it varied from 1 to 10, domains of each size having been distributed equally. From the appearance of these curves, it is difficult to infer that the sample contains 50% I/S having a high tendency to ordering. Such a conclusion, perhaps, could be drawn

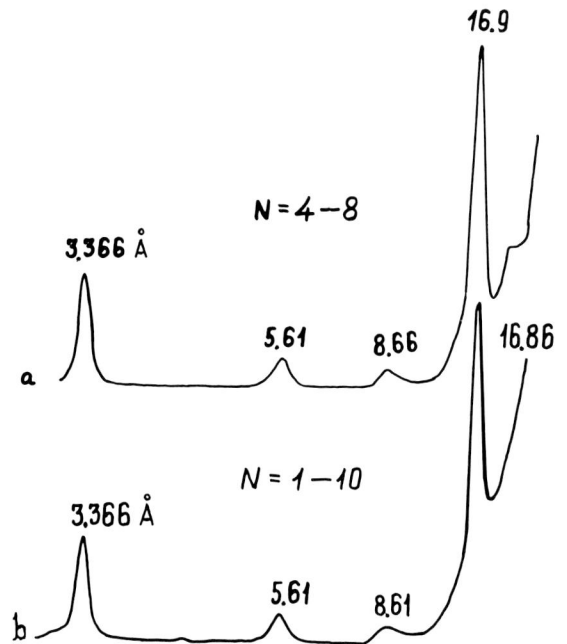

in Figure 6a from the poorly resolved superorder maximum in the low-angle region of the XRD pattern (d ≈ 29 Å). This peak, however, is absent in curve b, and therefore the true structure of the sample cannot be inferred.

It should be stressed that a mixture of smectite and rectorite in any proportion can be identified unambiguously using XRD data. The above considerations of calculated XRD patterns are in agreement with the experiments of Gradusov (1971) who examined artificial mixtures of interstratified minerals.

Homogeneous mixed-layer models. Statistical models in which the crystals have the same composition but differ in the distribution of layer types may be also envisaged. These models have been termed homogeneous models (Drits *et al.*, 1984). The problem is to compare XRD curves calculated for homogeneous and quasihomogeneous models with the same contents of layer types and the pattern in their sequence.

Figure 7 shows histograms for the distribution of compositions of coherent domains in two random mixed-layer samples having the same mean portion of A layers (W_A = 70%), one of them (the lower histogram) being homogeneous and the other (the upper histogram), quasihomogeneous. The mean number of layers in a coherent domain was assumed to be 10 for the quasihomogeneous sample. In the same figure, XRD curves calculated for I/S samples saturated by glycerol are shown. The two upper curves correspond to quasihomogeneous and homogeneous I/S samples, respectively, where $W_A = 70\%$, $R = 0$, $d_A = 10$ and $d_B = 17.8$ Å. In both samples the total number of layers in coherent domains, N, was assumed to range from 7 to 13, with domains of different thicknesses being distributed equally. For the homogeneous sample the contents of A layers, N_A/N, in different crystals were assumed to be 0.65–0.75 (see Figure 7). The comparison of the curves shows that they are similar with respect to the intensity distribution, but not identical in terms of shapes and positions of basal reflections. Thus, theoretically, these two models may be distinguished using XRD. It is not always easy, however, to do so because of the ambiguity in the precise determination of thicknesses of individual layers and coherent domains. For example, Figure 7c shows an XRD curve calculated for a quasihomogeneous I/S sample having $W_A = 65\%$, $W_B = 35\%$, $R = 0$, $d_A = 9.9$ Å and $d_B = 17.5$ Å. This curve is similar to the one in Figure 7b calculated for a random homogeneous I/S sample having $W_A = 70\%$, $d_A = 10$ Å and $d_B = 17.8$ Å and the same distribution of thicknesses of coherent domains. Very precise studies are therefore required to reveal the existence of homogeneous, quasihomogeneous, and heterogeneous mixed-layer samples.

Combination of direct and indirect methods. The most complete information on the structure of mixed-layer minerals may be obtained by a combination of direct and indirect methods. This combination is especially useful for the study of metabentonites and hydrothermal alteration products because these materials are usually either monomineralic or contain minerals that can be easily separated from or that do not interfere with the I/S. What does the initial use of the direct method provide? As has been shown, many different models may be proposed, and their number increases sharply with an increase in R. Application of the direct method narrows the possibilities for the correct model. First, data may be obtained immediately on the number of different layers and their thicknesses; such data are not at all easy to determine *a priori*. Second, the value of R as well as the layer combinations that are absent in the crystals may be determined directly. For example, Diakonov and Volosnykh (1979) showed, using the direct method, that a set of probability parameters with $R \geq 4$ is required to describe the pattern in the sequence of layer types in some I/S samples from hydrothermal environments. Such a pattern in the sequence of layer types is practically impossible to determine solely by the indirect method. At the same time, certain theoretical limitations are also present in the method of Diakonov. Therefore, the data obtained by the direct method should be refined by the indirect

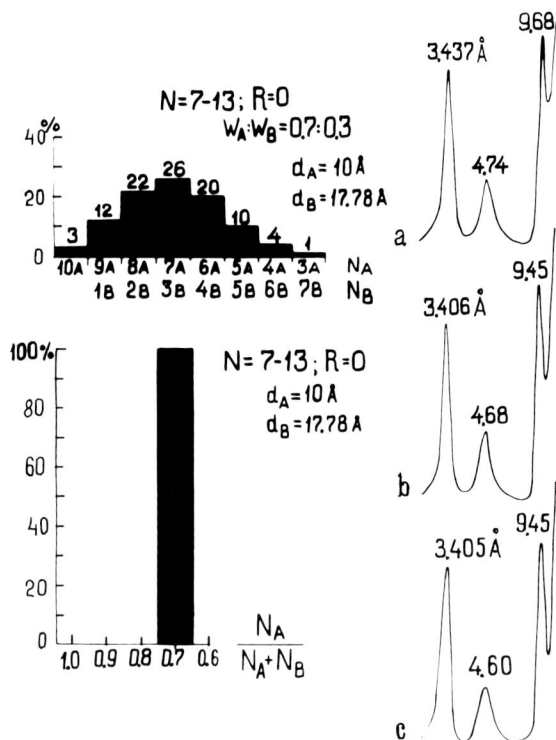

Figure 7. Comparison of histograms and diffraction curves calculated for random quasihomogeneous and homogeneous illite/smectite (I/S) saturated by glycerol. The upper histogram shows the contents of coherent domains containing N_A layers A and N_B layers B. The lower histogram shows that the homogeneous I/S contains domains in which the contents of A layers lie within the interval 65–75%. The upper diffraction curve corresponds to a quasihomogeneous I/S having $W_A = 70\%$, $R = 0$, the middle one to a homogeneous I/S having $N_A/N = 0.70 \pm 0.05$, the lower one to a quasihomogeneous I/S having $W_A = 65\%$, $R = 0$, $d_A = 9.9$ Å and $d_B = 17.5$ Å. Domains of thicknesses ranging from $N = 7$ to $N = 13$ are equally distributed for all the samples.

methods if the fine structural features of the interstratified sample under study are needed.

SELECTED AREA ELECTRON DIFFRACTION (SAD)

Previously, SAD has not been used as an independent means of structure analysis of poorly crystallized minerals, including mixed-layer minerals. Intensities of reflections in point electron diffraction patterns can be used for the determination of structural models of poorly crystallized minerals (Drits, 1981). SAD studies of mixed-layer minerals is based on the analysis of basal reflections obtained from bent edges and folds of particles (Gorshkov, 1970). Investigations of interstratified manganese minerals from Fe-Mn nodules demonstrate vividly the unique possibilities of SAD. The crystallinity of these minerals is so low that XRD was not effective for determining structural models or even for reliable identification. Therefore, it is not surprising that the problem of what minerals comprise oceanic and continental Fe-Mn nodules is still under discussion.

Studies of Fe-Mn nodules by a combination of SAD, X-ray dispersive analysis, and other methods have revealed a new world of minerals (Chukhrov et al., 1980a, 1980b, 1982, 1983a, 1983b, 1984; Drits et al., 1985). Most of these manganese minerals have layer structures and in many respects resemble phyllosilicates (Drits et al., 1985). In particular, ordered and random interstratified minerals are present. Mixed-layer minerals having an ordered alternation of layer types bear the general name of asbolanes (Chukhrov et al., 1980a, 1980b, 1982, 1983). Two essentially different groups of asbolanes have been found. In minerals of the first group, cations in both layer types are coordinated octahedrally (Chukhrov et al., 1980a, 1980b). In minerals of the second group, cations in different layer types have different coordination, i.e., octahedral and tetrahedral (Chukhrov et al., 1982, 1983). Structures of the first group consist of octahedral layers having the idealized composition MnO_2 alternating with octahedral layers having the composition $M(OH)_2$, where M may be Ni, Co, Mn^{2+}, or Mn^{3+}, and differences in the size of cations in different layers may lead to misfits in the basal plane of the layers. The second group consists of minerals that contain Mn^{4+} coordinated octahedrally and Co coordinated tetrahedrally, as well as of minerals having Mn^{4+} in tetrahedral layers alternating with Al, Ni, and Zn in octahedral layers. Different layer types have different parameters for their hexagonal unit cell. The general feature of all the asbolanes is that one of the layer types always has an island-like, discontinuous structure.

In oceanic Fe-Mn nodules, randomly interstratified asbolane-buserite is especially abundant (Chukhrov et al., 1983a). Its structure consists of Mn^{4+} octahedral layers sandwiching either water molecules and exchangeable cations (buserite interlayers) or $M(OH)_2$ layers (asbolane interlayers), having a random distribution.

STRUCTURAL ASPECTS OF THE FORMATION OF DIOCTAHEDRAL INTERSTRATIFIED MICA/SMECTITE

Numerous papers in the literature deal with the mechanism of formation of both natural and synthetic mixed-layer phases (Eberl, 1984; Eberl and Hower, 1976; Frank-Kamenetsky et al., 1983; Inoue, 1983; Roberson and Lahann, 1981; Środoń and Eberl, 1984). Nevertheless, many aspects of this problem are still unclear. The existence of a complete series from smectite to mica via interstratified mica/smectite may imply solid-phase transformations of smectite into mica. In other words, mixed-layer phases may preserve the initial matrix, which here is smectite crystals. In such materials, the main structural mechanism of the transformation is the fixation of K cations in an increasing number of interlayers of smectite particles accompanied by substitutions in the 2:1 layers that increase layer charge. Mica/smectite may also form by a dissolution-reprecipitation mechanism. In particular, in modern sediments in the rift zones of the Red Sea, mixed-layer nontronite/celadonite forms by subsequent recrystallization of nontronite. The change of physicochemical conditions and kinetic factors apparently led to the growth of new crystals from the dissolution of predecessors. In other words, the matrix of the nontronite crystals was not inherited (Butuzova et al., 1978, 1983).

The solid-phase transformation mechanism for the smectite-to-illite conversion in a diagenetic environment via intermediate mixed-layer phases is a widely accepted hypothesis. Much data confirm or at least do not contradict this point of view (Shutov et al., 1969; Eberl, 1984; Roberson and Lahann, 1981; Schultz, 1978; Środoń and Eberl, 1984). On the other hand, structural studies of I/S from diagenetic-hydrothermal environments do not completely support this hypothesis. The main structural features of I/S from areas of hydrothermal alteration may be summarized in brief as follows (see Diakonov and Volosnykh, 1979; Inoue and Utada, 1983). If the proportion of illite layers (A layers) exceeds 50%, BB pairs of smectite layers never occur in layer sequences. A strong tendency towards ordered alternation of A and B layers is observed, as among the possible layer subsequences BA^nB in which a certain n usually prevails (Figure 8). The most regular distribution of B layers among A layers is achieved if $0.3 \leq W(B) \leq 0.5$. If $W(B) > 0.5$, a strong tendency towards disordered layer alternation is observed. Interestingly, smectite layers, if fewer than 20%, are usually separated from each other by three or four illite layers, and the pattern in the sequence of layer types can be described only by $R \geq 4$.

I/S from rocks altered during diagenesis show similar structural features. Mixed-layer minerals containing predominantly smectite-like layers are usually characterized by a random alternation of layer types. On the contrary, a strong tendency towards ordering exists in mixed-layer minerals in which mica-like layers predominate. Note that structures have been found in metabentonites having BAAB as the prevalent subsequence, i.e., layers alternate with $R = 2$ (Sokolova et al., 1978). In the light of these data, a consequent series of solid phase transformations from mixed-layer structures having $W_B > W_A$ to those having $W_A > W_B$ is difficult to imagine. For example, replacement of B layers by A layers in a structure having $W_A = 0.55$, $W_B = 0.45$, $P_{BB} = 0$ should minimize the probability of occurrence for the subsequence BAAB which excludes mixed-layer structures having R = 2. If the sub-sequence BA^nB having a given n prevails in a mixed-layer structure, a solid-phase substitution of A layers for B layers could lead to mixed-layer phase having the prevalent subsequence $BA^{2n+1}B$, and never $BA^{n+1}B$. Thus, solid-phase transformations should impose certain limitations on the probability of occurrence of layer subsequences.

A possible explanation for the structural peculiarities of I/S having different contents of illite-like and smectite-like layers follows: at comparatively low temperatures, disordered I/S having $W_B > W_A$ is formed by solid-phase transformation, whereas at higher temperatures mixed-layer I/S having $W_A > W_B$ and $R > 1$ are formed by a dissolution-reprecipitation mechanism.

Additional research is obviously needed to solve the genesis problem. Note in this connection the results of the study of dioctahedral smectites saturated by K cations and then subjected to wetting and drying (WD) cycles according to the technique of Mamy and Gaultier (1976). Samples thus treated were studied by oblique texture electron diffraction (Tsipursky and Drits, 1984). Different smectites were found to have different patterns of distribution of the octahedral cations in *cis*- and *trans*-octahedra. In dioctahedral montmorillonites the *trans*-octahedra are usually occupied by cations at a probability of 70–100%. Simultaneously, in illites, including those containing 15–20% expandable layers, cations occupy *cis*-octahedra only, whereas *trans*-octahedra are vacant. Tsipursky and Drits (unpublished data) examined the samples of smectite and I/S described by Shutov *et al.* (1969) from a single diagenetic series having just such different distribution of cations over *cis*- and *trans*-octahedral sites. The transformation of montmorillonite to illite was accompanied by a redistribution of octahedral cations as well as by an increase in Al-for-Si substitution in tetrahedra in non-expandable layers. The latter effect is necessary to increase the negative layer charge (Eberl and Hower, 1976, 1977; Inoue and Utada, 1983; Środoń and Eberl, 1984).

The increase in the proportion of nonexpandable layers in I/S minerals should be accompanied by structural transformation of the 2:1 layers. The problem is to prove that this rearrangement can occur in the solid state. Another aspect of this problem is connected with factors favoring the irreversible fixation of K in the smectite interlayers. Besides the well-known role of the layer charge, temperature, and other factors, the mutual orientation of layers is essential. What is the role of wetting-drying (WD) cycles, after which the amount of K fixed in the smectite interlayers increases dramatically? During these cycles energy is spent reorienting adjacent layers, which leads to the formation of mica-like interlayers. The irreversible K fixation is possible only in those interlayers having centers of tetrahedral rings coinciding in projection in the (001) plane.

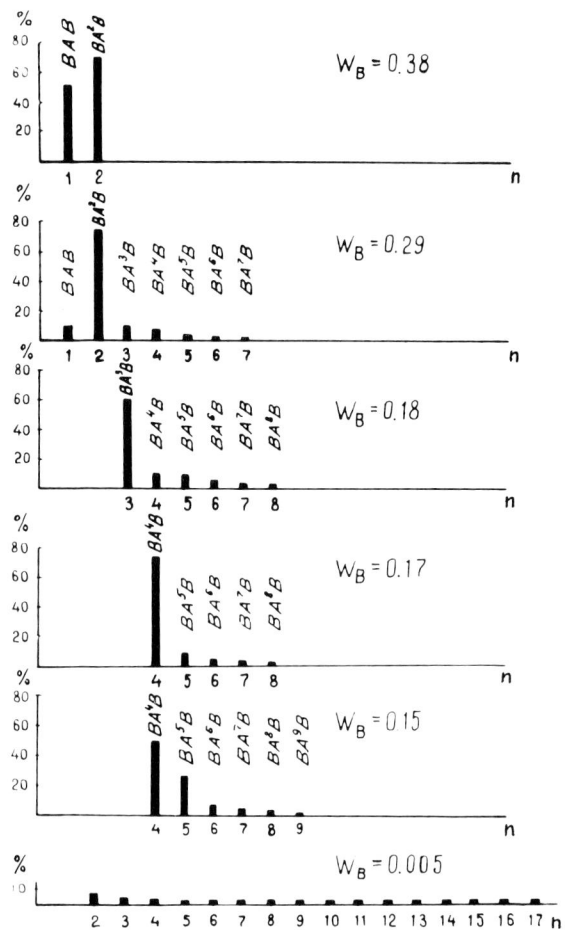

Figure 8. Histograms illustrating the contents of subsequences BA^nB in illite/smectite samples found in hydrothermal environments (after Diakonov and Volosnykh, 1974). W_B is the proportion of smectite-like layers. The height of each column corresponds to the contents of the subsequence BA^nB for the given value of n.

If this arrangement does not exist, K cations will remain exchangeable regardless of layer charge. Even small deviation of the azimuthal orientation of the adjacent 2:1 layers leads to significant consequences. Interlayer K cations will have an "uncomfortable" anionic environment, and the interlayers will not lose the ability to swell. On the contrary, if the hexagonal rings in interlayers face each other exactly, K fixation will take place even if the layer charge is relatively low. For example, after saturation by K, certain nontronites lose 80% of their ability to swell with glycerol even without WD cycles (Butuzova *et al.*, 1978). Apparently the lath-like form of nontronite crystals causes the unique azimuthal orientation for all the 2:1 layers favorable for fixation of K.

On the other hand, globular glauconitic material exists in which all the interlayers swell with glycerol despite a relatively high K_2O content (5%) (V. Muraviev, Geological Institute, Academy of Sciences, Moscow, personal communication). This swelling is due to different azimuthal orientations of 2:1 layers in microcrystals which results in the absence of "mica-like" interlayers.

To solve the problem of the solid-phase mechanism of the transformation of montmorillonites to illite, it is important to study the distribution of azimuthal orientations of 2:1 layers in mixed-layer phases and illites. Many illites are known to be of the 1*Md* polytype. If, for example, the distribution of layer orientations in I/S is the same as in the final illite, this would be a strong argument in favor of the solid-phase mechanism.

ACKNOWLEDGMENTS

I thank D. D. Eberl, L. G. Schultz, J. Środoń, and F. A. Mumpton for helpful criticism and for improving the English, B. Sakharov for calculation of the diffraction curves, and Bella Smoliar for translating the manuscript.

REFERENCES

Butuzova, G. Yu., Drits, V. A., Lisitsina, N. A., and Tsipursky, S. I. (1978) The dynamics of the formation of clay minerals in ore-bearing sediments of Atlantis II (Red Sea): *Litologiya i Poleznye Iskopayemye* **2**, 30–42.

Butuzova, G. Yu., Drits, V. A., Lisitsina, N. A., and Tsipursky, S. I. (1983) New data about the authigenic layer silicates in the metal-bearing sediments of Atlantis II (Red Sea): *Litologiya i Poleznye Iskopayemye* **5**, 82–88.

Chukhrov, F. V., Gorshkov, A. I., Vitovskaya, E. S., and Drits, V. A. (1980a) Crystal-chemical nature of Co-Ni asbolane: *Izv. Akad. Nauk S.S.S.R., Ser. Geol.* **6**, 73–81.

Chukhrov, F. V., Gorshkov, A. I., Vitovskaya, E. S., and Drits, V. A. (1980b) About crystal-chemical nature of Ni-asbolane: *Izv. Akad. Nauk S.S.S.R., Ser. Geol.* **9**, 108–120.

Chukhrov, F. V., Gorshkov, A. I., Drits, V. A., and Sivzev, A. V. (1982) New structural variation of asbolane: *Izv. Akad. Nauk S.S.S.R., Ser. Geol.* **6**, 69–77.

Chukhrov, F. V., Gorshkov, A. I., Drits, V. A., Shterenberg, L. E., and Sakharov, B. A. (1983a) Mixed-layer asbolane-buserite and asbolane minerals in the oceanic iron-manganese concretions: *Izv. Akad. Nauk S.S.S.R., Ser. Geol.* **5**, 91–100.

Chukhrov, F. V., Gorshkov, A. I., Drits, V. A., Finko, V. I., and Sivzev, A. V. (1983b) Structurally disordered asbolanes with the tetrahedral coordination of manganese: *Izv. Akad. Nauk S.S.S.R., Ser. Geol.* **12**, 85–95.

Chukhrov, F. V., Gorshkov, A. I., and Drits, V. A. (1984) Structural models and methods of study of buserite: *Izv. Akad. Nauk S.S.S.R., Ser. Geol.* **12**, 6–30.

Diakonov, Yu. S. (1961) About the application of Fourier analysis method to the interpretation of X-ray patterns of layer silicates with mixed-layer structure: *Crystallogr.* **7**, 624–625.

Diakonov, Yu. S. (1962) About the direct interpretation of X-ray patterns of mixed-layer minerals with the help of Fourier transform method: in *X-ray Study of Mineral Raw Materials, Vol. 1*, G. A. Sidorenko, ed., Gosgeoltechisdat, Moscow, 97–107.

Diakonov, Yu. S. (1974) The development of the direct method of interpretation of mixed-layer structures: in *Crystallochemistry and Structure of the Minerals*, G. A. Sidorenko, ed., Nauka, Leningrad, 33–43.

Diakonov, Yu. S. (1977) About the technique of calculation of layer distribution curves in the direct method of interpretation of mixed-layer structures: in *X-ray Study of Mineral Raw Materials, Vol. 11*, G. A. Sidorenko, ed., Nedra, Moscow, 75–80.

Diakonov, Yu. S. (1978) About the technique of interpretation of layer distribution curves in the direct method of interpretation of mixed-layer curves: in *X-ray Study of Mineral Raw Materials*, G. A. Sidorenko, ed., Nedra, Moscow, 7–14.

Diakonov, Yu. S. (1981) New data about the variety and identification of hydrobiotites: in *Crystallochemistry of Minerals*, V. A. Frank-Kamenetsky, ed., Nauka, Leningrad, 39–46.

Diakonov, Yu. S. (1983) Mixed-layer minerals: in *X-ray Study of the Basic Types of Rock-Forming Minerals*, V. A. Frank-Kamenetsky, ed., Nedra, Leningrad, 177–244.

Diakonov, Yu. S. and Volosnykh, G. T. (1979) Structural characteristics of mixed-layer illite-montmorillonites from the argillitized rocks close to ores: in *Crystallochemistry and Structural Mineralogy*, V. A. Frank-Kamenetsky, ed., Nauka, Leningrad, 69–82.

Drits, V. A. (1981) *The Structural Studies of Minerals by the Methods of Electron Microdiffraction and Electron Microscopy of High Resolution:* Nauka, Moscow, 350 pp.

Drits, V. A., Petrova, V. V., Gorshkov, A. I., Svalnov, V. M., and Sokolova, A. L. (1985) Manganese minerals from the Fe-Mn microconcretions in the sediments of the central part of Pacific Ocean and their postsedimentational transformations: *Litologiya i Poleznye Iskopayemye* **5**, 82–88.

Drits, V. A. and Sakharov, B. A. (1976) *X-ray Analysis of Mixed-Layer Minerals:* Nauka, Moscow, 256 pp.

Drits, V. A., Sakharov, B. A., Plançon, A., and Ben-Brahim, J. (1984) The distribution of layers in mixed-layer crystals of identical composition: *Crystallogr.* **29**, 350–355.

Eberl, D. D. (1984) Clay minerals formation and transformation in rocks and soils: *Phil. Trans. Royal Soc. Lond. A* **311**, 241–257.

Eberl, D. and Hower, J. (1976) Kinetics of illite formation: *Geol. Soc. Amer. Bull.* **87**, 1326–1330.

Eberl, D. and Hower, J. (1977) The hydrothermal transformation of sodium and potassium smectite into mixed-layer clay: *Clays & Clay Minerals* **25**, 215–227.

Frank-Kamenetsky, V. A., Kotov, N. N., and Goilo, E. A. (1983) *Transformation of Layer Silicates at the Increased P-T Parameters:* Nedra, Leningrad, 152 pp.

Gorshkov, A. I. (1970) The application of electron microdiffraction to obtaining the basal reflections of mica-layer silicates: *Izv. Akad. Nauk S.S.S.R., Ser. Geol.* **3**, 133–138.

Gradusov, B. P. (1971) X-ray patterns and direct Fourier-transformation of diffraction patterns of mixtures of illite, mixed layer mica-montmorillonite and montmorillonite: *Izv. Akad. Nauk S.S.S.R., Ser. Geol.* **12**, 86–92.

Gradusov, B. P. (1976) *Minerals with Mixed-Layer Structure in Soils:* Nauka, Moscow, 128 pp.

Inoue, A. (1983) Potassium fixation by clay minerals during hydrothermal treatment: *Clays & Clay Minerals* **31**, 81–91.

Inoue, A. and Utada, M. (1983) Further investigations of a conversion series of dioctahedral mica-smectites in the Shinzan hydrothermal alteration area, northeast Japan: *Clays & Clay Minerals* **31**, 401–412.

Jagodzinski, H. (1949) Eindimensionale Fehlordnung in Kristallen und ihr Einfluss auf die Rontgeninterferenzen. I. Berechnung des Fehlordnungsgrades aus der Rontgenintensitaten: *Acta Crystallogr.* **2**, 201–207.

Kakinoki, J. and Komura, Y. (1954a) Intensity of X-ray

diffraction by one-dimensionally disordered crystal. General derivation in the case of correlation range S ≥ 2: *J. Phys. Soc. Japan* **9**, 169–176.

Kakinoki, J. and Komura, Y. (1954b) Intensity of X-ray diffraction by one dimensionally disordered crystal. The close packed structure: *J. Phys. Soc. Japan* **9**, 177–183.

Kakinoki, J. and Komura, Y. (1965) Diffraction by a one-dimensionally disordered crystal. The intensity equation: *Acta Crystallogr.* **19**, 137–147.

MacEwan, D. M. C. (1956) Fourier transform methods for studying scattering from lamellar system. I. A direct method for analysing interstratified mixtures: *Kolloidzeitschrift* **149**, 96–108.

MacEwan, D. M. C. (1958) Fourier transform methods for studying X-ray scattering from lamellar systems. II. The calculation of X-ray diffraction effects for various types of interstratification: *Kolloidzeitschrift* **156**, 61–67.

Méring, J. (1949) L'interférence des rayons X dans les systèmes à stratification désordonnée: *Acta Crystallogr.* **2**, 371–377.

Mamy, J. and Gaultier, J. P. (1976) Les phenomènes de diffraction des rayonnements X et électroniques par les réseaux atomiques. Application à l'étude de l'ordre cristallin dans les minéraux argileux. II. Evolution structurale de la montmorillonite associée au phenomène de fixation irreversible du potassium: *Ann. Agron.* **27**, 1–16.

Plançon, A. (1981) Diffraction by layer structures containing different kinds of layers and stacking faults: *J. Appl. Cryst.* **14**, 300–304.

Plançon, A., Drits, V. A., Sakharov, B. A., Gilan, Z. I., and Ben-Brahim, J. (1983) Powder diffraction by layered minerals containing layers and/or stacking defects. Comparison between Markovian and non-Markovian models: *J. Appl. Cryst.* **16**, 62–69.

Plançon, A., Sakharov, B. A., and Drits, V. A. (1984) Diffractional effects from homogeneous mixed-layer crystals: *Crystallogr.* **29**, 657–662.

Reynolds, R. C. (1967) Interstratified clay systems: calculation of a total one-dimensional diffraction functions: *Amer. Mineral.* **52**, 661–672.

Reynolds, R. C. (1980) Interstratified clay minerals: in *Crystal Structures of Clay Minerals and their X-ray Identification,* G. W. Brindley and G. Brown, Mineralogical Society, London, 249–303.

Reynolds, R. C. and Hower, J. (1970) The nature of interlayering in mixed-layer illite-montmorillonite: *Clays & Clay Minerals* **18**, 25–36.

Roberson, H. E. and Lahann, R. W. (1981) Smectite-to-illite conversion rates: effects of solution chemistry: *Clays & Clay Minerals* **29**, 129–135.

Sakharov, B. A., Naumov, A. S., and Drits, V. A. (1982a) X-ray diffraction by mixed-layer structures with random distribution of stacking faults: *Dokl. Akad. Nauk S.S.S.R.* **265**, 339–343.

Sakharov, B. A., Naumov, A. S., and Drits, V. A. (1982b) X-ray intensities scattered by layer structure with short-range ordering parameters S ≥ 1 and G ≥ 1: *Dokl. Akad. Nauk S.S.S.R.* **265**, 871–874.

Sakharov, B. A., Naumov, A. S., and Drits, V. A. (1983) X-ray scattering by defect layer structures: *Crystallogr.* **28**, 951–958.

Schultz, L. G. (1978) Mixed-layer clays in the Pierre Shale and equivalent rocks, northern Great Plains region: *U.S. Geol. Surv. Prof. Pap.* **1064A**, 28 pp.

Shutov, V. D., Drits, V. A., and Sakharov, B. A. (1969) On the mechanism of a postsedimentary transformation of montmorillonite into hydromica: in *Proc. Int. Clay Conf., Tokyo, 1969, Vol. 1,* L. Heller and A. Weiss, eds., Israel Prog. Sci. Transl., Jerusalem, 523–532.

Sokolova, T. N., Drits, V. A., and Sokolova, A. L. (1976) Structural mineralogical characteristics and conditions of formation of leucophyllite from salt-bearing deposits of Inder Dome: *Litologiya i Poleznye Iskopayemye* **6**, 80–92.

Sokolova, T. N., Sakharov, B. A., and Drits, V. A. (1978) Mixed-layer leucophyllite-montmorillonite minerals: *Litologiya i Poleznye Iskopayemye* **6**, 87–101.

Środoń, J. (1980) Precise identification of illite/smectite interstratifications by X-ray powder diffraction: *Clays & Clay Minerals* **28**, 401–411.

Środoń, J. and Eberl, D. (1984) Illite: in *Micas, Reviews in Mineralogy 13,* S. W. Bailey, ed., Mineral. Soc. America, Washington, D.C., 495–544.

Tsipursky, S. I. and Drits, V. A. (1984) The distribution of octahedral cations in the 2:1 layers of dioctahedral smectites studied by oblique-texture electron diffraction: *Clay Miner.* **19**, 177–193.

Proceedings of the International Clay Conference, Denver, 1985, L. G. Schultz, H. van Olphen, and F. A. Mumpton, eds.,
The Clay Minerals Society, Bloomington, Indiana, 46–52 (1987).

HYDRATION ENERGIES OF SMECTITES: A MODEL FOR GLAUCONITE, ILLITE, AND CORRENSITE FORMATION

Y. Tardy and O. Touret

Institut de Géologie, Université Louis Pasteur and Centre de Sédimentologie et de Géochimie de la Surface
CNRS, 1, rue Blessig, 67 084 Strasbourg, France

Abstract—Dehydration isotherms of five smectites and one vermiculite saturated with Na, K, Mg, and Ca were used to calculate the hydration energies of clay minerals in different water-activity and cation-exchange conditions. As the water activity decreased, differences in hydration energies appeared to be dependent on the nature of the exchangeable cation and on the type of clay mineral involved. These differences in hydration energies were used to evaluate the variations of the apparent cation-exchange constants as a function of the water activity. As the water activity decreased, the less hydrated cations (i.e., K^+ for all samples and Na^+ or Mg^{2+} for some) were strongly preferred to the more hydrated cations (i.e., Ca^{2+} for all samples and Na^+ or Mg^{2+} for most). As the water activity decreased, nontronite showed a significant selectivity for K^+, and hectorite had a strong preference for Mg^{2+}. From these data, a model for the transformation of nontronite into glauconite, montmorillonite into illite, and stevensite into corrensite or chlorite during sediment compaction and burial diagenesis has been developed.

Key Words—Cation exchange, Chlorite, Corrensite, Dehydration, Diagenesis, Glauconite, Hydration energy, Illite, Montmorillonite, Water.

INTRODUCTION

Smectites transform during compaction and burial diagenesis in at least three different manners: beidellites or montmorillonites → illites, saponites or stevensites → corrensite or chlorites, and nontronites → glauconites. Chang and MacKenzie (1986) recently reviewed the conditions of clay diagenesis and compared the evolution of dioctahedral and trioctahedral smectites subjected to burial diagenesis. They noted that as temperature increases, dioctahedral smectites (beidellites or montmorillonites) transform into illites, whereas trioctahedral smectites (saponites or stevensites) transform into corrensite or chlorite.

The compositional changes accompanying the smectite-to-illite transformation, as deduced from published chemical analyses, were summarized by Hower and Mowatt (1966), Garrels and Mackenzie (1974), and Hower (1981). With increasing number of illite layers, the chief chemical changes are a gain of interlayer K, increasing substitution of Al for Si in the tetrahedral layer, and loss of octahedral Mg or Fe (Dunoyer de Segonzac, 1970). The compositional changes accompanying the saponite-to-chlorite transformation are a loss of interlayer cations, an increase in the substitution of Al for Si in the tetrahedral layer, and an overall increase in the total Fe and Mg contents, followed by a fixation of Mg in the brucitic layer. Chang and Mackenzie (1986) concluded that the ordering of illite/smectite (I/S) takes place at 90°–115°C, whereas the ordering of chlorite/saponite (corrensite) occurs at 60°–70°C. These authors also showed that each of the two different transformations may occur in the same

sediment volume, that is, in the same chemical environment.

Glauconite commonly forms in compacted Fe-bearing marine sediments during the earliest stage of diagenesis. In such environments, glauconite may form directly, but dioctahedral, Fe-rich smectite (nontronite) may also transform into mixed-layered nontronite/glauconite (Porrenga, 1967; Lamboy, 1967; Giresse and Odin, 1973). The compositional changes accompanying these transformations result in a gain of K and a gain of Al substituting for tetrahedral Si, and a loss of water (Burst, 1958; Thompson and Hower, 1975).

Most micaceous minerals, such as I/S, illite, and chlorite, form at elevated temperature and pressure. Glauconite, however, is unique in that it forms at low temperature and pressure. In marine environments, suspended clay materials, especially smectites, are not especially enriched in K or Mg, but simply select the seawater cations in the following amounts, as carefully measured by Sayles and Mangelsdorf (1977, 1979):

Na (51%), Mg (30%), Ca (12%), K (7%).

Fritz (1981), on the basis of an ideal solid solution model and using the cation-exchange constants of Tardy and Garrels (1974), calculated the amounts as:

Na (60%), Mg (22%), Ca (11%), K (7%).

Considering the above information, two major questions can be asked: (1) Why in seawater environments and at low temperature and pressure does glauconite form directly, and why does nontronite show such a

Table 1. Structural formulas of the clays studied.[1]

Hectorite
$(Si_{3.93}Al_{0.07})(Al_{0.04}Fe^{3+}_{0.02}Li_{0.17}Mg_{2.65})O_{10}(OH)_2 \cdot Na_{0.42}$

Wyoming montmorillonite
$(Si_{3.97}Al_{0.03})(Al_{1.51}Fe_{0.20}Mg_{0.22}Ti_{0.015})O_{10}(OH)_2 \cdot Na_{0.37}K_{0.03}$

Camp Berteau montmorillonite
$(Si_{3.93}Al_{0.07})(Al_{1.38}Fe^{3+}_{0.16}Fe^{2+}_{0.01}Ti_{0.02}Mg_{0.37})O_{10}(OH)_2 \cdot Na_{0.61}$

Nontronite
$(Si_{3.46}Al_{0.54})(Al_{0.17}Fe^{3+}_{1.88}Fe^{2+}_{0.01}Mg_{0.01})O_{10}(OH)_2 \cdot Na_{0.35}$

Beidellite
$(Si_{3.32}Al_{0.68})(Al_{1.40}Fe^{3+}_{0.47}Ti_{0.03}Mg_{0.20})O_{10}(OH)_2 \cdot Na_{0.55}$

Vermiculite
$(Si_{2.75}Al_{1.25})(Al_{0.19}Fe^{3+}_{0.25}Fe^{2+}_{0.02}Ti_{0.02}Mg_{2.49})O_{10}(OH)_2 \cdot Na_{0.83}$

[1] From Kehres (1983).

great affinity for K, whereas other smectites do not? (2) Why in the same diagenesis environment will aluminous montmorillonite select K and transform into illite, where stevensite or a saponite will select Mg and transform into corrensite or chlorite? We suggest that, perhaps the cation-exchange constants, i.e., the selectivities of smectites for the different cations, may change as a function of their chemical compositions (nontronite vs. beidellite vs. montmorillonite vs. saponite vs. stevensite), the degree of compaction, and temperature. In support of this idea, at low temperatures K seems strongly preferred by Fe-rich clay materials—glauconite forms readily from seawater solutions. At higher temperatures interlayer K seems to be preferentially selected by aluminous smectites, but Mg is preferred by Mg-rich smectites. K-fixation tends to form illite, and Mg-fixation tends to form corrensite or chlorite.

Because interlayer cations are hydrated, the pore size

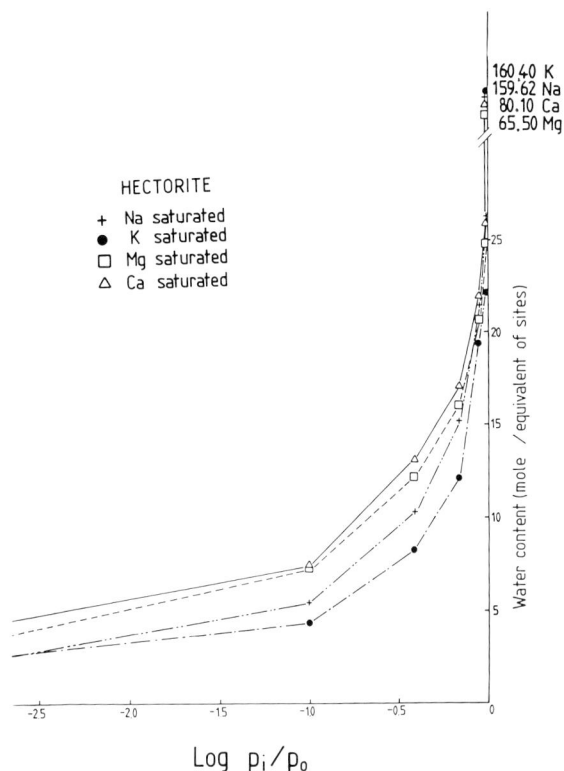

Figure 1. Dehydration isotherms for Na-, K-, Mg-, and Ca-saturated hectorites from Hector, California.

(degree of compaction) or the activity of water in which smectite particles equilibrate should have an important effect on the evolution of the cation selectivity. The present report attempts to evaluate the changes of the apparent cation-exchange constants for smectites of

Table 2. Water content (moles H_2O/eq of sites) as function of log p_i/p_0.

Mineral name[1]		Layer charge (eq/ $O_{10}(OH)_2$)	log p_i/p_0						
			0.00	−0.02	−0.05	−0.16	−0.41	−1.01	−2.70
Hectorite	Na	0.42	159.62	26.26	21.45	15.19	10.31	5.48	2.57
	K	0.42	160.84	22.12	19.40	12.07	8.26	4.31	2.60
	Mg	0.42	65.50	24.79	20.64	16.00	12.17	7.29	3.60
	Ca	0.42	80.10	26.93	21.98	17.10	13.19	7.43	4.69
Wyoming montmorillonite	Na	0.43	202.09	19.35	15.44	10.74	6.44	2.88	0.91
	Mg	0.43	81.63	21.91	18.21	13.98	10.63	6.63	2.81
Camp Berteau montmorillonite	Na	0.61	67.92	18.28	15.28	10.48	7.21	4.11	1.93
	Mg	0.61	63.77	17.92	14.56	11.13	8.67	5.64	2.67
	Ca	0.61	68.15	20.08	16.34	11.36	8.85	5.75	3.08
Nontronite	Na	0.35	129.71	28.43	24.77	19.00	15.06	9.49	4.77
	K	0.35	89.97	21.97	20.94	13.60	10.57	6.69	4.97
	Ca	0.35	129.03	33.17	28.46	22.74	18.66	12.31	6.74
Beidellite	Mg	0.55	46.65	13.62	11.65	7.89	6.02	4.24	2.45
	Ca	0.55	43.07	14.38	11.36	8.58	6.29	4.16	2.47
Vermiculite	K	0.83	19.51	3.54	3.30	2.00	1.60	1.23	0.89
	Ca	0.83	38.06	12.04	10.35	8.34	7.13	4.93	3.22

[1] Saturated with Na, K, Mg, or Ca.

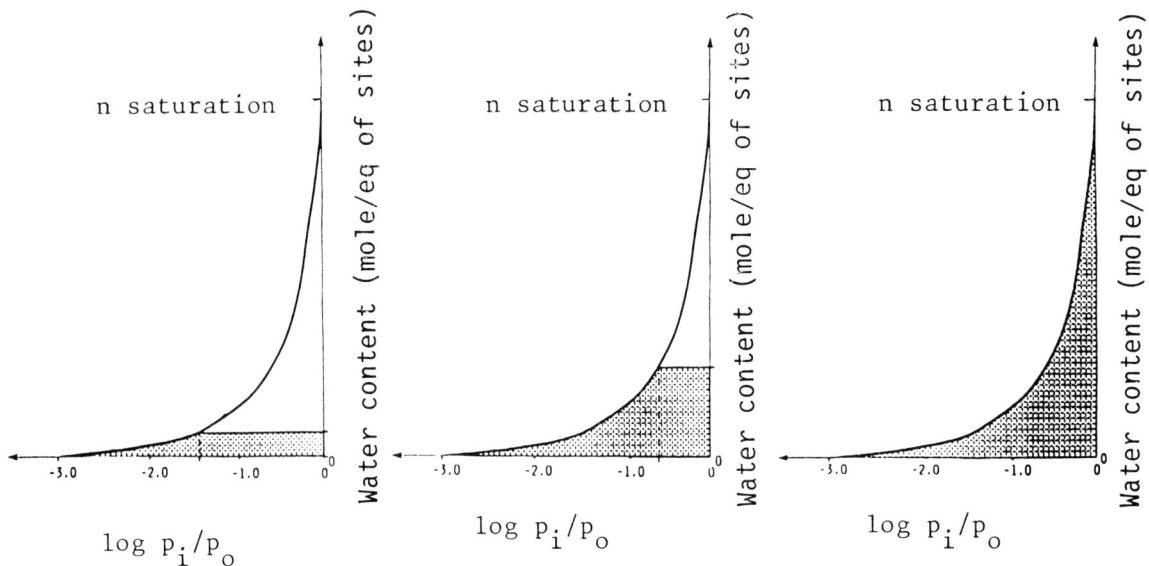

Figure 2. Schematic representations of the calculation of the Gibbs free energies of hydration, for different water activities ($a_{w,i} = p_i/p_0$), by integration of the dehydration isotherms.

different compositions in contact with water of different activities. These changes are evaluated from the dehydration isotherms.

DEHYDRATION ISOTHERMS

Dehydration isotherms were drawn from experimental data for seven clay samples (Table 1) reported by Kehres (1983). Each of the following clays had been saturated with Na, K, Ca, and Mg: hectorite, Wyoming montmorillonite, Camp Berteau montmorillonite, nontronite, beidellite, vermiculite, and kaolinite.

The activity of water a_w was defined by Garrels and Christ (1965) as the ratio of the vapor pressure, p, to the vapor pressure at saturation, p_0:

$$a_w = p/p_0 < 1.$$

The experimental procedure of dehydration was as follows: about 1 g of clay powder was first submitted to a liquid water imbibition until saturation was reached, i.e., until no more water was taken up after five weeks. Starting from this saturation point, defined by $a_w = p/p_0 = 1$, the samples were dehydrated in a stepwise manner in dry air, at different water activities ($a_{w,i} = p_i/p_0 < 1$). At each step, subscribed (i), the relative humidity of the atmosphere ($p_i/p_0 < 1$) was held constant by contact with a mixture of H_2O-H_2SO_4 (Tardy et al., 1980). The results obtained for hectorite, for example, are reported in Figure 1, where p_0 and p_i are, respectively, the partial vapor pressure at saturation with liquid free water and the partial vapor pressure at any stage (i) of hydration. Hydration or dehydration isotherms were drawn by plotting the water content n (moles) adsorbed on clays as a function of p_i/p_0.

Data for all the clay minerals studied are reported in Table 2, in which the number of moles of water given are calculated per one equivalent of exchangeable cation.

RESULTS AND DISCUSSION

Hydration energies

The Gibbs free energy of formation of a given hydrated clay mineral can be calculated as follows (Barshad, 1955; Tardy et al., 1980; Garrels and Tardy, 1982):

$$\Delta G^\circ_f(\text{hydrated clay}) = \Delta G^\circ_f(\text{dry clay})$$
$$+ \Delta G(\text{hydration})$$
$$+ n\Delta G^\circ_f(H_2O). \quad (1)$$

ΔG(hydration) is obtained by integrating along the corresponding isotherm curve (Figure 2):

$$\Delta G(\text{hydration}) = RT \int_0^{n_i} \ln a_w \cdot dn, \quad (2)$$

where n is the number of moles of water adsorbed on the clay, and n_i is the maximum number of moles adsorbed at equilibrium with a given activity of water $a_{w,i}$.

For each clay mineral saturated with one of the four cations (Na, K, Mg, Ca), hydration energies were determined at different stages of hydration ($a_{w,i} = p_i/p_0$) and listed in Table 3.

By comparing the hydration energy variations, either for a given clay mineral but between two different cations, or for a given pair of cations but between two different clay minerals, some important differences are

Table 3. Hydration energies of clay minerals at different stages of hydration (kJ/eq).

Mineral name		\multicolumn{7}{c}{log p_i/p_0}
		0.00	−0.02	−0.05	−0.16	−0.41	−1.01	−2.70
Hectorite	Na	−109.53	−102.05	−101.11	−97.01	−89.57	−71.37	−41.13
	K	−91.54	−83.85	−83.32	−79.90	−74.02	−59.36	−41.51
	Mg	−126.14	−123.86	−123.05	−119.96	−114.14	−95.53	−57.51
	Ca	−138.01	−135.03	−134.06	−131.22	−125.12	−103.53	−75.02
Wyoming montmorillonite	Na	−68.79	−58.54	−57.77	−55.08	−48.68	−35.09	−14.51
	Mg	−111.32	−107.97	−107.24	−104.65	−99.52	−84.72	−45.01
Camp Berteau montmorillonite	Na	−76.19	−73.41	−72.82	−70.04	−65.17	−53.64	−30.94
	Mg	−94.06	−91.48	−90.82	−88.54	−84.88	−73.63	−42.74
	Ca	−99.04	−96.34	−95.61	−92.74	−88.97	−77.11	−49.30
Nontronite	Na	−158.14	−152.61	−151.91	−148.63	−143.10	−121.91	−74.20
	K	−120.88	−117.17	−116.98	−113.60	−109.15	−94.66	−77.31
	Ca	−200.24	−195.01	−194.11	−191.05	−184.95	−161.25	−104.86
Beidellite	Mg	−71.67	−69.82	−69.43	−67.17	−64.42	−57.81	−39.26
	Ca	−72.33	−70.72	−70.13	−68.61	−65.32	−57.15	−39.55
Vermiculite	K	−21.03	−20.15	−20.10	−19.51	−18.94	−17.56	−4.09
	Ca	−82.05	−80.61	−80.28	−79.16	−77.34	−68.44	−50.84

apparent. Along dehydration paths, the relative stabilization or destabilization of a pair of cations (cation-exchange reaction) changes from one clay to another.

Exchange reactions involving hydrated cations

Cation-exchange reactions in soil materials or sediments involving, for example, Na^+ and Ca^{2+}, are commonly written as follows:

$$XNa + 0.5Ca^{2+}_{(aq)} \rightleftharpoons XCa_{0.5} + Na^+_{(aq)}, \quad (3)$$

in which X stands for a clay framework that is characterized by an interlayer charge of 1 eq, and $Ca^{2+}_{(aq)}$ and $Na^+_{(aq)}$ are aqueous cations. K_{ex}, the exchange constant of reaction (3), is defined by the expression:

$$K_{ex} = \frac{[XCa_{0.5}][Na^+_{(aq)}]}{[XNa][Ca^{2+}_{(aq)}]^{0.5}}, \quad (4)$$

in which $[XCa_{0.5}]$ and $[XNa]$ stand for the activities of solid end members, and $[Na^+_{(aq)}]$ and $[Ca^{2+}_{(aq)}]$ are the activities of the aqueous ions.

Exchangeable cations in the interlayer sites of clay minerals are in fact hydrated. In exchange processes in the laboratory or under natural conditions, water transfers from minerals to solutions take place, and several authors have considered the activity of water in the calculation of cation-exchange equilibria (Laudelout and Thomas, 1965; Laudelout *et al.*, 1968, 1972; Dufey and Laudelout, 1976; Laudelout, 1980). Taking such water transfers into consideration, Eq. (3) should be written as follows:

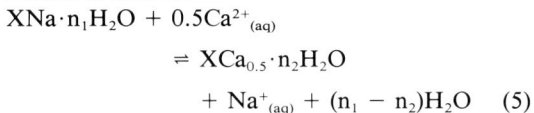

$$XNa \cdot n_1H_2O + 0.5Ca^{2+}_{(aq)}$$
$$\rightleftharpoons XCa_{0.5} \cdot n_2H_2O$$
$$+ Na^+_{(aq)} + (n_1 - n_2)H_2O \quad (5)$$

in which n_1 and n_2 stand, respectively, for the number of moles of water involved in the reaction.

The formulation of the cation-exchange constant, K_{ex}, for reaction (5) also must take into consideration the activity of the hydrated solid end members, respectively $[XCa_{0.5} \cdot n_2H_2O]$ and $[XNa \cdot n_1H_2O]$, the activity of the aqueous ions, respectively $[Na^+_{(aq)}]$ and $[Ca^{2+}_{(aq)}]$, as well as the activity of water as follows:

$$K_{ex} = \frac{[XCa_{0.5} \cdot n_2H_2O][Na^+_{(aq)}][H_2O]^{(n_1-n_2)}}{[XNa \cdot n_1H_2O][Ca^{2+}_{(aq)}]^{0.5}}. \quad (6)$$

By definition, the thermodynamic cation-exchange constant is constant and independent of the variations of water activity. This constant is easily measured in diluted conditions, i.e., in diluted aqueous solutions and in high water/clay proportions. Furthermore, it will be useful to calculate what was called K'_{ex}, the apparent cation-exchange constant:

$$K'_{ex} = \frac{[XCa_{0.5} \cdot n_2H_2O][Na^+_{(aq)}]}{[XNa \cdot n_1H_2O][Ca^{2+}_{(aq)}]^{0.5}}. \quad (7)$$

Obviously $K'_{ex} = K_{ex}$ only if the water activity $[H_2O] = 1$, but K'_{ex} is not constant and changes if the activity of water varies. The variations of K'_{ex} may allow an estimation of what would be, at a fixed ion-activity ratio in solution $[Na^+]/[Ca^{2+}]^{0.5}$, the variations of the exchangeable cation occupancies in the clay as the water activity decreases:

$$\frac{[XNa \cdot n_1H_2O]}{[XCa_{0.5} \cdot n_2H_2O]} = \frac{[Na^+_{(aq)}]}{[Ca^{2+}_{(aq)}]^{0.5}K'_{ex}}. \quad (8)$$

Variations of the apparent cation-exchange constants as a function of the activity of water

The Gibbs free energy of an exchange reaction, if the activity of water is smaller than the unity, ($a_{w,i} < 1$), can be written as follows:

$$\Delta Gr_{a_{w,i}} = \Delta G_{a_{w,i}} XCa_{0.5} \cdot n_2 H_2O$$
$$+ \Delta G_{a_{w,i}} Na^+_{(aq)}$$
$$+ (n_1 - n_2)\Delta G_{a_{w,i}} H_2O$$
$$- \Delta G_{a_{w,i}} XNa \cdot n_1 H_2O$$
$$- 0.5\Delta G_{a_{w,i}} Ca^{2+}_{(aq)}. \quad (9)$$

In fact,

$$\Delta G_{a_{w,i}} H_2O = \Delta G°_f H_2O + RT \ln a_{w,i}. \quad (10)$$

Introducing the activities of the different terms gives:

$$\Delta Gr_{a_{w,i}} = \Delta G°_f XCa_{0.5} \cdot n_2 H_2O$$
$$+ RT \ln[XCa_{0.5} \cdot n_2 H_2O]_{a_{w,i}}$$
$$+ \Delta G°_f Na^+_{(aq)} + RT \ln[Na^+_{(aq)}]_{a_{w,i}}$$
$$+ (n_1 - n_2)\Delta G°_f H_2O$$
$$+ (n_1 - n_2)RT \ln a_{w,i}$$
$$- \Delta G°_f XNa \cdot n_1 H_2O$$
$$- RT \ln[XNa \cdot n_1 H_2O]_{a_{w,i}}$$
$$- 0.5\Delta G°_f Ca^{2+}_{(aq)}$$
$$- 0.5RT \ln[Ca^{2+}_{(aq)}]_{a_{w,i}} . \quad (11)$$

Recasting Eq. (1), gives:

$$\Delta G°_f XCa_{0.5} \cdot n_2 H_2O = \Delta G°_f XCa_{0.5}(dry)$$
$$+ [\Delta G(hydration)]^{Ca}_{a_{w,i}}$$
$$+ n_2\Delta G°_f H_2O, \quad (12)$$

and

$$\Delta G°_f XNa \cdot n_1 H_2O = \Delta G°_f XNa(dry)$$
$$+ [\Delta G(hydration)]^{Na}_{a_{w,i}}$$
$$+ n_1\Delta G°_f H_2O. \quad (13)$$

The free energy of the exchange reaction at $a_{w,i}$ is then written as:

$$\Delta Gr_{a_{w,i}} = \Delta G°_f XCa_{0.5}(dry)$$
$$+ \Delta G°_f Na^+_{(aq)}$$
$$- \Delta G°_f XNa(dry)$$
$$- 0.5\Delta G°_f Ca^{2+}_{(aq)}$$
$$+ [\Delta G(hydration)]^{Ca}_{a_{w,i}}$$
$$- [\Delta G(hydration)]^{Na}_{a_{w,i}}$$
$$+ (n_1 - n_2)RT \ln a_{w,i}$$
$$+ RT \ln K'_{ex} \quad (14)$$

with:

$$K'_{ex} = \frac{[XCa_{0.5} \cdot n_2 H_2O]_{a_{w,i}}[Na^+_{(aq)}]_{a_{w,i}}}{[XNa \cdot n_1 H_2O]_{a_{w,i}}[Ca^{2+}_{(aq)}]_{a_{w,i}}}. \quad (15)$$

In free water ($a_w = 1$), the same kind of calculation yields:

$$\Delta Gr_{a_w=1} = \Delta G°_f XCa_{0.5} \cdot n'_2 H_2O$$
$$+ RT \ln[XCa_{0.5} \cdot n'_2 H_2O]_{a_w=1}$$
$$+ \Delta G°_f Na^+_{(aq)}$$
$$+ RT \ln[Na^+_{(aq)}]_{a_w=1}$$
$$+ (n'_1 - n'_2)\Delta G°_f H_2O$$
$$- \Delta G°_f XNa \cdot n'_1 H_2O$$
$$- RT \ln[XNa \cdot n'_1 H_2O]_{a_w=1}$$
$$- 0.5\Delta G°_f Ca^{2+}_{(aq)}$$
$$- 0.5RT \ln[Ca^{2+}_{(aq)}]_{a_w=1} \quad (16)$$

and

$$\Delta Gr_{a_w=1} = \Delta G°_f XCa_{0.5}(dry)$$
$$+ \Delta G°_f Na^+_{(aq)}$$
$$- \Delta G°_f XNa(dry)$$
$$- 0.5\Delta G°_f Ca^{2+}_{(aq)}$$
$$- [\Delta G(hydration)]^{Na}_{a_w=1}$$
$$+ [\Delta G(hydration)]^{Ca}_{a_w=1}$$
$$+ RT \ln K_{ex}, \quad (17)$$

in which K_{ex}, the thermodynamic cation-exchange constant, at $a_w = 1$, is:

$$K_{ex} = \frac{[XCa_{0.5} \cdot n_2 H_2O]_{a_w=1}[Na^+_{(aq)}]_{a_w=1}}{[XNa \cdot n_1 H_2O]_{a_w=1}[Ca^{2+}_{(aq)}]_{a_w=1}} . \quad (18)$$

If the equilibrium is reached in pure water, $\Delta Gr_{a_w=1} = 0$ and Eq. (17) becomes:

$$\Delta G°_f XCa_{0.5}(dry) + \Delta G°_f Na^+_{(aq)}$$
$$- \Delta G°_f XNa(dry) - 0.5\Delta G°_f Ca^{2+}_{(aq)}$$
$$= [\Delta G(hydration)]^{Na}_{a_w=1}$$
$$- [\Delta G(hydration)]^{Ca}_{a_w=1} - RT \ln K_{ex}. \quad (19)$$

In the same way, if the equilibrium is reached in brines where the activity of water is less than the unity ($a_{w,i} < 1$), Eq. (14) becomes:

$$\Delta G°_f XCa_{0.5}(dry) + \Delta G°_f Na^+_{(aq)}$$
$$- \Delta G°_f XNa(dry) - 0.5\Delta G°_f Ca^{2+}_{(aq)}$$
$$= [\Delta G(hydration)]^{Na}_{a_{w,i}} - [\Delta G(hydration)]^{Ca}_{a_{w,i}}$$
$$- (n_1 - n_2)RT \ln a_{w,i} - RT \ln K'_{ex}. \quad (20)$$

Combining Eqs. (19) and (20) yields the relation that gives the apparent cation-exchange constant for any activity of water, $a_{w,i}$:

Table 4. Calculated variations of the apparent cation-exchange constant (log K'_{ex}) for various types of clays as a function of water activity (log $a_{w,i}$ = log p_i/p_0).

Exchange reactions	log $a_{w,i}$ = log p_i/p_0						
	0.00^1	-0.02	-0.05	-0.16	-0.41	-1.01	-2.70
XNa → XK							
Hectorite	0.80	0.83	0.92	1.28	2.06	3.03	3.99
Nontronite	0.80	1.24	1.39	1.99	3.08	5.19	7.49
XMg → XCa							
Hectorite	0.00	-0.17	-0.22	-0.28	-0.58	-0.90	-1.95
Camp Berteau	0.00	-0.06	-0.12	-0.18	-0.23	-0.38	-0.82
Beidellite	0.00	0.03	0.02	0.03	-0.07	-0.16	-0.11
XK → XCa$_{0.5}$							
Hectorite	-0.72	0.04	-0.16	-0.80	-1.90	-4.27	-8.67
Nontronite	-0.72	-1.21	-1.49	-2.49	-4.58	-8.53	-14.90
Vermiculite	-0.72	-0.93	-1.22	-1.97	-3.43	-6.22	-11.25
XNa → XMg$_{0.5}$							
Hectorite	0.07	1.03	1.07	1.08	0.73	-0.34	-2.73
Wyoming	0.07	1.25	1.17	0.81	-0.16	-2.44	-7.22
Camp Berteau	0.07	0.12	0.13	0.08	-0.19	-1.09	-2.99

[1] For log p_i/p_0 = 0, log K'_{ex} = log K_{ex}.

$$\log K'_{ex} = \log K_{ex} - (n_1 - n_2)\log a_{w,i}$$
$$+ (1/RT \ln 10)\{([\Delta G(\text{hydration})]^{Ca}_{a_w=1}$$
$$- [\Delta G(\text{hydration})]^{Ca}_{a_{w,i}})$$
$$- ([\Delta G(\text{hydration})]^{Na}_{a_w=1}$$
$$- [\Delta G(\text{hydration})]^{Na}_{a_{w,i}})\}. \quad (21)$$

The corresponding results are listed in Table 4, in which the log K_{ex} values for the cation exchanges occurring in dilute conditions (a_w = 1, log a_w = 0) are those of Tardy and Garrels (1974):

$$\log K_{ex}(XNa \rightarrow XK) = 0.8,$$
$$\log K_{ex}(XNa \rightarrow XMg_{0.5}) = 0.075,$$
$$\log K_{ex}(XCa \rightarrow XMg) = 0.00, \text{ and}$$
$$\log K_{ex}(XK \rightarrow XCa_{0.5}) = -0.725.$$

As the activity of water decreases, K is strongly preferred to Na and to Ca, but Na is preferred to Mg and Mg to Ca. Furthermore, selectivities of nontronite for K and of hectorite for Mg are clearly different in dry conditions.

These results seem reasonable if one compares, for example, the apparent exchange constant K'_{ex}(XNa → XK) for hectorite at log p_i/p_0 = -2.70 (dry conditions) with the one calculated by the difference of log K_{sp} at 298.15 K of paragonite and muscovite (Helgeson et al., 1978):

$$\log K'_{ex}(XNa \rightarrow XK)\text{hectorite} = 3.99$$
$$\log K_{sp}(NaAl_2AlSi_3O_{10}(OH)_2)$$
$$- \log K_{sp}(KAl_2AlSi_3O_{10}(OH)_2) = 3.91.$$

Clearly, the apparent cation-exchange constants K'_{ex} (XNa → XK) are higher in dry than in wet conditions. At low water activities, Na is much less selected than K which is strongly fixed. At high water activities, the high hydration energy of Na tends to stabilize this cation in the interlayer sites of the clay minerals.

CONCLUSION

The data and calculations performed on hydration of clay minerals permit three major conclusions.

1. The cation-exchange constants measured in diluted conditions (a_w = 1) are considerably altered as the activity of water decreases and especially as the clay layers become dry (a_w = 0). Clearly, the exchange constants evolve such that the dehydrated cation is strongly preferred and selected compared with the initially more hydrated cation. If conditions are wet (a_w = 1), hydrated cations tend to be stabilized; if conditions are dry ($a_w \ll 1$) the dehydrated cations tend to be more stable.

2. The change in selectivity, as water activity decreases, is strongly dependent on the nature of the layer in which the exchange takes place. Nontronite shows a strong affinity for K and stevensite a strong affinity for Mg. K seems to be preferred by the dry ferric layers more than by the dry aluminous ones. Mg is preferred by the dry trioctahedral layers.

3. Assuming that a decrease in water activity has the same qualitative effect as an increase in the degree of compaction and an increase of temperature, nontronite will transform easily into glauconite and stevensite into corrensite if slight compaction or increase in temperature of the sediments takes place,

but more compaction and transformation of montmorillonite into illite will require higher temperature.

ACKNOWLEDGMENTS

The authors are deeply indebted to R. M. Garrels who had the original idea to derive the apparent exchange constants from the hydration energies and who spent a large part of his considerable energy to discuss the manuscript. Many thanks are also due to B. F. Jones for his meticulous correction of the manuscript. This work was supported by the Swedish Nuclear Fuel Supply (S.K.B) at Stockholm, Sweden, contract 85/907/180, and the Commissariat à l'Energie Atomique (C.E.A) at Fontenay aux Roses, France, contract no. BC-3133.

REFERENCES

Barshad, I. (1955) Adsorption and swelling properties of clay-water systems: in *Clays and Clay Technology,* J. A. Pask and M. D. Turner, eds., *Calif. Div. Mines Bull.* **169,** 70–71.

Burst, J. F. (1958) Glauconite pellets, their mineral nature and applications for stratigraphic interpretation: *Bull. Amer. Assoc. Petrol. Geologists* **42,** 310–327.

Chang, H. K. and Mackenzie, F. T. (1986) Comparisons between the diagenesis of dioctahedral and trioctahedral smectite diagenesis. Brazilian offshore basins: *Clays & Clay Minerals* **34** 407–423.

Dufey, J. E. and Laudelout, H. G. (1976) Hydration numbers of sodium calcium-motmorillonite: *Soil Sci.* **121,** 72–75.

Dunoyer de Segonzac, G. (1970) The transformation of clay minerals during diagenesis and low-grade metamorphism: a review: *Sedimentology* **15,** 281–346.

Fritz, B. (1981) Etude thermodynamique et modélisation des réactions hydrothermales et diagénétiques: *Mém. Sci. Géol.* **65,** 197 pp.

Garrels, R. M. and Christ, C. (1965) *Solutions, Minerals and Equilibria:* Harper and Row, New York, 450 pp.

Garrels, R. M. and Mackenzie, F. T. (1974) Chemical history of the oceans deduced from post depositional changes in sedimentary rocks: *Studies in Paleoceanography, Spec. Pub., 20,* Soc. Econ. Paleontol. Mineral., Tulsa, Oklahoma, 193–204.

Garrels, R. M. and Tardy, Y. (1982) Born-Haber cycles for interlayer cations of micas: in *Proc. Int. Clay Conf., Bologna and Pavia, 1981,* H. van Olphen and F. Veniale, eds., Elsevier, Amsterdam, 423–440.

Giresse, P. and Odin, G. (1973) Nature minéralogique et origines des glauconies du plateau continental du Gabon et du Congo: *Sedimentology* **20,** 457–488.

Helgeson, H. C., Delany, J. M., Nesbitt, H. W., and Bird, D. K. (1978) Summary and critic of the thermodynamic properties of rock-forming minerals: *Amer. J. Sci.* **278-A,** 1–229.

Hower, J. (1981) Shale diagenesis: in *Clays and the Resource Geologist,* F. Longstaffe, ed., Mineral. Assoc. of Canada Short Course Handbook **7,** 60–80.

Hower, J. and Mowatt, T. C. (1966) The mineralogy of illites and mixed layer illite-montmorillonites: *Amer. Mineral.* **51,** 825–854.

Kehres, A. (1983) Isothermes de déshydratation des argiles. Energies d'hydratation. Diagrammes de pores. Surfaces internes et externes: Thèse Sci., Univ. Toulouse, Toulouse, France, 163 pp.

Lamboy, M. (1967) Répartition de la glauconite sur le plateau continental de la Galice et des Asturies (Espagne): *C.R. Acad. Sci. Paris* **265,** 855–857.

Laudelout, H. G. (1980) L'échange d'ions dans les argiles: in *Géochimie des Interactions entre les Eaux, les Minéraux et les Roches,* Y. Tardy, ed., Elements, Tarbes, France, 7–25.

Laudelout, H. G. and Thomas, H. C. (1965) The effect of water activity on ion exchange selectivity: *J. Phys. Chem.* **69,** 339–341.

Laudelout, H. G., Van Bladel, R., Gilbert, M., and Cremers, A. (1968) Physical chemistry of cation exchange in clays: in *Proc. 9th. Int. Cong. Soil Science, 1968,* Adelaide, Australia, 565–576.

Laudelout, H. G., Van Bladel, R., and Robeyns, J. (1972) Hydration of cations adsorbed on a clay surface, from the effect of water activity on ions exchange selectivity: *Soil Sci. Amer. Proc.* **36,** 30–34.

Porrenga, D. H. (1967) Glauconite and chamosite as depth indicators in the marine environment: *Mar. Geol.* **5,** 495–501.

Sayles, F. L. and Mangelsdorf, P. C. (1977) The equilibration of clay minerals with sea water: exchange reactions: *Geochim. Cosmochim. Acta* **41,** 951–960.

Sayles, F. L. and Mangelsdorf, P. C. (1979) Cation exchange characteristics of Amazon river suspended sediments and its reaction with sea water: *Geochim. Cosmochim. Acta* **43,** 767–779.

Tardy, Y. and Garrels R. M. (1974) A method of estimating the Gibbs energies of formation of layer silicates: *Geochim. Cosmochim. Acta* **38,** 1101–1116.

Tardy, Y., Lesniak, P., Duplay, J., and Prost, R. (1980) Energie d'hydratation de l'hectorite: *Bull. Mineral.* **103,** 217–223.

Thompson, G. R. and Hower, J. (1975) The mineralogy of glauconite: *Clays & Clay Minerals* **23,** 289–300.

Proceedings of the International Clay Conference, Denver, 1985, L. G. Schultz, H. van Olphen, and F. A. Mumpton, eds.,
The Clay Minerals Society, Bloomington, Indiana, 53–58 (1987).

FORMATION OF PRIMITIVE CLAY PRECURSORS ON K-FELDSPAR UNDER EXTREME LEACHING CONDITIONS

KAZUE TAZAKI AND W. S. FYFE

Department of Geology, The University of Western Ontario
London, Ontario N6A 5B7, Canada

Abstract—Perthitic K-feldspar in Brazilian alkaline rocks covered with a thick weathering rind shows evidence for the formation of iron-rich transitional precursors of clay minerals. High-resolution transmission electron microscopy, energy dispersive X-ray analyses (EDX), and electron diffraction patterns clearly show the decomposition of K-feldspar and growth of these transitional products. They occur as long, curled fiber forms or circular forms on the altered feldspar surface. The electron diffraction patterns show diffuse rings at 4.41, 2.65, 1.56, and 1.38 Å, suggesting poor crystallinity or random orientation. These circular structures, typically 150–200 Å in diameter, contain duplicate to quintuplicate 7-, 14-, and 20-Å lattice images. Step scanning EDX analysis shows that the concentrations of Si, Al, and K tend to decrease and Fe increase from unaltered parts to altered parts of the clay precursors. Crystalline halloysite(7Å) appears to form from them.

Key Words—Clay precursors, Electron diffraction, Halloysite, High-resolution transmission electron microscopy, Potassium feldspar, Weathering.

INTRODUCTION

Studies of naturally weathered or artificially altered feldspars have shown the formation of various clay minerals, such as halloysite(7Å) and other kaolin-group minerals, montmorillonite, mica minerals, gibbsite, and imogolite (Guilbert and Sloane, 1968; Tazaki, 1976, 1978, 1979a, 1979b, 1981; Berner and Holdren, 1977, 1979; Eggleton and Buseck, 1980; Anand *et al.*, 1985). Using electron microscopy, numerous workers have observed etch pits (Wilson, 1975; Tazaki, 1976, 1978; Dearman and Baynes, 1979; Gilkes *et al.*, 1980; Velbel, 1983; Knauss and Wolery, 1983), clay coatings (Berner and Holdren, 1977, 1979) and a noncrystalline aluminosilicate (Guilbert and Sloane, 1968) on weathered feldspar during early stages of weathering.

Eggleton and Buseck (1980) observed that microscopic dissolution voids in K-feldspar from Australia were lined with 10-Å clays; they reported that the microenvironment in the voids may have been flushed sufficiently slowly so that dissolved K, Al, and Si reached the saturation level of 2:1 clay minerals. The feldspar structure breaks down completely during weathering, and clay weathering products are reconstituted from these breakdown products via dissolved or noncrystalline intermediates (Eggleton and Smith, 1983). The present study investigates these clay precursors and for the first time reports 7-, 14-, and 20-Å lattice spacings from such early weathering products.

MATERIALS AND METHODS

Weathered K-feldspar from an alkaline igneous complex, dated at 81 Ma, was collected from Ilha Bela, southern Brazil (Schobbenhaus *et al.*, 1984). The rock,

Figure 1. Transmission electron micrograph of slightly altered K-feldspar showing earliest-formed fibrous transitional materials.

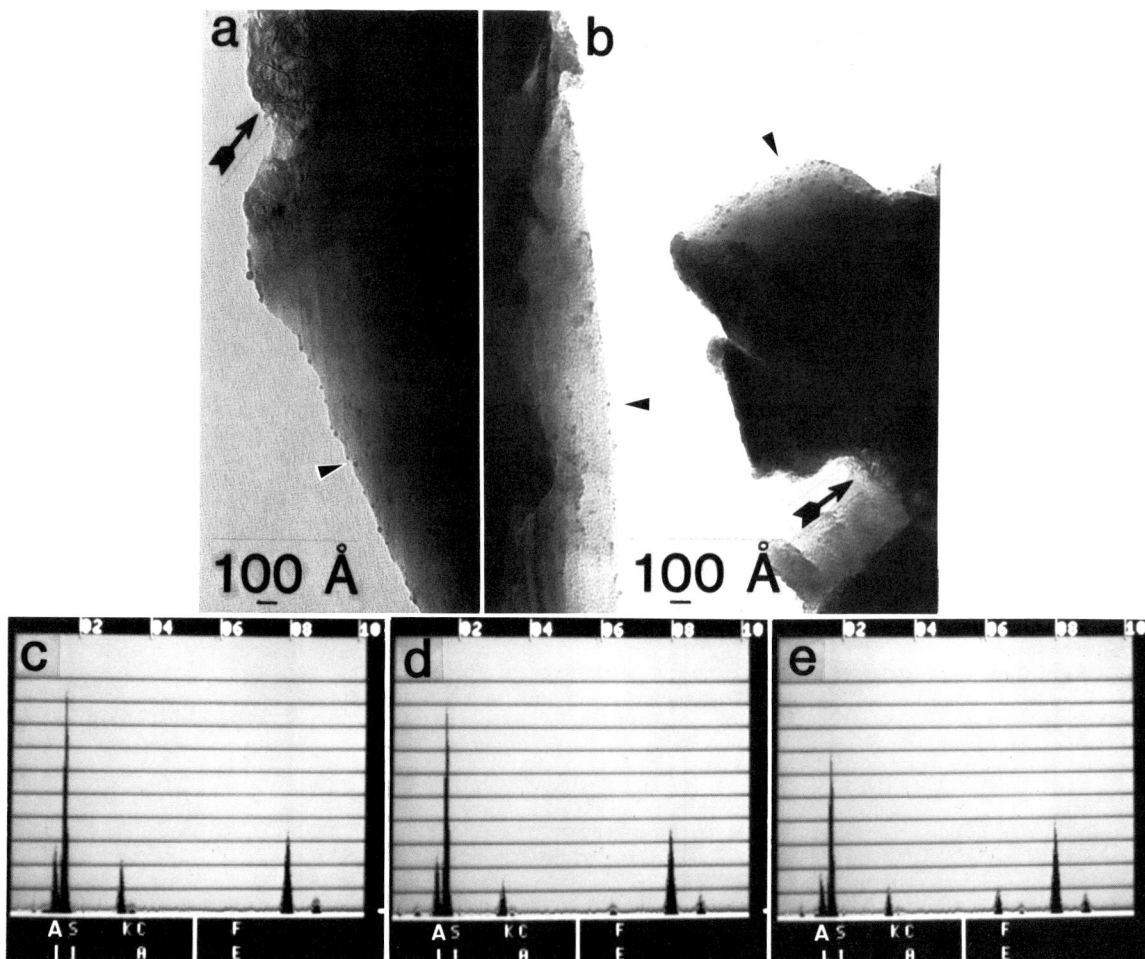

Figure 2. Transmission electron micrograph of a slightly altered surface of K-feldspar showing hydrated iron oxide spots (▲), clay fibers (←), and energy dispersive analyses of an unaltered surface (c), a spotted surface (d), and the earliest-formed fibrous material (e).

consisting mainly of perthitic K-feldspar (~90%), is covered with a thick reddish rind which contains well-developed gibbsite and tubular halloysite(7Å) formed under extreme leaching conditions. Unaltered microcline in this rock has hard, smooth, grey surfaces and strongly developed perthitic texture. Partly altered K-feldspar grains were handpicked from the bulk samples or separated by hydraulic elutriation methods. Grains <2 μm in diameter were collected for transmission electron microscopy (TEM) and energy dispersive X-ray analyses (EDX).

X-ray powder diffraction (XRD) patterns of the <2-μm size fraction showed strong peaks of albite and microcline only, indicating that transitional clay precursors could not be distinguished by XRD methods. These materials were observed by TEM, however, using a JEOL JEM 100C instrument, a Philips EM 400 STEM system equipped with EDX facilities, and a JEM-200 CS instrument for high resolution.

RESULTS

Figures 1–6 illustrate stages in the morphological transformation of feldspar to clay. Such intermediate materials are herein called transition materials. The surfaces of unaltered K-feldspar grains are flat and smooth, whereas the surfaces of altered K-feldspar initially show randomly oriented fibers (Figure 1). Such surfaces are also spotted with hydrated iron oxides (Figure 2). Electron diffraction patterns of these grains showed typical feldspar spacings and diffuse rings. EDX analyses showed that the unaltered surface contained no Fe (Figure 2c), but that spotted surfaces showed a trace of Fe (Figure 2d), and the earliest-formed fibers contained a slightly larger amount of Fe (Figure 2e). The Fe peak in Figure 2d is small because the iron oxide spots are much smaller than the electron beam. (The unaltered feldspar, an Fe spotted area, and the fibers correspond to 1, 2, and 3, respectively, in Figure

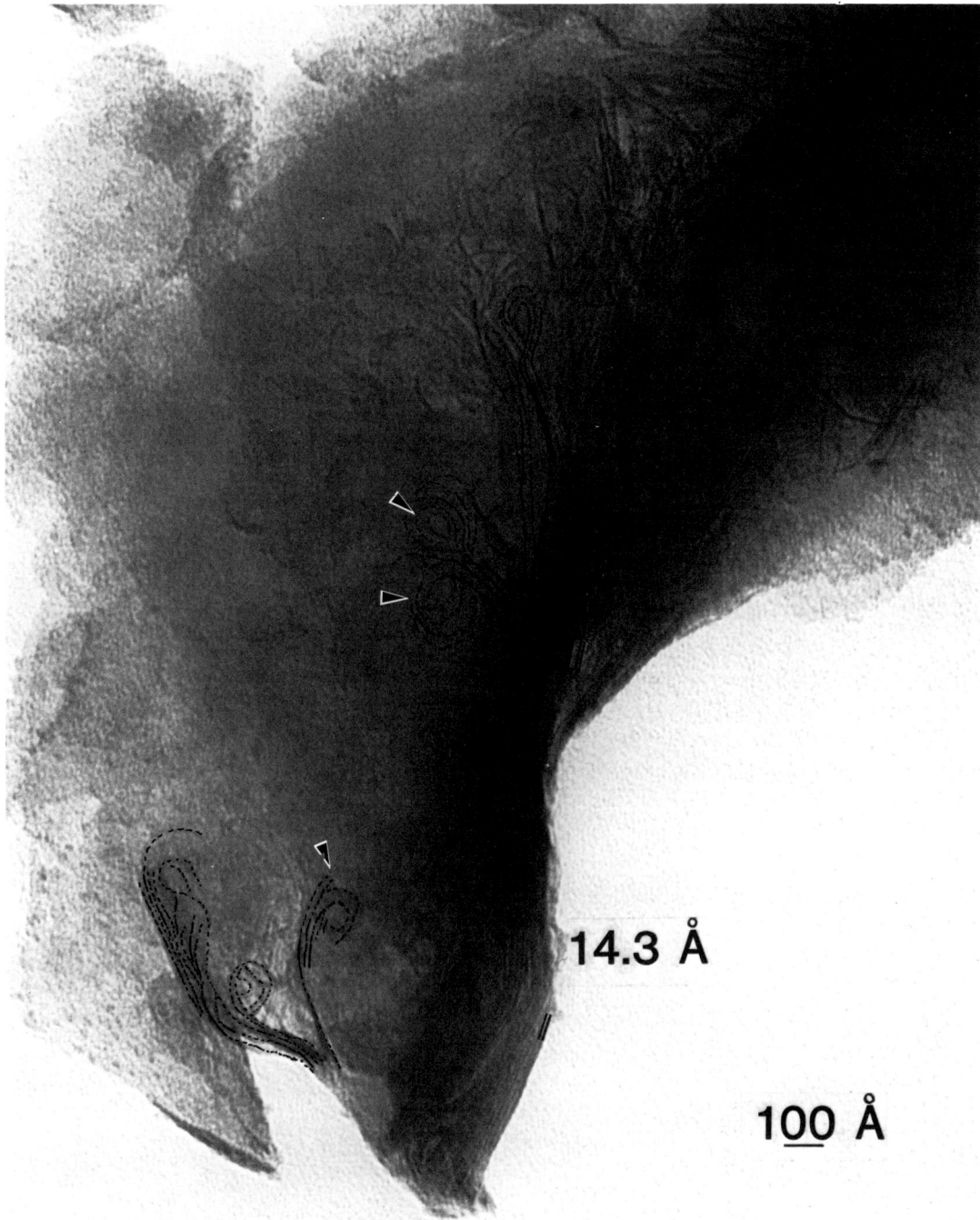

Figure 3.   Transmission electron micrograph of altered K-feldspar and bundled and curled fibers. Arrows show the end of curled fibers in circular forms. Note that 14.3-Å lattice images can be seen with bundled fibers.

7.) With increased weathering, the randomly oriented fibrous materials tended to be parallel and to form bundles of fibers or layers having 14.3-Å lattice images (Figure 3, arrows). The ends of the bundles tended to curve and curl inward forming roughly circular structures. Well-curled fibers or circular structures then de-

veloped; these materials gave diffuse electron diffraction rings at 4.41, 2.65, 1.56, and 1.38 Å (Figure 4). The diffuseness of the pattern suggests poor crystallinity or random orientation. Such well-developed transitional materials occurring with 150–200-Å circular forms and displaying 14–20-Å lattice images are shown

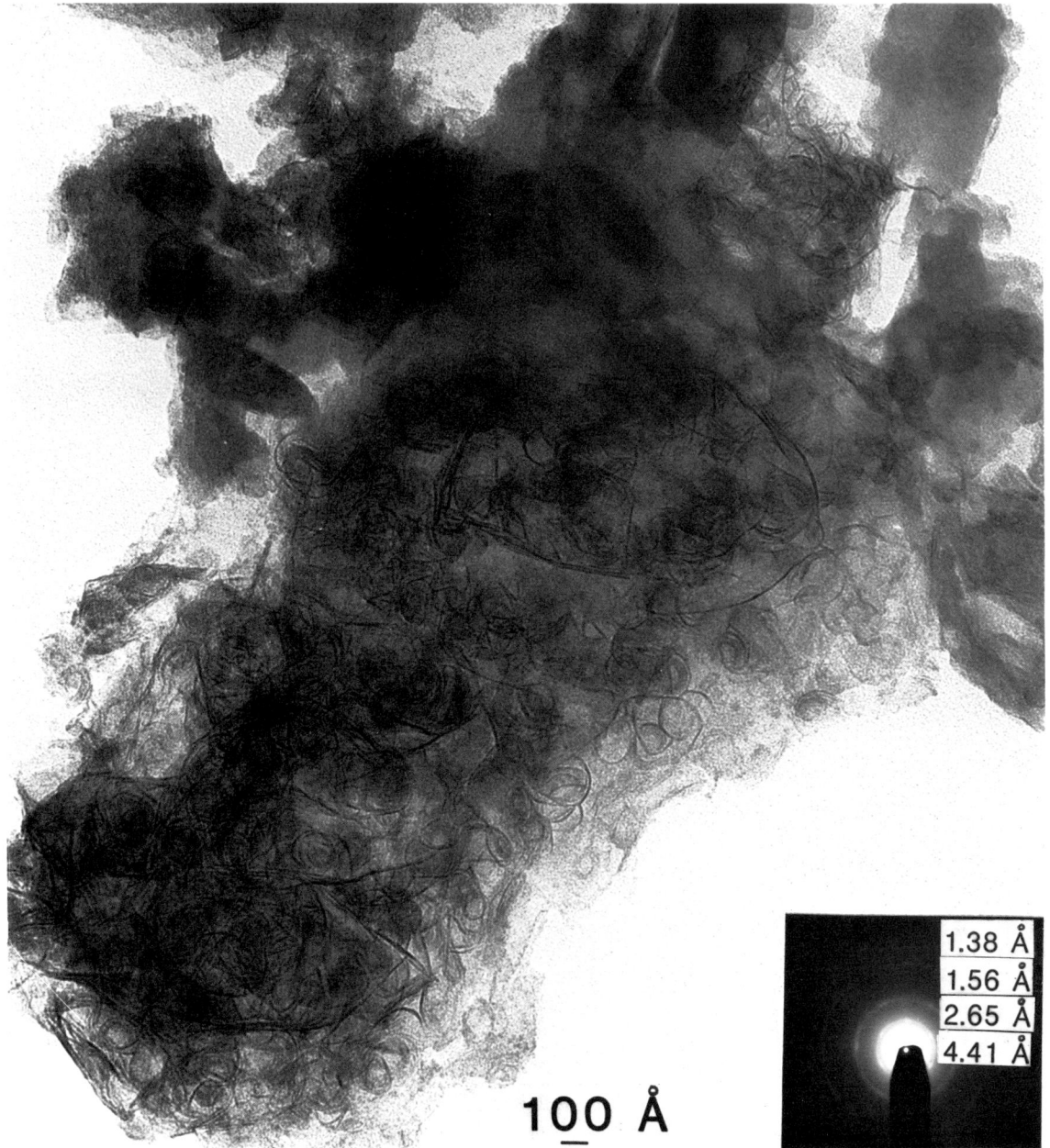

Figure 4. Transmission electron micrograph of the well-curled fibers or circular structures and an electron diffraction pattern showing diffuse rings at 4.41, 2.65, and 1.56 Å. A ring at 1.38 Å does not reproduce in the figure. Flakes of unaltered feldspar occur in the upper part. The surface has been cleaned ultrasonically.

in Figure 5a; these circular structures commonly contain duplicate, triplicate, and even quadruplicate lattice images. EDX showed that the transitional materials consisted mostly of Si and Fe and small amounts of Al, K, and Mn (Figure 5b). In the most weathered material, under the high-resolution transmission electron microscope, a 7-Å quintuplicate lattice image was noted in the transitional materials (Figure 6). In even a more intense stage of weathering, not illustrated here,

well-crystallized halloysite(7Å) in the thick weathered rind appears to have formed from the transitional materials that give the 7-Å lattice images and the circular forms.

EDX analyses, starting at unaltered feldspar and stepping toward more altered material at intervals of 400 Å (Figure 7), show the general trend in Si, Al, and Fe contents. The analyses of fresh microcline show only Si and Al (and K, which is not plotted). As explained

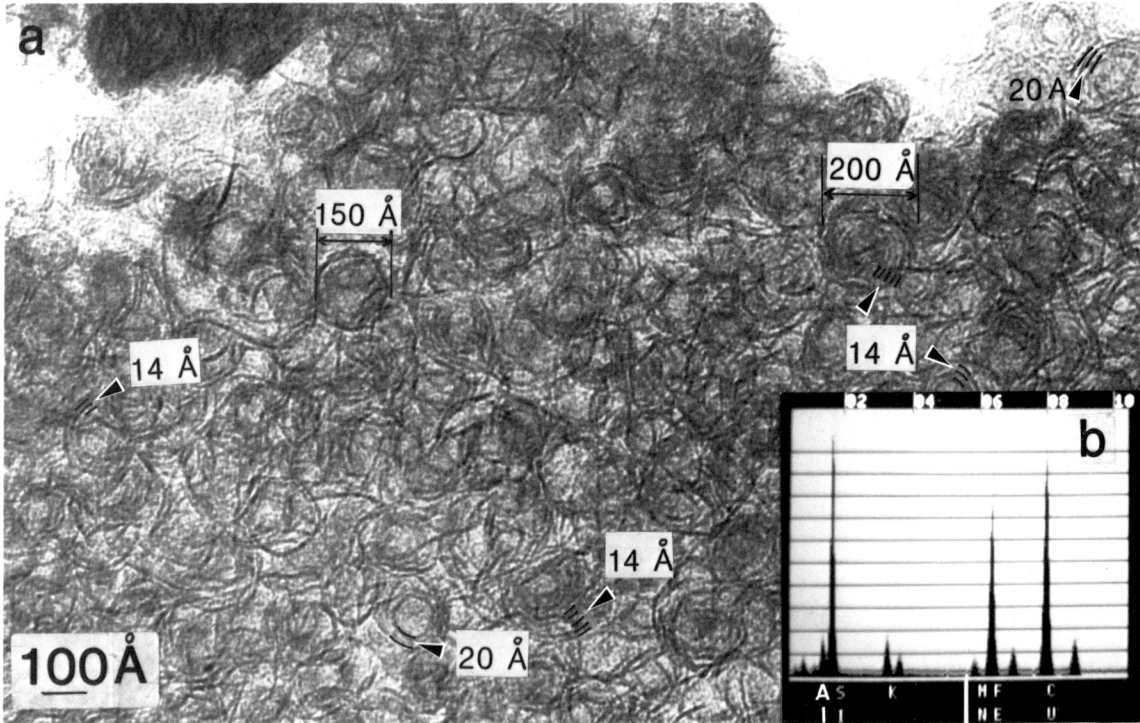

Figure 5. Transmission electron micrographs of the well-developed transitional products. (a) Sheet structures showing a 150–200-Å circular form and 14–20-Å lattice images; (b) Energy dispersive analysis of material in (a) which consists of Si and Fe with small amounts of Al, K, and Mn.

above, analyses of the two least altered products, i.e., the surface containing noncrystalline hydrated Fe oxide spots and the earliest formed fibrous products contain only small amounts of Fe. Fe increases progressively to a maximum in more highly weathered products, represented by material in Figures 4–6.

Secondary ion mass spectrometer (SIMS) (Cameca IMS-3F) data confirmed this compositional change and showed a decrease of Na, K, Rb, Sr, and Pb and an

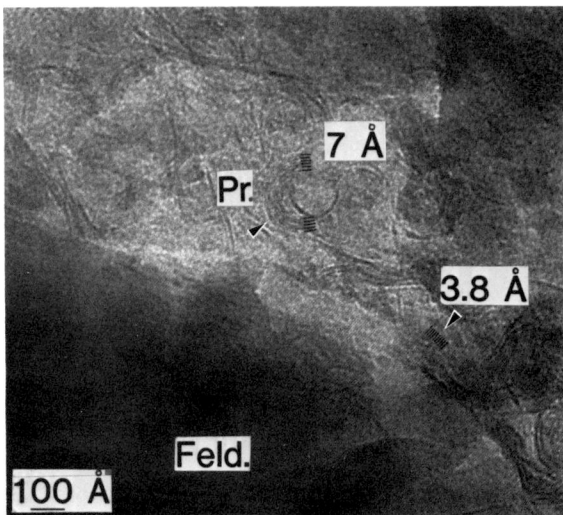

Figure 6. High-resolution electron micrograph of altered surface area with sheets and fibers. Arrows show the 3.8-Å spacings of feldspar (130) planes and the 7-Å spacings of primitive sheet structures (Pr).

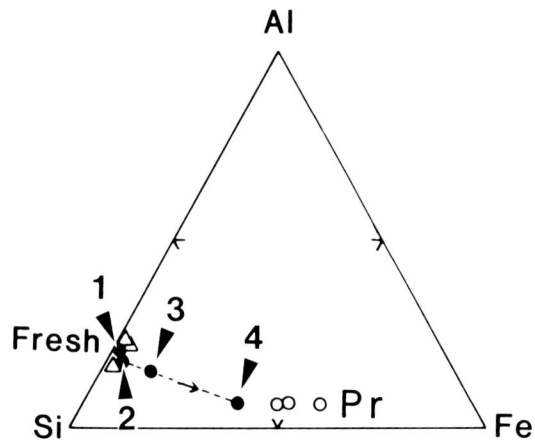

Figure 7. Al-Si-Fe diagram (counts per second ratios) of unaltered surface (△), primitive clay precursors (Pr) (○), and step-scanning intervals of 400 Å, using energy dispersive analysis. The filled circles ●1, ●2, ●3, and ●4 are related to the alteration stages discussed in text. Arrow shows the trend of development of products.

increase of Mg, P, Cl, Ca, Ti, Mn, Fe, Ni, Co, Zn, Ga, and rare earth elements, suggesting some substitution of OH and Fe for Na and K during residual and replacement processes.

DISCUSSION AND CONCLUSIONS

The present work describes intermediate formation of poorly crystalline material transitional to clays during the earliest weathering stages of K-feldspar.

TEM observations showed that alteration of microcline to various products proceeds in steps as follows:

1. Smooth surfaces (unaltered feldspar).
2. Randomly oriented fibers (nucleation) and tiny Fe spots (Figures 1 and 2).
3. Bundles of straight, curved, or curled fibers (Figure 3).
4. Circular forms (Figure 4).
5. Well-developed, circular sheet structures (Figure 5).

The feldspar surfaces of step 2 give an electron diffraction pattern of feldspar and diffuse rings of the earliest transitional products; EDX showed only a small amount of Fe. Fe also occurs as separate spots of noncrystalline hydrated iron oxide. As the transitional products transformed into curled and circular forms in steps 3 and 4, they developed some rudimentary crystallinity, as indicated by the electron diffraction pattern (Figure 4) and by the 14–20-Å lattice images. Their structure apparently was able to accommodate progressively larger amounts of Fe dissolved from the Fe spots of step 2, in agreement with the findings of Allcock (1985). Further development (step 5) gives 7-Å lattice images and circular primitive sheet structures reminiscent of globular halloysite. Indeed, such products must be the immediate precursor of Fe-rich spheroidal and squat cylinders of well-crystallized halloysite(7Å) found in the outermost parts (4 cm) of the thick weathered rinds. Fe-rich compositions are typical of such halloysites (Kirkman, 1977; Tazaki, 1981).

ACKNOWLEDGMENTS

We thank A. Melfi (University of São Paulo) and B. I. Kronberg (University of Western Ontario) for providing materials and R. Humphrey (University of Guelph) and M. Tsuji (University of Kyoto) for helping with analytical techniques. This project was partially supported by a grant from Atomic Energy Canada Ltd.

REFERENCES

Allcock, H. R. (1985) Developments at the interface of inorganic, organic, and polymer chemistry: *Chem. & Engineering News,* March 18, 1985, 22–36.

Anand, R. R., Gilkes, R. J., Armitage, T. M., and Hillyer, J. W. (1985) Feldspar weathering in lateritic saprolite: *Clays & Clay Minerals* **33**, 31–42.

Berner, R. A. and Holdren, G. R. (1977) Mechanism of feldspar weathering; some observation evidence: *Geology* **5**, 369–372.

Berner, R. A. and Holdren, G. R. (1979) Mechanism of feldspar weathering. II. Observations of feldspars from soils: *Geochim. Cosmochim. Acta* **43**, 1173–1186.

Dearman, W. R. and Baynes, F. J. (1979) Etch-pit weathering of feldspars. *Proc. Ussher Soc.* **4**, part 3, 390–401.

Eggleton, R. A. and Buseck, P. R. (1980) High-resolution electron microscopy of feldspar weathering: *Clays & Clay Minerals* **28**, 173–178.

Eggleton, R. A. and Smith, K. L. (1983) Silicate hydration mechanisms: in *Int. Coll. CNRS Petrology of Weathering and Soils,* D. Nahon and Y. Noack, eds., CNRS, Paris, France, p. 50 (abstract).

Gilkes, R. J., Suddhiprakarn, A., and Armitage, T. M. (1980) Scanning electron microscope morphology of deeply weathered granite: *Clays & Clay Minerals* **28**, 29–34.

Guilbert, J. M. and Sloane, R. L. (1968) Electron-optical study of hydro-thermal fringe alteration of plagioclase in quartz monzonite, Butte district, Montana: *Clays & Clay Minerals* **16**, 215–221.

Kirkman, J. H. (1977) Possible structure of halloysite disks and cylinders observed in some New Zealand rhyolitic tephras: *Clay Miner.* **12**, 199–216.

Knauss, K. G. and Wolery, T. J. (1983) Dependence of albite dissolution kinetics on pH and time at 70°C and 25°C. *Geol. Soc. Amer. Abst. with Programs* **15**, 616.

Schobbenhaus, C., Campos, D. A., Derze, G. R., and Asmus, H. E. (1984) Geologia do Brasil: *Ministerio das Minas e Energia,* DNPM, p. 323.

Tazaki, K. (1976) Scanning electron microscopic study of formation of gibbsite from plagioclase: *Inst. Thermal Spring Res., Okayama Univ. Paper* **45**, 11–24.

Tazaki, K. (1978) Micromorphology of plagioclase surface at incipient stage of weathering: *Earth Sci. (Chikyu Kagaku)* **32**, 8–12.

Tazaki, K. (1979a) Micromorphology of halloysite produced by weathering of plagioclase in volcanic ash: in *Proc. Int. Clay Conf., Oxford, 1978,* M. M. Mortland and V. C. Farmer, eds., Elsevier, Amsterdam, 415–422.

Tazaki, K. (1979b) Scanning electron microscopic study of imogolite formation from plagioclase: *Clays & Clay Minerals* **27**, 209–212.

Tazaki, K. (1981) Analytical electron microscopic studies of halloysite formation process—morphology and composition of halloysite: in *Proc. Int. Clay Conf., Bologna, Pavia, 1981,* H. van Olphen and F. Veniale, eds., Elsevier, Amsterdam, 573–584.

Velbel, M. A. (1983) A dissolution-reprecipitation mechanism for the pseudomorphous replacement of plagioclase feldspar by clay minerals during weathering: in *Pétrologie des Altérations et des Sols, Vol. I,* D. Nahon, and Y. Noack, eds., *Mem. Sci. Géol.* **71**, 139–147.

Wilson, M. J. (1975) Chemical weathering of some primary rock-forming minerals: *Soil Sci.* **119**, 349–355.

Proceedings of the International Clay Conference, Denver, 1985, L. G. Schultz, H. van Olphen, and F. A. Mumpton, eds.,
The Clay Minerals Society, Bloomington, Indiana, 59–65 (1987).

INTERSTRATIFIED CHLORITE/SMECTITE ("METAMORPHIC VERMICULITE") IN THE UPPER PRECAMBRIAN GREYWACKES OF ROUEZ, SARTHE, FRANCE

Daniel Beaufort

Laboratoire de Pétrologie des Altérations Hydrothermales
Université de Poitiers, U.A. 721 du C.N.R.S.
40, avenue du Recteur Pineau, 86022 Poitiers Cédex, France

Abstract—Investigations of what was previously called "metamorphic vermiculite" from the Upper Precambrian metagreywackes surrounding the massive sulfide deposit of Rouez, France, have led to the characterization of this material as an interstratified chlorite/smectite (C/S). X-ray powder diffraction data obtained from oriented mounts of small quantities of material (<0.001 mg) drilled from thin sections confirmed the presence of smectite layers (<20%) randomly interstratified with chlorite. This interstratified C/S has a composition that differs from chlorite in that it contains more titanium and alkali elements; however, the absolute amounts of these elements vary from grain to grain. Mössbauer spectrometry indicated a low Fe^{3+} content for this material.

In thin section, the C/S occurs as disseminated, brown flakes, optically similar to biotite or stilpnomelane. The C/S is apparently authigenic, and has been partly replaced by the assemblage chlorite + phengite during low-grade Hercynian regional metamorphism.

Key Words—Chlorite/smectite, Interstratified, Metamorphism, Titanium, Vermiculite, X-ray powder diffraction.

INTRODUCTION

Some recent studies of individual phyllosilicate minerals in metamorphic and hydrothermally altered rocks have mentioned the occurrence of a vermiculite-like mineral of high-temperature and high-pressure origin (Brown, 1967; Kerrick and Cotten, 1971; Black, 1975; Mac Dowell and Elders, 1980; Meunier, 1980; Meunier and Velde, 1982; Nicot, 1981; Beaufort and Meunier, 1983). In a compilation of optical and chemical data, Velde (1978) proposed "metamorphic vermiculite" for lack of a better name for these enigmatic minerals. Their chief diagnostic features seemed to be their optical properties, which are intermediate between those of biotite (or stilpnomelane) and chlorite, and their wide range of chemical compositions from grain to grain, which contrast with the properties of other coexisting ferromagnesian phyllosilicates.

The stability of these minerals compared with ferromagnesian phyllosilicates has not been clearly defined. They seem to be transition materials between the chlorite facies and other ferromagnesian facies in pelitic rocks (Velde, 1978; Mac Dowell and Elders, 1980), but they have also been found coexisting with randomly interstratified illite/smectite and kaolinite at temperatures slightly greater the 100°C, as a replacement of chlorite in hydrothermally altered porphyritic granite (Beaufort and Meunier, 1983).

The purpose of the present study was to define the interstratified chlorite/smectite (C/S) mineral in what was previously called "metamorphic vermiculite" that occurs in metagreywackes of the Upper Precambrian formations surrounding the massive sulfide deposit at Rouez, Sarthe, France. The variations in chemistry of this mineral were investigated, and Mössbauer spectra and trace element analyses were used to elucidate the relations between these minerals and accompanying chlorites, which are ubiquitous in the metagreywackes of Rouez.

FIELD OCCURRENCE AND LITHOLOGY

The Upper Precambrian formations surrounding the massive sulfide deposit of Rouez (Figure 1) are sequences of pelitic rocks and metagreywackes metamorphosed under greenschist facies conditions during the Hercynian orogeny (Icart and Safa, 1981). Mineralogical and petrological studies (Beaufort et al., 1985) of the metagreywackes showed the presence of a disseminated, brown phyllosilicate mineral, optically similar to biotite, which had been deformed and fractured by slaty cleavage and locally partly replaced by the metamorphic assemblage phengite + chlorite (Figure 2). This mineral normally occurs as authigenetic crystals with local poikiloblastic habits and shows a pronounced similarity to the so-called "metamorphic vermiculite" of the literature (vide supra).

Petrologic observations were made on 15 samples from drill cores that contained both interstratified C/S ("metamorphic vermiculites") and chlorites. The cores

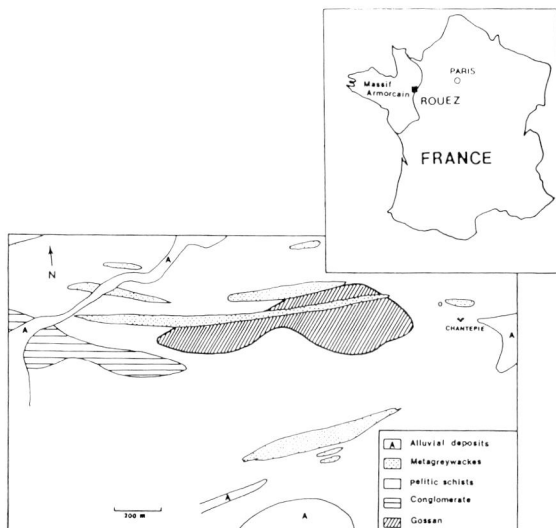

Figure 1. Geology of the Rouez district, Sarthe, France.

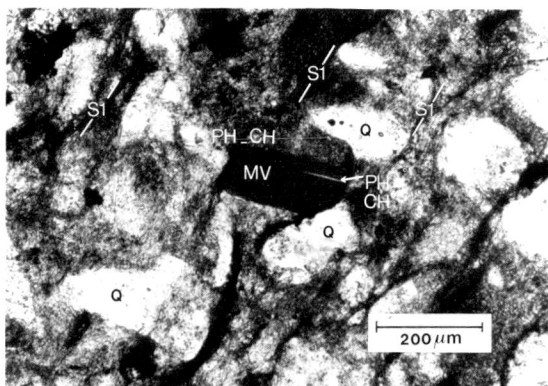

Figure 3. Photomicrograph of microdrilled area in a grain of interstratified chlorite/smectite (sample RZ8-277).

were from five boreholes located within 150 m of the massive sulfide deposit of Rouez and were obtained courtesy of The Société National Elf Aquitaine. All the samples selected for investigation were fresh and unaffected by weathering. Modal analyses of thin sections showed less than 1% of the interstratified C/S material.

EXPERIMENTAL

The identification of what was previously called "metamorphic vermiculites" by XRD was difficult because of its small abundance in the rocks and because of the ubiquitous presence of chlorite in the same samples. In the Rouez metagreywackes, interstratified C/S, distinguishable only in thin section, occurs as particles 100–300 μm in length. The paucity of XRD data in the literature concerning "metamorphic vermiculite"

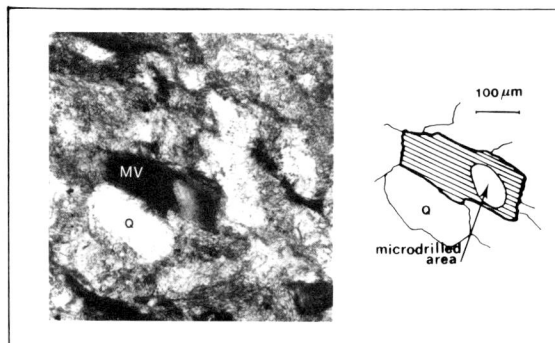

Figure 2. Photomicrograph of interstratified chlorite/smectite ("metamorphic vermiculite") in thin section of Rouez metagreywackes. S1 = slaty cleavage; PH-CH = phengite-chlorite assemblage of metamorphic origin; MV = interstratified chlorite/smectite ("metamorphic vermiculite"); Q = quartz.

(except Black, 1975) is due to the difficulty of mineral separation. In the present study, five interstratified C/S grains were microdrilled (Beaufort et al., 1983) from thin sections, and the oriented flakes were examined by XRD using the step-scan method described by Meunier and Velde (1982). The same grains were also analyzed by electron microprobe. The microdrilled volumes typically averaged 100 × 100 × 30 μm (Figure 3). The XRD analyses were carried out using a Philips PW 1730 diffractometer operated at 40 kV and 40 mA and Fe-filtered CoKα radiation. The diffractometer was equipped with 1° emergence, 0.1-mm receiving, and 1° antiscatter slits. Steps of 0.05°2θ and a counting time of 200 s were used. For quantities of matter (about 0.001 mg) at the lower level of discrimination, a counting time of 400 s/step was used. XRD patterns were made of individual grains in the natural state, after ethylene glycol saturation, and after they had been heated to 550° for 2 hr.

Chemical compositions of the grains were obtained with an electron microprobe (Cameca MS 46) equipped with an energy-dispersive X-ray system (ORTEC EDS system). The analytical conditions used in this study were similar to those suggested by Velde (1984) for highly reliable clay mineral analyses, particularly for the alkali elements. The conditions were: acceleration voltage, 15 kV; counting time, 120 s; beam intensity, 0.001 μA; spot size, 3–5 μm. The analytical reproducibility averaged 1.5% of the amount present for the nine elements considered. One hundred microprobe analyses of interstratified C/S were performed in the 15 thin sections for an accurate characterization of the composition range.

Chlorite and interstratified C/S were separated from each sample. The chlorite fraction was almost monomineralic by XRD examination, but the interstratified fraction contained as much as 20% quartz and phengite and, possibly, minor chlorite. Mössbauer spectroscopy was used to determine Fe^{3+}/Fe^{2+} ratios in both fractions. Mössbauer spectra at room temperature were obtained using the transmission method and an Elscint AME 30 spectrometer equipped with a ^{57}Co source in

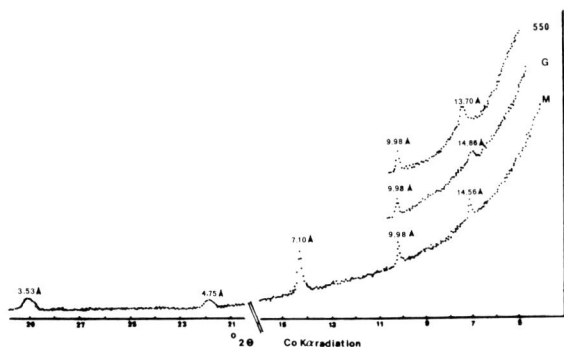

Figure 4. X-ray powder diffraction patterns of the material of the microdrilled area displayed in Figure 3 (sample RZ8-277); N = natural; G = ethylene-glycol saturated; 550 = after heating for 2 hr at 550°C.

a rhodium matrix (25 mCi). Analytical spectra were fitted by means of a least-squares computer program to a doublet, some having Lorentzian line shapes. About 100 mg of each fraction was analyzed by UV spectrometry for trace elements.

RESULTS

X-ray powder diffraction data

Table 1 summarizes the XRD data for interstratified C/S from five samples of Rouez metagreywackes; Figure 4 illustrates XRD patterns of a representative sample. The XRD patterns obtained on untreated samples displayed a moderately intense, broad reflection at about 14.40 Å, a strong, sharp reflection at 7.10–7.12 Å, and two weak reflections near 4.74–4.75 and 3.53 Å. The 001 and 002 basal reflections, respectively, shifted to about 14.60–14.80 and 7.12–7.14 Å after ethylene-glycol treatment and collapsed to about 13.60–13.70 and 7.08–7.09 Å after the sample had been heated to 550°C for 2 hr. The reflections at 9.98, 4.99, and 3.33

Table 2. Average chemical composition of interstratified chlorite/smectite and chlorite particles analyzed in 15 samples of Rouez metagreywacke.

Oxide	Interstratified chlorite/smectite (100)		Chlorite (53)	
	X̄	Sd	X̄	Sd
SiO₂	26.76	1.20	25.45	1.51
Al₂O₃	18.53	1.02	21.67	1.62
MgO	11.63	1.65	12.53	1.14
FeO	25.32	3.35	25.56	1.96
TiO₂	1.75	0.96	0.07	0.03
MnO	0.23	0.15	0.22	0.13
CaO	0.12	0.12	0.04	0.02
Na₂O	0.16	0.15	0.00	0.00
K₂O	0.23	0.09	0.01	0.01
Σ oxides wt. %	84.73	1.57	85.55	1.28

X̄ = average composition in wt. %. Sd = Standard deviation.
( ) = Numbers of microprobe analyses for each mineral species.

Table 1. X-ray powder diffraction data for five grains of interstratified chlorite/smectite.

	Sample RZ8-27			Sample RZ8-277			Sample RZ8-290			Sample RZ8-165			Sample MI1-123		
	Nat. d (Å)	EG d (Å)	H 550 d (Å)	Nat. d (Å)	EG d (Å)	H 550 d (Å)	Nat. d (Å)	EG d (Å)	H 550 d (Å)	Nat. d (Å)	EG d (Å)	H 550 d (Å)	Nat. d (Å)	EG d (Å)	H 550 d (Å)
	14.45 m	14.78 w	13.60 m	14.56 m	14.86 w	13.70 m	14.60 m	14.85 m	13.70 m	14.55 m	14.75 m	13.65 m	14.60 m	14.80 w	13.85 m
	7.10 s	7.12 s	7.08 s	7.10 s	7.12 s	7.09 s	7.12 s	7.14 s	7.09 s	7.10 s	7.12 s	7.09 s	7.10 s	7.12 s	7.08 s
	4.74 vw	na	na	4.75 vw	na	na	4.75 vw	na	na	4.74 vw	na	na	4.75 vw	na	na
	3.53 w	na	na	3.53 w	na	na	3.54 w	na	na	3.53 w	na	na	3.53 w	na	na

Nat = natural; EG = ethylene-glycol treated; H 550 = heated 2 hr at 550°C; na = not analyzed; m = moderate; s = strong; vw = very weak; w = weak.

Table 3. Microprobe analyses of individual grains of interstratified chlorite/smectite in a single thin section.

| Oxides | Sample RZ8-27 | | | | | | | | | |
| | 1 (4) | | 2 (3) | | 3 (6) | | 4 (4) | | 5 (5) | |
	X̄	Sd	X̄	Sd	X̄	Sd	X̄	Sd	X̄	Sd
SiO₂	27.97	0.30	26.70	0.11	26.58	0.45	25.61	0.44	26.97	0.43
Al₂O₃	18.86	0.32	19.08	0.21	18.19	0.40	18.78	0.39	18.95	0.50
MgO	9.56	0.33	10.59	0.08	10.34	0.30	10.30	0.18	10.64	0.25
FeO	26.55	0.28	26.88	0.46	27.76	0.42	27.67	0.49	27.53	0.68
TiO₂	3.13	0.23	1.27	0.09	1.24	0.09	0.67	0.16	1.31	0.08
MnO	0.24	0.03	0.15	0.05	0.15	0.10	0.10	0.06	0.24	0.12
CaO	0.07	0.03	0.02	0.02	0.05	0.01	0.06	0.02	0.03	0.02
Na₂O	0.19	0.10	0.20	0.14	0.21	0.20	0.18	0.11	0.25	0.12
K₂O	0.23	0.06	0.27	0.05	0.21	0.03	0.22	0.06	0.26	0.05
Σ oxides wt. %	84.89	0.40	85.21	0.60	84.87	1.47	83.28	0.80	86.12	1.56

X̄ = average composition in oxide wt. %. Sd = Standard deviation. ( ) = Number of microprobe analyses for each grain.

Å were probably due to <10-μm particles of optically observable phengite, which were impossible to separate from the interstratified C/S. The XRD patterns of the chlorite that partly replaced grains of interstratified C/S displayed sharp reflections at 14–15, 7.07, 4.73, and 3.52 Å. Its basal reflections did not shift after ethylene-glycol treatment; the 001 basal reflection collapsed slightly to 13.90 Å after the sample was heated at 550°C for 2 hr.

Microprobe analyses

Compared with chlorite from the same thin sections (Table 2), the average chemical analyses of interstratified C/S contained slightly more SiO₂, less Al₂O₃, slightly less MgO and FeO, and more CaO, Na₂O, and K₂O; the TiO₂ concentration (1.75 wt. %) was much greater. Interstratified C/S differed from chlorite by the wide range of TiO₂ and alkali element concentration between grains in each sample and between samples; however, in a single thin section (Table 3), the composition measured at different points inside each grain was relatively consistent.

On the basis of microprobe analyses performed in the 15 thin sections (Table 2), TiO₂ varies widely (from 0.25 to 4.50 wt. %) and can be correlated with variations of colors. Grains containing <1 wt. % TiO₂ are pale brown, and grains containing >3 wt. % TiO₂ are dark brown. X-ray micrographs of samples excited by the electron probe confirmed the homogeneous distribution of titanium in the grains of interstratified C/S. Total K₂O + Na₂O + CaO varies from 0.03 to more than 1%; K₂O generally predominates, however it is

lacking in chlorites. The Na₂O and CaO concentrations are more variable, typically at the detection limit of the microprobe. Locally, however, the Na₂O and CaO contents are more significant and predominate over K₂O. The FeO and MgO contents of chlorite, as well as the interstratified C/S, vary widely. No relation was found between compositions of either mineral from a given thin section. Trace elements (analyzed in sample RZ8-27) and were not identically distributed; chlorite contained more Ni and Li, whereas interstratified C/S contained more Cr, Co, and Zr (Table 4); the Co/Ni ratio was near 0.1 in the former and 5 in the latter.

Mössbauer spectroscopy

Representative Mössbauer spectra of chlorite and interstratified C/S separated from sample RZ8-27 are illustrated in Figures 5a and 5b; computed parameters are listed in Table 5. These results show that 75% of the Fe is in the ferrous state in both materials. The computed parameters for chlorite are similar to those for the iron chlorite reported by Goodman and Bain (1979), in which about 76% of the iron was reported to be in the ferrous state. They assigned the Fe^{2+} to two components and the Fe^{3+} to a single component. Computed parameters of Mössbauer spectra of interstratified C/S show that 80% of the Fe is in the ferrous state, being assigned to two components; the Fe^{3+} can be assigned to two components—a minor component similar to the Fe^{3+} component of the chlorite and a major component having an unusual isomer shift and larger half-height line-width values than those generally obtained for chlorite (Goodman and Bain, 1979),

Table 4. Trace element analyses of interstratified chlorite/smectite from sample RZ8-27 (analyzed by UV spectrometry).

| Mineral | Trace elements (ppm) | | | | | | | | | | |
	Pb	Zn	Cu	V	Cr	Ni	Co	Sr	Zr	La	Li
Interstratified chlorite/smectite	22	363	106	106	719	119	544	75	835	180	132
Chlorite	32	420	53	180	88	180	19	28	156	40	567

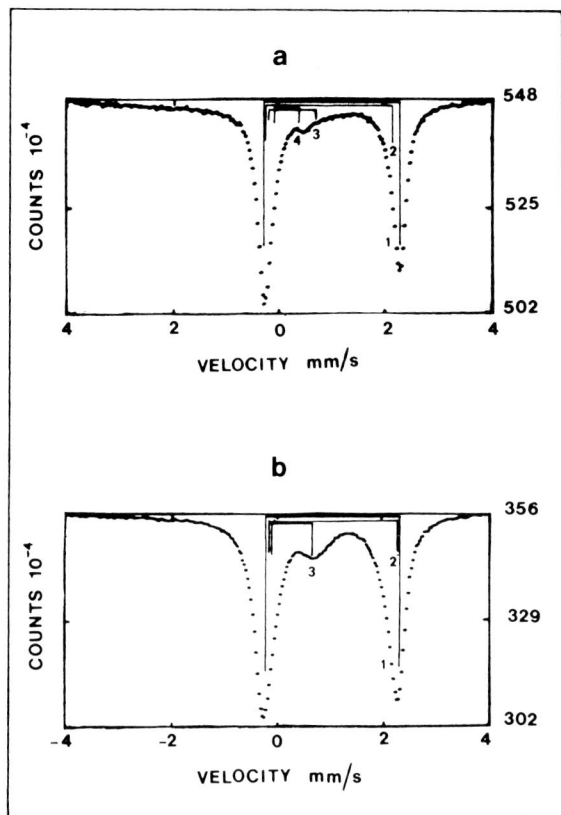

Figure 5. Mössbauer spectrum and quadrupole doublets of interstratified chlorite/smectite (a) and chlorite (b) at room temperature (sample RZ8-27).

Table 6. Average structural formula of interstratified chlorite/smectite (calculated on the basis of 28 oxygens) from sample RZ8-27.

	1	2
Si	5.74	5.77
Al^{IV}	2.26	2.23
Al^{VI}	2.56	2.62
Fe^{3+}	0.89	0.45
Fe^{2+}	3.94	4.47
Mg	3.32	3.34
Ti	0.29	0.29
Mn	0.04	0.04
Ca	0.01	0.01
Na	0.07	0.07
K	0.07	0.07

1 = both ferric doublets are included in the structure.
2 = the unusual ferric doublet is not included in the structure.

true vermiculite (Taylor *et al.*, 1968), or biotites (Bagin *et al.*, 1980). Because of the impure nature of the interstratified C/S fraction, this Fe^{3+} assignment must be treated with some reservation.

DISCUSSION

Optical and chemical data obtained for interstratified C/S in the present study are close to those from the literature for vermiculite-like minerals of high-temperature and high-pressure origin (Black, 1975; Velde, 1978, 1985; Nicot, 1981). The swelling properties and

the collapse on heating shown by the structure whose XRD data are listed in Table 1 are not consistent with those of a chlorite *sensu stricto* that contains no swelling layers; they also differ from those of soil vermiculites (Brindley and Brown, 1980), which generally expand to about 16–17 Å after saturation with ethylene glycol and collapse to 10 Å after being heated at 550°C for 2 hr. The XRD properties of the mineral identified in the Rouez metagreywackes apparently are due to a random interstratification of predominately chlorite layers and minor expandable smectitic layers (<20%).

Chemical analyses and Mössbauer spectrometry data for this interstratified C/S are consistent with a predominance of such a chlorite-like structure. Considering that the expandable trioctahedral smectites are

Table 5. Peak parameters from room-temperature Mössbauer spectra of interstratified chlorite/smectite and chlorite from sample RZ8-27.

	Interstratified chlorite/smectite				Chlorite			
	Δ	δ	Γ	%	Δ	δ	Γ	%
P$_1$	2.67	1.10	0.13	60	2.69	1.12	0.16	60
P$_2$	2.46	1.04	0.19	20	2.36	1.06	0.23	16
P$_3$	0.99	0.54	0.61	11	0.89	0.38	0.35	24
P$_4$	0.62	0.26	0.26	9				

P$_1$, P$_2$, P$_3$, P$_4$ represent quadrupole doublets; quadrupole splitting (Δ), isomer shift (δ), half-height line width (Γ); Δ, δ, and Γ are expressed in mm/s.

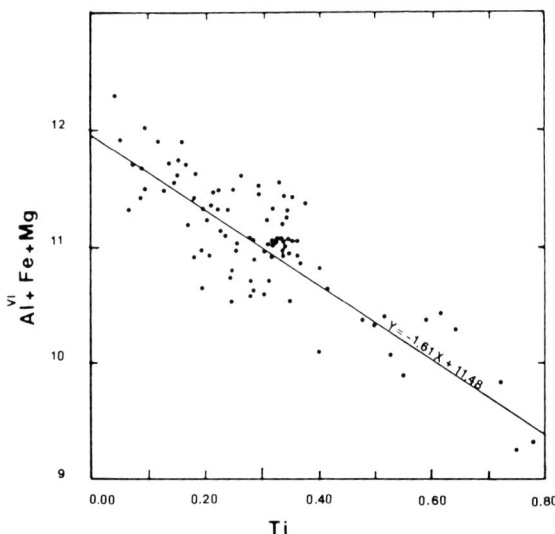

Figure 6. Ti vs. Al^{VI} + Fe + Mg contents of interstratified chlorite/smectite expressed in atomic proportions (calculated on the basis of 28 oxygens); negative correlation coefficient is .85.

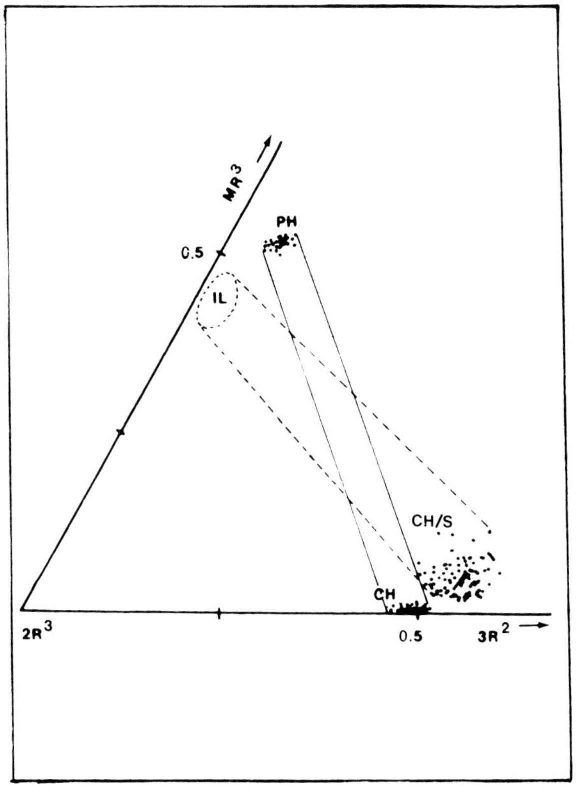

Figure 7. Composition field of interstratified chlorite/smectite, chlorites, and phengites analyzed with electron microprobe and tentative phase assemblage relation in Rouez metagreywackes. CH/S = interstratified chlorite/smectite; IL = hypothetical illite; PH = phengite; CH = chlorite. Dashed lines are tie lines joining CH/S stable with hypothetical illites in the diagenetic assemblage. Solid lines are tie lines joining CH to PH in the Hercynian metamorphic assemblage MR^3 = K + Na + 2Ca; $2R^3$ = (Al − MR^3)/3; $3R^2$ = (Fe + Mg + Mn)/2.

generally Fe^{3+} rich (Foster, 1963; Bain and Russell, 1981), the very low Fe^{3+} content of the mineral considered in this study confirms the few expandable layers present.

Considering the Fe^{3+}/Fe^{2+} ratio indicated by Mössbauer spectrometry, two tentative structural formulae of the Rouez interstratified C/S are suggested. Because of the predominance of chlorite layers, the formulae have been calculated arbitrarily on the basis of 28 oxygens (Table 6). In one formula, both ferric iron doublets have been considered to be from the C/S mineral; in the other formula, the ferric doublet having unusual parameters has been considered to be from an undetermined mineral impurity. The assignment of Ti to the octahedral site has been deduced from the very pronounced negative correlation between Al^{IV} + Fe + Mg and Ti in all the microprobe analyses (Figure 6). Compared with chlorite, the composition field of the Rouez interstratified C/S, plotted in MR^3-$2R^3$-$3R^2$ co-

ordinates (Velde, 1977), is slightly above and shifted to the right towards the biotite area (Figure 7).

The variability of major elements in the Rouez C/S can be summarized as follows: (1) Si and Al contents are almost invariable; (2) Ti, Ca, Na, and K vary from grain to grain in each sample; and (3) the variability of Fe and Mg is probably controlled chiefly by the geological environment, i.e., the Fe/Mg ratio varies between the different samples of metagreywackes. Slight variations of the Fe + Mg contents from grain to grain apparently result mainly from Ti substitutions in octahedral sites.

The origin of "metamorphic vermiculites" and their relation to other ferromagnesium phyllosilicates has been discussed by several authors (e.g., Black, 1975; Velde, 1978, 1985). In the Rouez metagreywackes such material has been partly replaced by a chlorite + phengite assemblage during the only metamorphic event (very low grade facies) recognized in the area (Icart and Safa, 1981). The different Fe and Mg distribution, the trace element content, and the Fe^{3+} content of the interstratified C/S and the chlorites suggest different pressure and temperature conditions for the formation of these materials, because the bulk rock chemistry probably did not change during such low-grade metamorphism. Studies of "metamorphic vermiculites" in pelitic rocks (Velde, 1978; Nicot, 1981) have shown that they occur at the transition between the chlorite (chlorite + illite) facies and the biotite facies in deeply buried assemblages: illite-1 + chlorite → illite-2 + "metamorphic vermiculite" (illite-2 being more aluminous that illite-1).

In the Rouez metagreywackes, the illitic phase in equilibrium with interstratified C/S could not be identified because of its general replacement by a phengite + chlorite metamorphic assemblage during the Hercynian orogenesis. Figure 7 illustrates the phase relations during the regional metamorphism. Consequently, interstratified C/S can be considered as a relict of a diagenetic assemblage which was mostly obliterated during the Hercynian orogenesis.

SUMMARY AND CONCLUSIONS

XRD investigation of what was previously called "metamorphic vermiculite" from metagreywackes surrounding the massive sulfide deposit of Rouez have shown the material to be a chlorite containing a few randomly interstratified expandable layers. One hundred microprobe analyses have defined a relatively wide range of compositions for these minerals. The Rouez interstatified C/S contains unusually large amounts of Ti, which varies from grain to grain and which changes the color of the grains from light to dark brown. The <20% expandable layers have a profound effect on the optical properties of the C/S, which closely resemble those of biotite, rather than chlorite.

REFERENCES

Bagin, V. I., Gendler, T. S., Dainyak, L. G., and Kuz'min, R. B. (1980) Mössbauer thermomagnetic, and X-ray study of cation ordering and high-temperature decomposition in biotite: *Clays & Clay Minerals* **28,** 188–196.

Bain, D. C. and Russell, J. D. (1981) Swelling minerals in a basalt and its weathering products from Morvern, Scotland: II. Swelling chlorite: *Clay Miner.* **16,** 203–212.

Beaufort, D., Champanhet, J. M., Meunier, A., Safa, P., and Sauvan, P. (1985) Les "vermiculites métamorphiques" des métasédiments encaissant l'amas sulfuré de Rouez (Sarthe, France): *Bull. Minéral.* **108,** 801–812.

Beaufort, D., Dudoignon, P., Proust, D., Parneix, J. C., and Meunier, A. (1983) Microdrilling in thin section: A useful method for the identification of clay minerals *in situ: Clay Miner.* **18,** 219–222.

Beaufort, D. and Meunier, A. (1983) Petrographic characterization of an argillic hydrothermal alteration containing illite, K-rectorite, K-beidellite, kaolinite and carbonates in a cupromolybdenic porphyry at Sibert (Rhône, France): *Bull. Minéral.* **106,** 535–551.

Black, P. M. (1975) Mineralogy of New Caledonian metamorphic rocks. IV. Sheet silicates from Ouegoa district: *Contrib. Mineral. Petrol.* **49,** 269–284.

Brindley, G. M. and Brown, G., ed. (1980) *Crystal Structures of Clay Minerals and their X-ray Identification*: Mineralogical Society, London, 495 pp.

Brown, E. H. (1967) The greenschist facies in part of Eastern Otayo, New Zealand: *Contrib. Mineral. Petrol.* **14,** 259–292.

Foster, M. D. (1963) Interpretation of the composition of vermiculites and hydrobiotites: in *Clays and Clay Minerals, Proc. 10th Natl. Conf., Austin, Texas, 1961,* Ada Swineford and P. F. Franks, eds., Pergamon, New York, 70–89.

Goodman, B. A. and Bain, D. C. (1979) Mössbauer spectra of chlorites and their decomposition products: in *Proc. Int. Clay Conf., Oxford, 1978,* M. M. Mortland and V. C. Farmer, eds., Elsevier, Amsterdam, 65–74.

Icart, J. C. and Safa, P. (1981) Rouez: Environnement géologique et minéralisation: *Chron. Rech. Minière* **458,** 12–32.

Kerrick, D. M. and Cotton, W. R. (1971) Stability relations of jadeite pyroxene in Franciscan metagreywackes near San Jose, California: *Amer. J. Sci.* **271,** 350–369.

Mac Dowell, D. M. C. and Elders, W. A. (1980) Authigenetic layer silicate minerals in Borehole Elmore 1, Salton Sea geothermal field, California, U.S.A.: *Contrib. Mineral. Petrol.* **74,** 293–310.

Meunier, A. (1980) Les mécanismes de l'altération des granites et le rôle des microsystèmes. Etude des arènes du massif granitique de Parthenay (Deux-Sèvres): *Mém. Soc. Géol. Fr.* **140,** 80 pp.

Meunier, A. and Velde, B. (1982) Phengitization, sericitization and potassium beidellite in a hydrothermally altered granite: *Clay Miner.* **17,** 285–299.

Nicot, E. (1981) Les phyllosilicates des terrains précambriens du Nord-Ouest du Montana (U.S.A.) dans la transition anchizone-épizone: *Bull. Minéral.* **105,** 615–624.

Taylor, G. L., Ruotsala, A. P., and Keeling, R. O., Jr. (1968) Analysis of iron in layer silicates by Mössbauer spectroscopy: *Clays & Clay Minerals* **16,** 381–391.

Velde, B. (1977) *Clays and Clay Minerals in Natural and Synthetic Systems:* Elsevier, Amsterdam, 218 pp.

Velde, B. (1978) High temperature or metamorphic vermiculites: *Contrib. Mineral. Petrol.* **66,** 319–323.

Velde, B. (1984) Electron microprobe analysis of clay minerals: *Clay Miner.* **19,** 243–247.

Velde, B. (1985) *Clay Minerals. A Physico-Chemical Explanation of their Occurrence:* Elsevier, Amsterdam, 427 pp.

Proceedings of the International Clay Conference, Denver, 1985, L. G. Schultz, H. van Olphen, and F. A. Mumpton, eds.,
The Clay Minerals Society, Bloomington, Indiana, 66–70 (1987).

FORMATION OF QUARTZ-TYPE PHASES DURING HIGH-TEMPERATURE REACTIONS OF MONTMORILLONITES

Yasuaki Uno,[1] Norihiko Kohyama,[2] Mitsuo Sato,[3] and Hideo Takeshi[4]

[1] College of General Education, Osaka University, Toyonaka, Osaka, Japan

[2] National Institute of Industrial Health, 6-21-1 Nagao, Tamaku, Kawasaki, Japan

[3] Faculty of Technology, Gunma University, Kiryu, Gunma, Japan

[4] Naruto University of Teacher Education, Naruto, Tokushima, Japan

Abstract—The phase transformations of homoionic montmorillonites and the compositions of high-temperature products have been studied by X-ray powder diffraction and electron microscopy. Quartz-type phases crystallized as high-quartz-like phases that were metastable because of their conversion into other compounds, such as cristobalite, spinel, and cordierite, on prolonged heating. Using the analytical electron microscope, the composition of high-quartz-type phases was found not to be pure SiO_2, but instead showed a structural formula close to $MgAl_2Si_5O_{14}$. Cristobalite-type phases were also noted and contained Mg and Al, although the amount was less than in the high-quartz-type phases. The mode of formation and decomposition of high-temperature products was affected by bulk chemical compositions, especially the Mg/Al ratio of the original montmorillonites.

Key Words—Cristobalite, Montmorillonite, Quartz, Thermal treatment, X-ray powder diffraction.

INTRODUCTION

It is well known that quartz-type phases develop from montmorillonites by heating in the temperature range 900°–1000°C. The X-ray powder diffraction patterns of these phases are closely similar to that of quartz, but the products do not show a low–high transition. They slowly transform to a cristobalite-type phase at more elevated temperatures. Bradley and Grim (1951) stated that some high-temperature products from montmorillonites, when observed at room temperature, were structurally similar to low-quartz and some were similar to high-quartz. Grim and Kulbicki (1961) suggested the possibility of some ion stuffing in the structure of the high-quartz-type phase. The composition and structure of these phases, however, are not clear. The aim of the present investigation was to describe the mode of formation as well as chemical composition of these quartz-type phases.

EXPERIMENTAL PROCEDURES

Montmorillonites used in the present study were from the Cheto mine, Apache County, Arizona, collected by Tadahisa Nakazawa (Mizusawa Chemical Co.) and the Clay Spur bed, Colony, Wyoming, collected by Yoichi Shiraki (Tokyo Institute of Technology). The original montmorillonites were investigated by chemical analysis, differential thermal analysis (DTA), and X-ray powder diffraction (XRD). After heat treatment, high-temperature products of the montmorillonites were analyzed by XRD and analytical electron microscopy (AEM).

Chemical components of the original montmoril-

lonites were quantified by gravimetric, volumetric, and atomic absorption spectrometric analyses. The original Cheto sample has a much higher Mg content than the Clay Spur sample. Cation-exchange capacities (CEC) were determined by the method of Schollenberger and Simon (1945). To study the correlation between bulk chemical composition and phase transformation, the Cheto and Clay Spur montmorillonites were made homoionic by saturation with Mg and Ca before the heating experiments. In addition, one sample of the Clay Spur montmorillonite was prepared by adding excess magnesium acetate to the Mg-saturated sample until the Mg:Al ratio in the specimen was about 0.5, i.e., similar to that of Mg-saturated Cheto specimen.

DTA was carried out from room temperature to about 1200°C at a heating rate of 15°C/min. Also, quartz-type phases were formed by heating the homoionic montmorillonites in an electric muffle furnace for 0.5–100 hr and at temperatures of 800°–1200°C. After cooling, the mineral composition of all products was determined by XRD. Low–high transition of quartz-type phases was also studied in a high-temperature XRD apparatus, using mineral quartz as an internal standard. A Hitachi H-500 electron microscope and Kevex-5100 analytical apparatus were used for chemical analysis of heated specimens.

RESULTS

Chemical analysis of montmorillonites

Chemical compositions of the original montmorillonites are given in Table 1. The Cheto specimen appears to be pure montmorillonite, as no other minerals

Table 1. Chemical compositions of original montmorillonites.

	Cheto, Arizona	Clay Spur, Wyoming
SiO_2	48.24	48.76[1]
TiO_2	0.02	0.00
Al_2O_3	17.01	20.71
Fe_2O_3	0.81	2.1
FeO	0.15	0.54
MnO	0.00	0.00
MgO	4.05	2.57
CaO	3.03	1.20
Na_2O	0.34	1.61
K_2O	0.16	0.14
H_2O+	8.59	6.15
H_2O-	17.21	16.23
Total	99.61	100.03

[1] Corrected for 7.6% quartz.

were observed optically or by XRD. Some quartz (7.6%) impurity was estimated in the Clay Spur specimen by XRD using the addition method. That amount of SiO_2 was subtracted from the original chemical analysis to arrive at the analysis given in Table 1. CECs and the amount of exchange cations in the montmorillonite specimens are listed in Table 2.

Structural formulae of the montmorillonites calculated from Tables 1 and 2 are as follows;

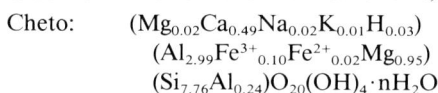

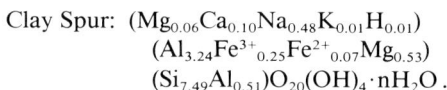

Cheto: $(Mg_{0.02}Ca_{0.49}Na_{0.02}K_{0.01}H_{0.03})$
$(Al_{2.99}Fe^{3+}_{0.10}Fe^{2+}_{0.02}Mg_{0.95})$
$(Si_{7.76}Al_{0.24})O_{20}(OH)_4 \cdot nH_2O$

Clay Spur: $(Mg_{0.06}Ca_{0.10}Na_{0.48}K_{0.01}H_{0.01})$
$(Al_{3.24}Fe^{3+}_{0.25}Fe^{2+}_{0.07}Mg_{0.53})$
$(Si_{7.49}Al_{0.51})O_{20}(OH)_4 \cdot nH_2O$.

Ca is the chief exchangeable cation in the Cheto sample, whereas Na is predominant in the Clay Spur sample. Nearly 25% of the octahedral sheet of the Cheto sample is occupied by Mg, whereas the Clay Spur sample is rich in Al compared with the Cheto sample.

Differential thermal analysis

Figure 1 shows DTA curves for Mg- and Ca-exchanged Cheto and Clay Spur montmorillonites. Three

Table 2. Cation-exchange capacity (CEC) and exchangeable cations.

	Cheto, Arizona	Clay Spur, Wyoming
CEC (meq/100 g)	101.5	84.0
Exchange cations (meq/100 g)		
Na^+	2.3	48.0
K^+	1.2	1.6
Mg^{2+}	4.8	13.3
Ca^{2+}	101.5	19.8
Al^{3+}	0.0	0.4
[1]H^+	2.7	0.9

[1] Calculated from CEC and total of exchange cations measured.

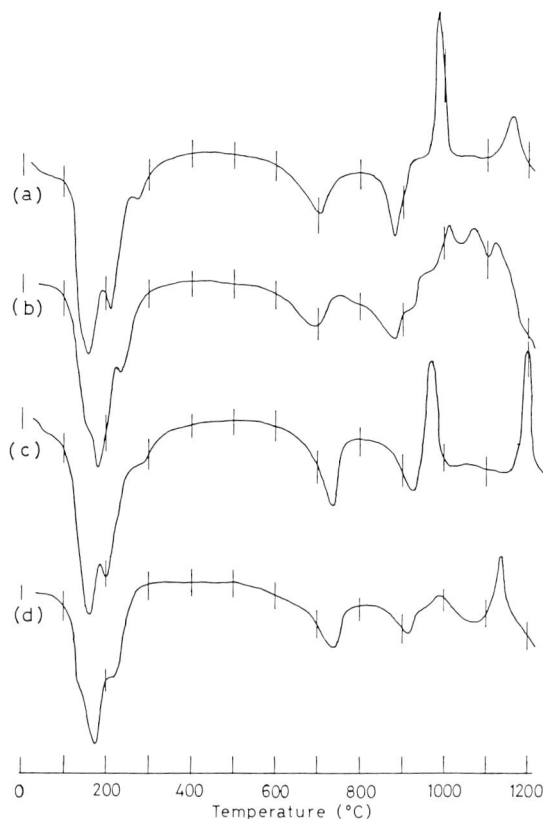

Figure 1. Differential thermal analysis curves of: (a) Cheto, Arizona, montmorillonite saturated with Mg; (b) Cheto sample saturated with Ca; (c) Clay Spur, Colony, Wyoming, montmorillonite saturated with Mg; (d) Clay Spur sample saturated with Ca.

endothermic peaks having different intensities are present: the first endothermic peak between 100° and 250°C, typically a doublet, is due to dehydration; the second endothermic reaction in the range 500°–700°C is due to dehydroxydation; and the third endothermic reaction at about 800°–900°C is due to the final breakdown of the montmorillonite as indicated by the X-ray-amorphous nature of the sample. The third endothermic peak is followed by two or three exothermic effects in the range of 950°–1200°C. As will be seen, the exothermic peaks are caused by the crystallization of new minerals.

Table 3. Unit-cell dimensions and silica content of high-temperature products.

	Quartz-type phase			Cristobalite phase	
	a_0 (Å)	c_0 (Å)	SiO_2 (%)	a_0 (Å)	c_0 (Å)
Cheto-Mg	5.13	5.38	67	5.00	7.07
Cheto-Ca	5.13	5.38	67	5.02	7.10
Clay Spur-Mg	—	—	—	5.01	7.09
Clay Spur-Ca	5.01	5.39	89	5.02	7.09
Clay Spur-excess Mg	5.13	5.38	67	5.01	7.08

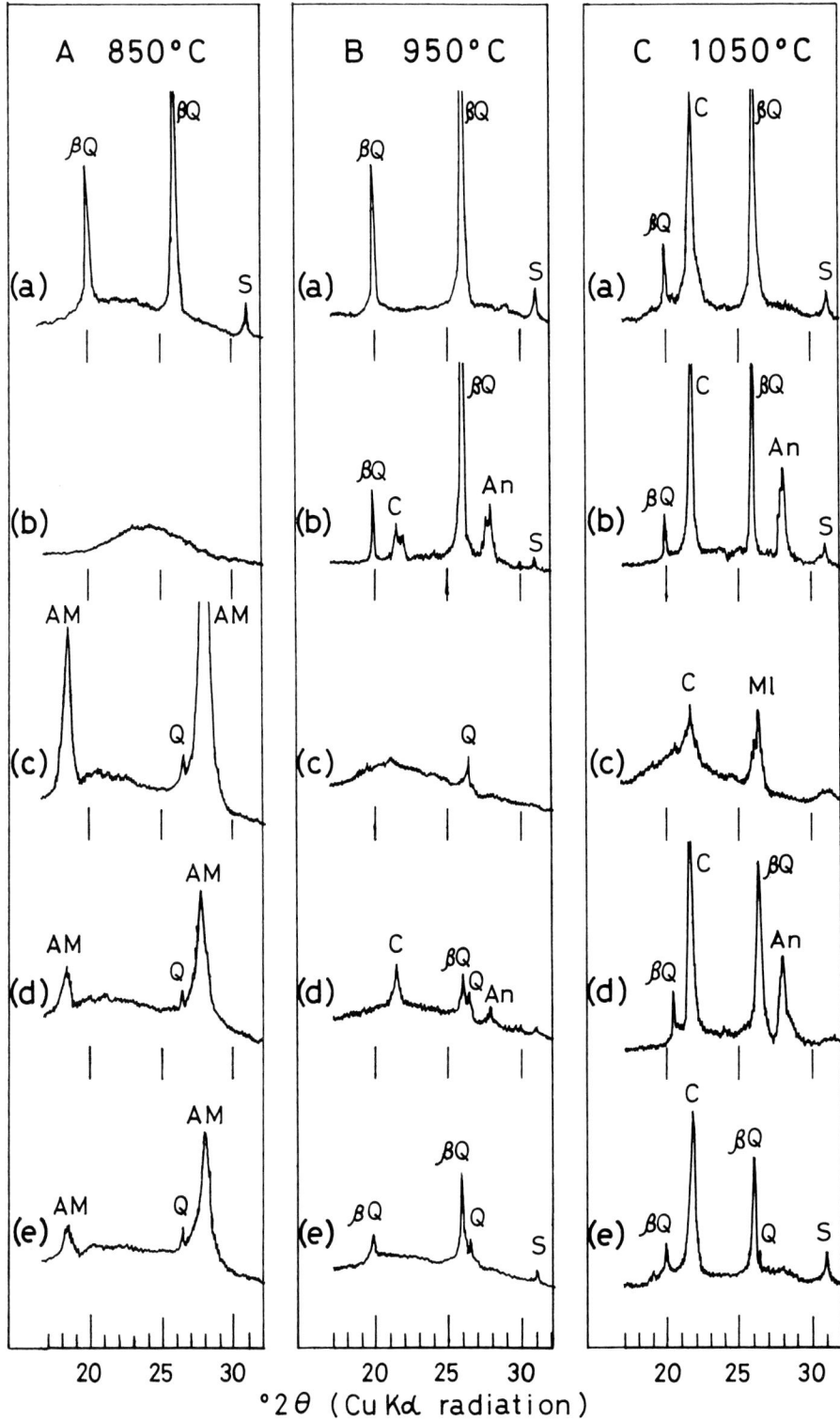

Figure 2.   X-ray powder diffraction patterns of samples heated for 30 min. A = at 850°C; B = at 950°C; C = 1050°C; (a) Cheto, Arizona, Mg-saturated; (b) Cheto, Ca-saturated; (c) Clay Spur, Colony, Wyoming, Mg-saturated; (d) Clay Spur, Ca-saturated; (e) Clay Spur, having excess Mg-ion. βQ = high-quartz-type phase; Q = quartz impurity in original montmorillonite; C = cristobalite phase; S = spinel; Ml = mullite; An = anorthite; AM = anhydrous montmorillonite.

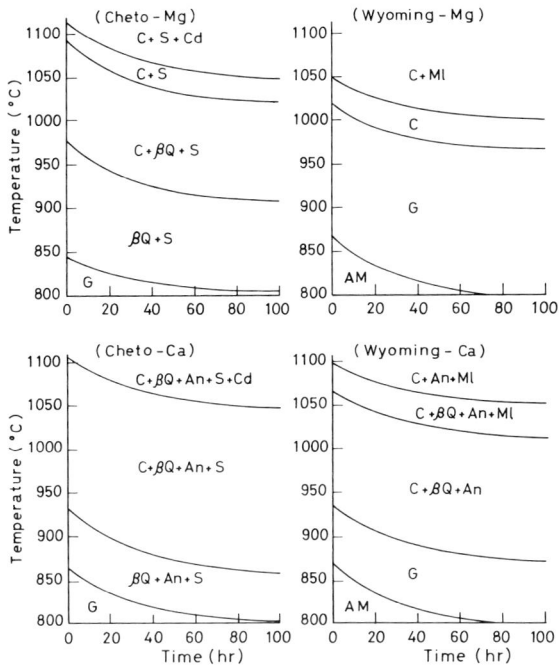

Figure 3. High-temperature phase development of Cheto, Arizona, and Clay Spur, Wyoming, samples with time and temperature. G = glass; Cd = cordierite. Other abbreviations are the same as those in Figure 2.

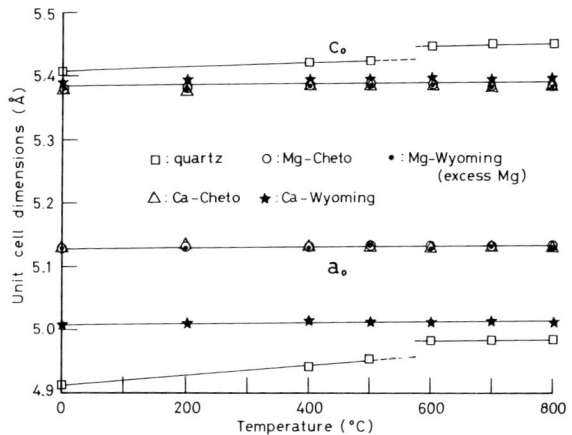

Figure 4. Variation of cell dimensions of mineral quartz and quartz-type phases derived from montmorillonites as determined on a high-temperature X-ray diffraction apparatus. Square = mineral quartz; open circle = quartz-type phase derived from Mg-saturated Cheto, Arizona, sample; solid circle = from Clay Spur, Wyoming, sample having excess Mg; triangle = from Ca-saturated Cheto sample; star = from Ca-saturated Clay Spur sample.

X-ray powder diffraction

XRD patterns of high-temperature products are shown in Figure 2. The high-quartz-type phase was formed in Mg-saturated Cheto montmorillonite above 800°C, but not in the Mg-saturated Clay Spur sample at any temperature except for the sample having excess Mg. Ca-saturated samples of the Cheto and Clay Spur materials developed quartz-type phases coexisting with anorthite, but the phase in the Clay Spur sample was closer to low-quartz in that its a_0 cell dimension was smaller than that from the Cheto sample. The cell dimensions of the quartz-type phases and cristobalite phases are shown in Table 3. Silica contents of quartz-type phases were estimated using the unit-cell variation diagram of quartz solid solution after Schreyer and Schairer (1961). Cell dimensions of cristobalite phases were slightly larger than the values of pure cristobalite.

Figure 3 shows the regions of phase development in the heated samples. High-quartz type phases were metastable because they converted into other compounds, such as cristobalite, spinel, and cordierite, with prolonged heating. The amount of quartz-type phases was related to heating time as well as to temperature or thermodynamic equilibrium.

Figure 4 shows the variation of cell dimensions of pure quartz and quartz-type phases between room temperature and 800°C. Although pure quartz showed the expected low–high transition at 573°C, the quartz-type

phases showed no such transition. They have larger a_0 and smaller c_0 cell dimensions compared with pure quartz, and showed no appreciable expansion of the unit cell with rising temperature.

Electron microscopic study

Chemical compositions of the quartz-type phase and the cristobalite phase in Mg-saturated Cheto sample

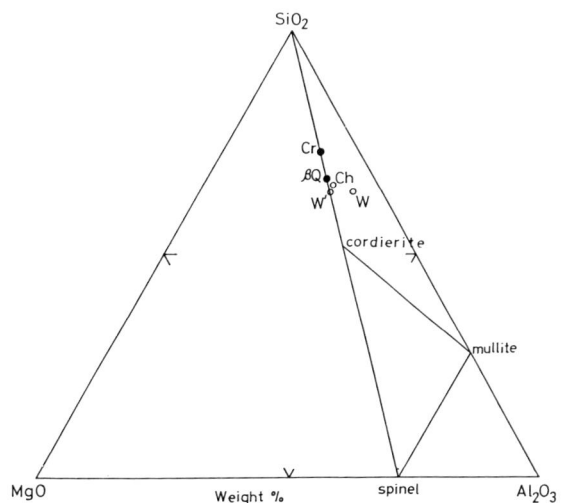

Figure 5. Chemical compositions of samples in the system MgO-Al₂O₃-SiO₂. Ch = Mg-saturated Cheto, Arizona, sample; W = Mg-saturated Clay Spur, Wyoming, sample; W = Clay Spur sample having excess Mg; βQ = high-quartz type phase derived from Mg-saturated Cheto sample; Cr = cristobalite phase derived from Mg-saturated Cheto sample.

determined by AEM are as follows: (1) quartz-type phase, $Mg_{1.00}Al_{1.98}Fe_{0.09}Si_{4.95}O_{14}$; (2) cristobalite phase, $Mg_{1.02}Al_{1.87}Fe_{0.24}Si_{6.91}O_{18}$. The silica content of the quartz-type phases derived from the Cheto sample agrees with the value estimated by XRD methods (Table 3). Chemial analyses of high-temperature products from other samples were not successful because the products were intimate mixtures of quartz-type and other phases.

DISCUSSION

The chemical composition of the high-quartz-type phases formed by heating montmorillonite was found not to be pure SiO_2, but close to $MgAl_2Si_5O_{14}$. Schreyer and Schairer (1961) reported the synthesis of various members of metastable quartz solid solutions from glass in the system $MgO-Al_2O_3-SiO_2$. High-quartz-type phases in the present study corresponded to one of them—$MgO:Al_2O_3:SiO_2 = 1:1:5$. Schulz et al. (1971a, 1971b) studied Mg-Al silicates having high-quartz-like structure and stated that an ordering of the (Si,Al) and Mg atoms results in a superstructure. All atoms take part in the formation of the superstructure by small displacements from the ideal high-quartz positions. Weak superstructure XRD patterns for the quartz-type phases derived from the Cheto sample were considered to be caused by similar displacements.

Figure 5 shows chemical compositions of the Mg-saturated Cheto and Clay Spur montmorillonites and the high-temperature products derived from the Cheto sample. The Mg:Al ratio of the Mg-saturated Cheto montmorillonite is about 1:2, whereas the Mg:Al ratio of the Mg-saturated Clay Spur montmorillonite is about 1:4. The Mg-saturated Clay Spur sample developed no quartz-type phases until excess Mg was added to raise the Mg:Al ratio to that of the Cheto sample. The Ca-saturated Cheto and Clay Spur samples were able to form quartz-type phases coexisting with anorthite evidently because the Mg:Al ratio of the anhydrous montmorillonite increased with the crystallization of anorthite ($CaAl_2Si_2O_8$). Evidently the Mg:Al ratio is one of the most important factors for the formation of quartz-type phases.

REFERENCES

Bradley, W. F. and Grim, R. E. (1951) High temperature thermal effects of clay and related materials: *Amer. Mineral.* **36,** 182–201.

Grim, R. E. and Kulbicki, G. (1961) Montmorillonite: high temperature reactions and classification: *Amer. Mineral.* **46,** 1329–1369.

Schollenberger, G. J. and Simon, R. N. (1945) Determination of exchange capacity and exchangeable bases in soil—ammonium acetate method: *Soil Sci.* **59,** 13–24.

Schreyer, W. and Schairer, J. F. (1961) Metastable solid solution with quartz-type structures on join SiO_2-$MgAl_2O_4$: *Z. Kristallogr.* **116,** 62–82.

Schulz, H., Muchow, G. M., Hoffmann, W., and Bayer, G. (1971a) X-ray study of Mg-Al silicate high-quartz phases: *Z. Kristallogr.* **133,** 91–109.

Schulz, H., Hoffmann, W., and Muchow, G. M. (1971b) The average structure of $Mg(Al_2Si_3O_{10})$, a stuffed derivative of the high-quartz structure: *Z. Kristallogr.* **134,** 1–27.

Proceedings of the International Clay Conference, Denver, 1985, L. G. Schultz, H. van Olphen, and F. A. Mumpton, eds., The Clay Minerals Society, Bloomington, Indiana, 71–72 (1987).

HIGHLY CRYSTALLINE MONTMORILLONITE OBTAINED BY TREATMENT WITH TiCl₄ AQUEOUS SOLUTIONS

Shoji Yamanaka, Tsuguo Ishihara, and Makoto Hattori

Department of Applied Chemistry, Faculty of Engineering
Hiroshima University, Higashi-Hiroshima 724, Japan

Abstract—After dispersion in 1–0.5 M $TiCl_4$ aqueous solutions and carefully drying, Na-montmorillonite showed a sharp X-ray powder diffraction (XRD) pattern. Such a pattern was obtained only when the $TiCl_4$-treated montmorillonite was dried quickly in a stream of dry air without rinsing with water. If the sample was washed with water before drying or treated with dilute $TiCl_4$ solutions (<0.05 M), only the broad diffraction pattern typical of montmorillonite was obtained. The use of $TiOSO_4$ instead of $TiCl_4$ also resulted in a broad pattern. Most of the XRD peaks of the pattern were indexed on the basis of a monoclinic cell with $a = 5.22(1)$ Å, $b = 9.13(2)$ Å, $c = 11.37(2)$ Å, and $\beta = 93.7(1)°$. The sharpening of the XRD peaks is probably caused by hydrogen bonded interlayer water molecules coordinated with the titanium ions, which restrict the relative orientation of the adjacent silicate layers, resulting in a highly crystalline product.

Key Words—Crystallinity, Interlayer water, Montmorillonite, $TiCl_4$, X-ray powder diffraction.

INTRODUCTION

In $ZrOCl_2 \cdot 8 H_2O$ aqueous solutions, zirconium is present in the form of tetrameric hydroxy ions $[Zr_4(OH)_8 \cdot n H_2O]^{8+}$ (Muha and Vaughan, 1960). When Na-montmorilonite is dispersed in this solution, the interlayer ions are exchanged with the hydroxy-zirconium ions, which are dehydrated to stable zirconium oxide pillars on heating. The pillars keep the silicate layers apart and form micropores between the layers (Yamanaka and Brindley, 1979).

Inasmuch as Ti and Zr are in the same IVB group and possess similar chemical properties, an attempt was made to introduce titanium oxide pillars instead of zirconium oxide pillars by a similar ion-exchange reaction, using $TiCl_4$ aqueous solutions. As described below, although such titanium oxide pillars could not have been formed between the silicate layers, the montmorillonite treated with $TiCl_4$ solutions showed an unusually sharp X-ray powder diffraction (XRD) pattern.

EXPERIMENTAL

Concentrated $TiCl_4$ solution (about 4 M) was prepared by dissolving anhydrous $TiCl_4$ into distilled water. The solution was diluted to the desired concentration with water prior to use. Another titanium solution was prepared by dissolving $TiOSO_4 \cdot 2 H_2O$ in distilled water.

The Na-montmorillonite used in this study was from Tsukinuno district in Yamagata Prefecture, Japan, the same sample used in a previous study (Yamanaka *et al.*, 1984). The structural formula of the montmorillonite is $(Na_{0.35}K_{0.01}Ca_{0.02})(Si_{3.89}Al_{0.11})(Al_{1.60}Fe_{0.09}Mg_{0.32})O_{10}(OH)_2 \cdot n H_2O$; the cation-exchange capacity (CEC) is 100 meq/100 g. Different kinds of homoionic (K⁺,

NH₄⁺, Mg^{2+}, and Ca^{2+}) montmorillonites were prepared by ion-exchange in the respective chloride solutions.

The homoionic montmorillonite (50 mg) was dispersed in 5 ml of distilled water and mixed with 5 ml of titanium solution of a concentration ranging from 2 to 0.1 M. After stirring for various periods of time at different temperatures, the samples were separated by centrifugation. The XRD pattern of the sample was measured after drying on a glass slide in a stream of dry air.

RESULTS AND DISCUSSION

Na-montmorillonite was dispersed in a 0.5 M $TiCl_4$ aqueous solution and separated by centrifugation. When

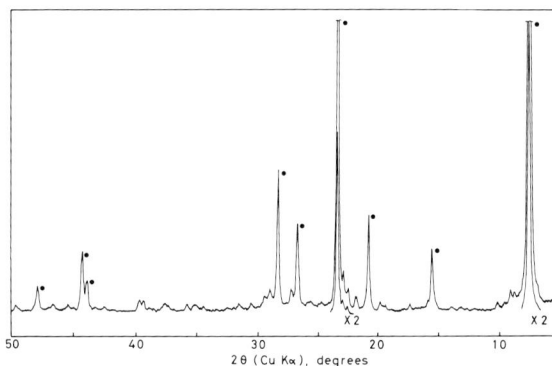

Figure 1. X-ray powder diffraction pattern of Na-montmorillonite treated with 0.5 M $TiCl_4$ solution and dried in a stream of dry air soon after the separation without washing with water.

71

Table 1. X-ray powder diffraction data of montmorillonite treated with 0.5 M $TiCl_4$ and dried in a stream of dry air soon after the separation without washing with water.

hkl	$d_{calc.}$ (Å)	$d_{obs.}$ (Å)	I
001	11.51	11.35	vs
002	5.69	5.67	m
021	4.26	4.24	m
003	3.78	3.78	vs
12$\bar{1}$	3.34	3.33	m
10$\bar{3}$	3.15	3.16	s
222	2.06	2.06	w
141	2.04	2.04	m
23$\bar{2}$	1.90	1.90	w

Only the reflections designated by (●) in Figure 1 are listed. vs = very strong; s = strong; m = medium; w = weak.

the separated sample was dried in air after washing with water, the sample showed a typical broad XRD pattern for montmorillonite having a basal spacing of 14.2 Å. When the sample was dried in a stream of dry air or nitrogen gas *without* washing soon after it was separated from the solution, it showed an unusually sharp XRD pattern, as shown in Figure 1. The other homoionic montmorillonites also gave similar sharp XRD patterns when treated in a similar manner. Most of the XRD reflections designated by (●) in the illustration were indexed on the basis of a monoclinic cell with $a = 5.22 \pm 0.01$ Å, $b = 9.13 \pm 0.02$ Å, $c = 11.37 \pm 0.02$ Å, and $\beta = 93.7 \pm 0.1°$ (Table 1). The lattice dimensions for a and b are comparable with those reported for typical montmorillonites (Brindley, 1980). The XRD pattern showed that the reaction was complete in 30 min at temperatures ranging from 10° to 30°C. If the sample was rinsed with water and then dried, it produced the broad pattern typical of a montmorillonite. The use of a more dilute $TiCl_4$ solution (<0.05 M) or a $TiOSO_4$ solution also resulted in the typical broad XRD pattern.

To obtain a sharp XRD pattern as in Figure 1, two precautions were necessary: (1) the montmorillonite treated with the $TiCl_4$ solutions had to be separated without rinsing with water and dried quickly in a stream of dry air or nitrogen gas; (2) the XRD pattern had to be obtained in a dry atmosphere, because in an atmosphere having a relative humidity >60%, the sample adsorbed water, the basal spacing increased, and the XRD pattern became broad.

It is difficult to explain why a highly crystalline montmorillonite having a sharp XRD pattern was formed, inasmuch as the montmorillonite which gave such a pattern was stable only if it was dried without washing. Consequently little information is available about its actual interlayer compositions; however the interlayer water molecules were apparently instrumental in forming the highly crystalline montmorillonite. Water molecules coordinated with titanium ions probably formed hydrogen bonded monomolecular interlayers, which restricted the relative orientation of the adjacent silicate layers. The repetition of this kind of stacking of the silicate layers may have resulted in the highly crystalline product.

ACKNOWLEDGMENT

This work was supported in part by a Grant-in-Aid for Scientific Research (No. 58850169) from the Ministry of Education, Science and Culture, Japan.

REFERENCES

Brindley, G. W. (1980) Order-disorder in clay mineral structures: in *Crystal Structures of Clay Minerals and their X-Ray Identification*, G. W. Brindley and G. W. Brown, eds., Mineralogical Society, London, 125–195.

Muha, J. M. and Vaughan, Ph.-A. (1960) Structure of the complex ion in aqueous solutions of zirconyl and hafnyl oxyhalides: *J. Chem. Phys.* **33,** 194–199.

Yamanaka, S. and Brindley, G. W. (1979) High surface area solids obtained by reaction of montmorillonite with zirconyl chloride: *Clays & Clay Minerals* **27,** 119–124.

Yamanaka, S., Doi, T., Sako, S., and Hattori, M. (1984) High surface area solids obtained by intercalation of iron oxide pillars in montmorillonite: *Mat. Res. Bull.* **19,** 161–168.

Proceedings of the International Clay Conference, Denver, 1985, L. G. Schultz, H. van Olphen, and F. A. Mumpton, eds.,
The Clay Minerals Society, Bloomington, Indiana, 73–77 (1987).

OCTAHEDRAL COMPOSITIONS OF INDIVIDUAL PARTICLES IN SMECTITE-PALYGORSKITE AND SMECTITE-SEPIOLITE ASSEMBLAGES

HÉLÈNE PAQUET,[1] JOËLLE DUPLAY,[1] MARIE-MADELEINE VALLERON-BLANC,[2]
AND GEORGES MILLOT[1]

[1] Centre de Sédimentologie et de Géochimie de la Surface, Institut de Géologie
1 rue Blessig, 67084 Strasbourg Cedex, France

[2] Laboratoirc de Géologie du Muséum National d'Histoirc Naturelle
43 rue de Buffon, 75231 Paris Cedex 05, France

Abstract—Chemical analyses were carried out on individual particles in <2-μm size fractions to study smectite-palygorskite and smectite-sepiolite assemblages extracted from calcareous and dolomitic continental formations in Provence, southeastern France. A plot of the octahedral composition of 145 individual particles of each clay mineral species considered on a (Al + Fe^{3+}) vs. Mg diagram shows a great variability for smectites and fibrous clays. The octahedral composition fields of the smectites and fibrous clays partly overlap. The sepiolite field is clearly in the trioctahedral domain, whereas the palygorskite field is both in the dioctahedral domain as well as between the dioctahedral and trioctahedral domains of smectites. Similar to smectites, the structure of palygorskite appears to be accommodating for octahedral substitutions; however, this characteristic cannot be inferred from chemical analyses of whole <2-μm fractions, which represent the average composition of several particles.

Key Words—Chemical analysis, Octahedral composition, Palygorskite, Sepiolite, Smectite.

INTRODUCTION

Palygorskites in calcareous formations and meteoric calcretes are generally mixed with smectites, and the composition of each type of clay mineral cannot be accurately determined. Typical chemical analyses of palygorskites from such rocks give chemical compositions similar to those of dioctahedral smectites of the Al-Mg-Fe^{3+}-beidellite type (Paquet, 1983). The individual particle analytical technique of Duplay (1982, 1984), however, is useful for smectites and fibrous clay minerals which commonly are intimately admixed in the clay fraction of carbonate rocks. The present paper presents the results of such a study of particles in <2-μm size fractions isolated from continental formations from southeastern France.

METHODS AND MATERIALS

Methods

Sample preparation. X-ray powder diffraction and chemical analyses are usually carried out on size-fractionated samples that are made up of several thousand particles of different clay minerals. As such, these analyses give average mineralogical or chemical compositions of the whole fraction. In the present study the <2-μm size fraction was first studied by means of standard X-ray powder diffraction (XRD) (Brindley and Brown, 1980) to identify the various mineral species present. Then, particles in aqueous suspension were carefully selected in a transmission electron microscope for individual chemical analyses. They consisted of clean and as-thin-as-possible individual smectite particles ranging from 1 to 2 μm in diameter and of ~1-μm thick bundles of palygorskite and sepiolite fibers.

Analytical techniques. Elemental microanalyses of selected individual particles were carried out using a Si(Li) energy-dispersive (EDX) spectrometer (Link system) coupled to a scanning-transmission electron microscope (STEM) having a <1-μm beam. The same grid served for the selection of individual particles from a given aqueous suspension with transmission electron microscope and, later, for the elemental analysis. X-ray emission spectra were obtained for each individual particle selected in the smectite-palygorskite and smectite-sepiolite assemblages studied. The various element contents were then calculated from peak areas, applying the Castaing relationship (Tixier, 1978), with reference to a standard similar to the analyzed mineral, as follows:

$$k_A = (I_A)sample/(I_A)standard$$
$$= [(C_A)sample/(C_A)standard]$$
$$\cdot [(\rho x)sample/(\rho x)standard],$$

where k_A is the relative intensity of an element A, I is the measured intensity, C is mass concentration, ρx is mass thickness, and

$$k_A/k_B = [(C_A/C_B)sample]/[(C_A/C_B)standard],$$

where C_A, C_B ... C_N are the unknown amounts of elements A, B ... N and $C_A + C_B ... + C_N$ is equal to 100.

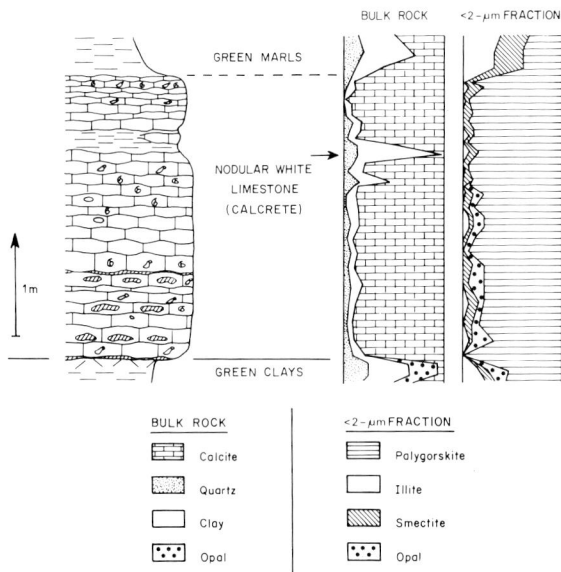

Figure 1. Schematic section of Laval St. Roman calcrete and bulk rock and <2-μm fraction mineral compositions (Valleron, 1981). Arrow indicates location of sample studied.

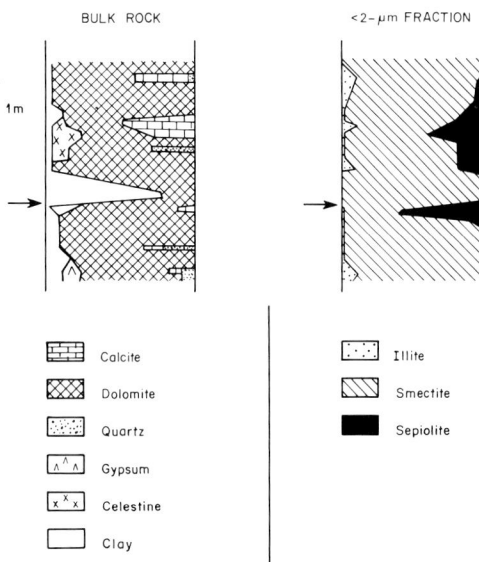

Figure 3. Schematic representation of bulk rock and of <2-μm fraction mineral compositions from Blauvac dolomite (Triat and Truc, 1972; Trauth, 1977).

This technique permitted the analysis of individual particles having areas of 1 to 2 μm^2 and thickness $\geqslant 1000$ Å. The relative error of the measurement of the peak intensities was about 1–2% for the major elements and ranged from 10 to 20% for the minor elements. Thus, only the first number after the decimal point is meaningful.

Materials

The samples analyzed were collected from three Late Eocene calcareous and dolomitic formations in southeastern France (Triat and Truc, 1972; Trauth, 1977; Valleron, 1981). A paleocalcrete sample from Laval St. Roman, Gard, contained palygorskite, smectite, chalcedony, and opal-CT (Valleron, 1981) (Figure 1). The limestone sample from Jocas in the Mormoiron basin, Vaucluse, was a calcareous diagenetic sediment containing palygorskite, smectite, and opal-CT (Truc *et al.,* 1984) (Figure 2). The sample extracted from the Blauvac evaporitic formation, also in the Mormoiron basin, consisted of dolomite, gypsum, celestite, and

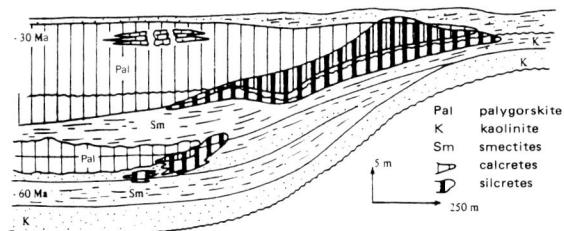

Figure 2. Section of the Jocas limestone showing the clay mineral distribution (Truc *et al.,* 1984).

clay beds composed of sepiolite and Mg-smectite of the saponite-stevensite type (Triat and Truc, 1972; Trauth, 1977) (Figure 3).

Electron photomicrographs illustrate typical smectite-sepiolite assemblages from Blauvac (Figure 4) and a studied smectite particle (Figure 5) and a bundle of sepiolite fibers (Figure 6) from the assemblage. The elemental analyses were made on such individual particles of smectite and fibrous clay minerals.

RESULTS

Calculation of structural formulae

Structural formulae were calculated from the analyses of the 145 individual particles of smectite, palygorskite, and sepiolite. The octahedral compositions obtained were based on the following conventions: (1) Octahedral compositions were calculated on the basis of the number of Al^{3+}, Fe^{3+}, and Mg^{2+} per six positive charges per half unit cell. They can thus be compared with each other despite the structural differences between smectites and fibrous clay minerals. (2) Total structural iron was considered as Fe^{3+} because the elemental analysis technique does not distinguish between Fe^{2+} and Fe^{3+}. Little or no Fe^{2+}, however, is likely present because chemical analyses carried out on bulk <2-μm size fractions typically showed very little Fe^{2+}. (3) All Al was assumed to be Al^{VI}. As will be seen, Figure 7 indicates the validity of this assumption.

Graphical analysis

The 145 octahedral compositions are plotted in Figure 7, in which the ordinate and abscissa axes represent

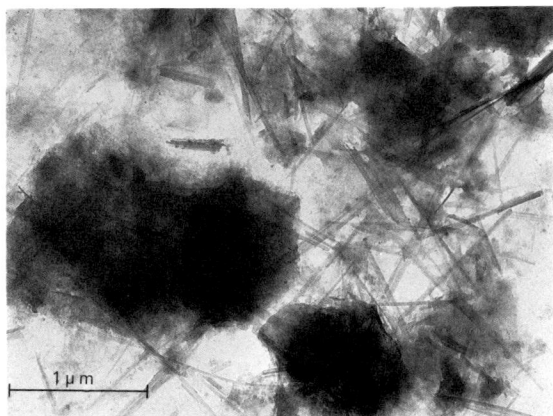

Figure 4. Transmission electron micrograph of the <2-μm fraction from the Blauvac dolomite, showing admixed smectites and sepiolites.

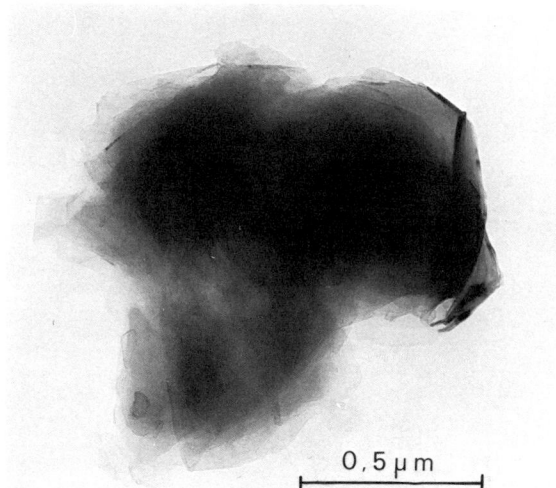

Figure 5. Transmission electron micrograph of the <2-μm fraction from the Blauvac dolomite, showing analyzed particle of smectite.

$(Al + Fe^{3+})^{VI}$ and the $(Mg^{2+})^{VI}$, respectively. The octahedral compositions are scattered close to the straight line $y = 2 - 2x/3$, where $y = (Al + Fe^{3+})^{VI}$ and $x = Mg^{VI}$. The ideally expectable relationship would occur only if almost all of the Al is in the octahedral layer. A scatter is also present within the size different groups of particles, as shown in Table 1.

Conventional limits

The usually accepted limits between dioctahedral and trioctahedral smectites are also shown in Figure 7. According to Weaver and Pollard (1973), the total trivalent cations must exceed 1.3 per half unit cell in dioctahedral smectites, whereas the total divalent cations must exceed 1.83 per half unit cell in trioctahedral phases (Foster, 1960). According to Martin-Vivaldi and Cano-Ruiz (1955), Weaver and Pollard (1973), and Valleron (1981), the octahedral compositions of palygorskites lie in a composition gap between dioctahedral and trioctahedral smectites.

Distribution of compositions

The compositions of individual smectite particles from the Jocas limestone fall within the dioctahedral domain, i.e., $(Al + Fe^{3+})$ is >1.3. The composition of all individual sepiolite particles from the Blauvac dolomite fall within the trioctahedral domain, i.e., Mg is >1.83, as expected for sepiolites. The compositions of the four other groups of individual particles, however, extend beyond these theoretical limits. Thirteen of the 23 smectite particles from Laval St. Roman are dioctahedral smectites, but the composition of 10 fall between the dioctahedral and trioctahedral domains. Seventeen of the 24 smectites from Blauvac correspond to trioctahedral smectites, one appears to be a dioctahedral smectite, and the compositions of six fall in the intermediate zone. Thirteen of the 27 palygorskite par-

ticles from Laval St. Roman have compositions in the dioctahedral domain, one in the trioctahedral domain, and 13 in the intermediate zone. Thirteen of the 25 palygorskites from Jocas have compositions in the dioctahedral domain, and 12 are in the intermediate zone. Thus, the octahedral compositions of the smectites appear to be intermediate between the dioctahedral and trioctahedral domains, whereas, the octahedral compositions of the palygorskites extend beyond this intermediate zone. In contrast, the *average* compositions

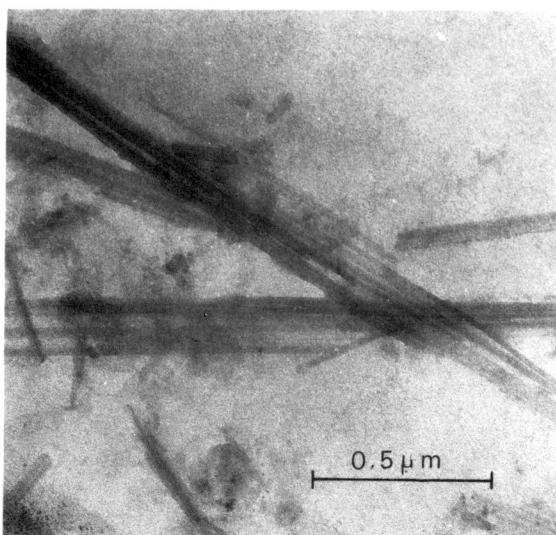

Figure 6. Transmission electron micrograph of the <2-μm fraction from the Blauvac dolomite, showing analyzed bundle of sepiolite fibers.

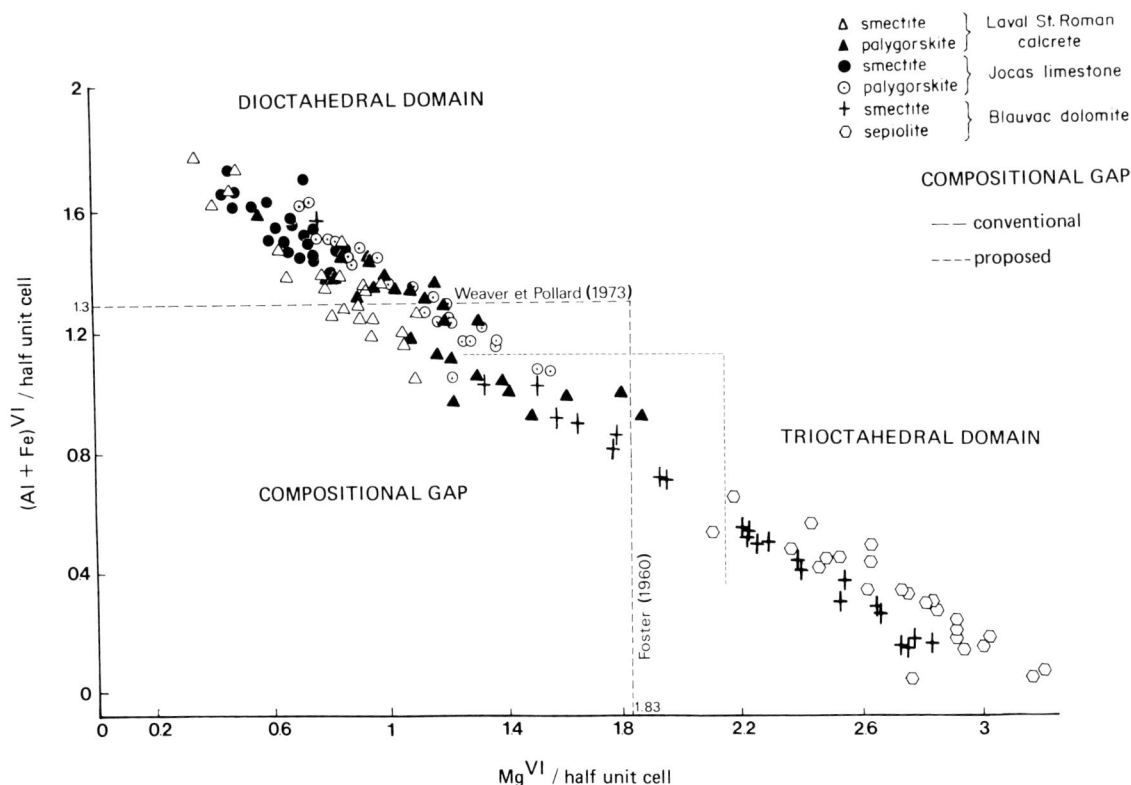

Figure 7.   Octahedral composition of 145 individual particles. $(Al + Fe^{3+})^{VI}$ vs. Mg.

of the 6 groups of particles are within the conventional limits of these phases, as shown by Figure 8.

INTERPRETATION

Variability of octahedral compositions

As discussed above, the elemental analyses of a large number of individual particles in a given clay mineral assemblage indicate a large variability of octahedral composition for each clay mineral species. These data corroborate previous studies by Paquet et al. (1982, 1983) and Duplay (1984) and also show that the octahedral compositions of smectites and fibrous clay minerals overlap considerably, with a preferential uptake of Mg by the fibrous clay minerals. The possibility of such an overlap among octahedral compositions was suggested earlier by Mackenzie et al. (1984).

Chemical analyses of whole <2-μm fractions may therefore be misleading because they represent only average values (Figure 8), and do not allow different admixed clay mineral species to be distinguished. Neither do they reflect the large variety of octahedral compositions in a given clay mineral assemblage.

Dioctahedral and trioctahedral domains

Of the 145 individual particles analyzed, the compositions of 71% clearly lie either in the dioctahedral or in the trioctahedral domains. The compositions of the other 29% fall between these two domains. The traditional limits of these two domains were established on the basis of chemical analyses carried out on the whole <2-μm fractions. These limits are no longer valid if the elemental analyses of individual particles

Table 1.   Range of $(Al + Fe^{3+})^{VI}$ and Mg values in the octahedral composition of the six groups of individual particles.

		$(Al + Fe^{3+}) = R^3$	$Mg = R^2$
Laval St. Roman calcrete	23 smectites	$1.8 > R^3 > 1$	$1.1 > Mg > 0.3$
	27 palygorskites	$1.6 > R^3 > 0.9$	$1.85 > Mg > 0.5$
Jocas limestone	22 smectites	$1.8 > R^3 > 1.4$	$0.9 > Mg > 0.4$
	25 palygorskites	$1.6 > R^3 > 1$	$1.6 > Mg > 0.7$
Blauvac dolomite	24 smectites	$1 > R^3 > 0.2$	$2.8 > Mg > 1.3$
	24 sepiolites	$0.7 > R^3 > 0$	$3.2 > Mg > 2.1$

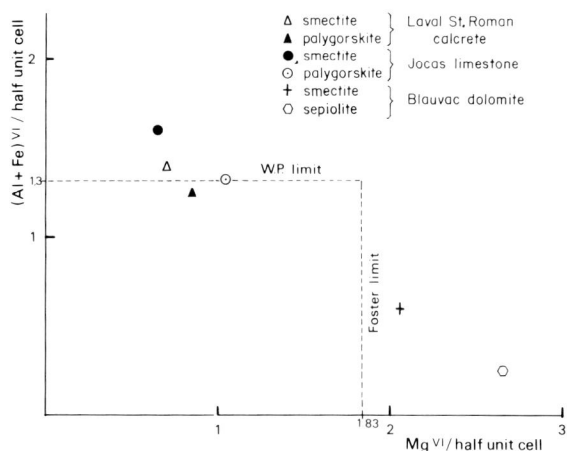

Figure 8. Average octahedral composition of 6 groups of individual particles. $(Al + Fe^{3+})^{VI}$ vs. Mg.

are considered. In fact, on the basis of the octahedral compositions of individual particles, the limits of the dioctahedral and trioctahedral compositional domains must also be modified.

Revised compositional gap

Numerous particles studied here have compositions between the dioctahedral and trioctahedral domains; that is, in an intermediate zone that is considered to be the smectite compositional gap. Sixteen of these particles are smectites. Although such samples are fewer in number than those falling in the dioctahedral and trioctahedral domains, the precise limits of the compositional gap should probably be changed. About half of the palygorskite particles (25 of 52) lie within the compositional gap, whereas 26 fall in the dioctahedral domain and 1 falls in the trioctahedral domain. Thus, it must be emphasized that, like smectites, palygorskites contain more octahedral substitution than is usually thought.

Finally, chemical analyses of whole <2-μm clay fractions do not show such compositional variability. These analyses represent average values and are usually depicted by single compositional points, in contrast to the analyses of individual particles which show considerable scatter.

ACKNOWLEDGMENTS

This work was supported by the grant CNRS-ATP "Paragenèses et associations minérales" (1983–1985). We are indebted to José Honnorez for his very useful contribution to the English translation and for his helpful discussion of the manuscript.

REFERENCES

Brindley, G. W. and Brown, G. (1980) *Crystal Structures of Clay Minerals and their X-ray Identification:* Mineralogical Society, London, 495 pp.

Duplay, J. (1982) Populations de monoparticules d'argiles: analyse chimique par microsonde électronique: Thèse 3e cycle, Univ. Poitiers, Poitiers, France, 110 pp.

Duplay, J. (1984) Analyses chimiques ponctuelles d'argiles. Relations entre variations de compositions dans une population de particules et température de formation: *Sci. Géol. Bull. Strasbourg* **37,** 307–317.

Foster, M. (1960) Interpretation of the composition of trioctahedral micas: *U.S. Geol. Surv. Prof. Pap.* **354-B,** 11–50.

Mackenzie, R. C., Wilson, M. J., and Mashhady, A. S. (1984) Origin of palygorskite in some soils of the Arabian Peninsula: in *Palygorskite-Sepiolite: Occurrence, Genesis and Uses:* A. Singer and E. Galan, eds., Elsevier, Amsterdam, 177–186.

Martin-Vivaldi, J. L. and Cano-Ruiz, J. (1955) Contribution to the study of sepiolite: some considerations regarding the mineralogical formula: in *Clays and Clay Minerals, Proc. 4th Natl. Conf., University Park, Pennsylvania, 1955,* Ada Swineford, ed., *Natl. Acad. Sci. Natl. Res. Counc. Publ.* **456,** Washington, D.C., 173–176.

Paquet, H. (1983) Stability, instability and significance of attapulgite in the calcretes of Mediterranean and tropical areas with marked dry seasons: *Sci. Géol. Mém. Strasbourg* **72,** 131–140.

Paquet, H., Duplay, J., and Nahon, D. (1982) Variations in the compositions of phyllosilicate monoparticles in a weathering profile on ultrabasic rocks: in *Proc. Int. Clay Conf., Bologna, Pavia, 1981,* H. van Olphen and F. Veniale, eds., Elsevier, Amsterdam, 595–603.

Paquet, H., Duplay, I., Nahon, D., Tardy, Y., and Millot, G. (1983) Analyses chimiques de particules isolées dans des populations de minéraux argileux. Passage des smectites magésiennes trioctaédriques aux smectites ferrifères dioctaédriques au cours de l'altération des roches ultrabasiques: *C.R. Acad. Sci. Paris* **296,** 699–704.

Tixier, R. (1978) Microanalyse sur échantillons minces: in *Microanalyse et Microscope à Balayage,* Ch. 5, Les Editions de Physique, Orsay, France, 433–448.

Trauth, N. (1977) Argiles évaporitiques dans la sédimentation carbonatée continentale et épicontinentale Tertiaire: *Sciences Géol. Mém.* **49,** 195 pp.

Triat, J. M. and Truc, G. (1972) L'Oligocène du bassin de Mormoiron (Vaucluse). Etude paléontologique et sédimentologique: *Document Laboratoire de Géologie, Faculté des Sciences de Lyon, France* **49,** 27–52.

Truc, G., Millot, G., Paquet, H., Sassi, S., and Triat, J. M. (1984) Remplacement isovolume dans les encroûtements carbonatés des roches meubles: exemples Quaternaires, Pliocènes et Éocènes du Maghreb, de l'Espagne et du Sud-Est de la France: in *Proc. 5th European Cong. Sedimentology, Marseille, France,* 384–385.

Valleron, M. M. (1981) Les faciès calcaires du Lutétien à *Planorbis pseudoammonius.* Argilogenèse et silicifications associées aux encroûtements calcaires: Thèse 3e cycle, Univ. Louis Pasteur, Strasbourg, France, 108 pp.

Weaver, C. E. and Pollard, L. D. (1973) *The Chemistry of Clay Minerals:* Elsevier, Amsterdam, 213 pp.

Proceedings of the International Clay Conference, Denver, 1985, L. G. Schultz, H. van Olphen, and F. A. Mumpton, eds.,
The Clay Minerals Society, Bloomington, Indiana, 78–84 (1987).

MINERALOGICAL CHARACTERIZATION OF 2:1 CLAYS
IN SOILS: IMPORTANCE OF THE CLAY TEXTURE

Daniel Tessier and Georges Pédro

Station de Science du Sol, INRA, 7800 Versailles, France

Abstract—A correlation was found between the net layer charge and the percentage of nonexchangeable potassium in 2:1 clay minerals having negative charges per half unit cell >0.45. This charge also marks the limit between low-charge smectites having entirely exchangeable cations and other 2:1 clay minerals. Furthermore, on samples prepared at 0.032 bar suction pressure, water retention decreased as the layer charge increased. Transmission electron micrographs, carried out so as to preserve the microstructure of the clay, also showed a relationship between layer charge, stability, and texture. For low-charge smectites (<0.45/half unit cell) exchanged with Na, clay particles were mainly individual crystals. Ca-exchanged clay particles, however, were arranged face-to-face in the form of "quasi-crystals." High-charge illites (layer charge = 0.8/half unit cell) occurred as stable aggregates that appeared to be characteristic of these minerals. An intermediate behavior was noted for smectites having charges between 0.45 and 0.6/half unit cell.

Key Words—Cation-exchange capacity, Illite, Layer charge, Potassium, Smectite, Texture, Transmission electron microscopy, Water retention.

INTRODUCTION

Since the International Clay Conference in Stockholm in 1963, the system adopted by the Nomenclature Committee of AIPEA has led to the classification of 2:1 clay minerals on the basis of a single criterion, i.e., layer charge (Mackenzie, 1965; Brindley, 1966; Pédro, 1967). Thus, a layer charge of x = 0.6 (per half unit cell of $(Si,Al)_4O_{10}$) has been selected as the dividing line between the two major groups of 2:1 phyllosilicates: for smectites, x < 0.6; for illites, x > 0.6.

Numerous studies carried out during the last four decades have identified the main factors that determine hydration and swelling processes of clays on the scale of interlayer spacings. For a comprehensive characterization of clay minerals, however, it is necessary to describe 2:1 clays not only on the scale of interlayer spacings, but also at different levels of their structural organization, particularly with respect to their behavior in water. Such information can be obtained by transmission electron microscopy (TEM) of samples prepared with different cations and at defined physical stresses.

The aim of the present paper is to specify for 2:1 clay minerals the relationships between layer charge and cation-exchange capacity, hydration, and texture.

MATERIALS AND METHODS

Ten dioctahedral and one trioctahedral 2:1 clay minerals were selected for study (Table 1). The Santa Rita, Lorena, Belle Fourche, and Cameron samples were studied by Schultz (1969). Hectorite from Hector, California (Prost, 1975), Greek and Wyoming montmorillonites (Ben Rhaiem et al., 1986), Le Puy illite (Robert, 1972), and Cormes glauconite (Robert and Barshad, 1972) were also investigated. For the Béthonvilliers smectite and Lichères illite, structural formulae were calculated on the basis of 11 oxygens per unit cell from the chemical composition of <2-μm fractions. To establish the structural formulae of the Lichères illite, iron oxides were extracted and determined by the method of De Endredy (1963); a small quantity of kaolinite also was estimated by thermogravimetry and deducted from the chemical composition. Cation-exchange capacity (CEC) was measured by chemical analysis of each material saturated with Ba (Gillman, 1979). Structural formulae are listed in Table 1 and illustrated in Figure 1.

For all materials except the Le Puy illite and the Cormes glauconite, X-ray powder diffraction (XRD) patterns of Mg-saturated samples prepared in the presence of glycerol showed several peaks at about 17 Å. For the Lichères illite, the 17-Å peak was less intense than the 10-Å peak. For the Béthonvilliers smectite and Lichères illite, XRD analysis of air-dry, oriented, Mg-saturated specimens showed a broadened 001 reflection of smectite that was masked by low-angle scatter.

The <2-μm fractions of the initially Na-saturated clays were prepared by suspending the clay in distilled water, followed by repeated cycles of sedimentation, siphoning, and resuspension. Na- and Ca-saturated clay suspensions were then prepared in the presence of a given salt concentration and homogenized by mechanical stirring. Water retention at 0.032 bar suction pressure was obtained in a filtration cell, as described by

Table 1. Cation formulae of the 2:1 clay minerals studied.

Sample and symbol	Formula
Hectorite (H) Hector, California	$Ca_{0.15}(Mg_{2.71}Li_{0.29})(Si_4)$
Smectite (BF) Belle Fouche, South Dakota	$Ca_{0.15}K_{0.03}(Al_{1.55}Fe^{3+}_{0.17}Fe^{2+}_{0.01}Mg_{0.29})(Si_{3.90}Al_{0.10})$
Smectite (W) Wyoming	$Ca_{0.17}K_{0.01}(Al_{1.53}Fe^{3+}_{0.18}Mg_{0.26}Ti_{0.01})(Si_{3.96}Al_{0.04})$
Smectite (G) Greece	$Ca_{0.18}Na_{0.03}(Al_{1.55}Fe^{3+}_{0.04}Fe^{2+}_{0.04}Mg_{0.38}Ti_{0.02})(Si_{3.92}Al_{0.08})$
Smectite (L) Lorena	$Ca_{0.22}K_{0.06}(Al_{1.54}Fe^{3+}_{0.20}Fe^{2+}_{0.02}Mg_{0.23})(Si_{3.82}Al_{0.18})$
Smectite (SR) Santa Rita, New Mexico	$Ca_{0.25}K_{0.01}(Al_{1.48}Fe^{3+}_{0.05}Mg_{0.45})(Si_{3.97}Al_{0.02})$
Smectite (B) Béthonvilliers	$Ca_{0.23}K_{0.11}Na_{0.01}(Al_{1.32}Fe^{3+}_{0.39}Mg_{0.23})(Si_{3.82}Al_{0.18})$
Smectite (C) Cameron, Arizona	$Ca_{0.20}K_{0.21}(Al_{1.36}Fe^{3+}_{0.42}Fe^{2+}_{0.02}Mg_{0.15})(Si_{3.70}Al_{0.30})$
Illite (Li) Lichères	$Ca_{0.14}K_{0.35}(Al_{1.48}Fe^{3+}_{0.33}Mg_{0.23})(Si_{3.47}Al_{0.53})$
Illite (LP) Le Puy	$Ca_{0.07}K_{0.66}(Al_{1.24}Fe^{3+}_{0.32}Fe^{2+}_{0.02}Mg_{0.38}Ti_{0.04})(Si_{3.48}Al_{0.52})$
Glauconite (CO) Cormes	$Ca_{0.08}K_{0.68}Na_{0.03}(Al_{0.45}Fe^{3+}_{1.14}Fe^{2+}_{0.04}Mg_{0.38})(Si_{3.63}Al_{0.37})$

Tessier and Berrier (1979). When equilibrium was reached at this suction pressure, the water content was measured by weighing the sample before and after heating to 150°C. Particle densities were estimated by recording the pressure exerted by the water-saturated samples in kerosene (Tessier, 1984). Because clay minerals have different particle densities, to compare the several clay minerals, the water volumes and CECs must be expressed relative to the solid volume of the particles, i.e., as a water ratio (θ) (I.S.S.S., 1976) for water-retention measurements, and as meq/cm^3 for CEC (Tessier, 1984).

TEM observations were carried out so as to avoid disrupting the structural organization of the sample. The interstitial solution was in turn displaced by propylene oxide and by an epoxy resin, as described by Spurr (1969). After the resin hardened, microtome sections about 500-Å thick were cut with a diamond knife and examined in a JEOL 100 CX electron microscope.

RESULTS

Relation between layer charge and cation-exchange capacity

The samples studied show a range of layer charges from 0.29 to 0.87/half unit cell. The variation of CEC as a function of layer charge (Figure 2) suggests that two main types of 2:1 clays can be distinguished, depending on whether the layer charge is greater or less than 0.45. For x < 0.45, all interlayer cations are exchangeable; this is true for smectites, such as the Wyoming, Belle Fourche, and Lorena smectites, and for hectorite (Figure 2). On the other hand, for x > 0.45, the CEC is always less than the layer charge, implying

that some of the compensating cations are not exchangeable. Here, K is fixed between layers, and the amount of interlayer K increases as the layer charge increases. Thus, a threshold exists at x = 0.45, and the various clay minerals studied here having layer charges of 0.45–0.60 are in fact interstratified (irregular) structures, as is also suggested by the XRD data discussed above.

Relation between layer charge and hydration

Water-retention measurements made at the very low suction pressure (0.032 bar) are plotted against layer charge in Figure 3. At x < 0.6, water retention de-

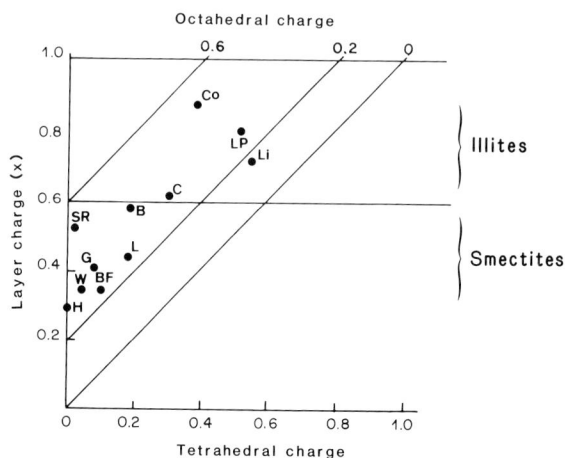

Figure 1. Net layer charge, tetrahedral layer charge, and octahedral position (diagonal lines) for the 2:1 clay minerals studied. (See Table 1 for symbols.)

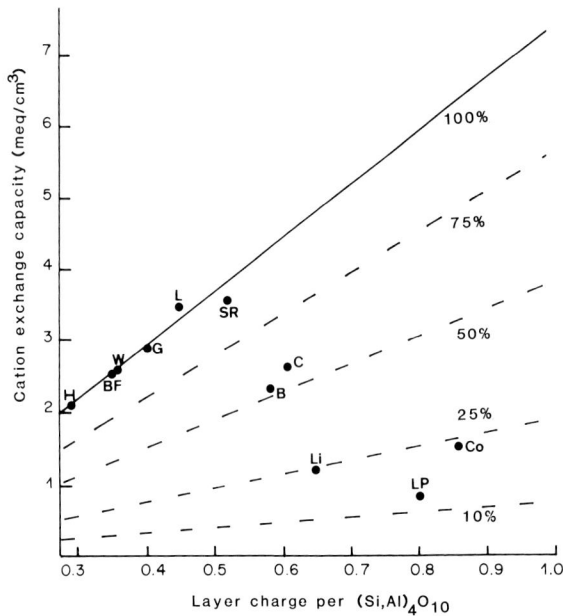

Figure 2. Cation-exchange capacity and layer-charge deficiency of the 2:1 clay minerals in this study. Diagonal lines indicate the percentage of exchangeable interlayer cations. (See Table 1 for symbols.)

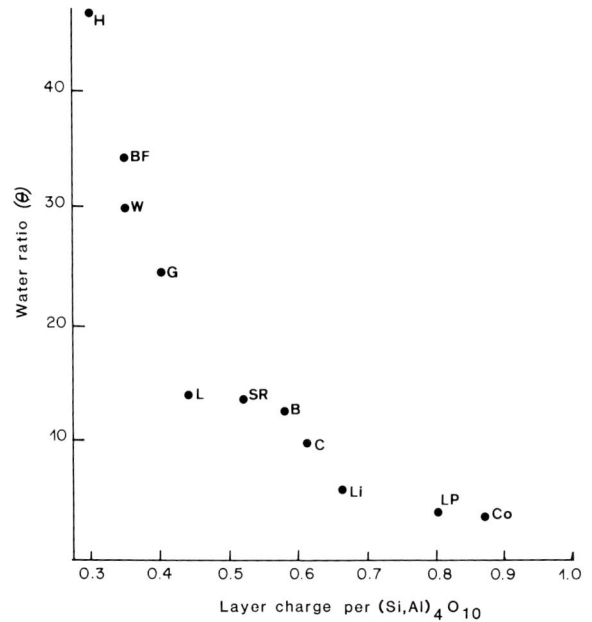

Figure 3. Water ratio of the 2:1 clay minerals prepared in a 10^{-3} M NaCl solution under a suction pressure of 0.032 bar as function of layer charge. (See Table 1 for symbols.)

creases as x increases. As indicated by a comparison of Figures 2 and 3, as water retention decreases, CEC increases.

At x > 0.6, the hydration level is low and nearly constant, irrespective of the layer charge. By taking the ratio of the water ratio to the CEC (Table 2), however, a break in the rate of change can be noted here also at x ≈ 0.45; at x < 0.45 (actually, <0.44), the ratio in-

creases rapidly as x decreases, whereas at x > 0.45, the ratio is fairly constant. This result, which confirms the conclusions of Foster (1955), leads to the conclusion that at x ≤ 0.45, the total water associated with exchangeable cations decreases as the layer charge increases, and all interlayer cations are exchangeable. For such materials, the surface area exposed to water amounts to the entire layer surface, i.e., about 2084 × 10^4 cm²/cm³ (Tessier, 1984) for dioctahedral clay min-

Table 2. Layer charge, water content, water ratio at 0.032 bar suction pressure, cation-exchange capacity (CEC), and particle density of 2:1 clays.

Clay	Layer charge[1] (x)	Tetrahedral charge[1]	Water content (wt. %)	Water ratio (θ)	CEC (meq/cm³)	CEC/θ	Particle density (g/cm³)
Hectorite	0.29	—	17.0	45.4	2.14	21.2	2.67
Belle Fourche smectite	0.35	0.10	12.5	34.6	2.55	13.6	2.77
Wyoming smectite	0.35	0.04	11.2	30.2	2.60	11.6	2.77
Greek smectite	0.40	0.08	9.3	24.6	2.96	8.31	2.74
Lorena smectite	0.44	0.46	5.1	14.2	3.53	4.02	2.78
Santa Rita smectite	0.52	0.01	5.0	13.8	3.63	3.80	2.75
Béthonvilliers smectite	0.58	0.18	4.8	12.9	2.17	5.94	2.71
Cameron smectite	0.61	0.30	3.6	10.0	2.54	3.94	2.79
Lichères illite	0.66	0.53	1.6	4.5	0.89	5.05	2.80
Le Puy illite	0.80	0.52	1.4	3.9	0.84	4.64	2.80
Cormes glauconite	0.87	0.37	1.2	3.6	1.35	2.66	3.00

[1] Per half unit cell.

$\longrightarrow$

Figure 4. Transmission electron microscopy observations of samples prepared at 0.032 bar suction pressure: (a) Na-hectorite prepared in a 10^{-3} NaCl M solution; (b) Na-Greek montmorillonite, 10^{-3} M NaCl; (c) Ca-Greek montmorillonite, 10^{-3} M CaCl₂; (d) Mg-Wyoming montmorillonite, 10^{-3} M MgCl₂. Micrographs by C. Clinard.

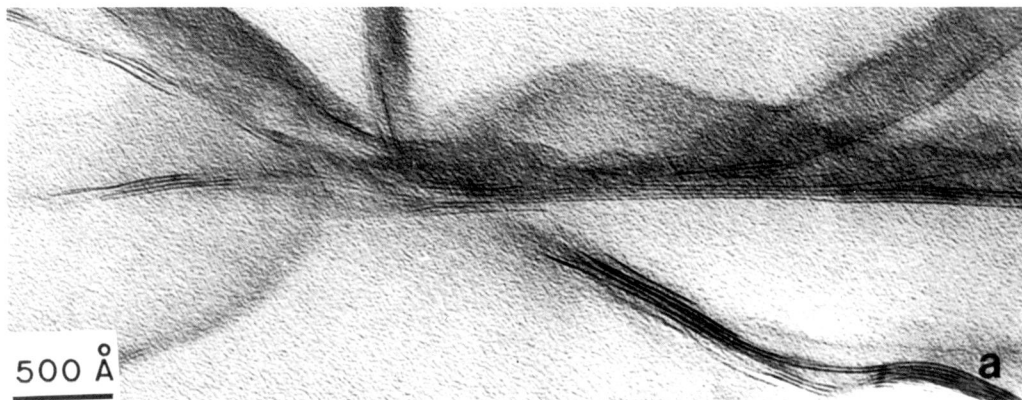

500 Å

a

500 Å

b

500 Å

c

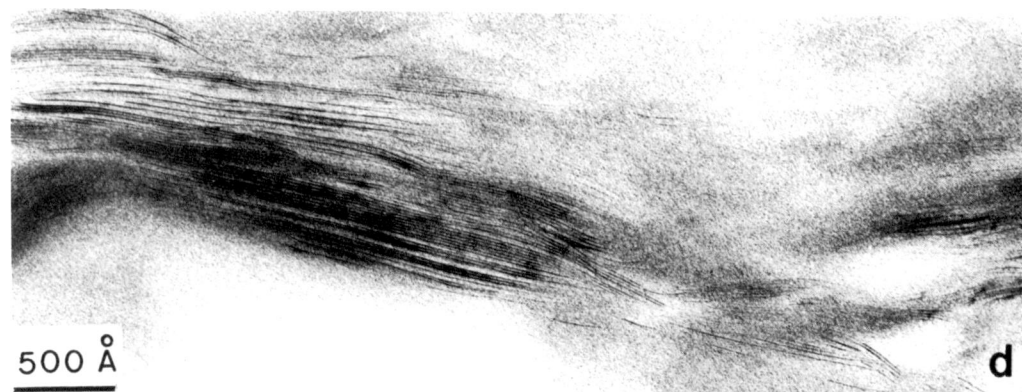

500 Å

d

erals. In clay minerals having a layer charge >0.45, however, nonexchangeable K must be present partly between layers and partly between particles. Thus, the behavior of the water ratio and CEC can be interpreted with respect to layer charge by considering the water is between layers and also between crystals or layer assemblages.

Relation between elementary layer charge and particle texture

Three types of clay particles have been noted: (1) crystals that consist of several layers having exact parallel orientation of *a-b* planes; these crystals are the same as the "primary particles" described by Méring and Oberlin (1971); (2) quasi-crystals that are nearly parallel face-to-face groupings of crystals or extensions due to overlap of crystals (Quirk and Aylmore, 1971) that have long-range order in the *a-b* directions; and (3) aggregates made up of crystals or quasi-crystals that are not easily disaggregated and that contain voids or internal discontinuities between them; they may also be oriented in a more-or-less parallel manner.

Layer charge <0.45 (e.g., true smectites). Na-smectites having a layer charge <0.45 in the presence of a dilute solution at low suction pressure (0.032 bar) consist chiefly of crystals containing only a few layers. This is true for a low-charge smectite, such as hectorite (Figure 4a, x = 0.29), in which crystals consist of five layers. Larger crystals are extended considerably in the *a-b* plane (~5000 Å). Similar crystals, but containing more layers, have been observed as the layer charge increases. For example, crystals of the Na-saturated Greek montmorillonite (Figure 4b, x = 0.4) contain from 3–4 to about 20 layers, and average about 10 layers (Figure 4b).

The effect of exchangeable cation on particle size and arrangement is shown in Figure 4c. Ca-saturated Greek montmorillonite occurs as quasi-crystals containing about 50 parallel layers laterally extended due to face-to-face bonding and overlap. Scanning electron microscopic observations of identical materials also show quasi-crystals (Tessier, 1984) as are also found in Mg-smectites (Figure 4d) and in Na-smectites prepared with a strong salt solution (Tessier and Pédro, 1982).

Layer charge >0.6. Na-saturated illites, e.g., the Le Puy illite (x = 0.8), even under optimal chemical dispersion conditions, consists chiefly of aggregate particles (Figure 5a). The aggregates are generally about 1000 Å thick and about 5000 Å in diameter in the *a-b* plane. Within the aggregates, the crystals usually consist of 3–6 layers; thicker particles may contain as many as 15 layers. Lateral extension in the *a-b* plane is usually 200–500 Å, but may reach 800 Å. Interlayer spacings within the crystals are 10 Å. TEM observations by Tessier (1984) of Ca-saturated Le Puy illite showed the same kind of aggregates. Apparently, illites and

low-charge smectites are basically different. Illites, especially those having high layer charge, are not likely to form the quasi-crystals that characterize low-charge smectites; instead, illites form aggregate particles.

Layer charge 0.45 < x < 0.60. Smectites having layer charges between 0.45 and 0.60 are common in soils. Na-clays, e.g., the Béthonvilliers smectite (Figure 5b, x = 0.58), occur as aggregates similar to those formed by illite (Figure 5a). Within these aggregates the crystals contain about 10 layers. Lateral extension in the *a-b* plane is usually 1000–1500 Å, and the interlayer spacing is 10 or 15 Å, even in the same crystal (Figure 5b), thus confirming the CEC data and the nonexchangeable K content, in that the clay is a mixed-layer phase.

As in Ca-saturated, low-charge smectites (x < 0.45), quasi-crystals having lateral extension have been identified (Figure 5c). Quasi-crystals appear to be characteristic of smectites, including mixed-layer smectites containing divalent cations, irrespective of their layer charge. For smectites having a layer charge between 0.45 and 0.60, however, exchange from Na to Ca cations results in the regrouping of individual crystals as in low-charge smectites (e.g., hectorite).

DISCUSSION AND CONCLUSIONS

Results obtained in this investigation suggest that layer charge is an essential factor in identifying, characterizing, and classifying the major 2:1 clay minerals. The real limit between low-charge smectites containing entirely exchangeable cations and other 2:1 clay minerals appears to be 0.45 negative charges/half unit cell, rather than 0.60, as previously thought. At a layer charge >0.45, the level of nonexchangeable K increases progressively as the CEC decreases. TEM observations show essentially a segregation of 10-Å and 15.6-Å crystals, rather than real mixed-layer crystals. This distinction between clays having entirely exchangeable cations (x < 0.45) and the others is important in that it implies that sample properties may be generalized to the entire surface of the system, i.e., the surface of all layers.

For Ca-smectites, however, cohesion between the particles that comprise quasi-crystals results in intercrystal spacings that are identical with interlayer spacings of crystals. Under such conditions, the face-to-face overlapping of crystals and quasi-crystals may lead to the formation of structures having substantial lateral extensions in *a-b* planes. If Ca is replaced by Na, extensive swelling is dependent on the development of diffuse double layers, which in turn is governed by the sodium salt concentration of the solution. This extensive interlayer disrupts the quasi-crystals (Quirk, 1968).

For intermediate-charge smectites (0.45 < x < 0.60), the level of nonexchangeable K increases with increasing charge, but it remains low enough to provide some flexibility to the structure; quasi-crystals can still form

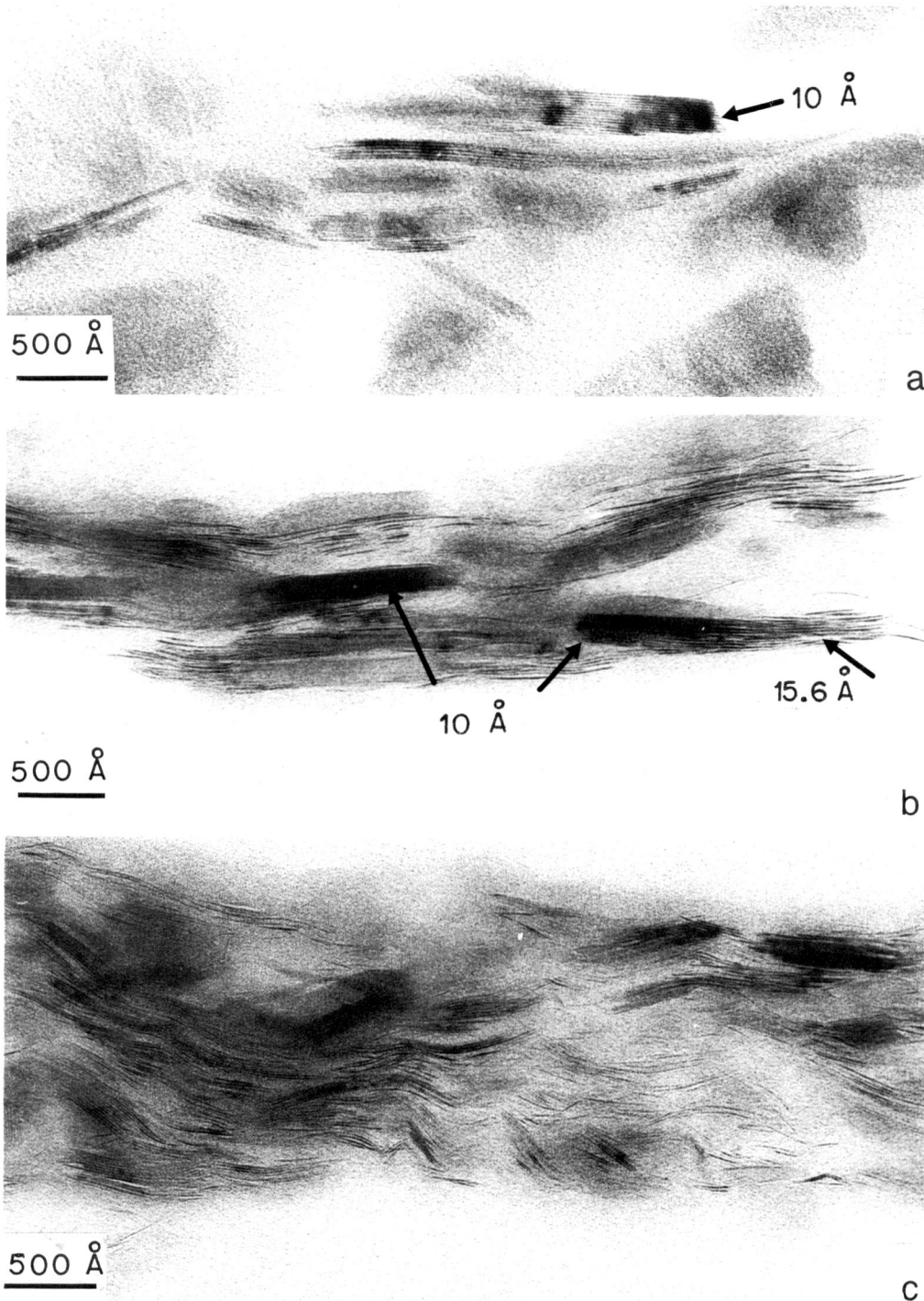

Figure 5.   Transmission electron microscopy observations of samples prepared at 0.032 bar suction pressure: (a) Na-Le Puy illite prepared in a 10^{-3} M NaCl solution; (b) Na-Béthonvilliers smectite, 10^{-3} M NaCl; (c) Ca-Béthonvilliers smectite, 10^{-3} M $CaCl_2$. Micrographs by C. Clinard.

by face-to-face bonding. The charge is also still low enough to allow a diffuse double layer to form on Na-saturation. The clay's behavior, however, no longer corresponds to the ideal behavior of low-charge smec-

tites because of the permanent presence of aggregate particles.

For 2:1 clays having a layer charge >0.6, especially high-charge illites ($x \approx 0.8$), the diffuse double layer may be involved between individual crystals if their layer charge, charge location, and compensating cation composition are favorable; however, the assemblage of rigid individual crystals and rigid aggregate particles is difficult to distort. Properties of these materials may be interpreted by considering a porous system of rigid particles are held together by cohesion forces that are necessarily weak due to the rigidity at joints.

Hence, to understand the relationship between the behavior of clay minerals and their crystal structure, it is necessary to take into account not only layer charge, location of charge, and layer species, but also the texture of the clay system. The assemblages of clay mineral particles in a soil will be characteristic of their physicochemical environment.

REFERENCES

Ben Rhaiem, H., Tessier, D., and Pons, C. H. (1986) Comportement hydrique et évolution structurale et texturale des montmorillonites au cours d'un cycle de dessication-humectation. Parte I. Cas des montmorillonites calciques: *Clay Miner.* **21**, 9–29.

Brindley, G. W. (1966) Discussions and recommendations concerning the nomenclature of clay minerals and related phyllosilicates: in *Clays and Clay Minerals, Proc. 14th Natl. Conf., Berkeley, California, 1965,* S. W. Bailey, ed., Pergamon Press, New York, 27–34.

De Endredy, A. D. (1963) Estimation of free iron oxides in soils and by a photolytic method: *Clay Min. Bull.* **29**, 209–217.

Foster, M. D. (1955) The relation between composition and swelling in clays: in *Clays and Clay Minerals, Proc. 3rd Natl. Conf., Houston, Texas, 1954,* W. O. Milligan, ed., *Natl. Acad. Sci. Natl. Res. Counc. Publ.* **395**, Washington, D.C., 296–316.

Gillman, G. P. (1979) A proposed method for the measurements of exchange properties of highly weathered soils: *Austral. J. Soil Res.* **17**, 129–139.

I.S.S.S. (1976) International Soil Science Society—soil physics terminology: *I.S.S.S. Bull.* **48**, 16–22.

Mackenzie, R. C. (1965) Nomenclature subcommittee of AIPEA: *Clay Min. Bull.* **6**, 123–126.

Méring, J. and Oberlin, A. (1971) The smectites: in *The Electron-Optical Investigation of Clays,* Mineralogical Society, London, 135–145.

Pédro, G. (1967) Commentaires sur la classification et la nomenclature des minéraux argileux: *Bull. Gr. Fr. Argiles* **19**, 69–86.

Prost, R. (1975) Etude de l'hydration des argiles: Interaction eau-minéral et mécanisme de la rétention de l'eau: *Ann. Agron.* **26**, 401–461, 463–535.

Quirk, J. P. (1968) Particle interaction and soil swelling: *Israel J. Chem.* **6**, 213–234.

Quirk, J. P. and Aylmore, L. A. G. (1971) Domains and quasi-crystalline regions in clay systems: *Soil Sci. Soc. Amer. Proc.* **35**, 652–654.

Robert, M. (1972) Transformation expérimentale de glauconite et d'illite en smectite: *C.R. Acad. Sci. Paris D* **275**, 1319–1322.

Robert, M. and Barshad, I. (1972) Transformation expérimentale des micas en vermiculites ou smectites. Propriétés des smectites de transformation: *Bull. Gr. Fr. Argiles* **24**, 137–151.

Schultz, L. G. (1969) Lithium and potassium adsorption, dehydroxylation temperature, and structural water content of aluminous smectites: *Clays & Clay Minerals* **17**, 115–149.

Spurr, A. R. (1969) A low-viscosity epoxy resin embedding medium for electron microscopy: *Ultrastructure Research* **26**, 31–43.

Tessier, D. (1984) Etude expérimentale de l'organisation des matériaux argileux: Dr. Science Thesis, INRA publ., 361 pp.

Tessier, D. and Berrier, J. (1979) Utilisation de la microscopie électronique à balayage dans l'étude des sols. Observations de sols humides à différents pF: *Science du Sol* **1**, 67–82.

Tessier, D. and Pédro, G. (1982) Electron microscopy of Na-smectites. Role of layer charge, salt concentration and suction parameters: in *Proc. Int. Clay Conf., Bologna, Pavia, 1981,* H. van Olphen and F. Veniale, eds., Elsevier, Amsterdam, 165–176.

Proceedings of the International Clay Conference, Denver, 1985, L. G. Schultz, H. van Olphen, and F. A. Mumpton, eds.,
The Clay Minerals Society, Bloomington, Indiana, 85–93 (1987).

MIXED-LAYER CHLORITE/SMECTITES FROM
A PENNSYLVANIAN EVAPORITE CYCLE,
GRAND COUNTY, UTAH

MARC W. BODINE, JR. AND BETH M. MADSEN

United States Geological Survey, MS 939, Denver Federal Center
Denver, Colorado 80225

Abstract—The typical salt cycle in the Paradox Member of the Hermosa Formation (Middle Pennsylvanian) in southeastern Utah consists of basal penesaline beds (dolomitic limestone, dolomite, anhydrite) and overlying bedded rock salt with or without intercalated potash salt beds. Each cycle is thought to represent an episode of seawater recharge followed by progressive evaporation and salt precipitation. The clay minerals identified in drill core through the penesaline interval of a single cycle are discrete chlorite (clinochlore), several mixed-layer clinochlore/trioctahedral smectites (including corrensite), talc, and illite. The illite contains negligible expandable interlayers. The progression from discrete clinochlore to corrensite parallels increasing salinity through the cycle; carbonate lithologies characteristically contain chlorite/smectites having as much as 15% expandable layers, whereas chlorite/smectites in the sulfate rocks typically contain 20–50% expandable layers. Based on data from Recent saline lakes and chemical considerations the transformation to the chlorite/smectites took place during burial diagenesis of precursor, authigenic, Mg-rich smectites that formed in response to evaporite brine compositions.

Key Words—Chlorite/smectite, Corrensite, Diagenesis, Evaporite, Illite, Mg-smectite, Talc.

INTRODUCTION

The unique character of clay-mineral assemblages in marine evaporite rocks has been recognized since Füchtbauer and Goldschmidt's (1959) study of clay minerals in the German Zechstein salts. They described the abundance of chlorite, corrensite, and talc in these rocks coexisting with ubiquitous, locally abundant, illite as the only dioctahedral clay. Subsequent findings by Braitsch (1960), Niemann (1960), Dreizler (1962), Echle (1961), Reinold (1965), Pundeer (1969), Lippmann and Savascin (1969), and Bodine (1971) confirmed these observations with additional data from the Zechstein, the German Keuper, and the Austrian Haselgebirge. Grim *et al.* (1960), Fournier (1961), Lounsbury (1963), and Bodine (1985a, 1985b) reported similar assemblages in several North American Paleozoic evaporites.

The chemical and mineralogic character of the clays in marine evaporites, when contrasted with typical argillaceous marine and continental detritus, suggests that many if not most clay minerals in the marine evaporites are authigenic, having formed by the interaction between detrital clays and magnesium-rich evaporite brine (Füchtbauer and Goldschmidt, 1959; Braitsch, 1971; Bodine, 1985a, 1985b). Therefore, changing salinity accompanying progressive evaporation may have resulted in systematic variation of coexisting clay mineral compositions that may have survived subsequent burial diagenesis.

Some mineralogical relations with salinity have been previously reported. In Zechstein rocks, for example, talc and corrensite occur in halite-rich zones, but are minor or absent in the potash-bearing zones and in the overlying claystone. Discrete chlorite, however, is abundant in the potash and claystone facies, but it is generally a minor clay constituent in the potash-free halites (Braitsch, 1971). Bodine and Rueger (1984) identified abundant talc and corrensite in halite rocks but noted their absence in less saline, chlorite-rich "black shales" and most dolomites in a 750-m-thick section of the Paradox Member of the Hermosa Formation in southeastern Utah. In the present paper the extent of smectite interlayering with chlorite in a penesaline (carbonate and sulfate minerals precipitated from halite-undersaturated brine) succession at the base of a single cycle of progressive evaporation in the Paradox Member is examined in more detail than previously reported.

GEOLOGIC SETTING

The Paradox basin in southeastern Utah (Figure 1) is a northwest–southeast trending structure that formed during the late Paleozoic along the southwestern margin of the emerging Uncompahgre uplift. A thick succession of Middle Pennsylvanian marine evaporites, the Paradox Member of the Hermosa Formation, was deposited in the form of 29 depositional cycles, the salt beds in which, numbered 1 to 29 from top to bottom, can be correlated throughout the basin (Hite, 1960).

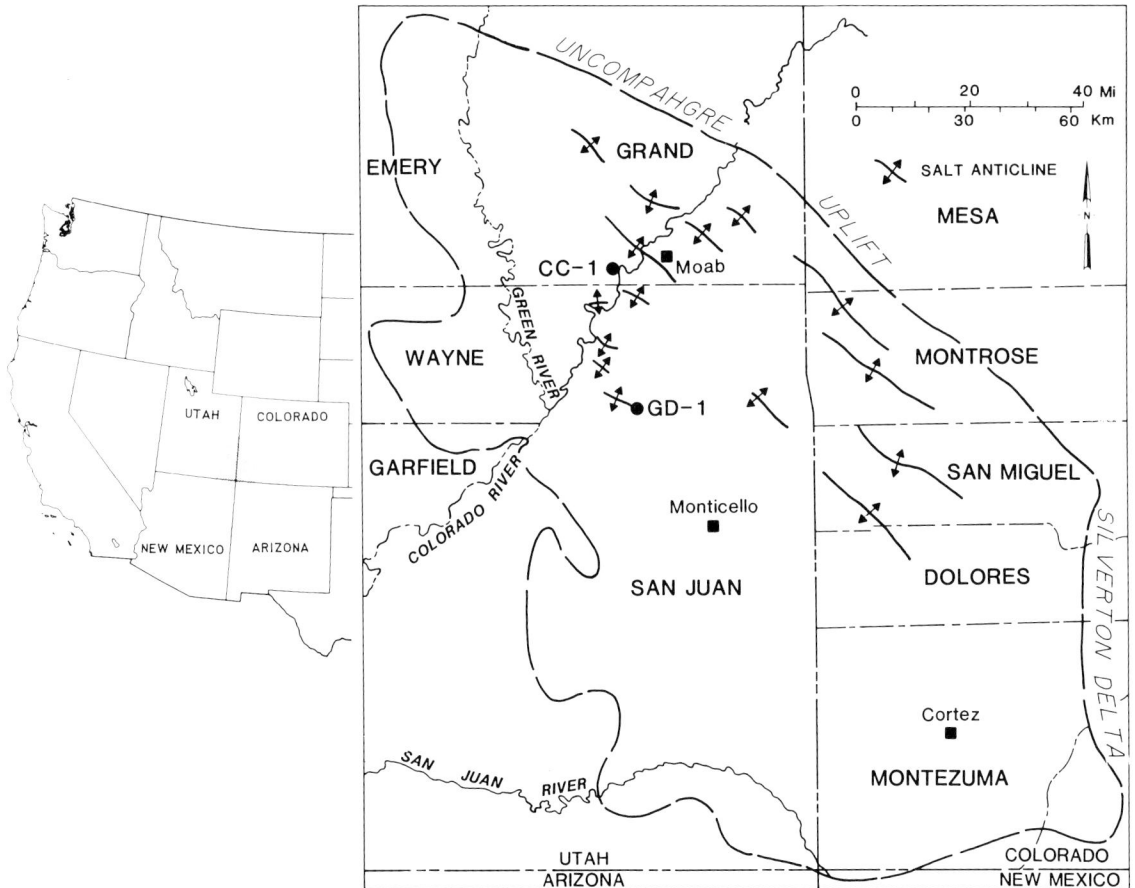

Figure 1.   Location map of the Cane Creek No. 1 (CC-1) and Gibson Dome No. 1 (GD-1) drill holes in the Paradox basin, Utah; approximate limit of halite deposition given by dashed boundary (modified from Hite *et al.,* 1984).

A typical cycle, illustrated in Figure 2, consists of a lower sequence of penesaline beds and an upper halite bed; in addition, beds containing potassium salts may occur locally toward the top of the cycle. The basal anhydrite of the penesaline sequence forms a sharp disconformable contact with the uppermost salts of the underlying unit. This anhydrite is successively overlain by silty dolomite, a "black shale" (i.e., silty, argillaceous, dolomitic limestone to calcareous, argillaceous, sparsely dolomitic siltstone containing locally abundant organic material (Hite *et al.,* 1984)), an upper dolomite, and an anhydrite bed. Hite and Buckner (1981) suggested that the cyclic sequences were the result of a sea level rise accompanied by abrupt basin recharge and residual brine dilution that resulted in dissolution of halite and deposition of calcium sulfate (point X in Figure 2); continued transgression and dilution resulted in the deposition of silty dolomite and "black shale." Marine transgression culminated at some point during "black shale" deposition (point Y in Figure 2) and was followed by an extended period of regression. Progressive evaporative concentration took

place throughout the remainder of the cycle, along with deposition of the remaining part of the characteristic marine evaporite succession of dolomite, anhydrite, and thick beds of salt. Further evidence of the integrity of the cycles and minimal postdepositional recrystallization of salt was reported by Raup (1966). He observed consistent increases of bromine content upwards through the halite in each of several cycles, with values reaching maxima of 150–250 ppm in the upper part of the cycles (Figure 2); the bromine content of salt increases as the salinity of the brine from which it crystallized increases (Braitsch, 1971), and if the salt had recrystallized the bromine content would have been homogenized.

MATERIALS AND METHODS

Our samples are from the Delhi-Taylor Oil Corporation's Cane Creek No. 1 drill hole (sec. 25, T.26S., R.20E., Grand County, Utah) on the northeastern flank of the Cane Creek anticline (Figure 1). Nineteen representative samples were collected from the penesaline sequence under salt 2. The interval is 30-m thick, and

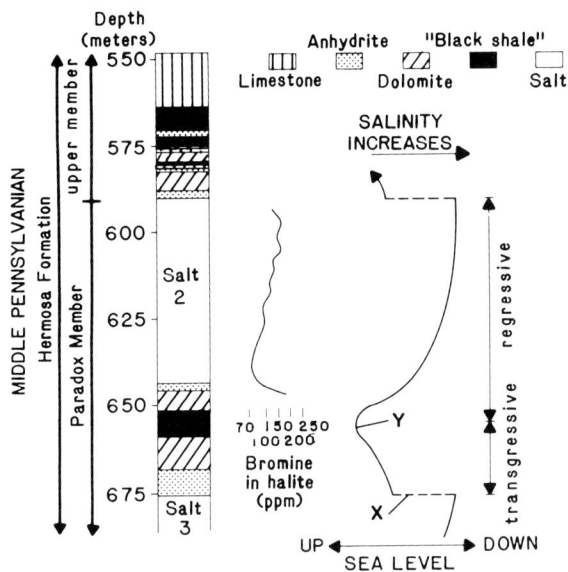

Figure 2. Stratigraphic section of salt 2 and its underlying penesaline interval in the Paradox Member of the Hermosa Formation in the Cane Creek No. 1 drill core; salt 2 overlain by lowermost beds of the upper member of the Hermosa Formation. Bromine content in salt 2 from Raup (1966); suggested salinity and sea level fluctuations from Hite and Buckner (1981). Evaporation of and salt precipitation from residual brines expelled from salt 3 suggested by Raup (1966) to explain anomalously high bromine content at the base of salt 2.

the complete sequence of penesaline lithologies (Figure 2) is well preserved.

The relative abundances of minerals in whole-rock samples and their clay-size insoluble fractions (Table 1) were estimated on the basis of the relative intensity of the principal X-ray powder diffraction reflection of each mineral from back-packed, randomly oriented, powder mounts. We recognize the errors inherent in this procedure and carry it no further than suggesting the qualitative relative abundances from the relative intensities in Table 1.

Clay-mineral separation and analysis

Ground, whole-rock samples (20–50 g) from the core studied earlier by Raup (1966) were used in this investigation. Clay-size fractions (<2 μm effective spherical diameter) of insoluble constituents were obtained by leaching the samples with water to dissolve water-soluble salts, and then with boiling, slightly alkaline 0.2 M Na_4-EDTA (versene) solution (Bodine and Fernalld, 1973) to extract "acid-soluble" salts (Ca-Mg-carbonates and Ca-sulfates) by chelating the divalent cations. The residue was disaggregated and dispersed in water with an immersible ultrasonic probe, washed repeatedly by centrifuging, decanting, and discarding the supernatant, and redispersed in distilled water until

it no longer flocculated. The suspended <2-μm fraction was then isolated by centrifugation.

Oriented X-ray powder diffraction (XRD) mounts were prepared from <2-μm fractions using Pollastro's (1982) filter-membrane peel technique and analyzed with CuKα radiation from 2° to 52°2θ with automated Picker and Siemens diffractometers. A series of three oriented mounts were examined that were untreated (air-dried), saturated with ethylene glycol, heated to 350°, and heated to 550°C. If sufficient sample remained, the suspension was dried at 60°C, and a randomly oriented mount was prepared to obtain 060 spacings. Unfortunately, the relatively small amounts of whole-rock sample, coupled with the low abundance of clay minerals in many samples, precluded any analysis of sample PCC-16, chemical analysis of any <2-μm fraction, separation of any specific size-fractions <2 μm, and 060 measurements of some samples.

The XRD results were evaluated using Brown and Brindley's (1980) determinative data and Reynolds' (1980) and Whitney's (1979) criteria for characterizing mixed-layer clays. Because Whitney's data apply to synthetic MgO-Al_2O_3-SiO_2-H_2O iron-free clay minerals devoid of the other minor components present in natural samples, and because minor quantities of interstratified high-charge layers (e.g., vermiculite) were found in some samples, errors for the proportions of expandable layers in mixed-layer chlorite/smectite (C/S) are at least ±5% absolute expandable layers.

Clay mineral characteristics

Clay minerals in the 18 <2-μm insoluble fractions (Table 1) were found to be illite, chlorite, interstratified chlorite/trioctahedral smectites (C/S), and talc. The C/S compositions range from chlorite-rich, randomly interstratified varieties to the 1:1 regularly interstratified clay mineral corrensite. Reversals in relative clay-mineral abundances between whole-rock and <2-μm fractions (Table 1) suggest that much of the talc in sample PCC-1 and significant proportions of the illite throughout the section occur as particles having a >2-μm effective spherical diameter. Basal reflections (Figures 3–5) from oriented mounts and 060 spacings (Figure 6) from random mounts characterize these assemblages.

Talc occurs only in sample PCC-1, the upper anhydrite unit immediately underlying the bedded halite of salt 2 (Figure 2). It is the most abundant phyllosilicate in the whole-rock sample (Table 1), and in the <2-μm fraction it exhibits sharp, diagnostic 001, 002, and 003 reflections (9.4, 18.9, and 28.6°2θ in Figure 5).

Illite is present in all size fractions of all samples. It is most abundant in "black shales" and overlying dolomites and least abundant in anhydrites (Table 1). The illite is remarkably free of expandable interlayers; the symmetry of the 001 spacing (8.8°2θ), the lack of a

Table 1. Relative X-ray powder diffraction intensities from minerals in penesaline lithologies.[1]

Sample	Depth (m)	Whole-rock[2]	<2-μm insoluble fraction[3]
PCC-1	643.6	Anhydrite ≫ halite > dolomite > talc > K-feldspar > quartz > C/S[4]	C/S (50%)[4] ≫ talc > illite
PCC-2	644.5	Anhydrite = quartz ≫ dolomite > illite > C/S > K-feldspar	C/S (40%) ≫ illite
PCC-3	645.5	Anhydrite ≫> dolomite > quartz > illite > C/S	C/S (30%) ≫ illite
PCC-4	645.9	Anhydrite ≫ dolomite ≫ quartz > K-feldspar > illite > C/S	illite ≫ C/S (<5%)
PCC-5	648.2	Dolomite ≫ quartz ≫ anhydrite > illite > K-feldspar > C/S	illite > C/S (<5%)
PCC-6	650.7	Dolomite ≫ quartz > calcite > illite > C/S = K-feldspar	illite > C/S (10%)
PCC-7	654.0	Calcite = quartz > dolomite = illite > C/S > K-feldspar	illite > C/S (<5%)
PCC-8	655.9	Calcite = quartz ≫ dolomite = illite > C/S > K-feldspar	illite > C/S (<5%)
PCC-9	657.1	Quartz > calcite ≫ dolomite = illite > C/S	illite > C/S (<5%)
PCC-10	659.2	Dolomite > calcite > quartz ≫ K-feldspar > illite > C/S	illite = C/S (15%)
PCC-11	659.7	Dolomite = calcite = quartz ≫> illite ≫ C/S = K-feldspar	illite > C/S (10%)
PCC-12	661.4	Quartz ≫ halite = calcite = dolomite > K-feldspar ≫ illite = C/S	C/S (10%) > illite
PCC-13	663.8	Quartz ≫ halite = K-feldspar > dolomite = calcite ≫ illite ≫ C/S	C/S (5%) = illite
PCC-14	665.7	Quartz ≫ dolomite = K-feldspar > calcite = halite ≫ illite > C/S	C/S (10%) > illite
PCC-15	669.1	Quartz = dolomite ≫> illite = K-feldspar ≫ C/S = anhydrite	C/S (<5%) = illite
PCC-16	669.2	Anhydrite ≫>> dolomite ≫ quartz ≫ K-feldspar > illite(?)	Not analyzed
PCC-17	670.1	Anhydrite ≫>> dolomite ≫ quartz = K-feldspar > illite	C/S (30%) > illite
PCC-18	672.0	Anhydrite ≫>> dolomite ≫ quartz > illite > C/S	C/S (20%) > illite
PCC-19	673.2	Anhydrite ≫>> dolomite ≫ quartz = K-feldspar > illite	C/S (15%) = illite

[1] Underlying salt 2 of the Paradox Member of the Hermosa Formation from the Cane Creek No. 1 drill core.

[2] Lithologic interval from Raup (1966). Upper anhydrite = PCC 1–3; upper dolomite = PCC 4–6; "black shale" = PCC 7–9; lower dolomite = PCC 10–14; lower anhydrite = PCC 15–19.

[3] Excludes minor quantities of quartz and K-feldspar.

[4] C/S = interstratified chlorite/smectite showing percentage of expandable (smectite) layers (±5%).

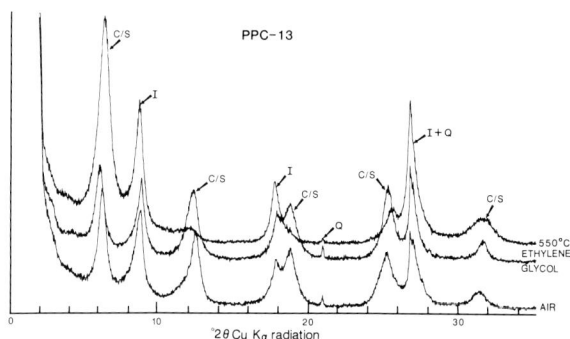

Figure 3. X-ray powder diffractometer traces from oriented mounts of the <2-μm fraction from sample PCC-13. Mixed-layer chlorite/smectite with 5 ± 5% expandable layers is dominant; peak labels: mixed-layer chlorite/smectite (C/S), quartz (Q), and illite (I).

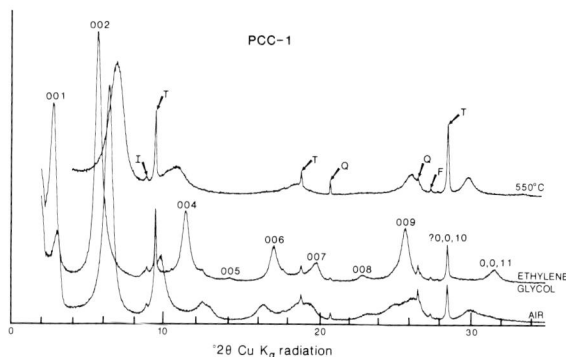

Figure 5. X-ray powder diffractometer traces from oriented mounts of the <2-μm fraction from sample PCC-1. Mostly corrensite with 50 ± 5% expandable layers. Peak labels: corrensite (Miller indices on ethylene glycol trace), talc (T), quartz (Q), K-feldspar (F), and illite (I). Anomalous high intensity of the 20.9°2θ peak, the 100 spacing of quartz, when compared with quartz's most intense peak, the 101 spacing at 26.6°2θ, due to euhedral character of the quartz with crystals oriented on their prism faces on the diffraction mount.

measurable shift in spacing or the lack of distortion of peak geometry with ethylene glycol saturation or heating, and the integral 10-Å periodicity of the 001, 002, and 003 spacings regardless of sample treatment (8.8, 17.8, and 26.7°2θ in Figures 3 and 4) document its discrete mica-clay character. The 060 reflection in the 61.5°–62.0°2θ region (~1.50 Å) in Figure 6 is attributed to the dioctahedral mica clay.

Discrete chlorites or chlorites with negligible (<5%) expandable layers show little or no expansion (<0.1°2θ) after ethylene glycol saturation; the (00l) for both air-dried and glycol-saturated samples define a c_0 dimension of 14.2–14.3 Å. An 060 spacing (sample PCC-8 in Figure 6) of ~1.54 Å indicates a trioctahedral species. Chlorite with 5% expandable (smectite) layers (sample PCC-13, Figure 3) shows minor expansion after ethylene glycol saturation; the 001 spacing, for example, expands from 14.22 to 14.57 Å. At 550°C, the 001 reflection is enhanced, but has collapsed to 13.39 Å.

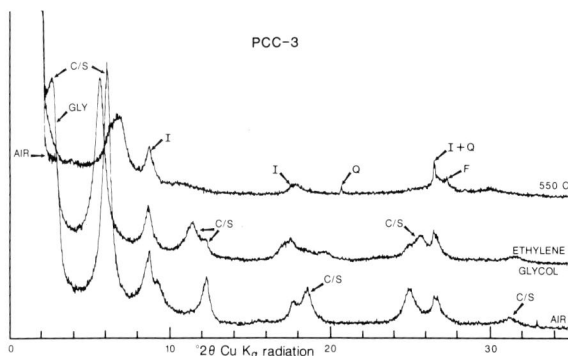

Figure 4. X-ray powder diffractometer traces from oriented mounts of the <2-μm fraction from sample PCC-3. Mixed-layer chlorite/smectite with 30 ± 5% expandable layers; peak labels: mixed-layer chlorite/smectite (C/S), illite (I), quartz (Q), and K-feldspar (F).

The 060 spacing (sample PCC-3, Figure 6) is about the same as that of sample PCC-8.

The air-dried mount of an interstratified clay with 30% expandable layers (sample PCC-3, Figure 4), yields a "001" reflection at 14.24 Å that expands to 15.24 Å after ethylene glycol saturation; the spacing collapses to 13.00 Å after heating to 550°C. A poorly developed superlattice spacing in the 2.5–3.0°2θ region indicates incipient regular stacking order, and makes the 001 designation of the 14 Å reflection questionable. The assignment of 30% expandable layers is based on the "001" spacing of the ethylene glycol-saturated clino-chlore/smectite from Whitney (1979). The 060 reflection (sample PCC-3, Figure 6) is diagnostic of trioctahedral clays. XRD data suggest that minor amounts of high-charge layers, e.g., vermiculite, are interstratified in some of the C/S. This is because the extent of collapse after heating is greater than can be attributed to both the anticipated collapse of a discrete sedimentary chlorite to ~13.8 Å and the collapse of the smectite estimated from the extent of expansion after ethylene glycol saturation.

The C/S having the most expandable layers (50%) was found in sample PCC-1 and was identified as corrensite. Data from the ethylene glycol-saturated sample (Table 2 and Figure 5) show 11 measurable 00l spacings (possibly 12, the 00,10 peak at 28.6°2θ was masked by the 003 reflection of talc). The spacings are integral and indicate both odd and even orders that define a mean c_0 dimension of 31.143 Å with a coefficient of variation of 0.70. If the 001 spacing (because of Lorentz polarization effects) and the 00,17 spacing (because of its low intensity and questionable identification as a 00l reflection of the mixed-layer clay) are excluded, the coefficient of variation is reduced to 0.20. The 00l spacings

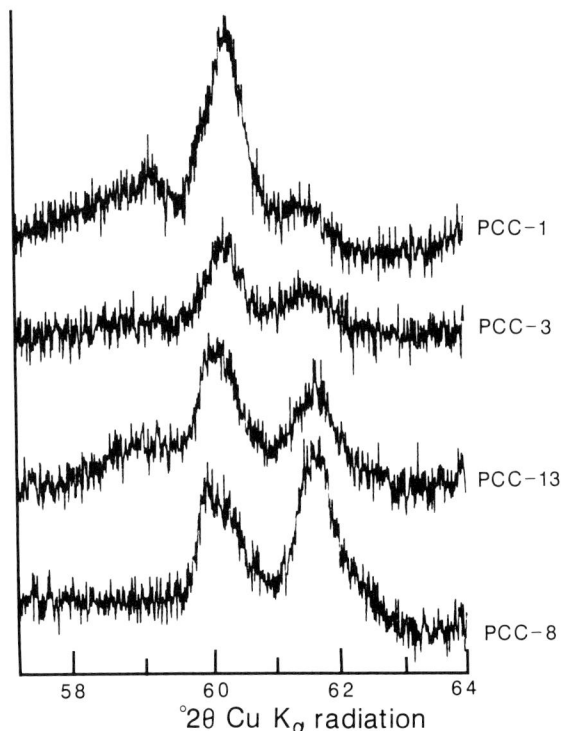

Figure 6. X-ray powder diffractometer traces from randomly oriented air-dried mounts of the <2-μm fraction showing 060 spacings in clay minerals from the penesaline interval underlying salt 2. Percent expandable layers in mixed-layer chlorite/smectite: PCC-1, 50 ± 5%; PCC-3, 30 ± 5%; PCC-13, 5 ± 5%, and PCC-8, <5%.

Table 2. X-ray powder diffractometer basal spacings and calculated c_0 for corrensite.[1]

hkl	I/I_0	d (Å)	c_0 (Å)
001	70	31.8	31.8
002	100	15.51	31.02
003	—	—	—
004	25	7.784	31.136
005	1	6.236	31.180
006	13	5.176	31.056
007	6	4.459	31.213
008	1	3.876	31.008
009	19	3.453	31.077
00,10[2]	14	3.123	31.230
00,11	5	2.825	31.075
00,12	—		
00,13	—		
00,14	—		
00,15	<1	2.0709	31.0635
00,16	—		
00,17	<1	1.8203	30.9451

$$\bar{c}_0 = 31.140 \text{ Å}$$
$$CV = 0.704$$

[1] Relative intensity (I/I_0 of interplanar spacing (d) and calculated c_0 dimension for 001 reflections from sample PCC-1 (<2-μm size fraction, ethylene glycol-saturated, oriented) from upper anhydrite below salt 2 in the Paradox Member of the Hermosa Formation in the Cane Creek No. 1 core; and the mean c_0 dimension ($\bar{c}_0$) and coefficient of variation (CV) for 11 reflections.
[2] Principally the 003 reflection from talc; not included in calculation of $\bar{c}_0$ and coefficient of variation.

of the ethylene glycol-saturated sample are greater than those of the air-dried mount, e.g., d(002) expands to 15.51 Å from 14.32 Å, whereas those of the 550°C mount collapsed, e.g., d(002) = 12.81 Å (Figure 5). These data identify the expandable layer as smectite rather than vermiculite. Based on the intense 060 spacing of ~1.54 Å (sample PCC-1, Figure 6), both the smectite and chlorite layers are trioctahedral. Our data conform to the AIPEA Nomenclature Committee's recommended criteria (Bailey, 1982) for assigning specific mineral names to regular interstratifications; hence, the clay is corrensite.

Although insufficient sample prevented chemical analysis of the Cane Creek No. 1 clay-size fractions, their compositions are undoubtedly similar to the compositions of clay-size fractions having identical mineralogies from the Paradox Formation in the Gibson Dome No. 1 core (Bodine and Rueger, 1984), 38 km to the south (Figure 1). In Figure 7, a plot of analyses of nine clay-size fractions from Gibson Dome fall below the muscovite-clinochlore join toward the MgO apex. These materials are similar in composition to clays extracted from several European and North American marine evaporites (Bodine, 1985a, 1985b).

The abundance of Mg in selected Gibson Dome assemblages from penesaline lithologies (Table 3) strongly suggest a clinochlore composition for the Cane Creek No. 1 chlorites (Mg/Fe mole ratios = 6.4, 6.5, 10.6, and 30.4). Insufficient Al (or Al + Fe—analysis 4, Table 3) precludes dioctahedral smectite (either montmorillonite or nontronitic montmorillonite) as the expandable layer in the Gibson Dome No. 1 corrensite, in agreement with the 060 spacing from the Cane Creek No. 1 corrensite.

Stratigraphic variation of mixed layering

The percentage of expandable layers in the C/S appears to be related to host lithology in the penesaline interval (Figure 8). The C/S in the lower three of the four samples from the lower anhydrite contains 15–30% expandable layers; only in the upper sample (PCC-15) in which dolomite is more abundant than anhydrite (Table 1) does the clay contain fewer expandable layers (essentially a discrete chlorite). The clay in the lower dolomite consistently contains 5–15% expandable layers. In the "black shale" and upper dolomite, the clay (with one exception, sample PCC-6, the lowest sample from the upper dolomite, with 10% expandable layers) is essentially discrete chlorite containing <5% expandable layers. In the upper anhydrite the clay ranges from 30% expandable layers toward the base of the

Table 3. Chemical analyses (wt. %) of <2-μm insoluble fractions from evaporites in the Paradox Member of the Hermosa Formation, Gibson Dome No. 1 drill core.

	1	2	3	4
SiO$_2$	47.2	41.6	39.6	42.9
TiO$_2$	0.48	0.37	0.27	0.15
Al$_2$O$_3$	20.1	15.9	15.8	9.31
Fe$_2$O$_3$[1]	3.57	5.16	4.44	2.00
MnO	<0.02	<0.02	<0.02	<0.02
MgO	11.6	17.0	23.7	30.7
CaO	0.69	0.80	0.35	<0.02
Na$_2$O	0.28	0.68	0.43	0.79
K$_2$O	5.37	3.58	2.07	0.11
P$_2$O$_5$	0.20	0.57	0.16	<0.05
LOI[2]	10.1	13.6	12.1	13.2
Total	99.5	98.6	98.9	99.2

Analyst: J. E. Taggert, Jr., U.S. Geological Survey Analytical Laboratory, Denver.

1. Illite and a chlorite with minor interstratified smectite in calcareous siltstone under salt 19 (Bodine and Rueger, 1984).
2. Chlorite with minor interstratified smectite, and illite in dolomitic siltstone under salt 8 (Bodine and Rueger, 1984).
3. Chlorite-rich interstratified chlorite/smectite and illite in silty dolomite under salt 6 (Bodine, 1985a).
4. Nearly monomineralic corrensite with minor talc and barely detectable illite in carnallite-bearing halite in salt 6 (Bodine and Rueger, 1984).

[1] Total Fe as Fe$_2$O$_3$.
[2] Loss on ignition at 900°C.

unit to 50% expandable layers (corrensite) near the contact with the overlying bedded halite. In general, discrete chlorite or chlorite containing negligible expandable layers characterizes the least saline lithologies ("black shale" and upper dolomite) of the succession, whereas C/S containing $\geq$20% expandable layers is restricted to the more saline anhydrite intervals.

DISCUSSION

The compositions of the C/S and the host evaporite (Figure 8) appear to be related to the salinity of the coexisting fluid during salt deposition and early diagenesis. Neither discrete clinochlore nor C/S, however, have been reported as authigenic minerals in modern, saline, sedimentary environments. Hence, these clays are probably products of burial diagenesis. The origin of these clays, therefore, appears to be a two-step process: (1) early authigenic formation of Mg-rich precursor clays whose composition reflects fluid composition; and (2) burial diagenetic transformation that preserves the diagnostic compositional signature. Bodine (1985b) suggested that the difference between the coexisting fluid:clay mass ratios of the depositional-early diagenetic environment and that of the subsequent burial diagenetic environment played a major role. In the former, the fluid:clay ratio is large; mass transfer affects clay compositions with negligible impact on the fluid. Alternatively, in the latter, the same ratio is markedly reduced, and mass transfer, particularly of

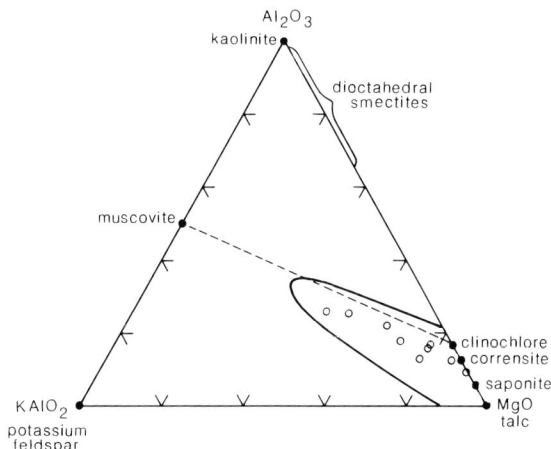

Figure 7. Molar proportions (Al$_2$O$_3$-MgO-KAlO$_2$, where Al$_2$O$_3$ = Al$_2$O$_3$ − [Na$_2$O + K$_2$O]) in clay-size fractions (O) in saline rocks in the Gibson Dome No. 1 drill core (Bodine and Rueger, 1984; Bodine, 1985a, unpublished). Sedimentary silicate mineral compositions (●) and dashed clinochlore-muscovite join shown for reference; heavy boundary encloses German Zechstein, Austrian Haselgebirge, Salina Group, and Salado Formation clay-size fraction compositions (Bodine, 1985a, 1985b).

hydrogen ion, results in the clays buffering the coexisting fluid, with little effect upon clay composition.

Deposition and early diagenesis

Aluminous dioctahedral clay minerals, abundant in terrigenous detritus, are not in equilibrium with sea water, much less with marine evaporite brines; [Mg^{2+}]/[H$^+$]^2 and [K$^+$]/[H$^+$] activity ratios in the brines are

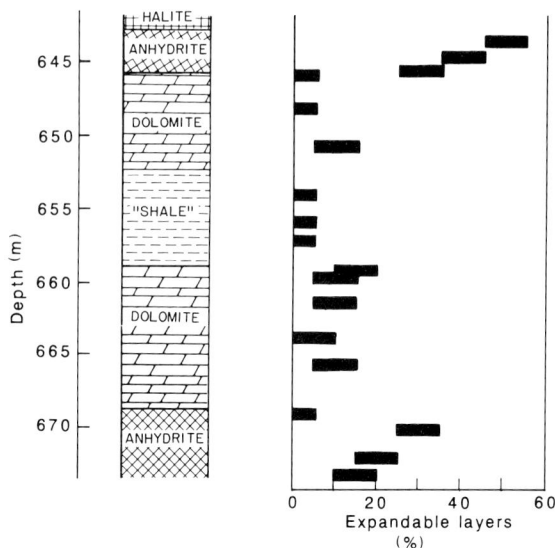

Figure 8. Percentage of expandable layers in mixed-layer chlorite/smectite as a function of evaporite lithology through the penesaline interval underlying salt 2.

Table 4. Idealized clay-mineral transformations as a function of salinity ($[a_{Mg^{2+}}]/[a_{H^+}]^2$) in depositionally early diagenetic environments of marine evaporites and subsequent isochemical burial diagenesis.

Penesaline sedimentation and early diagenesis

Anhydrite undersaturation

(1a) $Mg_{0.2}(Al_{1.6}Mg_{0.4})Si_4O_{10}(OH)_2 + 4.533SiO_2 + 3.4Mg_{aq}^{2+} + 5.533H_2O$
          detrital montmorillonite   quartz

$$\rightarrow 2.133Mg_{0.2}(Al_{0.75}Mg_{1.675})Si_4O_{10}(OH)_2 + 6.8H_{aq}^+$$
authigenic Mg-rich smectite $(2.425)^1$

Approaching or at halite saturation

(1b) $Mg_{0.2}(Al_{1.6}Mg_{0.4})Si_4O_{10}(OH)_2 + 6.492SiO_2 + 4.869Mg_{aq}^{2+} + 6.492H_2O$
          detrital montmorillonite   quartz

$$\rightarrow 2.623Mg_{0.2}(Al_{0.61}Mg_{1.885})Si_4O_{10}(OH)_2 + 9.738H_{aq}^+$$
authigenic Mg-rich smectite $(2.495)^1$

Burial diagenesis

(2a) $2.667Mg_{0.2}(Al_{0.72}Mg_{1.675})Si_4O_{10}(OH)_2 + 1.333H_2O \rightarrow (Mg_5Al)(Si_3Al)O_{10}(OH)_8 + 7.667SiO_2$
          Mg-rich smectite $(2.425)^1$ clinochlore   quartz

(2b) $3.933Mg_{0.2}(Al_{0.61}Mg_{1.885})Si_4O_{10}(OH)_2 + 1.067H_2O \rightarrow Mg_{0.2}(Mg_8Al)(Si_{6.6}Al_{1.4})O_{20}(OH)_{10} + 9.133SiO_2$
          Mg-rich smectite $(2.495)^1$ corrensite   quartz

---

^{1} Number of cations occupying three octahedral sites.

too high for equilibrium with kaolinite and dioctahedral smectites (Garrels, 1984; Bodine, 1985b). Primary authigenic Mg-enriched clays do form in saline, depositionally early diagenetic environments, and solute activities, thus salinity, should control resultant clay composition. Detrital dioctahedral smectites in Lake Abert, an alkaline, saline lake in Oregon, were reported by Jones and Weir (1983) to be replaced by smectites having substantially increased octahedral Mg and structural formulae midway between dioctahedral (montmorillonite) and trioctahedral (saponite) end members. Jones and Spencer (1985) described similar Mg-enriched, intermediate dioctahedral-trioctahedral smectites in the Great Salt Lake (Utah) Holocene sediments. Hypothetical Mg-enrichment of detrital dioctahedral smectite and its approach toward the trioctahedral structure is illustrated as a function of $[Mg^{2+}]/[H^+]^2$ activity in Table 4. Reaction (1a) represents transformation in a relatively dilute penesaline brine, and reaction (1b) represents transformation in a more concentrated penesaline brine. The equilibrium for reaction (1b) is defined by a greater $[Mg^{2+}]/[H^+]^2$ activity ratio than for reaction (1a); thus, reaction (1b) yields an authigenic smectite with greater Mg content and more trioctahedral character than does reaction (1a).

Burial diagenesis

Under burial conditions accompanied by an increase in temperature during an extended period of geologic time, the early authigenic MG-enriched clays were apparently replaced by the clays observed today. An idealized model for the essentially isochemical transformations allows no exchange with the pore fluid, except for the transfer of water, and preserves the com-

positional features imposed in the surficial environment. Reactions (2a) and (2b) in Table 4 produce the discrete clinochlore and corrensite from the Mg-rich smectites. The authigenic MG-rich smectites that effect these reactions show a relatively narrow compositional range, and their compositions and the degree of their trioctahedral character fall within the range observed by Jones and Weir (1983) for the Lake Abert clays. Note also that quartz is a product of reactions (2a) and (2b), and that euhedral, presumably authigenic quartz occurs in some Cane Creek No. 1 samples (Figure 5).

Neither the surficial nor the later burial processes inherent in this mechanism are considered to be proven for the origin of these clays. The mechanism appears reasonable, however, and imparts compositional signatures to the clays that we attribute to depositional and early diagenetic processes and preserves them through burial diagenesis.

ACKNOWLEDGMENTS

We are grateful to K. J. Esposito for assisting with some of the sample preparation and for X-ray powder diffraction analysis. This paper benefited from extended discussions with O. B. Raup and C. G. Whitney, although we alone take full responsibility for its contents. H. R. Northrop, O. B. Raup, and L. G. Schultz reviewed the manuscript and offered many useful suggestions.

REFERENCES

Bailey, S. W. (1982) Nomenclature for regular interstratifications: *Amer. Miner.* **67,** 394–398.
Bodine, M. W., Jr. (1971) Alteration of basic volcanic rocks by marine hypersaline brines, Hallstadt, Upper Austria: *Geol. Soc. Amer. Abstracts* **3,** 509.

Bodine, M. W., Jr. (1985a) Trioctahedral clay assemblages in Paleozoic marine evaporites: in *6th International Symposium on Salt, Vol. 2,* B. C. Schreiber and H. L. Harner, eds., Salt Institute, Arlington, Virginia, 267–284.

Bodine, M. W., Jr. (1985b) Clay mineralogy of insoluble residues in marine evaporites: in *Mineralogy V: Applications to the Mineral Industry,* D. M. Hausen and O. C. Kopp, eds., Soc. Min. Eng. Amer. Inst. Min. Metall. Petrol. Eng., New York, 133–156.

Bodine, M. W., Jr. and Fernalld, T. H. (1973) EDTA dissolution of gypsum, anhydrite, and Ca-Mg carbonates: *J. Sediment. Petrol.* **43,** 1152–1156.

Bodine, M. W., Jr. and Rueger, B. F. (1984) Progress report on clay-mineral assemblages in the Gibson Dome No. 1 drill core, Paradox Basin, Utah: *U.S. Geol. Surv. Open-File Rep.* **84-165,** 18 pp.

Braitsch, Otto (1960) Mineralparagenesis und Petrologie der Stassfurtsalze in Reyerhausen: *Kali Steinsalz* **3,** 1–14.

Braitsch, Otto (1971) *Salt Deposits: Their Origin and Composition:* Springer-Verlag, New York, 297 pp.

Brown, George and Brindley, G. W. (1980) X-ray diffraction procedures for clay mineral identification: in *Crystal Structures of Clay Minerals and their X-ray Identification,* G. W. Brindley and G. Brown, eds., Mineralogical Society, London, 305–359.

Dreizler, Ingo (1962) Mineralogische Untersuchungen an zwei Gipsvorkommen der Werraserie (Zechstein): *Beitr. Mineral. Petrogr.* **8,** 323–338.

Echle, Wolfram (1961) Mineralogische Untersuchungen an Sedimenten des Steinmergelkeupers und der Roten Wand aus der Umgebung von Göttingen: *Beitr. Mineral. Petrogr.* **8,** 28–59.

Fournier, R. O. (1961) Regular interstratified chlorite-vermiculite in evaporites of the Salado Formation, New Mexico: *U.S. Geol. Surv. Prof. Pap.* **424-D,** 323–327.

Füchtbauer, Hans and Goldschmidt, Hertha (1959) Die Tonminerale der Zechsteinformation: *Beitr. Mineral. Petrogr.* **6,** 320–345.

Garrels, R. M. (1984) Montmorillonite/illite stability diagrams: *Clays & Clay Minerals* **32,** 161–166.

Grim, R. E., Droste, J. B., and Bradley, W. F. (1960) A mixed-layer clay mineral associated with an evaporite: in *Clays and Clay Minerals, Proc. 8th Natl. Conf., Norman, Oklahoma, 1959,* Ada Swineford, ed., Pergamon, New York, 228–236.

Hite, R. J. (1960) Stratigraphy of the saline facies of the Paradox Member of the Hermosa Formation of southeastern Utah: in *Geology of the Paradox Fold and Fault Belt, 3rd Field Conference Guidebook,* K. G. Smith, ed., Four Corners Geological Society, 86–89.

Hite, R. J. Anders, D. E., and Ging, T. G. (1984) Organic-rich source rocks of Pennsylvanian age in the Paradox basin of Utah and Colorado: in *Hydrocarbon Source Rocks of the Greater Rocky Mountain Region,* Jane Woodward, F. F. Meissner, and J. L. Clayton, eds., Rocky Mountain Association of Geologists, Denver, Colorado, 255–274.

Hite, R. J. and Buckner, D. H. (1981) Stratigraphic correlations, facies concepts, and cyclicity in Pennsylvanian rocks of the Paradox basin: in *Geology of the Paradox Basin, 1981 Field Conference,* D. L. Wiegand, ed., Rocky Mountain Association Geologists, Denver, Colorado, 147–159.

Jones, B. F. and Spencer, R. J. (1985) Clay minerals in the Great Salt Lake basin: *Program and Abstracts, 1985 International Clay Conference,* Denver, Colorado, 114.

Jones, B. F. and Weir, A. H. (1983) Clay minerals of Lake Abert, an alkaline, saline lake: *Clays & Clay Minerals* **31,** 161–172.

Lippmann, F. and Savascin, M. Y. (1969) Mineralogische Untersuchungen an Lösungsrückständen eines württembergischen Keupergipsvorkommens: *Tschermaks Mineral. Petrogr. Mitt.* **13,** 165–190.

Lounsbury, R. W. (1963) Clay mineralogy of the Salina Formation, Detroit, Michigan: in *Symposium on Salt,* A. C. Bersticker, ed., Northern Ohio Geol. Soc., Cleveland, Ohio, 56–63.

Niemann, Hans (1960) Untersuchungen am Grauen Salzton der Grube "Königshall-Hindenburg," Reyerhausen bei Göttingen: *Beitr. Mineral. Petrogr.* **7,** 137–165.

Pollastro, R. M. (1982) A recommended procedure for the preparation of oriented clay mineral specimens for X-ray diffraction analysis: modifications to Drever's filter-membrane peel technique: *U.S. Geol. Surv. Open-File Rep.* **82-71,** 10 pp.

Pundeer, G. S. (1969) Mineralogy, genesis and diagenesis of a brecciated shaly clay from the Zechstein evaporite series of Germany: *Contrib. Mineral. Petrol.* **23,** 65–85.

Raup, O. B. (1966) Bromine distribution in some halite rocks of the Paradox Member, Hermosa Formation, in Utah: in *Second Symposium on Salt,* Jon Rau, ed., Northern Ohio Geol. Soc., Cleveland, Ohio, 236–247.

Reinold, P. (1965) Über das Vorkommen von Chlorit im alpinen Salinar: *Tschermaks Mineral. Petrogr. Mitt.* **9,** 195–201.

Reynolds, R. C. (1980) Interstratified clay minerals: in *Crystal Structures of Clay Minerals and their X-ray Identification,* G. W. Brindley and G. Brown, eds., Mineralogical Society, London, 249–303.

Whitney, C. G. (1979) The paragenesis of synthetic phyllosilicates on the talc-phlogopite join: Ph.D. dissertation, Univ. Illinois, Urbana-Champaign, Illinois, 221 pp.

GEOLOGY
AND
DIAGENESIS

Proceedings of the International Clay Conference, Denver, 1985, L. G. Schultz, H. van Olphen, and F. A. Mumpton, eds.,
The Clay Minerals Society, Bloomington, Indiana, 97–104 (1987).

ORIGIN OF CLAY MINERALS IN PENNSYLVANIAN STRATA
OF THE ILLINOIS BASIN

RANDALL E. HUGHES, PHILIP J. DEMARIS, W. ARTHUR WHITE, AND DOUGLAS K. COWIN

Illinois State Geological Survey, 615 East Peabody Drive
Champaign, Illinois 61820

Abstract—Clay mineral analyses of Pennsylvanian sediments from the Illinois basin indicate that three major processes account for the wide range of clay mineral composition in these sediments: preservation of source material, soil formation, and diagenesis. Most shales, underclays in areas of rapid sedimentation, and insoluble residues of limestones contain detrital illite and lesser amounts of chlorite, kaolinite, and mixed-layer illite/smectite (I/S). Source material composition does not seem to have varied significantly during the Pennsylvanian era. Most underclays, clays associated with coal, and some shales and claystones consist of source materials modified by soil-forming processes. Variation in clay mineral composition results from alteration in two distinct soil-forming environments. In areas of relatively low subsidence and sedimentation rate, the prolonged action of plant roots and/or percolation of water resulted in the formation of underclays and claystones containing poorly crystallized "soil" kaolinite, mixed-layer kaolinite/smectite (K/S), and minor I/S and vermiculite. The extraction of K^+ and Mg^{2+} by growing plants under saturated soil conditions (i.e., in underclays, coals, shales, and claystones) in other areas of the basin resulted in the formation of I/S and well-crystallized kaolinite with calcite, ferroan dolomite, siderite, pyrite, and/or marcasite. Sandstones commonly contain abundant authigenic kaolinite formed by post-depositional alteration of source materials. The lack of evidence of diagenetic formation of illite and chlorite suggests that the source detritus was derived by mechanical erosion in highland areas and that, similar to modern tropical basins, the intervening lowlands were protected from erosion by lush vegetation.

Key Words—Chlorite, Gley, Illite, Kaolinite, Sandstone, Shale, Smectite, Soil, Underclay.

INTRODUCTION

Parham (1964) described the composition of the Pennsylvanian underclays (i.e., claystones below coals) of the eastern and central part of the United States as three types of compositional facies: (1) a composition dominated by poorly crystalline kaolinite and minor illite, mixed-layer illite/smectite (I/S), and vermiculite; (2) one consisting of predominantly illite and minor kaolinite, chlorite, and I/S; and (3) one containing abundant I/S, minor illite and chlorite, and highly crystalline kaolinite. The occurrence of authigenic kaolinite in sandstones and siltstones (Glass *et al.*, 1956) is possibly a fourth type of facies. Therefore, the clay mineral composition of all clays, shales, limestones, coals, siltstones, and sandstones can be conveniently described. On the basis of a large body of data accumulated over several decades by the Illinois Geological Survey, the composition of the clay and nonclay minerals found with each of the four facies is reported below along with an origin for the occurrences.

REVIEW OF LITERATURE

Pennsylvanian strata in the Illinois basin are composed of shales, claystones, sandstones, siltstones, coals, and many thin but widespread limestones (Wanless *et al.*, 1963 and Figure 1). Basal units typically overlie an unconformity with as much as 150 m relief on strata varying in age from Mississippian to Ordovician (How-

ard, 1979). The oldest deposits are sandy and mostly nonmarine. The deposits become finer and include more marine units upward in the stratigraphic section (Wanless *et al.*, 1963 and Figure 1). Wanless and Weller (1932) observed the cyclic nature of the sediments and proposed the name cyclothem (Figure 2) for repetitive sequences in the Desmoinesian Series. Rarely are all 10 units of the composite cyclothem (Figure 2) present at one location, but more than 50 cyclothems have been named in the Illinois basin (Kosanke *et al.,* 1960). The western and northern parts of the basin were stable areas of relatively little subsidence; the eastern and southern parts were regions of more rapid subsidence. The latter cyclothems are thick and well developed, but in stable areas they are commonly thin and contain fewer members.

The climate during Pennsylvanian time in the Illinois basin was probably humid tropical. Based on the distribution of plant and animal fossils, Wanless and Weller (1932) proposed a change in water chemistry from fresh to marine that paralleled the change in cyclothemic units from bottom to top (Figure 2). Members of some cyclothems can be traced through several basins (Wanless *et al.*, 1963).

The clay mineral composition of the cyclothem units (Figure 2) was investigated by Grim and Allen (1938), Grim *et al.* (1957), Potter and Glass (1958), Glass (1958), and Bradbury *et al.* (1962). Underclays (Figure

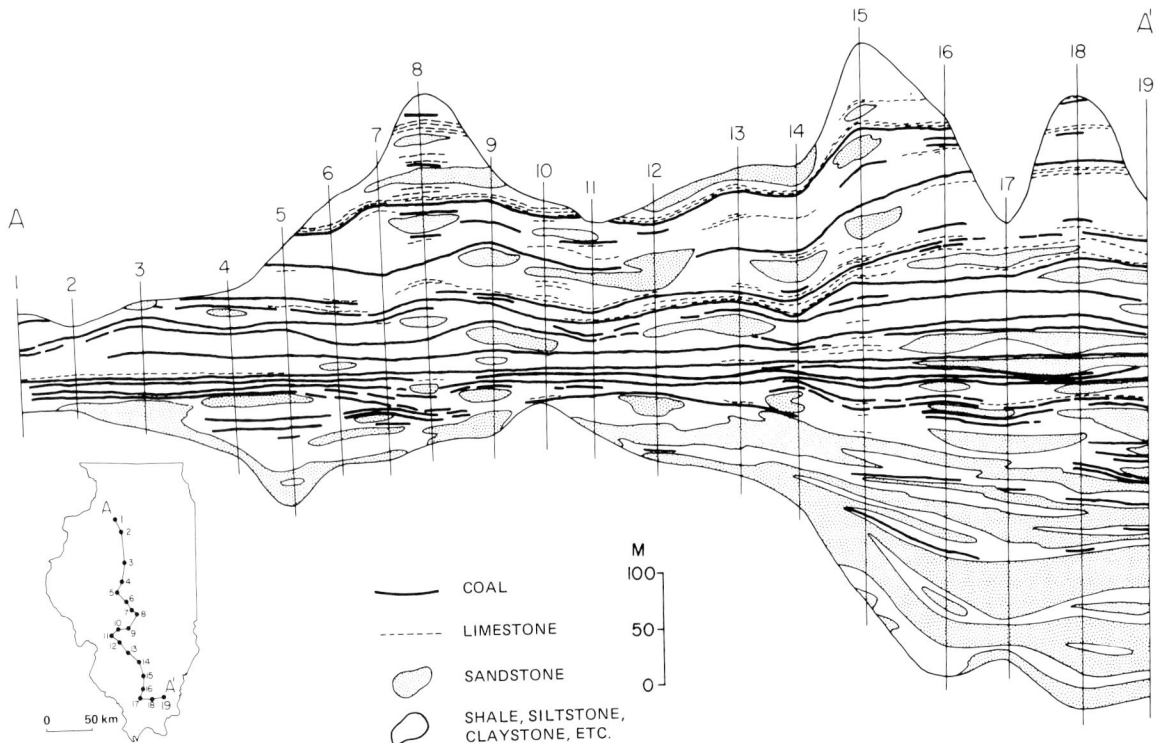

Figure 1.   North-south generalized cross section of Pennsylvanian strata in Illinois. Control datum is the Colchester Coal. Numbers (1, 2, etc.) mark the location of well-log control points. (Modified from unpublished figure of R. J. Jacobson and M. E. Hopkins, Illinois Geological Survey, Champaign, Illinois.)

2, unit 4) are widespread, show great variation in composition, and have been extensively studied (Grim and Allen, 1938; Glass, 1958; Schultz, 1958; Parham, 1964; Odom and Parham, 1968; Hughes, 1971; and Rimmer and Eberl, 1982). Summaries of theories of the origin of Pennsylvanian underclays in the basin were published in Parham (1964) and Rimmer and Eberl (1982). The disseminated mineral matter in coal (unit 5) was investigated by Hughes (1971), Rao and Gluskoter (1973), Ward (1977), and Harvey *et al.* (1983). The clay mineral composition of the marine part of the cyclothem (units 6–10) was investigated by Webb (1961) and Odom (1963).

SAMPLES AND EXPERIMENTAL PROCEDURE

The data base reviewed includes several thousand core and outcrop samples, mostly from underclays and roof shales (units 4 and 6), of many of the cyclothems in Illinois. The mineral content of the clay-size fraction was reported by Grim and Allen (1938), Grim *et al.* (1957), Webb (1961), Odom (1963), Parham (1964), and Odom and Parham (1968). Representative samples of shales and underclays from these studies and new samples were analyzed by whole-sample X-ray powder diffraction (XRD) smear techniques to eliminate compositional variations due to size separation

and to determine the associated nonclay minerals. All smear samples were analyzed by <2-μm sedimented-slide methods, and several other samples were also analyzed as <2-μm smear slides. Earlier studies used mainly composite channel samples, but data derived during the last twenty years are from multiple samplings from much smaller vertical intervals—commonly 5–25 cm.

XRD data were obtained using General Electric and Philips Norelco diffractometers with Cu radiation at a 2°2θ/min scan rate. Whole-sample and <2-μm smear slides and <2-μm sedimented samples were used to analyze mineral composition. Smear slides were made by smearing a wet paste of finely ground material (or the <2-μm fraction) on a glass slide with a microspatula. The <2-μm sedimented slides were prepared from dispersed clay after setting for 20 min, and pipeting about 2 ml from the top ~0.5 cm of the suspension to a glass petrographic slide. The samples were scanned after solvation with ethylene glycol and after heating to 375°C for 1 hr. The percentage of smectite layers in I/S was determined by the method given in Rimmer and Eberl (1982).

Kaolinite/smectite (K/S) was identified by its characteristic peak or shoulder on the diffractogram of the ethylene-glycol-solvated sample at spacings of 7.3–8 Å (11°–12°2θ) (Figure 3b, upper trace) and by the shift to

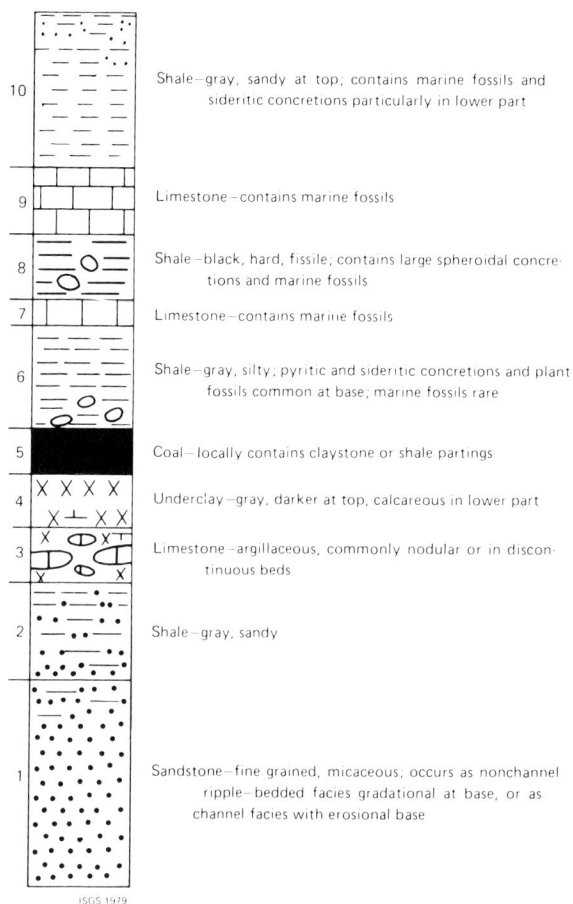

Figure 2. Complete cyclothem (after Willman and Payne, 1942). This cyclothem is a composite model; all ten units are only rarely present in the strata.

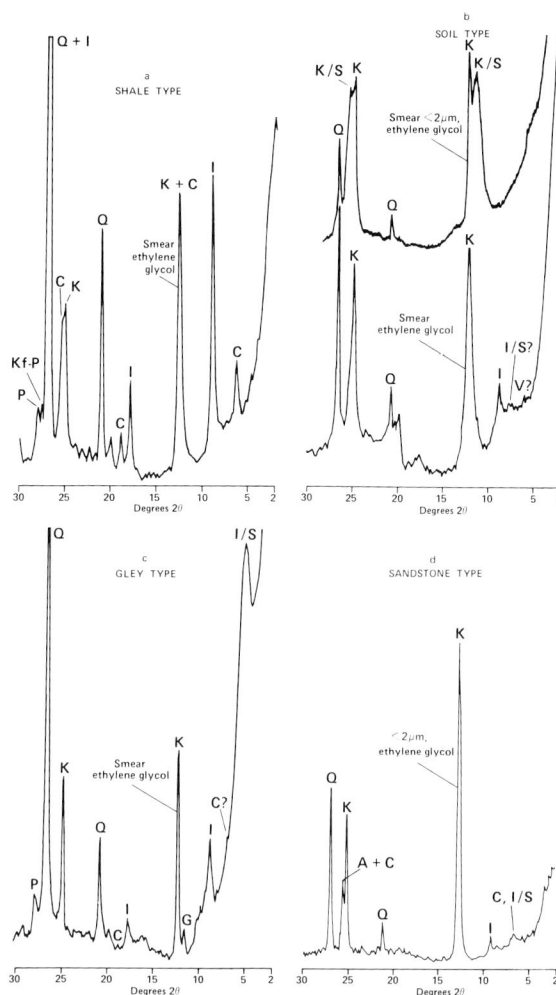

Figure 3. X-ray powder diffraction patterns of four clay mineral suites observed in the Illinois basin: (a) shale-type suite (source material); (b) soil-type suite (upper trace = K/S, lower trace = poorly crystalline kaolinite); (c) gley-type suite; (d) sandstone-type suite. I = illite; C = chlorite; K = kaolinite; K/S = kaolinite/smectite; I/S = illite/smectite; V = vermiculite; Q = quartz; KF = K-feldspar; P = plagioclase; G = gypsum; A = anatase. CuKα radiation.

larger spacings on heating at 375°C. The presence of K/S was also detected by an elevated background intensity on traces of both the glycolated and heated samples between 10 and 7 Å (8.8° and 12.3°2θ). Based on computer simulations of the diffractograms for the K/S compositional series (Reynolds, 1980), most of the K/S minerals we have observed have compositions between 80 and 95% kaolinite layers. Intercalation compounds, such as N-methyl formamide, will not penetrate the kaolinite layers in K/S, but discrete kaolinite can be expanded to 11.4 Å, and the K/S peak can be observed free of interference.

CLAY MINERAL SUITES

Four distinct suites were distinguished in Pennsylvanian rocks (Table 1 and Figure 3): (1) shale-type—illite with minor kaolinite, chlorite, I/S, orthoclase, and plagioclase (Figure 3a); (2) soil-type—poorly crystalline kaolinite and mixed-layer kaolinite/smectite (K/S) with minor illite, I/S, and vermiculite (Figure 3b); (3) gley-type—mixed-layer illite/smectite (I/S) and highly crystalline kaolinite, with minor illite, chlorite, and plagioclase (Figure 3c); and (4) sandstone-type—highly crystalline kaolinite and minor illite, chlorite, I/S, and anatase (Figure 3d). The soil-, gley-, and sandstone-type suites were derived by weathering or diagenesis from, and grade into, the shale-type composition. Parham (1964) showed a lateral gradation between almost 100% kaolinite (soil-type) and a shale-type suite, and from a shale-type to almost 100% smectite (gley-type).

Gradational changes in clay mineral composition were observed both laterally and vertically. The distance required for lateral change varied from a few hundred meters to a few hundred kilometers. Soil-type suites resulted from greater duration and intensity of weathering, and, therefore, larger lateral or vertical distances were observed for the soil- to shale-type transition. Vertical variation was usually observed over a few centimeters to a few meters. Vertical transitions from one cyclothem unit (Figure 2) to another typically resulted in abrupt changes in suite type. Soil- and gley-type suites were observed to change abruptly from one to another only if the rock type changed. For example, the composition of the clay suite in coals was usually a gley-type, whereas claystone partings in coal were soil-type (Figure 3b, lower trace).

Basal Pennsylvanian units were found to be sandy and to contain a highly altered clay mineral suite (sandstone-type). On the other hand, basal Pennsylvanian limestone and shale units contained a detrital, shale-type suite of clay minerals.

In many samples consisting of a shale-type suite of clay minerals, it was difficult to determine whether the minor associated kaolinite was detrital or authigenic. The XRD peak for kaolinite never indicated K/S or poorly crystalline kaolinite, but whether the peak was as sharp as that in gley-type kaolinite was obscured by overlap with chlorite. Resolution of this question awaits detailed scanning electron microscopic and energy-dispersive X-ray analyses of several shale-type samples. Microscopic examination of the sample in Figure 3a showed several euhedral kaolinite "books," which suggest little detrital kaolinite. Until the lack of detrital kaolinite can be demonstrated, however, the shale-type suite must be assumed to include minor but significant detrital kaolinite.

CLAY MINERAL SUITES IN CYCLOTHEM UNITS

Sandstones and siltstones

Sandstones at the unit 1 position in Figure 2, as well as sandstone and siltstone facies of other units, typically contain a sandstone-type clay mineral suite (Figure 3d). The occurrence of a sandstone-type suite with either a soil- or gley-type suite depends on whether the sediments were altered by soil processes or preserved by rapid burial.

Underclays and similar claystones

Underclays (unit 4 in Figure 2) and the underlying sandy-shale and nonmarine limestone (units 2 and 3) are both usually transitional with the underclay. *Underclay* is defined here as a non-bedded claystone present beneath coal (unit 5) that commonly contains *Stigmaria*, slickensides, and coalified-plant fragments. Where underclays are absent and coal overlies shale, sandstone, or limestone, the term *seat rock* is used.

Underclays (Figure 4) show the greatest variation in clay mineral suite and composition of all the units of

a cyclothem, including the following three clay mineral suites:

1. *Shale-type suite*—detrital source material (illite, minor chlorite, kaolinite, and I/S with orthoclase, plagioclase, and quartz) preserved by rapid burial (Figures 4a, 1–2.5 m; 4b, ~1 m; 4c, 1.5–2.5 m; and 4d, 1–1.4 m).
2. *Soil-type suite*—shale-type source material altered by plants and/or vadose water which extract K^+, Mg^{2+}, and SiO_2. Illite and chlorite are altered to K/S, poorly crystalline kaolinite, I/S, and vermiculite. Feldspars are altered to highly crystalline kaolinite (Figures 4b, 0–0.75 m and 1.25–2 m; and 4c, 0–1 m).
3. *Gley-type suite*—shale-type source material altered by K^+-extracting plants in a saturated soil environment. Illite is altered to I/S, and orthoclase is altered to highly crystalline kaolinite (Figure 4d, 0–1 m). Detrital chlorite and plagioclase are present, except where H^+-for-K^+ exchange has lowered the pH sufficiently to cause their alteration (Figure 4a, 0–0.75 m). With sufficient time, shale-type source materials alter to a "semi-flint clay" composition (Hughes and White, 1969).

Coals and claystone partings in coal

Disseminated clay minerals in coal (unit 5 in Figure 2) altered to a gley-type suite, although more kaolinite and illite are usually present than in gley-type underclays (Hughes, 1971; Rao and Gluskoter, 1973; Ward, 1977; and Harvey *et al.*, 1983). On the other hand, claystone partings in coal resulted from the alteration of shale-type source materials to a soil-type suite containing poorly crystalline kaolinite (Figure 3b, lower trace).

The widespread "blue band" of the Herrin Coal Member is a claystone parting typically 2–5-cm thick and containing a soil-type suite of clay minerals (Figure 3b, lower trace). The "blue band" near the Walshville channel, however, contemporaneous with the deposition of the Herrin Coal, progressively thickens to a meter or more (DeMaris *et al.*, 1983) and it has a shale-type composition.

Shales and limestones above coals

The unit immediately above the coal (unit 6 in Figure 2) is typically a gray or a black shale. The gray shale, which occurs locally along and near the axis of major paleorivers and tributary or distributary paleochannels, is considered a nonmarine deposit (Gluskoter and Simon, 1968). Black shale or limestone is typically present in areas between channels. Locally a limestone, sandstone, or siltstone overlies the coal. The black shale and limestone of unit 6 and the overlying shales and limestones (units 7–9) are marine. The lower

Table 1. Typical compositions of clay mineral suites in Pennsylvanian strata of Illinois.

Suite name	Minerals[1]								
	I (%)	K[2] (%)	C (%)	I/S, V (%)	Q	KF	P	Ca[3]	Py, Ma
Shale-type	70	10	10	10	c	m–c	m–c	m	m
Soil-type	10	80	0	10	c	–	–	r	r–m
Gley-type	15	15	5	65	c	–	m–c	m–c	m–c
Sandstone-type	10	75	5	10	a	r	r	r?	r–m

[1] I = illite; K = kaolinite; C = chlorite; I/S, V = illite/smectite and vermiculite; Q = quartz; KF = potassium feldspar; P = plagioclase feldspar; Ca = calcite, ferroan dolomite, and siderite; Py, Ma = pyrite and marcasite; a = abundant; c = common; m = minor; r = rare; – = absent.

[2] Kaolinite in soil-type suite is partly K/S or poorly crystalline; kaolinite in shale-, gley-, and sandstone-type suites is highly crystalline and euhedral.

[3] Distribution in clastic units (excluding limestone and dolomite).

part of unit 10 is typically marine, and the rest of the unit is typically nonmarine, transitional with the sandstone (unit 1) of the next cyclothem or terminated by an unconformity.

The grey, nonmarine roof shale and marine shales and limestones contain the least-altered source materials (shale-type suite). Authigenic kaolinite, however, is common in concretions that contain plant remains. Shales immediately above coals contain somewhat more I/S than overlying marine units, but illite and chlorite still make up >75% of the sample, and the percentage of hydrated layers in I/S is low— 10–30%.

Other strata

Hughes and White (1969) described flint clays from the interval between the Springfield Coal and the Herrin Coal that formed by the alteration of clay minerals in abandoned channels, oxbow lakes, and similar environments. The flint clays contain clasts consisting of smectite, highly crystalline kaolinite, and minor illite. Quartz, siderite, calcite, and dolomite are present. The flint clays also represent an example of a mineral suite that probably formed as a result of alteration by plant growth. K[+] and SiO_2 removal were contemporaneous with the crystallization of carbonate minerals.

Clay dikes in coals formed as the result of a downward flow of younger plastic sediments into the peat before coalification took place (Krausse and Damberger, 1979). Clay dikes contain a shale-type suite of clay minerals, but more authigenic kaolinite than the overlying marine shales and limestones. The detrital origin of the illite-chlorite suite was particularly evident in clay dikes, because fluids released during coalification apparently produced authigenic kaolinite and lacked K[+] and Mg^{2+} for the diagenetic formation of illite and chlorite.

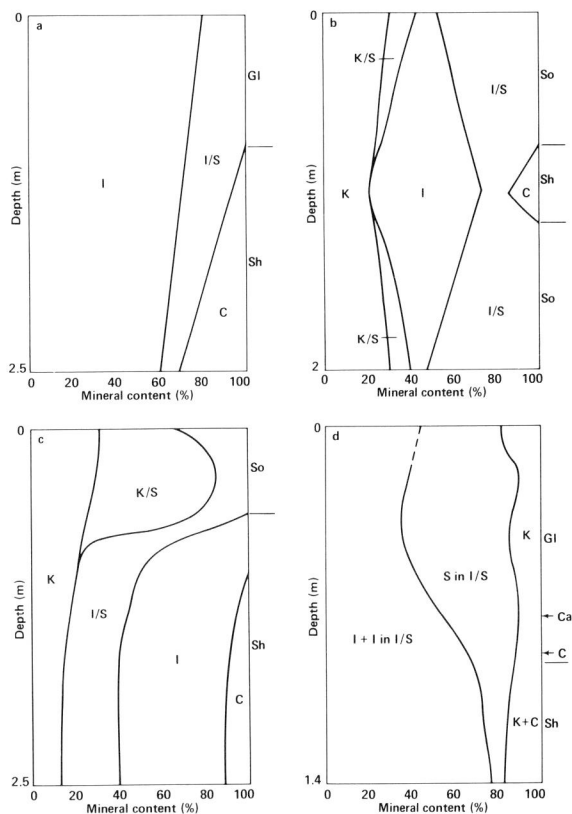

Figure 4. Typical underclay profiles: (a) shale-type suite with minor I/S development and absence of chlorite in the upper zone (from underclay of Houchin Creek (formerly Summum) Coal); (b) soil-type suite with an upper and lower zone of increased kaolinite, K/S, and I/S (from underclay of Houchin Creek Coal); (c) soil-type suite with a zone of well-developed K/S (from underclay of Houchin Creek Coal); (d) gley-type suite modified from Rimmer and Eberl (1982) showing strong development of I/S in the top third of the profile and the first detection of calcite (Ca) and chlorite (C) in the bottom half of the profile. I = illite; C = chlorite; K = kaolinite, I/S = illite/smectite, K/S = kaolinite/smectite, S = % smectite layers in I/S; Gl = gley-type suite; Sh = shale-type suite; So = soil-type suite.

EVIDENCE OF SUITE GENESIS

Lovering (1959) reviewed the ability of plants to extract and accumulate various inorganic elements and demonstrated the complex manner in which the growth and decay of plants rapidly alters minerals. He suggested that plant-induced alteration of minerals may be three or four orders of magnitude faster than inorganic alteration processes. Such an alteration mechanism must obviously be considered in any discussion of the origin of clay minerals in Pennsylvanian strata, but associations between type of plant, mechanism of mineral alteration, and the resulting changes in both mineral and chemical composition are only beginning to be investigated. Plant growth is probably a principal factor in the formation of poorly crystalline kaolinite,

K/S, flint clays, tonsteins, and bauxites. The I/S and authigenic kaolinite coexisting with carbonate zones in underclays is best explained by such a mechanism.

Two facts strongly support the hypothesis that underclays acted as soils for contemporaneous plants: (1) the presence of multiple zones of alteration within a single underclay (Figure 4b, 0–0.75 m and 1.25–2 m); and (2) the occurrence of gley- and soil-type suites in claystones (underclays) with no evidence of coal formation. In addition, the primitive nature of Pennsylvanian plants, which had mainly lateral root systems, suggests that the peat-forming plants caused little alteration of the minerals in the underclay, except, perhaps, near its contact with coal.

Wanless *et al.* (1963) showed the extensive distribution of underclays and suggested that some contain lacustrine facies. The widespread lateral distribution of individual underclays is best explained as resulting from periodic floods by major rivers into and over lowland areas.

The amount of quartz in disseminated mineral matter in coals, claystone partings, underclays, and shales is similar. The distributions of rare-earth elements in disseminated mineral matter in coal and claystone partings are also comparable (DeMaris *et al.,* 1983). These similarities suggest that the clay minerals in coal formed by alteration of typical, detrital, shale-type source material.

Recent work on carbonaceous roof shales (DeMaris *et al.,* 1983) indicates that claystone partings, such as the "blue band," may be produced by flooding similar to those which ended peat deposition and perhaps to those which produced underclays. The shale-type composition of partings near channels represents preservation of source material by rapid deposition and implies that claystone partings in some coals result mainly from fluvial deposition and are not tonsteins (i.e., kaolinitized volcanic ash falls) which are common in Pennsylvanian coals in other parts of the world.

Clay partings are accompanied by increased floral diversity (Phillips and DiMichele, 1981, Figure 7.6), which probably results from an increase in available nutrients. This sequence suggests that lycopods dominated in the low-nutrient environments between sediment floods, and pteridosperms and other plants became common during floral re-establishment on nutrient-rich clay partings. Therefore, highly crystalline kaolinite in disseminated mineral matter in coal and poorly crystalline kaolinite in clay partings may have resulted from differences in floral type and diversity.

SUMMARY AND CONCLUSIONS

Our ideas on the origin of the several clay mineral suites are therefore based on the following interrelated observations:

1. The shale-type suite of clay minerals occurs in underclays, clay dikes, near-channel clay partings in coal, and gray roof shales. These units were deposited in freshwater, weakly acidic environments and were unlikely sites for the formation of diagenetic illite and chlorite. Therefore, the shale-type suite was probably detrital.

2. The crystallinity and euhedral morphology of kaolinite in the gley- and sandstone-type suites and the concretions indicate that the kaolinite is authigenic. The anhedral morphology of illite and chlorite are typical of detrital materials.

3. The composition of I/S varies from about 10 to 90% smectite layers. XRD indicates a narrow range of percent hydration for individual samples, which strongly implies that the I/S originated *in situ.*

4. The soil-type suite completely lacks chlorite and feldspar (Table 1). If the suite had resulted from differential sorting, it should have included minor amounts of these minerals.

5. K-feldspar occurs only in the shale-type suite (Table 1), whereas the gley-type suite contains only plagioclase. The absence of K-feldspar in the gley-type suite is apparently the result of K^+ extraction by growing plants.

6. I/S, chlorite, and plagioclase coexist in the gley-type suite (Table 1). The coexistence of these minerals and the absence of K-feldspar is not adequately explained by hydrogen ion diffusion from the acid reservoir of the coal swamp. The well-known, rapid alteration of chlorite and plagioclase in acid solutions should have resulted in their removal before illite could be altered to I/S or K-feldspar to kaolinite. This alteration can best be explained by selective K^+ removal by growing plants.

7. Underclays usually show a vertical alteration sequence of the clay minerals (Figure 4), although commonly not as well developed as those described by Rimmer and Eberl (1982) (Figure 4d). Alteration sequences occur in underclays that contain soil- or gley-type suites and are best explained as the result of soil-forming processes.

8. Many Pennsylvanian sediments contain abundant plant remains and carbonaceous matter. Plant growth must, therefore, have been a significant factor in the alteration of detrital sediments.

9. The illite in Pennsylvanian sediments is mainly the 2*M* polytype (personal communication, H. D. Glass, 1981, Illinois State Geological Survey, Champaign, Illinois). These sediments are considered to have been buried no deeper than about 1.5 km (Damberger, 1971). This mica polymorph has been synthesized under conditions of high temperature and pressure (Yoder and Eugster, 1955), whereas 1*M* mica structures form experimentally at lower temperature and pressure (Yoder and

Eugster, 1955) and are known to be diagenetic in nature (e.g., glauconite). Thus, the $2M$ structure of illite observed in the Pennsylvanian sediments indicates a detrital origin.

10. A diagenetic formation of illite and chlorite in Pennsylvanian sediments is improbable, because no known source of sufficient K^+ and Mg^{2+} exists for the formation of these clay minerals.

The origin of the mineral suites observed in Pennsylvanian strata of the Illinois basin can be summarized as follows:

During sandstone deposition at the start of each cyclothem (Figure 2, unit 1), the detrital shale-type suite altered by soil processes. After burial, porous sandstones, siltstones, and sandy shales were principally altered by ground-water processes, and may have been further altered in outcrops after exposure. During underclay development (units 2–4, Figure 2), freshwater floods of sediment containing shale-type suites of clay minerals were altered or preserved depending on deposition rate. Sediments were altered by the action of growing plants to soil-type or gley-type suites of clay minerals, dependent upon whether the sediments were primarily emergent or submergent, and upon the type of plants present. During coal formation (unit 5, Figure 2), disseminated source material containing shale-type suites of clay minerals was altered by plant growth to a gley-type suite. Highly crystalline kaolinite (and possibly quartz) formed by the crystallization of plant-dissolved Al^{3+} and SiO_2. Clay partings in coal resulted from the alteration of shale-type source material to a soil-type suite.

During the initial phase of the transgression that formed the upper part of the cyclothem (units 6–10, Figure 2), shale-type source material was slightly altered by plant growth that diminished as the environment became more marine and/or as the rate of deposition increased. At the end of each cyclothem as conditions became nonmarine, the uppermost units (units 6–10, Figure 2) were locally altered by growing plants and/or vadose water. If the sediments during this interval were fine grained or if the sandy part of the overlying cyclothem (units 1 and 2, Figure 2) was absent, the underclay of the overlying cyclothem may appear transitional with the shale or limestone of the next lower cyclothem.

ACKNOWLEDGMENTS

The authors are indebted to H. D. Glass for critical review of the manuscript and many helpful suggestions and insights. We also gratefully acknowledge the assistance of John M. Fox and Robin L. Warren with sample preparation and analysis, the Center for Electron Microscopy at the University of Illinois, Urbana-Champaign, for the use of its facilities, and the Illinois Coal Development Board for funding.

REFERENCES

Bradbury, J. C., Ostrom, M. E., and Lamar, J. E. (1962) Chemical and physical character of the Pennsylvanian sandstones in central Illinois: *Illinois Geol. Surv. Circ.* **331**, 43 pp.

Damberger, H. H. (1971) Coalification pattern of the Illinois basin: *Econ. Geol.* **66**, 488–494.

DeMaris, P. J., Bauer, R. A., Cahill, R. A., and Damberger, H. H. (1983) Geologic investigation of roof and floor strata: Longwall demonstration, Old Ben Mine No. 24. Prediction of coal balls in the Herrin Coal: *Illinois Geol. Surv. Contract/Grant Rept.* **1983-2, Part 2**, 69 pp. (NTIS DOE/ET/1217-2).

Glass, H. D. (1958) Clay mineralogy of Pennsylvanian sediments in southern Illinois: in *Clays and Clay Minerals, Proc. 5th Natl. Conf. Urbana, Illinois, 1956*, A. Swineford, ed., *Natl. Acad. Sci.-Natl. Res. Counc. Publ.* **566**, 227–241.

Glass, H. D., Potter, P. E., and Siever, R. (1956) Clay mineralogy of some basal Pennsylvanian sandstones, clays, and shales: *Bull. Amer. Assoc. Petrol. Geol.* **40**, 751–754.

Gluskoter, H. J. and Simon, J. A. (1968) Sulfur in Illinois coals: *Illinois Geol. Surv. Circ.* **432**, 28 pp.

Grim, R. E. and Allen, V. T. (1938) Petrology of Pennsylvanian underclays of Illinois: *Geol. Soc. Amer. Bull.* **49**, 1485–1514.

Grim, R. E., Bradley, W. F., and White, W. A. (1957) Petrology of the Paleozoic shales of Illinois: *Illinois Geol. Surv. Rept. Invest.* **203**, 35 pp.

Harvey, R. D., Cahill, R. A., Chou, C.-L., and Steele, J. D. (1983) Mineral matter and trace elements in the Herrin and Springfield Coals, Illinois basin Coal Field: *Illinois Geol. Surv. Contract/Grant Rept.* **1983-4**, 161 pp.

Howard, R. H. (1979) Depositional history of the Pennsylvanian System in the Illinois basin—the Mississippian-Pennsylvanian unconformity in the Illinois basin—old and new thinking: in *Depositional and Structural History of the Pennsylvanian System of the Illinois Basin, Part 2*, J. E. Palmer and R. R. Dutcher, eds., *Illinois Geol. Surv. Guidebook Ser.* **15a**, 34–43.

Hughes, R. E. (1971) Mineral matter associated with Illinois coals: Ph.D. Thesis, University of Illinois, Urbana, Illinois, 145 pp.

Hughes, R. E. and White, W. A. (1969) A flint clay in Sangamon County, Illinois: in *Proc. Int. Clay Conf. Tokyo, 1969, Vol. 1*, L. Heller, ed., Israel Univ. Press, Jerusalem, 291–303.

Kosanke, R. M., Simon, J. A., Wanless, H. R., and Willman, H. B., eds. (1960) Classification of the Pennsylvanian strata of Illinois: *Illinois Geol. Surv. Rept. Invest.* **214**, 84 pp.

Krause, H. F. and Damberger, H. H. (1979) Clay-like faults and associated structures in coal bearing strata—deformation during diagenesis: in Abstracts of Papers, *9th Int. Cong. Carboniferous Stratigraphy and Geology, Univ. Ill., Urbana, Ill., 1979*, p. 111.

Lovering, T. S. (1959) Significance of accumulator plants in rock weathering: *Geol. Soc. Amer. Bull.* **70**, 781–800.

Odom, I. E. (1963) Clay mineralogy and clay mineral orientation of shales and claystones overlying coal seams in Illinois: Ph.D. Thesis, Univ. Illinois, Urbana, Illinois, 143 pp.

Odom, I. E. and Parham, W. E. (1968) Petrography of Pennsylvanian underclays in Illinois and their application to some mineral industries: *Illinois Geol. Surv. Circ.* **429**, 36 pp.

Parham, W. E. (1964) Lateral clay mineral variations in certain Pennsylvanian underclays: in *Clays and Clay Minerals, Proc. 12th Natl. Conf., Atlanta, Georgia, 1963*, W. F. Bradley, ed., Pergamon Press, New York, 581–602.

Phillips, T. L. and DiMichele, W. A. (1981) Paleoecology of Middle Pennsylvanian age coal swamps in southern Illinois—Herrin Coal Member at Sahara Mine No. 6: in *Paleobotany, Paleoecology and Evolution,* K. J. Niklas, ed., Praeger, New York, 231–284.

Potter, P. E. and Glass, H. D. (1958) Petrology and sedimentation of the Pennsylvanian sediments in southern Illinois—a vertical profile: *Illinois Geol. Surv. Rept. Invest.* **204,** 60 pp.

Rao, C. P. and Gluskoter, H. J. (1973) Occurrence and distribution of minerals in Illinois coals: *Illinois Geol. Surv. Circ.* **476,** 56 pp.

Reynolds, R. C. (1980) Interstratified clay minerals: in *Crystal Structures of Clay Minerals and Their X-ray Identification,* G. W. Brindley and G. Brown, eds., Mineralogical Society, London, 249–303.

Rimmer, S. M. and Eberl, D. D. (1982) Origin of an underclay as revealed by vertical variations in mineralogy and chemistry: *Clays & Clay Minerals* **30,** 422–430.

Schultz, L. G. (1958) Petrology of underclays: *Geol. Soc. Amer. Bull.* **69,** 363–402.

Wanless, H. R., Tubb, J. B., Jr., Gednetz, D. E., and Weiner, J. L. (1963) Mapping sedimentary environments of Pennsylvanian cycles: *Geol. Soc. Amer. Bull.* **74,** 437–486.

Wanless, H. R. and Weller, J. M. (1932) Correlation and extent of Pennsylvanian cyclothems: *Geol. Soc. Amer. Bull.* **43,** 1003–1016.

Ward, C. R. (1977) Mineral matter in the Springfield-Harrisburg (No. 5) Coal Member in the Illinois basin: *Illinois Geol. Surv. Circ.* **498,** 35 pp.

Webb, D. K., Jr. (1961) Vertical variations in the clay mineralogy of sandstone, shale, and underclay members of Pennsylvanian cyclothems: Ph.D. Thesis, University of Illinois, Urbana, Illinois, 107 pp.

Willman, H. B. and Payne, J. N. (1942) Geology and mineral resources of the Marseilles, Ottawa, and Streator Quadrangles: *Illinois State Geol. Surv. Bull.* **66,** 388 pp.

Yoder, H. S. and Eugster, H. P. (1955) Synthetic and natural muscovites: *Geochim. Cosmochim. Acta* **8,** 225–280.

Proceedings of the International Clay Conference, Denver, 1985, L. G. Schultz, H. van Olphen, and F. A. Mumpton, eds.,
The Clay Minerals Society, Bloomington, Indiana, 105–110 (1987).

CLAY MINERALS AND SEDIMENTARY FACIES IN THE UPPER CARBONIFEROUS PENNINE BASIN, ENGLAND: A REVIEW

D. A. SPEARS

Department of Geology, University of Sheffield
Mappin Street, Sheffield S1 3JD, United Kingdom

Abstract—The Pennine basin evolved during Upper Carboniferous times from a deep-water basin in which fine-grained sediments and turbidites accumulated via prograding delta complexes to a fluviatile-dominated environment with extensive coal swamps. A detailed study of a marine incursion (*G. subcrenatum* Marine Band) into the fluviatile environment has shown a decrease in kaolinite upwards through the section from nonmarine to marine and increases in: (1) the sum of illite + mixed-layer illite/smectite (I/S); (b) the illite to I/S clay ratio; and (3) the sum of chlorite + vermiculite. These changes reflect a shift in the balance between locally derived sediment and primary, less altered sediment from the rocks in the hinterland. The same explanation is advanced for similar changes noted through much of the Namurian (Tansley borehole). Very slow rates of sedimentation apparently led to metal enrichment in some of the marine shales. The influence of sorting was detected in turbidite sandstones and in the associated mudrocks; kaolinite is proportionally more important and illite and mixed-layer clay less so, the coarser the sediment. In the coal-swamp environment the fine-grained sediment available in suspension was found to be locally derived and comparable with underclay mineralogy. This material was noted dispersed through coal seams and concentrated in mudrock partings within coals. Kaolinite beds (tonsteins) have been recorded associated with the coals and a volcanic origin proved. In addition, diagenetic kaolinite was found as void and fracture infills.

Key Words—Coal beds, Illite, Illite/smectite, Kaolinite, Pennine basin, Sedimentary facies, Tonstein.

INTRODUCTION

Carboniferous sedimentation in the north of England was greatly influenced by the presence of uplifted blocks and basins. The Pennine basin (Figure 1) evolved from a 'rift' to a 'sag' basin (Leeder, 1982) during Upper Carboniferous time. The early Namurian (Table 1) sediments in the basin are black shales (Edale Shales) overlain by sediments from a prograding delta complex (Kinderscout Grit delta), which includes distal and proximal turbidites. This delta is one of several deltaic advances which eliminated in stages the relatively deep water of the basin (Jones, 1980) and led to the establishment of fluviatile facies and extensive coal formation in the Westphalian Coal Measures (Table 1). The paleontological control on the sequence is excellent, both for biostratigraphy and depositional environments (Calver, 1969; Ramsbottom, 1969).

In the Pennine basin the first fine-grained sediments to be chemically analyzed were the economically important fireclays (Ennos and Scott, 1924), which were later the subject of pioneering X-ray powder diffraction studies by Brindley and Robinson (1947). The search for uranium reserves (Ponsford, 1955) and the introduction of γ-ray geophysical logs provided a further stimulus for research into fine-grained Coal Measures rocks. Nicholls and Loring (1962) were interested in the abundance and distribution of elements. Possible links between element concentrations in shales and paleosalinity were sought in Coal Measures rocks by

Curtis (1964) and Spears (1965) and, although their success was limited, some progress was subsequently made in understanding depositional and diagenetic processes (Spears and Sezgin, 1985). Nevertheless, much still remains to be learned about the fine-grained sediments in the Pennine basin. An economic stimulus for such research currently exists in that hydrocarbon reservoirs and source rocks are known, and the East Pennine coalfields (Yorkshire, Nottinghamshire and Derbyshire) have an assured future.

This paper aims to relate the mudrock clay mineralogy and geochemistry to the varied sedimentary environments of the Pennine basin during Upper Carboniferous time. Such relationships may be of general application to other basins because the evolution of the Pennine basin followed a common pattern (Leeder, 1982). The emphasis in the present paper is on the primary dispersion of clay minerals rather than their diagenetic modifications.

MARINE AND NONMARINE MUDROCKS

Clay mineralogy

The Tansley borehole (Figure 1), 34 km south of Sheffield, was drilled to prove the Carboniferous succession. The Namurian mudrock geochemistry and clay mineralogy were described by Spears and Amin (1981a). The marine shales contain more illite and illite/smectite (I/S) and less kaolinite and chlorite than the nonmarine mudrocks (Table 2). The higher illite +

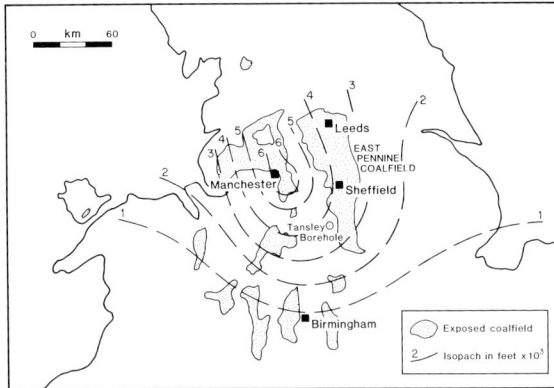

Figure 1. The Pennine basin, England, shown by the pattern of Westphalian A/B isopachs (after Leeder, 1982). Isopachs are at intervals of 1000 feet (~305 m).

size fractions. Figure 2, however, demonstrates that the difference in clay composition between marine and nonmarine mudrocks is not related to quartz content and thus to grain size, because quartz content and grain size are related.

The role of differential flocculation was considered in a study of the *G. subcrenatum* Marine Band and associated sediments (Table 1) by Spears and Sezgin (1985). The upwards sequence is underclay, coal, nonmarine shales, marine shales, and finally nonmarine shales. Through this sequence the kaolinite content decreases from bottom to top; the sum of illite + I/S, the ratio of illite to I/S, and the chlorite and vermiculite(?) contents, however, increase upwards. These clay changes, however, were gradual, unlike the paleosalinity changes based on the fauna, and therefore differential flocculation was eliminated as a control on the clay variation. The preferred explanation (Spears and Sezgin, 1985) is that of a progressive decrease in the amount of clay material derived from the mature soils in a flood-plain environment and an increase in the amount derived from a more distant and less-weathered upland source. The differences observed between nonmarine and marine mudrocks from the Tansley borehole could have a similar explanation of proximity to mature soils in the fluviatile environment for the nonmarine shales and of a greater contribution from the primary, upland source for the marine shales. One difficulty with this explanation is that the anticipated inverse relationship between kaolinite and chlorite was not observed (Spears and Amin, 1981a). Chlorite is not, however, a major component and in the marine shales the diagenetic formation of pyrite may have

I/S to kaolinite ratio for the marine shales is shown in Figure 2. The nonmarine mudrocks grade into siltstones and fine-grained sandstones (the mudrocks contain <30% quartz). Although nonmarine mudrocks are more abundant higher in the sequence, sufficient interdigitation of samples exists to demonstrate that the clay variation is related to depositional environment and is not a function of time. Differential flocculation is a possible explanation for the clay variation (Spears and Amin, 1981a). A grain size control is another possibility because in Namurian (Spears and Amin, 1981b) and Westphalian sediments (Spears and Taylor, 1972) the coarser grain-size fractions contain more kaolinite and chlorite and less illite and I/S than the finer grain-

Table 1. Outline classification of the Upper Carboniferous in the Pennine basin.[1]

		Northwest Europe		U.S.A.
		D		
		—		
		C		
	Westphalian	—		
	(Coal Measures)	B		Pennsylvanian
		—		
Upper		A		
			G. subcrenatum	
			Marine Band	
Carboniferous		G_1		
(Silesian)		—	Kinderscout Grit	
		R_2		
		R_1		
	Namurian	—	Mam Tor Beds	
	(Millstone Grit	H_2	Edale Shales	Mississippian
	Series)	H_1		
		—		
		E_2		
		E_1		

[1] Beds referred to in the text are shown. Letters subdividing the Westphalian are the stages and their names; those subdividing the Namurian are the goniatite index zones which define the stages but not their names. Stage names such as Yeadonian = G_1 are excluded for simplicity. From Spears and Amin (1981b).

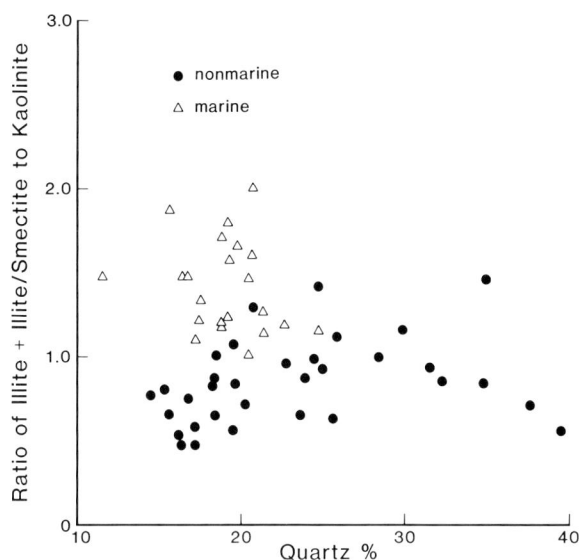

Figure 2. The ratio of illite + mixed-layer illite/smectite (I/S) to kaolinite vs. quartz concentrations in Namurian mudrocks and siltstones. The siltstones contain >30% quartz. All samples from the Tansley borehole.

Figure 3. Frequency distribution of multivariate means (13 trace elements as variables) plotted on a discriminant function line. Data for marine and nonmarine Namurian mudrocks and Mam Tor shales are plotted.

involved some destruction of chlorite (Ashby and Pearson, 1979).

Geochemistry

Marine and nonmarine Namurian mudrocks (Tansley borehole) can be distinguished by their geochemistry (Spears and Amin, 1981b). Pb and Cu make the greatest contribution to the difference between multivariate trace element means. The multivariate means are plotted on Figure 3 as a frequency distribution to show the separation between marine and nonmarine groups. The enrichment in marine rocks is due to paleosalinity, slow sedimentation rates, and the presence of reactive phases, including organic matter and oxyhydroxide material. Although complete discrimination between marine and nonmarine samples was achieved by Spears and Amin (1981a), this discrimination does not necessarily lead to a reliable indicator of paleosalinity, because paleosalinity was not the only control on metal enrichment. This point is illustrated by the data for the Mam Tor shales (from Amin, 1979) plotted

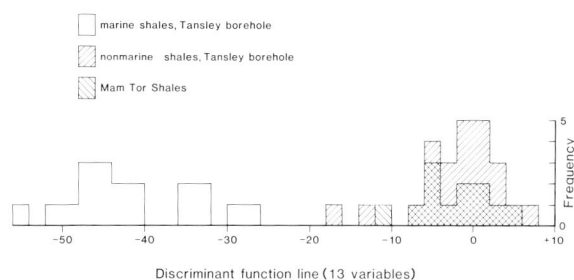

in Figure 3. Although the Mam Tor shales (Table 1) contain a marine fauna (Stevenson *et al.*, 1971), the samples plot with the nonmarine samples on Figure 3. The lack of detectable metal enrichment in the Mam Tor shales is attributed to an insufficiently slow rate of sedimentation. A comparison by Spears and Amin (1981a) of the Tansley samples with other black shales and modern anoxic marine muds led to the conclusions that the Tansley marine shales were deposited at a rate of about 0.03 cm/yr and that the rates of transfer of Pb and Cu from seawater into sediment was probably exceptionally high.

Comparing Mam Tor shales and Tansley nonmarine shales in a discriminant function analysis using trace elements (Amin, 1979) shows that significant differences exist, particularly for an element such as Cr, which is detrital in character and may therefore reflect differences in source rocks. Part of the Cr is likely to be in resistate minerals, the composition and distribution of which are largely unknown in mudrocks in spite of their greater preservation potential than in sandstones (Blatt, 1985). In the Pennine basin very little is known about feldspars in mudrocks apart from a record of their presence in whole-rock X-ray powder diffraction traces and in statistical analyses of geochemical data, which consistently reveal a positive relationship between the percentages of Na_2O (albite) and free silica (quartz). Geochemical relationships involving TiO_2 also led to a greater understanding of anatase and rutile distributions in the mudrocks (Spears and Kanaris-Sotiriou, 1979). Nevertheless, as Blatt (1985)

Table 2. Clay composition of some Namurian mudrocks (%).[1]

	Tansley borehole[1]		Mam Tor Beds[2]	
	Marine shales	Nonmarine mudrocks	Shales	Sandstones
Illite + I/S	31.0 ± 5	24.0 ± 5	23.0 ± 3	6.0 ± 2
Kaolinite	20.8 ± 5	29.9 ± 8	36.0 ± 4	18.0 ± 5
Chlorite	3.9 ± 1	5.3 ± 2	7.0 ± 1	3.0 ± 2
n	19	25	11	14

[1] From Spears and Amin (1981a).

noted, non-clay minerals in mudrocks have not been greatly utilized for solving petrological problems.

TURBIDITE SANDSTONES AND ASSOCIATED SHALES

Turbidite sandstones in the Pennine basin were first recognized by Allen (1960). Later workers (Reading, 1964; Walker, 1966; Collinson, 1969; McCabe, 1978) established the facies relationships for the turbidites. More recently, Aitkenhead (1977) provided a detailed description of turbidite sequences from a basinal area in Namurian and underlying Lower Carboniferous strata (Duffield borehole, SK 3428 4217).

The turbidite sandstones and associated shales in the Mam Tor Beds were studied by Spears and Amin (1981b) using outcrop samples. The proportion of kaolinite was found to be higher in the turbidites than in the associated shales and the proportion of illite and I/S to be lower (Table 2). Here, the differences were attributed to a grain size control, with a higher proportion of coarse clay minerals being present in the turbidites. Plots of Al_2O_3 vs. TiO_2, Al_2O_3 vs. K_2O, and K_2O vs. TiO_2 (Spears and Amin, 1981b) showed linear relationships for the sandstones which were attributed to systematic variations in the proportions of coarse and fine sediment. The shales were found to be chemically distinct from the sandstones and were therefore thought to be hemipelagic rather than turbiditic in origin. Although hemipelagic may be a suitable description, the shales are associated with turbidite sandstones and were therefore deposited nearer to a sediment source than the marine shales in the Tansley borehole. The more proximal character of the Mam Tor shales is seen in the clay mineral composition which is comparable with the nonmarine mudrocks in the Tansley borehole (Table 2). The geochemical plots referred to above show surprisingly little scatter of the sandstone samples from the linear relationships. Although the proportions of coarse and fine sediment in each turbidite unit may be variable, the composition of each fraction appears to be fixed. Lack of compositional variation might conceivably be a feature of turbidity currents originating from pro-delta slope deposits rather than directly from river discharge.

In many previous studies of ancient and modern turbidites the emphasis was on the structures and textures which could be identified directly in outcrop and core samples. One exception is the work of Kranck (1984) in which grain-size analyses were used to prove that unconsolidated turbidite units were continuously deposited from the same source sediment. Use of grain-size analyses is more difficult for turbidite sandstones because of the problems of disaggregation and diagenetic modification of original grain size. The work of Spears and Amin (1981b), however, on the Mam Tor Beds demonstrates that the geochemical approach has

potential including the possibility of obtaining original grain-size information. Two few samples were taken by Spears and Amin (1981b) through each turbidite unit to demonstrate continuous deposition from the same source sediment. This problem remains an area for future study, as is a more detailed examination of the turbidite muds in the Pennine basin described by Aitkenhead (1977).

CLAY MINERALS IN COALS AND ASSOCIATED CLAYSTONES

In the Pennine basin the maximum development of the fluviatile facies, and, hence, of coals, was during the Westphalian (Table 1). Coexisting with the coals are detrital and diagenetic clay minerals. The diagenetic minerals originated both from the slow breakdown of normal sedimentary material, including the detrital clays, and from the rapid breakdown of unstable volcanic material. One problem in analyzing clay minerals intimately admixed with the coal is that of separation. This problem has been overcome using low-temperature plasma ashing (Gluskoter, 1965). Direct methods of investigation without clay separation, mainly scanning electron microscopy and transmission electron microscopy are also finding increasing application in coal petrography.

The clay mineralogy of the *G. subcrenatum* horizon was described above. In addition to marine shales, the section includes a thin coal and an underclay, the latter consisting of poorly crystalline kaolinite, I/S, and illite, in that order of abundance. This particular underclay represents an extreme case of alteration; generally underclays in the Pennine basin contain less kaolinite, more illite, and a lower I/S to illite ratio. This clay mineral assemblage is one which developed in the coal swamp environment from the primary clay detritus and, where reworked, was incorporated into the peat. High-ash coals contain this assemblage, which commonly increases in amount close to the coal contacts. A complex balance apparently existed between the rate of supply of detrital clay and the rates of organic production and destruction.

Associated with coal seams of all ages are tonsteins, beds of well-crystallized kaolinite having considerable lateral extent. Fine-grained tonsteins are typically hard and have a characteristic conchoidal fracture. Although these features may be lacking in coarser grained tonsteins, the clay-rich character and lack of 'grittiness' in abrasion tests are retained. Alteration of air-borne volcanic ash has been proved for many tonsteins (Spears and Duff, 1985). In the Westphalian strata in the Pennine basin, tonsteins represent both local, basic volcanism and more distant, acid volcanism associated with Plinean eruptions (Spears and Kanaris-Sotiriou, 1979). K-bentonites were recorded by Trewin and Holdsworth (1972) in Namurian marine shales. Francis (1970) related the widespread, intermittent British Carboniferous igneous activity to crustal movements along the margins of the stable blocks. This extrusive and intrusive activity in central England is characterized by basic alkaline rocks (Harrison, 1977). The transition from I/S in the marine shales to kaolinite in the nonmarine environment was apparently controlled by the composition of the pore fluids and the respective clay stability fields. In the nonmarine environment ionic activities in the pore fluids were low and remained so provided that an open system was maintained by slow rate of sedimentation and thin ash falls. In limnic, nonmarine coal basins, the lateral passage from kaolinite to mixed-layer I/S (smectite precursor) corresponds to a limiting water depth beyond which vegetation was not established (Bouroz et al., 1983).

Kaolinite also precipitated within the coal seams from solutions which must have been derived from the diagenetic alteration of the normal sedimentary components (Spears and Caswell, 1986). This kaolinite is as highly crystalline as is that in the tonsteins, and thus differs from both the kaolinite in underclays and the detrital kaolinite in coal. Kaolinite fills voids in the coal and also occurs in the closely spaced joints (cleats). The latter kaolinite may be linked to diagenetic reactions in the sequence as a whole. These reactions are not considered here, except to note that in addition to the well-documented illitization of smectite and the consequent release of cations to the pore fluids, degradation of organic matter may also involve the release of Si and Al into pore solutions. If so, this reaction could be quantitatively important in mudrocks as well as in coals.

CONCLUSIONS

The Pennine basin evolved during Upper Carboniferous times from a 'rift' basin, in which marine shales accumulated, with turbidites and pro-delta sediments nearer the sediment source, to a 'sag' basin dominated by fluviatile, coal-bearing sediments. From the clay mineralogy and geochemistry of the mudrocks from a number of environments, the following conclusions can be drawn:

1. In the *G. subcrenatum* section the clay mineral assemblage changed from the underclay upwards through the section into the marine shales. The kaolinite content decreased and the contents of illite + I/S, chlorite, and vermiculite(?) and the ratio of illite to I/S all increased. The changes were gradual, not closely related to those of paleosalinity, and are interpreted as a gradual decrease in the proportion of clay derived from mature soils in a flood plain environment and increase in the clay originating from a less weathered, upland source.

2. A similar variation in clay mineralogy was recorded for Namurian marine and nonmarine mudrocks from the Tansley borehole. Proximity of the nonmarine mudrocks to the fluviatile environment is thought to account for the clay variation.

3. Trace elements are enriched in the marine mudrocks from the Tansley borehole compared with the nonmarine mudrocks. Enrichment is thought to be due to a combination of paleosalinity, slow rates of sedimentation, and suitable reactive phases. The marine Mam Tor shales are not enriched in trace elements, probably because the sedimentation rate was insufficiently slow.

4. In the turbidites from the Mam Tor Beds the proportion of kaolinite was found to be higher than in the associated shales and that of illite and I/S lower. These differences are attributed to a grain-size variation with more coarse clay being present in the turbidites.

5. The geochemistry of the turbidite sandstones shows a number of linear relationships which are thought to be related to grain size and clay content. The Mam Tor shales are chemically distinct from the turbidite sandstones, and a hemipelagic origin is preferred.

6. Reworking of mature sediment within the fluviatile environment and incorporation into the developing peat is thought to account for the close similarities between detrital clay minerals in Westphalian coals and in the underclays.

7. Highly crystalline, diagenetic kaolinite occurs in tonsteins associated with the coals. In the Pennine basin tonsteins are believed to have originated from volcanic ash falls of basic and acid composition; a distant source is postulated for the latter.

8. Highly crystalline, diagenetic kaolinite also occurs in the coals as an infill of voids and fractures.

REFERENCES

Aitkenhead, N. (1977) The Institute of Geological Sciences borehole at Duffield, Derbyshire: *Bull. Geol. Surv. G.B.* **No. 59**, 1–38.

Allen, J. R. L. (1960) The Mam Tor Sandstones: a 'turbidite' facies of the Namurian deltas of Derbyshire, England: *J. Sedim. Petrol.* **30**, 193–208.

Amin, M. A. (1979) *Geochemistry and mineralogy of Namurian sediments in the Pennine Basin, England:* Ph.D. Thesis, University of Sheffield, Sheffield, United Kingdom, 188 pp.

Ashby, D. A. and Pearson, M. J. (1979) Mineral distributions in sediments associated with the Alton Marine Band near Penistone, South Yorkshire: in *Proc. 6th Int. Clay Conf., Oxford, 1978,* M. M. Mortland and V. C. Farmer, eds., Elsevier, Amsterdam, 311–321.

Blatt, H. (1985) Provenance studies and mudrocks: *J. Sedim. Petrol.* **55**, 69–75.

Bouroz, A., Spears, D. A., and Arbey, F. (1983) Essai de synthese des donnees acquises sur la genese et l'evolution des marqueurs petrographiques dans les bassins houillers: *Soc. Geol. du Nord, Mem.* **16**, 114 pp.

Brindley, G. W. and Robinson, K. (1947) X-ray study of some kaolinite fireclays: *Trans. Br. Ceram. Soc.* **46**, 49–62.

Calver, M. A. (1969) Westphalian of Britain: in *Proc. 6th Int. Carboniferous Congress, Sheffield, 1967, Vol. 1,* Ernest van Aelst, Maastricht, Netherlands, 233–254.

Collinson, J. D. (1969) The sedimentology of the Grindslow Shales and the Kinderscout Grit: a deltaic complex in the Namurian of northern England: *J. Sedim. Petrol.* **30,** 194–221.

Curtis, C. D. (1964) Studies on the use of boron as a paleoenvironmental indicator: *Geochim. Cosmochim. Acta* **28,** 1125–1137.

Ennos, F. R. and Scott, A. (1924) Refractory materials: fireclays. Analyses and physical tests: *Mem. Geol. Surv. Spec. Rep. Miner. Resour. Gt. Br.* **28,** 81 pp.

Francis, E. H. (1970) Review of Carboniferous volcanism in England and Wales: *J. Earth Sci.* **8,** 41–56.

Gluskoter, H. J. (1965) Electronic low-temperature ashing of bituminous coal: *Fuel* **44,** 285–291.

Harrison, R. K. (1977) Petrology of the intrusive igneous rocks in the Duffield Borehole, Derbyshire: *Bull. Geol. Surv. G.B.* **No. 59,** 41–59.

Jones, C. M. (1980) Deltaic sedimentation in the Roaches Grit and associated sediments (Namurian R_2b) in the southwest Pennines. *Proc. Yorks. Geol. Soc.* **43,** 39–67.

Kranck, K. (1984) Grain size characteristics of turbidites: in *Fine Grained Sediments: Deep Water Processes and Facies,* D. A. V. Stow and D. J. W. Piper, eds., *Geological Society Special Publication* **No. 15,** Blackwell, London, 83–92.

Leeder, M. R. (1982) Upper Palaeozoic basins of the British Isles—Caledonide inheritance versus Hercynian plate margin processes: *J. Geol. Soc. Lond.* **139,** 479–491.

McCabe, P. J. (1978) The Kinderscoutian delta (Carboniferous) of northern England, a slope influence by density currents: in *Sedimentation in Submarine Canyons, Fans and Trenches,* D. J. Stanley and G. Kelling, eds., Dowden, Hutchinson and Ross, Stroudsburg, Pennsylvania, 116–126.

Nicholls, G. D. and Loring, D. H. (1962) The geochemistry of some British Carboniferous sediments: *Geochim. Cosmochim. Acta* **26,** 181–222.

Ponsford, D. R. A. (1955) Radioactive studies of some British sedimentary rocks: *Bull. Geol. Surv. G.B.* **No. 10,** 24–44.

Ramsbottom, W. H. C. (1969) The Namurian of Britain: in *Proc. 6th Int. Carboniferous Congress, Sheffield, 1967 Vol. 1,* Ernest van Aelst, Maastricht, The Netherlands, 219–232.

Reading, H. G. (1964) A review of the factors affecting the sedimentation of the Millstone Grit (Namurian) in the Central Pennines: in *Deltaic and Shallow Marine Deposits,* L. M. J. U. Van Straaten, ed., Elsevier, New York, 26–34.

Spears, D. A. (1965) Boron in some British Carboniferous sedimentary rocks: *Geochim. Cosmochim. Acta* **29,** 315–328.

Spears, D. A. and Amin, M. A. (1981a) Geochemistry and mineralogy of marine and non-marine Namurian black shales from the Tansley Borehole, Derbyshire, *Sedimentology* **28,** 407–417.

Spears, D. A. and Amin, M. A. (1981b) A mineralogical and geochemical study of turbidite sandstones and interbedded shales, Mam Tor, Derbyshire, U.K.: *Clay Miner.* **16,** 333–345.

Spears, D. A. and Caswell, S. A. (1986) Mineral matter in coals: cleat minerals and their origin in some coals from the English Midlands: *Coal Geol.* (in press).

Spears, D. A. and Duff, P. McL. (1985) Symposium S8 report, cinerites and tonsteins: in *Proc. 10th Int. Carboniferous Congress, Madrid, 1983, Vol. 4,* Instituto Geologico y Minero de Espana, Madrid, 171–173.

Spears, D. A. and Kanaris-Sotiriou, R. (1979) A geochemical and mineralogical investigation of some British and other European tonsteins: *Sedimentology* **26,** 407–425.

Spears, D. A. and Sezgin, H. I. (1985) Mineralogy and geochemistry of the *G. subcrenatum* Marine Band and associated coal-bearing sediments, Langsett, South Yorkshire: *J. Sedim. Petrol.* **55,** 570–578.

Spears, D. A. and Taylor, R. K. (1972) The influence of weathering on the composition and engineering properties of in-situ Coal Measures rocks: *Int. J. Rock Mech. Min. Sci.* **9,** 729–756.

Stevenson, I. P., Gaunt, G. D., Mitchell, M. A., Ramsbottom, W. H. C., Calver, M. A., and Harrison, R. K. (1971) Geology of the country around Chapel-en-le-Frith: *Mem. Geol. Surv. Gt. Br.,* 444 pp.

Trewin, N. H. and Holdsworth, B. K. (1972) Further K-bentonites from the Namurian of Staffordshire: *Proc. Yorks. Geol. Soc.* **39,** 73–91.

Walker, R. G. (1966) Shale grit and Grindslow shales: transition from turbidite to shallow water sediments in the Upper Carboniferous of northern England: *J. Sedim. Petrol.* **36,** 90–114.

Proceedings of the International Clay Conference, Denver, 1985, L. G. Schultz, H. van Olphen, and F. A. Mumpton, eds.,
The Clay Minerals Society, Bloomington, Indiana, 111–120 (1987).

CLAY MINERALOGY OF A WEATHERED GRANOPHYRE FROM NORTH QUEENSLAND, AUSTRALIA

A. Shayan, C. J. Lancucki, and S. J. Way

Commonwealth Scientific and Industrial Research Organization
Division of Building Research
P.O. Box 56, Highett, Victoria 3190, Australia

Abstract—The clay mineralogy of a pink granophyre of early Permian age, overlain locally by outcrops of Tertiary arkose, has been examined using thin section petrography, X-ray powder diffraction (XRD), electron probe microanalysis, measurement of exchangeable cations, and methylene blue adsorption. The rock consists of quartz, potassium feldspar, and plagioclase feldspar as major, and mica and chlorite as minor components. The feldspars show an increasing amount of alteration in slightly weathered and moderately weathered rocks. XRD showed that chlorite had partly altered to smectite in the slightly weathered rock, probably through a vermiculite intermediate phase. In the moderately weathered rock, chlorite is completely altered to smectite. The increase of methylene blue adsorption, total exchangeable cations, and the Mg/Ca ratio with weathering supports a progressive alteration of chlorite to smectite. Electron microprobe analyses of the clay phases in these rocks were variable, but indicated a considerable depletion in Mg (maximum MgO = 2.70%).

Kaolinite occurs mainly in the completely weathered rock at the surface (0–0.6-m depth), in highly to moderately weathered rock to a depth of 1 m, in a horizontal joint at a depth of 3.5 m, and in some, but not all, deeper samples of the highly to moderately weathered rock. Although plagioclase is presumeably the source of most of the kaolinite in these rocks, some of the kaolinite in samples from 0–1-m depth could have been contributed by weathered arkose that once covered the granophyre.

Key Words—Chlorite, Concrete aggregate, Diagenesis, Hydrothermal, Kaolinite, Methylene blue, Smectite, Vermiculite, Weathering.

INTRODUCTION

An early Permian granophyre (locally known as microgranite) which forms hills about 150 km south of Townsville, Queensland (20°39'S, 147°8'E, Figure 1), was proposed by the Queensland Water Resources Commission (QWRC) as a source of aggregate for concrete to be used in the construction of the Burdekin Falls Dam in northern Queensland, Australia. An important factor in the durability of an aggregate and its suitability for use in concrete is the degree to which primary minerals have altered, particularly the amount of swelling clay minerals that have formed in the rock. The strength of the rock, its resistance to dimensional changes due to water absorption, and its resistance to abrasion and salt crystallization all decrease with an increase of the clay content of the rock. Deterioration of concrete structures and concrete specimens made in the laboratory, due to the presence of swelling clay in the aggregate, has been documented in several publications (Stutterheim, 1954; Roper, 1960; Snowden and Edwards, 1962; Roper, 1974). In addition, Cole and Beresford (1980) demonstrated the adverse effect of swelling clays in basalt aggregates on concrete durability.

A simple test that has been employed to assess the overall swelling-clay content of potential source rocks for such applications is the methylene blue adsorption (MBA) test, originally proposed for determining the bentonite content of drilling mud (Jones, 1964), and later adopted for assessment of rocks (Sameshima *et al.,* 1978; Shayan *et al.,* 1984; Stewart and McCullough, 1985; Hosking and Pike, 1985). This test, however, does not provide information on the nature of clay minerals present nor on their influence on mechanical properties of the rock.

In the present paper, the clay minerals developed in the granophyre were characterized by X-ray powder diffraction (XRD), electron probe microanalysis (EPMA), measurement of exchangeable cations, and methylene blue adsorption in an effort to relate the clay content to various engineering properties of the granophyre aggregate.

MATERIALS AND METHODS

The area of the granophyre hills has an average rainfall of 685 mm, about 75% of which falls in summer (November to March). The maximum daily temperature occasionally reaches 40°C in December and January; the minimum daily temperature is in the range 10°–20°C throughout the year. Frost is rare in the area.

Five diamond drill cores (40-m long) were taken over an area of 100 × 200 m on one of the granophyre hills (Figure 1). Visually different parts of the cores were classified by officers of the Geological Survey of

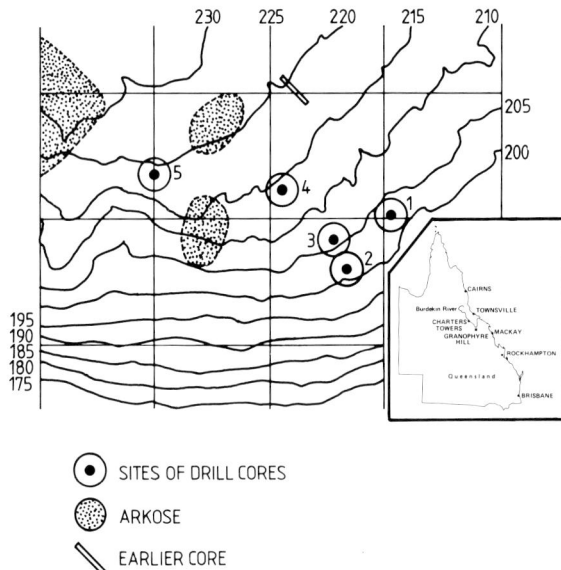

Figure 1. Contour map of the granophyre hill, showing the positions of the drill cores and arkose outcrops. The hill is south of the Burdekin river, as shown on the map of Queensland (insert). Contour numbers are elevations in meters; square grid is 50 × 50 m.

Queensland as fresh, slightly weathered, moderately weathered, highly weathered, and completely weathered, according to the definitions given in Appendix 1. Samples of these rock types were used by officers of the QWRC for routine engineering tests according to Australian Standard AS 1141 (see Appendix 2). Representative samples were examined by thin-section petrography; powders (<75 μm) were analyzed by X-ray powder diffraction (XRD).

XRD patterns were obtained in the range 2°–65°2θ using CuKα radiation at a scanning rate of ½°2θ/min. A separate pattern was obtained with the sample holder empty and served as a baseline above which the intensity of other peaks was measured at appropriate values of 2θ. The exchangeable cations were measured on the <75-μm-size rock powders by the method of Sameshima and Way (1982), which uses 0.1 M LaCl$_3$ at pH 7.7 as the extractant. The <2-μm-size fraction of the powdered samples was separated by sedimentation procedures for conventional XRD analyses of oriented specimens and for exchangeable-cation determinations.

To observe the distribution of the clay component, thin sections were stained with methylene blue by first etching them in concentrated orthophosphoric acid for 3 min and, after thoroughly washing them in water, immersing them in a 0.25% methylene blue solution for 15 s. Excess methylene blue was subsequently removed by washing the thin sections in a stream of water. Methylene blue adsorption (MBA) was measured quantitatively on rock powder (<75 μm) by the method of Jones (1964) and Sameshima et al. (1978), except that H$_2$O$_2$ and boiling H$_2$SO$_4$ treatments were unnecessary because the granophyre contained no organic matter. The MBA value (ml/g) is expressed as milliliters of 0.45% methylene blue solution sorbed by 1 g of rock powder.

Length-change measurements of granophyre rock prisms, exposed to wetting and drying conditions, were described by Shayan et al. (1984).

Specimens of the several rock types, embedded in resin, were polished for electron microprobe analysis. Due to the softness of the clay minerals, a good polish could not be achieved; consequently some error was introduced into the analysis. An Applied Research Laboratories EMX Microprobe equipped with an energy-dispersive spectrometer was employed and operated at 15-kV accelerating voltage, 1-nA beam current, and 100-s counting times. These conditions were even more favorable than those recommended by Velde (1984) for minimizing the loss of alkali elements from clay minerals due to interaction of the sample with the electron beam.

RESULTS

Petrography

The cores showed a considerable variation in the degree of alteration along the core length, particularly for those that intersected faults or shear zones; here, highly altered rock was found as deep as 36–37 m. The least altered rock was purple-pink; moderately to highly weathered rock was pale yellow-brown. Photomicrographs of thin sections of fresh, slightly weathered, and moderately weathered rocks are shown in Figure 2. On the basis of thin section examination, the unaltered granophyre is fine grained and consists of a micrographic intergrowth of quartz and orthoclase (55%), euhedral to subhedral albite (25%), discrete quartz grains (10%), and green, pleochroic chlorite having very low birefringence (5%). Some voids and opaque minerals are also present.

The XRD intensities of the basal reflections of the secondary minerals (Figure 3) illustrate the trend in mineralogy of core 1 with depth. Other cores showed more or less similar patterns, varying only in the exact positions of the several weathered zones. Samples mainly from core 1, therefore, were used for detailed clay mineralogy.

X-ray powder diffraction analyses

Representative XRD patterns of untreated and treated (glycerol, dimethyl sulfoxide, and heating at 580°C) samples are shown in Figures 4–8.

Electron microprobe analyses

Electron microprobe analyses and recalculated structural formulae of chlorites and smectites in the fresh

(36-m depth) and slightly weathered (28-m depth) rocks are listed in Table 1; those for smectites in the moderately weathered (1-m depth) rock and a relatively pure, green clay found in the joints of moderately weathered rock are listed in Table 2. Total iron is listed either as FeO (for chlorite) or Fe_2O_3 (for smectite); however, both oxidation states could have been present, particularly in the intermediate transition phase. The analyses in Tables 1 and 2 total less than 100% because the water content of these materials was not determined. The relatively rough nature of the surfaces of clay patches and the included voids could also have contributed to these low totals, but this contribution must have been minor because the deviation from 100% is generally close to the theoretical water content of the materials analyzed. The analysis of the green clay (Table 2), which totals 89% and which was made on a disk pressed under vacuum (thereby losing some absorbed water) supports this argument.

Other analyses

Exchangeable cations, determined on <75-μm rock powder from core 1, are plotted against depth in Figure 9; exchangeable cations, determined on the <2-μm fraction, are listed in Table 3. Methylene blue adsorption data were reported previously by Shayan et al. (1984).

Engineering test results, obtained by officers of QWRC according to the Australian Standard AS 1141, of aggregates prepared from rocks having different degrees of alteration are listed in Table 4. Length-change data of granophyre rock prisms exposed to wetting-drying cycles were also reported previously by Shayan et al. (1984).

DISCUSSION

Petrology

Photomicrographs in Figure 2 show some differences among fresh, slightly weathered, and moderately weathered rocks. Green pleochroic chlorite (Ch) in the fresh rock (Figure 2a) apparently formed by an earlier hydrothermal alteration of biotite. Remnants of largely altered biotite crystals are locally present. In this rock, the feldspars are cloudy, and the plagioclase crystals are partly altered to sericite (F + S). Discrete quartz grains (Q), nonpleochroic and birefringent, exfoliated (altered (?)) chlorite (Che), and a patch of brown carbonate (C) are also present.

In the slightly weathered rock (Figure 2b), the feldspar grains (F) are cloudier because of sericitic alterations, and a dusty, clay-like material coats some of the grains. Our examination of many thin sections showed that the proportion of exfoliated chlorite (vermiculite?) relative to the green, nonexfoliated chlorite is larger in the slightly weathered rock than in the fresh rock.

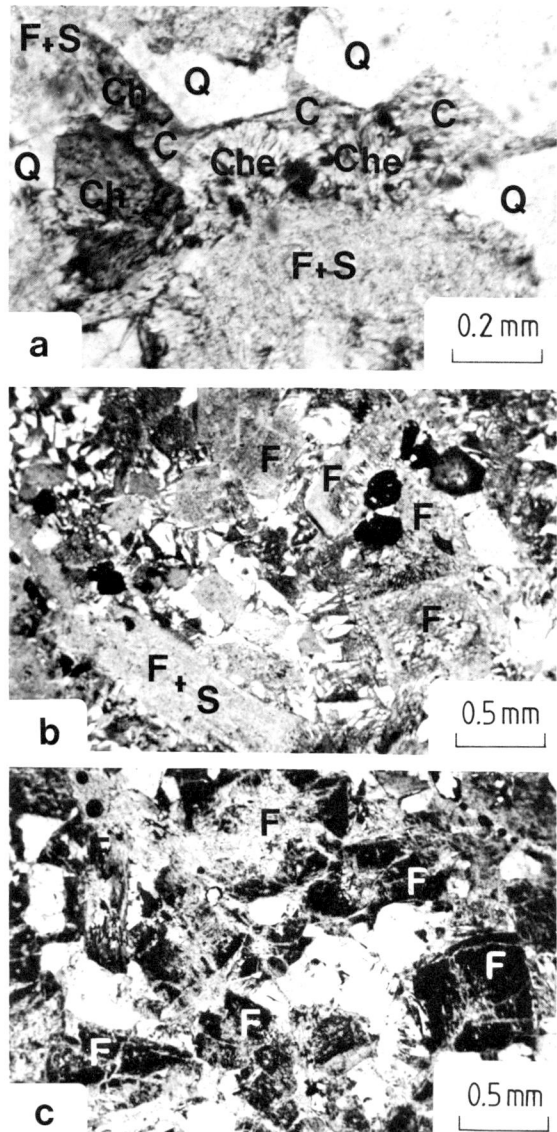

Figure 2. Photomicrographs taken with plane-polarized light showing thin sections of (a) fresh rock containing chlorite (Ch), exfoliated chlorite (Che), plagioclase feldspar with sericitic alterations (F + S), and discrete quartz grains (Q); (b) slightly weathered rock containing plagioclase crystals (F) which are more intensely altered and cloudier than in the fresh rock, and those with sericitic alteration (F + S); the white grains are quartz; (c) moderately weathered pale yellow-brown rock showing severely altered feldspar crystals (F), visibly altered at the grain boundaries and within twinning planes.

In the moderately weathered rock the chlorite is completely altered to smectite; feldspar grains also show extensive alterations (Figure 2c) at grain boundaries and within twinning planes. Some very fine grained opaque materials were apparently produced as a result of the feldspar alteration and oxidation of its small Fe content, giving the crystals a dark appearance under

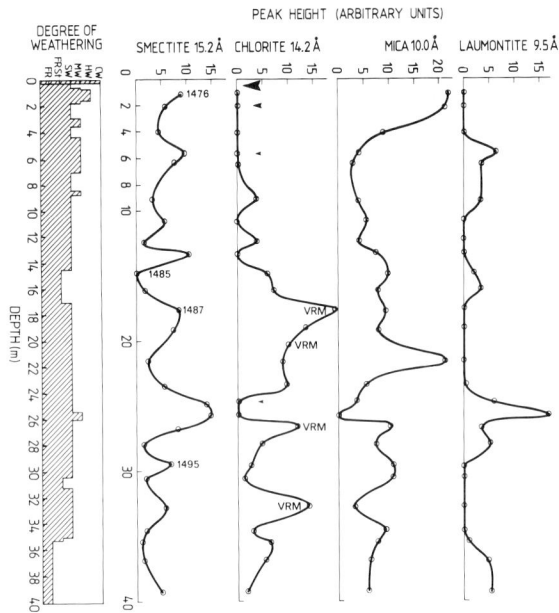

Figure 3. Plot of the intensity above the background of the basal reflection of the designated mineral against depth, for core 1, showing qualitative variation of the minerals with depth. Intense smectite and laumontite peaks are associated with moderately weathered rocks and fracture zones as shown by the borehole log at the left-hand side. Intense chlorite peaks (VRM) indicate presence of vermiculite as well as chlorite. Arrows at various depths indicate presence of kaolinite; relative abundance is shown by the size of arrow. The mica is largely sericite. Numbers on the smectite profile are sample numbers described in the text.

Figure 4. X-ray powder diffraction patterns of completely weathered rock from 0.4–0.6-m depth showing smectite (15.3 Å), mica (10 Å), and kaolinite (7.2 Å) in the air-dried sample. The d-value of kaolinite expands to an 11.2-Å on dimethyl sulfoxide treatment; 18-Å peak represents expanded smectite.

plane-polarized light. Regions of dusty, clay-like material are present in the highly altered feldspar. Staining the thin sections with methylene blue showed that these regions, surfaces, and fractures in the plagioclase strongly sorbed the dye because of the presence of the clay. The plagioclase crystals in the slightly weathered rock were also stained by the methylene blue to some extent, but hardly at all in the fresh rock. Powder of fresh, slightly, and moderately weathered rock (<75 μm) absorbed an average of 0.5, 1.0, and 5.0 ml/g methylene blue from a 0.45% solution, respectively, indicating the presence of more clay in the more extensively weathered rocks. Figure 2c illustrates the most severe alteration of the moderately weathered rock from a fault zone in which material has been extensively weathered. Other moderately weathered rocks were intermediate between pale brown and purple-pink. Their plagioclase crystals showed no internal cracking such as can be seen in Figure 2c; however, smectite was identified in these rocks (*vide infra*).

X-ray powder diffraction analysis

XRD indicated quartz, orthoclase, and albite as major components and muscovite, chlorite, and/or smec-

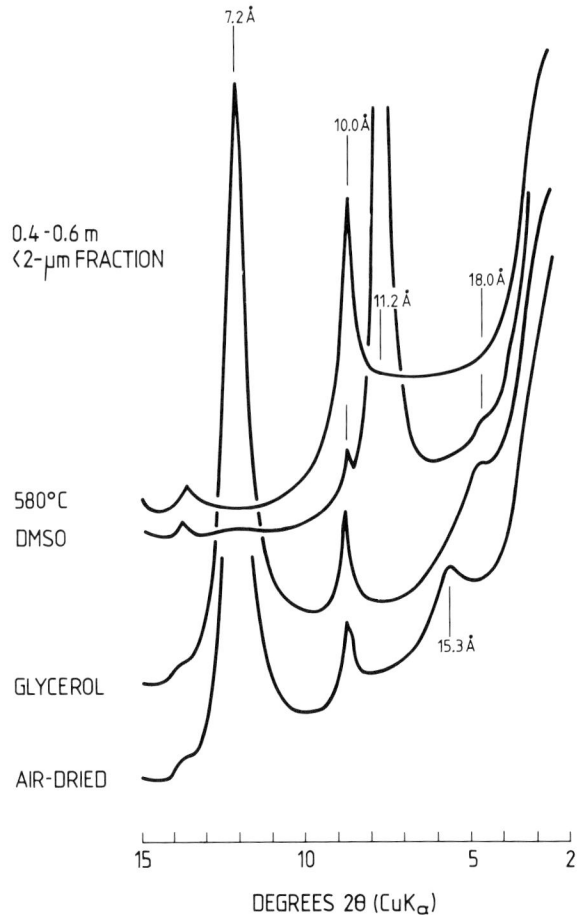

tite, and locally laumontite as minor components of the granophyre. Figure 3 suggests that close to the surface, chlorite has completely altered to smectite. The increased mica (sericite) content and the presence of kaolinite suggest feldspar alteration and weathering, respectively. Abrupt increases in the intensity of the XRD reflections of smectite and laumontite at particular depths reflect the presence of faults and shear zones, in which intense weathering has completely altered the chlorite to smectite. In such zones, smectite and laumontite may be present as narrow veins; laumontite veins 2–3 mm wide are also present in some large masses of tuff surrounding the granophyre hills. Both forms of laumontite could have been introduced from extraneous sources, perhaps during some later hydrothermal activity. Samples from intervals labeled VRM (Figure 3) and that show intense 14.2-Å peaks are vermiculite-rich (*vide infra*).

The clay mineralogy of the completely weathered

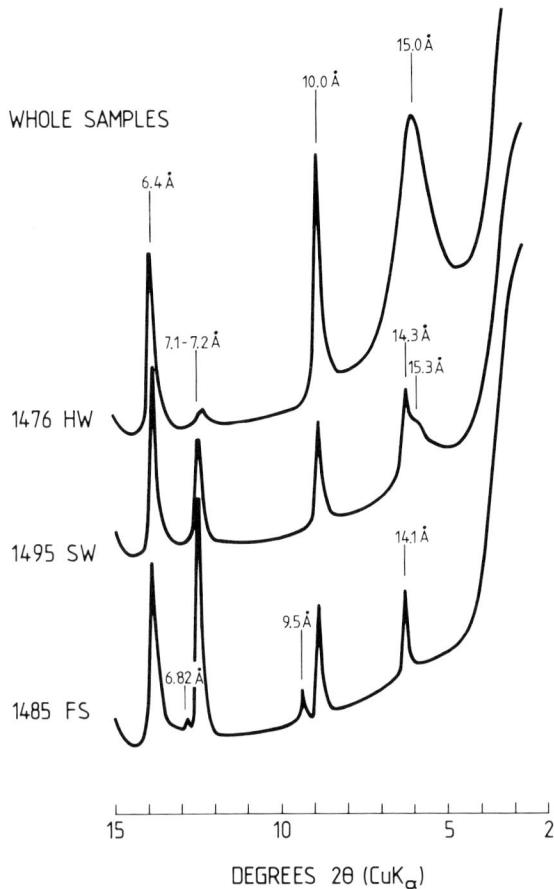

Figure 5. X-ray powder diffraction patterns of fresh rock (FS) containing fresh chlorite; slightly weathered rock (SW) showing development of smectite (15.3 Å); highly weathered rock (HW) containing smectite and probably kaolinite (7.1–7.2 Å). Mica (10 Å) and feldspar (6.4 Å) are also present. The small amount of laumontite (9.5 and 6.8 Å) in the fresh rock is largely in fine cracks.

Figure 6. X-ray powder diffraction patterns of a slightly weathered rock containing smectite and chlorite and some mica (see text).

rock at 0–0.6-m depth is similar to that of the rock at 0.4–0.6-m depth (Figure 4). Kaolinite (7.2 Å) is the dominant clay mineral, and its basal spacing expands to 11.2 Å with dimethyl sulfoxide (DMSO) treatment (Garcia and Camazano, 1968). Some smectite and mica are also present. The amount of kaolinite in the highly weathered, but still coherent, rock at 1-m depth is much less than that in the completely weathered surface material; at 6-m depth only a trace of kaolinite, just detectable by XRD, is present. A severely fractured zone at 24–26-m depth is the only other region to contain a trace of kaolinite. Tertiary arkose which once covered extensive areas in the region is still present as isolated outcrops. It contains large amounts of smectite and kaolinite and may have contributed these minerals to the surface layer of the granophyre, in addition to that which formed by the alteration of feldspar and chlorite, respectively. XRD patterns of the slightly weathered

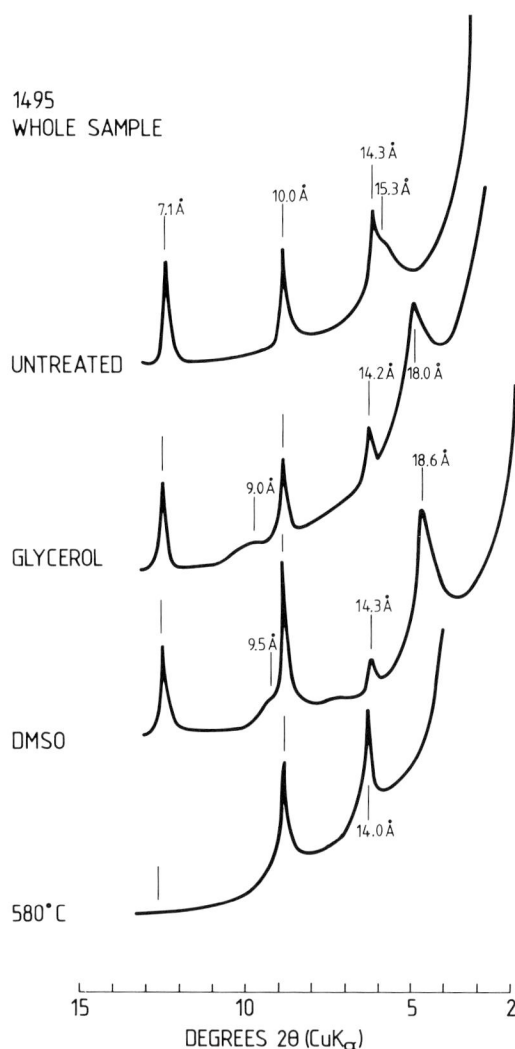

rock (Figure 2b) show a trace of kaolinite; those of the moderately weathered rock (Figure 2c) (both taken from the earlier core located in Figure 1) show moderate amounts of kaolinite and large amounts of smectite.

The influence of weathering on the clay mineralogy of the rocks is evident from Figure 5. The fresh rock (sample 1485 FS) contains Fe-rich chlorite for which the intensity of the 002 reflection (7.1 Å) is much greater than that of the 001 reflection (14.1 Å). Some mica (10 Å) and a small amount of laumontite (9.5 and 6.8 Å) are also present. The peak at 6.4 Å is due to feldspar. In the slightly weathered rock (sample 1495 SW), a shoulder at 15.3 Å indicates the presence of a small amount of smectite and/or vermiculite. On further weathering (sample 1476 HW), the chlorite has apparently completely altered to smectite. The weak peak

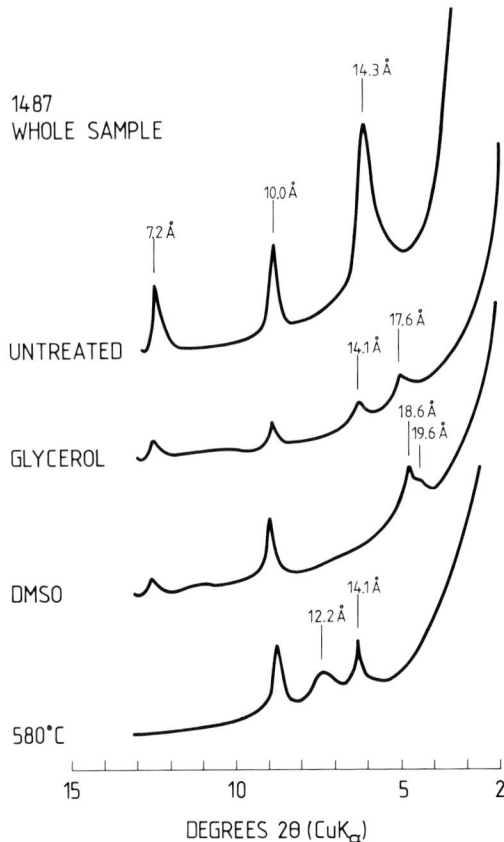

Figure 7. X-ray powder diffraction patterns of a slightly weathered rock that shows an intense 14.3-Å and a weak 7.2-Å spacing for the untreated sample. This sample contains smectite, chlorite, and a random interstratification of chlorite/vermiculite (see text).

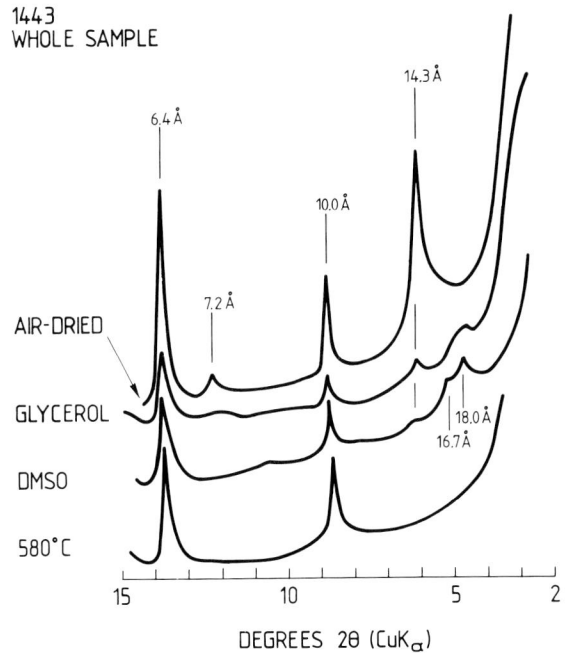

Figure 8. X-ray powder diffraction patterns of a slightly weathered sample that contained vermiculite and smectite, as shown in Figure 7, but without chlorite. The 6.4-Å spacing is due to feldspar.

at 7.1–7.2 Å in the XRD pattern of this sample is due to kaolinite, because the spacing expanded to 11.2 Å with DMSO treatment (not shown). The 14.1- and 7.1-Å spacings of the chlorite were not affected by this treatment nor by a glycerol treatment. The presence of smectite in the slightly weathered rock (sample 1495 SW) is indicated in the XRD patterns of the untreated and the glycerol- and DMSO-treated samples shown in Figure 6. The characteristic 15.3-Å spacing, in the pattern of the untreated sample, expanded to 18 and 18.6 Å by glycerol and DMSO treatments, respectively; the chlorite component was unaffected. Vermiculite, like smectite, can also be intercalated with DMSO (expanding to 17–18 Å), but not with glycerol. Therefore, the weaker 14.3-Å spacing in the DMSO-treated sample relative to that in the glycerol-treated sample likely resulted from the expansion of small amounts of vermiculite during the DMSO treatment. Heating at 580°C collapsed the smectite and vermiculite components (10 Å) and destroyed the 7.1-Å spacing of the Fe-rich chlorite.

The transition from chlorite to smectite, thus, appears to have taken place through a vermiculite intermediate phase. The 14.2-Å peak in the XRD patterns of samples taken from those intervals in Figure 3 that showed both a large 14.2-Å peak and a large amount of smectite were found to be due both to vermiculite and to chlorite. These samples were mainly from depths of 17–20, ~27, and ~33 m, although some vermiculite was also found in samples from depths of 35 and 39 m. These samples, unlike those containing fresh chlorite, gave XRD patterns with a weak 7.2-Å spacing, indicating the presence of small amounts of chlorite. Figure 7 illustrates this feature for the untreated sample and also shows XRD patterns of treated sample 1487 from a depth of ~18 m. In this sample the absence of kaolinite is indicated by the lack of an 11.2-Å peak in the XRD pattern of the DMSO-treated material. The presence of chlorite is indicated by the 14.1- and 7.2-Å peaks in the XRD pattern of the glycerol-treated sample, as well as by the 7.2-Å peak in the DMSO-treated and the 14.1-Å peak in the 580°C heat-treated sample. The 14.1-Å peak of chlorite must have been too small in the pattern of the DMSO-treated material to be registered. Inasmuch as vermiculite does not expand by glycerol treatment, the 17.6-Å spacing in the pattern of the glycerol-treated sample must be due to smectite. The expanded spacings at 19.6 and 18.6 Å in the pattern of the DMSO-treated sample are prob-

Table 1. Electron probe analysis of chlorites and smectites in fresh and slightly weathered granophyre.

	Fresh chlorites		Altered chlorite	Smectite in vein	Smectite in vesicle	
SiO$_2$	25.69	30.40	35.41	44.82	53.16	51.78
TiO$_2$	0.04	0.12	0.21	0.05	0.13	0.12
Al$_2$O$_3$	18.45	17.23	16.09	15.87	19.40	19.70
Fe$_2$O$_3$[1]	—	—	13.34	21.78	4.54	8.60
			(or)			
FeO[1]	21.69	16.99	12.00	—	—	—
MnO	0.89	0.70	0.55	—	0.02	—
MgO	14.16	13.28	6.40	1.58	0.94	1.53
CaO	—	0.85	0.87	2.47[2]	0.84	1.60
Na$_2$O	1.32	0.61	2.92[2]	0.75	1.65	1.38
K$_2$O	—	0.14	0.74	0.16	0.52	0.17
Total	82.24	80.32	75.19	85.30	81.20	84.02
Recalculated formulae: 22 oxygens for smectite and 28 oxygens for chlorite						
Si	5.74	6.67	6.05	6.65	7.79	7.41
Al	2.26	1.33	1.95	1.35	0.21	0.59
Total	8.0	8.0	8.0	8.0	8.0	8.0
Ti	0.01	0.02	0.03	0.01	0.01	0.01
Al	2.60	3.13	1.28	1.43	3.14	2.73
Fe	4.04	3.12	1.71	2.43	0.50	0.93
Mn	0.17	0.13	0.08	—	—	—
Mg	4.71	4.34	0.90	0.13	0.21	0.33
Total	11.53	10.14	4.0	4.0	3.86	4.0
Mg	—	—	0.73	0.22	—	—
Ca	—	0.20	0.16	0.39	0.13	0.25
Na	0.57	0.26	0.97	0.21	0.47	0.38
K	—	0.04	0.16	0.03	0.10	0.03
Total	12.11	11.24				
Σ exch. cations	—	—	2.02	0.85	0.70	0.66

[1] Total Fe as FeO in the chlorites and as Fe$_2$O$_3$ in the smectites.
[2] High value probably from contamination with albite (Na$_2$O) and laumontite (CaO).

ably due to smectite and vermiculite, respectively. Although the Na-saturated vermiculite used by Kirkman (1974) showed no spacing greater than 14.7 Å after DMSO treatment, the spacing of a macrocrystalline vermiculite ground to <75 μm and treated with DMSO (Garcia and Camazano, 1968) expanded from 14.3 to 17.9 Å. The basal spacing of the smectites treated with DMSO by Kirkman (1974) expanded to 18.8 and 20.5 Å and, like those in the DMSO pattern of Figure 7, were larger than those obtained by glycerol treatment. Brindley (1980) showed that the basal spacing of DMSO-treated smectite depended on the type of interlayer cation, which may be the reason for different d-spacings obtained by different workers using different smectites.

An interesting feature of the XRD pattern of the sample heated to 580°C (Figure 7) is the presence of a 12.2-Å reflection, which is probably the result of an interaction of the 14.1-Å basal spacing of chlorite and the ~10-Å basal spacing of heated vermiculite in a random interstratification. A similar spacing, however, can result from a random interstratification of chlorite/smectite (C/S), but the latter would show a 16-Å spacing (002) in the XRD pattern of a glycerol-treated sample. Because no 12.2-Å peak was observed in the XRD

patterns of samples which contained chlorite and smectite alone, e.g., those similar to sample 1495, the former explanation is more likely. Vermiculite-containing samples from 27-m- and 33-m-depths also showed a 12.2-Å spacing after they had been treated to 580°C.

Transformation of chlorite to regularly (and subsequently randomly) interstratified chlorite/vermiculite (C/V) has been reported from various geological environments. For example, Nakamuta (1981) found C/V to have formed by the preferential removal of alternate hydroxide sheets from chlorite during the weathering of a talc-chlorite vein. Similarly, Senkayi et al. (1981) found that the artificial weathering of Fe-rich chlorite produced C/V and C/S intermediates and finally smectite. The transformation of chlorite to C/V in soils formed on "glacialmarine" drift was also reported by Pevear et al. (1984). In their study of soil profiles derived from mafic rocks, Rice et al. (1985) found that the weathering of chlorite produced C/V which, in turn, altered to randomly interstratified C/V.

In the XRD pattern of a more typical vermiculite, from drill core 4 (Figure 8), the intensities of the strong 14.3-Å and the very weak 7.2-Å peaks are reversed compared with the corresponding peaks of the Fe-rich chlorite. A small amount of smectite is indicated in

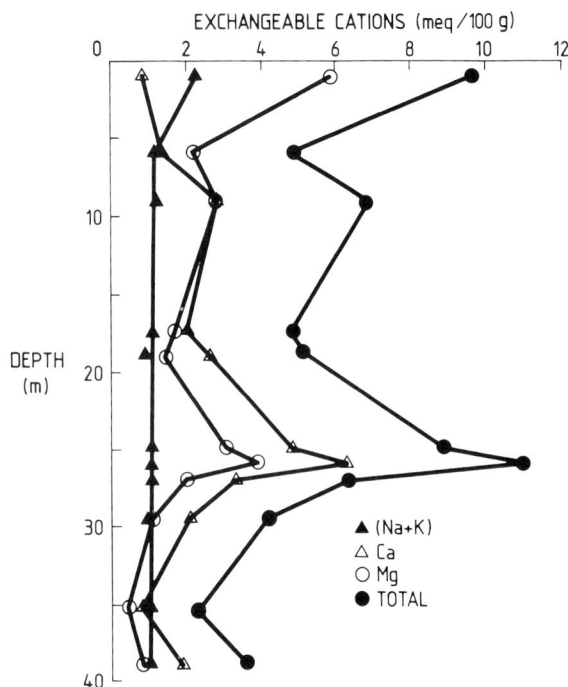

Figure 9. Exchangeable cations vs. depth for drill core 1, showing a larger Mg/Ca ratio above 9-m depth. The greater amount of exchangeable Mg near the surface suggests a more extensive breakdown of chlorite than at depth.

Table 2. Electron probe analysis of smectites in moderately weathered granophyre.

	Beidellite		Nontronite		Green clay from joints[2]
	1	2	3	4	
SiO_2	51.78	45.42	42.18	46.73	51.62
TiO_2	–	–	–	–	0.04
Al_2O_3	19.53	27.89	11.35	16.18	14.33
Fe_2O_3[1]	10.18	8.39	29.28	19.89	19.12
MnO	–	–	0.07	–	0.02
MgO	2.73	0.52	1.16	1.41	2.04
CaO	0.87	0.36	0.94	0.63	1.48
Na_2O	0.50	0.61	0.95	0.73	0.48
K_2O	–	0.12	–	0.13	0.06
Total	85.59	83.31	86.31	85.70	89.19
Recalculated formula on the basis of 22 oxygens					
Si	7.35	6.60	6.57	6.94	7.31
Al	0.65	1.40	1.43	1.06	0.69
Total	8.00	8.00	8.00	8.00	8.00
Ti	–	–	–	–	–
Al	2.61	3.37	0.66	1.78	1.70
Fe	1.09	0.92	3.43	2.22	2.04
Mn	–	–	0.01	–	–
Mg	0.30	–	–	–	0.26
Total	4.00	4.29	4.10	4.00	4.00
Mg	0.28	0.11	0.27	0.31	0.18
Ca	0.13	0.06	0.16	0.10	0.22
Na	0.14	0.17	0.29	0.21	0.13
K	–	0.02	–	0.02	0.01
Total	0.55	0.36	0.72	0.64	0.54

[1] Total Fe as Fe_2O_3.
[2] Clay pressed under vacuum into a disk for electron probe microanalysis.

the pattern of the glycerol-treated sample. Although the XRD pattern of DMSO-treated sample shows a very weak peak at about 14.3 Å, the lack of a 7.2-Å reflection in this pattern and of a 14.3–14.0-Å reflection in the 580°C pattern indicate that the sample contained vermiculite and smectite but no chlorite.

Electron microprobe analyses

Fresh chlorite containing about 25% SiO_2 (Table 1) is not common; most chlorites analyzed contained about 30% SiO_2. The enrichment of SiO_2 was apparently accompanied by a loss of Fe and, to some extent, a loss of Al and Mg. These compositions are similar to those of chlorites that replace igneous biotite (Parry and Downey, 1982). Although the chlorite compositions in Table 1 are poorer in Mg than the biotite-derived chlorites, the ratio of Mg/(Fe + Mg) is in the same range of 0.52–0.60 for both types of material, indicating a structural similarity between them.

The altered chlorite listed in Table 1 is from a sample, the XRD pattern of which showed an asymmetrical 001 reflection at about 14.2 Å. The composition of this chlorite shows a further enrichment of SiO_2 and a depletion of Fe and Mg relative to the composition of the fresh chlorites (Table 1). Whereas the compositions listed for the fresh chlorites yield structural formulae compatible with a chlorite structure, the analysis of the

altered chlorite does not, nor does it yield a realistic structural formula for smectite (high value for the exchange cations). The analysis probably is of a mixture of phases, or randomly interstratified C/V. Table 1 also shows that even in the slightly weathered rock chlorite has completely altered to smectite. Moreover, Mg is highly depleted in these smectites. The structural formulae for the smectites in the vein and in the vesicle indicate the dioctahedral nature of these clays.

Moderately weathered rock (6-m depth) that contains smectite as the sole clay mineral shows a range of composition for the clay material (Table 2). Analyses 1 and 4 and that of the green clay, however, yield reasonable structural formulae, compatible with dioctahedral smectite; the ratio Fe/Al varies considerably, suggesting Fe-rich beidellite (analysis 1) and nontronite (analyses 4 and green clay). Analyses 2 and 3 yield excess cations for the octahedral layer, suggesting the probable presence of separate alumina and iron oxide phases, respectively. The presence of such phases reflects the severity of alteration in the moderately weathered rock and local differences in the chemical environment during the alteration process.

Table 3. Exchangeable cations (meq/100 g) in the <2-µm fraction of rock powder.

Rock type	Depth (m)	Na	K	Ca	Mg	Total (CEC)
Moderately	6.0	1.38	1.24	9.71	14.00	26.3
weathered	26.0	0.58	0.84	16.20	10.00	27.6
Slightly	9.0	0.94	1.18	8.99	9.27	20.4
weathered	19.0	1.20	1.00	7.29	6.93	16.4
	29.5	1.26	1.11	9.05	6.12	17.5
Fresh	28.0	0.75	1.06	6.60	3.58	12.0
	35.5	1.05	1.17	5.38	4.00	11.7

Table 4. Results of aggregates testing[1] on bulk samples of the granophyre.

Weathering grade	Bulk density (kg/m³)	Water adsorption (%)	ACV[2] (% loss)	LA[3] (% loss)	Soundness test[4] (% loss)
Moderately weathered	2435	2.5	22.4	35	37
Slightly weathered	2495	1.8	17	24	27
Fresh	2510	1.5	—	22	29

[1] Obtained by the Queensland Water Resources Commission according to the Australian Standard AS 1141 (see Appendix 2).
[2] Aggregate crushing value.
[3] Los Angeles abrasion test.
[4] Sodium sulfate soundness test.

Exchangeable cations

Figure 9 shows that the amount of exchangeable Na and K is small and almost constant throughout core 1. Whereas the amount of exchangeable Mg exceeds that of Ca above a depth of 9 m, the situation is reversed deeper in the profile, and Ca dominates the exchangeable cations. Where chlorite has completely weathered to smectite (Figure 3), exchangeable Mg exceeds exchangeable Ca. The weathering of chlorite to Mg-poor smectite (Table 2) probably provided sufficient Mg to dominate the exchange sites. The source of the exchangeable Ca was likely the solution from which laumontite formed. The largest amount of exchangeable Ca corresponds to the fracture zone at about 26 m depth (Figure 3), where the largest amount of laumontite also was noted. Although laumontite, and zeolites in general, have large cation-exchange capacities (CEC), they do not contribute significantly to the measured exchangeable cations if 0.1 M $LaCl_3$ is used as the extracting solution, apparently due to the energy of dehydration and diffusion of La^{3+} into zeolite structures (Sameshima and Way, 1982). In the method employed in the present study, soluble salts were also removed with 10% ethylene glycol in ethanol before the extraction of exchangeable cations. Thus, the measured exchangeable Ca is that extracted from the exchange sites on the clay mineral.

Table 3 lists typical exchangeable-cation contents of the <2-µm fraction of the several rock types. The CEC values increase with increased weathering and show that the ratio of exchangeable Mg + Ca is greater at the surface because of chlorite transformation into Mg-poor smectite. The relatively high CEC of the fresh rock (which contains chlorite) is probably due to the preferential removal of chlorite and the concentration of small amounts of sericite in the <2-µm fraction during the sedimentation procedure.

Implications of clay development in the rock

The data in Table 4 illustrate the adverse effects of weathering on the engineering properties of aggregates made from these several types of weathered rocks. Bulk density decreases and water absorption increases be-

cause of increasing clay content with weathering. Resistance to crushing (ACV), abrasion (LA), and salt crystallization (soundness test) decrease (higher losses) with increased weathering. Large decreases in these last properties obtained for the moderately weathered rock indicated that it is unsuitable for use as an aggregate in concrete.

As noted above, powders of the moderately weathered rocks sorbed more methylene blue dye (5.0 ml/g) than those of the slightly weathered (1.0 ml/g) and fresh (0.5 ml/g) rocks. Shayan et al. (1984) derived the following regression equation for 171 samples of the granophyre:

$$Z = -2.26 + 6.10\chi, \tag{1}$$

where Z = smectite XRD peak height at 15.2 Å, and χ = ml/g of methylene blue sorbed from a 0.45% solution. This equation had a correlation coefficient of .946, highly significant at the 0.1% level. Shayan et al. (1984) also derived a second regression equation for the relationship between rock shrinkage (on drying wet rock prisms) and methylene blue dye sorption to serve as a measure of the clay content for 123 samples of the granophyre:

$$Y = -0.003 + 0.019\chi, \tag{2}$$

where Y = rock shrinkage (%) and χ = methylene blue adsorption (ml/g). This equation has a correlation coefficient of .78 and is significant at the 0.1% level.

Rocks having drying shrinkage values >0.03% are dimensionally unstable and unsuitable for use in structural concrete because they induce cracking. From Eq. (2), a drying shrinkage (Y) of 0.03% corresponds to a value of 1.74 ml/g of methylene blue. All the moderately weathered rocks sorbed larger amounts of methylene blue, indicating their dimensional instability, whereas the fresh and the slightly weathered rocks sorbed <1.5 ml/g methylene blue and were dimensionally stable and thus suitable for use in concrete.

ACKNOWLEDGMENTS

The authors thank the Queensland Water Resources Commission for financial support and Peter Westgate, David Ritchie, and Russell Diggins for technical assistance.

REFERENCES

Australian Standard 1141 (1983) Sampling and testing aggregates: Standards Association of Australia.

Brindley, G. W. (1980) Intracrystalline swelling of montmorillonites in water-dimethyl sulfoxide systems: *Clays & Clay Minerals* **28**, 369–372.

Cole, W. F. and Beresford, F. D. (1980) Influence of basalt aggregate on concrete durability: *Amer. Soc. Test. Mat. Spec. Tech. Pub.* **691**, 617–628.

Garcia, G. S. and Camazano, M. S. (1968) Differentiation of kaolinite from chlorite by treatment with dimethyl-sulfoxide: *Clay Miner.* **7**, 447–450.

Hosking, J. R. and Pike, D. C. (1985) The methylene blue dye adsorption test in relation to aggregate drying shrinkage: *J. Chem. Tech. Biotechnol.* **35A**, 185–194.

Jones, F. O. (1964) New, fast accurate test measures bentonite in drilling mud: *Oil and Gas J.* **62**, 76–78.

Kirkman, J. H. (1974) Dimethyl sulfoxide treatment of standard clay minerals and clay fractions of selected New Zealand soils: *New Zealand J. Science* **17**, 503–509.

Nakamuta, Y. (1981) A regularly interstratified chlorite/vermiculite in a talc-chlorite vein: *Mem. Fac. Sci. Kyushu Univ., Ser. D, Geol.* **24**, 253–279.

Parry, W. T. and Downey, L. M. (1982) Geochemistry of hydrothermal chlorite replacing igneous biotite: *Clays & Clay Minerals* **30**, 81–90.

Pevear, D. R., Goldin, A., and Sprague, J. W. (1984) Mineral transformations in soils formed in glacialmarine drift, northwestern Washington: *Soil Sci. Soc. Amer. J.* **48**, 208–216.

Rice, T. J., Jr., Buol, S. W., and Weed, S. B. (1985) Soil-saprolite profiles derived from mafic rocks in the North Carolina Piedmont: I. Chemical, morphological, and mineralogical characteristics and transformations: *Soil Sci. Soc. Amer. J.* **49**, 171–178.

Roper, H. (1960) Volume change of concrete affected by aggregate type: *J. Portland Cement Assoc. Res. Devel. Lab.* **2**, 13–19.

Roper, H. (1974) The properties of concrete manufactured with some coarse aggregates of the Sydney area: *Aust. Road Res.* **5**, 40–50.

Sameshima, T., Black, P. M., and Heming, R. F. (1978) Hydrochemical degradation of greywacke road aggregate: *Proc. 9th Aust. Road Res. Board Conf. Pt. 15,* 21–26.

Sameshima, T. and Way, S. J. (1982) Exchangeable bases of swelling clays in Melbourne basalts: *Aust. Road Res.* **12**, 166–172.

Senkayi, A. L., Dixon, J. B., and Hossner, L. R. (1981) Transformation of chlorite to smectite through regularly interstratified intermediates: *Soil Sci. Soc. Amer. J.* **45**, 650–656.

Shayan, A., Lancucki, C. J., and Way, S. J. (1984) Assessment of a microgranite source rock for use in concrete: *Bull. Int. Assoc. Eng. Geol.* **29**, 433–435.

Snowdon, L. C. and Edwards, A. G. (1962) The moisture movement of natural aggregates and its effect on concrete: *Mag. Conc. Res.* **14**, 109–116.

Stewart, E. T. and McCullough, L. M. (1985) The use of the methylene blue test to indicate the soundness of road aggregates: *J. Chem. Tech. Biotechnol.* **35A**, 161–167.

Stutterheim, N. (1954) Excessive shrinkage of aggregates as a cause of deterioration of concrete structures in South Africa: *Trans. S. Afr. Inst. Civ. Eng.* **4**, 351–367.

Velde, B. (1984) Electron microprobe analysis of clay minerals. *Clay Miner.* **19**, 243–247.

APPENDIX 1

ROCK WEATHERING DEFINITIONS ADOPTED BY THE GEOLOGICAL SURVEY OF QUEENSLAND

Completely weathered rock

Rock which retains most of the original rock texture (fabric), but the bond between its mineral constituents is weakened by chemical weathering to the extent that the rock will disintegrate when immersed and gently shaken in water. In engineering usage, this is a soil.

Highly weathered rock

Rock which is weakened by chemical weathering to the extent that dry pieces about the size of a 50-mm-diameter drill core can be broken by hand across the rock fabric. Highly weathered rock does not readily disintegrate when immersed in water.

Moderately weathered rock

Rock which exhibits considerable evidence of chemical weathering, such as discoloration and loss of strength to prevent dry pieces about the size of a 50-mm-diameter drill core (of inherently hard rock) being broken by hand across the rock fabric. Moderately weathered rock does not ring when struck with a hammer.

Slightly weathered rock

Rock which exhibits some evidence of chemical weathering, such as discoloration, but which has suffered little reduction in strength. Except for some inherently soft rocks, slightly weathered rock rings when struck with a hammer.

Fresh rock

Rock which exhibits no evidence of chemical weathering. Joint faces may be clean or coated with clay, calcite, chlorite, or other minerals.

APPENDIX 2

LIST OF AUSTRALIAN STANDARD TEST METHODS (AS 1141) USED FOR EVALUATION OF THE GRANOPHYRE

Test	AS 1141 section
Bulk density and water absorption of coarse aggregate	Section 6
Aggregate crushing value	Section 21
Los Angeles (abrasion) value	Section 23
Sodium sulfate soundness	Section 24

Proceedings of the International Clay Conference, Denver, 1985, L. G. Schultz, H. van Olphen, and F. A. Mumpton, eds.,
The Clay Minerals Society, Bloomington, Indiana, 121–127 (1987).

MICROTEXTURE OF CLAY-RICH SEDIMENTS FROM
THE OSLOFJORD, NORWAY

RAY E. FERRELL, JR.

Basin Research Institute and Department of Geology
Louisiana State University, Baton Rouge, Louisiana 70803

Abstract—Marine clays derived from glacial debris in the Bonnefjord, the most easterly portion of the Oslofjord, Norway, are characterized by distinctive microscopic textures that originated at the time of deposition and are similar to the particle-to-particle arrangements reported in studies of "quick clays." Scanning electron micrographs show that these clays consist chiefly of flocculated domains of face-to-face particles arranged in an edge-to-face, card-house, random pattern. Pelletization by organisms has caused partial collapse of some of the floccules and has produced a rough parallelism of clay domains around silt-size particles. Other bioturbation structures, such as burrows, tubes, and feeding traces, are not associated with any changes in the microtexture of the sediment. The card-house structure is the characteristic textural arrangement of the clays deposited in this quiet, anoxic environment.

Key Words—Bioturbation, Card-house structure, Marine clays, Quick clays, Scanning electron microscopy, Sediments.

INTRODUCTION

The sizes, shapes, and fabric of sedimentary particles are the key factors in the successful analysis of ancient sedimentary environments. The correct assessment of variations in these textural parameters due to source rock characteristics, processes of transportation and deposition, and diagenetic modifications has been largely responsible for the progress in modern sedimentology, especially in the areas of sandstone and carbonate petrology. Progress has been somewhat slower in the area of clay petrology, because the small particle sizes of the constituent grains make primary textural data extremely difficult to obtain. Therefore, knowledge of sedimentary features and their paleonvironmental significance in the most volumetrically significant sedimentary rock types, those composed chiefly of clay minerals, is limited.

In a review of shale fabric, O'Brien (1981) summarized the small amount of work that had been done in the field and illustrated the utility of the transmission electron microscope (TEM) and the scanning electron microscope (SEM) in textural studies. Evidence from studies of turbidites (O'Brien *et al.,* 1980) suggests that fissility in the hemipelagic units is a direct consequence of flake orientation that took place at the time of deposition. Studies of other clay-rich materials have related fissility to lamination with quartz (Spears, 1976) and other mineral segregations (Curtis *et al.,* 1980) or to accumulations of organic matter (Odom, 1967).

Studies of modern sediments and other clay-rich rocks lacking fissility also have demonstrated the importance of electron microscopic techniques for evaluating particle-to-particle arrangements. For example, prodelta clays of the modern Mississippi River were deposited as flocculated sediment (Bennett *et al.,* 1979) composed of randomly oriented stacks of clay platelets. Keller (1978) observed that massive-appearing kaolins commonly exhibit microscopic structures that reflect their geologic origins. He suggested that small, tightly intergrown crystals of kaolinite are indicative of growth from solution and large, vermicular aggregates are often the result of deposition by rivers and streams.

In the present report, the microscopic textures of modern sediments from the Oslofjord, Norway, are described and the processes responsible for their formation are discussed. The location is an ideal one because the source of sediments is well known, the sediments contain only traces of smectite, and limited evidence exists of reworking by waves, coastal currents, and organisms. Thus, the observed textures should directly record the clay particle arrangements that are produced when the materials are deposited in the marine environment.

MATERIALS AND METHODS

Shallow piston cores (6 cm diameter) were collected at four locations in Bonnefjord, the relatively isolated eastern terminus of the Oslofjord, Norway (Figure 1). About 1 m of organic-rich, gray to black clay was recovered at each site. Water depth at the sample stations ranged from 45 to 156 m, but no apparent textural differences associated with sample depth were noted. The sediments were typical of those derived from glacial debris, and the dominant mineral components were

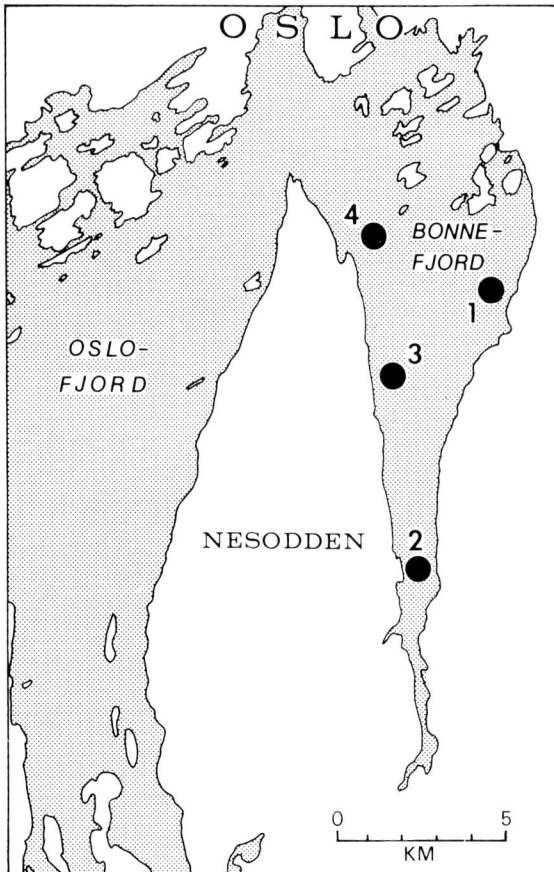

Figure 1. Sketch map of the section of Norway south of the city of Oslo, illustrating approximate sample locations.

A 2-cm-diameter cork borer was used to remove plugs from representative slabs for examination in the SEM. Each plug was trimmed with a razor to a small cube (<1 cm) and dried. Some samples were dried by the critical-point method (Naymik, 1974), but most were frozen by immersion in liquid freon and then dried under vacuum. After drying, the cubes were fractured to reveal internal surfaces, mounted on specimen support stubs, and sputter-coated with gold. About 20 specimens were examined in detail with a JEOL-200 SEM, operated at 15 kV and at magnifications ranging from 2000 to 10,000 times. An energy dispersive X-ray analyzer was employed to help identify the mineral particles.

RESULTS

The typical macroscopic structures of the sediments occurring in this part of the Oslofjord are illustrated in the X-radiographs of Figure 2. Each core segment is 10-cm long. Wavy, irregular bedding, pyrite accumulations, and bioturbation structures are obvious. Larger bed forms are not apparent because of the limited size of the core.

Areas of pyrite concentration appear as the darkest black features shown in the X-radiographs. They are especially common in Figures 2C–2F, i.e., samples from depths greater than 50 cm below the sediment–water interface. Pyrite occurs as threads, a few tenths of a millimeter in cross section and as much as 1 cm long, and as spherical and irregular aggregates about 1 mm in diameter. Branching, thread-like bodies (PT), are common in Figures 2D and 2E. The small circular masses in Figures 2C and 2E are framboidal aggregates of pyrite crystals. A larger mass of pyrite appears to have crystallized around an open crack in the upper part of Figure 2F.

A burrow, or feeding trace of an organism (B), is the large, irregular feature trending essentially horizontally across Figure 2D. The cross section of these structures may be as large as 2 cm, and their length usually exceeds the dimensions of the core slab. A second burrow occurs in the top of Figure 2D, and another partly inclined burrow occurs in Figure 2B. Most burrows are oriented sub-parallel to faint horizontal bedding. The interiors of burrows are generally lighter (less dense) than the surrounding clay, and some swirling or irregular banding may be seen perpendicular to their axes.

Apparent horizontal layering was only observed in a few samples. The layers are only poorly discernible, and most are wavy and irregular. A typical example

illite (40–50 wt. %) and chlorite (40–50 wt. %) mixed with clay-sized rock and quartz detritus (5–15 wt. %). Present-day bottom conditions are saline and anoxic.

Each core was examined by X-ray radiographic techniques similar to those used by Jorgensen *et al.* (1981). A thin, longitudinal slab, about 1-cm thick, was removed from each core by sawing through the diameter of the core barrel and then trimming the excess material from one half with a wire or osmotic knife. Radiographs were produced in a FACIT unit operated at 45 kV and 60 mA. All illustrations were prepared by making photographic prints of the X-ray negatives; hence, dark areas represent dense materials in the cores. The X-radiographs were used to analyze sedimentary structures and to serve as a guide to sampling for microscopic analyses.

→

Figure 2. X-ray radiographs of slabs of sediment showing typical macroscopic structures. The only primary sedimentary structures observed are the wavy bed (WB) forms in Figure 2B. The other structures are produced by organisms (B = burrows, FF = forams or fecal pellets, and T = tubes) or the alteration of organic matter (GB = gas bubbles and PT = pyrite threads). Station numbers and depths are indicated. Each core segment is 10 cm long.

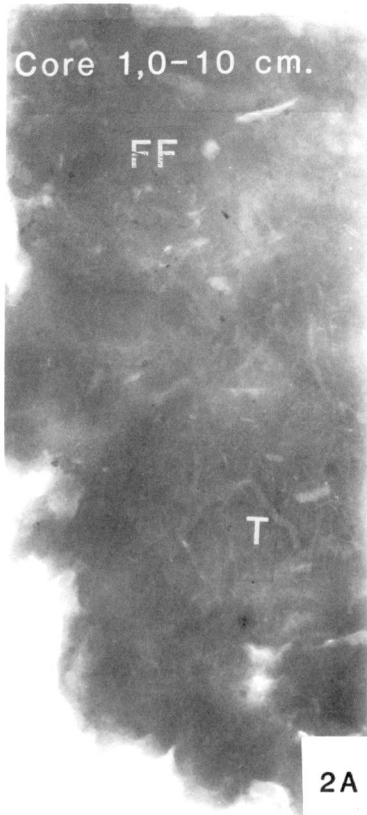

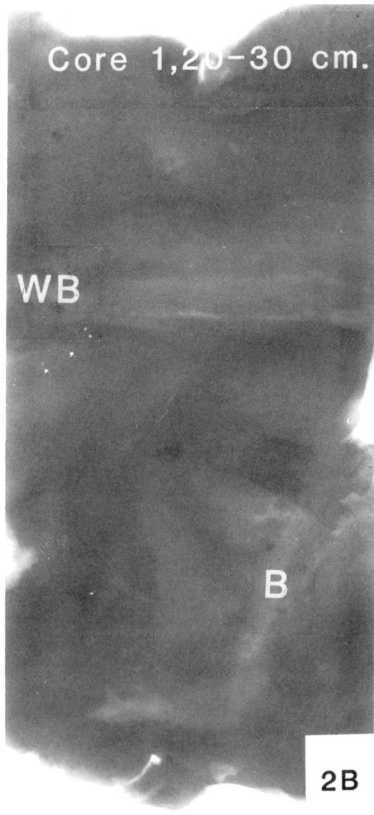

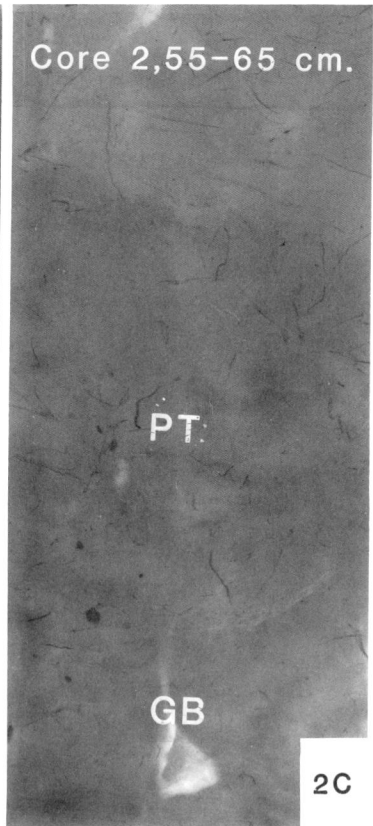

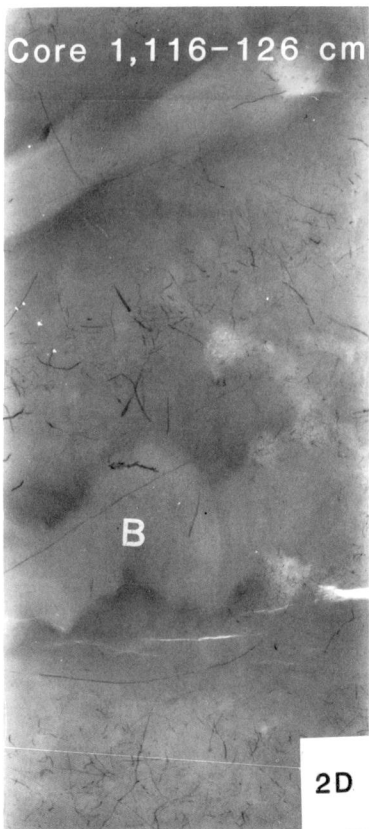

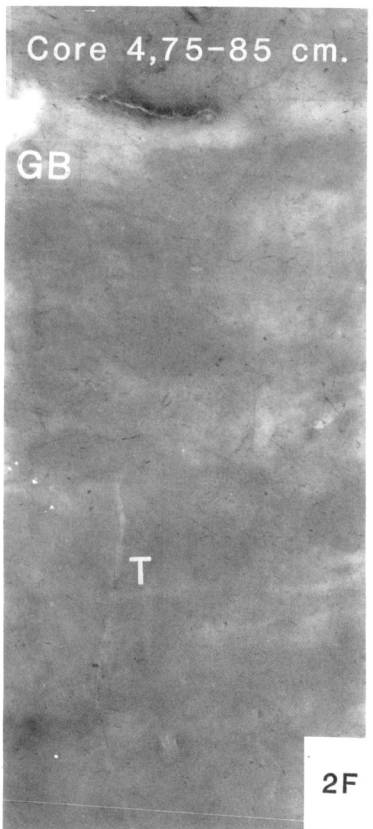

124   Ferrell

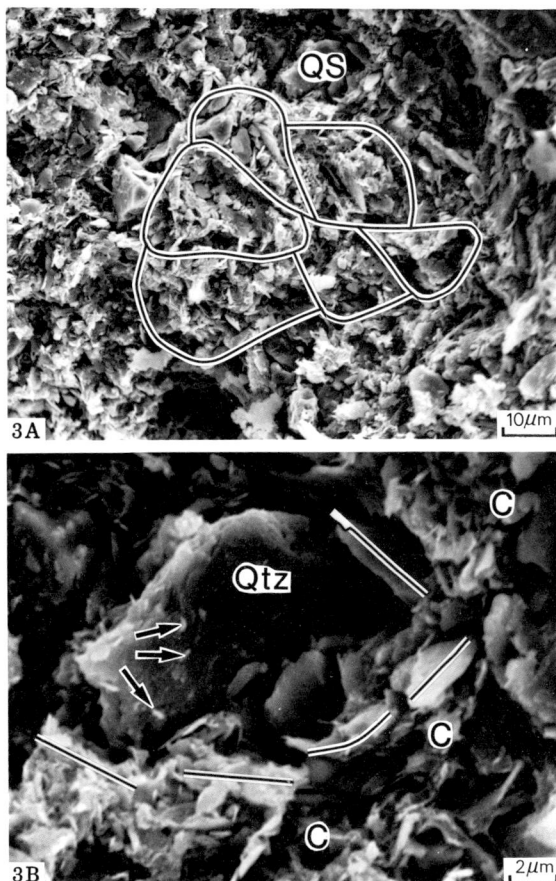

Figure 3. Scanning electron micrographs of burrowed area
of Figure 2D illustrating random orientation of clay particles
outlined in Figure 3A and semi-parallelism of clays to large
grain surfaces (QS). Pelletization by organisms is responsible
for the swirling effect evident in the enlarged view (Figure
3B). The line segments indicate the long axes of clay particles
roughly parallel to a quartz grain (Qtz). Areas of bridging
particles and chains (C) and small clay flakes (arrows) adhering
to the quartz grain are evident.

(WB) is shown in the upper part of Figure 2B. Individual bands are more or less lens shaped, and in some,
the lenses appear to be truncated at one, or both ends.
Some cores exhibit other types of features. Light-colored circular particles can be seen near the top of Figure
2A (FF) and in Figures 2E and 2F. Small, light-colored
tubes, 0.5–4.0 cm long and about 1 mm in diameter
are present in Figures 2A and 2F (T). Irregular, or
circular, light- and dark-toned patches as large as 0.5
cm occur in Figures 2C, 2E, and 2F.

The origins of these features are problematic, but
most are probably deformation structures resulting from
bioturbation. The wavy bedded layers in Figure 2B are
the only ones that are primary sedimentary structures.
The organically produced structures have been identified by comparison with those described by Jorgensen
et al. (1981) from clay-rich sediments recovered from

the Skagerrak south of the Oslofjord. Echinoids, or
other organisms, produced the large irregular feeding
traces or burrows. The small hollow tubes are characteristic of Trichichnus or Chondrites, and the thread-
like pyrite is probably a pseudomorph of these tubes.
Forams and fecal pellets have produced the light circular objects and gas bubbles are responsible for the
larger, irregular, light-colored areas. The dark circular
masses of pyrite formed by replacing forams and fecal
pellets or as inorganically precipitated spherulites, similar to those observed by Love (1963).

The microscopic features of four typical specimens
are illustrated in the SEM photos of Figures 3–5. The
photomicrographs of Figure 3 are of specimens from
the bioturbated zone (feeding trace or burrow) shown
in Figure 2D. The photomicrographs of Figures 4 and
5 are typical of areas that appeared structureless in the
X-radiographs of other samples. Energy dispersive
X-ray spectra confirmed that most of the <5-μm particles are clay minerals and that the larger particles are
fragments and slivers of quartz. The general appearance of the texture is one of randomly oriented aggregates of clay flakes and other particles. The average
porosity estimated by visual comparisons of the relative abundances of mineral grains and dark intergranular voids is 40–50%. The random pattern with high
porosity is a result of flocculation.

The randomness of the microscopic texture of marine clays deposited in this section of the Oslofjord is
evident in Figures 3A and 4A. These lower magnification micrographs illustrate that most of the grains
are aggregates of individual flakes that form thicker
books and that they are joined to form clusters similar
to the card-house structures described by Ingles (1968).
Some typical clusters are outlined by solid lines. The
average book is <0.5-μm thick and about 5-μm long.
Clusters may be 10–20 μm in diameter. The microscopic fabric is essentially the same in bioturbated (Figure 3A) and massive (Figure 4A) zones of the cores.

In some regions, clusters of clay flakes are oriented
tangentially to the >10-μm quartz grains, and the clays
create a swirling effect that may extend 20–30 μm from
the grain (QS, Figure 3A). Figure 3B is an enlarged
view of the area in Figure 3A in which the apparent
swirling can be observed. Many of the complex clay
aggregates have their long axes (short solid lines) oriented roughly parallel to the surface of the quartz grain
(Qtz). Small clusters (C) of clay books form chains or
bridges between the larger units. The quartz grain has
<1-μm clay crystallites adhering to its surface (arrows),
and many similar particles also can be seen on the
larger aggregates of clays in Figure 4B. This partially
parallel pattern is thought to be the result of fecal pellet
formation by organisms.

Samples from other cores generally have the same
texture as that shown in Figure 3A. SEMs of samples
from 111-cm (Figures 4A and B), 97-cm (Figure 5A)

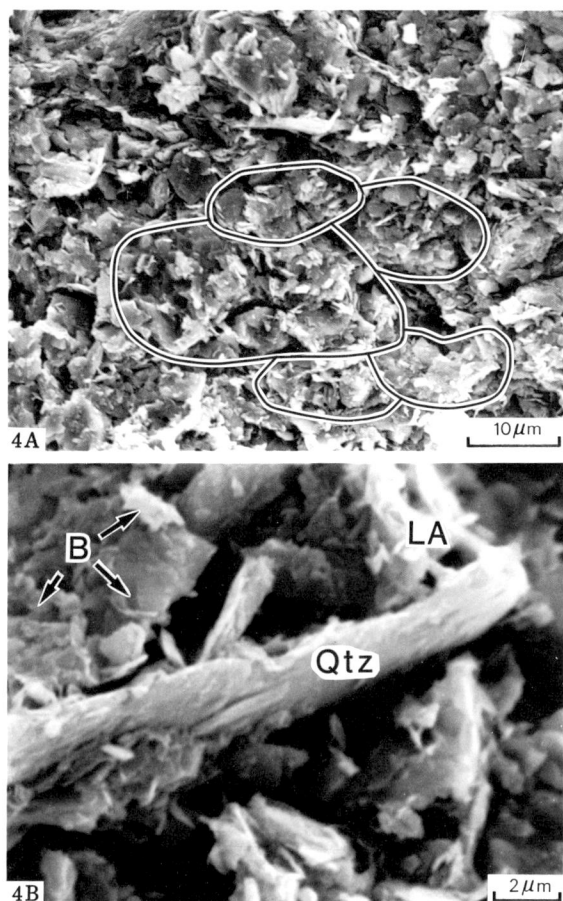

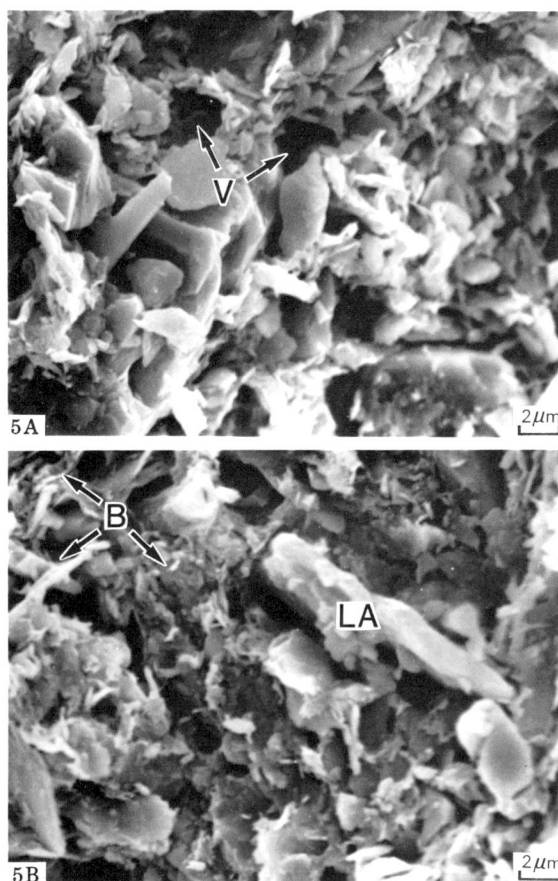

Figure 4. Scanning electron micrographs of homogeneous area 1.0 cm above burrow shown in Figure 2D. Randomly oriented clusters of clay particles are outlined in Figure 4A. At higher magnification, some clay flakes form bridges and chains (B) linking the larger aggregates (LA) and a sliver of quartz (Qtz) to other clusters in Figure 4B.

Figure 5. Scanning electron micrographs of samples from core 2 and core 3 at depths of 97 cm and 35 cm, respectively. Void spaces (V) are almost as large as the clusters of clay flakes comprising the well developed random microtexture of Figure 5A. Additional examples of a large aggregate (LA) and bridging particles (B) are shown in Figure 5B.

and 35-cm (Figure 5B) depths illustrate the similarities among samples and that the microscopic fabric does not change within about the first meter of sediment. Visual estimates of apparent porosities also are similar.

The clays in the Oslofjord occur as books of flakes; the books are randomly oriented in an edge-to-face manner. Figure 5A is the best high-magnification example of the particle-to-particle arrangement most frequently detected in the Oslofjord sediments. Void spaces between clay aggregates (V) are almost as large as the grains observed in Figure 5A. This fabric remains essentially the same even though larger aggregates (LA, Figures 4B and 5B) than those in Figure 5A are present. At higher magnifications, the principal particle arrangement, especially near the large aggregates and quartz particles, appears to be reinforced by smaller clusters of bridging clay flakes. These finer structures (B) are evident in the upper left parts of Figures 4B and 5B.

DISCUSSION

The spatial arrangement of particles in the marine clays of the Oslofjord appears to be fairly simple, and it does not appear to have been modified by bioturbation. The effects of various kinds of organisms are limited to the macroscopic scale features revealed in the X-radiographs and the swirling of clay aggregates associated with fecal pellets. The microtexture of homogeneous, massive sections of the uppermost one meter of sediment is essentially the same as the microtexture of clays in burrows and feeding traces. Organisms have not changed the primary clay fabric. The textures described here are essentially the same as those initially described by Rosenqvist (1959) for glacio-marine clays of the Oslo area and later modified by Pusch (1970). In these earlier studies of exposed "quick clays," the basic card-house structures cataloged by Lambe (1953) were detected and used to explain the physical failure of uplifted marine clays. Pusch (1970) detected a more porous structure than Rosenqvist and described the existence of dense aggregates of clay flakes connected by linking chains of particles. He measured intergranular void spaces with diameters 2–5 times greater than the particle clusters. The present study provides evidence for the existence of the card-house clusters and linking chains, but the chains are not as prominently developed and the void spaces are not as large as those reported by Pusch. The higher porosity values reported by Pusch are probably the consequence of the preparation procedure, and the critical-point or rapid-freezing techniques employed herein more reliably preserve the *in situ* textures of these illite-chlorite clays.

Two processes are apparently responsible for the observed microtextures of the Oslofjord sediments. The dominant process is flocculation. The salinity conditions creating the floccules could not be established, but the clay particles seem to have agglomerated initially in a face-to-face manner to form books; these and other detrital components then combined in an edge-to-face fashion to produce the commonly observed card-house, randomly oriented texture. Pelletization in the gastrointestinal tract of organisms is the second process which locally influenced the clay arrangements. Clay floccules have been partially oriented parallel to the surfaces of detrital grains creating a local swirling effect. The orientation was accomplished by rotating some of the components of the card-house and did not completely eliminate the edge-to-face arrangement. The clay pellets are physically mixed with the flocculated clays in the massive and bioturbated zones of the cores.

Minor quantities of clays are attached to the larger aggregates and form linked chains of smaller flakes or clusters. The chains are not as volumetrically significant as those described by Bennett *et al.* (1979) and Bohlke and Bennett (1980) in recently deposited Mississippi prodelta sediments. The greater smectite content of the Mississippi sediments and the different sedimentological conditions may be responsible for the differences in the character of the flocculated sediments in these two areas. Many of the smaller chains observed in the Oslofjord sediments may have formed by the settling of dispersed clays on the pore walls during the specimen preparation procedure.

This study establishes the pattern of the clay texture in a fairly simple geologic environment. It suggests that random clay fabrics in ancient rocks could be a consequence of deposition in some marine environments. The microscopic textures described above are assumed to be typical of those produced when glacial debris accumulates by flocculation and/or pelletization and settles through marine waters in a protected fjord. Anoxic bottom-water conditions retard extensive bioturbation and the original sedimentary textures are preserved. It remains for future studies to determine whether consolidation during burial will produce oriented textures similar to those observed by Odom and Mesri (1970) or Bennett *et al.* (1979). It will also be important to establish how representative these textures are of marine clays in general and of sediments with larger populations of in-fauna, greater wave and current resuspension and reworking, or higher percentages of non-clay materials and smectite.

ACKNOWLEDGMENTS

This work was accomplished during the author's appointment as a Senior Fulbright Researcher at the University of Oslo. I thank the U.S. Educational Foundation in Norway and the University of Oslo for their support in Norway and Louisiana State University for granting a sabbatical leave. Many stimulating discussions of this topic were held with H. Dypvik, I. Rosenqvist, and P. Aagaard. Tor Mellom of Saga Petroleum A/S assisted with the scanning electron microscope analyses.

REFERENCES

Bennett, R. H., Bryant, W. R., and Keller, G. H. (1979) Clay fabric and geotechnical properties of selected submarine sediment cores from the Mississippi delta: *U.S. Dept. Commerce, NOAA Prof. Paper* **9**, 86 pp.

Bohlke, B. M. and Bennett, R. H. (1980) Mississippi prodelta crusts: a clay fabric and geotechnical analysis: *Mar. Geotechnology* **4**, 55–82.

Curtis, C. D., Lipshie, S. R., Oertel, G., and Pearson, M. J. (1980) Clay orientation in some Upper Carboniferous mudrocks, its relationship to quartz content and some inferences about fissility, porosity, and compactional history: *Sedimentology* **27**, 333–339.

Ingles, O. G. (1968) Soil chemistry relevant to the engineering of soils: in *Soil Mechanics—Selected Topics,* I. K. Lee, ed., Butterworth's, London, 1–57.

Jorgensen, P., Erlenkeuser, H., Lange, H., Nagy, J., Rumohrs, J., and Werner, F. (1981) Sedimentological and stratigraphical studies of two cores from the Skagerrak: in *Holocene Marine Sedimentation in the North Sea Basin,* S-D. Nio, T. E. Shuttenhelm, and Tj. C. E. van Weering, eds., *Spec. Pub. Int. Assoc. Sediment.* **5**, 397–414.

Keller, W. D. (1978) Classification of kaolins exemplified by their textures in scan electron micrographs: *Clays & Clay Minerals* **26**, 1–20.

Lambe, T. W. (1953) The structure of inorganic soil: *Amer. Soc. Civil Engr. Proc.* **79**, 1–49.

Love, L. G. (1963) Pyrite spheres in sediments: in *Biochemistry of Sulphur Isotopes,* M. L. Jensen, ed., *Proc. Nat. Sci. Found. Symp., Yale, New Haven, Conn., 1962,* 121–143.

Naymik, T. G. (1974) The effects of drying techniques on clay-rich soil texture: in *Proc. 32nd Ann. Mtg. Electron Microscopy Society of America,* C. J. Arceneaux, ed., Claitor's Inc., Baton Rouge, Louisiana, 466–467.

O'Brien, N. R. (1981) SEM study of shale fabric—a review: in *Proc. 14th Ann. Scanning Electron Microscopy Symposium,* O. Johari, ed., IITRI, Chicago, 569–575.

O'Brien, N. R., Nakazava, K., and Tokuhashi, S. (1980) Use of clay fabric to distinguish turbidite and hemipelagic siltstones and silts: *Sedimentology* **27**, 47–61.

Odom, I. E. (1967) Clay fabric and its relation to structural properties in mid-continent Pennsylvanian sediments: *J. Sed. Petrol.* **37,** 610–623.

Odom, I. E. and Mesri, G. (1970) Mechanisms controlling compressibility of clays: *J. Soil Mech. Found. Div., Proc. Amer. Soc. Civ. Eng.* **96,** 1863–1878.

Pusch, R. (1970) Microstructural changes in soft quick clay failure: *Canadian Geotech. J.* **7,** 1–7.

Rosenqvist, I. Th. (1959) Physio-chemical properties of soils: soil-water systems: *J. Soil Mech. and Found. Div., Proc. Amer. Soc. Civ. Eng.* **85,** 31–53.

Spears, D. A. (1976) The fissility of some Carboniferous shales: *Sedimentology* **23,** 721–725.

Proceedings of the International Clay Conference, Denver, 1985, L. G. Schultz, H. van Olphen, and F. A. Mumpton, eds.,
The Clay Minerals Society, Bloomington, Indiana, 128–134 (1987).

GENESIS OF SOUTH AFRICAN RESIDUAL KAOLINS
FROM SEDIMENTARY ROCKS

R. O. HECKROODT

Department of Materials Engineering, University of Cape Town
Rondebosch, Republic of South Africa

D. BÜHMANN

Department of Geology, University of Natal, Pietermaritzburg
Republic of South Africa

Abstract—The Grahamstown kaolin deposits in the Republic of South Africa occur associated with two distinct geomorphological features, the Grahamstown peneplane (at about 650 m elevation and of Miocene age) and the Coastal plain (at about 520 m elevation and of some later Tertiary age). The parent rocks consist of three lithologically different shales and a diamictite of Devonian to Permian age. The clay occurs in irregularly shaped deposits having no relationship to the structure and folding or to the stratigraphic position within the sedimentary assemblages. The clay deposits are covered by thick silcrete horizons, suggesting that the deposits formed under semi-arid conditions in very flat areas having a fluctuating water table. Pyrophyllite formed prior to kaolinite in those clay deposits that developed from sediments containing both $2M_1$ and $1M$ micas as the only silicate minerals. At the beginning of weathering the less stable $1M$ mica appears to have favored the formation of pyrophyllite, and only at a later stage of weathering did kaolinite become more stable than pyrophyllite. The recent discovery of alunite, natroalunite, and wardite veins indicates that, beside supergene weathering, some low- to moderate-temperature hydrothermal activity may have contributed to the genesis of the kaolinitic clay deposits.

Key Words—Alunite, Genesis, Hydrothermal, Kaolin, Mica polytypes, Pyrophyllite, Weathering.

INTRODUCTION

Residual kaolin deposits derived from sedimentary rocks are particularly well developed in the surroundings of Grahamstown (33°S, 26°E), about 115 km northeast of Port Elizabeth in the Eastern Cape Province, Republic of South Africa. These white-firing clay materials are extensively used in the ceramics industry and as filler material where low abrasiveness is not a critical requirement. Some of these deposits were utilized as early as 1875. The earliest systematic mineralogical investigation of the clay deposits was carried out by Blignaut (1928), who described the clay as consisting of quartz, feldspar, mica, and some "amorphous" material. Since then, numerous studies of these materials have been undertaken (e.g., Mountain, 1931, 1946; Smuts, 1983), but some problems regarding the genesis of these kaolinitic clays remain unresolved. Early in 1985, a mineralogical study of the deposits was initiated by the authors with the aim to examine anew the origin of the deposits.

GEOLOGY

Formations found in the Grahamstown region are entirely sedimentary and represent the upper part of the Cape Supergroup (Devonian and Carboniferous) and the lower part of the Karoo Sequence (Carboniferous and Permian). The lithostratigraphy of the formations is given in Table 1. Large areas are overlain locally by much younger Tertiary silcretes. The oldest formation in this area belongs to the Bokkeveld Group of the Cape Supergroup, and the youngest formations form part of the lower Ecca Group of the Karoo Sequence. The Tertiary rocks occur locally and have not been assigned formational rank.

Physiography

The region is characterized by long ridges and escarpments trending ESE–WNW, parallel to the strike of the broad synclines and anticlines of the Cape folded belt, which had its origin in early Jurassic time. Important physiographical features of the area are the remains of the Grahamstown peneplane and the Coastal plain. The peneplane is at an altitude of about 650 m and extends as a broad area, roughly 2 × 8 km, along the fold axis. The peneplane is flanked on the north and south by prominent quartzite ridges, the southern ridge forming an escarpment between the Coastal plain and the interior. The altitude of the Coastal plain in this area is 500–530 m.

Stratigraphy

The general stratigraphy and lithology of the formations found in the Grahamstown area were described by Mountain (1946, 1975), but only those formations associated with the kaolin occurrences are considered here. The sediments of the Cape Supergroup in the Grahamstown region—both the Bokkeveld and the Witteberg Groups—are probably of brackish origin, because only plant fossils and brachiopods have been found. The structural features of the series of synclines and anticlines can be clearly observed because of the well-developed bedding of the sediments. Some of the beds are locally highly contorted and even faulted.

Table 1. Lithostratigraphy of the Cape Supergroup and the lower part of the Karoo Sequence in the Eastern Cape Province.[1]

	System	Group	Subgroup	Formation	Dominant lithology(ies)
Karoo Sequence	Jurassic	Lebombo			
	Triassic		Tarkastad		
		Beaufort	Adelaide		
	Permian			Fort Brown	Shale
				Ripon	Sandstone, shale
		Ecca		Collingham	Shale (volcanic ashes)
				Whitehill	Shale (white band)
				Prince Albert	Shale (with phosphorites)
				Dwyka	Diamictite, tillite
	Carboniferous		Kommadagga		Sandstone, shale, diamictite
Cape Supergroup		Witteberg	Lake Mentz		Shale
					Sandstone
	Devonian		Traka		Shale, siltstone
		Bokkeveld	Ceres		Shale, sandstone
	Silurian Ordovician	Table Mountain			Sandstone, shale

(unconformity)

Sediments dealt with in this study

[1] South African Committee for Stratigraphy, 1980.

The sediments of the Bokkeveld Group are predominantly sandy micaceous shales and vary considerably in grain-size distribution and color. Most of the shales are dark colored and carbonaceous, but on exposure turn yellow, red, or purple. The sandstone beds, as thick as 15 m, are dark colored, well bedded, and generally not highly silicified.

The older parts of the Witteberg Group are almost exclusively arenaceous. These sediments consist mainly of a pale-grey sandstone (rarely massive) in which the bedding is usually very distinct. Locally, the sandstone grades into highly micaceous flagstone. Although the argillaceous horizons in the lower part of the Witteberg Group are not uncommon, they are rarely more than 6 m thick. They are micaceous and in places sandy or carbonaceous. Somewhat carbonaceous shales that predominate in the younger parts of the Witteberg Group are blue-grey when fresh and weather to a yellow-brown color. The shales contain a small percentage of phosphate, as well as pyrite that is locally well crystallized. Near the base of the Witteberg shales, well-formed gypsum crystals occur which are most probably derived from the oxidation of pyrite. Thinly bedded sandstone bands, as thick as 15 m, are also present in the younger parts of the group.

The Dwyka Tillite of the Karoo Sequence unconformably overlies the sediments of the Witteberg Group, although the unconformity is not obvious at first sight (Truswell, 1977). The fresh tillite is generally a massive blue-grey siltstone and contains unsorted angular fragments and rounded boulders as large as 1 m in diameter. The groundmass shows no stratification, but the formation shows some highly imperfect cleavage and jointing.

Only the oldest unit of the Ecca Group, the Prince Albert Formation, is preserved in the Grahamstown region. It consists of easily erodable argillaceous sediments, typically olive-green shale or fine-grained sandstone. The sediments show well-developed bedding planes and cleavage along steeply dipping planes. Phosphate beds are known in the formation, as are two horizons of chert.

Silcretes, possibly of Miocene age (Smuts, 1983), are the youngest formations of the Grahamstown peneplane. Their texture varies considerably, and Mountain (1946) described the most common variety as a fine-grained, massive, cream-colored rock. The silcrete near the margin of the original peneplane, however, consists of subangular rubble of Witteberg quartzitic sandstone cemented with fine-grained silica. The silcrete shows no stratification and is regarded as a silicified soil or subsoil. The Coastal plane silcrete, thought to be younger than the silcrete of the Grahamstown peneplane and probably of later Tertiary age, occurs mainly in isolated outcrops and small patches, although pebbles and nodules of the silcrete are widespread on the plain and locally form continuous sheets of gravel and rubble.

MATERIALS AND METHODS

Unweathered Cape and Karoo sediments were collected. Seven samples of Bokkeveld and Witteberg shales were obtained from a recent road cut along the national road N2 southeast of Grahamstown, and 20 samples representing Dwyka Tillite and Ecca shales were obtained from the Prince Albert, Whitehill, Collingham, and Ripon Formations along the regional road R67 northeast of Grahamstown, in the area known as the Ecca pass. Weathered Cape and Karoo sediments were sampled at 11 open pit kaolin mines in the vicinity of Grahamstown, including 24 clay samples that formed from the Witteberg and Bokkeveld shales and 23 from the Dwyka Tillite and Ecca shales.

Air-dried material of the whole rock was examined by X-ray powder diffraction (XRD). Finely ground powder was mounted into aluminum frames, the powder being pressed against filter paper covering the surface in order to achieve as random an orientation as possible. For clay mineralogical analysis the samples were dispersed ultrasonically, and the <2-μm fraction was separated by sedimentation. Orientation of the clay fraction was achieved by the smear method (Gibbs, 1965). Expansion tests were performed by solvation with ethylene glycol (in vacuo overnight), hydrazine (in vacuo for 24 hr, the boiling point of hydrazine having been reached), and formamide (a few drops added to the air-dried specimen on the glass slide, which was then X-rayed 10 min after the formamide addition). Samples were also heated at 500°C for 4 hr.

X-ray diffractometry data were obtained using a Philips X-ray diffractometer equipped with a graphite monochromator and CoKα radiation, generated at 40 mA and 40 kV. Whole-rock, random powder patterns were scanned from 3° to 75°2θ (oriented specimens from 2° to 35°2θ) at a scanning rate of 1°2θ/min. For the semiquantitative evaluation of the mineralogy, heights of selected peaks of the various minerals present were measured and their sum equated to 100%. The proportions of the various heights are reported in Table 2. Subgroup and polytype identification was based on the 060 and hkl reflections, respectively.

THE CLAY DEPOSITS

Smuts (1983), among others, showed that the clay deposits in the Grahamstown region are distributed in broad belts at specific altitudes and that they are similar in appearance, although they comprise four different lithologies. The kaolinization is remarkably deep, generally about 25 m in all the deposits, although the usable clay does not always extend to this depth. The deposits are commonly capped by thick silcrete. Two distinct silcrete horizons have been noted, separated from each other by a lateritic layer, which locally is composed of a poor quality clay in its lower part.

The deposits are not stratabound and occur on both the Grahamstown peneplane and the Coastal plain. Most of the known deposits are at the present-day edges of the two plains and have been exposed by the erosional incisions of the original plains. The presence of highly weathered material away from the edges of the plains is well documented from boreholes, but the hardness and thickness of the overlying silcrete horizons make exploitation currently unattractive.

Although the properties of the clay materials and structures of all the deposits in the Grahamstown area are superficially similar, the quality of the deposits varies considerably and is controlled by the lithology of the parent rock. Detailed descriptions of many of the deposits have been reported elsewhere (e.g., Smuts, 1983), and data on the chemical, physical, and ceramic

Table 2. Minerals in unweathered sediments and weathered clay materials developed on them in the Grahamstown area.

	Whole rock																<2-μm fraction					
	Halite	Hematite	Goethite	Quartz	Calcite	Alunite	Natroalunite	Wardite	Apatite	Plagioclase	K-feldspar	Pyrophyllite	Kaolinite	Mica	Chlorite	Illite/Smectite	Chlorite	Kaolinite	Pyrophyllite	Mica	Smectite	Number of samples analyzed
Clay deposits																						
Karoo																						
Ecca shales				55			4	7					22	12 (2M₁)			0	57	0	43	0	12
Cape																						
Dwyka tillite				32									67				0	97	0	3	0	11
Witteberg shales		2		38		4					3	12	28	12 (2M₁)		1	0	54	8	38	0	24
Unweathered sediments																						
Karoo																						
Ripon Formation¹				44	26					21				5 (2M₁)	4		28	0	0	72	0	3
Cullingham Formation¹				55	10				1	6				23 (2M₁)	4	1	17	0	0	83	0	11
Whitehill Formation¹	2	2		78										10 (2M₁)	8		18	0	0	80	2	3
Prince Albert Formation¹				83					2					10 (2M₁)	3	2	15	8	0	69	8	2
Dwyka tillite		5		34						49				4 (2M₁)	13		67	0	0	33	0	1
Cape																						
Witteberg and Bokkeveld shales				53									10	31 (1M + 2M₁)	1		0	22	0	78	0	7

¹ Ecca shales.

properties of the clay materials can be found in the CSIR Technical Notes Series X/BOU-KER.

Mineralogy of the parent sediments

Until the present study, very little had been reported on the mineralogy of the sediments in this area. Heckroodt (1968) found that unweathered samples of Ecca shales, taken from shallow pits alongside the road leading through the Ecca pass near Grahamstown, contained abundant quartz, substantial quantities of illite, and minor to trace amounts of chlorite or vermiculite. Traces of kaolinite and montmorillonite may possibly have been present in some samples. Smuts (1983) reported chemical analyses of relatively fresh, unweathered sediments, but not their mineralogical composition.

The mineralogical analyses of 27 samples of unweathered rock collected during the present study are given in Table 2. The mineralogy of the Karoo sediments is significantly different from that of the Cape sediments. The Witteberg and Bokkeveld shales consist of quartz (40–60%), kaolinite (5–25%), micaceous minerals (30–40%), and varying proportions of $1M$ and $2M_1$ mica polytypes. Most Karoo sediments, on the other hand, contain substantial amounts of plagioclase and quartz, and the Dwyka Tillite generally contains more plagioclase and less quartz than the Ecca shales. Both the tillite and the Ecca shales contain about 15–35% of $2M_1$ mica + chlorite, but kaolinite is rare.

Mineralogy of the clay materials

The mineralogical analyses of 47 samples of weathered Cape and Karoo rocks are presented in Table 2. The clay materials formed from the Witteberg and Bokkefeld shales are characterized by the presence of pyrophyllite (as much as 35%) and about 5% K-feldspar. The quartz content ranges roughly between 30 and 60%, and the kaolinite content may be as high as 70%, but rarely less than 20%. The micaceous mineral, typically comprising as much as 25% of the clay material, consists only of a $2M_1$ mica polytype. Very small amounts of a smectite/illite have been found in some samples, but never chlorite. Alunite was also detected in these clays for the first time during the present study.

The Dwyka Tillite clay materials contain no pyrophyllite, chlorite, plagioclase, or K-feldspar. Kaolinite is present in amounts as high as 70%, and the quartz content varies from 20 to 60%. Illite is typically sparse in the clay. Some of the pebbles in the original tillite have been altered to a material consisting of as much as 80% kaolinite. In some vein fillings the kaolinite content is as high as 95%.

The clay materials derived from the Ecca shales also contain no pyrophyllite, chlorite, or feldspar. The kaolinite and quartz contents are generally 20–35% and about 55%, respectively. The micaceous mineral ($2M_1$ mica polytype only) constitutes about 10% of the clay material. In one deposit, appreciable amounts of natroalunite and wardite were also detected in some layers.

GENESIS OF THE CLAY MATERIALS

The generally held opinion about the genesis of the Grahamstown clay deposits is that they are residual and were formed by the weathering of the Cape and Karoo sediments under unusual conditions. Mountain (1931) ascribed the extensive weathering of the shales to the extraordinary complete peneplanation that took place in the Grahamstown area. According to him, poor drainage allowed surface water to soak deep into the underlying formations and to leach the soluble constituents. During the dry seasons the leached silica was transported to the surface and precipitated in the form of silcrete, commonly cementing rubble material. It is certain that the deposits associated with a specific peneplane are all of the same age.

Smuts (1983) confirmed the ideas of Mountain and used loss-gain diagrams and Barth plots to show that the alumina content remained essentially constant and that the alkali and alkaline earths were leached at an early stage and iron at a later stage in the alteration process. He contended that the process involved hydrolysis by acidic surface water, solution of K, Na, Mg, and Ca, and an increase in alkalinity. This increase in alkalinity then increased the solubility of iron and silica, particularly during dry periods when no fresh water was introduced into the system. The alkaline water would then have been drawn by evaporation and capillary action to the surface during the periodic dry spells. When it came into contact with the more acidic surface water, the dissolved iron and silica precipitated. Iron is now found as oxides along the joint and bedding surfaces. Likewise, the dissolved silica precipitated as quartz in fracture zones and as the extensive silcrete cappings now present above the leached areas.

Thus, leaching caused by weathering appears to have played the major role in the genesis of the deposits. The presence of pyrophyllite, K-feldspar, alunite, natroalunite, and wardite in some of the deposits, even though the parent rocks are devoid of these minerals, needs to be explained. One of the major difficulties in establishing the genesis of the deposits is the variability of the parent rocks, particularly over short distances. Thus, the original composition of the sediment from which a particular clay material was derived cannot be inferred with confidence. Pyrophyllite appears to occur only in those deposits derived from the older sediments, i.e., the Cape sediments. Only these sediments contain the $1M$ polytype of mica.

An assemblage of $1M$ and $2M_1$ micas indicates a nonequilibrium condition, because under leaching conditions the $1M$ mica is less stable than the $2M_1$ polymorph (Velde, 1965). Lippmann (1979) postulated the formation of pyrophyllite during the early stages of weathering as follows:

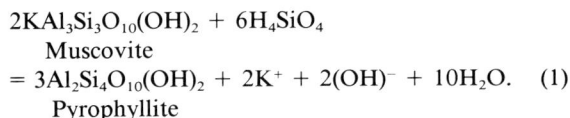

$$2KAl_3Si_3O_{10}(OH)_2 + 6H_4SiO_4$$
Muscovite
$$= 3Al_2Si_4O_{10}(OH)_2 + 2K^+ + 2(OH)^- + 10H_2O. \quad (1)$$
Pyrophyllite

He argued that the stability of pyrophyllite decreases during later stages of weathering, whereas the stability of kaolinite increases, as follows:

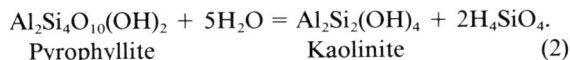

$$Al_2Si_4O_{10}(OH)_2 + 5H_2O = Al_2Si_2(OH)_4 + 2H_4SiO_4.$$
Pyrophyllite $\quad\quad\quad$ Kaolinite $\quad\quad\quad$ (2)

How much of the $2M_1$ mica polytype transformed into kaolinite could not be determined in the present study, but the variable amount of pyrophyllite present in the clay materials seems to reflect the variable amount of $1M$ polytype mica in the parent rock. The mica:pyrophyllite:kaolinite ratios of the clay materials derived from Witteberg shale, as estimated for the whole rock and the <2-μm fractions, are different, namely 24:23:53 and 38:8:54, respectively (see Table 2). These ratios suggest that most of the pyrophyllite in the weathered profile is in the >2-μm fraction and that the mica tended to break down to fine particles during weathering. The Dwyka tillite and Ecca shales, on the other hand, are devoid of the $1M$ mica polytype, and no pyrophyllite has been found in the clay materials derived from these sediments. All the plagioclase has been changed into kaolinite, and the micaceous mineral now found in the clay material is probably the degraded remnant of the $2M_1$ polytype mica that was present in the original sediments.

Until now, hydrothermal action and low-level metamorphism have been discounted as influencing the formation of these deposits. Mountain (1946) pointed out that, although overfolding took place, this region shows no evidence of major faulting, even though the beds are highly contorted on a local scale, with both normal and thrust faulting being common. Also, Smuts (1983) reported no evidence to suggest that the clay deposits were influenced by metamorphism, major faulting, or hydrothermal activity. Furthermore, as far as igneous activity is concerned, the southern limit of the Karoo dolerite intrusions is far to the north.

The presence of alunite, natroalunite, and wardite, however, represent strong evidence for hydrothermal activity in this area. Alunite is generally a product of rock alteration in volcanic regions, caused by the presence of sulfurous waters (Höller, 1967; Sharma, 1970; Kromer, 1975). In this area alunite has been found only in those clay deposits derived from Witteberg shale, and its formation appears to have been the result of the breakdown of mica, which provided the necessary potassium. A small but persistent amount of

K-feldspar is also present in the clay materials from these deposits. This K-feldspar seems to be authigenic, because no K-feldspar was detected in the parent rocks. It might have formed during the weathering process when $K^+ + (OH)^-$ (released from mica during reaction (1), above) and H_4SiO_4 (released during reaction (2), above) reacted with part of the remaining mica, according to reaction (3), below:

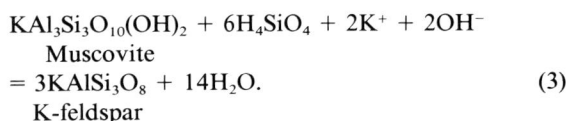

$$KAl_3Si_3O_{10}(OH)_2 + 6H_4SiO_4 + 2K^+ + 2OH^-$$
Muscovite
$$= 3KAlSi_3O_8 + 14H_2O. \quad (3)$$
K-feldspar

The presence of this K-feldspar may also indicate low-temperature hydrothermal activity (Kastner and Siever, 1979), a condition which, like alunite mentioned above, is also suggested by the presence of natroalunite and wardite in the Ecca sediments. The later minerals are regarded as being formed by the hydrothermal reaction of sulfurous water with plagioclase and apatite, respectively (Höller, 1967; Sharma, 1970; Kromer, 1975).

Low- to moderate-temperature hydrothermal action was apparently confined to the Grahamstown peneplane and the Coastal plain clay deposits, irrespective of their parent rocks. To date, hydrothermal activity has not been found in unweathered sediments in the Grahamstown region, although hydrothermal vein deposits of fluorite, calcite, and quartz occur in Karoo sediments along the coasts of the Eastern Cape Province, Transkei, and Natal. These veins cut the youngest Karoo sediments and are therefore Late Jurassic or younger in age. Whether these hydrothermal veins do or do not cut the silcrete has not been determined.

The hydrothermal action on the sedimentary rocks in the Grahamstown area should not be overemphasised, but, in addition to surface weathering, it definitely contributed to the formation of the kaolinitic clay deposits.

ACKNOWLEDGMENTS

The guidance given by Johann Smuts during the fieldwork, the discussions with Eric Hammerbeck, and the financial assistance by the Geological Survey and the Foundation for Research Development of the South African Council for Scientific and Industrial Research are gratefully acknowledged.

REFERENCES

Blignaut, J. J. G. (1928) Clays derived from the Lower Dwyka shales: M.Sc. thesis, Rhodes University, Grahamstown, Republic of South Africa, 63 pp. (unpublished).

CSIR Technical Note Series (X/BOU-KER) (19XX) Ceramic raw materials of Southern Africa. C.S.I.R., Pretoria, Republic of South Africa, 00 pp.

Gibbs, R. J. (1965) Error due to segregation in quantitative clay mineral X-ray diffraction mounting techniques: *Amer. Min.* **50,** 741–751.

Heckroodt, R. O. (1968) The mineralogy of the Ecca Series: *Ann. Geol. Surv. S. Afr.* **7**, 95–98.

Höller, H. (1967) Experimentelle Bildung von Alumit-Jarosit durch die Einwirkung von Schwefelsäure auf Mineralien und Gesteine: *Contr. Mineral. Petrol.* **15**, 309–329.

Hurlbut, C., Jr. (1952) Wardite from Beryl Mountain, New Hampshire: *Amer. Mineral.* **37**, 849–852.

Kastner, M. and Siever, R. (1979) Low temperature feldspar in sedimentary rocks: *Amer. J. Sci.* **279**, 435–479.

Kromer, H. (1975) Geochemical aspects of genesis of kaolinite, alunite and silica minerals in the vicinity of the trass-deposit near Gleichenberg: *Styria. Miner. Deposita.* **10**, 249–253.

Lindberg, M. L. (1957) Relationship of the minerals avelinoite, cyrilovite, and wardite: *Amer. Mineral* **42**, 204–213.

Lippmann, F. (1979) Stabilitätsbeziehungen der Tonminerale: *N. Jb. Miner. Abh.* **136**, 287–309.

Mountain, E. D. (1931) The Grahamstown ceramic industry: *S. Afr. J. Sci.* **28**, 135–139.

Mountain, E. D. (1946) The geology of an area east of Grahamstown: explanation of sheet No. 136 (Grahamstown): *Geol. Surv. S. Afr.,* 56 pp.

Mountain, E. D. (1975) The geology of the Upper Dwyka Stage in Grahamstown: *Trans. Geol. Soc. S. Afr.* **78**, 161–165.

Sharma, R. R. (1970) Mineralbestand und Bildungsbedingungen der Alunit-Kaolinit-Vorkommen im Ostteil der Insel Milos (Griechenland): *Gueppinger Akad. Beit.* **3**, 1–53.

Smuts, J. (1983) The geology, mineralogy and chemistry of the Grahamstown clay deposits: M.Sc. thesis, Rhodes University, Grahamstown, Republic of South Africa, 100 pp. (unpublished).

South African Committee for Stratigraphy (1980) Stratigraphy of South Africa. Part I: in *Lithostratigraphy of the Republic of South Africa, South West Africa/Namibia, and the Republics of Bophuthatswana, Transkei and Venda,* L. E. Kent, compiler, *Handb. Geol. Surv. S. Afr.* **8**, 690 pp.

Truswell, J. F. (1977) *The Geological Evolution of South Africa*: Purwell, Cape Town, 218 pp.

Velde, B. (1965) Experimental determination of muscovite polymorph stabilities. *Amer. Miner.* **50**, 436–499.

Proceedings of the International Clay Conference, Denver, 1985, L. G. Schultz, H. van Olphen, and F. A. Mumpton, eds.,
The Clay Minerals Society, Bloomington, Indiana, 135–143 (1987).

DIAGENESIS IN SHALES: EVIDENCE FROM BACKSCATTERED ELECTRON MICROSCOPY AND ELECTRON MICROPROBE ANALYSES

T. J. Primmer [1] and H. F. Shaw

Department of Geology, Imperial College of Science and Technology
London SW7 2BP, United Kingdom

Abstract—Information gained from standard X-ray powder diffraction (XRD) analyses of Gulf Coast Tertiary and North Sea Mesozoic shales are compared with backscattered electron microscopy and electron microprobe data to illustrate the potential of backscattered electron imagery (BEI) for diagenetic studies of shales. The XRD analyses of Gulf Coast shales showed that one of the principal changes is the increased illitization of mixed-layer clays with depth. The electron microprobe data confirmed this change, but the BEI data revealed that closed microenvironments also exist within the shales, in which other diagenetic processes have taken place, e.g., the formation of authigenic kaolinite and chlorite. The XRD analyses of Mesozoic North Sea shales showed abrupt changes in the nature of illite/smectite compositions with stratigraphic position, suggesting that the compositional changes could be due to differences in detrital source materials. The BEI and probe analyses of these North Sea samples indicate a variety of diagenetic reactions in different microenvironments including dissolution, precipitation, and alteration involving phyllosilicate, feldspar, carbonate, and sulfide phases. The diagenetic reactions were found to vary significantly depending on the nature of the original sediments.

Key Words—Backscattered electron microscopy, Chlorite, Diagenesis, Electron microprobe analysis, Illite/smectite, Kaolinite, Shale.

INTRODUCTION

To date, most interpretations of diagenetic processes in argillaceous rocks have been based on mineralogical and geochemical analyses of bulk samples (e.g., Hower *et al.,* 1976). Scanning electron microscopic (SEM) techniques have also been used to study diagenesis in sedimentary rocks, principally sandstones (e.g., Scholle and Schluger, 1979; Welton, 1984). More recently, backscattered electron imagery (BEI) has been introduced as a further means of investigating such diagenetic assemblages. The ability of BEI to contrast differences in mineral compositions (a distinct advantage over standard, secondary electron techniques) is especially useful in the study of fine-grained lithologies (Pye and Krinsley, 1983; Krinsley *et al.,* 1983). The usefulness of BEI is greatly enhanced by energy- and wavelength-dispersive X-ray spectrometers which allow the electron microscope to be used as a microprobe and thus yield elemental analyses of mineral phases (White *et al.,* 1984, 1985).

The present study examines specimens from Tertiary Gulf Coast sediments and Mesozoic shales from the North Sea to demonstrate how a detailed view of shale diagenesis can be made using BEI and microprobe analyses. Differences between the two areas are discussed with respect to their burial history and the influence of different sediment provenance materials and pore-fluid chemistry.

MATERIALS AND METHODS

The materials studied were shale cuttings of Oligocene-Miocene age from a Gulf Coast well, offshore Texas, and a mixture of cuttings and core specimens (see Table 2) of deeply buried Lower Cretaceous-Upper Jurassic shales from the Brae and Magnus fields in the North Sea (U.K. sector).

The samples were examined by standard X-ray powder diffraction (XRD) methods. Analyses were made of random powder mounts for whole-rock mineral identification and of oriented <2-μm and <0.5-μm size fractions saturated with Mg^{2+} for specific clay mineral identification. The XRD data from the Mg-saturated, <0.5-μm size fractions after solvation with ethylene glycol were used to determine the "end-member" compositions of the mixed-layer illite/smectites (I/S) following the method of Reynolds and Hower (1970).

The principles involved in producing high-resolution backscattered electron images have been discussed in detail elsewhere (see, e.g., White *et al.,* 1984). Suffice it to say that the proportion of backscattered electrons is a function of the mean atomic number of the mineral; thus, the higher the mean atomic number, the brighter the image on the SEM screen, enabling chemically different mineral grains to be distinguished. The BEI contrast is maximized by the use of flat, polished specimens. The specimens were prepared by mounting the cuttings or core slices in resin, cutting them flat, and

---

[1] Present address: British Petroleum p.l.c., BP Research Centre, Chertsey Road, Sunbury-on-Thames, Middlesex TW16 7LN, United Kingdom.

135

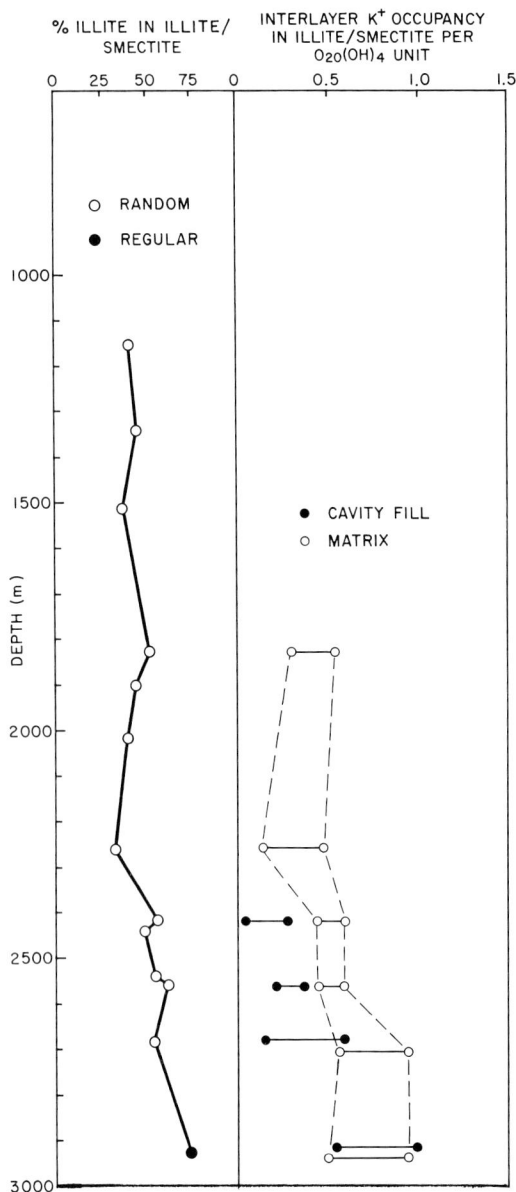

Figure 1. Percentage of illite layers in illite/smectite (I/S) from Gulf Coast shale cuttings vs. depth. A parallel plot (right) illustrates microprobe determination of interlayer K^+ content in the I/S from different chemical microenvironments (for clarity, analyses at depth of 2694 m and 2940 m are slightly offset for each microenvironment).

polishing them prior to BEI analysis. In all analytical microprobe work, flat, polished specimens are essential to produce representative chemical analyses.

RESULTS

Gulf Coast samples

XRD analysis revealed similar mineralogical trends with increased depths of burial to those reported elsewhere (e.g., Perry and Hower, 1970; Hower *et al.*, 1976).

These include an increased proportion of illite layers in the dominant I/S (Figure 1) and the gradual appearance of chlorite at depths >2800 m. No apparent change, however, was detected in the relative proportion of kaolinite with increasing depth of burial. Increased illitization is supported by microprobe analyses of the I/S, which showed greater amounts of K in the I/S with increased depth of burial (Figure 1 and Table 1). The data presented in Table 1 also indicate that the Al contents of the I/S phases in the matrix and cavity environments decreased with increasing depth of burial. These results contrast with some previous studies in which the illitization of the I/S was found to be accompanied by increased Al contents (Hower *et al.*, 1976), by the reaction:

$$Smectite + K^+ + Al^{3+} \rightarrow Illite + Si^{4+}.$$

The data of the present study suggest that illitization was accompanied here by loss of Al^{3+} and the incorporation of Mg/Fe^{2+} into the structure to provide the increased layer charge required for illite formation.

From the BEI/probe analyses of these specimens, the details of in situ minerals, associations, and microstructures were revealed (Figure 2) as follows:

1. The sediments are relatively uncompacted throughout the section and show only minor evidence of grain fracture (e.g., microfossil test in Figure 2c) or pressure dissolution between quartz and K-feldspar (Figure 2d). Although some evidence of compaction on a broad scale exists, Figures 2e and 2f show that the relatively non-oriented nature of the matrix clay is preserved.
2. The dissolution of K-feldspar is followed by the neoformation of phyllosilicates (kaolinite, chlorite, I/S) in most of the solution cavities, as shown in Figures 2a, 2c, 2e, 2f, and 2g.
3. Figures 2c and 2e attest to the existence and maintenance of chemical equilibrium on a local scale within preserved microfossil tests during burial. Both figures illustrate an early chlorite lining the test which, as Figure 3 suggests, is consistent in chemical composition throughout the depth range studied. Crystallization of kaolinite or I/S appears to have followed the formation of the chlorite in most of the specimens where this association was recorded, surrounding the chlorite and filling any remaining void space in the test. The composition of the I/S in the cavity environment is distinctly K-poor with respect to matrix I/S at shallower depths (Figure 1 and Table 1), but with increasing depths of burial, the difference between matrix and cavity compositions diminishes.

North Sea samples

Some recent studies have suggested parallels between shale diagenesis and burial history in the North

Table 1. Typical microprobe analyses and calculated formulae of two Gulf Coast samples (wt. %).[1]

			2430-m depth			2940-m depth		
			Chlorite	Illite/smectite		Chlorite		Illite/smectite	
			Cavity fill	Cavity fill	Matrix	Cavity fill	Matrix	Cavity fill	Matrix
SiO_2			32.67	59.10	52.56	27.08	31.12	58.29	58.14
TiO_2			0.33	0.00	0.62	0.08	0.03	0.48	0.60
Al_2O_3			20.98	22.70	22.21	21.05	20.83	18.94	17.90
FeO^2			28.35	1.90	5.03	31.07	30.04	3.84	4.46
MgO			6.40	3.13	2.32	6.87	6.88	3.03	3.09
CaO			0.35	0.44	1.02	0.67	0.18	0.18	0.57
Na_2O			0.64	1.50	0.86	0.32	1.01	1.53	1.07
K_2O			0.36	1.39	3.30	0.12	0.87	6.30	4.59
Total			90.08	90.16	87.92	87.26	90.96	92.59	90.42
			(a)	(b)	(b)	(a)	(a)	(b)	(b)
Si	}	IV	3.33	7.77	7.37	2.94	3.19	7.79	7.90
Al	}		0.67	0.23	0.63	1.06	0.81	0.21	0.10
Al	}		1.85	3.29	3.04	1.64	1.71	2.78	2.76
Ti	}	VI	0.03	–	0.07	–	–	0.04	0.07
Fe	}		2.42	0.20	0.59	2.83	2.58	0.44	0.51
Mg	}		0.97	0.62	0.48	1.12	1.05	0.67	0.62
Ca			0.04	0.07	0.15	0.08	0.01	0.02	0.09
Na			0.12	0.37	0.23	0.07	0.20	0.40	0.29
K			0.05	0.24	0.59	0.01	0.11	1.08	0.79

[1] Formulae recalculated on the basis of: (a) 14 equivalent oxygen atoms for chlorite and (b) 22 equivalent oxygen atoms for illite/smectite.

[2] Total Fe expressed as FeO.

Sea with those determined for Gulf Coast samples, but these studies have been based entirely on mineralogical analyses of bulk samples (Pearson et al., 1982, 1983). Even within a regional geological context, such a direct comparison should be viewed with caution. Clearly, the Gulf Coast basin (as reviewed by Jackson and Galloway, 1984) has undergone a much simpler and more rapid subsidence than the North Sea basin (see Kent, 1975; Ziegler, 1978). Such differences may manifest themselves in the diagenetic history of the sediments.

Variations in the nature of the I/S and accompanying clay minerals as identified by XRD are summarized in Table 2. Figure 4 illustrates XRD patterns of typical samples listed in Table 2. In the samples from the Brae field, a change can be seen from smectite-rich, random I/S in the Lower cretaceous samples to ordered I/S in the Upper Jurassic samples. These variations are broadly similar to those reported by Pearson et al. (1983) for samples from the South Viking graben of the North Sea; however, in the samples from the Magnus field, random I/S phases were found in the Upper Cretaceous and Upper Jurassic samples, but illite-rich, ordered I/S was found in the Lower Cretaceous samples.

Such changes in the I/S composition related to stratigraphic position are marked by recorded unconformities in the succession and imply that the variations may be due chiefly to differences in detrital source materials rather than due solely to illitization reactions during burial diagenesis.

Figure 5 is a series of BEI micrographs of Upper Jurassic Kimmeridge Clay samples from different wells buried to >3 km. The four micrographs demonstrate the development of a well-defined compaction fabric as indicated by a planar, parallel arrangement of mineral flakes around the larger grains. Figure 5a shows an early (precompactional) growth of framboidal pyrite, and Figure 5b, a later (postcompactional) growth of euhedral pyrite. Figure 5b also illustrates the nature of detrital mica that has been partially transformed to kaolinite along the mica cleavage plane to produce extremely fine-scale mica-kaolinite intergrowths.

A common feature of these shales is the different degree of alteration of the Na- and K-feldspars. Whereas Figure 5c illustrates dissolution of K-feldspar, Figure 5d illustrates preferential dissolution of albite (possibly having previously replaced adjacent K-feldspar), with an infilling of I/S in the cavities produced.

Another feature that will affect the burial diagenetic sequence is the mineralogy and chemistry of the original sediment. Figure 6 shows BEI micrographs of a Lower Cretaceous marly shale (Cromer Knoll Group) that is significantly more calcareous and less feldspathic and micaceous than the Kimmeridge Clay discussed above. As a result of this difference in mineralogy, the shales have developed significantly different diagenetic mineralogies. From Figure 6 the following diagenetic reactions can be inferred: (1) dissolution of albite and replacement by calcite; (2) dissolution of calcite and subsequent formation of a chlorite rim; (3) partial al-

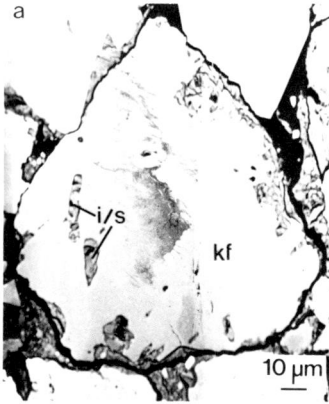

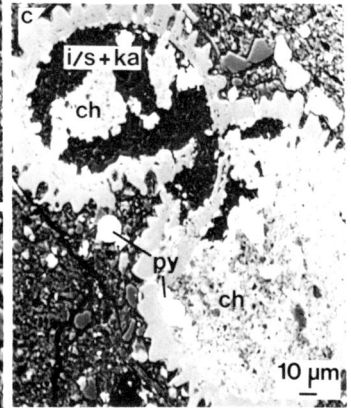

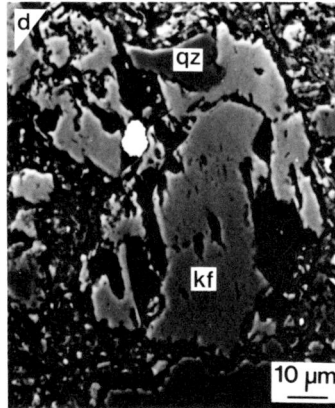

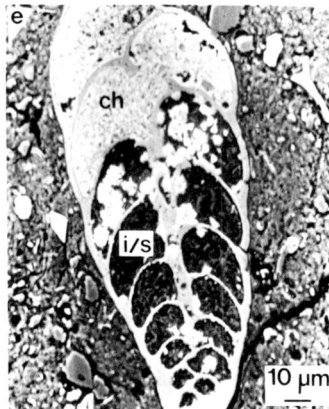

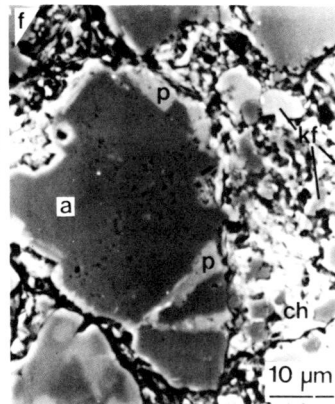

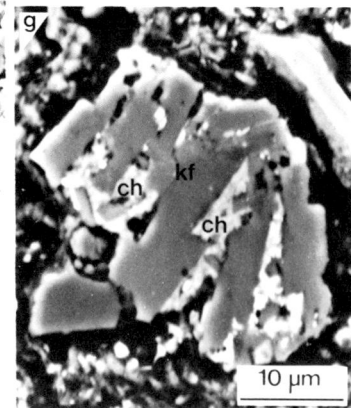

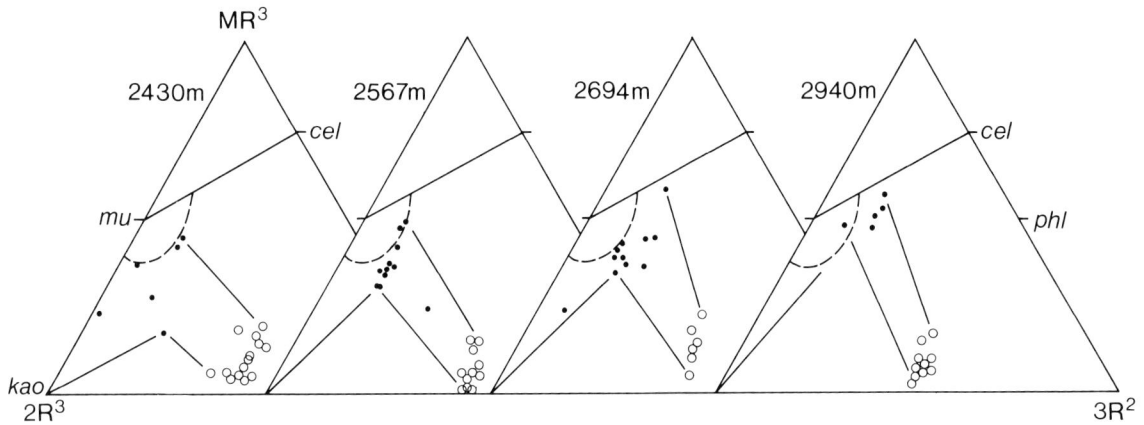

Figure 3. Chemical compositions of cavity clay fillings plotted on Velde (1977) MR^3–$2R^3$–$3R^2$ diagrams. Open circles = chlorite compositions; solid circles = illite/smectite (I/S) compositions. Although composition of the I/S varies with depth of burial, chlorite compositions are tightly grouped. Dashed field is the range of compositions of naturally occurring illites: Kao = kaolinite; mu = muscovite; cel = celadonite; phl = phlogopite.

Table 2. Variations in the compositions of illite/smectite (I/S) in Mesozoic shales from the Brae and Magnus fields of the North Sea.

Stratigraphic epoch	Depth (m)	Nature of I/S phase	
		Ordering	Smectite content (%)
Brae (Kimmeridge Clay Formation)			
Lower Cretaceous	3569	Random	70
	3630	Random	63
Upper Jurassic	3678	Ordered	20
	3685		
	3692		
	3697		
	3700		
	3704		
	3707		
	3710	↓	↓
	3715	Ordered	15
Magnus			
Upper Cretaceous (Shetland Group)	3110	Random	41
Lower Cretaceous (Cromer Knoll Group)	3118	Ordered	19
Upper Jurassic (Kimmeridge Clay Formation)	3129	Random	45

←

Figure 2. Backscattered electron image micrographs of Gulf Coast shale cuttings. Sample depths correspond to depths plotted on left-hand side of Figure 1. (a) Grain edge and surface dissolution of K-feldspar (kf) with illite/smectite (i/s) filling the cavities (1830 m). (b) Undercompacted fabric; little evidence of preferred matrix mineral orientation or pressure dissolution and/or distortion of detrital grains. Note later state *in situ* overgrowth of fine-grained pyrite (py) on skeletal carbonate (2276 m). (c) Authigenic clay fillings in skeletal carbonate remains. Only minor distortion of the skeletal remains due to compaction is noted. Chlorite (ch) is postdated by a mixture of kaolinite (ka) and I/S (i/s) (2430 m). (d) Increased degree of K-feldspar dissolution, creating extensive microporosity in the shale. Note also evidence of minor pressure dissolution in the center of the K-feldspar due to impingement of darker quartz grain (2567 m). (e) Complete carbonate skeleton filled with neoformed phyllosilicates illustrating the extent of localized chemical environments. Ch = chlorite, i/s = I/S (2694 m). (f) K-feldspar (kf) showing extensive dissolution and local chlorite (ch) development in surrounding matrix adjacent to plagioclase apparently composed of an almost wholly albitized core (a) and remnant calcic plagioclase (p) mantle (2940 m). (g) Chlorite (ch) authigenesis within cavities produced by K-feldspar (kf) dissolution. This texture is typical of the onset of widespread chlorite formation (2490 m).

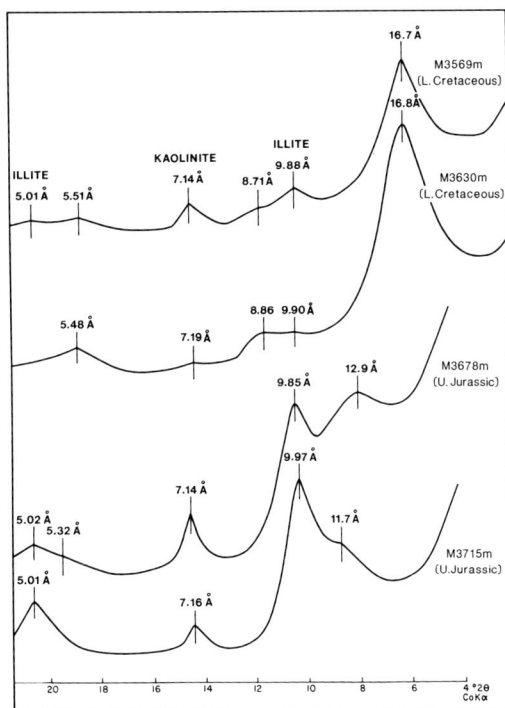

Figure 4.   Smoothed X-ray powder diffraction traces of <0.5-μm, Mg²⁺-saturated, ethylene glycol-solvated, oriented specimens of samples from Lower Cretaceous and Upper Jurassic of the Brae field in the North Sea. Traces show the presence of random I/S phases in the Lower Cretaceous samples and ordered I/S phases in the Upper Jurassic samples (for details, see Table 2) plus illite and kaolinite in all samples.

teration of K-mica to kaolinite containing local intergrowths of chlorite. Note, however, the relative stability of the K-feldspar and the lack of evidence for the neoformation of an illitic clay, in contrast to the Kimmeridge Clay Formation samples (Figure 5d).

Skeletal fossil debris in Kimmeridge Clay samples show a variety of authigenic mineral fillings (Figure 7). Early precompactional calcite cements were the first phases to precipitate (Figures 7a and 7d). Some of these early calcite cements are associated with and locally

replaced by chlorite (Figure 7d), or, more commonly, microcrystalline quartz (Figure 7a) in response to diagenetic changes in fluid chemistry. Subsequent precipitation of kaolinite of the type shown in Figure 7b or replacement of the original carbonate test by pyrite with a cavity infill of illite (Figure 7c) indicate further changes in the postcompactional pore-fluid chemistry.

DISCUSSION

By means of XRD analysis, BEI micrographs, and electron microprobe analysis, shales from both the Gulf Coast and North Sea appear to be complex, multiphase systems rather than the homogeneous sediments often implied from studies of their bulk mineralogy and geochemistry.

Diagenetic processes in the Gulf Coast involved release of K and Al from the feldspars. Most of the K released probably migrated out of the local microenvironment, although some may have been incorporated into the neoformed I/S within the dissolution cavities and, thus, was likely to have been less mobile than the K outside this microenvironment. The relative mobility of K and immobility of Al with respect to other ions have been proposed as major factors in controlling patterns of mineral authigenesis (Boles and Franks, 1979). The observations of the present study confirm this interpretation; the authigenesis process is well illustrated by the BEI techniques described.

Similarly, recognition of the existence of chemical microenvironments within preserved microfossil tests suggests an equilibrium buffering similar to that proposed by Hutcheon (1981). Here, the controlling factor was the K supply into the pore fluid of the microenvironment, which buffered the crystallization of either kaolinite or I/S as the pore-fluid chemistry changed. The depth-related influence of K mobility as evidenced by the changing chemical composition of I/S in cavity and matrix environments with increasing depth of burial is shown in Figure 1. An approach towards more widespread chemical equilibrium with increased depth of burial is suggested by the convergence of the K content in the I/S from the cavities and from the matrix.

→

Figure 5.   Backscattered electron image micrographs of North Sea shale core and cuttings. Sample depths correspond to those listed in Table 2. (a) Early framboidal pyrite (py) predating primary compaction of sediment (Magnus 3129 m). (b) Late-state euhedral pyrite (py) postdating compaction and overgrowing detrital mica (m) that has been altered to thin intergrowths of kaolinite (ka) (Magnus 3129 m). (c) Extensive K-feldspar dissolution (kf) postdating main compaction; note the relative stability of nearby quartz (qz) (Brae 3704 m). (d) Dissolution of albite (a) with illite/smectite (i/s) filling the cavities; note the relative stability of K-feldspar (kf) with respect to albite (Brae 3621 m).

Figure 6.   Backscattered electron image micrographs of North Sea shale core. Sample depths correspond to those listed in Table 2. (a) Perthitic feldspar illustrating the relative instability of albite (a), dissolved and replaced by calcite (ca), with respect to K-feldspar (kf). The light color of the matrix is due to abundant carbonate (Magnus 3118 m). (b) Detrital mica (m) altered to kaolinite (ka) with large carbonate plate (ca) rimmed by Fe-rich chlorite (ch) (Magnus 3118 m). (c) Albite (a) much dissolved and replaced by calcite (ca), itself rimmed by Fe-rich chlorite (ch) (Magnus 3118 m). (d) Detrital micas showing almost complete alteration to kaolinite (ka—center) and partial alteration to kaolinite (ka—lower left) having intergrowths of chlorite (ch) (Magnus 3118 m).

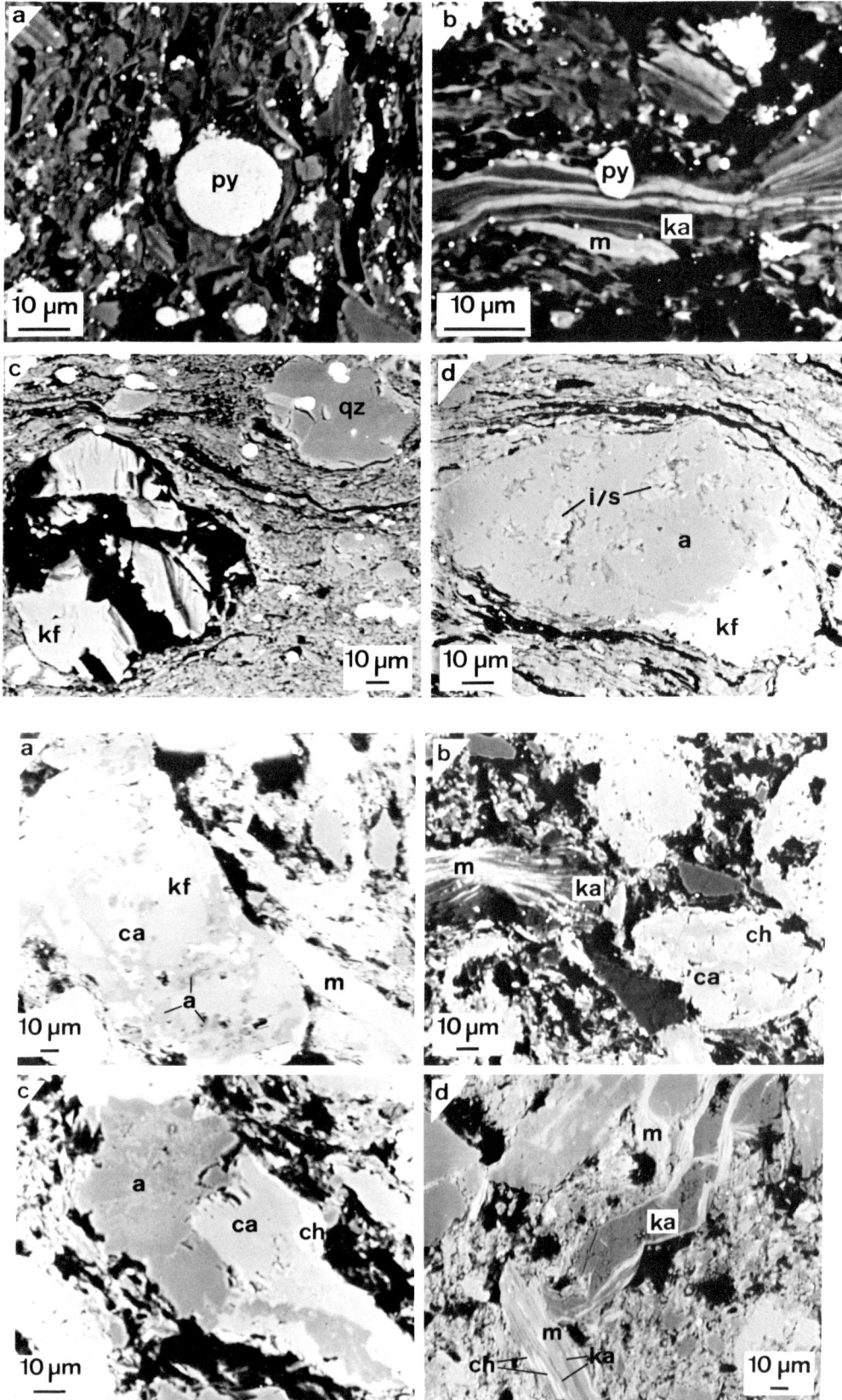

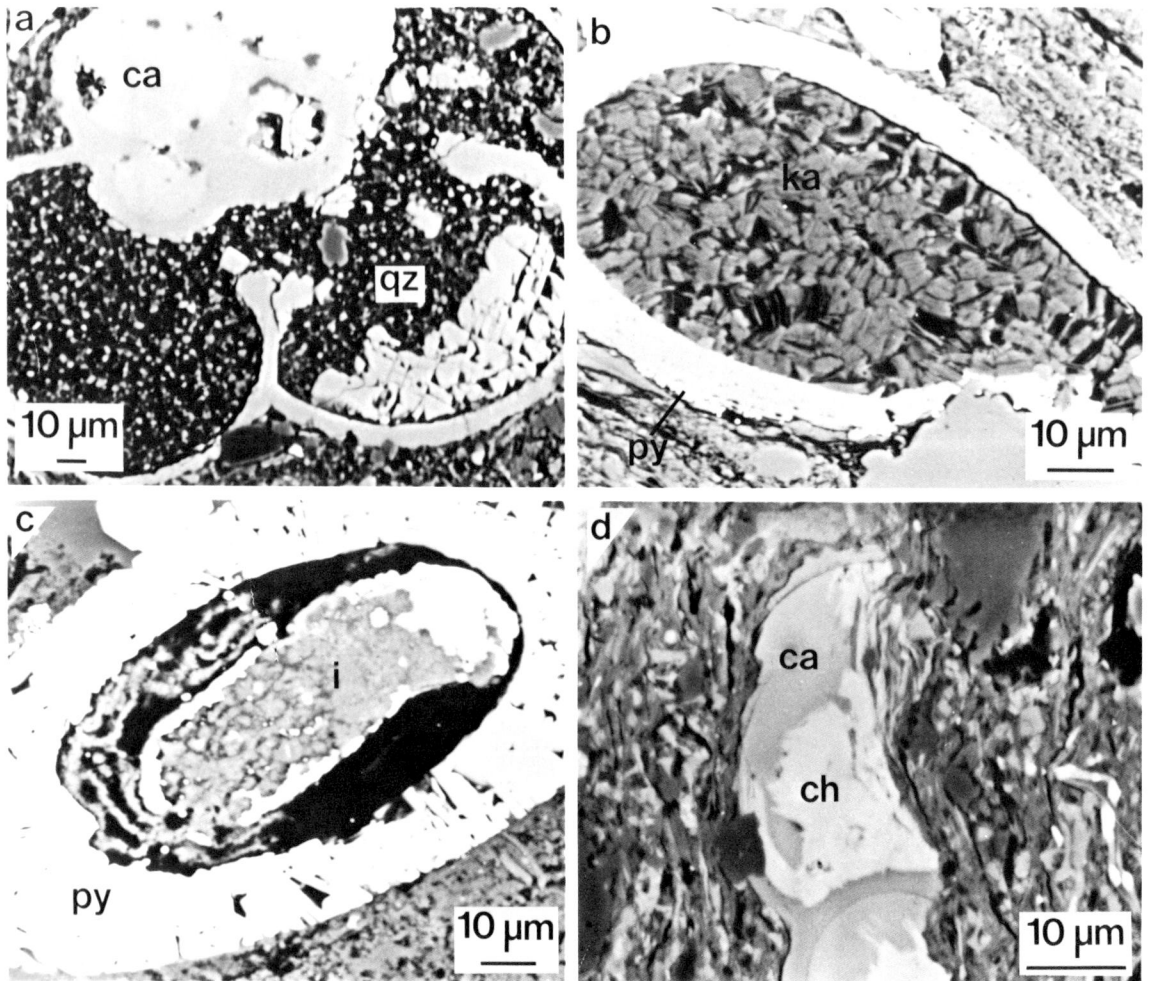

Figure 7.   Backscattered electron image micrographs of North Sea shale core and cuttings. Sample depths correspond to those listed in Table 2. (a) Early calcite (ca) cement filling partially dissolved and replaced by microcrystalline quartz (qz) (Brae 3569 m). (b) Blocky kaolinite (ka) infilling and incipient pyritization (py) of carbonate test (Brae 3707 m). (c) Complete pyritization (py) of carbonate skeletal remains (postcompaction) and a subsequent cavity filling of illite (i) (Brae 3710 m). (d) Early (precompaction) calcite (ca) lining carbonate skeleton and subsequent chlorite (ch) fillings replaced by matrix after compaction: note indurated quartz (qz) grain penetrating bottom of test and fill (Brae 3631 m).

The North Sea examples do not show the simple depth-related changes in bulk I/S composition, as calculated from XRD analyses, that were found for the Gulf Coast samples. These North Sea samples show variations in the I/S compositions that apparently reflect changes in detrital source materials rather than the result of illitization reactions. Hurst (1982) also proposed that an ordered I/S in some Jurassic shales from the North Sea province could be detrital in origin and that its presence may reflect source-provenance changes. These conclusions disagree with the interpretations of Pearson *et al.* (1982, 1983), who proposed that changes in the nature of the I/S phases with increased depth of burial and/or geological age in Mesozoic shales of the North Sea were diagenetic in origin. Differences in source materials and their possible in-

fluence on the evolution of pore fluids could also account for the effects of diagenesis being significantly different in the adjacent lithologies of the Cromer Knoll Group and the Kimmeridge Clay Formation (Figures 5 and 6).

The variety of authigenic minerals filling the skeletal components in the Kimmeridge Clay samples (Figure 7) testify to various diagenetic changes in pore-water chemistry. The petrographic evidence suggests: (1) connate marine waters and early precipitation of calcite followed by later chlorite and microcrystalline quartz (Figures 7a and 7d); (2) a subsequent stage of dilute, acidic pore fluids that dissolved calcite and precipitated kaolinite (Figure 7b); in some samples the kaolinite replaces microcrystalline quartz and thus postdates its formation; and (3) a postcompactional stage of alkaline fluids caused the formation of illite and, in some samples, pyrite, suggesting anoxic conditions (Figure 7c). Note, however, that the various stages of flushing were not pervasive because remnants of each stage survived.

The Mesozoic shales from the North Sea appear to have undergone variable and complex diagenetic changes that differ from those that affected samples from the Gulf Coast. Variable detrital mineralogy and fluctuating pore-fluid chemistry, dependent on the interaction of formations with deep circulating ground waters in the basin (Bjørlykke *et al.*, 1979) produced a more complicated diagenetic mineralogy than that observed in the Gulf Coast.

SUMMARY

The petrologies of shales from the Gulf Coast and North Sea are significantly different and are related to the different depositional lithologies and different diagenetic histories of these regions. The principal differences are summarized below:

North Sea Samples	*Gulf Coast samples*
Well-compacted fabrics	No well-developed compaction fabrics
Variable stability of feldspars	K-feldspars are unstable
Transformation of detrital micas to kaolinite	
No pervasive carbonate cements, but recrystallization and filling of carbonate tests and later replacement of carbonate by quartz, kaolinite, chlorite, pyrite, and illite	Restricted, but pervasive carbonate cements that subsequently dissolved, but which show no evidence of later replacement by other phases
Microenviroments in which precipitation and dissolution took place according to fluctuations in porewater chemistry	Microenvironments showing a simple, two-stage precipitation history (chlorite, followed by either kaolinite or I/S), indicating restrictions on fluid flow and subsequent mobility of components

ACKNOWLEDGMENTS

This research was funded by Schlumberger Cambridge Research. We thank the British Geological Survey, British Petroleum, and Marathon Oil (U.K.) Ltd. for providing the samples.

REFERENCES

Boles, J. R. and Franks, S. G. (1979) Clay diagenesis in Wilcox sandstones of southwest Texas: implications of smectite diagenesis on sandstone cementation: *J. Sediment. Petrol.* **49**, 55–70.

Bjørlykke, K., Elverhoi, A., and Malm, A. O. (1979) Diagenesis in Mesozoic sandstones from Spitzbergen and the North Sea—a comparison: *Geol. Rundsch.* **68**, 1152–1171.

Hower, J., Eslinger, E. V., Hower, M. E., and Perry, E. A., Jr. (1976) Mechanism of burial metamorphism of argillaceous sediment: 1. Mineralogical and chemical evidence: *Geol. Soc. Amer. Bull.* **87**, 725–737.

Hurst, A. (1982) The clay mineralogy of Jurassic shales from Brora, N. E. Scotland: in *Proc. Int. Clay Conf., Bologna, Pavia, 1981,* H. van Olphen and F. Veniale, eds., Elsevier, Amsterdam, 677–684.

Hutcheon, I. (1981) Applications of thermodynamics to clay minerals and authigenic mineral equilibria: in *Clays and the Resource Geologist,* F. J. Longstaffe, ed., Short Course Handbook, Mineral Assoc. Can., Toronto, Ontario, 169–193.

Jackson, M. P. A. and Galloway, W. E. (1984) *Structural and Depositional Styles of Gulf Coast Tertiary Continental Margins: Application to Hydrocarbon Exploration:* Continuing Education Course Notes Series, No. 25, Amer. Assoc. Pet. Geol., Tulsa, Oklahoma, 226 pp.

Kent, P. E. (1975) Review of North Sea basin development: *J. Geol. Soc. London* **131**, 435–468.

Krinsley, D. H., Pye, K., and Kearsley, A. T. (1983) Application of backscattered electron microscopy in shale petrology: *Geol. Mag.* **120**: 109–114.

Pearson, M. J., Watkins, D., Pitton, J. L., Caston, D., and Small, J. S. (1983) Aspects of burial diagenesis, organic maturation and palaeothermal history in an area in the South Viking graben, North Sea: in *Petroleum Geochemistry and Exploration of Europe,* J. Brooks, ed., *Spec. Publ. Geol. Soc. London,* Blackwells, Oxford, 161–173.

Pearson, M. J., Watkins, D., and Small, J. S. (1982) Clay diagenesis and organic maturation in northern North Sea sediments: in *Proc. Int. Clay Conf., Bologna, Pavia, 1981,* H. van Olphen and F. Veniale, eds., Elsevier, Amsterdam, 665–676.

Perry, E. A., Jr. and Hower, J. (1970) Burial diagenesis in Gulf Coast pelitic sediments: *Clays & Clay Minerals* **18**, 165–177.

Pye, K. and Krinsley, D. A. (1983) Inter-layered clay stacks in Jurassic shales: *Nature (London)* **304**, 618–620.

Reynolds, R. C. and Hower, Jr. (1970) The nature of interlayering in mixed layer illite-montmorillonites: *Clays & Clay Minerals* **18**, 25–36.

Scholle, P. A. and Schluger, P. R. (1979) *Aspects of Diagenesis: Spec. Publ.* **26**, Soc. Econ. Paleontol. Mineral., Tulsa, Oklahoma, 443 pp.

Velde, B. (1977) *Clays and Clay Minerals in Natural and Synthetic Systems:* Elsevier, Amsterdam, 218 pp.

Welton, J. E. (1984) *SEM Petrology Atlas:* Methods in Exploration Series, No. 4, Amer. Assoc. Pet. Geol., Tulsa, Oklahoma, 237 pp.

White, S. H., Huggett, J. M., and Shaw, H. F. (1985) Electron-optical studies of phyllosilicate intergrowths in sedimentary and metamorphic rocks: *Miner. Mag.* **49**, 413–423.

White, S. H., Shaw, H. F., and Huggett, J. M. (1984) The use of backscattered electron imaging for the petrographic study of sandstones and shales: *J. Sediment. Petrol.* **54**, 487–494.

Ziegler, P. A. (1978) Northwestern Europe: tectonics and basin development: *Geol. Mijnbouw* **53**, 43–50.

Proceedings of the International Clay Conference, Denver, 1985, L. G. Schultz, H. van Olphen, and F. A. Mumpton, eds.,
The Clay Minerals Society, Bloomington, Indiana, 144–150 (1987).

INFLUENCE OF SHALE FABRIC ON ILLITE/SMECTITE DIAGENESIS IN THE OLIGOCENE FRIO FORMATION, SOUTH TEXAS

JAMES J. HOWARD

Schlumberger-Doll Research, Ridgefield, Connecticut 06877

Abstract—Variations in mixed-layer illite/smectite (I/S) expandability within a 60-m interval of shale from the lower Frio Formation of south Texas are much greater than expected from a simple temperature-dependent burial diagenesis model. This variation is associated with differences in shale fabric that appear to have influenced rates of diagenesis rather than with differences in original composition. The section investigated consists of two distinct shale fabric types interbedded with several-meter thick beds of fine-grained sandstone. Laminated and massive shales differ by the presence of either discrete laminations of silt or dispersed grains of silt, respectively.

The shales are composed of 65–80% clay minerals, with quartz, plagioclase, and calcite as the other major constituents. The predominant clay mineral is an I/S that ranges in expandability from 45 to 100%. Massive shales contain I/S that ranges in expandability from 60 to 95%. The I/S in the massive shales is on the high side of the range of expandability observed in other Gulf Coast wells of similar depth and age. In contrast, the I/S in the laminated shale contains 45 to 80% expandable layers.

Both massive and laminated shales contain little K-feldspar and nearly equal amounts of K. The entire section has about the same total K content as other shallow Tertiary Gulf Coast sections. The difference in I/S expandability between the massive and laminated shales is not reflected in their bulk K content. I/S expandability therefore does not appear to be controlled simply by the shale's bulk composition. Shale fabric, as measured by its effect upon fluid flow properties, is a critical factor in understanding the difference in the relative extent of reaction, as measured by I/S expandability. The thin silt laminations in the laminated shale act as fluid conduits, providing both more reactants and more sites available for reaction.

Key Words—Diagenesis, Fabric, Illite/smectite, Interstratification, Shale, Smectite.

INTRODUCTION

Interest in interstratified clays, especially illite/smectites (I/S), has extended to their application as geothermometers and as potential sources of enhanced fluid pressures in basin modeling (Hoffman and Hower, 1979; Burtner and Warner, 1983; Keith and Rimstidt, 1985). The degree of illitization as an indicator of the extent of reaction reflects the time-temperature history of a basin and thereby provides a potential inorganic measure of source rock maturation (Foscolos *et al.,* 1976; Dypvik, 1983). The degree of illitization also may be a measure of fluid pressure and composition conditions in different lithologies. Proponents of both equilibrium and kinetic interpretations of I/S compositions have applied clay mineral diagenesis results to basin models, either by applying maximum burial temperatures as assumed in equilibrium models or a time-temperature integral that reflects kinetic considerations (Morton, 1985). The timing of the onset and the subsequent rate of smectite dewatering during I/S alteration may also play a role in the formation of overpressured zones (Foscolos, 1984; Bruce, 1984; Keith and Rimstidt, 1985).

These efforts in clay mineral geothermometry are based on studies of I/S diagenesis in the Gulf Coast Tertiary section (Hower *et al.,* 1976; Perry and Hower, 1970, 1972; Boles and Franks, 1979; Freed, 1982; Weaver and Beck, 1971). These studies showed a gradual monotonic decrease in I/S expandability from >80% smectite layers to <20% expandable layers over a depth range of several kilometers. The absolute depths correspond to present-day temperatures of 80°–100°C at the top of the illitization zone to 130°–170°C at the base of the "active" diagenesis interval. The overlap of the interval of maximum clay diagenesis with the intervals related to oil generation and to over-pressured zones suggests that clay mineral diagenesis is intertwined with both hydrocarbon generation and migration and the creation of elevated fluid pressures.

This commonly observed distribution of depth/temperature-related expandabilities can be evaluated as a first-order kinetic reaction, providing activation energies and pre-exponential factors that are applicable in thermal history modeling (Keith and Rimstidt, 1985). In both kinetic and equilibrium evaluations of I/S distributions, temperature is the dominant factor that influences the observed extent of reaction. Experiments have demonstrated that fluid composition is also im-

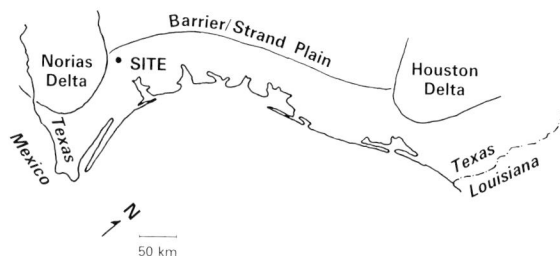

Figure 1. Depositional setting of lower Frio Formation taken from regional study of cores and logs (Galloway *et al.*, 1983). Samples in this study were taken from a well proximal to the Norias delta complex.

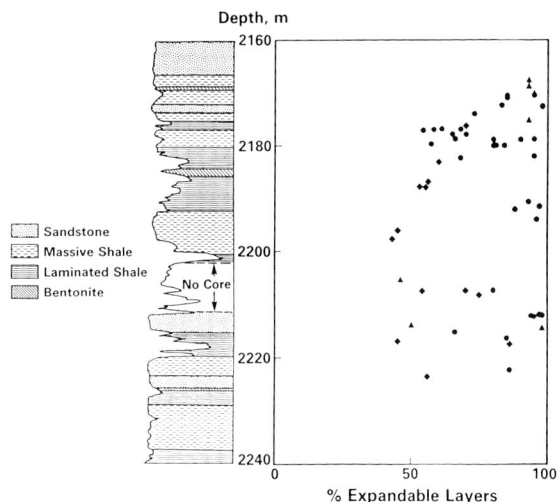

Figure 2. Stratigraphic section illustrating relative distribution of major shale lithologies plus sandstones and thin bentonite zones. Caliper log emphasizes difference in texture and strength of massive and laminated shales. Percentage of expandable layers in I/S within interval for massive shales ●, laminated shales ◆, and sandstones ▲. Estimated error for individual expandabilities ±5%. Depths are correct to within 10 cm.

portant in determining the extent of illitization (Roberson and Lahann, 1981; Howard and Roy, 1985). The application of these experimental results to shale diagenesis is hindered by the absence of fluid composition data from deeply buried sections. The recognition of different amounts of I/S expandability in proximal sandstones and shales, however, indicates that fluid composition differences in natural settings can influence diagenesis (Boles and Franks, 1979; Howard, 1981). The effect of fluid pressures on mineral stability or reaction rates is less well understood, though the key variable appears to be the ratio of fluid to lithostatic pressures (Colten, 1985).

A principal purpose of the present study is to illustrate that over a relatively short depth interval I/S expandability varies markedly. The study also examines the relationship between I/S expandability and the lithology and fabric of different shales. A third purpose of the study is to characterize the source of the different shale lithologies in the hope of determining whether the differences in I/S expandability are due to differences in original composition or in reaction rates and mechanisms of reaction.

MATERIALS AND METHODS

The cored interval investigated in this study was obtained from a well in south Texas that penetrated the Oligocene Frio Formation. The samples were from the lower Frio, at a depth of 2200 m, in a well from a transitional depositional setting between shallow marine and the nearshore (Figure 1). Paleontological evidence for this interval indicated brackish to shallow marine conditions (P. Basan, ERCO, Houston, Texas, personal communication, 1984). The source of sediments for this locality during deposition of the Frio was the Norias delta complex, which is characterized by greater volcanic input than that associated with the Houston delta complex (Galloway *et al.*, 1983).

The 60-m section of core examined here consisted of thick shale intervals along with several thin sandstones that comprised less than 20% of the total section (Figure 2). The shales are characterized by two dom-

inant lithologies that differ in texture and that were identified in hand specimen. Both lithologies contain about 10% silt-size quartz. The first lithology is a massive, well-indurated shale that displays no bedding and has its silt dispersed throughout. The second shale lithology contains distinct, 0.1- to 10.0-mm thick, layers of silt. Thin section petrography of these two types of shale revealed no large or systematic difference in the orientation of clay particles. These two major shale lithologies were distinguished with a caliper log in this well, because the laminated shales are more susceptible to borehole failure than the massive shales. The sandstones are mostly fine to very fine grained and contain few sedimentary structures other than laminations with reverse size-grading. Several zones, <50-cm thick, of very finely laminated clay-rich layers are also present and contain only a small percentage of silt. These thin zones comprise a third minor shale lithology type.

Ninety samples of shale were taken from the 60-m core at regular intervals, insofar as the core was preserved. To study vertical heterogeneity of the shale composition, several parts of the core were sampled at 1-cm spacings. Sedimentary structures and bedding features were described megascopically, and silt distribution and particle orientation were verified with thin section petrography.

Mineralogy was determined by X-ray powder diffraction (XRD) analysis of both the bulk samples, and the <1-μm size fractions. A Philips diffractometer and CuKα radiation was used. Quantitative estimates of

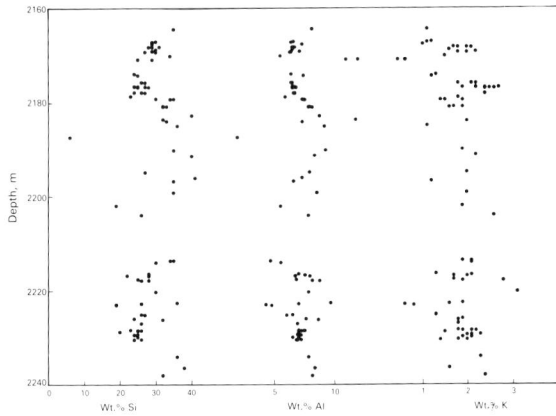

Figure 3.   Bulk Si, Al, and K composition of Frio shales and sandstones from 2165 to 2225 m.

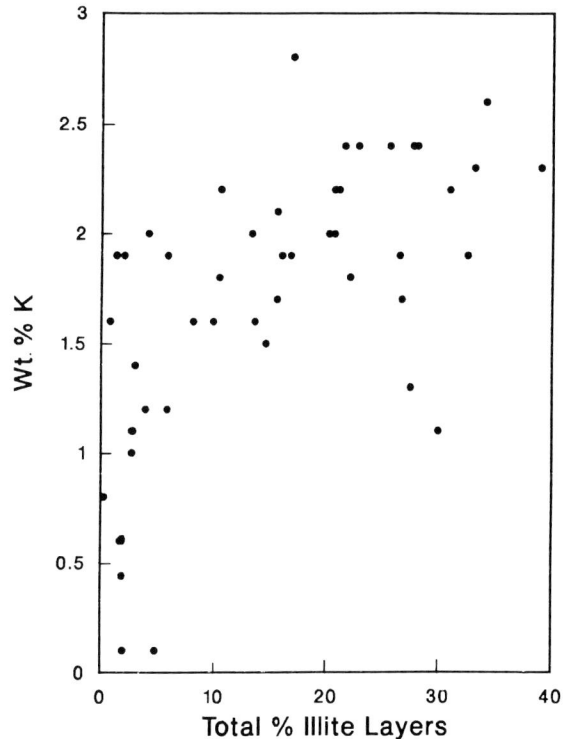

Figure 4.   Total percentage of illite layers in samples (% illite in I/S times the abundance of I/S plus any discrete illite) and their bulk K content. Trend for a 100% I/S clay shale is defined by line from origin to 2.7% K at 40% total illite layers.

bulk mineralogy were made using peak areas and a set of external standards. XRD patterns of the <1-μm fraction were obtained from oriented mounts of air-dried, ethylene-glycol solvated, and, for some K-saturated samples. The expandability of the I/S clays was determined by the positions of the $003_{10}/005_{17}$ and $004_{10}/006_{17}$ reflections, and, if necessary, the lower-angle $002_{10}/003_{17}$ reflection using experimentally established graphs (Środoń, 1980, 1981). The presence of only small amounts of discrete illite in these samples reduced the error in estimating the percentage of expandable layers in the I/S to less than $\pm5\%$ of the absolute value.

Chemical analyses of major and trace elements in the bulk samples were made by instrumental neutron activation analyses (INAA). The rare earth element (REE) values were normalized with respect to the North American shale composite (NASC) to facilitate comparison of this study with previous work (Gromet *et al.,* 1984).

RESULTS

The mineralogical and chemical composition of the two types of shales was quite uniform, making it difficult to discriminate lithology based upon bulk composition. The shales contained 65–80% total clay minerals, of which I/S was the dominant clay, with lesser amounts of discrete illite and kaolinite. Quartz comprised 10–25% of the total sample, and generally occurred as silt-size particles. Several weight percent plagioclase feldspar was present in all shale samples, but only a few samples contained detectable K-feldspar. All of the sandstones contained several weight percent K-feldspar, some of which had distinctive euhedral morphologies as observed by scanning electron microscopy. Many of the shales contained ~10% calcite, although several intervals contained no calcite. These

calcite-free zones were not restricted to a particular shale lithology, and the presence of fossil molds in several of these samples suggests that leaching was responsible for the absence of calcite, rather than its non-deposition. The third type of shale, i.e., the thin clay-rich zones, was composed almost entirely of a fully expandable smectite.

The average chemical composition of the shales from this interval was similar to the shallow samples from well CWRU No. 6 located in the Houston delta complex (Hower *et al.,* 1976). This composition reflects the presence of aluminosilicates and abundant calcite. The distribution patterns of Si and Al were similar with depth (Figure 3). The mean Si content for all the shales in this interval was $28.3 \pm 5.0\%$ whereas the Al mean was $7.3 \pm 1.1\%$. Several of the smectite-rich clay beds of the third type of shale were distinguished by Al contents $>10\%$. The calcareous shales averaged 5.5% Ca, a value that is comparable to the shallowest calcareous samples in well CWRU No. 6. The systematic increase of Si and Al at several intervals, e.g., between 2180 and 2190 m, represents a non-calcareous zone and, hence, increased aluminosilicate components, rather than a change in aluminosilicate mineralogy.

The studied interval had a mean whole-rock K con-

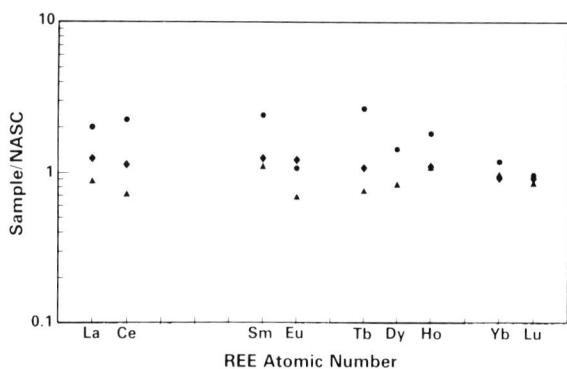

Figure 5. Rare earth element abundances normalized to North American shale composite (NASC). ▲ Average value for 84 shales sampled in this study. ● Average value for 6 samples from this interval characterized by a pure smectite content. These samples were interpreted to represent bentonite deposits. ◆ Average value for samples from an offshore Louisiana shale sequence (Chaudhuri and Cullers, 1979).

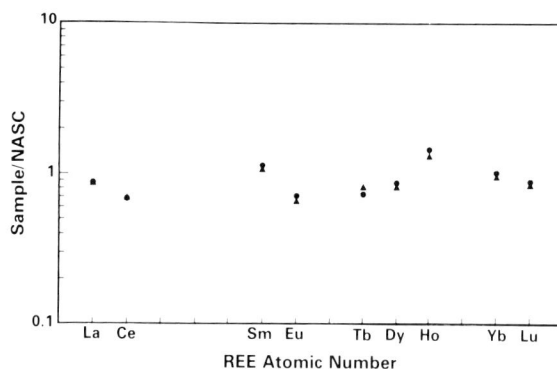

Figure 6. Rare earth abundances normalized to North American shale composite (NASC) for 61 massive shales ▲, and 23 laminated shales ●, demonstrate similar distribution patterns. La and Ce values for each lithology overlap.

tent of $1.9 \pm 0.6\%$, which is similar to values measured in shallow samples of well CWRU No. 6, despite the lack of K-feldspar observed in this well. The absence of other K-bearing phases allows a direct correlation of bulk K and the percentage of illitic layers in the entire sample (Figure 4). The total percentage of illite layers in a sample is the product of the I/S non-expandability times the abundance of I/S, plus the amount of discrete illite. Based on a K content of 6.7% for discrete illite, most of the samples contained more K than expected from the calculated total percentage of illite layers. Many of the samples that contained <10% total illite layers were from more sandy samples and contained only a small amount of highly expandable clay in addition to a small percentage of K-feldspar. The more illite-rich samples represented the shales, although no clear discrimination between massive and laminated lithologies could be made due to the variable amount of I/S in each sample. These shales still had more K than expected, based on a simple I/S contribution, indicating that either another source of K was present in the shale or that the estimates of either I/S expandability or abundance were wrong.

The major element composition of the bulk samples clearly cannot discriminate between the two major lithologic types of shale. Even the sandstone intervals contained enough clay minerals and K-feldspar to render their Al and Si composition indistinguishable from that of the shales.

The average concentrations of individual rare earth elements in these shales were similar to those of the NASC. Interpretations of source and chemical alteration often are based upon subtle differences in NASC-normalized REE abundances (Gromet et al., 1984). The difference in the abundances of the light rare earth

elements (LREE; La to Sm) specifically between samples from south Texas and offshore Louisiana was sufficient to suggest two separate sources (Figure 5). Several of the thin clay-rich zones contained rare earth elements two to three times more abundant than that of the Frio rocks, suggesting a different provenance than the rest of these shales (Figure 5). In contrast, the REE signatures offer no clear discrimination of the two major lithologic types of shales (Figure 6). Several highly calcareous samples, i.e., Ca > 15%, contained significantly lower rare earth abundances, most likely due to dilution of the very low rare earth contribution of the carbonates rather than due to another terrigenous source (Fleet, 1983).

The vertical distribution of the I/S clay was found to be the most important parameter for distinguishing this sedimentary sequence from the numerous others that are cited in the literature. Within the 60-m interval, the I/S ranged in expandability from 100% expandable layers in the several thin, clay-rich zones to <45% expandable layers in the laminated shales (Figure 2). No clear trend in expandability was evident with depth nor with any compositional variations, e.g., calcareous vs. noncalcareous shales. Expandability, however, could be correlated with shale lithology or fabric. The massive shales had an average expandability of 81% but the laminated shales contained an average of 59% expandable layers. K-saturation resulted in an uneven change in I/S expandability, and many samples exhibited a 5–15% reduction in the number of expandable layers. Expandability of other I/S clays was not reduced by K-saturation, and no readily apparent correlation was found between K-saturated behavior and lithology. In addition to their behavior upon K-saturation, a 001 basal spacing of the I/S of 16.9 Å and less suggested that some of the smectites had high-layer charges (Środoń, 1980).

DISCUSSION

Such a clear difference in I/S expandability in the two major lithologic types of shale suggests that two sources of different original clay composition contributed to this Frio stratigraphic section. Several lines of evidence, however, point to a single source for this particular interval. First, whereas the averages are distinct, a wide range of individual expandabilities was noted within each of the two types of shale (Figure 2). A second factor is that the site of this particular well is close to the Norias delta source, and therefore high sedimentation rates would probably have diluted contributions from more distant sources (Figure 1).

The most persuasive argument for a single source is the rare earth signature of the major shale lithologies (Figure 5). Rare earth distributions in shales traditionally have been ascribed to source characteristics; they appear to be only slightly influenced by post-depositional processes (Cullers *et al.,* 1979; Fleet, 1983). The small difference in LREE abundances for this south Texas well and for an offshore Louisiana well is sufficient to suggest the presence of two different source compositions (Figure 4, see also Chaudhuri and Cullers, 1979). South Texas was supplied by volcaniclastics from the west, whereas offshore Louisiana received sediment via the Mississippi River from the mid-continent. Because of the apparent sensitivity of REE in detecting sediment sources, the near similarity of the REE distributions in the massive and laminated shales in this interval strongly suggest only one source. Samples from the several thin zones of pure clay had REE compositions that were twice that of the NASC, and were comparable to values associated with silicic volcanic rocks (Fleet, 1983). A similarity in the REE composition of calareous and non-calcareous samples indicates that the diagenetic leaching of calcite from some of the shales did not alter the REE signature. This pattern indicates that original source rather than diagenesis played the more significant role in the REE composition of these shales.

The regional setting of south Texas Oligocene deposition suggests episodic volcanic events during a long period of humid weathering and rapid transport of volcanic material to depositional sinks (Galloway *et al.,* 1983). The thin clay layers with high rare earth element concentrations likely represent "bentonitic" ash falls that altered from glass to clay, whereas the weathering of volcanic debris at the surface produced a smectite-rich, rare earth-depleted sediment.

The above arguments suggest that the original clay mineral composition of this interval was the same throughout, but the question remains by what processes did illitization proceed within the various shale lithologies. Because so little is known experimentally about how fluid compositions and pressures influence both reaction kinetics and thermodynamic stability,

and because analyses of actual fluid compositions and pressures in shales are lacking, a definitive model for smectite diagenesis cannot be proposed at this time. If the extent of illitization considered reflects an equilibrium with environmental conditions, the range of observed expandabilities requires large variations in temperatures, pore fluid compositions, and pressures over this short vertical interval. A more plausible explanation is that the smectite alteration progressed at different rates within the several types of shales. The greater degree of illitization in the laminated shales indicates that the reaction proceeded faster here. Assuming that fluid compositions and temperatures in the various shales were constant within this 60-m interval, the critical variable is the rate at which material was supplied to and from the reaction site. Illitization should be faster in more permeable shales if the necessary materials are supplied faster than they can be removed. The importance of thin silt laminations in shales for enhancing permeability is illustrated by gas production in Devonian shales wherein only laminated shales have sufficient permeability for production (Nuhfer *et al.,* 1979). Similarly, in the Frio shales the laminations were effective conduits for material transported to and from the clay mineral reaction sites. The massive shale containing dispersed silt particles experienced slower rates of diagenesis because of its intrinsic lower permeability.

The question of whether the potassium distribution observed in the bulk samples of this section were due to different original compositions or a redistribution process cannot be resolved with the data of this study. Without chemical analyses of the <1-μm size fraction, redistribution of K from the coarse- to fine-sized fractions cannot be shown. The lack of a 1:1 correlation between bulk K and the calculated total percentage of illite emphasizes the difficulties in mass balancing in seemingly simple systems. Here, shifting the analyses of samples with a constant K content to a higher total percentage of illite layers to coincide with the ideal illite trend required an increase of 10% relative to the original estimate of total clay abundance. The presence of K-feldspar in the sandstones, some of which was authigenic, indicates that diagenetic conditions were not severe enough to destroy the feldspar, as is commonly seen in deeper Gulf Coast sections (Hower *et al.,* 1976).

It is not intuitively obvious why the laminated shale would have more available reaction sites or a higher effective surface area than a massive shale. Because the two types of shales contain the same amount of clay, similar surface areas might be expected. In terms of ability to supply material to a selected site, the more permeable laminated shales clearly have the advantage over the massive shales. Preliminary measurements in this laboratory of surface areas normalized to pore volume also indicate that the laminated shales have a higher effective surface area than the massive shales. Intuition is obscured because surface areas clearly are not simply a summation of total clay content, which is the same in the two lithologies. Here, surface area in consolidated rocks is more a measure of the size and distribution of the clays and of pores.

CONCLUSIONS

The range in I/S expandability observed in this 60-m interval of the lower Frio Formation from south Texas does not fit the pattern of gradual transformation of smectite to illite previously described in many studies. There was apparently strong lithological control on the distribution of I/S expandability in this interval. Massive shales tend to contain the more expandable I/S clays, whereas a greater degree of illitization was observed in the laminated shales. Lithological and geochemical evidence, primarily the similar rare earth element signatures of the different lithologies, suggest that the different shales had the same source. Thus, observed I/S differences were probably the result of different diagenetic reaction processes or rates. Although temperature, fluid pressure, and fluid composition may have controlled broad trends of diagenesis, other factors must have operated in the small interval sampled. In addition, the rate of I/S transformation in shales was apparently influenced by lithology-controlled fluid flow properties of the shales. The fine silt laminations acted as fluid conduits and increased the effective surface area available for reaction.

One broad implication of these results is that the application of I/S compositions to basin geothermic models must be seriously evaluated in terms of the various factors that control rates of transformation. The rate of material supply, as controlled by permeability and the composition of the fluids, probably plays a major role in smectite diagenesis, especially on the scale of individual beds. A second implication for future work is the importance in sampling from the same lithologic shale unit when attempting to define diagenetic trends with depth.

ACKNOWLEDGMENTS

The author acknowledges the laboratory assistance of M. Supp, L. McGowan, D. Neuberger, and P. Dryden and the editorial comments of G. R. Thompson, G. Whitney, L. G. Schultz, K. M. Gerety, and K. A. Gruebel.

REFERENCES

Boles, J. R. and Franks, S. G. (1979) Clay diagenesis in Wilcox sandstones of southwest Texas: implications of smectite diagenesis on sandstone cementation: *J. Sediment. Petrology* **49**, 55–70.

Bruce, C. H. (1984) Smectite dehydration: its relation to structural development and hydrocarbon accumulation in Northern Gulf of Mexico Basin: *Amer. Assoc. Petrol. Geol. Bull.* **68**, 673–683.

Burtner, R. L. and Warner, M. A. (1983) Illite/smectite diagenesis and hydrocarbon generation in Cretaceous Mowry and Skull Creek shales of Northern Rocky Mountains-Great Plans region: *Amer. Assoc. Petrol. Geol. Bull.* **67**, 434–435.

Chaudhuri, S. and Cullers, R. L. (1979) The distribution of rare-earth elements in deeply buried Gulf Coast sediments: *Chemical Geol.* **24**, 327–338.

Colten, V. A. (1985) Experimental determination of smectite hydration states under simulated diagenetic conditions: Ph.D. Dissertation, University of Illinois, Urbana, Illinois, 144 pp.

Cullers, R. L., Chaudhuri, S., Kilbane, N., and Koch, R. (1979) Rare earths in size fractions and sedimentary rocks of Pennsylvanian–Permian age from the mid-continent of the U.S.A.: *Geochim. Cosmochim. Acta* **43**, 1285–1301.

Dypvik, H. (1983) Clay mineral transformations in Tertiary and Mesozoic sediments from North Sea: *Amer. Assoc. Petrol. Geol. Bull.* **67**, 160–165.

Fleet, A. J. (1983) Aqueous and sedimentary geochemistry of the rare earth elements: in *Rare Earth Element Geochemistry*, P. Henderson, ed., Elsevier, Amsterdam, 343–374.

Foscolos, A. E. (1984) Catagenesis of argillaceous sedimentary rocks: *Geoscience Canada* **11**, 67–75.

Foscolos, A. E., Powell, T. G., and Gunther, P. R. (1976) The use of clay minerals, inorganic and organic geochemical indicators for evaluating the degree of diagenesis and oil generating potential of shales: *Geochim. Cosmochim. Acta* **40**, 953–960.

Freed, R. L. (1982) Clay mineralogy and depositional history of the Frio Formation in two geopressured wells, Brazoria County, Texas: *Gulf Coast Assoc. Geol. Soc. Trans.* **32**, 459–463.

Galloway, W. E., Hobday, D. K., and Magara, K. (1983) Frio Formation of Texas Gulf coastal plain: depositional systems, structural framework, and hydrocarbon distribution: *Amer. Assoc. Petrol. Geol. Bull.* **66**, 649–688.

Gromet, L. P., Dymek, R. F., Haskin, L. A., and Korotev, R. L. (1984) The "North American shale composite": its compilation, major and trace element characteristics: *Geochim. Cosmochim. Acta* **48**, 2469–2482.

Hoffman, J. and Hower, J. (1979) Clay mineral assemblages as low grade metamorphic geothermometers: application to the thrust-faulted disturbed belt of Montana: in *Aspects of Diagenesis*, P. A. Scholle and P. R. Schulger, eds., Soc. Econ. Paleontol. Mineral. Spec. Publ. **26**, 55–79.

Howard, J. J. (1981) Lithium and potassium saturation of illite/smectite clays from interlaminated shales and sandstones: *Clays & Clay Minerals* **29**, 136–142.

Howard, J. J. and Roy, D. M. (1985) Development of layer charge and kinetics of experimental smectite alteration: *Clays & Clay Minerals* **33**, 81–88.

Hower, J., Eslinger, E., Hower, M., and Perry, E. (1976) Mechanism of burial metamorphism of argillaceous sediment: I. Mineralogical and chemical evidence: *Bull. Geol. Soc. Amer.* **87**, 725–737.

Keith, L. A. and Rimstidt, J. D. (1985) A numerical compaction model of overpressuring in shales: *J. Math. Geol.* **17**, 115–135.

Morton, J. P. (1985) Rb-Sr evidence for punctuated illite/smectite diagenesis in the Oligocene Frio Formation, Texas Gulf Coast: *Bull. Geol. Soc. Amer.* **96**, 114–122.

Nuhfer, E. B., Vinopal, R. J., and Klanderman, D. S. (1979) X-radiograph atlas of lithotypes and other structures in the Devonian shale sequence of West Virginia and Virginia: *U.S. Dept. Energy Report METC/CR 79/27*, 45 pp.

Perry, E. and Hower, J. (1970) Burial diagenesis in Gulf

Coast pelitic sediments: *Clays & Clay Minerals* **18,** 165–178.

Perry, E. and Hower, J. (1972) Late-stage dehydration in deeply buried pelitic sediments: *Amer. Assoc. Petrol. Geol. Bull.* **56,** 2013–2021.

Roberson, H. E. and Lahann, R. W. (1981) Smectite to illite conversion rates: effects of solution chemistry: *Clays & Clay Minerals* **29,** 129–135.

Środoń, J. (1980) Precise identification of illite/smectite in-terstratification by X-ray powder diffraction: *Clays & Clay Minerals* **28,** 401–411.

Środoń, J. (1981) X-ray identification of randomly inter-stratified illite-smectite in mixtures with discrete illite: *Clay Miner.* **16,** 297–304.

Weaver, C. E. and Beck, K. (1971) Clay-water diagenesis during burial: how mud becomes gneiss: *Geol. Soc. Amer. Spec. Paper* **134,** 96 pp.

Proceedings of the International Clay Conference, Denver, 1985, L. G. Schultz, H. van Olphen, and F. A. Mumpton, eds.,
The Clay Minerals Society, Bloomington, Indiana, 151–157 (1987).

TRANSMISSION ELECTRON MICROSCOPIC STUDY OF THE DIAGENESIS OF KAOLINITE IN GULF COAST ARGILLACEOUS SEDIMENTS[1]

JUNG HO AHN[2] AND DONALD R. PEACOR

Department of Geological Sciences, The University of Michigan
Ann Arbor, Michigan 48109

Abstract—Transmission and analytical electron microscopy have revealed diagenetic relationships between kaolinite and other clay minerals in core samples from Tertiary Gulf Coast sediments. In all samples studied (depths of 1750, 2450, and 5500 m), kaolinite occurs as thin packets of layers contained exclusively within a matrix of smectite or illite and is typically interstratified with illite. Interstratified kaolinite and illite exhibit apparently coherent (001) grain boundaries. The thickness of the kaolinite packets appears to increase with depth. These data suggest that kaolinite was derived from smectite concomitant with the smectite-to-illite reaction in a locally K-deficient environment. The interlayer cation K in smectite may be important in determining the final proportions of kaolinite and illite.

Key Words—Diagenesis, Illite, Interstratification, Kaolinite, Lattice fringe image, Potassium, Transmission electron microscopy.

INTRODUCTION

Kaolinite formation in burial diagenetic environments has received little attention, because kaolinite commonly appears to disappear with increasing depth (e.g., Perry and Hower, 1970; Heling, 1978; Boles and Franks, 1979) or because the proportion of kaolinite may show no systematic trend with burial depth in argillaceous sediments (Perry and Hower, 1970; Weaver and Beck, 1971; Hower et al., 1976), implying that such kaolinite was not derived by diagenesis. Eberl (1971), however, proposed that kaolinite may form from smectite in K-deficient environments. Many hydrothermal experiments have shown that smectite can react to form both illite and kaolinite (Eberl and Hower, 1977; Eberl, 1978a, 1978b; Eberl et al., 1978). On the basis of hydrothermal experiments in which smectite was converted to illite, Eberl and Hower (1977) suggested that kaolinite formed from Al and Si derived from some dissolved smectite layers in excess of that required for transformation of smectite layers to illite (Eberl and Hower, 1977; Eberl, 1978a, 1978b).

The lack of detailed data concerning kaolinite in fine-grained argillaceous sediments is due in part to the technical difficulties inherent in accurate characterization and quantification of kaolinite; kaolinite occurs as a minor phase with chlorite and berthierine, both of which exhibit 00*l* reflections that overlap those of kaolinite in X-ray powder diffraction (XRD) patterns. Furthermore, no detailed textural and structural characteristics of kaolinite have been reported, partly because the actual grain size of kaolinite is usually too small to allow study by conventional techniques. In the present study, transmission electron microscopy (TEM) and analytical electron microscopy (AEM) have been used to overcome such difficulties and to characterize diagenesis involving kaolinite and other phyllosilicates in argillaceous sediments.

SPECIMENS AND EXPERIMENTAL TECHNIQUES

Shales from three different depths (1750, 2450, and 5500 m) from a Gulf Coast well (Case Western Reserve University Gulf Coast 6) were examined in this study. The specimens were provided by John Hower and previously investigated in detail by Hower et al. (1976). They reported an increase with depth in the illite content of mixed-layer illite/smectite (I/S) in these three samples of 20, 40, and 80%, increasing with depth. These samples have been used to study chlorite formed during diagenesis (Ahn and Peacor, 1985) and the transition of smectite to illite (Ahn and Peacor, 1986).

Ion-thinned specimens were prepared from petrographic thin sections in order to preserve the original structure of the sediments. Specimens were observed at 100 kV in a JEOL JEM-100CX scanning-transmission electron microscope (STEM) fitted with a solid-state detector for energy dispersive analysis. The detailed sample preparation procedures and experimental techniques were described by Lee et al. (1984) and Ahn and Peacor (1985, 1986).

---

[1] Contribution No. 407 from the Mineralogical Laboratory, Department of Geological Sciences, The University of Michigan, Ann Arbor, Michigan 48109.
[2] Current address: Department of Geology, Arizona State University, Tempe, Arizona 85287.

151

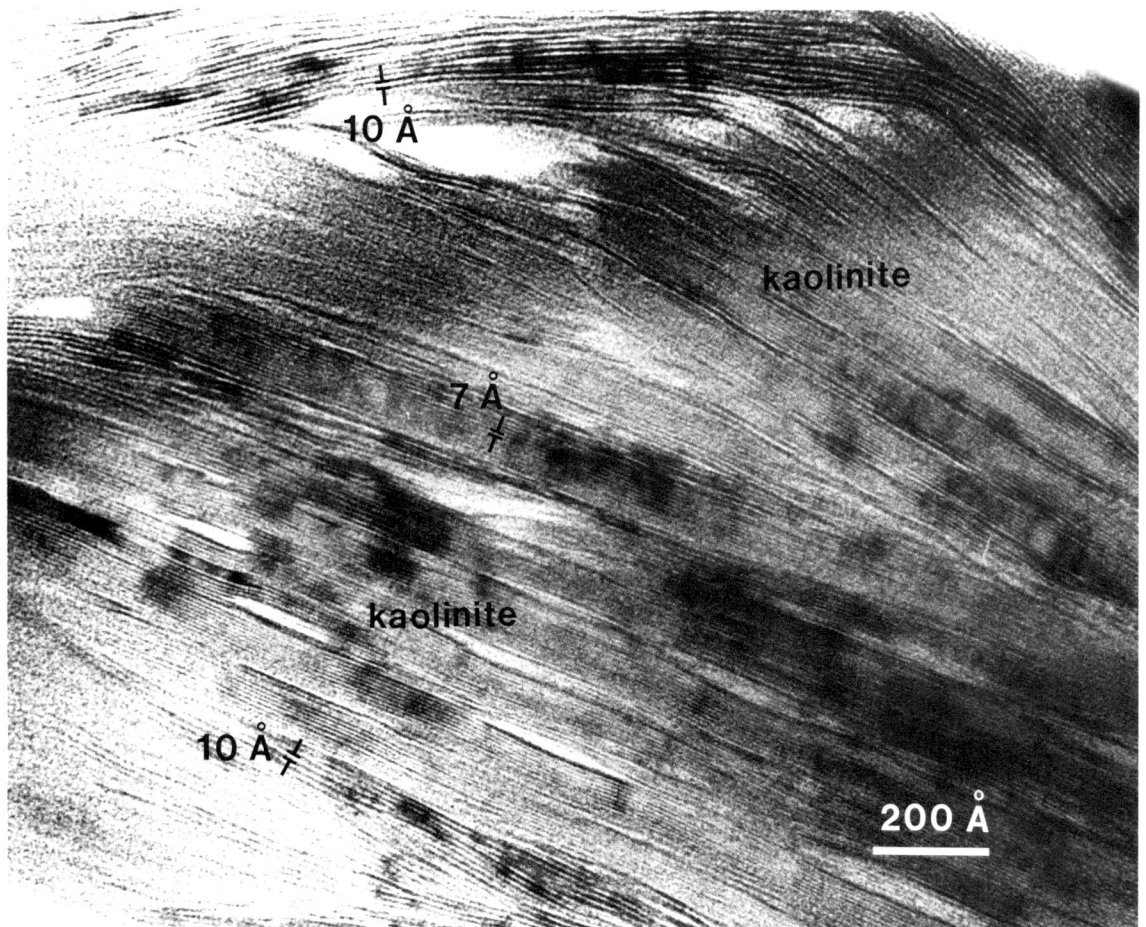

Figure 1.   Lattice fringe image of thick packets of kaolinite layers which are inferred to be detrital in origin (from the 1750-m-depth sample).

TEM study of kaolinite is especially difficult because kaolinite is markedly subject to beam damage. Lattice fringe images dissipate within a few seconds upon exposure of the specimen to a relatively intense electron beam. Kaolinite appears to be much less stable under electron beam exposure than other phyllosilicates, including smectite, illite, chlorite, and berthierine, all of which have been observed in the same samples. Indeed, unless care is taken, kaolinite may remain undetected due to beam damage even though it is present.

EXPERIMENTAL RESULTS

Kaolinite was characterized by a combination of electron diffraction patterns and resulting lattice fringe images that display straight lattice fringes of d = 7 Å and AEM data which show high Al and Si contents. AEM analysis was necessary because berthierine, an Fe-rich aluminous trioctahedral phase (Bailey, 1980; Brindley, 1982), occurs in the same samples (Ahn and Peacor, 1985) and displays 7-Å lattice fringe images similar to those of kaolinite. Berthierine, however, exclusively occurs interlayered with chlorite (Ahn and

Peacor, 1985), and kaolinite was never observed to be directly associated with chlorite.

Detrital kaolinite

Two different modes of occurrence of kaolinite were noted. In one, the grains are very large (as thick as 0.1 μm, perpendicular to (001)), even in samples from shallow depths (Figure 1). These kaolinite grains commonly exhibited imperfections: lattice fringes were locally curved and grain boundaries displayed no coherent relation (i.e., commonly were not parallel) to surrounding smectite or illite. Such grains were interpreted to be detrital in origin; the imperfections were inferred to be pre-diagenetic features. This interpretation is consistent with the XRD observations of Hower et al. (1976) that showed kaolinite to be concentrated principally in shallower samples (<3400 m), implying that such kaolinite was not a product of diagenesis. Furthermore, Hower et al. (1976) showed that kaolinite occurred primarily in the 0.1–0.5-μm size fractions, as opposed to the <0.1-μm size fractions. In the present study, a second population of packets (a "packet" of

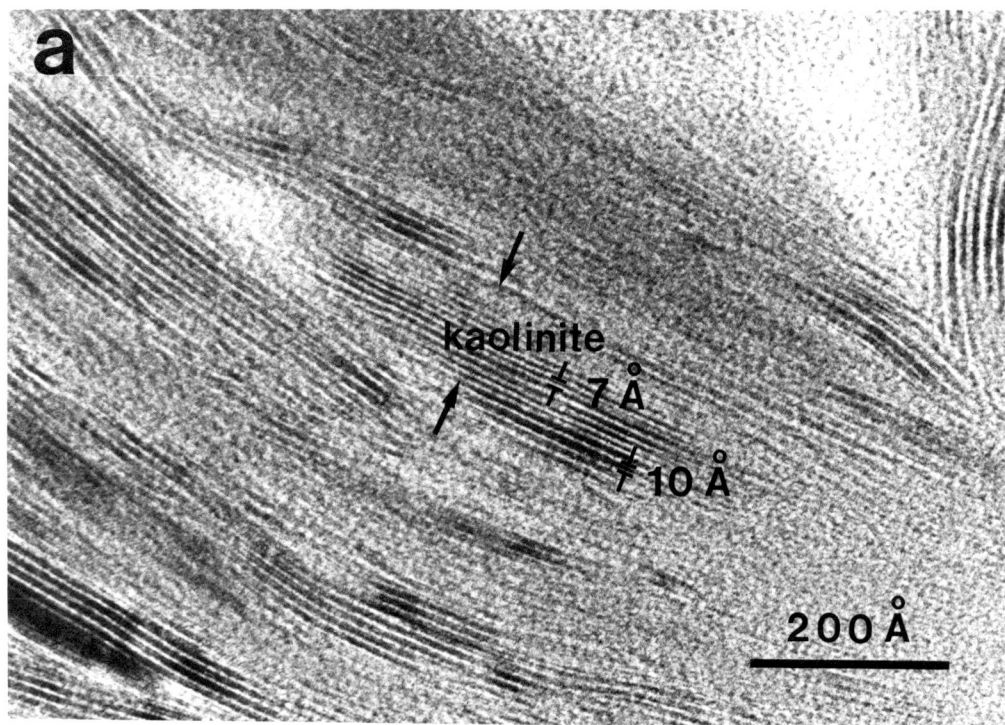

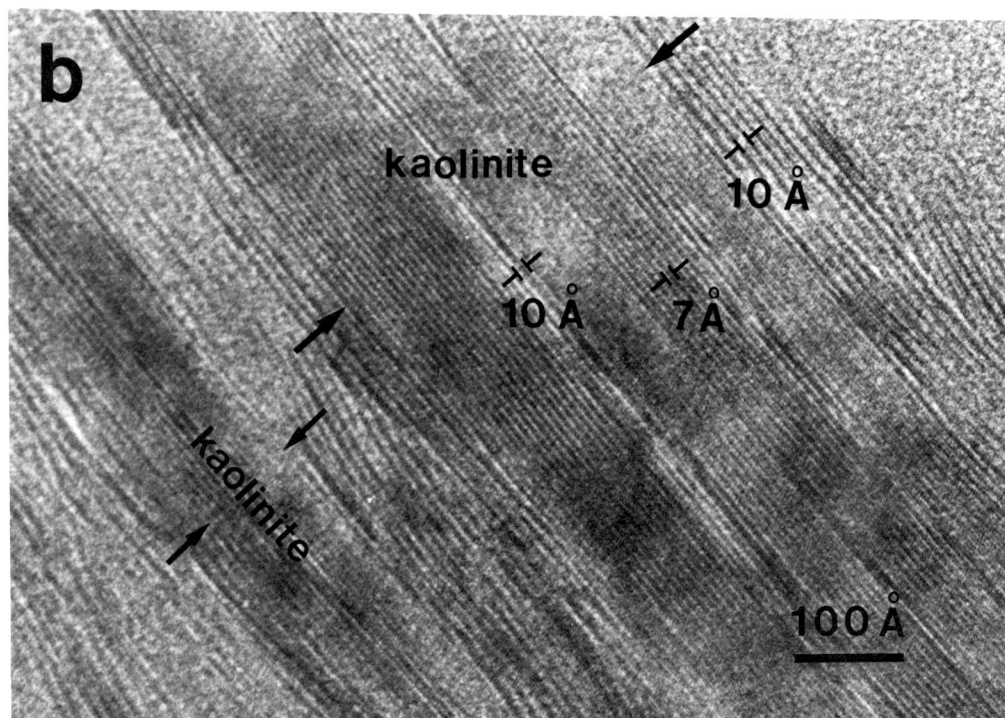

Figure 2. Lattice fringe images of: (a) thin kaolinite packets which appear to be in an early stage of formation (from the 1750-m-depth sample); straight 10-Å layers are apparently coherent with respect to 7-Å layers; (b) kaolinite packets showing textural relations to surrounding illite and smectite (from the 1750-m-depth sample).

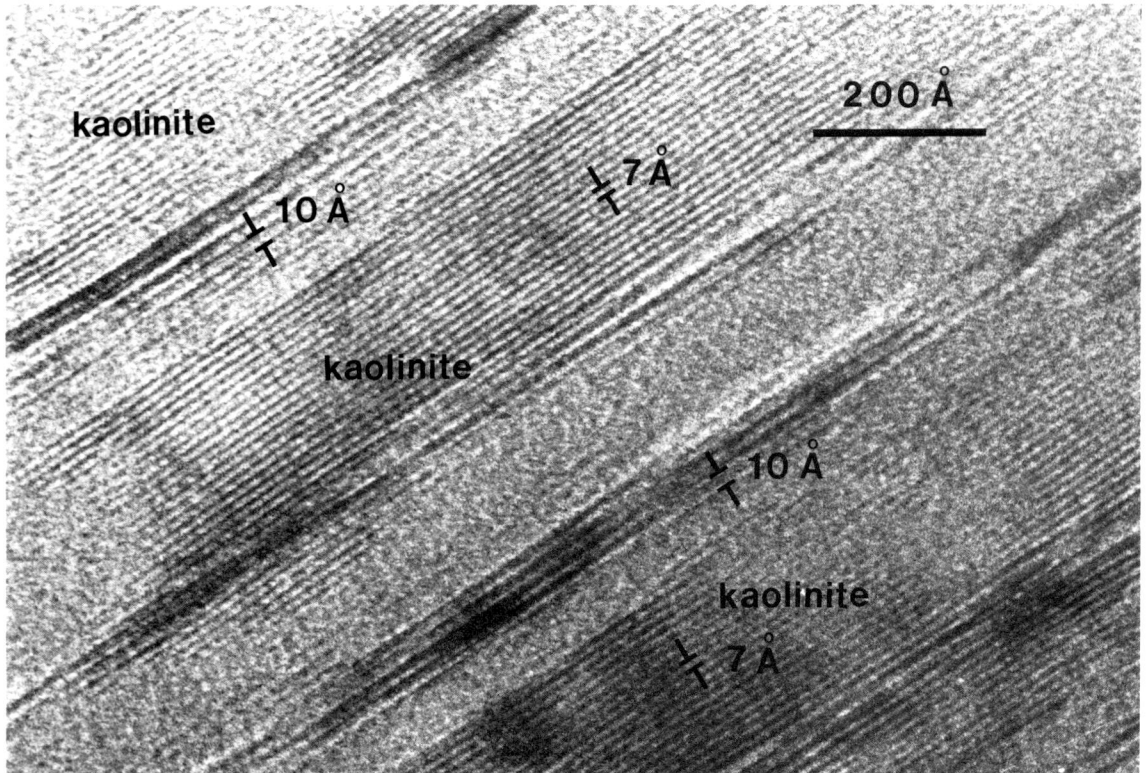

Figure 3.   Lattice fringe image of relatively thick packets of kaolinite layers from the 2450-m-depth sample. Kaolinite packets are interstratified with illite layers.

layers refers to a sequence of parallel, contiguous layers as seen in lattice fringe images, which has uniform characteristics) of kaolinite layers was observed that have fewer imperfections than the detrital kaolinite. They are less than a few hundred Ångstroms in thickness and are commonly intercalated with very straight 10-Å layers. These grains are interpreted as being authigenic, and they are exclusively the source of the data and conclusions described below.

Authigenic kaolinite

Figure 2a shows a thin packet of kaolinite layers 100 Å in thickness from the 1750-m depth sample. The straight 001 lattice fringes of kaolinite are markedly different than those of the surrounding materials, in which interlayer spacing ranges from 10 to 13 Å. The latter material is probably smectite that has dehydrated and partially collapsed in the TEM environment (Ahn and Peacor, 1986). Smectite layers are wavy and anastomosing with variable interplanar spacing and image contrast; they contain many edge dislocations, small-angle grain boundaries, and other defects (Ahn and Peacor, 1986). On the other hand, illite layers are characterized by straight, uniform lattice fringes having few edge dislocations (Ahn and Peacor, 1986). Figures 2a and 2b show thin packets of kaolinite layers interstrat-

ified with straight 10-Å layers. Packet thicknesses of about 100 Å are typical of the 1750-m depth sample. The association and textural relations of kaolinite and surrounding wavy 10-Å layers are similar to those of chlorite and smectite or illite in the same samples (see Ahn and Peacor, 1985, Figures 1 and 2). A ubiquitous feature is the apparent coherency or near-coherency of the (001) boundaries of packets of layers of kaolinite with those of surrounding smectite or illite. Furthermore, 10-Å layer units are commonly intercalated with kaolinite lattice fringes (Figures 2a and 2b).

In the 2450-m depth sample, the packets of kaolinite layers appear to be thicker and 10-Å layers were more frequently observed interstratified within kaolinite (Figure 3). These features are accompanied by an increased illite content (from 20 to 40%) in mixed-layer I/S caused by the smectite-to-illite transition with increasing depth, as inferred from XRD data (Hower *et al.*, 1976). AEM analyses of the area of kaolinite packets that includes thin packets of 10-Å layers show only small amounts of K (Figure 4). The association of K with relatively high Al contents indicates that the 10-Å layers interstratified with kaolinite are illite (as opposed to collapsed layers of smectite). In some areas, kaolinite layers were observed locally interlayered within thick illite packets (Figure 5).

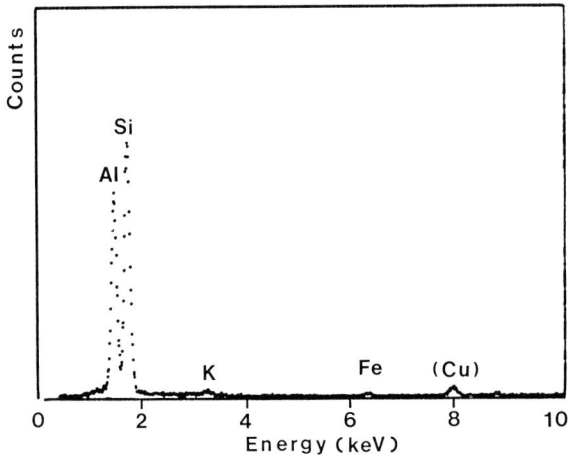

Figure 4.   Energy dispersive X-ray spectrum of the area shown in Figure 3. The K peak is due to interstratification of thin packets of illite. The small Cu peak is due to contamination.

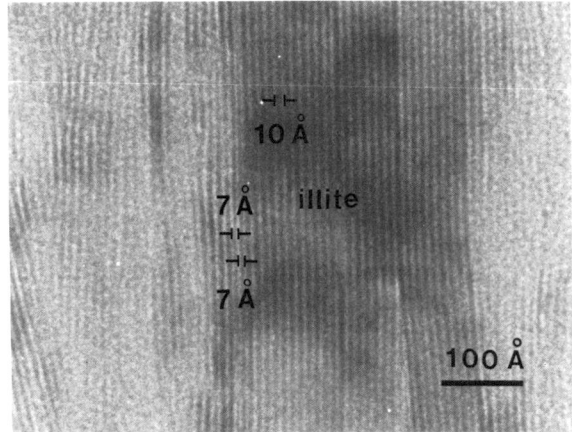

Figure 5.   Lattice fringe image of an illite packet in the 2450-m-depth sample. Two 7-Å layers (kaolinite) are interlayered within illite.

Kaolinite packets in the 5500-m depth sample are generally larger than those in shallower samples (Figure 6). They average about 300 Å in thickness. The layers of both illite and kaolinite represented by the lattice fringes are exceedingly straight and free of defects (Figure 6). In addition, interstratification of illite and kaolinite was more frequently observed, although discrete kaolinite free of interstratified illite is also present (Figure 7).

DISCUSSION AND CONCLUSIONS

Non-detrital kaolinite ubiquitously occurs as thin packets of layers within a matrix of smectite and/or illite in samples from all three depths. Kaolinite layers are generally subparallel to the surrounding smectite and/or illite layers. In addition, kaolinite packets commonly occur interstratified with thin illite packets. The interfaces between kaolinite and interstratified illite are apparently coherent. Where layers of kaolinite appear to terminate, they are discontinuous; i.e., there is no transition or reaction relationship between kaolinite and illite. These textural relations collectively suggest that the kaolinite and illite crystallized simultaneously at the expense of smectite. This origin is further supported by the increase in kaolinite packet thickness with depth. The gradual increase in illite and decrease of smectite with increasing depth is probably due to the smectite-to-illite reaction with further burial diagenesis (Hower et al., 1976). The observed increase in packet thickness of kaolinite with depth is compatible with the crystallization of kaolinite, utilizing components derived from smectite, which give rise primarily to illite.

The observed structural and textural relations imply that the formation of kaolinite involved the dissolution of smectite and the direct crystallization of kaolinite from solution without retention of major structural units of the smectite, such as combined octahedral and tetrahedral sheets. This process is entirely different from the transformation mechanism proposed by Altschuler et al. (1963), who suggested that kaolinite inherited one octahedral and one tetrahedral layer from a smectite layer through stripping off of interlayer cations and one of the silicate tetrahedral layers of smectites.

A dissolution-crystallization mechanism is compatible with the results of many hydrothermal experiments which show that kaolinite and illite are both produced by reactions starting with smectite as a reactant (e.g., Eberl and Hower, 1977; Eberl, 1978a, 1978b; Eberl et al., 1978). Eberl and Hower (1977) suggested that some smectite layers may dissolve to supply the Al necessary in the reaction of other smectite layers to illite layers and in the process may have generated excess Al and Si for kaolinite. Środoń (1980) suggested that the mixed-layer kaolinite/smectite synthesized from smectite formed by the dissolution of some smectite and subsequent crystallization of kaolinite in the interlayer space of unreacted smectite. Mixed-layer kaolinite/smectite, however, was not observed in the present samples, although it has often been reported to form at the expense of smectite (e.g., Altschuler et al., 1963; Schultz et al., 1971; Wiewiora, 1971; Brindley et al., 1983). Nevertheless, the syntheses, which were carried out under hydrothermal conditions, demonstrate the necessity of a dissolution-crystallization mechanism.

Eberl and Hower (1977) proposed a reaction for formation of illite and kaolinite from potassium smectite (K-smectite + H_2O = illite + kaolinite + SiO_2), based on the results of hydrothermal experiments. This reaction suggests that smectite itself can react to form both kaolinite and illite without a supply of K from an external source; i.e., such a reaction may occur if pore solutions are K-free. Hoffman and Hower (1979) pro-

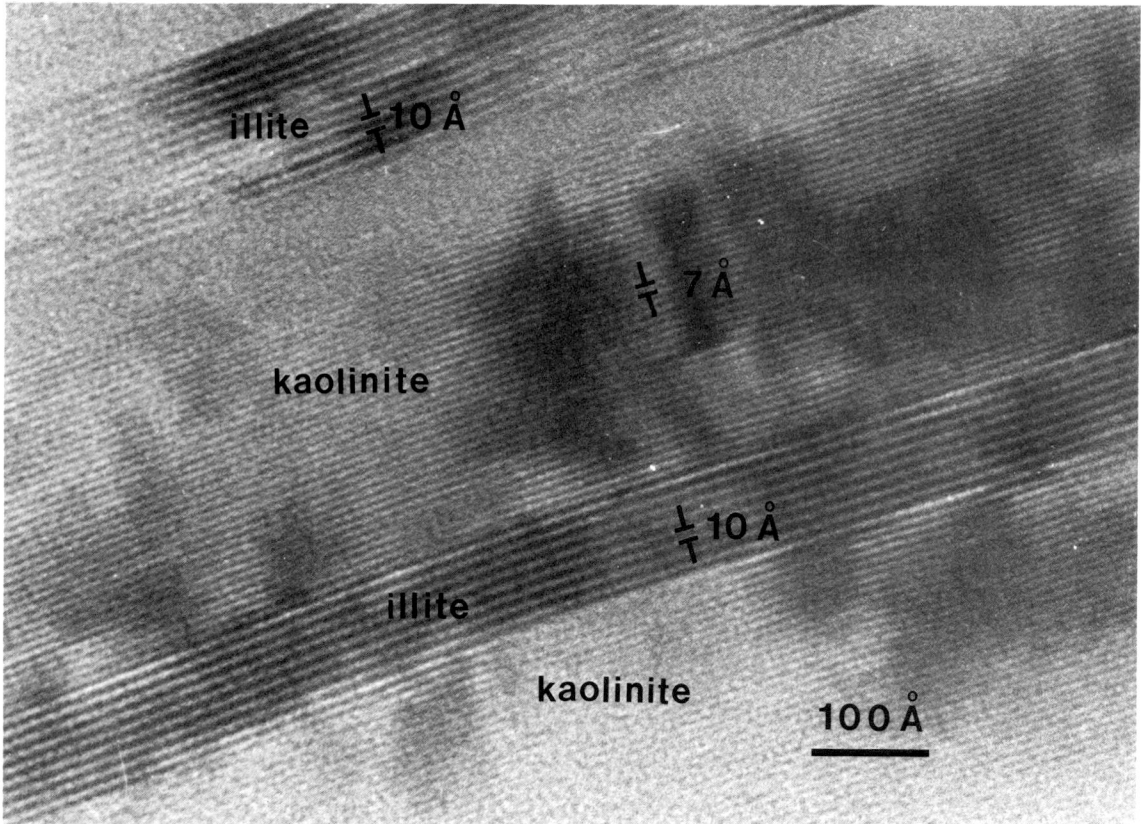

Figure 6.   Lattice fringe image of interstratified kaolinite and illite packets from the 5500-m-depth sample. Illite and kaolinite exhibit well-defined boundaries.

posed three reactions involving smectite as a reactant: (1) with no supply of K; (2) with a limited supply of K; and (3) with an unlimited supply of K. These reactions suggest that the ratio of kaolinite to illite is controlled by the activity of K in solution, such that only kaolinite and SiO_2 are produced from smectite in the absence of K. The interstratification of kaolinite and illite observed in this study appears to be the result of their simultaneous crystallization where smectite was a reactant and where the availability of K was limited. The smectite in the same samples was shown to contain significant interlayer K (Ahn and Peacor, 1986), although the composition of smectite appeared to vary substantially from grain to grain. The presence of K in smectite itself (as well as K from external sources) appears to be important in the reaction of smectite. As originally proposed by Eberl (1971), local formation of kaolinite thus occurs during the conversion of smectite to illite if the availability of K is limited.

Insofar as the K (and Al) necessary for the formation of illite during burial diagenesis is apparently derived from detrital minerals such as K-feldspar and/or micas (Perry and Hower, 1970; Hower et al., 1976), the activity of K in solution must be controlled by proximity to such grains in the relatively impermeable shales,

and may vary from point to point. The replacement of smectite by kaolinite (and illite and chlorite) with domains of reactants entirely enveloped within a matrix of reactant and other products, requires diffusion of ions through the phyllosilicate matrix. Ions are apparently transported by diffusion along dislocations and

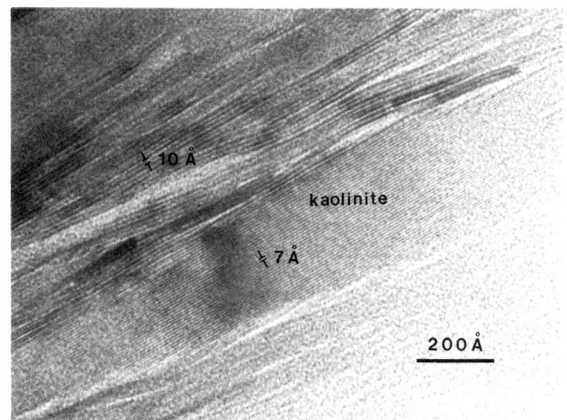

Figure 7.   Lattice fringe image of a packet of kaolinite layers free of interstratification with illite layers from the 5500-m-depth sample.

other defects (Yau *et al.,* 1984; Veblen, 1985), and the rate of transport must be relatively slow at low temperatures. These considerations are consistent with the observed heterogeneity in chemistry and mineralogy and, specifically, with the possible local variation in the activity of K.

The progressive reaction of smectite to illite and kaolinite with increasing burial depth is therefore envisioned as a complex process dependent on extreme heterogeneity in chemistry and structure in impermeable shale. Dissolution of smectite occurs at reaction interfaces, with Al, Si, Mg, and Fe going into solution and with at least local mass transport of those components over distances of as much as hundreds of Ångstroms. The pore solutions themselves may be heterogeneous with respect to a complex series of dissolution-crystallization processes that locally produce variations in the activities of K and Al resulting in the crystallization of kaolinite and illite, with significant amounts of Fe and Mg giving rise to chlorite (Ahn and Peacor, 1985). The product is a heterogeneous assemblage of reactant smectite locally replaced by domains of illite, chlorite/berthierine, and kaolinite.

ACKNOWLEDGMENTS

We are especially grateful to the late John Hower for providing us with samples, data, and encouragement. We thank E. J. Essene for his valuable comments on early versions of the manuscript. We are grateful to I. D. R. Mackinnon, F. A. Mumpton, J. Środoń, and L. G. Schultz for their reviews and helpful comments. J.H.A. is grateful to J. H. Lee for his help during this study. We thank W. C. Bigelow and the staff of the University of Michigan Electron Microbeam Laboratory for providing STEM facilities. This study was supported by NSF grants EAR-8107529 and EAR-8313236 to D. R. Peacor.

REFERENCES

Ahn, J. H. and Peacor, D. R. (1985) Transmission electron microscopic study of diagenetic chlorite in Gulf Coast argillaceous sediments: *Clays & Clay Minerals* **33**, 228–236.

Ahn, J. H. and Peacor, D. R. (1986) Transmission and analytical electron microscopy of the smectite-to-illite transition: *Clays & Clay Minerals* **34**, 165–179.

Altschuler, Z. S., Dwornik, E. J., and Kramer, H. (1963) Transformation of montmorillonite to kaolinite during weathering: *Science* **141**, 148–152.

Bailey, S. W. (1980) Structure of layer silicates: in *Crystal Structures of Clay Minerals and their X-ray Identification,* G. W. Brindley and G. Brown, eds., Mineralogical Society, London, 1–123.

Boles, J. R. and Franks, S. G. (1979) Clay diagenesis in Wilcox sandstones of southeast Texas: Implications of smectite diagenesis on sandstone cementation: *J. Sed. Petrol.* **49**, 55–70.

Brindley, G. W. (1982) Chemical compositions of berthierines—a review: *Clays & Clay Minerals* **30**, 153–155.

Brindley, G. W., Suzuki, T., and Thirty, T. (1983) Interstratified kaolinite/smectites from the Paris basin; correlations of layer proportions, chemical compositions and other data: *Bull. Mineral.* **106**, 403–410.

Eberl, D. D. (1971) Experimental diagenetic reactions involving clay minerals: Ph.D. thesis, Case Western Reserve Univ., Cleveland, Ohio, 145 pp.

Eberl, D. (1978a) The reaction of montmorillonite to mixed-layer clay: The effect of interlayer alkali and alkaline earth cations: *Geochim. Cosmochim. Acta* **42**, 1–7.

Eberl, D. (1978b) Reaction series for dioctahedral smectites: *Clays & Clay Minerals* **26**, 327–340.

Eberl, D. and Hower, J. (1977) The hydrothermal transformation of sodium and potassium smectite into mixed-layer clay: *Clays & Clay Minerals* **25**, 215–227.

Eberl, D., Whitney, G., and Khoury, H. (1978) Hydrothermal reactivity of smectite: *Amer. Mineral.* **63**, 401–409.

Heling, D. (1978) Diagenesis of illite in argillaceous sediments of the Rhinegraben: *Clay Miner.* **13**, 211–220.

Hoffman, J. and Hower, J (1979) Clay mineral assemblages as low-grade metamorphic geothermometers: Application to the thrust faulted Disturbed Belt of Montana, U.S.A.: *Soc. Econ. Paleontol. Mineral. Spec. Publ.* **26**, 55–79.

Hower, J., Eslinger, E. V., Hower, M. E., and Perry, E. A. (1976) Mechanism of burial metamorphism of argillaceous sediments: 1. Mineralogical and chemical evidence: *Geol. Soc. Amer. Bull.* **87**, 725–737.

Lee, J. H., Peacor, D. R., Lewis, D. D., and Wintsch, R. P. (1984) Chlorite-illite/muscovite interlayered and interstratified crystals: A TEM/STEM study: *Contrib. Mineral. Petrol.* **88**, 372–385.

Perry, E. and Hower, J. (1970) Burial diagenesis in Gulf Coast pelitic sediments: *Clays & Clay Minerals* **18**, 165–177.

Schultz, L. G., Shepard, A. O., Blackmon, P. D., and Starkey, H. C. (1971) Mixed-layer kaolinite-montmorillonite from the Yucatan Peninsula, Mexico: *Clays & Clay Minerals* **19**, 137–150.

Środoń, J. (1980) Synthesis of mixed-layer kaolinite/smectite: *Clays & Clay Minerals* **28**, 419–424.

Veblen, D. R. (1985) Extended defects and vacancy nonstoichiometry in rock-forming minerals: in *Point Defects in Minerals, Geophysical Monograph 31,* R. N. Shock, ed., American Geophysical Union, Washington, D.C., 122–131.

Weaver, C. E. and Beck, K. C. (1971) Clay-water diagenesis during burial: How mud becomes gneiss: *Geol. Soc. Amer. Spec. Pap.* **134**, 96 pp.

Wiewiora, A. (1971) A mixed-layer kaolinite-smectite from Lower Silesia, Poland: *Clays & Clay Minerals,* **19**, 415–416.

Yau, Y. C., Anovitz, L. M., Essene, E. J., and Peacor, D. R. (1984) Phlogopite-chlorite reaction mechanisms and physical conditions during retrograde reaction in the Marble Formation, Franklin, New Jersey: *Contrib. Mineral. Petrol.* **88**, 299–308.

Proceedings of the International Clay Conference, Denver, 1985, L. G. Schultz, H. van Olphen, and F. A. Mumpton, eds.,
The Clay Minerals Society, Bloomington, Indiana, 158–164 (1987).

CONVERSION OF SMECTITE TO CHLORITE BY HYDROTHERMAL AND DIAGENETIC ALTERATIONS, HOKUROKU KUROKO MINERALIZATION AREA, NORTHEAST JAPAN

Atsuyuki Inoue

Geological Institute, College of Arts & Sciences
Chiba University, Chiba 260, Japan

Abstract—The conversion of trioctahedral smectite to chlorite has been examined using drill core samples from the Hokuroku Kuroko mineralization area of Japan, where silicic pyroclastic rocks have undergone intensive hydrothermal and diagenetic alterations. The percentage of expandable layers in trioctahedral chlorite/smectite (C/S) decreases discontinuously with depth, with steps at 100–80%, 50–40%, and 15–0%; thus the C/S exhibits a trimodal frequency in expandability throughout a given drill hole. These three types of C/S, having expandabilities of about 80%, 50% (corrensite), and 15%, coexist over a depth range of about 200 m. This discontinuous change in expandability contrasts with the continuous change from 100 to 0% expandability found for illite/smectite. As smectite converts to chlorite, C/S increases in tetrahedral Al, octahedral Fe, and exchangeable Mg and Fe and decreases in tetrahedral Si and exchangeable Na and K.

Key Words—Chlorite, Diagenesis, Hydrothermal alteration, Interstratification, Kuroko, Smectite.

INTRODUCTION

Interstratified trioctahedral chlorite/smectite (C/S) is one of the most common mixed-layer minerals in nature. Although C/S of different expandabilities, including regularly interstratified corrensite, have been reported from various geologic settings, the overall process of conversion of trioctahedral smectite to chlorite through intermediate C/S is poorly understood, partly because the process is obscured by a close association between these minerals and dioctahedral smectite, illite, and illite/smectite (I/S).

In the Hokuroku district of Japan, which is the most important Kuroko (stratabound Cu-Pb-Zn sulfide-sulfate deposits) mining district of the country, C/S occurs extensively in hydrothermally and diagenetically altered, silicic pyroclastic sediments of late Tertiary age. In the present investigation, drill cores in the Hokuroku district were studied to determine the mode of occurrence and mineralogical properties of C/S formed by both processes.

SAMPLES AND ANALYTICAL PROCEDURES

Samples studied

Samples were obtained from 22 drill holes in the Ohdate area and from one drill hole in the Ohyu area of the Hokuroku district (Figure 1). In the Ohdate area, lower and upper Miocene formations, composed of silicic tuff, mudstone, and basaltic rocks, have been subjected to hydrothermal alterations related to Kuroko ore mineralization. In the Ohyu area, about 25 km east of Ohdate, a Mio-Pliocene formation composed of pumice tuff and tuff breccia has been altered mainly by diagenesis.

Analytical procedures

The whole-rock mineralogy of core samples was determined using X-ray powder diffraction (XRD) techniques. The clay fraction (<1 μm) was smeared on a glass slide to give an oriented film for XRD examination. XRD analysis was carried out on air-dried natural, ethylene-glycol-solvated (EG-solvated), K-saturated, and Mg-saturated specimens. The percentage of expandable layers in mixed-layer minerals was determined from XRD peak positions of Mg-saturated and EG-solvated specimens (Reynolds, 1980; Watanabe, 1981). The precision of the proportion of expandable layers determined was within 5%.

Thin sections of selected samples were prepared for petrographic examination and electron microprobe analysis. Following exposure of powdered samples to 0.1 N $SrCl_2$ solution, the Na, K, Mg, Ca, and Fe contents of the supernatant solution were determined as the exchangeable interlayer cations of the clays by atomic absorption spectroscopy. Cation-exchange capacities (CEC) of the clays were calculated from the amount of adsorbed Sr.

RESULTS AND DISCUSSION

Mode of occurrence of clay minerals

In the Ohdate area, an alteration halo envelops Kuroko ore deposits as shown in the cross section of

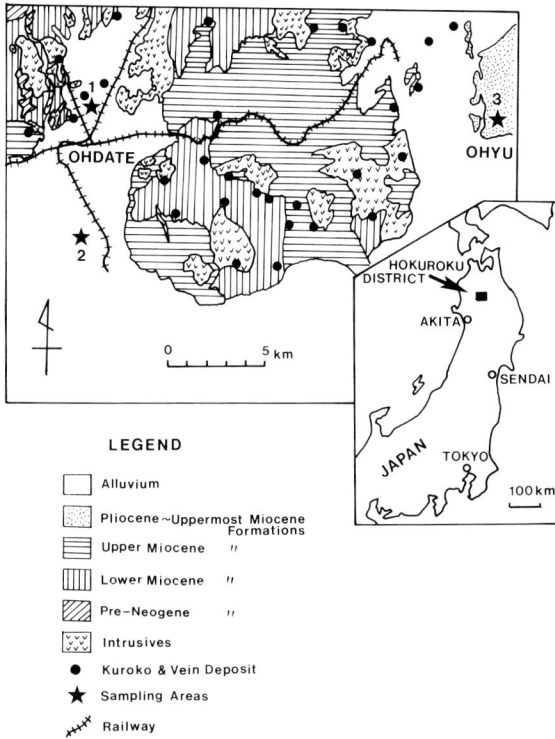

Figure 1. Geological map of Hokuroku Kuroko mineralization area (simplified from Sato *et al.,* 1974). Samples studied were collected from drill holes in the starred areas.

Figure 2. Typically, an illite-chlorite zone exists in the center of the hydrothermally altered mass and a smectite or zeolite zone at the margin. An intermediate zone consists of mixed-layer clay minerals, such as I/S and C/S. Dioctahedral smectite is a major authigenic phyl-

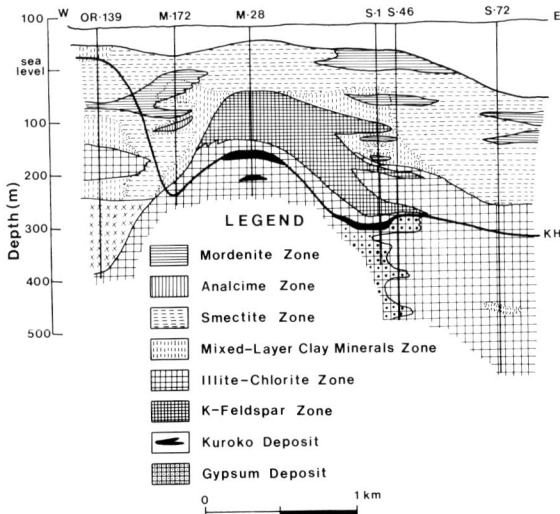

Figure 2. Cross section showing alteration zones around Kuroko deposits in northern part of Ohdate area. "KH" line indicates horizon of Kuroko ore deposits.

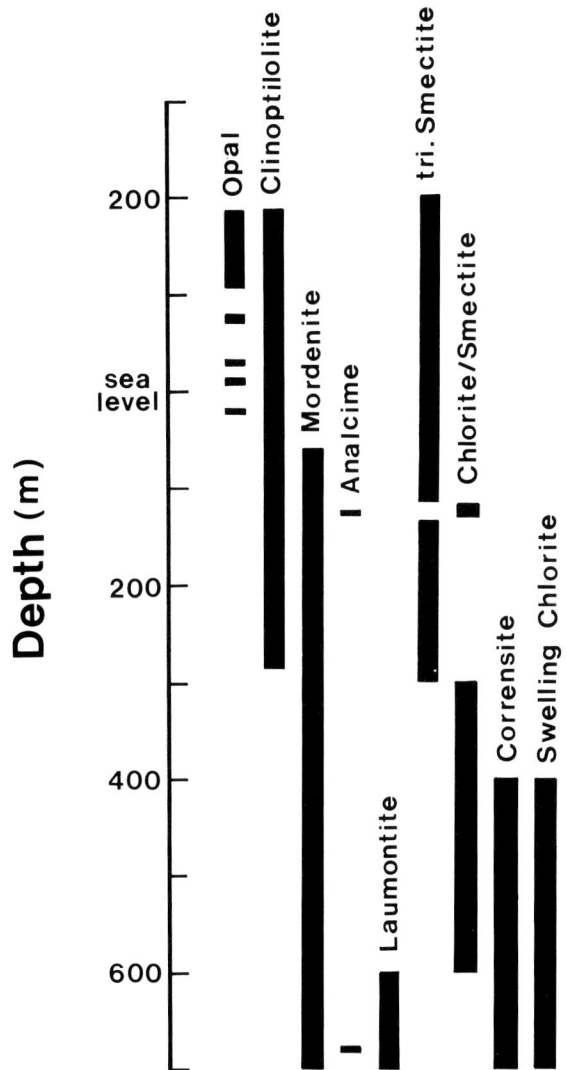

Figure 3. Diagram showing authigenic mineral assemblages in drill hole HT-42 of Ohyu diagenetic alteration area. "Chlorite/smectite" corresponds to that having expandabilities considerably more than 50%.

losilicate in the smectite and zeolite zones. Trioctahedral smectite (saponite) occurs exclusively in hyaloclastic basalt or in green patches of pumiceous tuff in the smectite zone. Trioctahedral C/S is concentrated in dark patches in the pumiceous tuff and in tuff and tuff breccia in the zone of mixed-layer clay minerals; it is also abundant in basaltic rocks where it coexists with laumontite. The mineral also coexists commonly with montmorillonite, I/S, illite, and celadonite. Detrital illite is common in the smectite zone.

Authigenic mineral assemblages of diagenetic origin in drill hole HT-42 in the Ohyu area are summarized in Figure 3. Dioctahederal smectite and I/S are rare throughout this drill hole, but trioctahedral smectite and C/S are abundant. Trioctahedral smectite (sapo-

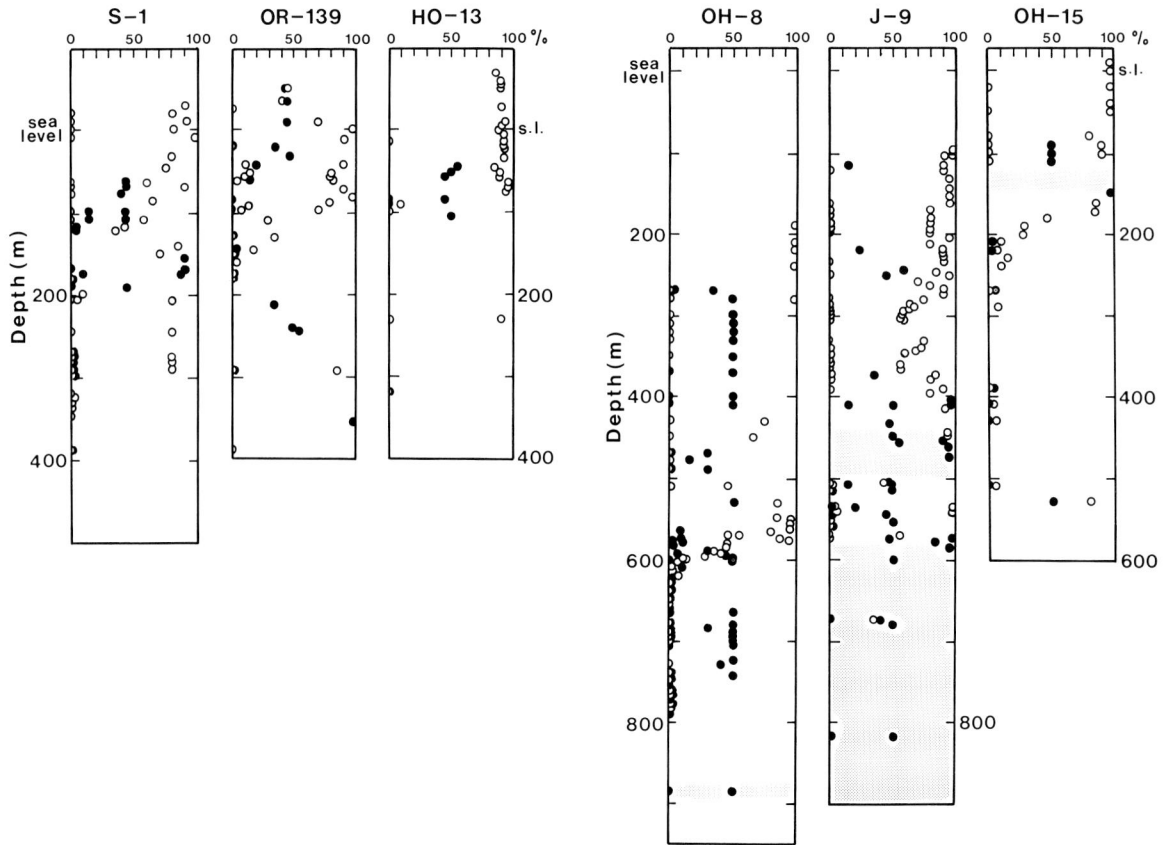

Figure 4. Variation of percentage of expandable layers in chlorite/smectite (●) and illite/smectite (○) from the northern part (drill holes S-1, OR-139, HO-13) and southern part (drill holes OH-8, J-9, OH-15) of Ohdate hydrothermal alteration area. Shadowed parts indicate basaltic rocks.

nite) occurs as a brown to pale-green aggregate which replaces silicic glass shards and pumice fragments in tuff or fills interstitial pores in tuff and tuff breccia. As shown in Figure 3, saponite commonly coexists with clinoptilolite and mordenite. Regularly interstratified C/S (corrensite) locally replaces pumice and glass at >420-m depth and is generally greener than the saponite. Corrensite commonly coexists with laumontite and mordenite and locally with analcime. Slightly expandable C/S (swelling chlorite) has been found with corrensite at >420-m depth. Primary augite phenocrysts are present throughout the drill core. Quartz phenocrysts are unaltered. Calcic plagioclase (An_{66}-An_{50}) coexisting with laumontite or calcite has been albitized.

Variation in expandability of chlorite/smectite

The variation of the percentage of expandable layers in I/S and C/S in the Ohdate hydrothermally altered area is shown as a function of depth in Figure 4. The percentage of expandable layers in I/S varies continuously from 100% (dioctahedral smectite) to 0% (illite) with increasing depth, as typically shown in drill holes

S-1, OR-139, OH-8, and OH-15, although the variation in expandability of I/S has been locally perturbed by repeated hydrothermal alterations. On the other hand, the variation in expandability of C/S shows a stepwise change and little or no relation to depth. The frequency of the percentage of expandable layers in C/S appears to be trimodal throughout a given drill hole, that is, ~100% (saponite), ~50% (corrensite), and ~0% (chlorite). As is shown below in Figure 7, such a stepwise change in the variation in expandability of C/S is even more evident for the diagenetic C/S in the Ohyu area.

Variation of the basal and 060 spacings of C/S in the Ohyu area after EG-solvation is shown in Figure 5. The d(060) values of the clay minerals are about 1.54 Å, consistent with a trioctahedral structure. Saponite from <300-m depth expanded to about 17 Å after EG-solvation. The mineral shrank to 12.8 Å after K-saturation (Inoue *et al.*, 1984b). A clay mineral from >420-m depth having a d value of 32 Å after EG-solvation shrank to 24 Å on heating at 500°C and to 27 Å after K-saturation, thereby identifying it as corrensite, composed of regularly interstratified chlorite

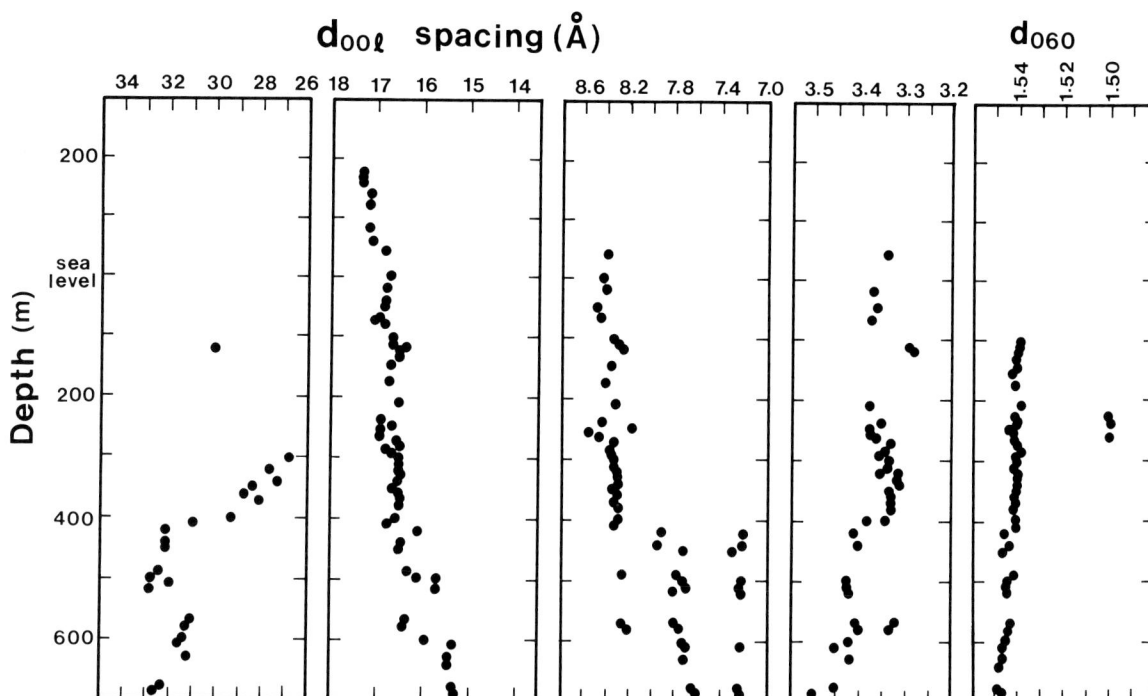

Figure 5. Variation of d values of basal reflections and d(060) value of chlorite/smectite after ethylene glycol-solvation in drill hole HT-42 in Ohyu diagenetic alteration area.

and saponite. The Mg-saturated corrensite had a mean d(001) value of 28.7 Å (air-dried with CV = 0.49), and EG-solvated corrensite had a mean d(001) value of 31.2 Å (CV = 0.42). CV values are the coefficients of variation of the d(00l) values as defined by Bailey *et al.* (1982); and mixed-layer minerals having CV values <0.75 merit the specific name, corrensite. The weak, long-spacing reflection ranging from 27 to 31 Å after EG-solvation in samples from 300- to 400-m depth (Figures 5 and 6a), are interpreted as somewhat imperfect precursors of corrensite containing <50% chlorite layers and having slightly irregular stratification. The clay minerals in samples from >420-m depth yielded a weak XRD 7.2-Å peak after EG-solvation, suggesting a chloritic mineral having only a few expandable layers (Figures 6b and 6c).

The variation in expandability of C/S of drill hole HT-42 shows that the percentage of expandable layers decreases discontinuously with increasing depth, with steps at 100–80%, 50–40%, and 15–10% (Figure 7). In addition, these three types of C/S, having expandabilities about 80%, 50% (corrensite), and 15%, coexist between 420 and 600 m. In the sample from 420.1-m depth, as shown in Figure 6b, three XRD peaks are present between 10° and 14°; 8.2-, 7.7-, and 7.2-Å peaks can be assigned to C/S phases containing 20, 55, and 85% chlorite layers, respectively. Such a stepwise decrease in expandability of C/S with depth has been observed in diagenetically altered pyroclastic sediments in other Green Tuff regions of Japan (Kimbara

and Sudo, 1973) and in hydrothermally altered basaltic rocks of Reykjanes, Iceland (Tomasson and Kristmannsdottir, 1972; Kristmannsdottir, 1976, 1983). Therefore, the stepwise decrease in expandability probably is a general change during the conversion of smectite to chlorite through interstratified minerals regardless of its genesis.

Variation in chemical composition of chlorite/smectite

The range of chemical composition of C/S from drill hole HT-42, as determined by electron microprobe analysis, is given in Figure 8. The structural formulae of all the clays were uniformly calculated on the basis of $O_{20}(OH)_4$ to facilitate comparison, although technically the formula of corrensite should be calculated on the basis of $O_{20}(OH)_{10}$ (Brigatti and Poppi, 1984). Iron was calculated as Fe^{2+}. The clays from 156.5- and 460.0-m depth, which contain exceptionally large amounts of Fe, have replaced pyroxene phenocrysts. The other clays have replaced pumice fragments or filled interstitial pores.

The substitution of Al for Si in the tetrahedral sheet generally increases with increasing depth (Figure 8a). The number of octahedral cations also increases with increasing depth (Figure 8b). Ideally, octahedral cations in excess of 6 represent the brucite layers of chlorite in the C/S, and the marked increase of number of octahedral cations in the samples from >420-m depth is caused by the appearance of corrensite and more

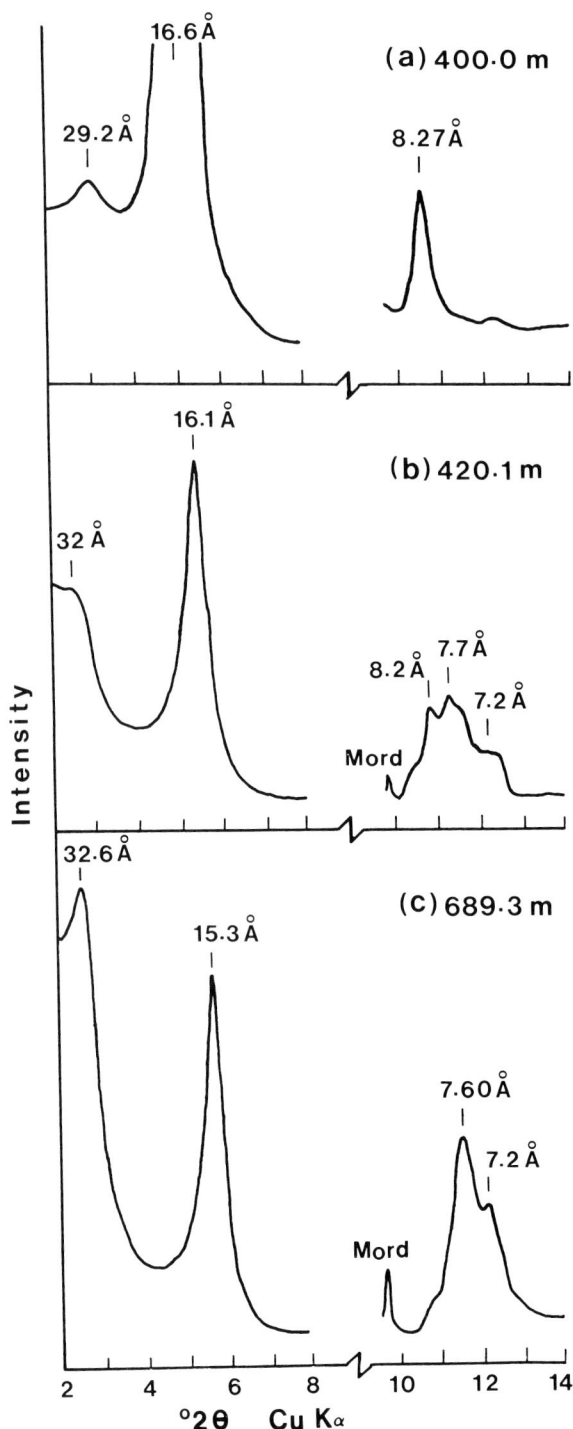

Figure 6. X-ray powder diffraction patterns of chlorite/smectite (C/S) after ethylene glycol-solvation measured in the range 2°–14°2θ, CuKα radiation. Numerical figures are d values (Å). (a) 20/80 C/S from 400.0-m depth, (b) mixture of 20/80 (8.2 Å), 55/45 (7.7 Å), and 85/15 (7.2 Å) C/S from 420.1-m depth, and (c) mixture of 60/40 (7.6 Å) and 85/15 (7.2 Å) C/S from 689.3-m depth. Mord = mordenite.

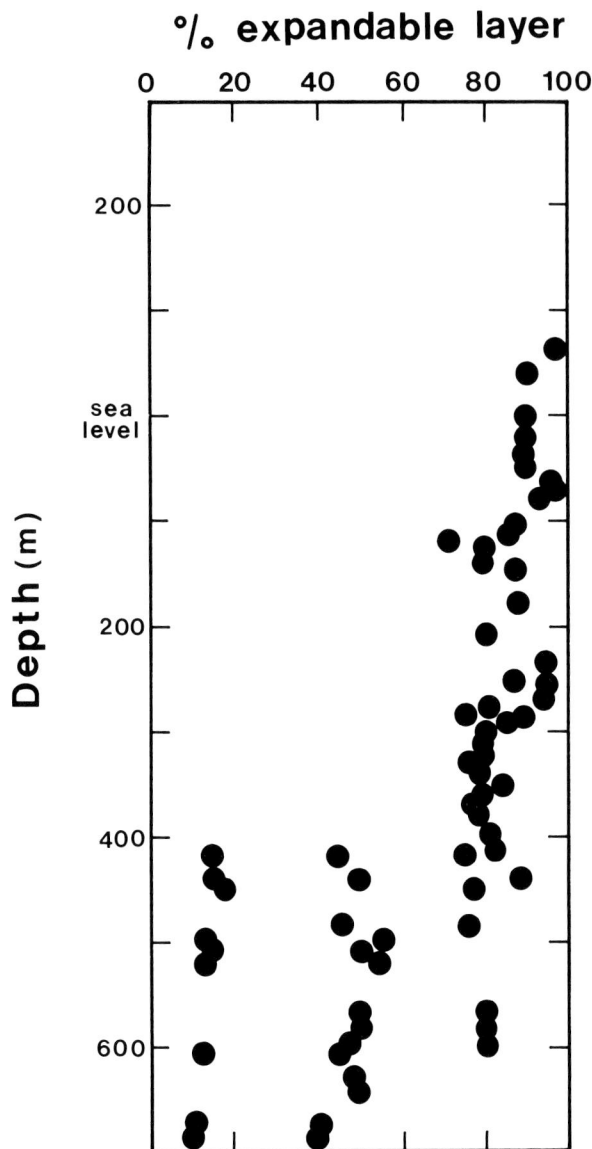

Figure 7. Variation of percentage of expandable layers in chlorite/smectite as function of depth in drill hole HT-42 from Ohyu diagenetic alteration area.

chloritic C/S (see Figure 7). As shown in Figure 8c, the octahedral Fe content increases slightly, but the Al content decreases only slightly with increasing depth. The increase of the Fe content in C/S with decreasing expandability is consistent with the downward increase of d(060) values from 1.539 to 1.547 Å, as shown in Figure 5. The saponite and corrensite from the Ohyu diagenetic area are richer in Fe than those reported from other localities (Yoshimura, 1983; Kristmanns-dottir, 1983; Brigatti and Poppi, 1984). Corrensite in the Kuroko deposits is usually Mg-rich (Shirozu et al., 1975); thus, the octahedral cation composition of C/S most likely depends on the physicochemical conditions

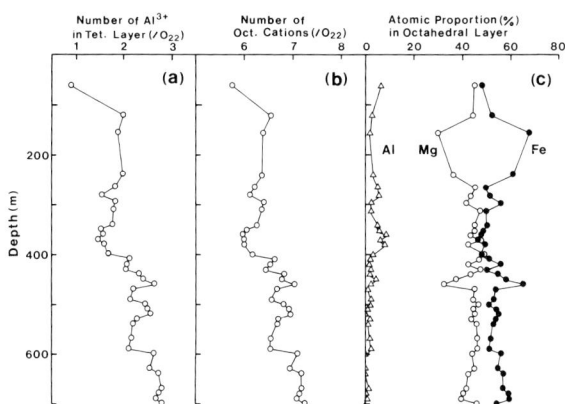

Figure 8. Variation of chemical composition in chlorite/smectite as function of depth in drill hole HT-42 from Ohyu diagenetic area: (a) number of Al in tetrahedral sheet, assuming 22 oxygens; (b) number of octahedral cations, assuming 22 oxygens; (c) atomic proportion (%) in octahedral sheet. $\triangle$ = Al; $\bigcirc$ = Mg; $\bullet$ = Fe. In the figure, Mn content is neglected.

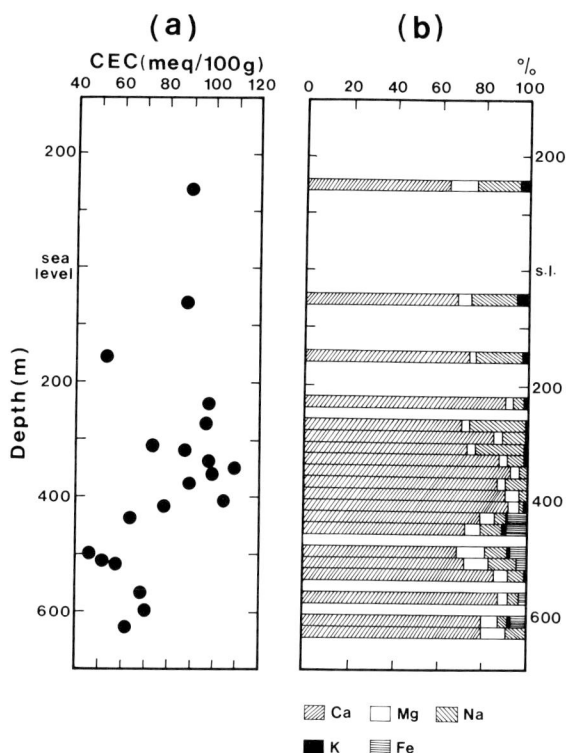

Figure 9. Variation of (a) cation-exchange capacity and (b) composition of exchangeable interlayer cations in the clays as function of depth in drill hole HT-42 from Ohyu diagenetic alteration area.

of the formation and the original rock composition. The change in the chemical composition of C/S in the Ohyu area is probably typical of the conversion of Fe-rich saponite to chlorite during diagenesis.

The CEC of C/S in drill hole HT-42 appears to decrease with decreasing expandability, although quite erratically so, from 90–110 meq/100 g for saponite to 40–50 meq/100 g for corrensite (Figure 9a). The erratic decrease in CEC is attributed to the overall increase in chlorite layers in the saponite-corrensite-chlorite mixture. The exchangeable interlayer cations of the clays (Figure 9b) are mainly Ca regardless of the expandabilities. Apart from the enrichment of Mg in the samples near the surface (Inoue *et al.*, 1984a), the percentages of exchangeable Na and K generally decrease and the percentage of exchangeable Mg increases with depth. Fe was detected as an exchangeable interlayer cation only in the specimens from >420-m depth. The increase of exchangeable Mg and Fe in C/S has been recognized in hydrothermally altered pyroclastics of the Ohdate area (Inoue *et al.*, 1984a).

Trioctahedral smectite-to-chlorite conversion processs

Smectite in hydrothermally altered silicic pyroclastics from the Ohdate area is largely dioctahedral; trioctahedral smectite is rare, except in basaltic rocks. Trioctahedral C/S commonly coexists with dioctahedral smectite, I/S, illite, and celadonite. It is not clear how trioctahedral C/S and chlorite were formed from dioctahedral smectite under either hydrothermal or diagenetic conditions. The conversion of trioctahedral smectite to chlorite, on the other hand, has been clarified by the present study. The characteristic features in the conversion of C/S in the Hokuroku district are summarized as follows:

In terms of the structural changes: (1) the decrease in expandability of C/S is discontinuous during the conversion of smectite to chlorite, and the three types of C/S coexist in the depth range of about 200 m; (2) somewhat imperfect precursors of corrensite containing <50% chlorite layers and slightly irregular interstratification appear prior to corrensite; and (3) corrensite occurs extensively over a wide range of depth due to both diagenetic and hydrothermal alterations. In terms of the chemical changes as the proportion of chlorite in C/S increases: (4) tetrahedral Al increases and tetrahedral Si decreases; (5) octahedral Fe increases and octahedral Al decreases; (6) exchangeable Na, K, and Ca decrease and exchangeable Mg and Fe increase in the interlayers; and (7) these chemical changes take place nearly continuously as smectite converts to chlorite.

Thus, the mechanism responsible for conversion of trioctahedral smectite to C/S and then to chlorite must explain two features that at first appear to be contradictory: the proportions of layers in the C/S of the alteration sequence is distinctly discontinuous and trimodal, whereas changes in the chemical composition, though somewhat irregular, apparently are at least roughly continuous. Two possible mechanisms must be considered: neoformation and transformation. Neoformation of three types of C/S in the same small volume of rock seems improbable. From the viewpoint of a transformation mechanism, however, a stepwise decrease in expandability with continuous chemical change could result from the much greater stability of corrensite and highly smectitic or chloritic C/S than of intermediate compositions, i.e., 70/30 or 30/70 C/S (in contrast to the relatively small stability preferences in the smectite-to-illite transformation, except at 20% smectite). Saponite hydrothermally and diagenetically altered from tuff particles of slightly different composition, to say nothing of that from basalt or augite crystals, probably had a wide range of composition and layer charge. Saponite crystals having relatively high layer charge would be more susceptible to the formation of interlayer brucite than others, and, once beyond the 20% chlorite limit, brucite would form rapidly in alternate adjacent interlayers to yield relatively stable corrensite; this reaction would be followed by saponite crystals having progressively smaller layer charges. Before the saponite-to-corrensite transformation was complete, however, some corrensite crystals having relatively large layer charges began to convert in a similarly rapid manner to expandable chlorite having ~15% expandable layers. Thus, the process would be discontinuous on a crystallite basis, as seen in XRD patterns (Figure 6), but by varying the relative abundances of the three types of C/S in samples from 420–600-m depths (Figure 7), chemical compositions of samples could vary continuously. In addition, the apparently continuous changes in the chemical combination of C/S may be due to the large chemical variability of both chlorite and corrensite (Brigatti and Poppi, 1984).

ACKNOWLEDGMENTS

The author thanks M. Utada, University of Tokyo, and K. Kanehira, Chiba University, for their valuable suggestions and critical reading of the manuscript, and D. D. Eberl, U.S. Geological Survey, for his critical review of the initial manuscript. I gratefully acknowledge the help of the Metal Mining Agency of Japan and Dowa Mining Company, Ltd., which supplied the core samples and allowed publication of this study.

REFERENCES

Bailey, S. W., Brindley, G. W., Kodama, H., and Martin, R. T. (1982) Nomenclature for regular interstratifications: Clays & Clay Minerals 30, 76–78.

Brigatti, M. F. and Poppi, L. (1984) Crystal chemistry of corrensite: a review: Clays & Clay Minerals 32, 391–399.

Inoue, A., Utada, M., and Kusakabe, H. (1984a) Clay mineral composition and their exchangeable interlayer cation composition from altered rocks around the Kuroko deposits in the Matsumine-Shakanai-Matsuki area of the Hokuroku district, Japan: Nendo Kagaku (J. Clay Sci. Soc. Japan) 24, 69–77 (in Japanese).

Inoue, A., Utada, M., Nagata, H., and Watanabe, T. (1984b) Conversion of trioctahedral smectite to interstratified chlorite/smectite in Pliocene acidic pyroclastic sediments of the Ohyu district, Akita Prefecture, Japan: Clay Sci. 6, 103–116.

Kimbara, K. and Sudo, T. (1973) Chloritic clay minerals in tuffaceous sandstone of the Miocene Green Tuff formation, Yamanaka district, Ishikawa Prefecture, Japan: J. Japan Assoc. Min. Petrol. Econ. Geol. 68, 246–258.

Kristmannsdottir, H. (1976) Types of clay minerals in hydrothermally altered basaltic rocks, Reykjanes, Iceland: Jokull 26, 30–39.

Kristmannsdottir, H. (1983) Chemical evidence from Icelandic geothermal systems as compared to submarine geothermal systems: in Hydrothermal Processes at Seafloor Spreading Centers, P. A. Rona, K. Bostrom, L. Laubier, and K. L. Smith, Jr., eds., NATO Conf. Series IV 12, Plenum, New York, 291–320.

Reynolds, R. C. (1980) Interstratified clay minerals: in Crystal Structures of Clay Minerals and Their X-ray Identification, G. W. Brindley and G. Brown, eds., Mineralogical Society, London, 249–303.

Sato, T., Tanimura, S., and Ohtagaki, T. (1974) Geology and ore deposits of the Hokuroku district, Akita Prefecture: Soc. Mining Geologists Japan, Spec. Issue 6, 11–18.

Shirozu, H., Sakasegawa, T., Katsumoto, N., and Ozaki, M. (1975) Mg-chlorite and interstratified Mg-chlorite/saponite associated with Kuroko deposits: Clay Sci. 4, 305–321.

Tomasson, J. and Kristmannsdottir, H. (1972) High temperature alteration minerals and thermal brines, Reykjanes, Iceland: Contrib. Miner. Petrol. 36, 123–134.

Watanabe, T. (1981) Identification of illite/montmorillonite interstratifications by X-ray powder diffraction: J. Miner. Soc. Japan, Spec. Issue 15, 32–41 (in Japanese).

Yoshimura, T. (1983) Neoformation and transformation of trioctahedral clay minerals in diagenetic process: J. Sediment. Soc. Japan, Spec. Issue 17/18/19, 177–185 (in Japanese).

SOILS

Proceedings of the International Clay Conference, Denver, 1985, L. G. Schultz, H. van Olphen, and F. A. Mumpton, eds.,
The Clay Minerals Society, Bloomington, Indiana, 167–173 (1987).

SOIL SMECTITES AND RELATED INTERSTRATIFIED MINERALS: RECENT DEVELOPMENTS

M. J. WILSON

Department of Mineral Soils, The Macaulay Institute for Soil Research
Craigiebuckler, Aberdeen AB9 2QJ, Scotland, United Kingdom

Abstract—Soil smectites differ from standard bentonite-type smectites in several respects. First, soil smectites generally have a composition varying between montmorillonite and beidellite. They commonly contain >20% octahedral Fe^{3+} (excluding magnesium) and can be described as iron-rich beidellites. Second, interstratified smectites formed in soils differ from interstratified smectites formed during diagenesis. Thus, interstratified soil smectites commonly form by means of transformation reactions involving relatively large crystals of mica or chlorite, whereas, according to a recently developed conceptual model, diagenetic interstratified clays consist of exceedingly fine particles some tens of Ångstrom units in thickness which are characterized by the phenomenon of interparticle diffraction. Third, soil smectites may be interlayered with non-exchangeable aluminous, organic, or other material, a phenomenon that may produce anomalously high basal spacings on X-ray powder diffraction patterns. Many soil smectites are, therefore, significantly different from bentonite-type smectites; such differences must be fully characterized if the properties and behavior of smectitic soils are to be understood.

Key Words—Beidellite, Interlayer complexes, Interstratification, Montmorillonite, Nontronite, Smectite, Soil clay.

INTRODUCTION

Smectites occur in many different soil types and environments, and the mineralogy of some soils may be dominated by smectite, e.g., Vertisols. The clay fractions of many other soil types contain appreciable quantities of smectite which strongly influence the overall properties and behavior of the soil. Soil smectites may differ in composition from the smectites found in sedimentary deposits, particularly bentonites. Based on earlier work by Marshall (1935), Ross and Hendricks (1945) stated that "beidellite with a high content of ferric iron is probably the predominant soil clay mineral of the montmorillonite group." This conclusion was at least partly substantiated by Sawhney and Jackson (1958) and Carson and Dixon (1972) who showed a continuous compositional variation in the montmorillonite-beidellite series in some soil smectites.

Despite such findings, a tendency still exists in the soil science literature to describe any expandable clay mineral encountered in soil clays either as montmorillonitic, implying a similarity with bentonitic clay, or as smectitic, implying uncertainty as to the real nature of the mineral. This uncertainty is understandable if one considers the difficulties faced in characterizing the expandable clay minerals in soils. Thus, soil smectites may be difficult to separate in a reasonably pure form, they may be interstratified with other layer silicates, and/or they may contain absorbed non-exchangeable inorganic or organic species. Nonetheless, these prob-

lems must be overcome because a better understanding of the differences in properties and behavior of smectitic soils, and of the diversity of pedogenic processes affecting these soils, requires that the smectite minerals themselves be characterized as fully as possible.

The present paper describes some recent developments in characterizing smectites from different soil types and weathered materials, with particular emphasis on the questions of (1) composition, (2) interstratification with other layer silicate minerals, and (3) interlayering, i.e., occupation of the interlamellar space by non-exchangeable aluminous, organic, or other material.

COMPOSITION OF SOIL SMECTITES

Because Vertisols are smectite-dominant, this clay mineral is relatively easy to separate and characterize from these soils. The most recent review of Vertisol smectites is that of Dixon (1982) who plotted the formula of fifteen such smectites on a diagram relating octahedral composition [Al/(Al + Fe)] and relative tetrahedral charge [$Al^{IV}/(Al^{IV} + Mg^{VI})$]. The smectites divided about equally between montmorillonitic and beidellitic compositions, but all showed appreciable amounts of octahedral Fe^{3+} (which commonly accounted for about 25% of the cations of the octahedral sheet), and most showed a relatively high tetrahedral charge compared with ideal montmorillonite. Certainly, the original observations of Paquet (1967) that Vertisol smectites are iron-rich (arbitrarily defined as >20%

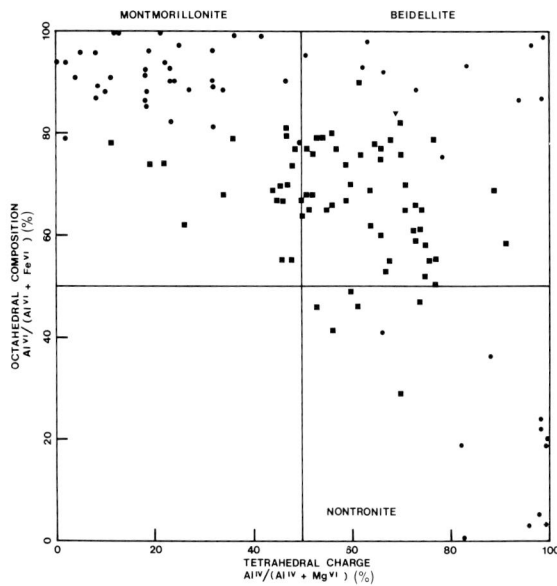

Figure 1. Octahedral composition and relative tetrahedral charge of authigenic smectites ● and soil smectites ■. Data for authigenic smectites are from Ross and Hendricks (1945). Data for soil smectites are from Carson and Dixon (1972), Güzel and Wilson (1981), Kantor and Schwertmann (1974), Ildefonse (1980), Özkan and Ross (1979), Paquet (1967), Proust and Velde (1978), Ross and Mortland (1966), Rossignol (1983), Sawhney and Jackson (1958), and Wilson and Mitchell (1979).

octahedral Fe^{3+} excluding magnesium) and are commonly beidellitic, or at least tend towards being beidellitic, now seem to be fully substantiated. Rossignol (1983), for example, found that the smectites in the lithomorphic Vertisols of northern Uruguay are mainly iron-rich beidellites, and Özkan and Ross (1979) and Güzel and Wilson (1981) came to a similar conclusion with regard to the smectite in Turkish Vertisols. The smectite in the Vertisols of the Blue Nile Plains of Sudan is also iron-rich and possesses a high tetrahedral charge; it does not quite fall, however, into the beidellite compositional field (Wilson and Mitchell, 1979). From the response of the Hofmann-Klemen or Greene-Kelley test of the smectites from the Blue Nile Plains soils, Adam *et al.* (1983) concluded that these clays were mixtures of montmorillonite and beidellite.

Iron-rich beidellite also seems to be widespread in other soil types. In Turkish soils, for example, Özkan and Ross (1979) identified this mineral in *terra rossa*, rendzina, and volcanic ash soils. Also, the smectites in reddish brown soils developed on basalt on the Malwa Plateau, Madhya Pradesh, India, are mainly iron-rich beidellites (Singh and Krishna-Murti, 1975). Other occurrences of this mineral, or at least a smectite approaching it, have been reported for paddy soils derived from marine alluvium in Japan (Egashira and Ohtsubo, 1983), in magnesium-affected and sodic soils

of Egypt (Rabie and El-Araby, 1979), in fersiallitic brown eutrophic soils of Senegal (Tobias and Janot, 1981), and in some French Andosols (Moinereau, 1977). Recently, Norrish and Pickering (1983) showed that the same smectite occurred in Australian soils that were formed on basic igneous parent material.

Other types of ferriferous smectite have also been reported from soils and weathered materials. For example, Wilson and Berrow (1978) found a pedogenic smectite in the basal horizons of serpentinite-derived soils in Scotland, although here the mineral was described as an iron-rich saponite. A iron-rich smectite was also described by Kodama and de Kimpe (1983) from a gabbro saprolite in Quebec. The mineral was reported to have a high tetrahedral charge and a total octahedral population of 2.69, indicating an intermediate di-trioctahedral character. The compositions of the smectites from the direct lateritic weathering of orthopyroxene in the Ivory Coast (Nahon and Colin, 1982) suggest solid solution between the nontronite-beidellite series and saponite.

Aluminous smectites (>80% octahedral Al) also occur widely in soils, although they are commonly directly or indirectly inherited from sedimentary parent materials. Truly pedogenic aluminous smectites have been found, however, in the A_2 horizons of some Spodosols. These clays were first characterized by Ross and Mortland (1966) from some Michigan Podzols and found to be distinctly beidellitic. The smectites in some New Zealand podzols are also beidellites, when assessed in terms of their response to lithium-saturation, heating, and glycerol treatment (Churchman, 1980). Beidellitic smectite has also been reported from Ultisols (Nash, 1979) and an Alfisol (Douglas, 1982). This kind of smectite is thought to originate by weathering of preexisting micaceous minerals (Robert, 1973), hence a beidellitic nature is probably not unexpected.

Figure 1 shows that the composition of smectites in soils and weathered saprolites tends to be different from that of smectites from other sources. Most of the nonpedogenic smectites falling into the montmorillonite field are aluminous and are from bentonites; the nonpedogenic smectites in the beidellite and nontronite fields are usually of hydrothermal origin. Soil smectites show something of a continuum between montmorillonite-beidellite-nontronite with respect to octahedral composition and charge distribution, but these compositions clearly tend to cluster around the iron-rich beidellite part of the diagram.

In general, the differences between soil and bentonite smectites probably arise because soil smectites commonly derive directly or indirectly from parent materials containing ferromagnesian primary minerals, whereas bentonites mostly form from acid or intermediate pyroclastic material. Paquet (1967) noted a positive correlation between the amount of iron in the smectite in Vertisols and that in the parent rock, a relationship most readily observed in lithomorphic Vertisols deriving directly from parent rock under conditions of poor drainage. For the topomorphic Vertisols, whose formation is controlled essentially by lateral drainage towards low-lying topography, this relationship is rather obscure. The work of Nahon and Colin (1982) suggests that the formation of iron-rich smectite from the weathering of orthopyroxene can take place via a noncrystalline phase or directly from the mineral itself.

In a tropical environment, these weathering products are distinctly transitory and are usually confined to the basal parts of the profile. They quickly give way to iron oxyhydroxides closer to the surface. In more temperate climates, detailed studies of weathering profiles show the stability of both aluminous and ferriferous beidellites (Proust and Velde, 1978; Ildefonse, 1980), suggesting that these minerals may exist in a range of soil environments. In a general way, the stability of smectites in soils appears largely dependent upon the Si activity and pH of the soil solution. The exact stability field, however, will depend on, amongst other things, the chemical composition of the particular smectite and the activities in solution of other ions, such as Mg and Fe, that make up the mineral structure. The composition of soil smectites strongly suggests that any consideration of the formation and stability of soil smectites must take into account the activity of iron in soil solutions, as pointed out earlier by Krishna-Murti and Satyanarayana (1969, 1970).

The compositional differences between soil smectites and bentonite smectites gives rise to significant differences in the physical and chemical properties of the clays. For example, Norrish and Tiller (1976) showed that the poor dispersion, low plasticity, and high aggregate stability of some Australian smectitic soils was due mainly to the high tetrahedral charge of the clay mineral. The physical behavior of these clay soils is significantly different from what might normally be expected of smectitic materials. Similarly, Egashira and Ohtsubo (1983) attributed the low-swelling character of an iron-rich smectite in some Japanese paddy soils to substitution of Fe^{2+} for Al^{3+} in the octahedral sheet. They suggested that the smectite became high-swelling after octahedral Fe^{2+} was oxidized to Fe^{3+}. Low-swelling behavior might also be expected of the "transformation" smectites found in the A horizons of Spodosols. Under the transmission electron microscope, these clays are morphologically similar to vermiculite. They occur as large relatively thick, platy particles, yielding a single-spot type of electron diffraction pattern that is indicative of a high degree of three-dimensional order (unpublished work of the author). In contrast, most other soil smectites occur as thin, poorly shaped particles that produce rotational turbostratic electron diffraction patterns indicative of a low degree of order between the layers.

Soil smectites having high tetrahedral charge may be able to fix potassium and ammonium ions more readily than their bentonitic counterparts with low tetrahedral charge, and there is evidence that this is indeed the case (Kantor and Schwertmann, 1974).

INTERSTRATIFICATION OF SOIL SMECTITES

Interstratified smectitic minerals are extremely widespread in both soils and sediments and have been studied intensively, especially by X-ray powder diffraction (XRD). The currently accepted concept of interstatified clays views them as consisting of physically separable crystallites that are made up of a mosaic of discrete domains, which themselves consist of fixed sequences of two or more types of silicate layers arranged in a regular, partially ordered, or random way (Reynolds, 1980). Most interstratified smectitic clays typically involve illite or chlorite as their non-expandable components; kaolinite-smectite (K/S) is relatively rare. Until recently, little discrimination was made between interstratified clays formed in soils and weathering environments and those formed during sediment diagenesis. Indeed, diagenesis of smectites is often regarded as weathering in reverse, but it now seems that this picture may be misleadingly simple and that fundamentally different types of interstratified clays are formed during weathering and diagenesis.

Wilson and Nadeau (1985) concluded that interstratified minerals actually forming in weathering environments do so by transformation reactions of relatively coarse-grained micas and chlorites, first to interstratified vermiculite and then, with further weathering, to interstratified smectite. This mechanism is, of course, in agreement with the generally accepted view, but it was also pointed out that many of the most frequently described interstratified minerals are not really characteristic of weathering environments at all, even though they are often taken to be, at least by implication. For example, there is no evidence that the partially ordered IIS and IIIS types of illite/smectite (I/S) form during weathering. Moreover, although some occurrences of rectorite-like and corrensite-like clays have been described from weathering profiles (Churchman, 1980; Johnson, 1964), in general they are more characteristic of diagenetic and/or hydrothermal environments. The same may be true, although more evidence is required on this point, of randomly interstratified I/S that is so characteristic of great thicknesses of sedimentary rocks. This mineral is, of course, common in soils, but it is commonly inherited from underlying argillaceous material (Kodama and Brydon, 1966; Weir and Rayner, 1974).

Wilson and Nadeau (1985) rationalized these observations in terms of a new conceptual model for interstratified clays previously proposed by Nadeau *et al.* (1984a, 1984b). In this model interstratified clays, particularly those in sediments, are viewed as consisting of aggregates of exceedingly fine-grained fundamental particles, whose interfaces within the aggregates are capable of adsorbing water and organic molecules and which are perceived by XRD as being interstratified smectitic clays. Thus, according to this concept, the interstratified character of many clays results from what is termed an interparticle diffraction phenomenon. Complementary transmission electron microscopy and XRD evidence were presented to show that regularly interstratified I/S (K-rectorite) and chlorite/smectite (C/S, corrensite) are made up of only one type of particle, namely 20-Å-thick "illite" and 24-Å-thick "chlorite" particles, respectively. Similarly, long-range interstratified I/S of the IIS and IIIS types are composed primarily of "illite" particles 20–50-Å thick. Randomly interstratified I/S and C/S were experimentally synthesized simply by mixing completely dispersed suspensions of 10-Å smectite with suspensions of 20- and 24-Å "illite" and "chlorite" particles, respectively. In these suspensions smectite consists of single silicate sheets and "illite" and "chlorite" consist of double silicate sheets coordinated by layers of K ions and brucite, respectively.

At present, however, this concept will probably not be of general applicability in the interpretation of interstratified smectitic clays in soils, except if the influence of inheritance is strong. In this connection, recent unpublished electron microscope studies in this laboratory on randomly interstratified I/S in gley soils in Great Britain (derived from Mesozoic sediments) tend to confirm that these clays are mechanical mixtures of smectite and thin "illite." Such "illite," which may show crystallographic faces, is thought to have formed during sediment burial by neoformation (Nadeau *et al.,* 1985) rather than during weathering. There is indeed little indication that illite (or chlorite) originate by neoformation in soils, although Norrish and Pickering (1983) recently presented evidence for the crystallization of ferriferous illite in Australian soils. Such illite could possibly be more widespread in arid soils than is presently realized.

Masshady *et al.* (1980) and Viani *et al.* (1983) reported a smectitic mineral in Saudi Arabian soils that yields an XRD pattern at least superficially similar to randomly interstratified I/S. The clay does not appear to have formed by the weathering of mica. In the author's experience the clays from many other soil types may be characterized by broad diffraction effects between 14 and 18 Å after glycerol or ethylene glycol solvation, but whether such material is interstratified I/S or should be interpreted in terms of interlayering remains to be assessed critically.

Although interstratified kaolinite/smectite (K/S) has not been widely reported in soils, it may be more widespread than is presently appreciated. For example, Norrish and Pickering (1983) identified K/S as the only clay mineral in about 40 Australian soils developed upon basic igneous rocks, and, on the basis of its weak XRD pattern, this interstratified material may have gone undetected in many other profiles where it was accompanied by more highly crystalline clays. K/S has also been described in some Vertisols from El Salvador (Yerima *et al.,* 1985), from a toposequence of some red-black tropical soils in Burundi (Herbillon *et al.,* 1981), and from some Arctic soils in the Yukon (Kodama *et al.,* 1976). Under the transmission electron microscope, K/S consists of very small, sub-rounded hexagonal particles, typically 0.05–0.10 μm in the *ab* plane and about 20–70 Å in thickness, that yield a single-spot type of electron diffraction pattern. Little evidence exists that the material consists of two discrete phases, as in randomly interstratified I/S. On the other hand, if a suspension of completely dispersed smectite is mixed with a fine suspension of K/S, the dried material yields a diffraction pattern characteristic of interstratified K/S but with more smectite as assessed by the calculated curves of Reynolds (1980), in proportion to the amount of montmorillonite added (P. H. Nadeau, The Macaulay Institute for Soil Research, Aberdeen AB9 2QJ, Scotland, United Kingdom, personal communication). Thus, the evidence is conflicting as to whether the concept of interparticle diffraction is applicable to interstratified K/S, and the physical character of this material remains to be more completely evaluated.

INTERLAYERING OF SOIL SMECTITES

Soil smectites, unlike most sedimentary smectites, are commonly interlayered with non-exchangeable aluminous material (Barnhisel, 1977). Various opinions exist regarding the exact nature of this material, the way in which soluble Al interacts with smectite, and the structure of aluminum clays. Most evidence, however, favors hydrolysis of Al to a basic polymeric cation, although the size of the polymer unit is not clear. Aluminum interlayering of smectites is an extremely important process as it bears directly on a variety of soil properties, including acidity, phosphate fixation, potassium fixation, and physical behavior. Such smectites have been reported from a large number of acidic, actively leached soil types, and have been widely reported, for example, in Alfisols, Ultisols (Juo, 1980) and Spodosols (Ross, 1980; Wilson *et al.,* 1984).

In general, the optimum conditions for the Al-interlayering of smectite would appear to be a pH of about 5.0, low organic matter content, and perhaps frequent wetting and drying of the soil (Rich, 1968).

Aluminum interlayering in expandable clays seems to be more prominent in the surface horizons of soils, but in some soils, e.g., those of the North Carolina coastal plain (Malcolm *et al.,* 1969) and on heavy textured New Zealand podzols (Lee *et al.,* 1985), it is at a maximum in B horizons where the pH is higher. The XRD characteristics of Al-interlayered smectite, in conjunction with auxiliary treatments such as alkali dissolution, enable identification to be made fairly readily. The properties of this material are intermediate between those of smectite and chlorite with respect to responses to glycerol/ethylene glycol and heat treatments, and it is usually (although not always) possible to extract the interlayer material by chemical treatment (Rich, 1968).

The identification of the interlayer material in smectites from alkaline soils is not so straightforward. Usually, this interlayering is manifested by a broad, diffuse basal XRD reflection both in the air-dry state and after glycerol/ethylene glycol solvation. Whittig (1959) described a smectite in a solodized solonetz which expanded to 20 Å after glycerol treatment. A similar mineral was reported from solodic and associated soils in Montana (Klages and Southard, 1968) and in some Bulgarian salt-affected soils (Behar *et al.,* 1972). In the latter soils the XRD pattern of the mineral was explained in terms of aluminum/silicic acid interlayering. On the other hand, a high-spacing smectitic mineral in some sodic soils in India was thought to involve three-component interstratification of illite, smectite, and chlorite (Kapoor *et al.,* 1981). Other recent examples of this enigmatic, high-spacing smectite in alkaline (but not sodic) soils were described from some Turkish Entisols and Inceptisols (Güzel and Wilson, 1983) and from some Indian alluvial soils (Tomar, 1985). Here, spacings of as much as ~19 Å were recorded for material in the air-dry state and of as much as ~22 Å after glycerol treatment. These observations were interpreted by Güzel and Wilson (1983) as being due to an interstratification/interlayering phenomenon, but so far no complete explanation has been proposed. In the author's experience, smectitic material with anomalously high spacings is common in Aridisols.

Finally, the smectites in many Vertisols also consistently show higher basal spacings than might normally be expected from Ca-saturated montmorillonite. Thus, a spacing of about 16 Å for material in the air-dry state and 18–20 Å after glycerol treatment is common. Such values were first observed in some American Vertisols (Johnson *et al.,* 1962) and have been shown since to be characteristic of these soils in other countries, such as Kenya (Kantor and Schwertmann, 1974), Republic of South Africa (Fitzpatrick and Le Roux, 1977), Turkey (Güzel and Wilson, 1981), and Uruguay (Rossignol, 1983). This high spacing is thought to be the result

of organic matter adsorbed in the interlamellar space, a concept that is well-established in the French soil science literature (Duchaufour, 1982) but not elsewhere. Perez-Rodriguez *et al.* (1977) have provided evidence for the existence of a smectite-organic complex in an Andalusian Vertisol, although here the clay mineral showed normal spacings and normal swelling behavior.

CONCLUSIONS

This review of the composition, interstratification, and interlayering of soil smectites shows that they are generally different from smectites in bentonites and other deposits. Soil smectites are commonly beidellitic and iron-rich, their interstratification characteristics differ from those of diagenetic clays, and they may contain aluminous, organic, or other interlayer material. These features impinge directly on chemical and physical soil properties and should be more thoroughly investigated. We need to know in much greater detail how soil smectites differ from those found in sediments.

REFERENCES

Adam, A. I., Anderson, W. B., and Dixon, J. B. (1983) Mineralogy of the major soils of the Gezira Scheme (Sudan): *Soil Sci. Soc. Amer. J.* **47**, 1233–1240.

Barnhisel, R. I. (1977) Chlorites and hydroxy interlayered vermiculite and smectite: in *Minerals in Soil Environments,* J. B. Dixon and S. B. Weed, eds., Soil Science Society of America, Madison, Wisconsin, 331–356.

Behar, A., Hubenov, G., and van der Marel, H. W. (1972) Highly interstratified clay minerals in salt-affected soils: *Contr. Mineral. Petrol.* **34**, 229–235.

Carson, C. D. and Dixon, J. B. (1972) Potassium selectivity in certain montmorillonitic soil clays: *Soil Sci. Soc. Amer. Proc.* **36**, 838–843.

Churchman, G. J. (1980) Clay minerals formed from micas and chlorites in some New Zealand soils: *Clay Miner.* **15**, 59–76.

Dixon, J. B. (1982) Mineralogy of Vertisols: in *Symposia Papers II, Vertisols and Rice Soils of the Tropics: Trans. 12th Internat. Congr. Soil Sci., New Delhi, India,* 48–60.

Douglas, L. A. (1982) Smectites in acidic soils: in *Proc. Int. Clay Conf., Bologna, Pavia, 1981,* H. van Olphen and F. Veniale, eds., Elsevier, Amsterdam, 635–640.

Duchaufour, P. (1982) *Pedology:* George Allen & Unwin, London, 448 pp. (Translated by J. R. Paton).

Egashira, K. and Ohtsubo, M. (1983) Swelling and mineralogy of smectites in paddy soils derived from marine alluvium, Japan: *Geoderma* **29**, 119–127.

Fitzpatrick, R. W. and Le Roux, J. (1977) Mineralogy and chemistry of a Transvaal black clay toposequence: *J. Soil Sci.* **28**, 165–179.

Güzel, N. and Wilson, M. J. (1981) Clay mineral studies of a soil chronosequence in southern Turkey: *Geoderma* **25**, 113–129.

Güzel, N. and Wilson, M. J. (1983) Chemical, physical and mineralogical characteristics of some Turkish soils derived from volcanic material: *Trans. Roy. Soc. Edinb. Earth Sci.* **74**, 153–163.

Herbillon, A. J., Frankart, R., and Vielvoye, L. (1981) An

occurrence of interstratified kaolinite-smectite minerals in a red-black soil toposequence: *Clay Miner.* **16**, 195–201.

Ildefonse, P. (1980) Mineral facies developed by weathering of a meta-gabbro, Loire-Atlantique (France): *Geoderma* **24**, 257–273.

Johnson, J. J. (1964) Occurrence of regularly interstratified chlorite-vermiculite as a weathering product of chlorite in a soil: *Amer. Mineral.* **49**, 556–572.

Johnson, W. M., Cady, J. G., and James, M. S. (1962) Characteristics of some brown Grumusols of Arizona: *Soil Sci. Soc. Amer. Proc.* **26**, 389–393.

Juo, A. S. R. (1980) Mineralogical characteristics of Alfisols and Ultisols: in *Soils with Variable Charge*, B. K. G. Theng, ed., New Zealand Society of Soil Science, Palmerston North, New Zealand, 69–86.

Kantor, W. and Schwertmann, U. (1974) Mineralogy and genesis of clays in red-black soil toposequences on basic igneous rocks in Kenya: *J. Soil Sci.* **25**, 63–78.

Kapoor, B. S., Singh, H. B., and Goswami, S. C. (1981) Three component interstratification in sodic soils: *J. Indian Soc. Soil Sci.* **129**, 123–124.

Klages, M. G. and Southard, A. R. (1968) Weathering of montmorillonite during formation of a solodic soil and associated soils: *Soil Sci.* **106**, 363–368.

Kodama, H. and Brydon, J. E. (1966) Interstratified montmorillonite-mica clays from sub-soils of the Prairie Provinces, western Canada: in *Clays and Clay Minerals, Proc. 13th Natl. Conf., Madison, Wisconsin, 1964*, W. F. Bradley and S. W. Bailey, eds., Pergamon Press, New York, 151–173.

Kodama, H. and de Kimpe, C. R. (1983) Ferruginous swelling clay minerals in a gabbro saprolite from Mount Megantic, Quebec: *Can. J. Soil Sci.* **63**, 143–148.

Kodama, H., Miles, N., Shimoda, S.., and Brydon, J. E. (1976) Mixed-layer kaolinite-montmorillonite from soils near Dawson, Yukon Territory: *Can. Mineral.* **14**, 159–163.

Krishna-Murti, G. S. R. and Satyanarayana, K. V. S. (1969) Significance of magnesium and iron in montmorillonite formation from basic igneous rocks: *Soil Sci.* **107**, 381–384.

Krishna-Murti, G. S. R. and Satyanarayana, K. V. S. (1970) Discussion on the significance of magnesium and iron in montmorillonite formation from basic igneous rocks: *Soil Sci.* **110**, 287–288.

Lee, R., Bache, B. W., Wilson, M. J., and Sharp, G. S. (1985) Aluminium release in relation to the determination of cation-exchange capacity of some podzolised New Zealand soils: *J. Soil Sci.* **36**, 239–253.

Malcolm, R. L., Nettleton, W. D., and McCracken, R. J. (1969) Pedogenic formation of montmorillonite from a 2:1–2:2 intergrade mineral: *Clays & Clay Minerals* **16**, 405–414.

Marshall, C. E. (1935) Layer lattices and base exchange clays: *Z. Kristallogr.* **91(A)**, 433–449.

Masshady, A. S., Reda, M., Wilson, M. J., and Mackenzie, R. C. (1980) Clay and silt mineralogy of some soils from Qasim, Saudi Arabia: *J. Soil Sci.* **31**, 101–115.

Moinereau, J. (1977) Altération des matériaux basaltiques et genèse des argiles en climat tempéré humide et milieu organique: *Cah. ORSTOM sér. Pédol.* **15**, 157–153.

Nadeau, P. H., Wilson, M. J., McHardy, W. J., and Tait, J. M. (1984a) Interstratified clays as fundamental particles: *Science* **225**, 923–925.

Nadeau, P. H., Wilson, M. J., McHardy, W. J., and Tait, J. M. (1984b) Interparticle diffraction: a new concept for interstratified clays: *Clay Miner.* **19**, 757–769.

Nadeau, P. H., Wilson, M. J., McHardy, W. J., and Tait, J. M. (1985) The conversion of smectite to illite during dia-

genesis: evidence from some illitic clays from bentonites and sandstones. *Mineral. Mag.* **49**, 393–400.

Nahon, D. B. and Colin, F. (1982) Chemical weathering of orthopyroxenes under lateritic conditions: *Amer. J. Sci.* **282**, 1232–1243.

Nash, V. E. (1979) Mineralogy of soils developed on Pliocene–Pleistocene terraces of the Tombigbee River in Mississippi: *Soil Sci. Soc. Amer. Proc.* **43**, 616–623.

Norrish, K. and Pickering, J. G. (1983) Clay minerals: in *Soils: an Australian Viewpoint*: CSIRO, Melbourne and Academic Press, London, 281–308.

Norrish, K. and Tiller, K. G. (1976) Subplasticity in soils. V. Factors involved and techniques of dispersion: *Aust. J. Soil Res.* **14**, 273–289.

Özkan, A. I. and Ross, G. J. (1979) Ferruginous beidellites in Turkish soils: *Soil Sci. Soc. Amer. J.* **43**, 1242–1248.

Paquet, H. (1967) Les montmorillonites des Vertisols; altération alcaline en milieu tropical: *Bull. Serv. Carte. geol. Als. Lorr.* **20**, 293–306.

Perez-Rodriguez, J. L., Weiss, A., and Lagaly, G. (1977) A natural clay organic complex from Andalusian black earth: *Clays & Clay Minerals* **25**, 243–251.

Proust, D. and Velde, B. (1978) Beidellite crystallization from plagioclase and amphibole precursors: local and long range equilibrium during weathering: *Clay Miner.* **13**, 199–209.

Rabie, F. and El-Araby, A. (1979) The mineralogy of magnesium-affected soils: *Egyptian J. Soil Sci.* **19**, 221–229.

Reynolds, R. C. (1980) Interstratified clay minerals: in *Crystal Structures of Clay Minerals and their X-ray Identification*, G. W. Brindley and G. Brown, eds., Mineralogical Society, London, 249–303.

Rich, C. I. (1968) Hydroxy interlayers in expansible layer silicates: *Clays & Clay Minerals* **16**, 15–30.

Robert, M. (1973) The experimental transformation of mica toward smectite: relative importance of total charge and tetrahedral substitution: *Clays & Clay Minerals* **21**, 167–174.

Ross, C. S. and Hendricks, S. B. (1945) Minerals of the montmorillonite group: their origin and relation to soil clays: *U.S. Geol. Surv. Prof. Pap.* **205B**, 23–79.

Ross, G. J. (1980) The mineralogy of Spodosols: in *Soils with Variable Charge*, B. K. G. Theng, ed., The New Zealand Society of Soil Science, Palmerston North, New Zealand, 127–143.

Ross, G. J. and Mortland, M. M. (1966) A soil beidellite: *Soil Sci. Soc. Amer. Proc.* **30**, 337–343.

Rossignol, J. P. (1983) Les Vertisols du nord de l'Uruguay: *Cah. ORSTOM sér. Pédol.* **20**, 271–291.

Sawhney, B. L. and Jackson, M. L. (1958) Soil montmorillonite formulas: *Soil Sci. Soc. Amer. Proc.* **22**, 115–118.

Singh, G. and Krishna-Murti, G. S. R. (1975) Charge distribution in smectites developed from basalts in Madhya Pradesh: *J. Indian Soc. Soil Sci.* **23**, 177–183.

Tobias, C. and Janot, C. (1981) L'évolution de la montmorillonite ferrifère, des oxydes et hydroxydes de fer dans une séquence fersiallitiques au Sénégal: *Cah. ORSTOM sér. Pédol.* **23**, 47–69.

Tomar, K. P. (1985) High spacing irregularly interstratified layer silicates in the alluvial soil clays of Meerut, India: *Clay Miner.* **20**, 115–124.

Viani, B. E., Al-Masshady, A. S., and Dixon, J. B. (1983) Mineralogy of Saudi Arabian soils: central alluvial basins: *Soil Sci. Soc. Amer. J.* **47**, 149–157.

Weir, A. H. and Rayner, J. H. (1974) An interstratified illite-smectite from Denchworth Series soil in weathered Oxford Clay: *Clay Miner.* **10**, 173–187.

Whittig, L. D. (1959) Characteristics and genesis of a solodised solonetz of California: *Soil Soc. Amer. Proc.* **23,** 469–473.

Wilson, M. J. and Berrow, M. L. (1978) The mineralogy and heavy metal content of some serpentinite soils in northeast Scotland: *Chem. Erde* **37,** 181–205.

Wilson, M. J. and Mitchell, B. D. (1979) Comparative study of a Vertisol and an Entisol from the Blue Nile plains of Sudan: *Egypt J. Soil Sci.* **19,** 207–220.

Wilson, M. J., Bain, D. C., and Duthie, D. M. L. (1984) The soil clays of Great Britain: II. Scotland: *Clay Miner.* **19,** 709–735.

Wilson, M. J. and Nadeau, P. H. (1985) Interstratified clay minerals and weathering processes: in *The Chemistry of Weathering,* J. I. Drever, ed., D. Reidel, Dordrecht, Holland, 97–118.

Yerima, B. P. K., Calhoun, F. G., Senkaji, A. L., and Dixon, J. B. (1985) Occurrence of interstratified kaolinite/smectite in El Salvador Vertisols: *Soil Sci. Soc. Amer. J.* **49,** 462–466.

Proceedings of the International Clay Conference, Denver, 1985, L. G. Schultz, H. van Olphen, and F. A. Mumpton, eds.,
The Clay Minerals Society, Bloomington, Indiana, 174–178 (1987).

CLASSIFICATION OF ANDISOLS IN JAPAN BASED ON PHYSICAL PROPERTIES

Takashi Maeda and Katsuyuki Soma

Faculty of Agriculture, Hokkaido University
Sapporo 060, Japan

Abstract—Japanese Andisols are volcanic ash soils that have characteristic physical properties such as low bulk density, high natural water content, high water content at 15 bars, and high liquid limit at the natural water content. They are classified as allophanic Andisols or crystalline Andisols according to the major clay minerals: chiefly allophane and imogolite for allophanic Andisols and Al-vermiculite, chlorite, small amounts of halloysite, and traces of gibbsite for crystalline Andisols.

The physical properties of Andisols are changed markedly and irreversibly on air-drying. Andisols can be classified into two types on the basis of the decrease in liquid limit on drying. A-type Andisols exhibit a large decrease in flow index, which is the slope of the shear strength-water content curve in the liquid limit test. B-type Andisols show no change in flow index with decreasing liquid limit on air-drying. Moreover, the decrease in liquid limit of A-type Andisols is larger than that of B-type Andisols.

Allophanic Andisols without a past history of drying and wetting, and a few crystalline Andisols with high organic matter content, belong to the A-type. Many crystalline Andisols and allophanic Andisols with a past history of drying and wetting belong to type B. Drying and wetting occur under natural conditions at the soil surface, but some subsoils may remain permanently wet.

The differences in physical properties between A- and B-type Andisols also show in the relationship between dry bulk density and natural water content, dry bulk density and water content at 15 bars, and plasticity index and liquid limit.

Key Words—Allophane, Andisol, Drying-wetting, Liquid limit, Plasticity, Volcanic ash soils.

INTRODUCTION

Much of the arable land in Japan is covered with soils derived from volcanic ash. These soils are classified as Andosols or Andepts, or more recently, Andisols. Andisols have characteristic physical and engineering properties that are related to the particular structure of allophane, the dominant clay material in many of these soils. Allophane has been found to consist of hollow spherules having many openings that permit the entry and exit of water molecules (Henmi and Wada, 1976). Some of the Andisols in Japan contain crystalline clay minerals such as Al-vermiculite or chlorite as the dominant clay materials (Masui and Shoji, 1967). These soils have also been described in the United States by Baham and Simonson (1985).

Therefore, Andisols can be classified as allophanic Andisols or crystalline Andisols according to the dominant clay mineral species. The clay minerals of allophanic Andisols consist chiefly of allophane and imogolite, whereas crystalline Andisols consist of Al-vermiculite, chlorite, small amounts of halloysite, and traces of gibbsite.

The physical properties that distinguish Andisols from other soils are low bulk density, high natural water content, high water content at 15 bars, and high liquid limit at natural water content (Maeda *et al.,* 1983). These properties are changed markedly and irreversibly by air-drying; this change is larger for allophanic Andisols than for crystalline ones.

Two characteristic physical properties have been proposed as criteria to distinguish volcanic ash soils in Soil Taxonomy by the Soil Survey Staff (1975), by G. D. Smith (Soil Conservation Service, U.S. Department of Agriculture, Washington, D.C., personal communication, 1978), and recently by Leamy (1983). These criteria are: (1) A bulk density of the fine earth fraction of <0.90 g/cm^3; and (2) Water retention of the undried fine earth of 40% or more at 15-bar soil water potential.

In this paper, we investigated the relationship between physical properties of Andisols and their mineralogy using samples at the natural water content and samples dried to lower water contents. Andisols were subsequently classified into two groups on the basis of their physical properties.

MATERIALS AND METHODS

Surface and subsoil samples of Andisols were collected from Hokkaido, Honshu (including Tohoku, Kanto, and Tokai regions), and Kyushu islands, Japan, and stored at their natural water content. Measurements were made at this *in situ* water content and on samples that were dried in the laboratory to varying water contents between the original and the air-dry state. Dry bulk density, defined as the mass of oven-dry soil divided by total volume of the wet soil mass, was measured on undisturbed soil samples taken with a core sampler of 5-cm diameter and 5-cm height.

Water retention was measured with a pressure plate

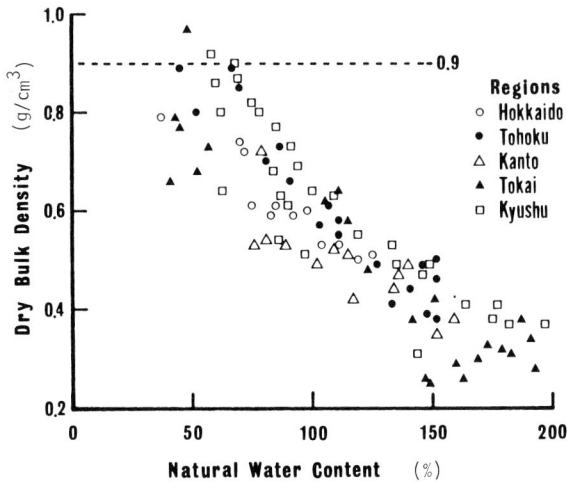

Figure 1. Relation between dry bulk density and natural water content of Japanese Andisols.

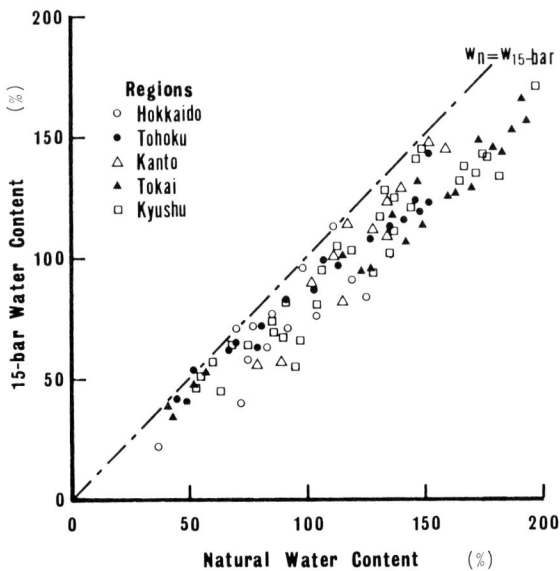

Figure 2. Relation between natural water content and 15-bar water content of Japanese Andisols.

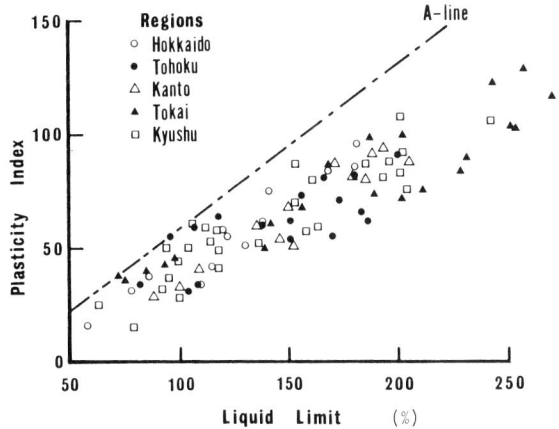

Figure 3. Position of Japanese Andisols on the Casagrande plasticity chart.

apparatus on undisturbed soil samples of 5-cm diameter and 2.5-cm height (Warkentin and Maeda, 1974). The samples were placed in an enclosed chamber on a water-saturated ceramic plate that had an air-entry value greater than 15 bar. The bottom of this plate was in contact with atmospheric pressure. As air pressure in the chamber was increased, water moved out of the samples through the ceramic plate, until equilibrium was reached between the confining air pressure and the energy with which water was retained in the soil sample. This pressure is defined as the soil water potential. The 15-bar water content is the water content of the soil in equilibrium with 15-bar air pressure, or 15-bar soil water potential.

Plasticity properties, the liquid and plastic limits, were measured according to Japanese Industrial Standards (JIS) A 1205 (equivalent to ASTM 423-66) and JIS A 1206 (equivalent to ASTM D 424-59), using <2-mm soil samples at varying initial water contents. The liquid limit is defined as the water content at which 25 blows in a special apparatus close a groove in the soil to a specified degree. Although they are empirical, the measured values are reproducible. The flow index is the slope of the water content vs. the log of the number of blows, which is usually linear. The liquid limit represents the water content at which soil behavior changes from plastic to liquid. The plastic limit is the water content where soil behavior changes from semi-solid to plastic. In the empirical standard method, soil at varying water contents is rolled into threads. The water content at which the soil can be rolled into ⅛-inch threads before breaking is the plastic limit.

RESULTS AND DISCUSSION

Physical properties of Andisols in Japan

In view of the changes in physical properties of Andisols on drying, it was first necessary to establish proper methods of measurement if these properties were to be useful in the classification of these soils. Moreover, these properties had to be measured on undisturbed soil samples. Andisols have a well-developed aggregate structure that consists of assemblages of microaggregates. Many physical properties, such as bulk density and water retention, are closely related to this soil structure.

Figure 1 shows dry bulk density vs. natural water content of undisturbed core samples of Japanese Andisols. Natural water content is the water content of the soil samples as they were taken in the field. Dry bulk density decreases with increasing natural water

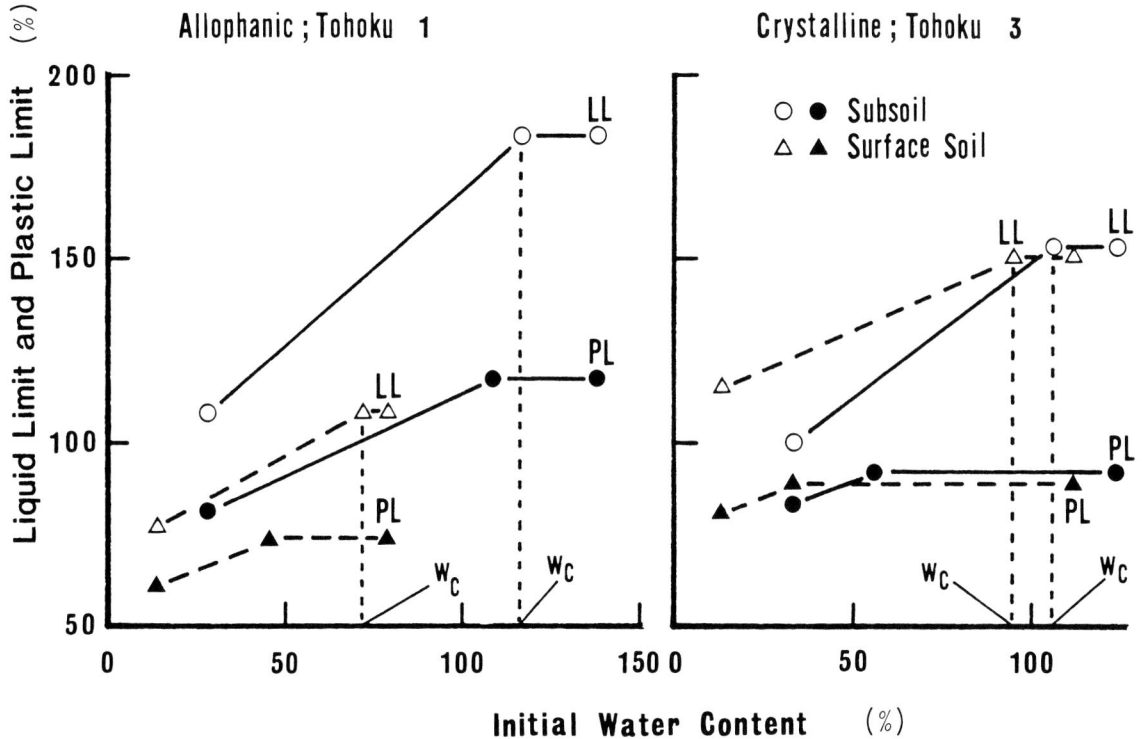

Figure 4.   Relation between liquid (LL) and plastic limit (PL) and initial water content of Japanese Andisols. w_c = critical water content $\cong 15$-bar water content.

content. Samples from different regions do not appear to be separated in the plot. The dry bulk density of Japanese Andisols is <0.90 g/cm³, except for some unweathered pumice soils containing heavy minerals.

The natural water content for undisturbed core samples generally exceeds the 15-bar water content, i.e., the amount of water retained at 15-bar soil water potential (Figure 2). In Japan, because of the wet climate, Andisols retain a high natural water content, and have apparently not been dried below the 15-bar water content. Again, the different regions do not separate out in the plot.

Plasticity is one of the physical properties that distinguish allophane from crystalline clay minerals. Many papers have pointed out that at natural water contents Andisols have high liquid and plastic limits and low plasticity indexes (Warkentin and Maeda, 1974; Maeda et al., 1977). Plasticity measurements for soils are usually plotted on a Casagrande plasticity chart as liquid limit vs. plasticity index (difference betwen liquid and plastic limit). The position of samples of Japanese Andisols at natural water content on the Casagrande plasticity chart is shown in Figure 3. The significance of the "A" line is that most inorganic soils with crystalline clay minerals have plasticity values around the line (Casagrande, 1948). The plasticity values of Andisols are distributed in an area below the A-line and

liquid limit >50%. The values of liquid limit and plasticity index for these Japanese Andisols are comparatively higher than those reported in the literature (Warkentin and Maeda, 1974; McNabb, 1979).

Changes in physical properties on drying

The liquid and plastic limits of Andisols decrease markedly and irreversibly on air-drying (Soma, 1978). Figure 4 shows the relation between liquid or plastic limits and initial water content for selected soil samples. Here, the initial water content is the water content to which the soil samples were dried before the plasticity measurement. If the soil samples at natural water content were air-dried gradually in the laboratory, the liquid limit at first remained constant with decreasing initial water content. If the initial water content fell below a certain value, however, the liquid limit decreased sharply with further decreases in initial water content. This particular water content was called the critical point by Soma and Maeda (1974). It is nearly equal to the 15-bar water content measured on an undisturbed core sample. The plastic limits also change on air-drying, but the change is smaller than for the liquid limits.

The decrease in the liquid and plastic limits caused by drying produced a change in the position of the plasticity values on the Casagrande plasticity chart.

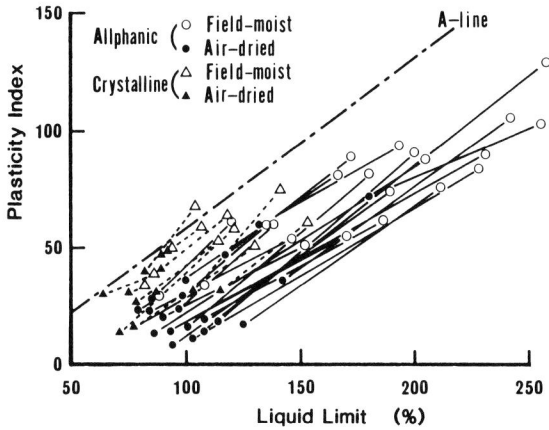

Figure 5. Change of plasticity of selected samples on air-drying.

Figure 5 shows the change from natural water content to air dry. The liquid limit decreased markedly, and the plasticity index decreased to a lesser degree. These changes were larger for allophanic than for crystalline Andisols.

The change in liquid limit caused by air-drying was studied in more detail by examining flow curves obtained from the liquid limit test for samples at natural water content and air-dried. Examples shown in Figures 6 and 7 indicate that the changes in liquid limit

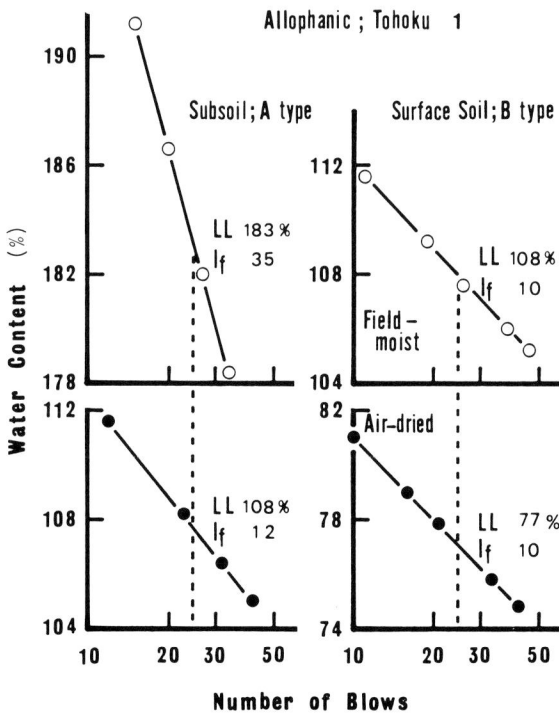

Figure 7. Flow-curves of crystalline Andisols.

on air drying can be divided in two groups, an A-type and a B-type. A-type changes show a large decrease in the slope of the flow curve (flow index) with decreasing liquid limit when air dried, whereas the B-type shows no change on air drying. Also, the decrease in liquid limit for A-type Andisols is comparatively larger than that for the B-type. Allophanic Andisols not subjected to previous drying and a few crystalline Andisols rich in organic matter belong to the A-type, whereas many crystalline Andisols and those allophanic Andisols with a past drying-wetting history belong to the B-type.

Classification of Andisols in Japan based on physical properties

On the basis of these measurements, Andisols in Japan can be classified into two types. Figure 8 shows the differentiation based upon dry bulk density vs. natural water content, dry bulk density vs. 15-bar water content, and plasticity index vs. liquid limit. The volume change of undisturbed Japanese Andisols on drying from natural water content to 15-bar soil water content is very small; hence, the dry bulk density does not change with this water content change. In Figure 8, therefore, dry bulk density (at natural water content) vs. 15-bar water content is also plotted as dry bulk density vs. natural water content.

A-type Andisols have natural water contents of 80–200%, dry bulk densities of 0.2–0.6 g/cm³, and 15-bar water contents of 70–170%. On the Casagrande plasticity chart, A-type Andisols occupy a broad area below the A-line, have liquid limits of 120–250% and plasticity indexes >40. B-type Andisols have natural water contents of 40–80%, dry bulk densities of 0.6–0.9 g/

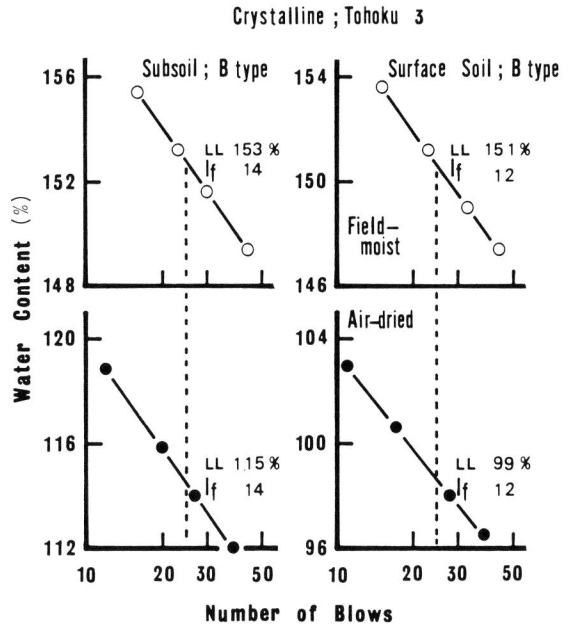

Figure 6. Flow-curves of allophanic Andisols. I_f = flow index; LL = liquid limit.

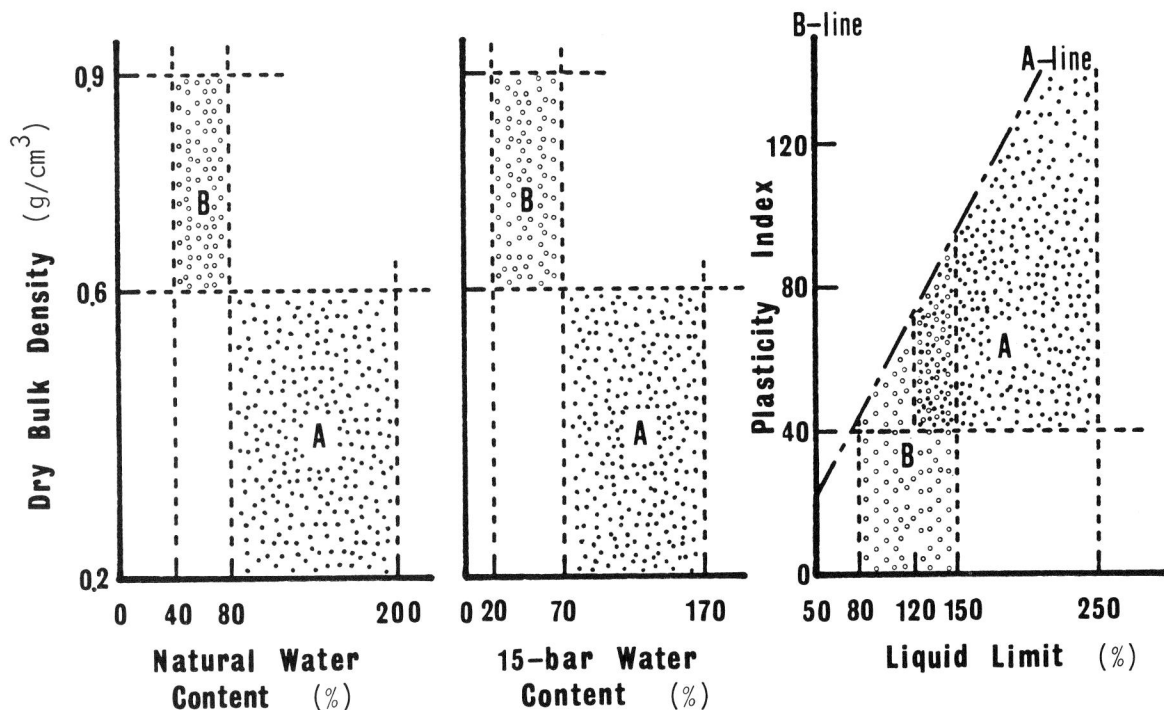

Figure 8.   Physical properties of A- and B-type Japanese Andisols.

cm³, and 15-bar water contents of 20–70%. On the Casagrande plasticity chart, B-type Andisols occupy a somewhat narrow area below the A-line and have liquid limits of 80–150%.

SUMMARY

Andisols, or volcanic ash soils, in Japan can be classified into two mineralogic groups, allophanic and crystalline Andisols. For the allophanic group, but not for the crystalline group, the flow index in the liquid limit test decreases remarkably on air-drying. The Andisols can be also classified into two types on the basis of their physical properties: A-type Andisols include the allophanic Andisols not subjected to a past dry-wet history and a few crystalline Andisols rich in organic matter; B-type Andisols include crystalline Andisols and allophanic Andisols subjected to past drying-wetting cycles.

ACKNOWLEDGMENTS

We thank K. Wada and C. Mizota, Department of Agricultural Chemistry, Kyushu University, and N. Yoshinaga, Department of Agricultural Chemistry, Ehime University, for permission to use their analytical data on mineralogy of the soil samples. We also thank B. P. Warkentin, Department of Soil Science, Oregon State University, for his help in revising the manuscript. This work was supported in part by a Grant-in-Aid for Research from the Ministry of Education, Japan.

REFERENCES

Baham, J. and Simonson, G. H. (1985) Classification of soils with andic properties from the Oregon coast: *Soil Sci. Soc. Amer. J.* **49,** 777–780.

Casagrande, A. (1948) Classification and identification of soils: *Trans. Amer. Soc. Civil Eng.* **113,** 901–903.

Henmi, T. and Wada, K. (1976) Morphology and composition of allophane: *Amer. Miner.* **61,** 379–390.

Leamy, M. L. (1983) Proposed revision of the Andisol proposal: *ICOMAND Circular Letter* **5,** 3–29.

Maeda, T., Takenaka, H., and Warkentin, B. P. (1977) Physical properties of allophane soils: *Adv. Agron.* **29,** 229–264.

Maeda, T., Soma, K., and Warkentin, B. P. (1983) Physical and engineering characteristics of volcanic ash soils in Japan compared with those in other countries: *Irrigation Engineering and Rural Planning* **3,** 16–31.

Masui, J. and Shoji, S. (1967) Some problems on clay minerals of volcanic ash soil: *Pedologist* **11,** 33–45.

McNabb, D. H. (1979) Correlation of soil plasticity with amorphous clay constituents: *Soil Sci. Soc. Amer. J.* **43,** 613–616.

Soil Survey Staff (1975) *Soil Taxonomy (A Basic System for Making and Interpreting Soil Surveys): U.S. Dept. Agr. Handbook* No. **436,** Washington, D.C., 230 pp.

Soma, K. (1978) Studies on the relationship between Atterberg limits and initial water content of soil: *Soil Physical Condition and Plant Growth, Japan* **38,** 16–22.

Soma, K. and Maeda, T. (1974) Studies on the relationship between liquid limit and initial moisture content of volcanic ash soil: *Trans. Japanese Soc. Irrigation, Drainage and Reclamation Eng.* **49,** 27–34.

Warkentin, B. P. and Maeda, T. (1974) Physical properties of allophane soils from the West Indies and Japan: *Soil Sci. Soc. Amer. Proc.* **38,** 372–377.

Proceedings of the International Clay Conference, Denver, 1985, L. G. Schultz, H. van Olphen, and F. A. Mumpton, eds.,
The Clay Minerals Society, Bloomington, Indiana, 179–185 (1987).

INFLUENCE OF NONCRYSTALLINE MATERIAL ON PHOSPHATE ADSORPTION BY KAOLIN AND BENTONITE CLAYS

K. P. C. Rao[1] and G. S. R. Krishna Murti[2]

Division of Agricultural Physics, Indian Agricultural Research Institute
New Delhi 110 012, India

Abstract—Phosphate adsorption by noncrystalline aluminosilicate gels and their mixtures with kaolin and bentonite was studied. The infrared and ion-exchange data indicated that the noncrystalline gels were chemically bonded to the crystalline minerals. The noncrystalline aluminosilicates showed a high phosphate adsorption capacity which increased with increasing Al in the gels. Phosphate adsorption was initially fast due to adsorption on Al-OH polymer surfaces and then slowed due to the breakdown of Al-OH polymers and/or displacement of structural silicate by phosphate groups. Similar studies of the clay-gel mixtures showed that noncrystalline material played a dominant role in phosphate adsorption by increasing the adsorption capacity by 8.5 and 3.8 times for kaolin and bentonite, respectively. Similar adsorption capacities of both clay-gel mixtures indicated that the phosphate reactive sites on the crystalline minerals appeared to be blocked by the noncrystalline material. In these mixtures phosphate adsorption appeared to be mainly on the noncrystalline aluminosilicate gels.

Fluoride ion was sorbed by the noncrystalline gels in a manner similar to that for the phosphate ion. The amount of hydroxyls released during the fluoride reaction varied linearly with the amount of phosphate adsorbed by the systems. This relationship may be useful in characterizing soils for their content of noncrystalline material and phosphate adsorption capacity.

Key Words—Adsorption, Aluminosilicate gel, Bentonite, Kaolin, Noncrystalline, Phosphate.

INTRODUCTION

Recently, noncrystalline inorganic constituents have been extensively reported in the colloidal fraction of soils developed on parent material other than volcanic ash (see, e.g., in Alfisols and Ultisols, Krishna Murti, 1982; Oxisols, Herbillon, 1980; Spodosols, Ross, 1980; Vertisols, Rao and Krishna Murti, 1984). These components are highly surface-reactive, and they have large cation-exchange capacities (Krishna Murti and Rao, 1984) and pH-dependent charges (Krishna Murti *et al.*, 1982).

Phosphate adsorption is one of the many important properties that characterize noncrystalline aluminosilicate gels. Soils containing large amounts of allophane are typically described as being "phosphate hungry" and are notorious for the amount of phosphate needed to maintain an adequate supply of this element to plants. Buol and Hole (1959) reported that the clays having noncrystalline oxide coatings adsorbed 186% more P than did normal soil. Similar observations by Soileau *et al.* (1964) and Khalifa and Buol (1969) underline the importance of noncrystalline material in plant nutrition.

Inasmuch as these noncrystalline components are intimately associated with crystalline aluminosilicate minerals, their surface properties can only be studied in synthetic systems. Synthetic aluminosilicate gels have been reported to adsorb greater amounts of phosphate as the Al/(Al + Si) molar ratio of the gels increases from 0.29 to 0.64 (Cloos *et al.*, 1968; Rajan and Perrott, 1975). Crystalline clay minerals also adsorb phosphate, although not of the same magnitude. Exchange characteristics of synthetic noncrystalline aluminosilicate gels of varying composition were reported earlier by Rao and Krishna Murti (1983). The cation selectivity (Ca or K over Mg) of kaolin and bentonite was reported to be significantly influenced by the nature of the clay and the aluminosilicate gel added (Rao and Krishna Murti, 1986).

Phosphate adsorption and availability are important parameters in plant nutrition. It is worthwhile to investigate how the phosphate adsorption characteristics of clays are affected by the presence of noncrystalline aluminosilicates, which are present in many soil clays in appreciable amounts. The present paper reports the influence of synthetic noncrystalline aluminosilicate gels of varying composition on the phosphate adsorption kinetics of kaolin and bentonite clays.

MATERIALS AND METHODS

Five synthetic, noncrystalline aluminosilicate gels of varying Al/(Al + Si) molar ratios (0.12, 0.21, 0.29,

[1] Present address: Dry Land Research Project, Saidabad, Hyderabad 500 659, India.

[2] Present address: Department of Soil Science, University of Saskatchewan, Saskatoon, Saskatchewan S7N 0W0, Canada.

Table 1. Characteristics of the samples used for study.

Samples	Chemical composition			Cation-exchange capacity (meq/100 g)	Surface area (m²/g)
	SiO_2 (%)	Al_2O_3 (%)	Al/(Al+Si) (molar)		
Noncrystalline aluminosilicate gels					
G1	58.25	6.90	0.12	165.3	480.7
G3	46.15	10.40	0.21	240.4	554.3
G5	41.60	14.50	0.29	282.3	424.3
G7	32.15	19.30	0.41	190.0	436.4
G9	29.40	22.85	0.48	159.3	590.9

Crystalline aluminosilicate minerals

	Mineralogy			
	Dominant	Accessory		
Bentonite (B)	Sm	Kl, Nc	82.3	509.9
Kaolin (K)	Kl		9.6	31.5

Sm = smectite; Kl = kaolinite; Nc = noncrystalline material (0.5-N-KOH extractable).

0.41, and 0.48) were used in the present study. The details of preparation, their structure, and their ion-exchange characteristics were reported by Rao and Krishna Murti (1983). After preparation, the gels were not allowed to dry, but held in an acetone-water solution. The kaolin and bentonite clays used in the present study are <2-μm fractions separated from clay deposits of Bihar, India, that contained more than 75% kaolinite and smectite, respectively, as determined by X-ray powder diffraction and cation-exchange capacity methods (Jackson, 1979). Data on a few characteristics of the samples are presented in Table 1.

The clays were mixed separately with four gels having Al/(Al + Si) molar ratios of 0.21, 0.29, 0.41, and 0.48 (termed, herein, G3, G5, G7, and G9, respectively) by addition of the corresponding gel suspension. The amount of gel added was adjusted to give an effective 10% gel (by weight) in the clay-gel mixture after evaporation. The gel suspension and the clay were mixed by shaking the clay-gel suspension vigorously

before the clay-gel mixture was allowed to dry. The bentonite-gel and kaolin-gel mixtures are referred to herein as BG3, BG5, BG7, and BG9 and KG3, KG5, KG7, and KG9, respectively.

Noncrystalline material from the clay-gel mixtures was extracted by two different methods, herein called the 0.5-N-KOH (Briner and Jackson, 1970) and the kinetic-dissolution (Segalen, 1968) techniques.

For the 0.5-N-KOH dissolution technique, 100 mg of the clay-gel mixture was transferred to a nickel beaker and boiled in 200 ml of 0.5 N KOH for exactly 2.5 min. The suspension was then rapidly cooled in a cold water bath. The extract was removed immediately by centrifugation and made up to a known volume (Briner and Jackson, 1970). For the kinetic-dissolution technique, about 400 mg of clay-gel mixture was subjected to eight repeated dissolution treatments as described by Segalen (1968). The Si and Al contents in the dissolved extracts were estimated colorimetrically by the yellow molybdosilicate (Jackson, 1958) and aluminon (Krishna Murti *et al.,* 1974) methods respectively.

The CEC (K/NH_4) of the noncrystalline gels and the clay-gel mixtures was determined following the method of Wada and Harada (1969). Potassium in the extract was determined by a flame photometer (ELICO Model C 22A). Infrared spectra of the samples were recorded in a Nujol medium using a Perkin-Elmer IR spectro-photometer, Model 399B. Surface area of the samples was estimated following the EGME method of Carter *et al.* (1965).

Phosphate adsorption was determined by the method of Cloos *et al.* (1968), after equilibrating the samples with 0.04 M KH_2PO_4 solution at pH 5 for 1, 3, 5, 10, 20, 25, and 30 days. The phosphorus content in the extract was determined colorimetrically by the molybdovanadate method (Jackson, 1958), using the extracts of the blanks to account for possible interference of silicon. Phosphate extractable by 0.02 N H_2SO_4 (pH 3.0, sample/solution ratio of 1/100) was assumed to

Table 2. Nature of the noncrystalline material extracted from the clay-gel mixtures.

Sample	Noncrystalline material extracted		Al/(Al + Si) molar ratio			CEC (meq/100 g)	
	B & J (%)	Segalen (%)	B & J	Segalen	Original gel (Table 1)	Before KOH dissolution	After KOH dissolution
B	3.0	5.7	0.10	0.19		82.3	103.3
BG3	10.0 (77)	12.8 (81)	0.20 (0.24)	0.24 (0.28)	0.21	90.3	102.5
BG5	10.0 (77)	12.4 (79)	0.25 (0.30)	0.32 (0.42)	0.29	98.0	102.9
BG7	9.3 (72)	12.4 (79)	0.32 (0.39)	0.39 (0.53)	0.41	86.3	102.5
BG9	8.7 (67)	11.4 (72)	0.37 (0.48)	0.45 (0.67)	0.48	87.3	101.4
K	0.8	1.2			0.41	9.6	9.8
KG3	9.7 (90)	11.8 (105)	0.19 (0.20)	0.23 (0.21)	0.21	25.8	9.2
KG5	8.2 (76)	9.3 (83)	0.26 (0.28)	0.31 (0.30)	0.29	33.9	9.3
KG7	8.0 (74)	9.4 (84)	0.37 (0.40)	0.41 (0.41)	0.41	25.5	9.1
KG9	8.2 (76)	10.5 (94)	0.46 (0.48)	0.47 (0.48)	0.48	23.0	10.0

B & J = extracted with 0.5 N KOH (Briner and Jackson, 1970); Segalen = kinetic dissolution (Segalen, 1968). Values in parentheses in columns 2 and 3 indicate percentage of the added gel that could be extracted; values in columns 4 and 5 are corrected for noncrystalline material present in the clays.

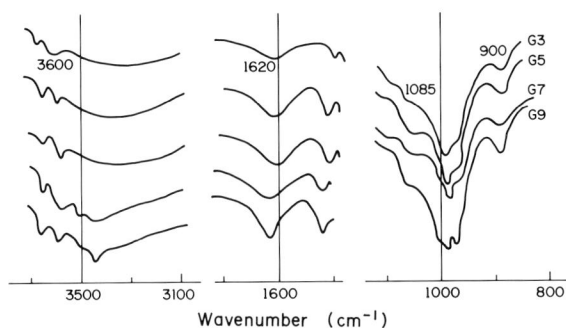

Figure 1. Infrared spectra of noncrystalline aluminosilicate gels.

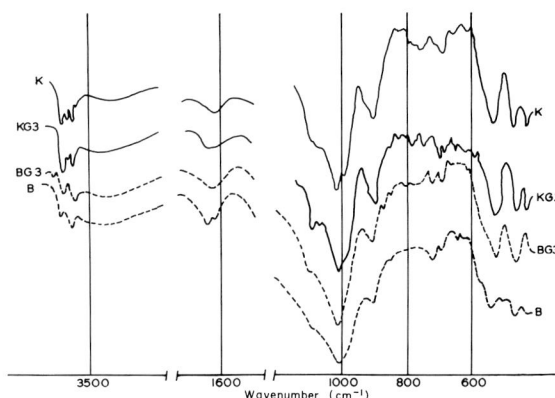

Figure 2. Infrared spectra of bentonite (B), kaolin (K), and their mixtures with noncrystalline aluminosilicate gels KG3 and BG3.

be available phosphate, as suggested by Troug (1930). Phosphorus in the extract was estimated as described above.

The rate of OH release from the samples when treated with 1.0 M NaF solution was measured using the procedure of Bracewell et al. (1970). Aluminum released during the fluoride reaction was determined on fluoride-free solutions (Huang and Jackson, 1965) by the aluminon method (Krishna Murti et al., 1974).

RESULTS AND DISCUSSION

Nature of clay-gel mixtures

The nature of noncrystalline material extracted (viz., the amounts of SiO_2 and Al_2O_3 extracted, the molar Al/(Al + Si) ratio of the extracted material and the CEC before and after extraction) from the clay-gel mixtures following the 0.5-N-KOH and the kinetic-dissolution methods is presented in Table 2. Neither of the two methods completely extracted the added noncrystalline gel; complete extraction was more nearly obtained from the kaolin-gel than from the bentonite-gel mixture (Table 2), the values ranging from 74–105%

and 67–81%, respectively. Similarly, the molar Al/(Al + Si) ratio of the extracted material was in closer agreement with the corresponding ratios of the added material (column 6, Table 2) for the kaolin than for the bentonite mixtures. The CEC values of the clay-gel mixtures before and after KOH dissolution suggest that the noncrystalline gels added were chemically bonded to the crystalline clay minerals. The increase in the CEC of bentonite and kaolin after KOH dissolution may be due to the removal of the noncrystalline material which was earlier occupying the exchange sites of the original, untreated clays (B and K, Table 2). The decrease of CEC of the clay-gel mixtures after KOH dissolution relative to the respective KOH-treated original clay samples (B and K, column 8, Table 2) may be due to the bonding of the gels to the minerals at the exchange sites (except sample KG9). This bonded gel represents, at least qualitatively, the difference between the percentage of gel removed and the amount originally added.

Table 3. Data on kinetics of phosphate adsorption by the aluminosilicate gels and the clays.

Sample[1]	Phosphate adsorbed (mmole/100 g) by the samples after:							
	1 day	3 days	5 days	10 days	15 days	20 days	25 days	30 days
G1	13.2 (6.0)	15.5 (6.1)	17.1 (6.1)	20.4 (4.7)	25.2 (4.4)	26.1 (4.7)	27.0 (4.7)	30.6 (4.3)
G3	34.2 (5.4)	40.3 (5.9)	51.5 (6.0)	56.5 (6.1)	58.1 (6.1)	64.4 (6.1)	66.4 (6.1)	68.4 (7.0)
G5	41.5 (6.8)	56.9 (7.2)	70.3 (7.9)	81.8 (7.2)	86.6 (7.5)	93.3 (6.9)	95.5 (6.9)	102.1 (7.3)
G7	93.3 (6.8)	117.8 (6.9)	126.1 (6.9)	129.4 (6.9)	135.3 (9.0)	150.0 (10.1)	152.0 (10.8)	160.5 (11.3)
G9	115.7 (6.7)	129.8 (6.8)	147.7 (6.8)	154.4 (6.8)	166.6 (9.8)	174.0 (10.5)	176.8 (10.6)	179.6 (12.1)
B	10.5	15.3	17.6	19.4	21.4	22.1	22.3	22.5
K	4.9	6.6	7.3	8.6	9.0	9.6	9.6	9.6

[1] G1, G3, G5, G7, and G9 are noncrystalline aluminosilicate gels with molar Al/(Al + Si) ratio of 0.12, 0.21, 0.29, 0.41, and 0.48, respectively; B and K are the bentonite and kaolin clays. The values in parentheses indicate the amount of available phosphate in mmole/100 g.

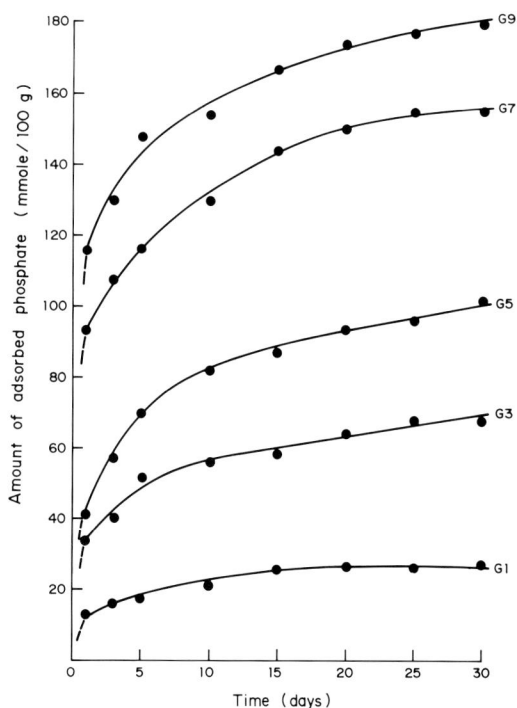

Figure 3.   Phosphate adsorption kinetics of the aluminosilicate gels.

The infrared spectra of the aluminosilicate gels and their mixtures with kaolin and bentonite are given in Figures 1 and 2. The noncrystalline aluminosilicate gels are characterized by an Si–O absorption maxima at about 1000 cm⁻¹ which shifts to lower frequencies with increasing Al content (Figure 1). This shift is probably due to an increasing substitution of Al for Si in the gels (Mitchell *et al.*, 1964; Kanno *et al.*, 1968). The characteristic absorption maxima at 1620 cm⁻¹ is probably due to adsorbed water in the gels (van der Marel and Beutelspacher, 1976). The absorption maxima in the range 3500–3800 cm⁻¹ are probably due to Si–OH and

Al–OH stretching vibrations (Fieldes *et al.*, 1972). The increasing number of absorption maxima observed with increasing Al content presumably indicates that the OH ions are held in more than one type of bonding. Rao and Krishna Murti (1983) showed that Al in excess of one in four substitution for Si in tetrahedral coordination, forms octahedrally coordinated hydroxyaluminum polymers.

The absorption spectra observed at about 3600, 1000, and 700 cm⁻¹ for kaolin and bentonite shifted significantly when the gels were mixed with the kaolin and bentonite clays (Figure 2). The maxima at 3694, 3620, and 920 cm⁻¹, attributed to Al–O–H, and the maxima at 1040 and 720 cm⁻¹, attributed to Si–O–Al (van der Marel and Beutelspacher, 1976), either shifted in frequency or additional maxima were noted in the region indicating that the noncrystalline gels were chemically bonded to the crystalline clays. The effect was most pronounced for the KG9 and BG9 samples.

Phosphate adsorption and release

Aluminosilicate gels. Phosphate adsorption data on the aluminosilicate gels are presented in Table 3. Phosphate adsorption was generally high compared with the crystalline clays and increased with increasing Al/(Al + Si) molar ratio of the gels. The kinetics of phosphate adsorption is shown in Figure 3. An initial rapid adsorption was followed by a slower adsorption, and the isotherms, instead of reaching a constant value, show a continuous increase in the rate of adsorption.

The adsorption of phosphate is associated mostly with Al–OH surfaces (Parfitt, 1978). With increasing proportion of Al in the gels, the Al–OH surfaces increase beyond a molar Al/(Al + Si) ratio of 0.25 due to increased formation of hydroxy-Al polymers (Rao and Krishna Murti, 1983), thereby giving rise to an increase in the phosphate adsorption value. The initial fast reaction is probably due to adsorption on exposed (Al–OH)-type surfaces by ligand-exchange reactions

Table 4.   Data on kinetics of phosphate adsorption by the clays and their mixtures with noncrystalline aluminosilicate gels.

Sample[1]	Phosphate adsorbed (mmoles/100 g) by the samples after:								OH released[2] by fluoride (meq/100 g)
	1 day	3 days	5 days	10 days	15 days	20 days	25 days	30 days	
B	10.5	15.3	17.6	19.4	21.4	22.1	22.3	22.5	44.6
BG3	16.7	18.1	21.5	23.2	25.5	26.9	28.2	28.2	125.7
BG5	27.0	29.7	31.7	33.8	37.0	40.3	42.5	43.9	228.5
BG7	30.6	38.8	44.1	48.3	50.9	52.6	55.0	56.6	278.4
BG9	39.6	42.3	48.1	53.8	56.6	59.5	62.5	64.0	318.7
K	4.9	6.6	7.3	8.6	9.0	9.6	9.6	9.6	14.9
KG3	13.2	15.1	17.5	20.2	23.5	27.6	28.6	28.6	108.5
KG5	20.1	25.2	29.7	35.2	40.3	41.5	42.8	43.0	236.7
KG7	30.6	35.2	42.9	44.3	47.0	49.6	50.5	50.9	250.2
KG9	41.5	46.6	50.5	50.8	55.3	56.6	60.5	62.4	292.7

[1] B and K are the bentonite and kaolin clays; BG3, BG5, BG7, BG9 and KG3, KG5, KG7, and KG9 are the bentonite-gel and kaolin-gel mixtures, respectively.

[2] Reaction time is 25 min.

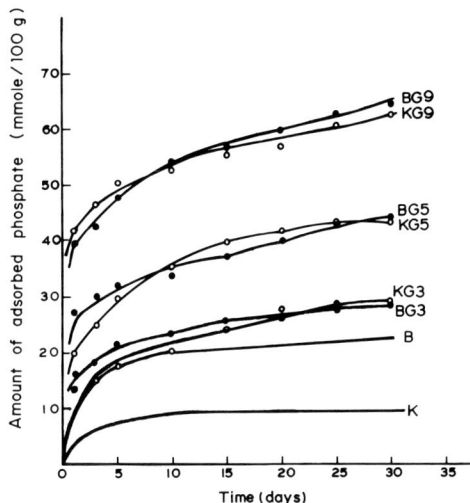

Figure 4. Phosphate adsorption kinetics of the bentonite, kaolin, and their mixtures with the aluminosilicate gels.

(Rajan and Perrott, 1975; Veith and Spossito, 1977). The subsequent slower reaction is probably due either to the surfaces exposed by progressive breakdown of the silicate structure and/or to the new sites formed by breakdown of hydroxy-Al polymers, as suggested by Rajan (1975a, 1975b) and Imai et al. (1981). The Al released during the breakdown of the structure was probably precipitated with phosphate in solution to form sparingly soluble aluminophosphates of the type $Al(OH)_2H_2PO_4$ and $Al(OH)Na(PO_4)$.

Most of the phosphate adsorbed by the gels was not in an available form (Table 3); the percentage of available phosphate decreased with increasing Al in the gels and with time (up to 10 days). Beyond ten days of equilibration, the percentage of available phosphate either decreased (G1, G3, and G5 gels) or showed a slight increase (G7 and G9 gels). Aluminum hydroxide and hydroxy-Al polymers are generally reported to be major components that are responsible for phosphate retention. If hydroxy-Al polymers form, the average net positive charge per Al ion decreases with increase in the size of the polymer and results in a weakening of the attraction for further OH ions. Phosphate ions adsorb on the surface only if the Al–OH–P attraction

is strong enough to remove such OH ions, resulting in a breakdown of the hydroxy-Al polymers and exposing new surfaces for phosphate adsorption (Hsu and Rennie, 1962).

Clay-gel mixtures. The kinetics data of phosphate adsorption on clay-gel mixtures are given in Table 4. The phosphate adsorption capacity of the clays increased markedly with the addition of noncrystalline gels, e.g., the kaolin and bentonite showed 8.5- and 3.8-fold increases in phosphate adsorption after treatment with gel G9 and one-day equilibration with KH_2PO_4 solution. The adsorption of phosphate on the crystalline minerals took place chiefly on edges where small proportions of positive charge occur through ligand exchange. The bentonite, having a higher surface area (both on faces and edges), contributed more such sites and gave rise to greater adsorption than kaolin.

The total amount of phosphate adsorbed by the two clays after treatment with the gels was almost the same (Figure 4), indicating that in mixed systems, phosphate adsorption was governed mainly by the amount of noncrystalline material in the system. As pointed out in the discussion of the exchange characteristics of the clay-gel mixtures, the noncrystalline material was apparently bonded to the exposed exchange sites of the crystalline minerals. Consequently, the phosphate reactive sites on the crystalline minerals were apparently blocked by the noncrystalline gel, and the only sites available for phosphate adsorption were those in the noncrystalline gel added to the crystalline minerals. The similar adsorption values for the mixtures (treated with same amount of gel), even though the phosphate adsorption capacity of the crystalline minerals were significantly different, support this conclusion.

The kinetics data of phosphate adsorption (Table 4) further indicate that the adsorption instead of reaching a constant value continued even after 30 days of reaction time, which can be attributed to the slower reaction due to breakdown of the structure of the aluminosilicate gel as discussed above.

Fluoride reactivity

Huang and Jackson (1965) showed that Al released from an aluminosilicate system during reaction with

Table 5. Fluoride reactivity of the aluminosilicate gels.

Sample[1]	Molar Al/(Al + Si) ratio	OH (meq/100 g) released after:					Al released[2] (meq/100 g)	OH/Al ratio
		5 min	10 min	15 min	20 min	25 min		
G1	0.12	136.2	189.1	428.2	484.0	508.0	337.2	4.52
G3	0.21	297.3	481.0	593.8	657.6	693.3	603.2	3.45
G5	0.29	496.0	649.7	752.7	811.5	827.9	773.8	3.21
G6	0.39	570.3	737.4	840.8	917.8	960.9	951.5	3.03
G7	0.41	590.2	798.3	906.3	1061.7	1079.1	1048.6	3.09
G9	0.48	531.9	751.3	817.8	875.0	918.8	906.2	3.04

[1] G1, G3, G5, G6, G7, and G9 are noncrystalline aluminosilicate gels.
[2] Reaction time = 25 min.

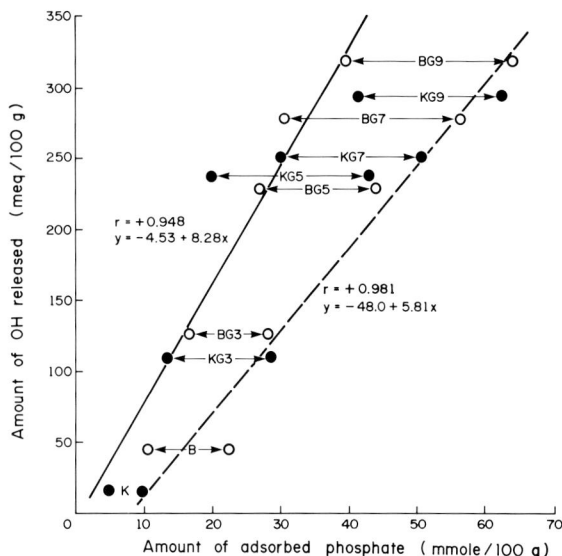

Figure 5. Relationship between fluoride reactivity and phosphate sorption of the clays and their mixtures with noncrystalline aluminosilicate gels. Phosphate sorption values after one-day (——) and 30-days (- - - -) equilibration time.

1.0 M NaF is equal to the equivalent amount of hydroxyls released into the system. Bracewell *et al.* (1970) and Perrott *et al.* (1976) suggested that OH ions released at constant pH can be a measure of the relative amounts of the noncrystalline components in the system.

Fluoride reactivity of the gels and their mixtures with clays was studied to look for a possible relation between phosphate adsorption capacity and fluoride reactivity. The results are summarized in Table 5. Introduction of fluoride solution resulted in an immediate increase in pH from the initial value of 6.8 to 10.0. The rate of release was fast during the first 15 min, and the amount of OH released increased with increasing amount of Al in the gel system to a value of molar Al/(Al + Si) = 0.41 (sample G7) and then decreased. The total amount of the OH ions released by the noncrystalline gels was much higher than for the crystalline minerals. The magnitude of OH ions released further indicates that the release of Al was mainly due to the disruption of the gel structure, as it is unlikely that such a large amount of OH was released by the adsorption reaction of fluoride alone. Table 5 also shows that the OH/Al ratio is higher in the gels with small amounts of Al, and that it was nearly constant at about 3.0 at a molar Al/(Al + Si) ratio of 0.39 (sample G6) and higher, indicating that Al was present in different forms in the two ranges. The magnitude of the OH ions released by fluoride from the clay-gel mixture as compared to crystalline clays (column 10, Table 4) indicates that noncrystalline gels alone contributed significantly to the fluoride reactivity of the system.

Because of the similarity of the reaction of fluoride

and phosphate with noncrystalline gels, viz., initial sorption due to ligand exchange followed by structural breakdown, a relationship was sought between the OH released by the fluoride reaction and phosphate adsorbed by the systems (Figure 5). The significant linear relationship observed indicates a possible method for rapidly assessing the phosphate adsorption capacity and the amount of noncrystalline components in the system.

CONCLUSIONS

The phosphate adsorption kinetics of noncrystalline aluminosilicate gels of varying composition indicated an initial fast sorption due to adsorption on Al–OH surfaces of the hydroxy-Al polymers of the gel. The subsequent slow adsorption observed was attributed to the breakdown of the aluminosilicate gel structure. The infrared spectra of the bentonite-gel and kaolin-gel mixtures indicated that the noncrystalline gels were chemically bonded to the crystalline clays. The similar adsorption values of the clay-gel mixtures indicate that phosphate adsorption was mainly on the noncrystalline gels. The fluoride reactivity, as measured by the amount of hydroxyls released during the fluoride reaction, varied linearly with phosphate adsorption capacity of the clay-gel mixtures. This relationship may be a useful measure of the noncrystalline material content and phosphate adsorption capacity in soils.

ACKNOWLEDGMENT

The work forms part of the Ph.D. thesis of the senior author (K. P. C. Rao), who is grateful to the Indian Agricultural Research Institute, New Delhi, India, for the award of senior fellowship during the course of the investigation.

REFERENCES

Bracewell, J. M., Campbell, A. S., and Mitchell, B. D. (1970) An assessment of some thermal and chemical techniques used in the study of the poorly ordered aluminosilicates in soil clays: *Clay Miner.* **8,** 325–335.

Briner, G. P. and Jackson, M. L. (1970) Mineralogical analysis of clays in soils developed from basalts in Australia: *Israel J. Chem.* **8,** 487–500.

Buol, S. W. and Hole, F. D. (1959) Some characteristics of clay skins on the peds in the B horizon of a gray brown podzolic soil: *Soil Sci. Soc. Amer. Proc.* **23,** 239–241.

Carter, D. L., Heilman, M. D., and Gonzalez, G. L. (1965) Ethylene glycol monoethyl ether for determining the surface area of silicate minerals: *Soil Sci.* **100,** 356–360.

Cloos, P., Herbillon, A. J., and Echeverria, J. (1968) Allophane-like synthetic silicoaluminas. Phosphate adsorption and availability: *Trans. 9th Int. Congr. Soil Sci., Adelaide, Australia* **2,** 733–745.

Fieldes, M., Furkert, R. J., and Wells, N. (1972) Rapid determination of constituents of the whole soils using infrared observation: *N.Z. J. Sci.* **15,** 615–627.

Herbillon, A. J. (1980) Mineralogy of oxisols and oxic materials: in *Soils with Variable Charge,* B. K. G. Theng, ed., New Zealand Soc. Soil Sci., Lower Hutt, New Zealand, 109–126.

Hsu, P. H. and Rennie, D. A. (1962) Reactions of phosphate in aluminum systems. I. Adsorption of phosphate by X-ray amorphous aluminum hydroxide: *Can. J. Soil Sci.* **42**, 197–209.

Huang, P. M. and Jackson, M. L. (1965) Mechanism of reaction of neutral fluoride solution with layer silicates and oxides of soils: *Soil Sci. Soc. Amer. Proc.* **29**, 661–665.

Imai, H., Goulding, K. W. T., and Talibudeen, O. (1981) Phosphate adsorption in allophanic soils: *J. Soil Sci.* **32**, 555–570.

Jackson, M. L. (1958) *Soil Chemical Analysis:* Prentice Hall, Englewood Cliffs, N.J., 498 pp.

Jackson, M. L. (1979) *Soil Chemical Analysis—Advanced Course:* 2nd ed., Published by the author, Dept. of Soil Science, Univ. of Wisconsin, Madison, Wisconsin, 895 pp.

Kanno, I., Onikura, Y., and Higashi, T. (1968) Weathering and clay mineralogical characteristics of volcanic ashes and pumices in Japan: *Trans. 9th Int. Congr. Soil Sci., Adelaide, Australia, Vol. 3,* 111–122.

Khalifa, E. M. and Buol, S. W. (1969) Studies of clay skins in Cecil (typic Hapludult) soil. II. Effect on plant growth and uptake: *Soil Sci. Soc. Amer. Proc.* **33**, 102–105.

Krishna Murti, G. S. R. (1982) Amorphous constituents in soil clays: in *Review of Soil Research—Part II,* Indian Soc. Soil Sci., New Delhi, India, 725–730.

Krishna Murti, G. S. R., Bhavanarayana, M., Rao, T. V., and Rao, K. P. C. (1982) pH-dependent charge of soil amorphous material: *Clay Res.* **1**, 57–62.

Krishna Murti, G. S. R. and Rao, T. V. (1984) A model system for soil amorphous material: *Agrochimica* **28**, 257–265.

Krishna Murti, G. S. R., Sarma, V. A. K., and Rengasamy, P. (1974) Spectrophotometric determination of aluminium with aluminon: *Indian J. Tech.* **12**, 270–271.

Mitchell, B. D., Farmer, V. C., and McHardy, W. J. (1964) Amorphous inorganic materials in soils: *Adv. Agron.* **16**, 327–383.

Parfitt, R. L. (1978) Anion adsorption by soils and soil materials: *Adv. Agron.* **30**, 1–50.

Perrott, K. W., Smith, B. F. L., and Mitchell, B. D. (1976) Effect of pH on the reaction of NaF with hydrous oxides of Si, Al, and Fe and with poorly ordered aluminosilicates: *J. Soil Sci.* **27**, 348–356.

Rajan, S. S. S. (1975a) The mechanism of phosphate sorption of allophane clays: *N.Z. J. Sci.* **18**, 93–101.

Rajan, S. S. S. (1975b) Phosphate adsorption and the displacement of structural silicon in allophane clay: *J. Soil Sci.* **26**, 250–256.

Rajan, S. S. S. and Perrott, K. W. (1975) Phosphate adsorption by synthetic amorphous aluminosilicates: *J. Soil Sci.* **26**, 257–266.

Rao, K. P. C. and Krishna Murti, G. S. R. (1983) Structure and exchange characteristics of amorphous aluminosilica gels: *Clay Res.* **2**, 10–15.

Rao, K. P. C. and Krishna Murti, G. S. R. (1986) Influence of amorphous silicoaluminas on the cation exchange selectivity of soil clays: *Clay Res.* **5** (in press).

Rao, T. V. and Krishna Murti, G. S. R. (1984) Mineralogy and genesis of a few red and black complex soils of Hyderabad, India: *Recent Res. Geology* **11**, 225–247.

Ross, G. J. (1980) Mineralogical, physical and chemical characteristics of amorphous constituents in some podzolic soils from British Columbia: *Can. J. Soil Sci.* **60**, 31–43.

Segalen, P. (1968) Note sur une méthode de détermination des produits minéraux amorphes dans certains sols à hydroxydes tropicaux: *Cah.-Orstom Ser. Pedol.* **6**, 105–125.

Soileua, G., Fackson, W. A., and McCracken, R. J. (1964) Cutans (clay films) and potassium availability to plants: *J. Soil Sci.* **15**, 117–123.

Troug, E. (1930) The determination of readily available phosphorus of soils: *J. Amer. Soc. Agron.* **22**, 874–882.

van der Marel, H. W. and Beutelspacher, H. (1976) *Atlas of Infrared Spectroscopy of Clay Minerals and their Admixtures:* Elsevier, Amsterdam, 396 pp.

Veith, J. A. and Sposito, G. (1977) On the use of the Langmuir equation in the interpretation of adsorption phenomena: *Soil Sci. Soc. Amer. J.* **41**, 697–702.

Wada, K. and Harada, Y. (1969) Effects of salt concentration and cation species on the measured cation exchange capacity of soils and clays: in *Proc. Int. Clay Conf., Tokyo, 1969, Vol. 1,* L. Heller, ed., Israel Univ. Press, Jerusalem, 561–571.

Proceedings of the International Clay Conference, Denver, 1985, L. G. Schultz, H. van Olphen, and F. A. Mumpton, eds.,
The Clay Minerals Society, Bloomington, Indiana, 186–194 (1987).

TEXTURAL AND MINERALOGICAL RELATIONSHIPS BETWEEN FERRUGINOUS NODULES AND SURROUNDING CLAYEY MATRICES IN A LATERITE FROM CAMEROON

JEAN-PIERRE MULLER

O.R.S.T.O.M., UR 605, and Laboratoire de Minéralogie et Cristallographie, UA CNRS 09
Universités Paris 6 et 7, 2 place Jussieu, 75251 Paris Cédex 05, France

GÉRARD BOCQUIER

Laboratoire de Pédologie, Université Paris 7, 2 place Jussieu
75251 Paris Cédex 05, France

Abstract—In lateritic profiles from Cameroon, the relationships between ferruginous nodules, in which the texture has been inherited from the parent rock, and surrounding red and yellow matrices, in which the original rock texture has disappeared, were investigated by optical and scanning electron microscopy and by electron microprobe, X-ray powder diffraction, infrared, and electron spin resonance analyses. An orderly succession of changes was found for kaolinite from the ferruginous nodules to the clayey matrices as follows: (1) successive generations of the clay, each of smaller particle size; (2) concomitant decrease in degree of crystallinity; (3) increase in amount of iron substitution; and (4) decrease in the degree of orientation relative to the foliate texture of the parent gneiss. Correlative changes were also observed for associated iron oxides: (1) progressive decrease in the overall content of iron oxides; (2) decrease in the content of hematite (containing a minor amount of Al substitution); and (3) an increase in the content of Al-substituted goethite. These data suggest successive transformations from the ferruginous nodules to the surrounding clayey matrices, although progressive recrystallization may have taken place due to variation of the geochemical conditions of weathering of the original rock.

Key Words—Aluminum, Goethite, Hematite, Iron, Kaolinite, Laterite, Nodules, Texture.

INTRODUCTION

Most laterite formations of central Africa that formed in a humid climate and under forest cover show a consistent pattern of development from the bottom to the top of the profiles (Stoops, 1967; Bocquier *et al.,* 1984) as follows (Figure 1A): (1) a lower saprolite; (2) an intermediate nodular horizon composed of indurated ferruginous nodules (according to Brewer, 1964) surrounded by loose and clayey red and yellow matrices; and (3) an upper horizon mainly composed of a yellow clayey matrix. Weathering of the parent rock apparently takes place with little textural change. Millot and Bonifas (1955) and numerous other authors have shown that the weathering products commonly replicate the texture and structure of the original rock in both the saprolite and in some of the ferruginous nodules in the overlying nodular horizon. The secondary products of these materials are chiefly kaolinite and iron oxides which coexist with resistant primary minerals, such as quartz and muscovite. In contrast, the clayey matrices are characterized by the absence of original rock texture and structure. The principal mineralogical components of these matrices are also kaolinite and iron oxides (together with stable primary minerals), but these minerals have different crystal chemical characteristics (structural order, substitution rate, etc.) than those found in the saprolite and in the nodules (Herbillon, 1980; Didier *et al.,* 1983; Cantinolle *et al.,* 1984).

The present report compares the textural and crystal chemical characteristics of ferruginous nodules whose texture and structure are inherited from a parent granite gneiss with the surrounding clayey matrices.

MATERIALS AND METHODS

Materials

About 100 weathering profiles derived from gneissic granite in a tropical rain forest in central Cameroon have been described in detail by Sarazin *et al.* (1982) and Muller and Bocquier (1986). Seven of them in which the original rock structure and texture are the best preserved in the soil profiles have been selected for the present study. About thirty samples were taken from each profile.

Ferruginous nodules in these profiles are of two chief types (Figure 1): (1) Large nodules are 20–80 mm in diameter, irregular in shape, have textures more or less inherited from original gneiss that grade into the surrounding matrix. They are the more abundant type of nodule in the profiles studied. (2) Small nodules are

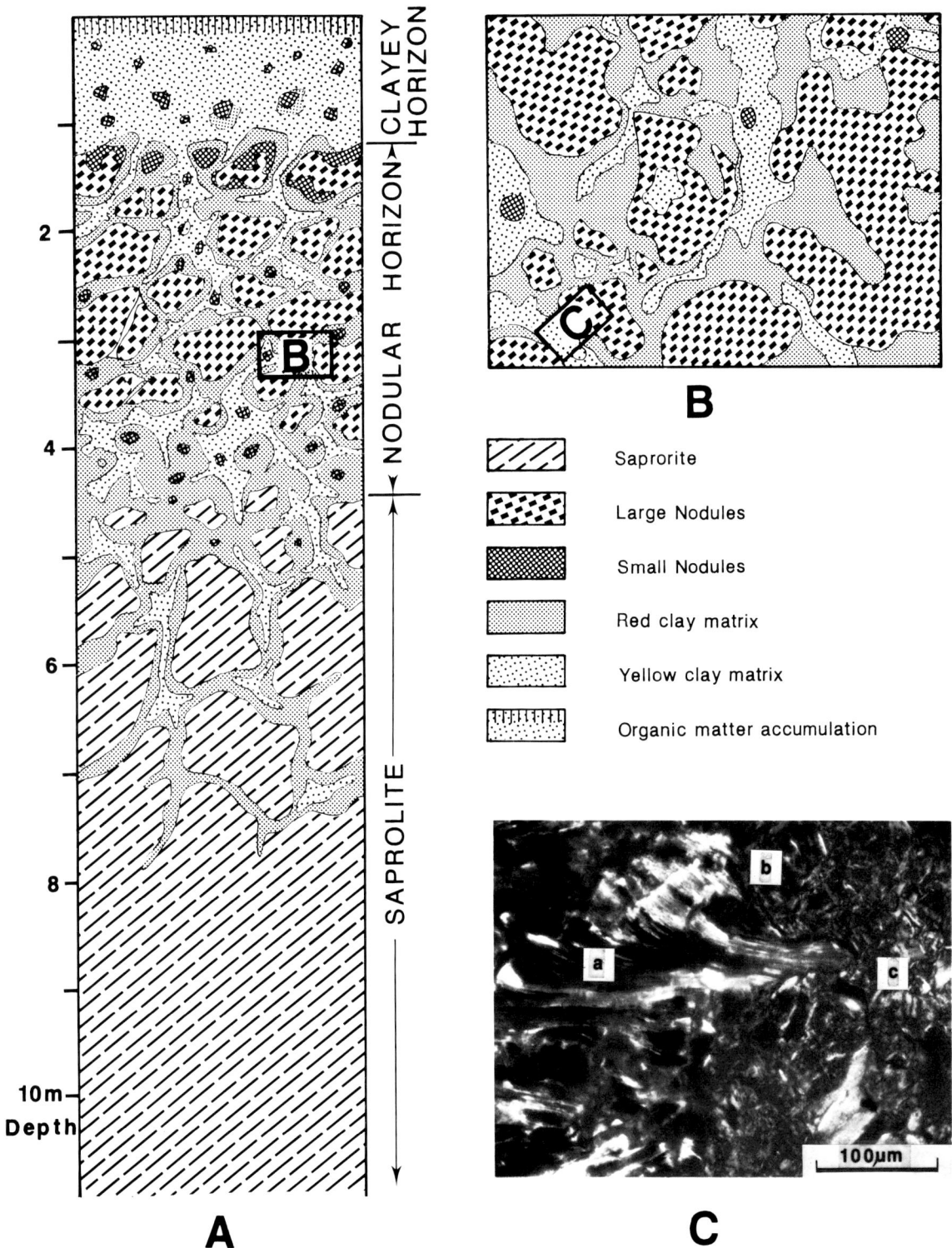

Figure 1. (A) Schematic representation of a laterite profile, Cameroon; (B) sketch of the relative distribution of matrices within an impregnated block of material from the nodular horizon; (C) optical micrograph showing the transition between a ferruginous nodule (note the inherited gneissic texture and the large crystallites of kaolinite) and clayey matrices (note the random texture and the smaller crystallites of kaolinite). a = ferruginous nodule; b = red clayey matrix; c = yellow clayey matrix.

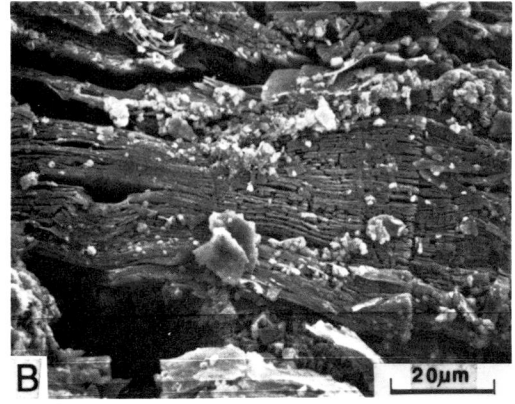

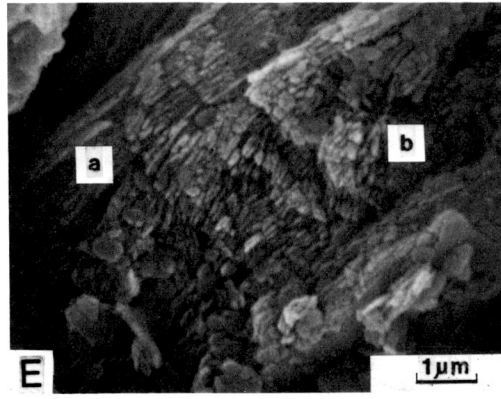

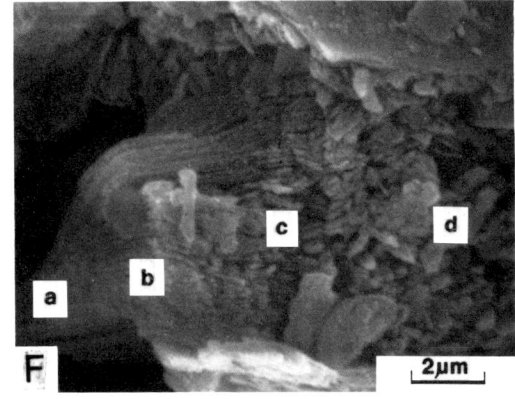

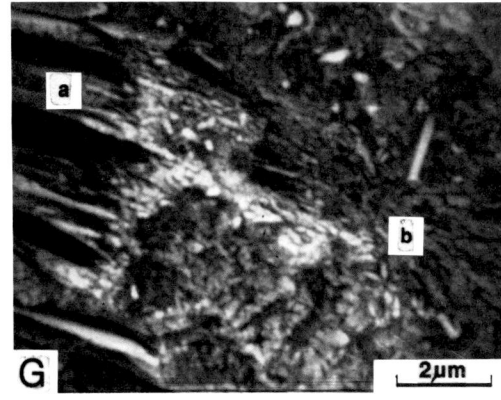

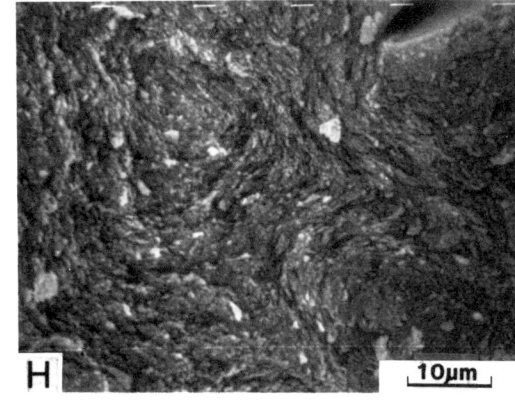

less than 20 mm in diameter, oval in shape, have sharp boundaries, and do not display the gneissic rock texture. Only the former are taken into account in this study. The smaller nodules were studied previously by Muller and Bocquier (1986).

The relationships between large nodules having inherited rock texture and surrounding clayey matrices were studied at different levels in the nodular horizon. Indeed, the nodules become more and more indurated and the original rock texture progressively less distinct upward in the horizon.

As shown schematically in Figures 1B and 1C, macroscopic and microscopic sections display an ordered relationship and gradual transition from the ferruginous nodules to a red clayey matrix and to a yellow clayey matrix. Textural, chemical, and mineralogical analyses were carried out on each of these three types of weathered materials.

Methods

Samples were impregnated with epoxy resin, and thin sections were prepared for petrographic examination. Fracture surfaces of clods were coated with gold and examined in a JEOL JSM 20 scanning electron microscope (SEM) at 20 kV. Quantitative elemental analyses were obtained on thin sections using an electron microprobe (Camebax) equipped with an EDS ORTEC multiline analyzer. Semiquantitative elemental analyses of clods were also obtained during SEM examination by energy dispersive X-ray techniques (EDX).

X-ray powder diffraction (XRD) data were obtained using CoKα radiation (40 kV, 40 mA) and a Philips PW 1730 vertical goniometer at scanning rates of 0.25 or 0.125°2θ/min. The degree of Al substitution in the goethite structure was estimated from 111 XRD line shifts using Vegard's rule. The degree of Al substitution in the hematite structure was calculated according to the quadratic equation of Perinet and Lafont (1972) after determining the unit-cell parameters by the method of William (1964).

Infrared (IR) absorption spectra were recorded in a FTIR Nicollet DX spectrograph on samples pressed in KBr. Peak intensities were calculated after an auto-matic baseline correction. Electron spin resonance (ESR) measurements were made at 298 K on an X-band Varian CSE 109 spectrometer, using 100 kHz modulation and 40 mW incident microwave power. Calibrated quartz tubes were filled with 50 mg of powdered samples which had been pretreated by the complexing procedure of De Endredy (1963) to remove Fe oxides.

RESULTS

Petrographic results

Ferruginous nodules. In the central part of the ferruginous nodules, especially the less indurated ones, textures of the parent gneiss were inherited at macroscopic, microscopic, and ultramicroscopic scales. For example, the SEM shown in Figure 2A illustrates the preservation of the original foliation of the gneiss, as defined by lithologic layering and the planar orientation of grain boundaries (Turner and Weiss, 1963). In Figure 2B a large crystallite of kaolinite (150 μm in diameter) appears to be pseudomorphous after mica flakes that originally were parallel to the rock cleavage, typical of *in situ* weathering as described by several authors (e.g., Bisdom *et al.,* 1982). Iron oxides occur at the boundaries between the large kaolinite booklets and interparticle voids.

Figure 2C shows another face-to-face packing of large crystallites of kaolinites. Kaolinite lamellae (40 μm in diameter) are continuous and close-packed at the edges of the aggregates, but they appear to be split into much smaller crystallites (2–5 μm in diameter) separated by cracks perpendicular to basal planes (best seen in Figure 2D) in the central part. Moreover, these smaller kaolinite crystals have been noted only where iron oxide has accumulated in the form of rosettes of intertwinned hematite platelets (see also mineralogy discussed above). Note, however, in Figure 2C that the original phyllitic texture inherited from mica is preserved.

At the periphery of the ferruginous nodules, particularly the more indurated ones, gneissic texture is macroscopically less distinct, and optically the large crystallites of kaolinite appear to be less abundant. In SEM, as in Figure 2E, even smaller kaolinite crystallites (0.5

Figure 2. Scanning electron micrographs of: (A) slightly indurated ferruginous nodule showing inherited gneissic texture (Q = quartz; K = pseudomorphs of kaolinite; O = iron oxides; V = voids); (B) detail of a large crystallite of kaolinite apparently pseudomorphous after mica; (C) large crystallites of kaolinite from a ferruginous nodule—low magnification micrograph showing close-packed lamellae of kaolinite at edges of aggregate; (D) detail of outlined rectangle in Figure 2C, showing smaller crystallites of kaolinite and rosettes of hematite (arrow); (E) a ferruginous nodule showing textures of different generations of kaolinite (a = preservation of original phyllitic texture by large lamellae of kaolinite, b = disruption of this texture by zones of smaller crystallites of kaolinite); (F) a ferruginous nodule showing spatial relationship between mica and successive generations of kaolinite (a = mica, b = exfoliated and large crystallite of kaolinite, c = smaller crystallites of kaolinite packed face-to-face (texture of precedent kaolinite is preserved), d = small crystallites of kaolinite in random texture); (G) the transition between (a) a ferruginous nodule and (b) the surrounding red clayey matrix showing same spatial relationship between large and small crystallites of kaolinites as in Figure 6F; (H) red clayey matrix showing "microaggregates" of small crystallites of kaolinites.

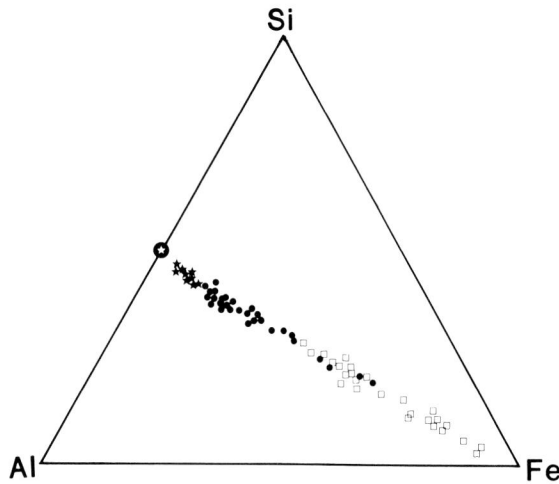

Figure 3. Chemical compositions in terms of Si, Al, and total Fe expressed as Fe^{3+}. Microprobe analyses. □ = ferruginous nodules; ● = red clayey matrix; ★ = yellow clayey matrix; ◉ = ideal kaolinite.

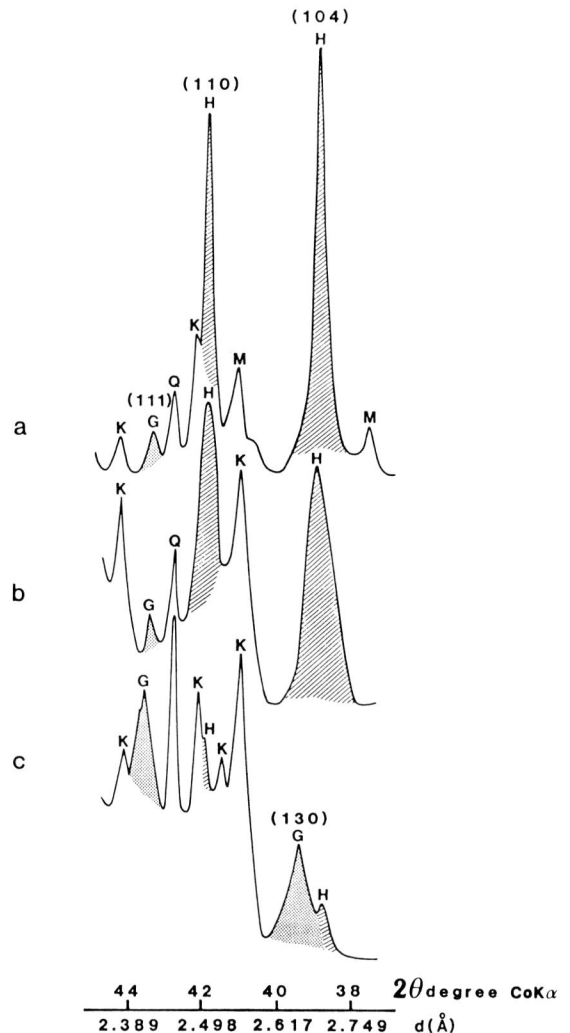

Figure 4. X-ray powder diffraction patterns of (a) ferruginous nodule; (b) red clayey matrix; (c) yellow clayey matrix. (111) = Miller indices of reflection. K = kaolinite; H = hematite; G = goethite; Q = quartz; M = muscovite-$2M_1$.

μm in diameter) are less oriented with respect to the larger lamellae of kaolinite (or relict mica). Figure 2F shows an orderly relationship between (a) close-packed lamellae of residual mica (identified by the presence of K in semi-quantitative analysis), (b) exfoliated large crystallites of kaolinite, (c) smaller crystallites of kaolinite, but which are still parallel to the lamellae of the exfoliated kaolinite, and (d) randomly oriented small crystallites of kaolinites. A similar sequence can be seen in Figure 2G that corresponds to the progressive macroscopic transition from ferruginous nodules to surrounding red clayey matrices shown in Figure 1C.

Clayey matrices. Both red and yellow clay matrices are characterized by a random arrangement of kaolinite crystallites, <1 μm in diameter, as has been reported elsewhere for similar materials (Eswaran, 1983). The crystal size of the kaolinites is so small in the yellow clayey matrices that these matrices appear almost optically isotropic; however, new textures are apparent, particularly in the red clayey matrices wherein kaolinite crystallites tend to be arranged concentrically (Figure 2H) in what has been called "microaggregates" (Muller, 1977; Chauvel et al., 1983).

Geochemical results

In situ quantitative microprobe analyses of the secondary products in thin sections are presented in Figure 3. From these data: (1) the materials appear to be a mixture of kaolinite (Al/Si ratio is constant) and variable amounts of iron oxides; (2) the amount of iron appears to decrease from the ferruginous nodules (40–90% Fe_2O_3) to the red clayey matrix (10–50% Fe_2O_3), to the yellow clayey matrix (<12% Fe_2O_3); and (3) the iron appears to be progressively homogenized from the

ferruginous nodules to the yellow clayey matrix, as is shown by a decrease in the scatter of the points in the diagram.

X-ray powder diffraction results on iron minerals

Hematite and goethite coexist in all the materials examined; however, as the iron content decreases and the color changes from red (5R) to yellow (7.5YR) (Munsell Color Chart, 1954), the iron oxide changes from almost exclusively hematite in the ferruginous nodules to predominantly goethite in the yellow clayey matrix.

Hematite. The XRD patterns of the hematite in the ferruginous nodules are different from those of hematites in the red clayey matrix (Figures 4a and 4b),

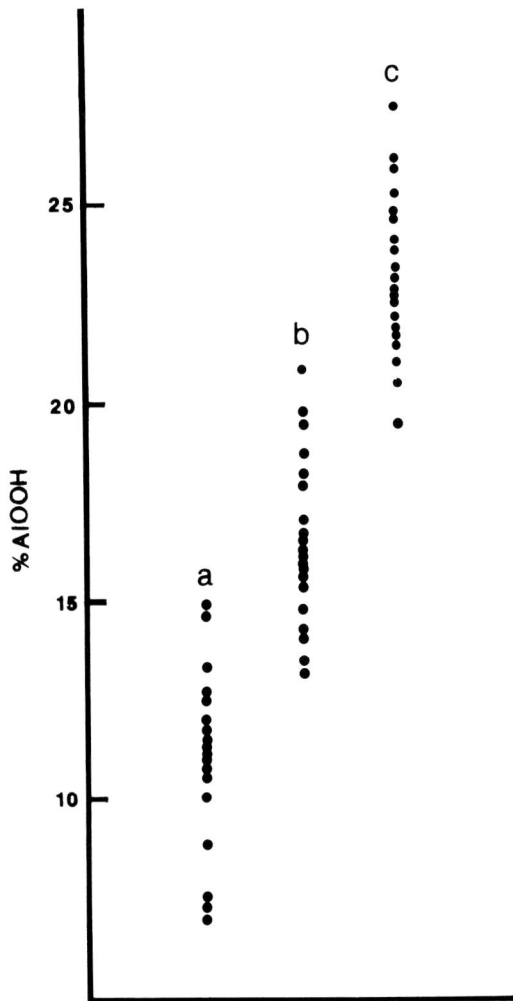

Figure 5. Plot of Al-substitution rate in goethites of (a) ferruginous nodules; (b) red clayey matrix; (c) yellow clayey matrix; expressed in % AlOOH.

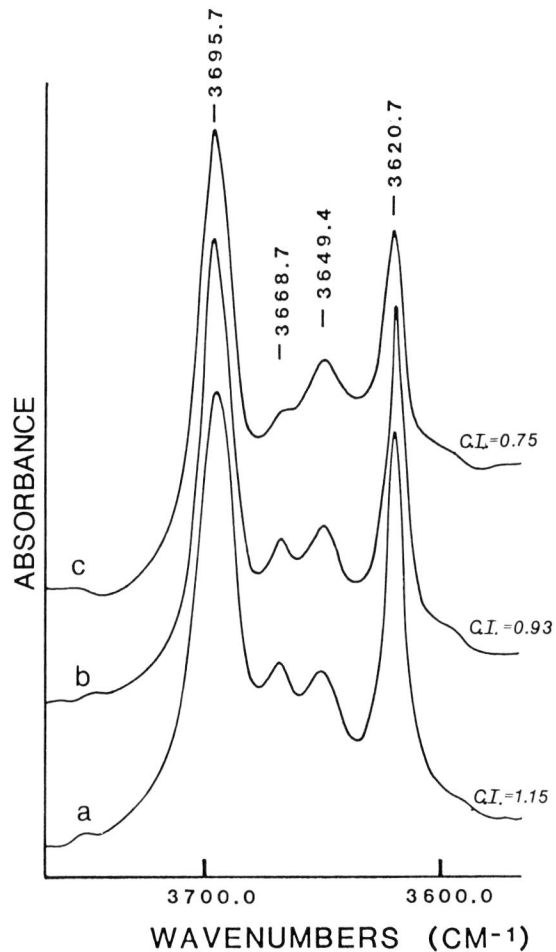

Figure 6. Infrared OH-stretching vibrations of selected kaolinites. (a) ferruginous nodules; (b) red clayey matrix; (c) yellow clayey matrix.

as follows: (1) a small shift in the 110 reflection towards lower d-values in the latter materials; (2) a broadening of the 104 and 110 reflections, attributed to a reduced growth of hematite crystals in the Z-direction in the presence of Al (Schwertmann *et al.,* 1979), but which can also be a result of decreased particle size and/or crystallinity; and (3) an inversion of the intensity ratios of the 104 and 110 reflections, which is probably due to a change in the amount of Al in the crystal structure (Schwertmann *et al.,* 1977). (Hematite in the yellow clayey matrices was not examined because of its low concentration and intimate occurrence with goethite and kaolinite.)

Using the XRD vs. Al-substitution data developed by Perinet and Lafont (1972), the degree of Al substitution calculated for 15 samples ranges from 3 to 6% in the ferruginous nodules and from 5 to 9% in the red clayey matrices. More quantitative analyses, how-

ever, needed to state precisely the scatter in the data points and the overlap of the different populations is in progress, using the procedure of Kampf and Schwertmann (1982) for the concentration of oxides.

Goethite. Although the differences in the properties of goethite may be related to both Al-for-Fe substitutions and structural defects (Schulze and Schwertmann, 1984), the amount of Al substitution may be estimated from the position of the 111 XRD reflection (Schulze, 1984). Data for Al content of goethite in about 20 samples of ferruginous nodules and red and yellow matrices are shown on Figure 5. The Al content of the goethite increases from the ferruginous nodules (7–15%) to the red clayey matrices (13–21%) to the yellow matrices (19–28%). These data agree with those reported by Fitzpatrick and Schwertmann (1982) who found that in lateritic environments Al substitution is much lower in red Plinthite than in yellow clayey matrices. Moreover, the comparison of the Fitzpatrick and Schwertmann (1982) data with our hematite data shows

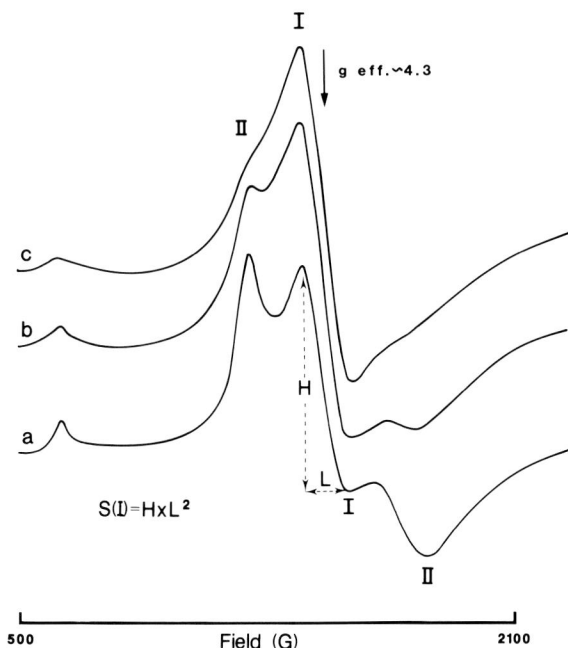

Figure 7. Electron spin resonance spectra of selected kaolinites. $g_{eff} = 4$-band splitting into their components I and
II. (a) = ferruginous nodule; (b) = red clayey matrix; (c) = yellow clayey matrix.

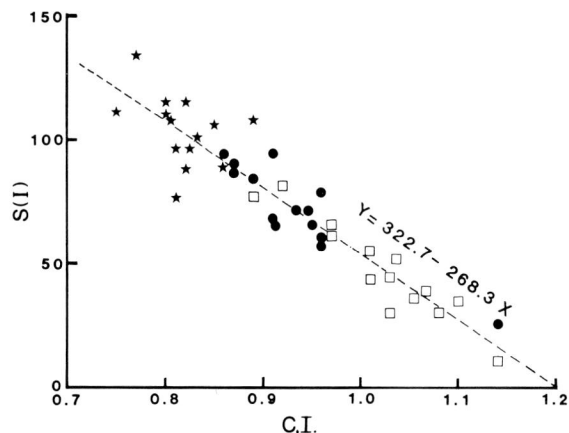

Figure 8. Correlation of infrared crystallinity index (C.I.)
and S(I), the area of the internal $g_{eff} = 4$ signal I, expressed in
arbitrary units. □ = ferruginous nodules; ● = red clayey matrix; ★ = yellow clayey matrix.

that in each type of material goethite is more Al rich
than the coexisting hematite. These results agree with
those reported by Nahon et al. (1977).

Infrared spectroscopic results for kaolinite

Most methods used to estimate the degree of order
or disorder in kaolinite are based on XRD data and
are somewhat limited for soil materials such as laterites, that are rich in iron oxides and quartz. IR techniques avoid such limitations (Parker, 1969). Cruz et
al. (1982) showed that interlamellar cohesion energy
increases with the degree of layer stacking disorder.
Giese and Datta (1973) showed that this disorder is
intimately related to the orientation of inner-surface
hydroxyls. Inner-surface hydroxyls, in turn, have been
found to give rise to four OH-stretching bands in IR
spectra (Rouxhet et al., 1977). Barrios et al. (1977)
found a wide distribution of both absolute and relative
intensities of the bands at 3669 and 3649 cm^{-1} as a
function of the degree of ordering of kaolinites, giving
rise to a crystallinity index, defined as C.I. = I(3669
cm^{-1})/I(3649 cm^{-1}). According to Cases et al. (1982),
this cristallinity index corresponds well with those based
on XRD data, and C.I. decreases with decreasing degree of order.

OH-stretching bands of kaolinites selected from the
different materials examined in the present work are
presented in Figure 6. The corresponding C.I. values
range from 1.15 for ferruginous nodules to 0.75 for

yellow clayey matrix. The C.I. of kaolinite from the
red clayey matrix is intermediate.

Electron spin resonance of Fe-bearing kaolinite

Angel and Hall (1973) and Meads and Malden (1975)
showed from ESR spectroscopy that small (<2.5%,
according to Mestdagh et al., 1980) and variable
amounts of iron substitute in the octahedral layer of
kaolinite. Such substitutions produce two components
near g = 4 that are related to Fe^{3+} occupying two distinct sites in the structure of the kaolinite (centers I
and II, Figure 7). The relative intensities of these two
components correlate with the crystallinity of the kaolinite as follows: (1) center I is associated with layer
stacking disorder; and (2) center II corresponds to domains of greater crystallinity and regular stacking (Jones
et al., 1974). The degree of "kaolinite perfection" and
the related amount of iron substitution in I sites are
closely related (Mestdagh et al., 1980).

From the ESR spectra for kaolinites from the various
materials examined here (Figure 7), all the kaolinites
appear to be iron bearing, in agreement with Herbillon
et al. (1976) and Mendelovici et al. (1979). Moreover,
the relative intensity of the signal from center I with
respect to center II increases from the ferruginous nodules (Figure 7a) to the yellow clayey matrices (Figure
7c), suggesting more Fe substitution in the latter. Figure
8 shows a reasonable agreement ($r^2 = .87$) between the
integrated area (intensity × square of linewidth peak-
to-peak, expressed in arbitrary units (Mestdagh et al.,
1980) of the signal from center I (S(I)) and the IR
crystallinity index (C.I.).

DISCUSSION AND CONCLUSIONS

The gradual transition analyzed at different scales
between some lateritic ferruginous nodules and their

Table 1. Comparative textural and crystal chemical characteristics of the ferruginous nodules and their surrounding clayey matrices in the nodular horizon.

	Textures	Ferruginous nodules	Red clayey matrices	Yellow clayey matrices
		Inherited rock texture	Soil texture	
Iron oxides	Fe$_2$O$_3$ (%)	>40	10–50	<12
	Mineralogy	Al-hematite (3–6% Al) + Al-goethite (7–15% Al)	Al-hematite (5–9% Al) + Al-goethite (13–21% Al)	Al-goethite (19–28% Al)
Kaolinites	I.R. Crystallinity index (C.I.)[1]	0.95–1.15	0.85–0.95	0.75–0.85
	E.S.R. S(I)[2]	25–75	50–90	75–130

[1] C.I. = I(3669 cm^{-1})/I(3649 cm^{-1}).

[2] S(I) = intensity × square of linewidth peak-to-peak of "I" E.S.R. signal near g = 4.3 (arbitrary units).

surrounding loose and clayey matrices corresponds to progressive and correlative changes in textures, geochemistry, mineralogy, and crystal chemistry of both kaolinites and iron oxides (Table 1). An orderly succession of changes was found for kaolinite from ferruginous nodules to clayey matrices as follows: (1) successive generations of the clay, each of smaller particle size; (2) concomitant decrease in degree of crystallinity; (3) increase in amount of iron substitution; and (4) decrease in the degree of orientation relative to the foliate texture of the parent gneiss. In the same manner, one observes for the iron oxides: (1) progressive decrease in their overall amount; (2) decrease in the content of hematite, containing a relatively small amount of Al substitution; and (3) an increase in the content of substituted goethite, containing a relatively large amount of Al.

Because of their inherited rock texture, i.e., the size and orientation of kaolinite particles are similar to the precursor, the ferruginous nodules seem to be an earlier stage of weathering than the surrounding matrices. If this is so, successive alterations from the nodule state to the red and then presumably also to the yellow clayey matrix state suggest an *in situ* transformation of minerals in the nodules to related minerals of a different age in the matrices. We have not proven, however, that the kaolin and iron oxide minerals in the nodules have not actually dissolved and reprecipitated in slightly different form and with different chemical compositions in the matrices, in which case, the minerals could be of similar ages and formed in response to different geochemical conditions that, with time, migrated generally inward toward centers of the nodules and downward in the soil profile.

ACKNOWLEDGMENTS

We are grateful to G. Calas for laboratory facilities, to MM. Locati and Morin for their technical assistance and to G. Calas, U. Schwertmann, and R. Fitzpatrick for helpful discussions.

REFERENCES

Angel, B. R. and Hall, P. L. (1973) Electron spin resonance studies of kaolinite: in *Proc. Int. Clay Conf., Madrid, 1972*, J. M. Serratosa, ed., Div. Ciencias C.S.I.C., Madrid, 47–60.

Barrios, J., Plançon, A., Cruz, M. J., and Tchoubar, C. (1977) Qualitative and quantitative study of stacking faults in a hydrazine treated kaolinite. Relationships with the infrared spectra: *Clays & Clay Minerals* 25, 422–429.

Bisdom, E. B. A., Stoops, G., Delvigne, J., Curmi, P., and Altemuller, H. J. (1982) Micromorphology of weathering biotite and its secondary products: *Pedologie* 32, 225–252.

Bocquier, G., Muller, J. P., and Boulangé, B. (1984) Les latérites. Connaissances et perspectives actuelles sur les mécanismes de leur différenciation: *Livre Jubilaire Cinquantenaire A.F.E.S.,* Paris, 123–138.

Brewer, R. (1964) *Fabric and Mineral Analysis of Soils:* Wiley, New York, 470 pp.

Cantinolle, P., Didier, P., Meunier, J. D., Parron, C., Guendon, J. L., Bocquier, G., and Nahon, D. (1984) Kaolinites ferrifères et oxyhydroxydes de fer et d'alumine dans les bauxites des Canonettes (S.E. de la France): *Clay Miner.* 19, 125–135.

Cases, J. M., Lietard, O., Yvon, J., and Delon, J. F. (1982) Etude des propriétés cristallochimiques, morphologiques, superficielles de kaolinites désordonnés: *Bull. Minéralogie* 105, 439–455.

Chauvel, A., Soubiès, F., and Melfi, A. (1983) Ferrallitic soils from Brazil. Formation and evolution of structure: *Sciences Geol.* 72, 37–46.

Cruz, M., Sow, C., and Fripiat, J. J. (1982) Etude par spectrométrie infrarouge des kaolinites désordonnées: *Bull. Minéralogie* 105, 493–498.

De Endredy, A. S. (1963) Estimation of free iron oxides in soils and clays by a photolytic method: *Clay Miner.* 5, 209–217.

Didier, P., Nahon, D., Fritz, B., and Tardy, Y. (1983) Activity of water as a geochemical controlling factor in ferricretes. A thermodynamic model in the system kaolinite Fe-Al oxihydroxides: *Sciences Geol.* 71, 35–44.

Eswaran, H. (1983) Characterization of domains with the scanning electron microscope: *Pedologie* 38, 41–54.

Fitzpatrick, R. W. and Schwertmann, U. (1982) Al-substituted goethite. An indicator of pedogenic and other weathering environments in South Africa: *Geoderma* 27, 335–347.

Giese, R. F. and Datta, P. (1973) Hydroxyl orientation in kaolinite, dickite, and nacrite: *Amer. Min.* 58, 471–479.

Herbillon, A. J. (1980) Mineralogy of oxisols and oxic materials: in *Soils With Variable Charge*, B. K. B. Theng, ed., N.Z. Soc. Soil Sci., Wellington, 109–126.

Herbillon, A. J., Mestdagh, M. M., Vielvoye, L., and Derouane, E. G. (1976) Iron in kaolinite with special reference to kaolinite from tropical soils: *Clay Miner.* **11**, 201–220.

Jones, J. P. D., Angel, B. R., and Hall, P. L. (1974) Electron spin resonance studies of doped synthetic kaolinites. II: *Clay Miner.* **10**, 257–269.

Kampf, N. and Schwertmann, U. (1982) The 5-M NaOH concentration treatment for iron oxides in soils: *Clays & Clay Minerals* **30**, 401–408.

Meads, R. E. and Malden, P. J. (1975) Electron spin resonance in natural kaolinites containing Fe^{3+} and other transition metal ions: *Clay Miner.* **10**, 313–345.

Mendelovici, E., Yariv, S. H., and Villalba, B. (1979) Iron bearing kaolinite in Venezuelan laterites. Infrared spectroscopy and chemical dissolution evidence: *Clay Miner.* **14**, 323–331.

Mestdagh, M. M., Vielvoye, L., and Herbillon, A. J. (1980) Iron in kaolinite. The relationship between kaolinite crystallinity and iron content: *Clay Miner.* **15**, 1–14.

Millot, G. and Bonifas, M. (1955) Transformations isovolumétriques dans les phénomènes de latérisation et de bauxitisation: *Bull. Service Carte Géologique Alsace Lorraine* **8**, 3–20.

Muller, J. P. (1977) Microstructuration des structichrons rouges ferrallitiques, à l'amont des modelés convexes, Centre-Cameroun: *Cahiers ORSTOM Pédologie* **15**, 25–44.

Muller, J. P. and Bocquier, G. (1986) Dissolution of kaolinites and accumulation of iron oxides in lateritic-ferruginous nodules: mineralogical and microstructural transformations: *Geoderma* **37**, 113–136.

Munsell Color Chart (1954) *Munsell Soil Color Charts:* Munsell Color Company, Baltimore, Maryland.

Nahon, D., Janot, C., Karpoff, A. M., Paquet, H., and Tardy, Y. (1977) Mineralogy, petrography and structures of iron crusts (ferricretes) developed on sandstones in the western part of Senegal: *Geoderma* **19**, 263–278.

Parker, T. W. (1969) A classification of kaolinites by infrared spectroscopy: *Clay Miner.* **8**, 135–141.

Perinet, G. and Lafont, R. (1972) Sur les paramètres cristallographiques des hématites alumineuses: *C.R. Acad. Sci. Paris* **275**, 1021–1024.

Rouxhet, P. G., Samudacheata, N., Jacobs, H., and Anton, O. (1977) Attribution of the OH stretching bands of kaolinite: *Clay Miner.* **12**, 171–179.

Sarazin, G., Ildefonse, Ph., and Muller, J. P. (1982) Contrôle de la solubilité du fer et de l'aluminium en milieu ferrallitique: *Geochim. Cosmochim. Acta* **46**, 1267–1279.

Schulze, D. G. (1984) The influence of aluminum on iron oxides. VIII. Unit-cell dimensions of Al-substituted goethites and estimation of Al from them: *Clays & Clay Minerals* **32**, 36–44.

Schulze, D. G. and Schwertmann, U. (1984) The influence of aluminium on iron oxides. X. Properties of Al-substituted goethites: *Clay Miner.* **19**, 521–539.

Schwertmann, U., Fitzpatrick, R. W., and Le Roux, J. (1977) Al substitution and differential disorder in soil hematites: *Clays & Clay Minerals* **25**, 373–374.

Schwertmann, U., Fitzpatrick, R. W., Taylor, R. M., and Lewis, D. G. (1979) The influence of aluminum on iron oxides. II. Preparation and properties of Al-substituted hematites: *Clays & Clay Minerals* **27**, 105–112.

Stoops, G. (1967) Le profil d'altération du Bas Congo (Kinshasa): *Pédologie* **17**, 60–105.

Turner, F. J. and Weiss, L. E. (1963) *Structural Analysis of Metamorphic Tectonites:* McGraw-Hill, New York, 525 pp.

William D. E. (1964) LCR2 Fortran lattice constant refinement program: U.S. Atom. Energy Comm. Report, **IS-1052**.

IRON AND ALUMINUM OXIDES

Proceedings of the International Clay Conference, Denver, 1985, L. G. Schultz, H. van Olphen, and F. A. Mumpton, eds.,
The Clay Minerals Society, Bloomington, Indiana, 197–204 (1987).

ACID DISSOLUTION OF SYNTHETIC METAL-CONTAINING GOETHITES AND HEMATITES

R. Lim-Nunez and R. J. Gilkes

Soil Science and Plant Nutrition Group, School of Agriculture
University of Western Australia, Nedlands, Western Australia 6009, Australia

Abstract—Goethites consisting of acicular and lathlike crystals and containing between 0 and 14 mole % Co, Cr, Ni, Mn, or Al were synthesized at high pH. Hematites were prepared by heating goethites at 300° and 800°C for 2 hr. The dissolution rates of these iron oxides in 1 M HCl differed greatly (from 0.03 to 21.8×10^{-2} g Fe/g/hr) and increased with increasing temperature. On the basis of unit surface area, Cr-goethites dissolved at about one-tenth the rate of unsubstituted, Ni- and Al-goethites (~ 3 g Fe/m^2/hr), whereas Co- and Mn-goethites dissolved at about twice this rate. The frequency factor for Cr-goethites ($\sim 10^{-16}$–10^{-14} g Fe/m^2/hr) was much larger than for other goethites, which may have been due to the extensive development of etchpits during dissolution; however, these large frequency factors were not sufficient to counteract the effect of the high activation energies (21–33 kcal/mol for Cr-goethites), and the dissolution rate for Cr-goethites was the lowest recorded.

Hematite produced by dehydroxylation of goethite at 300°C is microporous and dissolved much more quickly (3–107 times as fast) than corresponding goethites. Hematites produced at 800°C appeared sintered and massive; they mostly dissolved at similar rates (0.3–2.7 times) to goethites, although Cr-substituted hematite heated to 800°C dissolved much more rapidly (10 times) than the poorly soluble Cr-goethite.

Dissolution curves for both goethite and hematite heated to 300°C showed that Fe and other metals dissolved congruently and therefore were probably uniformly distributed within crystals. The only exceptions were for goethites containing Al and Ni which did not dissolve congruently; these elements were slightly concentrated towards the centers and rims of crystals respectively. For hematite heated to 800°C, Co dissolved much more rapidly than Fe, and Ni much more slowly than Fe, indicating that these elements were not present in the hematite structure. Cr-, Mn-, and Al-hematites (800°C), however, dissolved congruently, indicating that the nearly uniform substitution of these elements was preserved on heating to 800°C.

Key Words—Acid dissolution, Goethite, Hematite, Solid solution, Synthesis, Thermal treatment.

INTRODUCTION

Little is known about minor elements in most soil minerals. Iron oxides and oxyhydroxides in soils typically contain a higher proportion of minor elements than other soil constituents, but the nature of this substitution is poorly understood. The mode of occurrence of minor elements in soil minerals is of interest *inter alios* to geologists exploring weathered terrain for ore bodies, to metallurgists developing extractive techniques, and to soil chemists investigating their availability to plants. Electron microprobe analyses have shown that Ni, Cu, and Zn (Taylor *et al.,* 1964); Ni, Cr, and Al (Roorda and Queneau, 1973); and Ni and Cr (Schellmann, 1983) appear to be contained in iron oxides in soils; however, the fine particle size of most soil iron oxides (<1000 Å) restricts microprobe analysis to aggregates of crystals. It is therefore uncertain whether these metals are present as: (1) discrete micrometer-size mineral species; (2) ions adsorbed onto the very large surfaces of iron oxides (Forbes *et al.,* 1976; Padmanabham, 1975); or (3) ionic substituents for Fe in the structure of the iron oxides (Kuhnel *et al.,* 1975; Norrish, 1975). The third possibility can be

investigated by the synthesis of metal-substituted iron oxides.

Iron oxides and oxyhydroxides in soils can contain significant amounts of Al (Schwertmann and Taylor, 1977). Apart from an attempt by Nalovic and Pinta (1972) to synthesize an iron oxide from a solution containing Mn, Cr, V, Co, Ni, and Cu, little information is available on the ability of goethite and hematite to accommodate other metals or on the effects of such substituents on crystal morphology and solubility. Such a study of synthetic Co-, Cr-, Ni-, Mn-, and Al-substituted goethite and hematite is the subject of the present paper.

MATERIALS AND METHODS

Synthesis of goethites and hematites

Metal (Co, Cr, Ni, Mn, Al)-substituted and unsubstituted goethites were prepared at high pH using the method developed by Lewis and Schwertmann (1979). Mixed solutions of 1 M ferric nitrate and metal nitrate and an equal volume of 2 M NaOH were added simultaneously into 1-liter Polythene flasks. Additional 2 M NaOH was added to ensure that the final pH was

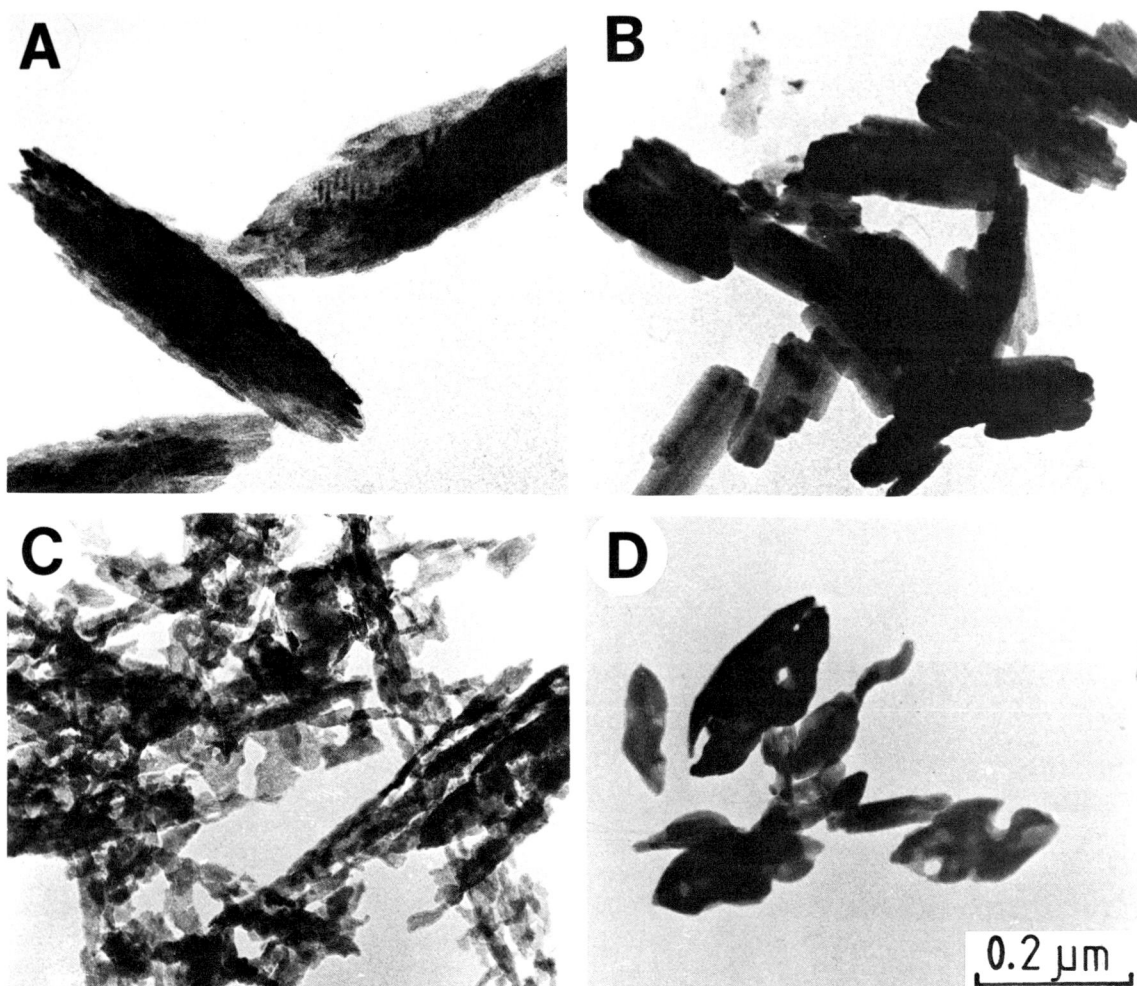

Figure 1. Transmission electron micrographs of synthetic goethites: (A) Goethite containing 12 mole % Cr; (B) goethite containing 8 mole % Al; (C) partially (35%) dissolved 12 mole % Cr-goethite; (D) partially (35%) dissolved 8 mole % Al-goethite.

greater than 12. The amounts of the metals added were equivalent to 5, 10, 15, and 20 mole % of the iron plus metal; however, as discussed below, the metals did not enter the goethite structure in these same proportions. The mixed solutions were stored at 70°C for 10 days with gentle shaking once every 24 hr. The precipitates were collected by centrifugation and extracted several times in acid ammonium oxalate solution (Schwertmann, 1964) to remove any Fe and metals not in the goethite. As many as ten extractions were required before the ratio of metal to iron in successive extracts remained constant. Generally, by the fifth extraction <1% of the sample was dissolved by each extraction. After repeated washing with deionized water, the goethites were dried at room temperature and stored in a desiccator. Duplicates of each metal-substituted goethite (A and B samples) were synthesized and triplicates for unsubstituted goethites (C1, C2, C3). The oxidation

status of the different metals presumably did not change due to incorporation into goethite.

Hematite was prepared from the most highly substituted and an unsubstituted goethite by heating them in an electric muffle furnace for 2 hr at 300° and 800°C. Such hematite products are referred to below as hematite (300°C) and hematite (800°C), respectively.

Sample characterization

All samples were examined by X-ray powder diffraction (XRD) using a Philips vertical goniometer with curved graphite crystal monochromator. For precise spacing measurements, quartz or halite was used as an internal standard. Samples were also examined by transmission electron microscopy (TEM) using a Hitachi HU11 instrument, infrared (IR) spectrometry, and thermal analysis. Surface areas of oxides were measured by the BET nitrogen-adsorption procedure.

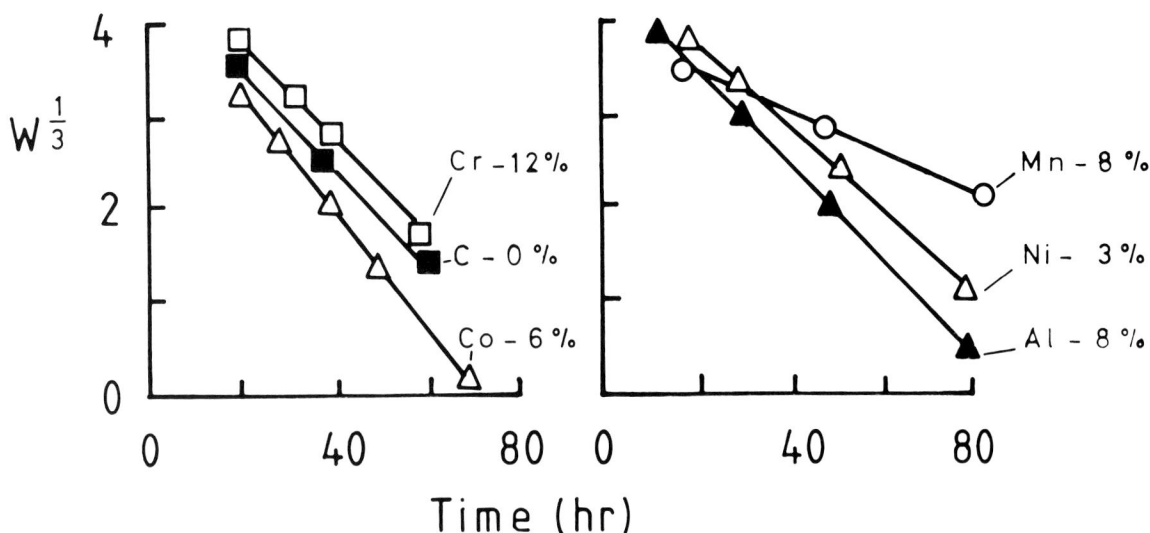

Figure 2. Plots of the cube root of the percentage goethite not dissolved ($W^{1/3}$) against period of dissolution for several goethites.

Acid-dissolution measurements

Separate 25-mg subsamples of each goethite and hematite were dissolved in 50 ml of 1 M HCl in Polythene bottles at 40°, 50°, 60°, and 70°C in a controlled-environment chamber. The bottles were continuously shaken. Aliquots of 2 ml were withdrawn at various intervals using a plastic syringe and passed through a 0.22-μm Millipore filter. The solutions were analyzed for Fe and metals using a Perkin Elmer 403 atomic absorption spectrophotometer. The partly dissolved residues were examined by TEM.

RESULTS AND DISCUSSION

Characterization of the goethites

For the three lowest levels of metal addition (0, 5, 10 mole % metal) the sole mineral present was goethite. For the addition of 15 and 20 mole % Ni and 15 mole % Cr, goethite was also the sole mineral detected. For the addition of 15 and 20 mole % Co, hematite was the only mineral to form; for the addition of 15 mole % Mn, goethite and a spinel phase were detected; and for the addition of 20 mole % Mn only a spinel phase formed. Subsequent investigations were restricted to those specimens consisting solely of goethite and which contained the levels of metal substitution shown in Figure 3.

Although substantial morphological differences between the various goethites were noted (to be described elsewhere), all specimens displayed the basic lathlike morphology that has been reported for this method of synthesis (Figues 1A and 1B) (Lewis and Schwertmann, 1979). TEM and XRD line-broadening measurements showed that the overall average crystal dimensions (length × width × height) were about 4000 × 500 ×

150 Å. These dimensions correspond to the crystallographic c, b, and a axes, respectively (Cornell et al., 1974), as was confirmed by electron diffraction analysis of single crystals of all products. Mottled contrast and extinction contours were common, and some crystals displayed a composite structure consisting of several lathlike domains parallel to the c axis (Smith and Eggleton, 1983; Schwertmann, 1984; Cornell et al., 1974; Cornell et al., 1983). Most of the surface area of the crystals was probably the (110) face rather than the (100) face (Schwertmann, 1984), and a domain may be a region of the crystal having a common (110) or (1$\bar{1}$0) face. Many crystals were etched, due to dissolution during ammonium oxalate extraction.

The specific surface area of the goethites ranged from 29 to 69 m^2/g and was inversely related to several crystal size parameters as measured by both XRD line broadening and TEM. For example, the relationship ($p < .10$) between BET surface area (A, m^2/g) and the average number of unit cells in each crystal (N) was A = 71.4 − 1.32 × 10^{-4}N; r = .78.

Dissolution of goethites

Dissolution rate increased sharply with increasing temperature, and a wide range of dissolution rates was observed for the various metal-substituted goethites. The shapes of these dissolution curves are similar to those for goethite measured by Cornell et al. (1975) and Sidhu et al. (1981). Dissolution closely followed the cube root law (Figure 2), which indicates dissolution of a single phase (Gastuche et al., 1960). The cube root law should not have been followed if the crystals dissolved incongruently or if two or more discrete minerals or particle types dissolved, as reported by Surana

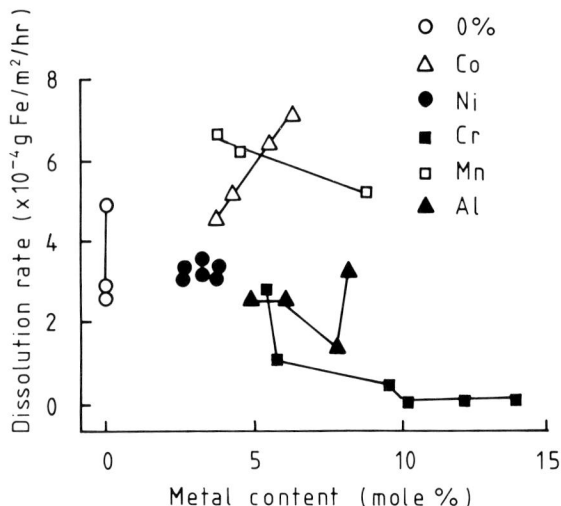

Figure 3. Plots of dissolution rate D ($\times 10^{-4}$ g Fe/m²/hr) vs. mole % metal content for all goethites.

and Warren (1969) and Warren and Roach (1971). The inhomogeneous dissolution evident in the electron micrographs of partly dissolved goethite is inconsistent with the uniform dissolution associated with the cube root law.

The dissolution rates of the goethites expressed on a unit surface-area basis were calculated from the initial slopes of the dissolution curves and the BET surface areas. The influence of the type and amount of metal substitution in goethite on initial dissolution rate at 60°C is shown in Figure 3. Two distinct features can be discerned from this mode of presentation. First, the dissolution rate varied significantly for goethites containing the same type and about the same level of metal substitution. For example, the dissolution rates at 60°C for unsubstituted goethites differed 1.7-fold (2.9–4.9 g Fe/m²/hr), 2.3-fold for goethites containing about 6 mole % Cr (1.2–2.8 g Fe/m²/hr), and 2.4-fold for goethites containing about 8 mole % Al (1.4–3.3 g Fe/m²/hr). Dissolution rate does not appear, therefore, to be simply related to the chemical composition of these goethites; it must be affected by other factors, such as crystal morphology, although no statistically significant relationship was noted with crystal shape parameters. Second, relative to dissolution rates for unsubstituted goethites, the dissolution rate decreased with increasing level of metal substitution for Cr-goethites, systematically increased for Co-goethites, first increased then decreased for Mn-goethites, and was not significantly affected for Al- and Ni-goethites. The different levels of substitution, however, and, in most experiments, the large difference in dissolution rate for similar goethites makes this interpretation rather speculative. For Cr-goethites, the effect of Cr-substitution was major, increasing the levels of substitution (from 0 to 14 mole % Cr) and decreasing the dissolution rate by about 90%

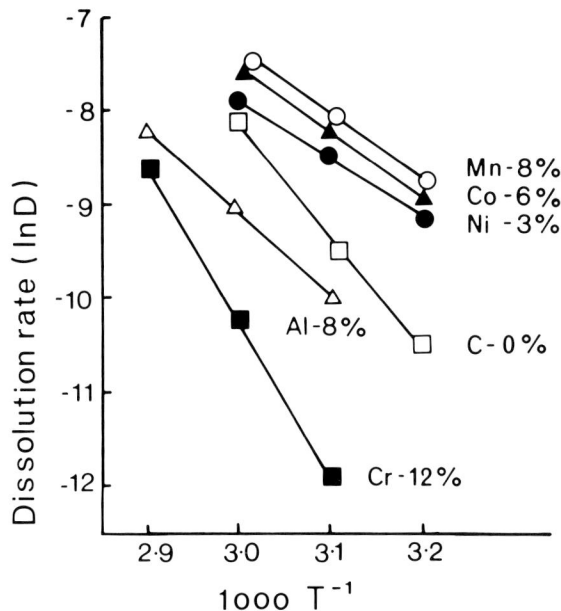

Figure 4. Plots of ln(dissolution rate, D) vs. $1000T^{-1}$ for several goethites.

(from 2.9 to 0.22 × 10⁻⁴ g Fe m²/hr). Evidently, incorporation of small amounts of Cr into goethite greatly decreased its solubility in acid. Mn- and Co-goethites probably dissolved in 1 M HCl at a faster rate (~2×) than unsubstituted goethites, and Ni- and Al-goethites dissolved at a similar rate to unsubstituted goethite. Schwertmann (1984) found that Al-substituted goethite dissolved in 6 M HCl much more slowly than unsubstituted goethite.

Each goethite was dissolved at three temperatures, and, although dissolution rates increased with increasing temperature, the trends for the different substitutions persisted. For each goethite, plots of the logarithm

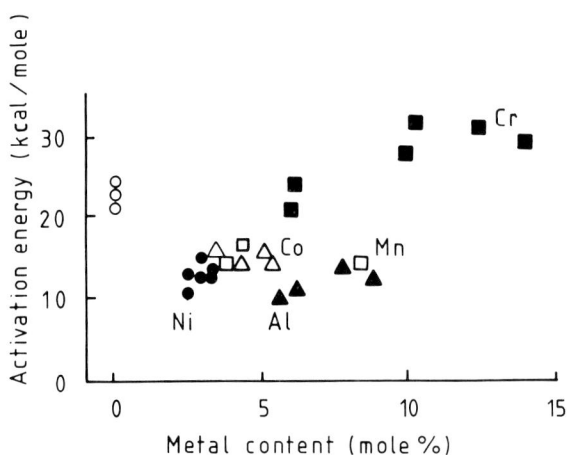

Figure 5. Plot of activation energy E (kcal/mole) vs. mole % metal content for all goethites.

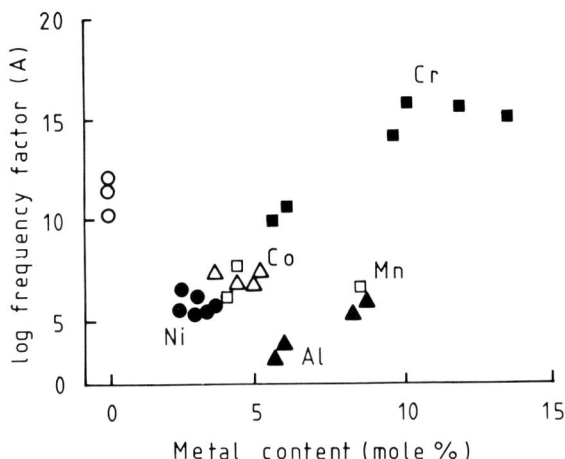

Figure 6.   Plot of log of frequency factor (A) vs. mole % metal content for all goethites.

of dissolution rate vs. reciprocal absolute temperature are linear (Figure 4), indicating that dissolution of goethite conformed to the Arrhenius equation (Cornell *et al.*, 1975; Sidhu *et al.*, 1981):

$$D = Ae^{-E/RT},$$

where D = dissolution rate (g Fe/m²/hr), A = frequency factor (g Fe/m²/hr), E = activation energy (kcal/mole), R = gas constant (1.987×10^{-3} kcal/degree/mole), and T = absolute temperature (K).

Activation energies and frequency factors obtained from the linear graphs are plotted against mole percent metal in Figures 5 and 6, respectively. The values of activation energy reported here for unsubstituted goethites are similar to those published by Surana and Warren (1969), Cornell *et al.* (1975), and Sidhu *et al.* (1981). The lowest values were for the Al-, Ni-, Co-, and Mn-goethites (11–16 kcal/mole); unsubstituted and Cr-goethites yielded systematically higher values. Activation energies for acid dissolution are believed to be related to the ionic properties of iron and substituting metals, including bond strength, electronegativity, and ionic potential (Glasstone *et al.*, 1941), but no consistent relationship with these parameters was observed in the present work.

The changes in dissolution rate with increasing metal-substitution described above (Figure 3) for Cr-, Co-, and Mn-goethites were evidently not simply a consequence of systematic changes in activation energy. The frequency factors for Ni-, Co-, Mn-, and Al-substituted goethites were systematically lower than for unsubstituted and Cr-goethites; no systematic change occurred with increasing substitution (Figure 6). The frequency factors reported here are higher (about 4 times) than those measured for goethite by Sidhu *et al.* (1981), probably because of the different acid strengths used. Although a frequency factor may be influenced by differences in crystal morphology, no significant correlation was observed between frequency factor and various measures of crystal size and shape (e.g., crystal length (r = .35 NS), width (r = −.04), surface area (r = .04), length to width ratio (r = .06), mean crystal dimension from broadening of the 110 reflection (r = −.29)). Thus, differences in frequency factor were probably due to different modes of acid attack on goethite crystals rather than to differences in crystal shape (i.e., exposure of different exterior crystallographic faces). Some support for this interpretation was obtained from TEMs of partly (~35%) dissolved goethite crystals. For example, Cr-goethite (12 mole % Cr) that had a high frequency factor developed a lacelike morphology due to the growth of etchpits that penetrated through the crystals along the *a* axis (Figure 1C). In contrast, Al-goethite (8 mole % Al) had a much lower frequency factor and developed fewer etchpits (Figure 1D). Thus, the high frequency factor for Cr-goethite may simply have been a consequence of a higher density of sites for acid dissolution on the (110) face. It should be pointed out that despite the markedly inhomogeneous nature of acid attack revealed in electronmicrographs, dissolution followed the cube root law, which assumes uniform acid attack of all faces of crystals.

Dissolution of hematites

Hematites produced by heating goethites at 300°C retained the morphology of the parent goethite crystals (Figure 7A), but developed a microporous fabric which appears as a fine granular texture in TEMs. This phenomenon was explained by Watari *et al.* (1979a, 1979b, 1983) who showed that each goethite crystal that dehydroxylated at 270°C (i.e., transformed to hematite) consisted of an array of small (50–100 Å) hematite grains in perfect parallel alignment, but separated by a network of planar voids. In marked contrast, hematite (800°C) consisted of massive, sintered aggregates of former goethite crystals (Figure 7B). The same topotaxial alteration of goethite (G) to hematite (H) apparently had occurred at 800°C as at 300°C (i.e., $c_H \parallel a_G$, $a_H \parallel b_G$, $hh0_H \parallel c_G$; Brown, 1980; Francombe and Rooksby, 1959). Diffusion of ions at 800°C, however, altered the shape of particles and eliminated the microporous fabric that existed in hematite (300°C).

Dissolution curves for hematites are not shown, but were similar in shape to those for goethites and conformed to the cube root law. Dissolution rates for goethites, hematites (300°C), and hematites (800°C) expressed on the basis of a unit weight of oxide are given in Table 1. It is evident that all hematites (300°C) dissolved at a much faster rate (3–107 times) than the corresponding goethites. This increase was most marked for those goethites that dissolved at a relatively slow rate (i.e., unsubstituted, Al-, Ni-, and Cr-substituted goethites). This increase was likely due to the micro-

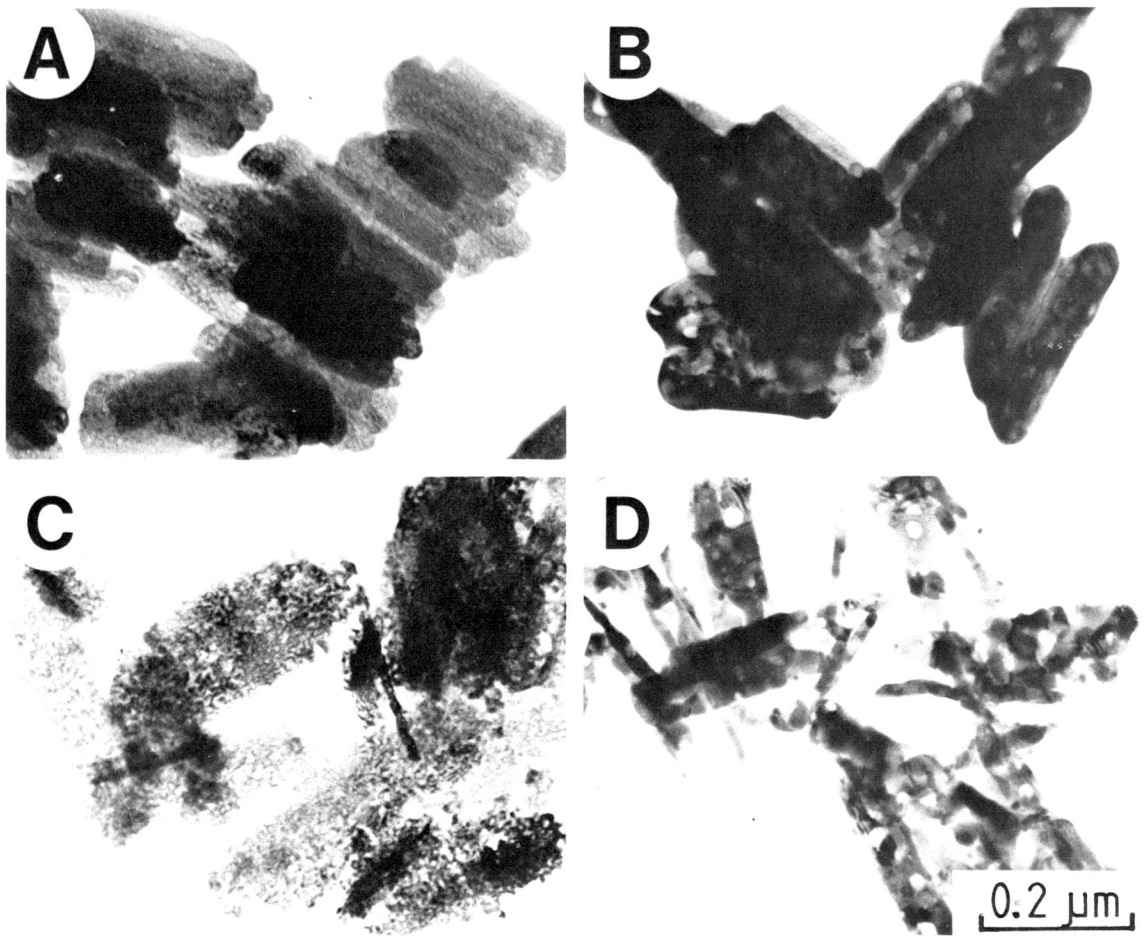

Figure 7.   Transmission electron micrographs of hematites formed by dehydroxylation of goethite: (A) Hematite formed at 300°C; (B) hematite formed at 800°C; (C) partly (35%) dissolved hematite (300°C); (D) partly (35%) dissolved hematite (800°C).

porous fabric of hematite (300°C) that provided many more sites for acid attack. Consequently, the partially (35%) dissolved residues of all hematites (300°C) consisted of highly etched, microporous grains (Figure 7C).

In marked contrast, dissolution rates for most hematites (800°C) (with the exception of Cr-hematite) were similar (0.3–2.7 times) to those for corresponding goethites (Table 1). Particles of hematite (800°C) did not have a microporous fabric that exposed grain interiors so that, as for goethite, acid attack was mostly

Table 1.   Dissolution rates ($\times 10^{-2}$ g Fe/g oxide/hr) for iron oxides.

Substitution	Goethite	Hematite (300°C)	Hematite (800°C)
Nil	0.5	21.8	1.0
6% Co	2.4	19.2	3.3
3% Ni	0.5	16.0	1.3
12% Cr	0.03	3.2	0.3
8% Mn	1.1	3.8	0.3
8% Al	0.5	6.4	0.6

of their external surface. Thus, partly (35%) dissolved particles of hematite (800°C) contained fewer but larger etchholes (Figure 7D). The large increase in dissolution rate for Cr-hematite (800°C) relative to unheated Cr-goethite mostly reflected the particularly slow dissolution of Cr-goethite, inasmuch as Cr-hematite (800°C) dissolved at a similar rate to most other hematites (800°C).

Distribution of metals in crystals of goethite and hematite

The goethites that contained high levels of Al (ionic radius 0.51 Å) exhibited shifts in unit-cell dimensions which nearly obeyed Vegard's Rule and which were similar in magnitude to those reported by Schulze (1984). No statistically significant ($p < .10$) systematic displacement of unit-cell dimensions from those for unsubstituted goethites was observed for Cr- and Ni-goethites. Hence, Cr and Ni could not be shown to be uniformly distributed through the goethite crystals. The same result was found for the hematites. The ionic

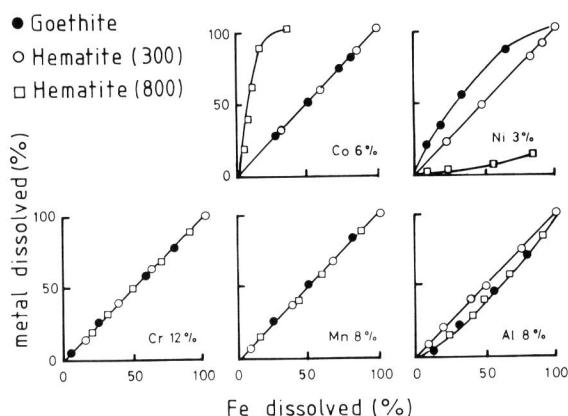

Figure 8. Plots of percentage metal dissolved vs. percentage Fe dissolved for the complete dissolution of goethite, hematite (300°C), and hematite (800°C).

radius of Cr^{3+} (0.63 Å) is almost identical to that of Fe^{3+} (0.64 Å); hence, despite the incorporation of as much as 13.8 mole % Cr in goethite, relatively little diffraction line displacement would be anticipated and none was observed. Bracewellite ($CrO \cdot OH$), the Cr analogue of goethite, has slightly smaller unit-cell dimensions than goethite, the a axis exhibiting the greatest (2%) reduction. The presence of significant levels (as much as 5.4 mole % Mn, 3 mole % Ni, 7 mole % Co) of divalent metals having substantially different ionic radii from Fe^{3+} (i.e., $Mn^{2+} = 0.80$ Å, $Ni^{2+} = 0.69$ Å, $Co^{2+} = 0.72$ Å) should have induced significant changes in unit-cell dimensions. Minor systematic displacements of reflections were observed for Mn and Co. For example, the statistically significant (p < .10) linear relationships between the c unit-cell dimension and level of substitution were: for Mn-goethite, c (Å) $=$ 3.0239 − 0.0011(mole % Mn), r = .70; for Co-goethite, c (Å) = 3.0256 − 0.0026(mol % Co), r = .85; and for Al-goethite, c (Å) = 3.0250 − 0.0012(mole % Al), r = .77.

The absence of a statistically significant relationship between unit-cell dimension and level of substitution of Ni might be taken to indicate that Ni did not enter the goethite structure, but that it was present either as a separate phase or adsorbed onto the surface of crystals. A proportion of the other metals could also be in these forms; however, TEM, XRD, IR, differential thermal analysis, thermogravimetric analysis, and electron diffraction data indicated that no compounds other than goethite or hematite were present in any of the specimens. Furthermore, any noncrystalline gels and coatings or adsorbed metals should have been removed by the repeated acid ammonium oxalate treatments (Schwertmann, 1964).

An indirect means of estimating the distribution of a substitution in crystals is to determine the relative dissolution rates of all constituents of the crystal. Con-

gruent dissolution at all stages of dissolution is to be expected if the substituents are distributed uniformly within the crystal. Alternatively, incongruent dissolution suggests that a separate phase is present or that substituents are concentrated towards the core or periphery of the crystals. This procedure has been applied to studies of natural and synthetic metal-substituted magnetites, maghemites, and hematites (Sidhu et al., 1980, 1981). Ideally, all crystals should be the same size and shape, and uniform dissolution of all faces must be assumed. As discussed above, none of these conditions applies exactly to the dissolution of the goethite and hematite specimens investigated in this work, but any major anisotropy in distribution of metals in crystals should still be revealed by such an analysis.

Dissolution data for Fe and substituent metals for the most highly substituted goethites and the corresponding hematites are shown in Figure 8 as plots of percentage metal dissolved vs. % Fe dissolved. The data for Co-, Cr-, and Mn-substituted goethites fall on lines having slopes of 1, indicating that these elements were uniformly distributed in the crystals. All metals were uniformly distributed in the hematites (300°C), but Ni dissolved more rapidly than Fe for goethite and therefore appeared to be slightly concentrated near the surfaces of crystals. Conversely, Al dissolved more slowly than Fe and thus appeared to be slightly concentrated towards the core of the goethite and hematite (800°C) crystals. Dissolution of Al-substituted goethite, however, was nearly congruent, which is consistent with the XRD data that indicates the uniform substitution of Al for Fe. The apparent change to congruent dissolution for Ni- and Al-hematites (300°C) may have been due to dissolution of the entire volume of the crystals via the many etchpits at the closely spaced, 50–100-Å subgrain boundaries identified by Watari et al. (1979a, 1979b, 1983). Goethites containing Ni and Al and Al-hematite (800°C) dissolved incongruently because most dissolution was from the external surfaces of crystals. Co dissolved from the hematite (800°C) product much more rapidly than Fe, whereas Ni dissolved much more slowly, indicating that these elements were present as discrete compounds. This different behavior of divalent Co and Ni from that of trivalent Cr, Mn, Al, and Fe may be a consequence of the inability of well-crystalline hematite to accommodate divalent metals. Cr-goethite developed a more highly etched appearance during dissolution than did other goethites, but the surface density of these etchpits (i.e., one about every 10^6 Å²) was much less than for all hematites (300°C), for which high-resolution TEM showed one pit was present for about 10^4 A² of crystal surface.

GENERAL DISCUSSION

A deficiency of the present work is that goethites were not prepared with a common range of mole %

metal for all metals. Thus, the comparisons of properties made here are to some extent comparisons among iron oxides with different levels of substitution. The work does not demonstrate conclusively the existence of Co, Cr, Ni, and Mn ionic substitutions in the structures of goethite and hematite. The dissolution data, however, are consistent with ionic substitution of these metals; dissolution rates were systematically affected by their presence. For Mn- and Co-goethites the systematic changes in unit-cell parameters are evidence for ionic substitutions of these elements. Iron oxides from soils should be investigated by these and other techniques to establish if associated minor elements are indeed present as ionic substitutions in the crystal structure. The role of iron oxides in sorbing phosphate in soils is well established and of agricultural significance; the potential of these minerals for incorporating micronutrient metals during crystallization, thereby making these metals unavailable to plants, deserves investigation.

ACKNOWLEDGMENTS

We thank Terry Armitage and Irene McKissock who provided the electron optical data, and Udo Schwertmann and Darrell Schulze for helpful criticism.

REFERENCES

Brown, G. (1980) Associated minerals: in *Crystal Structures of Clay Minerals and Their X-ray Identification,* G. W. Brindley and G. Brown, eds., Mineralogical Society, London, 361–410.

Cornell, R. M., Mann, S., and Skarnulis, A. J. (1983) A high resolution electron microscopy examination of domain boundaries in crystals of synthetic goethite: *J. Chem. Soc. Faraday Trans.* **1,** 2679–2684.

Cornell, R. M., Posner, A. M., and Quirk, J. P. (1974) Crystal morphology and the dissolution of goethite: *J. Inorg. Nucl. Chem.* **36,** 1937–1946.

Cornell, R. M., Posner, A. M., and Quirk, J. P. (1975) The complete dissolution of goethite: *J. Appl. Chem. Biotech.* **25,** 701–706.

Forbes, E. A., Posner, A. M., and Quirk, J. P. (1976) The specific adsorption of divalent Cd, Co, Cu, Pb and Zn on goethite: *J. Soil Sci.* **27,** 154–166.

Francombe, M. H. and Rooksby, H. P. (1959) Structure transformations affected by the dehydration of diaspore, goethite and delta ferric oxide: *Clay Miner. Bull.* **4,** 180–187.

Gastuche, M. C., Delman, B., and Vielvoye, L. (1960) La cinétique des réactions hétérogènes: ataque du réseau silico-aluminique des kaolinites par l'acide chlorohydrique: *Bull. Soc. Chem. France,* 60–70.

Glasstone, S., Laidler, K. J., and Eyring, H. (1941) *The Theory of Rate Processes:* McGraw Hill, New York, 661 pp.

Kuhnel, R. A., Roorda, H. J., and Steensma, J. J. (1975) The crystallinity of minerals—a new variable in pedogenetic processes: a study of goethite and associated silicates in laterites: *Clays & Clay Minerals* **23,** 349–354.

Lewis, D. G. and Schwertmann, U. (1979) The influence of

Al on iron oxides. Part III. Preparation of Al goethite in M KOH: *Clay Miner.* **14,** 115–126.

Nalovic, L. and Pinta, M. (1972) Compartiment du fer en présence des éléments de transition, étude expérimentale: lessivage des hydroxydes hydrates par de l'eau, à l'air libre: *C. R. Acad. Sci. Paris,* **275,** D153–D156.

Norrish, K. (1975) Geochemistry and mineralogy of trace elements: in *Proc. Waite Inst. Symp., Adelaide, 1974,* D. J. D. Nicholas and A. R. Egan, eds., Academic Press, New York, 55–81.

Padmanabham, M. (1975) The behaviour of heavy metal cations at the oxide solution interface: Ph.D. Thesis, Univ. Western Australia, Nedlands, Western Australia, 210 pp.

Roorda, H. J. and Queneau, P. E. (1973) Recovery of nickel and cobalt from ilmenites by aqueous chlorination in sea water: *Trans. Inst. Mining Met.* **C82,** 79–87.

Schellmann, W. (1983) Geochemical principles of lateritic nickel ore formation: in *Proc. Int. Sem. Lateritization Processes, São Paulo, Brazil, 1983,* A. J. Melfi and A. Carvalho, eds., Univ. São Paulo, São Paulo, Brazil, 119–135.

Schulze, D. G. (1984) The influence of aluminum on iron oxides. VIII. Unit-cell dimensions of Al-substituted goethites and estimation of Al from them: *Clays & Clay Minerals* **32,** 36–44.

Schwertmann, U. (1964) Differenzierung der Eisenoxide des Bodens durch Extraktion mit Ammoniumoxalat-Lösung: *Z. Pflanzenernaehr. Bodenk.* **105,** 194–202.

Schwertmann, U. (1984) The influence of aluminium on iron oxides: IX. Dissolution of Al-goethites in 6 M HCl: *Clay Miner.* **19,** 9–19.

Schwertmann, U. and Taylor, R. M. (1977) Iron oxides: in *Minerals in Soil Environments,* J. B. Dixon and S. B. Weed, eds., Soil Sci. Soc. Amer., Madison, Wisconsin, 145–180.

Sidhu, P. S., Gilkes, R. J., Cornell, R. M., Posner, A. M., and Quirk, J. P. (1981) Dissolution of iron oxides and oxyhydroxides in hydrochloric and perchloric acids: *Clays & Clay Minerals* **29,** 269–276.

Sidhu, P. S., Gilkes, R. J., and Posner, A. M. (1980) The behavior of Co, Ni, Zn, Cu, Mn, and Cr in magnetite during alteration to maghemite and hematite: *Soil Sci. Soc. Amer. J.* **44,** 135–138.

Smith, K. L. and Eggleton, R. A. (1983) Botryoidal goethite: a transmission electron microscopic study: *Clays & Clay Minerals* **31,** 392–396.

Surana, V. S. and Warren, I. H. (1969) The leaching of goethite. *Trans. Instn. Min. Metall.* **80,** C133–C139.

Taylor, R. M., MacKenzie, R. M., and Norrish, K. (1964) The mineralogy and chemistry of manganese in some Australian soils: *Aust. J. Soil Res.* **2,** 235–248.

Warren, I. H. and Roach, G. I. D. (1971) Physical aspects of the leaching of goethite and hematite: *Trans. Instn. Min. Metall.* **80,** C152–C155.

Watari, F., Delavignette, R., and Aleminekx, S. (1979a) Electron microscopic study of dehydration transformation. II. The formation of 'superstructures' on the dehydration of goethite and diaspore: *J. Solid State Chem.* **29,** 417–427.

Watari, F., Delavignette, R., and Amelinekx, S. (1983) Electron microscopic study of dehydration transformation. III. High resolution observation of the reaction process FeOOH → Fe$_2$O$_3$: *J. Solid State Chem.* **48,** 49–64.

Watari, F., Van Landuyt, J., Delavignette, P., and Amelinekx, S. (1979b) Electron microscopic study of dehydration transformation. I. Twin formation and mosaic structure in hematite derived from goethites: *J. Solid State Chem.* **29,** 137–150.

Proceedings of the International Clay Conference, Denver, 1985, L. G. Schultz, H. van Olphen, and F. A. Mumpton, eds.,
The Clay Minerals Society, Bloomington, Indiana, 205–211 (1987).

AN UNUSUAL OCCURRENCE OF MAGHEMITE IN SOILS DEVELOPED ON DOLOSTONES OF THE KNOX GROUP, OAK RIDGE, TENNESSEE

O. C. Kopp[1] AND S. Y. Lee[2]

Environmental Sciences Division, Oak Ridge National Laboratory[3]
Oak Ridge, Tennessee 37831

Abstract—Maghemite, γ-Fe_2O_3, was identified in soils developed from residuum of Knox Group dolostones (Cambro-Ordovician age) along Chestnut Ridge on the U.S. Department of Energy Oak Ridge Reservation in Oak Ridge, Tennessee. The parent dolostones in the area were free of maghemite and magnetite. Scanning electron micrographs and electron probe analyses indicated that the maghemite replaces oolites, cements siltstones, and occurs as massive to botryoidal material. The most complex replacement by maghemite was displayed by the oolitic material. Oolitic limestone was dolomitized, as revealed by the presence of numerous dolomite rhombohedra, during early diagenesis. During the next stage, dolomite rhombohedra were replaced by chert; much of the remaining oolites and carbonate cements also have been partly replaced by chert. After uplift and erosion brought these oolitic cherts into the upper weathering zone, the cherts were replaced by iron hydroxide. Massive chert and siltstone in all stages of replacement by iron oxides have been observed. The replaced iron phase was converted to maghemite and hematite during a late-stage of paragenesis, possibly a pedogenic process.

Key Words—Dolostone, Iron, Maghemite, Magnetite, Soil, Weathering.

INTRODUCTION

Maghemite, γ-Fe_2O_3, is isostructural with magnetite. It has been reported in soils rich in iron and organic matter, especially soil that has been heated by fire (Winchell, 1931; van der Marel, 1951). Bonifas (1959) attributed maghemite in soils to the aerial oxidation of detrital magnetite derived from igneous rocks. Harrison and Peterson (1965) reported a high concentration of a magnetic mineral near the top of two deep-sea cores from the Indian Ocean. The mineral was reported to be between magnetite and maghemite in structure and was considered to have formed in the deep-sea environment.

Schwertmann and Taylor (1977) stated that maghemite is most common in highly weathered soils formed under tropical to subtropical conditions, such as in Hawaii, Australia, India, and Africa, and also in some temperate regions, such as Japan, Holland, and Germany. In all occurrences, the soils were apparently derived from mafic igneous rocks. Fagan (1969) reported a mixture of hematite and magnetite in the oxidized zone of the Lost Creek barite deposit, but the iron minerals were formed by surface oxidation of pyrite in the breccia body.

In the present study, maghemite is described from soils developed on dolostones of the Cambrian-Lower Ordovician Knox Group. The study area has been farmed, and the upper soil horizons are rich in organic matter. This report characterizes the magnetic and nonmagnetic iron-rich nodules in the soils developed from residuum of the iron-poor dolostones and examines the mode of the maghemite formation in the soils.

FIELD DESCRIPTIONS AND MATERIALS

The Chestnut Ridge site is on the U.S. Department of Energy Oak Ridge Reservation (Figure 1), about 30 km southwest of the city of Oak Ridge, Tennessee. Soils have developed on top of the Knox Group dolostones (the Copper Ridge Formation of Cambrian age and the Chepultepec Formation of Lower Ordovician age). Knox Group strata in this region are mainly dolostones, but limestone, sandstone, and siltstone beds are also present. Cherts, both primary and secondary, are common in the dolostones. Most primary chert occurs as rounded to elliptical nodules; secondary cherts are generally irregular in shape. Secondary cherts commonly preserve the textures and structures of the carbonates, such as oolites and dolomite rhombohedra. The Knox Group carbonates are deeply weathered, and the depth to fresh bedrock may exceed 30 m. Present-day solution activity is a common feature revealed by

---

[1] Department of Geological Sciences, The University of Tennessee Knoxville, Tennessee 37916.

[2] Correspondence to S. Y. Lee.

[3] Operated by Martin Marietta Energy Systems, Inc., under Contract No. DE-AC05-84OR21400 with the U.S. Department of Energy. This article was supported by the Office of Defense Waste and Byproducts Management.

Copyright © 1987, The Clay Minerals Society

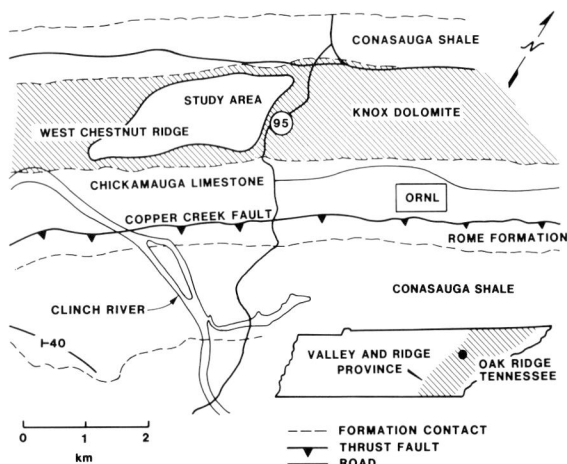

Figure 1. Location of the study area near Oak Ridge National Laboratory, Oak Ridge, Tennessee.

Table 1. X-ray powder diffraction data of magnetic fraction separated from soil samples and selected data of reference iron oxide minerals

Soil sample		Hematite[1]		Maghemite[2]		Magnetite[3]	
d (Å)	I/I_max	d (Å)	I/I_max	d (Å)	I/I_max	d (Å)	I/I_max
2.95	30			2.95	34	2.97	60
2.69	38	2.69	100				
2.51	100	2.52	70	2.51	100	2.53	100
2.20	21	2.21	17				
2.08	23			2.08	24	2.10	50
1.84	21	1.84	31				
1.69	31	1.69	36	1.70	12	1.71	40
1.60	26	1.60	8	1.60	33	1.62	60
1.48	33	1.49	22	1.47	53	1.48	70
1.45	22	1.45	21				

[1] JCPDS File, mineral powder diffraction card 24-72.
[2] JCPDS File, mineral powder diffraction card 24-81.
[3] JCPDS File, mineral powder diffraction card 7-322.

the presence of numerous sinkholes and subterranean drainage. Some karst features were formed during Middle Ordovician time when the Knox Group terrane was uplifted and exposed to subaerial weathering and erosion.

Soils at Chestnut Ridge developed in humid climates on very old, freely drained, stable land surfaces. The morphology of soils developed on top of the Knox Group rocks changes abruptly, both horizontally and vertically, because material derived from various levels in the Knox Group formations has been redistributed by alluvial and colluvial processes. Soil and residuum samples from several pits and cores were examined. Maghemite-containing nodules were observed only in the Paleudults but not in Entisols, which consist of undifferentiated Holocene alluvium. The clay fractions of soils contain varying amounts of kaolinite, hydroxy-Al-interlayered vermiculite (HIV), mica (illite), iron oxides, gibbsite, and quartz. Kaolinite is most abundant in subsurface horizons, and HIV is most abundant in the surface horizons (Lee *et al.,* 1984).

EXPERIMENTAL METHODS

Each air-dried soil and residuum sample was gently crushed with a plastic roller and passed through a 2-mm sieve. The gravel-size fraction (>2 mm) was washed with demineralized water, and the magnetic and non-magnetic iron oxide-rich nodules were separated by a horseshoe magnet for morphological and mineralogical analyses. Some of the iron oxide-rich nodules were impregnated with epoxy resin and polished until the cross-section surfaces of the nodules were exposed. The morphology of fractured surfaces of the nodules was examined by scanning electron microscopy (SEM) using a JEOL 35CF instrument. Elemental distribution maps and backscattered images of the polished surfaces were obtained by electron microprobe (EMP) analysis

using a JEOL SUPERPROBE 733 instrument. For X-ray powder diffraction (XRD) analysis, some of the magnetic nodules were gently crushed, and the fine fractions (<5 μm) were separated by centrifugation. The magnetic fractions were separated using a laboratory magnetic stirring bar. A Philips diffractometer equipped with a graphite monochromator and Cu target (35 kV, 20 mA) was used. Randomly oriented powder mounts were prepared using petrographic glass slides.

RESULTS

Occurrence and identification

The magnetic nodules were red (10 YR 4/8) to dark reddish brown (2.5 YR 3/2), similar to non-magnetic iron oxide nodules. The nodules from surface soils (A and E horizons) were darker in color, perhaps because of the presence of organic matter. A major component in the gravel fraction of the soils was chert, which occurred in massive, oolitic, and dolomoldic habits. Most cherts were highly weathered and partially replaced by and/or coated with iron oxides.

The iron oxide nodules were distributed throughout the soil profiles but were less abundant in surface horizons. A study of sand fractions in borehole core samples from the same area, however, found that the magnetic iron oxide particles were absent in the residua at depths greater than 3–5 m (H. C. Monger, University of Tennessee, 1985, personal communication). In the clay fraction from the B horizons of the soils, many well-defined cubic iron oxide crystallites were found by transmission electron microscopy (Lee *et al.,* 1984).

In soils, possible ferromagnetic minerals (i.e., attracted by a horseshoe magnet) are maghemite and/or magnetite. The XRD pattern of the magnetic fraction separated from gravel indicates that maghemite is the dominant iron mineral (Table 1). Hematite (α-Fe_2O_3) was also identified in the magnetic fraction, which suggests that the magnetic grains are aggregates of both minerals. The strongest peak, at 2.51 Å, of the sample is characteristic of both maghemite and hematite. Therefore, the 2.95-, 2.08-, and 1.60-Å peaks were used for maghemite identification and the 2.69-, 2.20-, and 1.84-Å peaks for hematite identification.

Differentiation between maghemite and magnetite by XRD is difficult because both have similar crystal structures (Table 1). For a graphical illustration, a powdered magnetite specimen was mixed with the magnetic fraction separated from a soil in a 1:2 ratio. The XRD pattern of the mixture shows the separation between maghemite and magnetite peaks: 2.95 vs. 2.97 Å, 2.51 vs. 2.53 Å, 2.08 vs. 2.10 Å, and 1.60 vs. 1.62 Å (Figure 2). The d value differences, particularly at higher diffraction angles, together with the line broadening and missing weak diffraction peaks that are typical for the small grain size of pedogenic maghemite, indicate that the magnetic mineral in the soil is maghemite rather than magnetite.

Paragenesis

Maghemite occurs in several textural forms in soils at West Chestnut Ridge, including various types of cherts that have been replaced by iron oxides and siltstone in which cement and pore-filling minerals have been replaced by iron oxides. A specimen illustrating multiple stages of replacement is shown in Figures 3A and 3B. The original rock was an oolitic limestone. The initial replacement occurred when some of the oolite cores were partially replaced by dolomite (in the form of rhombohedra, arrow r). Later, parts of the partially dolomitized oolitic limestone were replaced by chert. The dolomite rhombohedra appear to have been completely replaced, whereas the surrounding material was only partially replaced. Still later, the oolites were replaced by iron oxides.

Figures 3C and 3D illustrate magnetic grains in more altered states. In Figure 3C, the chert appears to be strongly altered, but it is better preserved in the region near the top of the photo (arrow c). The iron oxides appear to have been introduced along fractures and voids in the specimen (arrow i). At least two distinct bands were observed when the specimen was viewed in reflected light. Figure 3D illustrates a very light-colored, magnetic grain (arrow i). The chert is highly altered and shows less replacement by iron oxides. Some dolomite rhombohedra that were replaced by chert appear relatively fresh (arrow r).

Figures 3E and 3F show two ferruginous specimens that were not attracted to a horseshoe magnet. Many of these oolitic grains appear to be hollow. The iron oxides can be seen as surface coatings on chert grains, former ooid walls, etc. Figure 3F illustrates a sample in which most iron oxides are not magnetic. Grains of chert (arrow c), aluminum oxyhydroxides (arrow a), and mica (arrow m) are embedded in a relatively dense

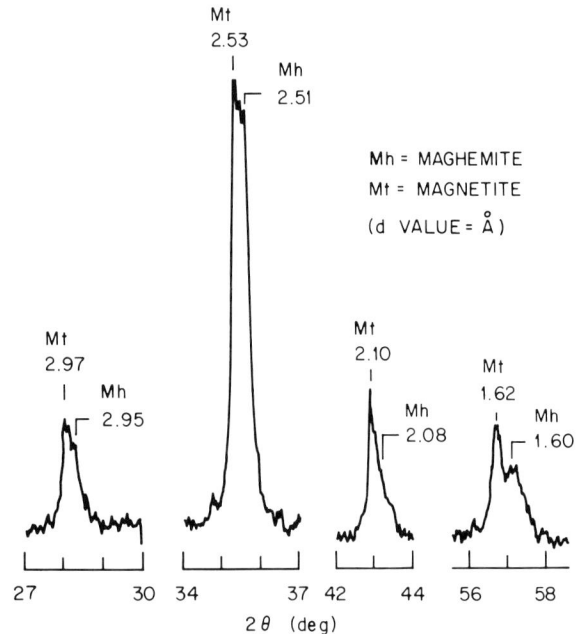

Figure 2. Selected (*hkl*) diffraction pattern of a 2:1 mixture of maghemite separated from magnetic nodules in West Chestnut Ridge soils and a reference magnetite (source unknown). CuKα radiation.

iron oxide matrix (arrow i). Relict grains of chert embedded in the iron oxides suggest that the oxide has replaced preexisting chert (possibly a porous variety).

Figure 4 illustrates an altered siltstone grain that is magnetic. A relict K-feldspar grain has somehow survived the replacement and weathering process (arrow f). The feldspar appears to have a kaolinite rind (arrow k) and some iron oxides (arrow i) along its outer surface. The silt-size quartz grains (arrow q) are relatively fresh, but have ragged edges. Iron oxides and kaolinite fill the spaces between the grains. Elemental maps for Al, Si, and K help elucidate the mineral distribution. The Al-rich areas represent kaolinite (arrow k in Figure 4B), the Si-rich area represents quartz (arrow q in Figure 4C), and the K-rich area represents feldspar (arrow f in Figure 4D). The original siltstone may have had a carbonate cement that was replaced by iron oxides. Figures 4E and 4F illustrate, respectively, the surfaces of a previously fractured and a newly fractured grain. The old fracture is filled with numerous small grains of silica (arrow s) which are coated with ferruginous materials and a few crystals of iron oxide minerals (arrow i). The newly fractured surface shows kaolinite (arrow k) replacing feldspar (arrow f) and small grains of iron oxide.

Why the replacement process is commonly associated with grains of chert and siltstone is not obvious. Where the chert is dense, as in the pseudomorphic dolomite rhombohedra, it appears to resist replacement. The replacement process may have begun in

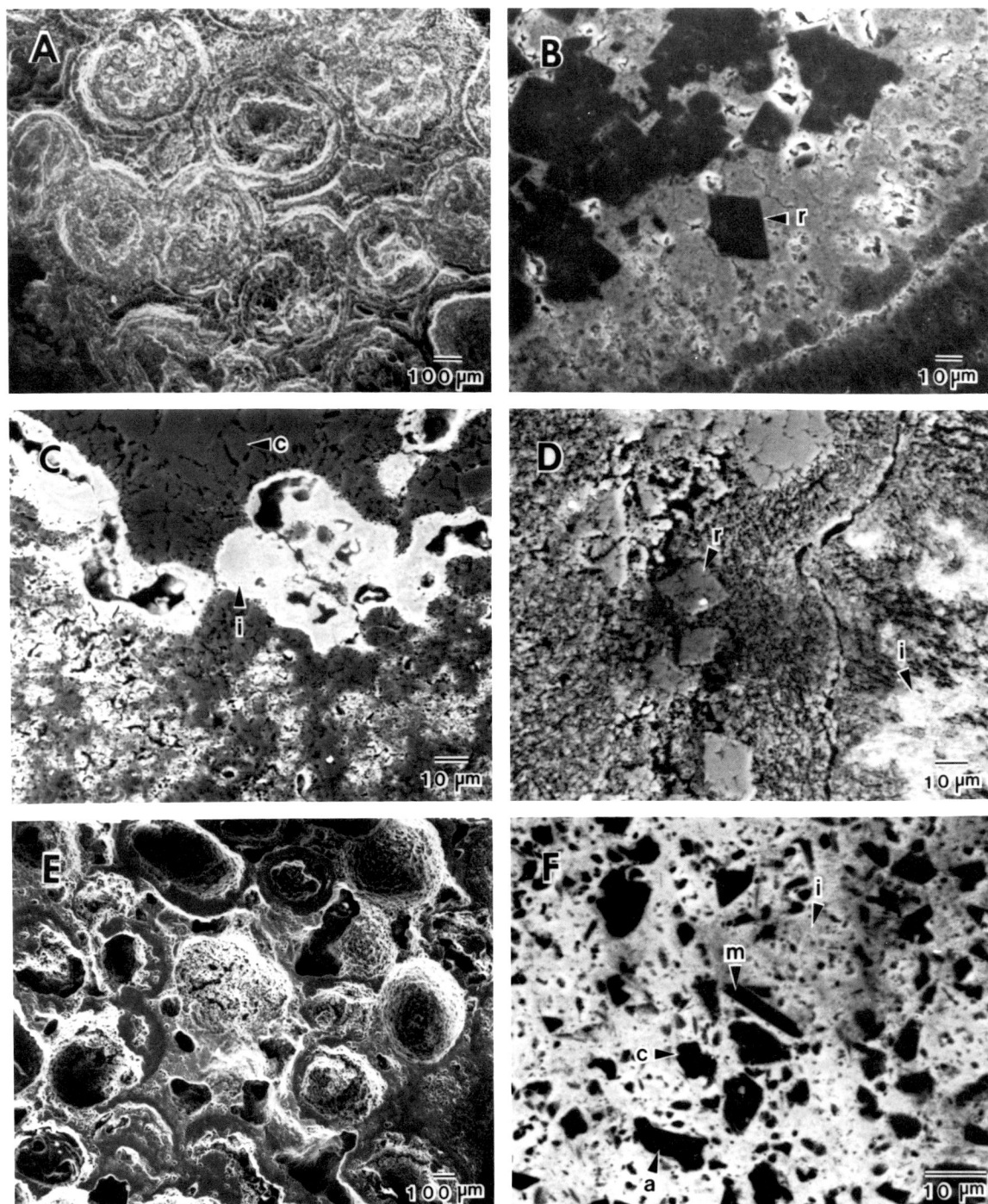

Figure 3.   Backscattered images of polished cross-section surface of magnetic (A, B, C, D) and nonmagnetic (E, F) iron oxide nodules in the soils studied (a = aluminum hydroxide, c = chert, i = iron oxide, m = mica).

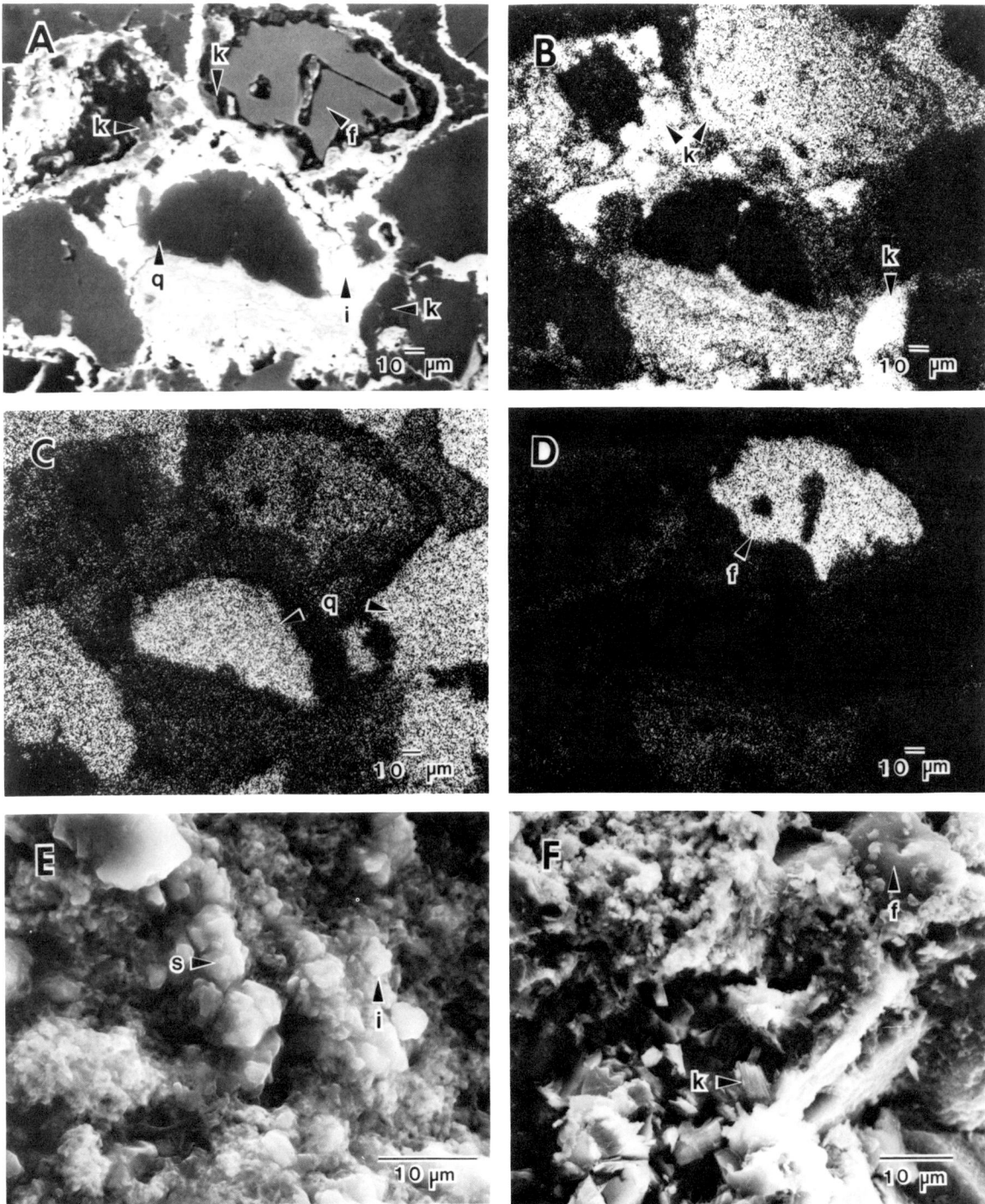

Figure 4. Backscattered image (A); elemental maps of Al (B), Si (C), and K (D) of polished cross-section surface; and scanning electron-microscopic images (E, F) of fractured surface of a weathered siltstone fragment in the soil studied (f = feldspar, i = iron oxide, k = kaolinite, q = quartz, s = silica).

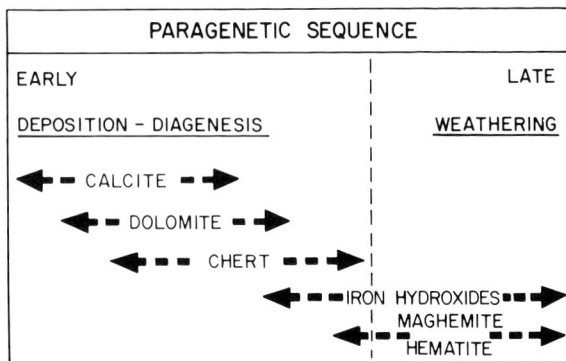

Figure 5. Proposed paragenetic sequence for iron oxide mineral formation.

porous chert grains which contained relict carbonate grains.

DISCUSSION

The morphological features and distribution of iron oxide-containing nodules in soil and residuum core samples indicate that the iron oxide minerals formed during a late stage of paragenesis, possibly a pedogenic process (Figure 5). As diagenesis progressed, calcite was replaced by dolomite, dolomite by chert, and chert by iron oxide minerals.

Bonifas (1959) and Fitzpatrick and Le Roux (1976) reported pedogenic maghemite that was formed by the oxidation of magnetite in soils derived from basic igneous rocks. Because of its absence in the original Knox Group dolostones, magnetite could not have been a precursor of the maghemite and hematite that occur in the soils. Thermal transformation of various iron oxides in the presence of organic matter as a result of burning is another possible mechanism of maghemite formation (Le Borgne, 1955; Graham and Scollar, 1976; Schwertmann and Fechter, 1984). Such a thermal influence is expected only in surface horizons of soils and certainly not 2 to 3 m below the surface, where higher amounts of maghemite nodules were found in this study.

In pedogenic environments, reducing conditions, which are required to dissolve iron, would be provided by microbial decomposition of organic matter in the soils (Mullins, 1977), and a partial oxidation of the reduced iron would promote precipitation of ferrous-ferric hydroxy compounds (Schwertmann and Taylor, 1977). Taylor and Schwertmann (1974) found that higher pH (pH 6–8) and temperature (20°–40°C) and slower oxidation rates were necessary to synthesize maghemite from a mixed ferrous and ferric solution. Maghemite was also the dominant mineral formed from a $FeSO_4$ solution after 7 days aging at pH 8.5 and 70°C (Lee and Bondietti, 1983). Thus, the maghemite in soils weathered from carbonate rocks might have been formed by slow oxidation and simultaneous dehydra-

tion of iron hydroxide precipitates within nodules or on the surface of soil components.

Hematite, the only other coexisting iron mineral in these soils, could have formed by the oxidation of the maghemite and/or through an independent path. In terms of environmental conditions, both hematite and maghemite appear to have formed at similar temperatures, soil pHs, and moisture conditions; however, the formation of hematite apparently required a constant oxidizing environment containing a small amount of organic matter, whereas the formation of maghemite required either repeated reduction-oxidation cycles or an ample supply of ferrous iron from weathering of primary minerals.

The replacement sequence of minerals in the nodules suggests that the maghemite was formed by: (1) dissolution of iron-bearing minerals such as carbonates and layer silicates, (2) replacement of leachable components in cherts and siltstones, and (3) crystallization or phase transformation of the replaced or precipitated iron. Maghemite, a common minor constituent in soils derived from the Knox Group dolostones, may be a more widely distributed, but overlooked phase, in other soils.

ACKNOWLEDGMENTS

The authors thank S. Owen and T. J. Henson for excellent technical support and D. A. Lietzke for soil characterization studies. Research was supported by the Office of Defense Waste and Byproducts Management, U.S. Department of Energy, under Contract No. DE-AC05-84OR21400 with Martin Marietta Energy Systems, Inc. Publication No. 2689, Environmental Sciences Division, Oak Ridge National Laboratory.

REFERENCES

Bonifas, M. (1959) Contribution a l'etude geochimique de l'alteration laterique: *Mem. Serv. Carte Geol. Alsace Lorraine* **17**, 159 pp.

Fagan, J. M. (1969) Geology of the Lost Creek barite mine: in *Papers on the Stratigraphy and Mine Geology of the Kingsport and Mascot Formations (Lower Ordovician) of East Tennessee, Tennessee Div. Geol., Rept. Invest.* **23**, 40–44.

Fitzpatrick, R. W. and Le Roux, J. (1976) Pedogenic and solid solution studies on iron titanium minerals: in *Proc. Int. Clay Conf., Mexico City, 1975,* S. W. Bailey, ed., Applied Publishing, Wilmette, Illinois, 585–599.

Graham, I. D. G. and Scollar, I. (1976) Limitations on magnetic prospecting in archaeology imposed by soil properties: *Archaeo-Physida* **6**, 1–124.

Harrison, C. G. A. and Peterson, M. N. A. (1965) A magnetic mineral from the Indian Ocean: *Amer. Mineral.* **50**, 704–712.

Le Borgne, E. (1955) Abnormal magnetic susceptibility of the top soil: *Annls. Geophys.* **11**, 399–419.

Lee, S. Y. and Bondietti, E. A. (1983) Technetium behavior in sulfide and ferrous iron solutions: in *Scientific Basis for Nuclear Waste Management,* D. G. Brookins, ed., *Mat. Res. Soc. Symp. Proc.* **5**, 315–322.

Lee, S. Y., Kopp, O. C., and Lietzke, D. A. (1984) Mineralogical characterization of West Chestnut Ridge soils: *Oak Ridge National Laboratory, Rept.* **ORNL/TM-9361,** 85 pp.

Mullins, C. E. (1977) Magnetic susceptibility of the soil and its significance in soil science—a review: *J. Soil Sci.* **28,** 223–246.

Schwertmann, U. and Fechter, H. (1984) The influence of aluminum on iron oxides: XI. Aluminum-substituted maghemite in soils and its formation: *Soil Sci. Soc. Amer. J.* **48,** 1462–1463.

Schwertmann, U. and Taylor, R. M. (1977) Iron oxides: in *Minerals in Soil Environments,* J. B. Dixon and S. B. Weed, eds., Soil Science Society of America, Madison, Wisconsin, 145–180.

Taylor, R. M. and Schwertmann, U. (1974) Maghemite in soils and its origin. II. Maghemite synthesis at ambient temperature and pH 7: *Clay Miner.* **10,** 299–310.

van der Marel, H. W. (1951) Gamma ferric oxide in sediments: *J. Sediment. Petrol.* **21,** 12–21.

Winchell, A. N. (1931) Maghemite or oxymagnite?: *Amer. Mineral.* **16,** 270–271.

Proceedings of the International Clay Conference, Denver, 1985, L. G. Schultz, H. van Olphen, and F. A. Mumpton, eds.,
The Clay Minerals Society, Bloomington, Indiana, 212–220 (1987).

MAGNETIC PROPERTIES OF SYNTHETIC FEROXYHITE (δ'-FeOOH)

C. J. W. Koch,[1] O. K. Borggaard,[1] M. B. Madsen,[2] and S. Mørup[2]

[1] Chemistry Department, Royal Veterinary and Agricultural University
DK-1871 Frederiksberg C, Denmark

[2] Laboratory of Applied Physics II, Technical University of Denmark
DK-2800 Lyngby, Denmark

Abstract—Mössbauer studies of the magnetic properties of synthetic feroxyhites (δ'-FeOOH) of varying crystallinity showed that large crystallites are ferrimagnetic, probably due to uncompensated spins associated with defects. Heating of these crystallites led to a reduction of their magnetic moment, presumably because the number of defects was reduced during heating. Small crystallites of feroxyhite were found to be speromagnetic, having a magnetic ordering temperature between 80 and 300 K, in contrast to a ferrimagnetic Curie temperature of about 450 K for large crystallites. A speromagnetic component was also found in the spectra of the large crystallites, probably due to the surface atoms.

Key Words—Defects, Feroxyhite, Magnetic properties, Mössbauer spectroscopy, Speromagnetism, Thermal treatment.

INTRODUCTION

Recent works by Chukhrov et al. (1977) and Carlson and Schwertmann (1980) have added feroxyhite (δ'-FeOOH) to the list of known naturally occurring ferric oxyhydroxides. Its presence in soils is of particular interest inasmuch as synthesis experiments have indicated that feroxyhite forms only by the rapid oxidation of an Fe^{2+}-containing solid in a weakly acid to alkaline suspension. These experimental findings indicate that the presence of feroxyhite could be used as an indicator mineral in geology and pedology for zones of rapid oxidation of Fe^{2+} (Carlson and Schwertmann, 1980).

Identification of feroxyhite by X-ray powder diffraction methods is difficult because of the close crystallographic similarity of feroxyhite, hematite (α-Fe_2O_3), and ferrihydrite (Carlson and Schwertmann, 1980), especially if these minerals occur as mixtures. To facilitate the identification of feroxyhite, it is useful to use other physical characteristics of the mineral, e.g., its magnetic properties. The magnetic properties of feroxyhite are not presently understood in detail; Mössbauer studies of natural and synthetic samples have shown magnetically split spectra at 4 K (work of E. Murad cited in Carlson and Schwertmann, 1980), but no information on the magnetic order of the samples was given. Pernet et al. (1984) concluded from neutron diffraction studies that δ-FeOOH ideally is antiferromagnetic, but that it shows ferrimagnetic order due to magnetic uncompensated spins in the crystallites. Madsen et al. (1985) found that the magnetic behavior of synthetic feroxyhite[3] is highly dependent on crystallite size; the largest crystallites are ferri- or ferromagnetic, whereas the smallest crystallites have only a small net magnetic moment. In the present communication a more detailed characterization of the magnetic properties of feroxyhite is reported along with the dependence of these properties on other characteristics of the samples.

MATERIALS AND METHODS

Samples

The samples were prepared by rapid oxidation by H_2O_2 of a suspension formed by precipitating an Fe^{2+} solution with NaOH. All chemicals were of analytical reagent grade, and the water was doubly ion exchanged. The details of the preparations are: 0.1 mole Ferrum reductum was dissolved in 30 cm^3 of 36% HCl by boiling it in a plastic beaker. The solution was diluted with water to make the Fe^{2+} concentration 0.27 or 0.10 M. During the dilution and the following steps the solution/suspension was flushed with Ar (>99.998% Ar, <0.002% O_2) and stirred vigorously, and the beaker was cooled on an ice bath. When the pH of the solution was adjusted to 6–11 with 5 M NaOH, a green precipitate formed. The suspension was oxidized by a dropwise addition of 30% H_2O_2 in excess, and simultaneously 5 M NaOH was added to keep the pH as constant as possible during oxidation. During the oxidation the temperature of the suspension increased from room temperature to 25–35°C. The suspensions were stored for a short period (<1 hr) at the "final pH"

---

[3] The studied samples are designated feroxyhite because of the absence of an observable 001 reflection in the X-ray powder diffraction pattern, as suggested by Chukhrov et al. (1977). The properties of feroxyhite, however, are not markedly different from material designated δ-FeOOH in the literature if this criterion is used for the distinction.

Table 1. Preparation conditions for feroxyhite.

| | Sample | | | |
	A	B	C	D
Initial [Fe^{2+}] (M)	0.1	0.27	0.1	0.1
pH during oxidation	8–9	6–8	4–8	3–8
Final pH	11.0	11.8	8.0	8.0

before they were centrifuged ($\sim$8000 g), washed five times with water, and dried at room temperature. Sample C was, however, washed once with water, twice with acetone, and dried at 60°C. The dried samples were gently crushed in an agate mortar. Table 1 summarizes the parameters for the preparation of the samples.

To study the effect of heating, the samples were heated at 105°C for 24 hr in air (indicated by adding suffix "v" to the sample code).

X-ray powder diffraction

X-ray powder diffraction (XRD) studies were performed on back-filled samples using CoKα radiation and a Philips PW 1050 diffractometer with a graphite diffracted-beam monochromator. The XRD pattern was obtained by line scanning using a goniometer speed of 0.125°2θ/min and fixed slits of 2°.

Transmission electron microscopy

Electron micrographs were obtained after ultrasonic dispersion of the samples in water using a JEOL JEM 100B instrument operating at 80 kV.

Mössbauer spectroscopy

Mössbauer spectra were obtained by means of a constant acceleration Mössbauer spectrometer with a 50 mCi source of ^{57}Co in Rh. The spectrometer was calibrated using a 12.5-μm foil of α-Fe at room temperature. Isomer shifts are given relative to the centroid of the spectrum of this absorber. Spectra of all the samples were obtained at room temperature and at 80 K without and with an applied magnetic field, B (between 0.5 and 1 T), applied perpendicular to the direction of propagation of the γ-rays. Spectra of samples A and D were also obtained at 5 K without and with magnetic fields (1 and 4 T) applied parallel to the propagation direction of the γ-rays.

Specific surface area and adsorbed H_2O

The specific surface area was determined by water adsorption at p/p$_0$ = 0.2, as suggested by Pyman and Posner (1978). From the weight loss of the samples after 4 weeks equilibration over P_2O_5, the amount of adsorbed H_2O at air humidity was determined.

Iron content

The iron content of the samples was determined by permanganate titration.

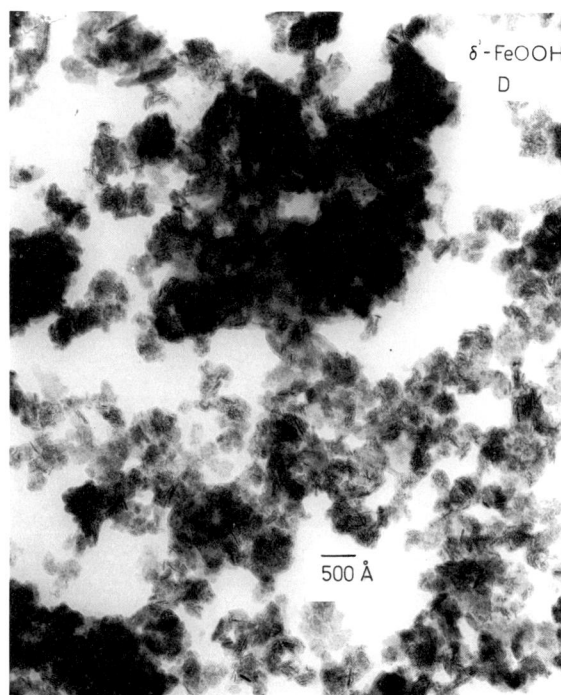

Figure 1. Transmission electron micrograph of sample D.

RESULTS

Iron content and surface areas

The iron content, after correction for adsorbed H_2O, and the specific surface area of the samples are listed in Table 2. The iron content of the samples is in reasonable agreement with the theoretical composition of FeOOH (88.85 wt. % Fe_2O_3).

Transmission electron micrographs

A transmission electron micrograph (TEM) of sample D is shown in Figure 1. Micrographs of samples A and C were presented by Madsen et al. (1985). Two different particle morphologies were observed in all samples: (1) translucent, rounded to hexagonal plates, and (2) electron-dense, needle-like particles. Particles of similar morphologies were also observed by Chu-

Table 2. Iron content, specific surface area, and particle dimensions obtained from transmission electron micrographs of feroxyhites.

| Sample[1] | Fe_2O_3 (wt. %) | Specific surface area (m^2/g) | Particle dimensions | |
			Thickness (Å)	Diameter (Å)
A	86.61	110	20–30	150–900
B	85.14	144	15–20	200–600
C	88.18	148	15–20	300–400
D	86.55	197	20–25	300–500

[1] See Table 1 for preparation conditions.

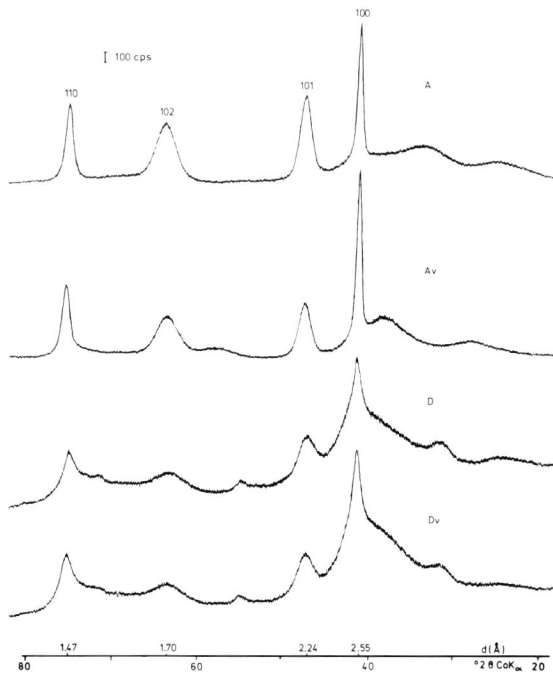

Figure 2. X-ray powder diffraction patterns of samples A, Av, D, and Dv.

Table 3. Mean crystallite size (Å) of feroxyhites obtained from X-ray powder diffractograms along different directions.

hkl	Sample[1]					
	A	Av	B	C	D	Dv
100	270	280	175	n.m.[2]	n.m.	n.m.
101	80	80	60	40	45	45
102	45	40	30	20	20	25
110	160	160	125	60	65	55

[1] See Table 1 for preparation conditions.
[2] n.m. = not measurable.

3). The changes in mean crystallite size after heating were estimated to be negligible within the limits of measuring error.

Mössbauer spectroscopy

Mössbauer spectra of magnetic microcrystals obtained in external magnetic fields are useful for determination of the magnetic structure (Mørup *et al.,* 1980, 1982). If the magnetization vectors are oriented at random in the material, the area ratios are 3:2:1:1:2:3. If the sample is magnetized perpendicular to the γ-ray direction, the area ratios are 3:4:1:1:4:3, whereas a magnetization parallel to the γ-ray direction results in area ratios of 3:0:1:1:0:3, i.e., lines 2 and 5 vanish. Mössbauer spectroscopy is also sensitive to superparamagnetic relaxation effects. For very small magnetic particles the relaxation time is commonly of the order of or smaller than 10^{-9} s, leading to a collapse of the magnetic hyperfine splitting. If the particles possess a magnetic moment, however, the application of an external magnetic field may restore the magnetic hyperfine splitting.

Samples A and Av. At room temperature the zero-field spectrum of samples A and Av consisted of a superposition of a six-line component with broad lines and a quadrupole doublet (Figure 3). Application of an external field of 1.0 T resulted in the disappearance of most of the quadrupole doublet and in a better resolution of the six-line component, giving an area ratio close to 3:4:1:1:4:3 (because of the different widths of the lines the intensity ratios deviate considerably from the area ratios). These results indicate that part of the microcrystals were superparamagnetic at this temperature and that the particles possessed a magnetic moment which was partly aligned by the applied magnetic field.

At 80 K the zero-field spectrum of sample A was magnetically split; the application of the external field resulted in alignment of the magnetization vectors, as seen from the change in the line area ratio (Figure 3). The asymmetry of the 80-K spectrum was due to the presence of more than one Fe^{3+} site in the structure.

The unheated sample, A, was also studied at 5 K (Figure 4). In zero applied field the spectrum consisted

khrov *et al.* (1977), Carlson and Schwertmann (1980), and Krakow *et al.* (1980). The last authors showed that the apparently different morphologies were the result of viewing platy particles at different angles. They also suggested that the thickness of the platy particles could be estimated from the smallest dimension of the needles. From the measurements listed in Table 2, it is apparent that all samples examined in the present study contained particles of different size.

No significant changes in the electron micrographs could be observed after the samples were heated to 105°C.

X-ray powder diffraction

XRD patterns of samples A–D were presented by Madsen *et al.* (1985). The effects of heating the samples are illustrated in Figure 2 by samples A, Av, D, and Dv; samples B and C show changes after heating similar to those in samples A and D, respectively. Indices in Figure 2 are based on a hexagonal unit cell having $a = 2.95$ and $c = 4.56$ Å (Patrat *et al.,* 1983). Heating of sample A resulted in changes in the diffuse scattering from the sample (Patrat *et al.,* 1983), indicating a change of the local structural order of the sample. The pattern of sample Dv showed no major changes compared to sample D. Assuming that the broadening of the peaks is due only to crystallite size, the mean crystallite size along different directions was calculated using the Scherrer formula (Klug and Alexander, 1974) (Table

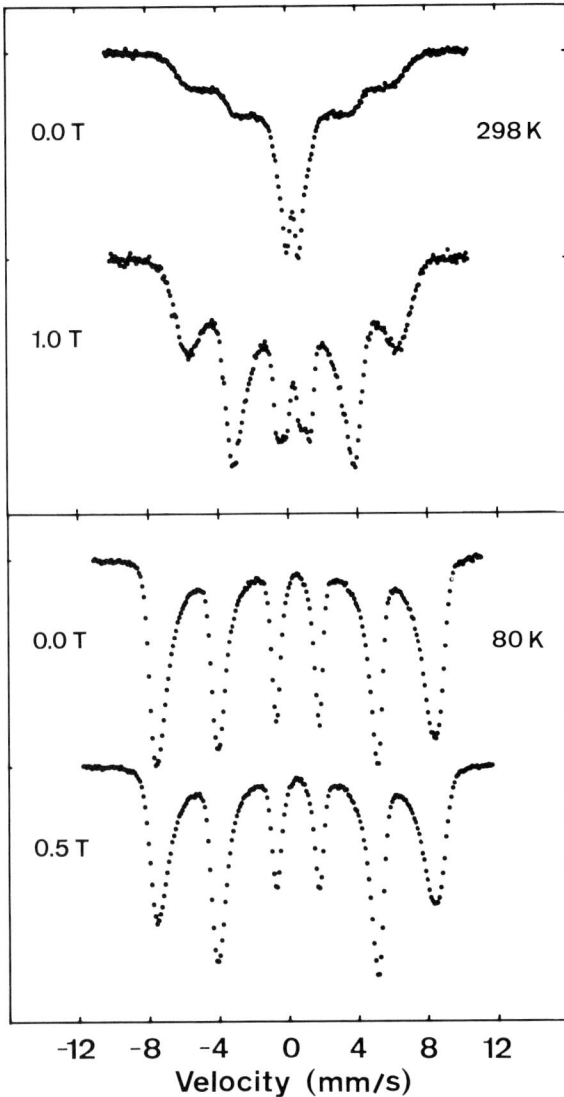

Figure 3. Mössbauer spectra of sample A obtained at 298 and 80 K. Magnetic fields were applied perpendicular to the γ-ray direction.

of a single sextet. At B = 1 T (applied parallel to the γ-ray direction) the intensities of lines 2 and 5 decreased drastically, indicating partial alignment of the magnetization. At 4.0 T the relative areas of lines 2 and 5 were further reduced compared with those seen in the 1.0-T spectrum. Moreover, lines 1 and 6 split, indicating the presence of (at least) two magnetically split components.

After heating, the zero-field, room-temperature spectrum contained only a magnetically split component, and the magnetic hyperfine splitting at 80 K was better resolved (Figure 5); line area ratios were only slightly affected by the applied magnetic field (Figure 5).

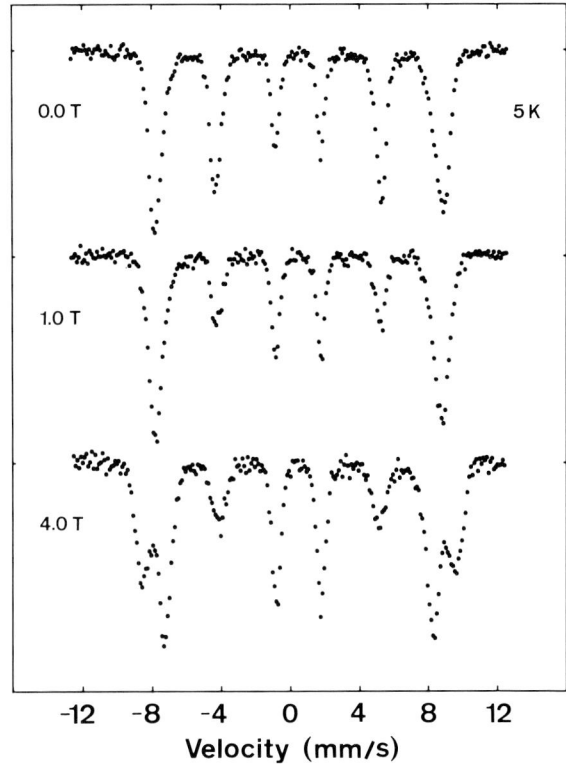

Figure 4. Mössbauer spectra of sample A obtained at 5 K. Magnetic fields were applied parallel to the γ-ray direction.

Samples B and Bv. The zero-field, room-temperature spectrum of sample B consisted of a quadrupole doublet only; however, in an applied field of 1.0 T, a magnetically split component was present together with the quadrupole doublet (Figure 6).

At 80 K the spectra of sample B were magnetically split (a quadrupole doublet of low intensity was present in the zero-field spectra). As was also noted for sample A, the intensity ratios of the lines were significantly affected by the external field, indicating that the magnetization was parallel to the applied magnetic field.

Heating of this sample introduced some remarkable effects. First, a six-line component appeared in the zero-field, room-temperature spectrum (Figure 6). Second, the area ratios of the six lines in the 80-K spectra were only slightly affected by an external magnetic field, as was also observed for sample Av.

Samples C and Cv. The room-temperature Mössbauer spectrum of sample C (Figure 7) consisted of a quadrupole doublet (isomer shift = 0.36 mm/s and quadrupole splitting = 0.68 mm/s). The spectrum was only slightly affected by application of an external magnetic field; line widths increased from about 0.54 to 0.60 mm/s and a broad, low-intensity component was present.

The spectra of sample C obtained at 80 K showed

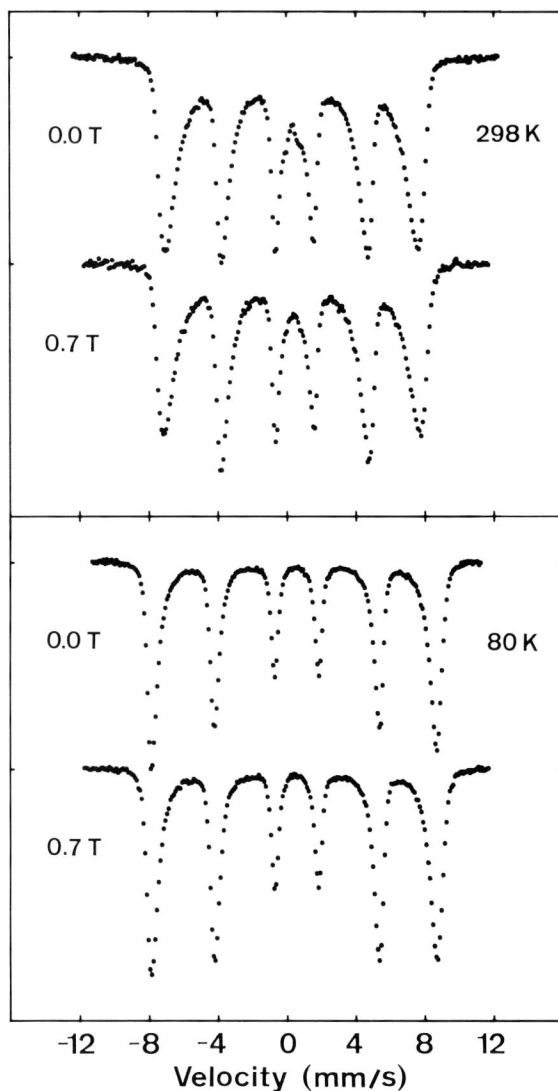

Figure 5. Mössbauer spectra of sample Av obtained at 298 and 80 K. Magnetic fields were applied perpendicular to the γ-ray direction.

Figure 6. Mössbauer spectra of samples B (top) and Bv (bottom) obtained at 298 K. Magnetic fields were applied perpendicular to the γ-ray direction.

magnetic hyperfine splitting with broad lines. Remarkably, the application of an external magnetic field to this sample did not significantly change the line area ratios (Figure 7), in contrast to the behavior of the unheated samples discussed above.

Heating of this sample did not change the room-temperature spectra, but at 80 K the magnetic hyperfine splitting was considerably better resolved. The line area ratios, however, were still unaffected by the external magnetic field.

Samples D and Dv. The room-temperature spectra of samples D and Dv were similar to those of samples C and Cv. At 80 K the zero-field spectrum of sample D (Figure 8) consisted of a superposition of a six-line

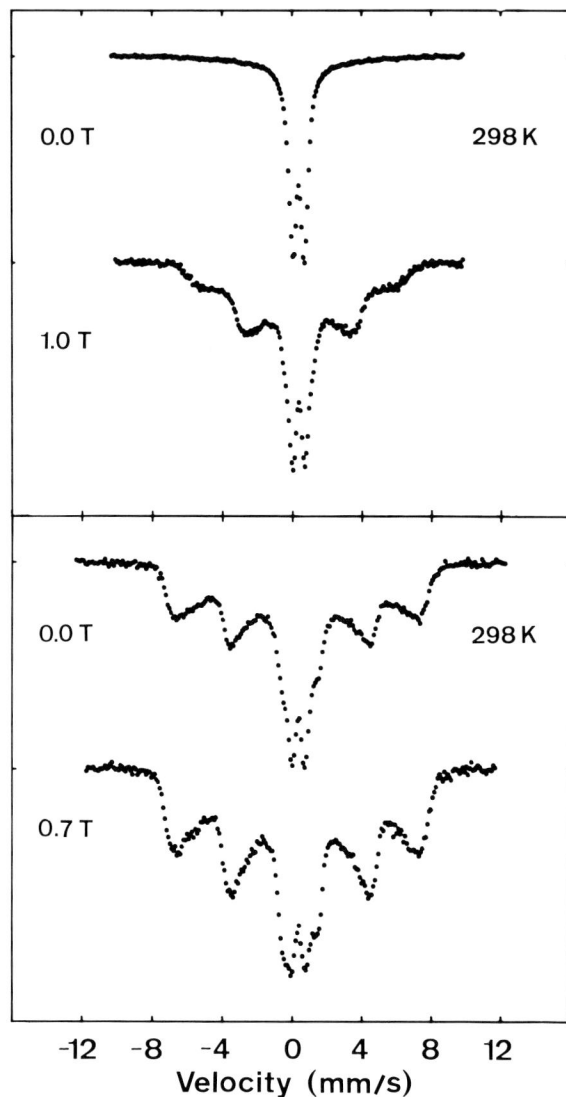

component and a quadrupole doublet. At this temperature, the application of an external magnetic field of 1.0 T resulted in substantial line broadening of the quadrupole doublet, indicating the presence of superparamagnetic particles.

At 5 K the spectrum was magnetically split with a hyperfine field of 49.6 ± 0.5 T and a negligible quadrupole shift. The application of an external field of 4.0 T resulted only in broadening of the lines (Figure 9).

The spectra of sample Dv obtained at 80 K were substantially different from those of sample D and consisted mainly of a magnetically split component. The application of a magnetic field of 0.5 T further reduced the relative intensity of the quadrupole doublet.

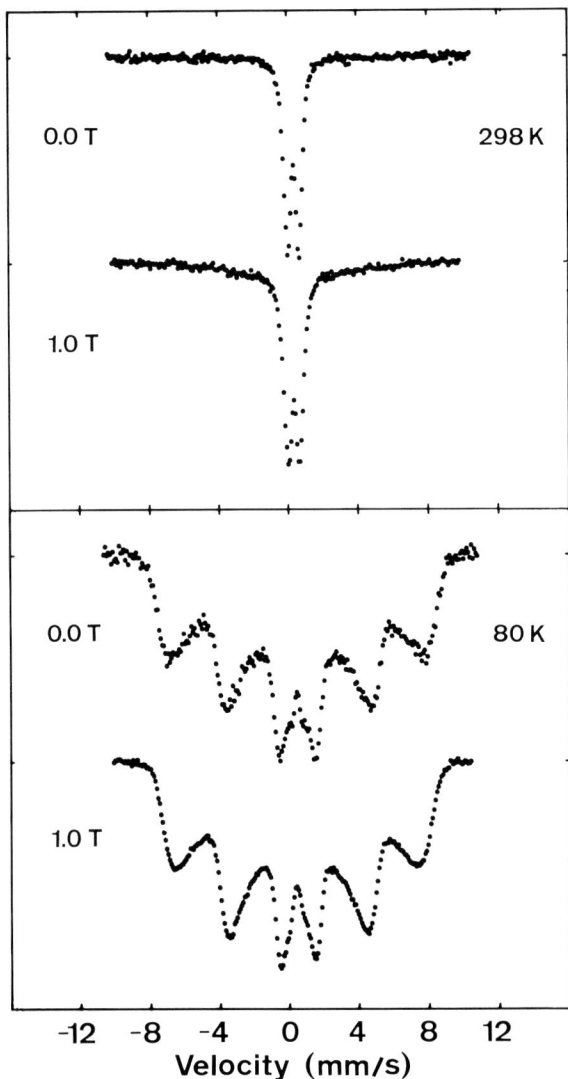

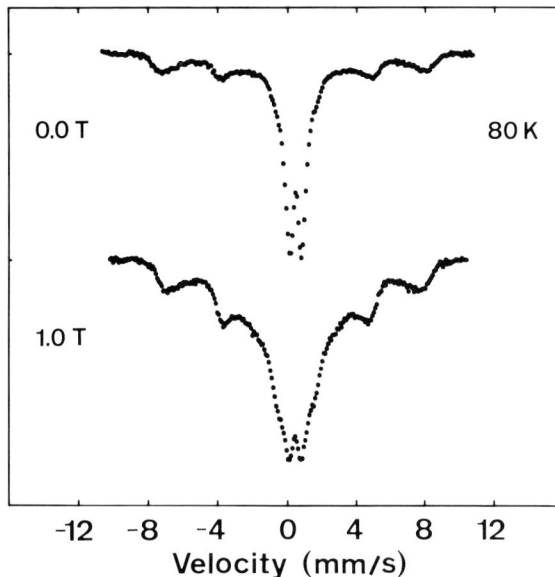

Figure 8. Mössbauer spectra of sample D obtained at 80 K. The magnetic field was applied perpendicular to the γ-ray direction.

Figure 7. Mössbauer spectra of sample C obtained at 298 and 80 K. Magnetic fields were applied perpendicular to the γ-ray direction.

DISCUSSION

Crystallographic properties and morphology

Patrat *et al.* (1983) reported that the crystal structure of δ-FeOOH is based on hexagonal close-packed layers of oxygen, wherein iron ions occupy half of the octahedral interstices. Furthermore, they noted that the structure was somewhat disordered. Although in the present study, the local order in the samples was not determined in detail, the samples were classified on the basis of the XRD data. Samples A and B appear to be well crystallized, but have more defects than samples Av and Bv; samples C, Cv, D, and Dv appear to be poorly crystallized.

From a comparison of the particle dimension ob-

served in the electron micrographs (Table 2) with the mean crystallite size obtained from XRD (along directions with $l \neq 0$) (Table 3) the c-axis appears to be parallel to the smallest dimension of the particles, i.e., the crystallites are only a few unit cells thick in the c-direction. Krakow *et al.* (1980) made similar observations by high-resolution electron microscopy. The data of Table 3 also suggest that the dimensions perpendicular to the c-direction of the crystallites of samples C and D are generally about half of those of samples A and B. Thus, the specific surface area of samples C and D should be much larger compared to that of samples A and B. As can be seen from Table 2, however, this is not true, probably because of aggregation. Moreover a comparison of the results in Tables 2 and 3 indicates that not all the particles observed in the electron micrographs are single crystallites.

Magnetic properties

The Mössbauer results clearly show that different samples of feroxyhite exhibited different magnetic ordering. The smallest crystallites (samples C and D) appeared to possess only a very small magnetic moment at 80 K, although they gave magnetically split spectra. As discussed by Madsen *et al.* (1985), these particles presumably had a disordered magnetic structure (spin-glass structure or speromagnetic order). This conclusion was based on the fact that the magnetic hyperfine fields could not be aligned by external fields as great as 1.0 T at 300 and 80 K. In the present study the magnetic hyperfine fields could not be aligned, even in external fields as strong as 4.0 T at 5 K (Figure 9).

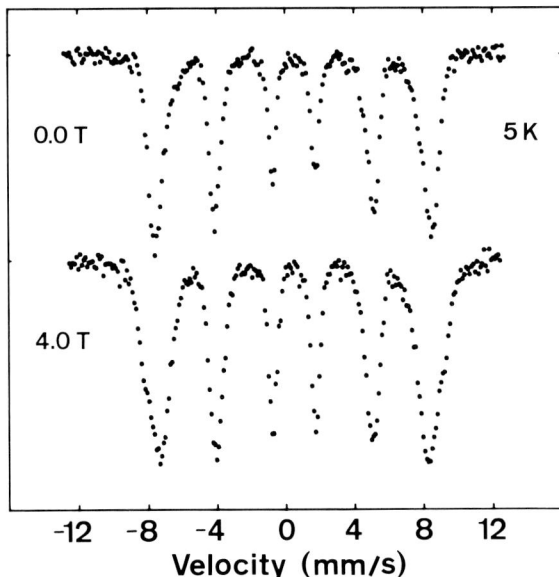

Figure 9. Mössbauer spectra of sample D obtained at 5 K. The magnetic field was applied parallel to the γ-ray direction.

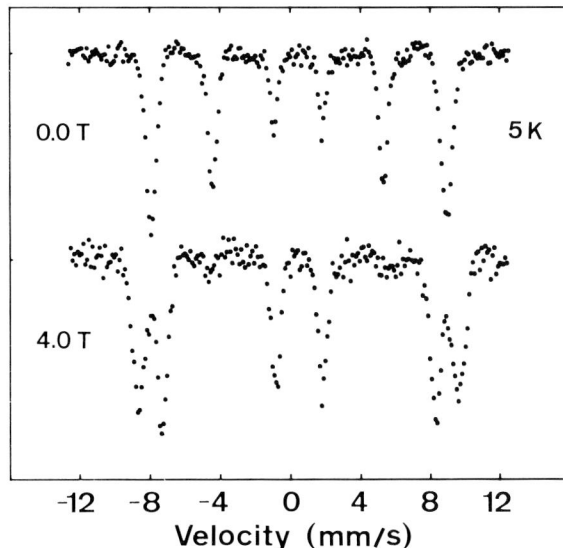

Figure 10. Mössbauer spectra of "bulk" feroxyhite at 5 K. Spectra were generated as differences between spectra of samples A and D. Magnetic fields were applied parallel to the γ-ray direction.

This result corroborates the conclusions of Madsen *et al.* (1985).

In the larger crystallites (samples A and B), the application of external fields resulted in a substantial change in the relative line intensities, indicating that the hyperfine fields, and thus also the magnetization of the crystallites, were aligned parallel to the external field. Therefore, the larger particles must have been ferro- or ferrimagnetic.

Detailed information about the magnetic structure of the larger crystallites was obtained from the spectra obtained at high external magnetic fields at 5 K (Figure 4). In these spectra the low intensities of lines 2 and 5 indicated a high degree of alignment of the magnetization at 1.0 T and an even higher degree of alignment at 4.0 T. Inasmuch as lines 2 and 5 did not vanish completely, some of the spins were not aligned even in a field of 4.0 T. A similar effect has been observed for microcrystals of maghemite (γ-Fe_2O_3), in which the non-collinearity in spin structure was due to surface effects, i.e., the surface spins were not parallel to the spins inside the particles (Coey, 1971; Morrish *et al.*, 1976).

The present results for the large crystallites of feroxyhite can probably be explained by a similar effect. As discussed above, the main difference between the well-crystallized samples (A and B) and the poorly crystallized samples (C and D) was the size of the crystallites. Measurements of crystallite size indicated that samples C and D were only a few unit cells thick in all directions (Table 3), suggesting that surface effects might have been significant for all atoms in the small crystallites.

The Mössbauer spectra of samples C and D show that the magnetically disordered surface atoms must have had slightly smaller hyperfine fields than the atoms inside the particles at 5 K, namely 49.6 T, whereas the bulk value was 52.5 T. Moreover, the surface component had broader Mössbauer lines. This component can be analyzed, for example, in the 5-K spectra of sample D shown in Figure 9.

Assuming that lines 2 and 5 in the 4.0-T spectrum of sample A (Figure 4) were due to a surface component with non-collinear spin structure, the lines 1, 3, 4, and 6, corresponding to lines 2 and 5 of the surface component, must also have been present in the spectrum of the sample A. To obtain Mössbauer spectra of the pure bulk structure, the spectra of sample D (Figure 9) were subtracted from those of sample A after multiplying the former by an appropriate factor such that lines 2 and 5 disappear in the difference spectrum at 4.0 T. The resulting spectra are shown in Figure 10. For the spectrum obtained in zero applied field, the subtraction of the surface component led to a substantial decrease of the line widths of the sextet. In the spectrum resulting from the subtraction of the spectra obtained at B = 4.0 T, some structure remained in the background at the positions of lines 2 and 5. This structure is caused by a small difference between the splitting of lines 2 and 5 in the spectra of samples A and D, respectively.

The hyperfine parameters of the bulk component determined by fitting the difference spectra shown in Figure 10 are given in Table 4. The two components of the 4.0-T spectrum had the same isomer shift of

Table 4. Hyperfine parameters for the "bulk" component of feroxyhite at 5 K.

B^1 (T)		B_{hf}^2 (T)	Isomer shift (mm/s)	Quadrupole shift (mm/s)	Relative area (%)
0.0		52.5 ± 0.5	0.48 ± 0.02	0.04 ± 0.03	100
4.0	$\{$(a)^3	57.0 ± 1.0	0.49 ± 0.03	~0	46 ± 2
	$\{$(b)	48.5 ± 1.0	0.49 ± 0.03	~0	54 ± 2

[1] B = the magnetic field applied to the sample.
[2] B_{hf} = the hyperfine field of the sample.
[3] a and b designate the two magnetic sublattices.

0.49 mm/s, indicating that the iron was octahedrally coordinated in both sites. The hyperfine fields were consistent with a sublattice magnetization completely parallel to the applied field. The changes in the hyperfine fields relative to the zero-field spectrum were $\Delta_a =$ 57.0 T − 52.5 T = 4.5 ± 1.0 T and $\Delta_b =$ 49.5 T − 52.5 T = −4.0 ± 1.0 T, which is equal to the applied field, B = 4.0 T, within the uncertainty. The presence of these two components shows that "bulk" feroxyhite is ferrimagnetic. From the area ratio of the two components 54% of the iron ions was found to belong to one sublattice and 46% to the other. Thus, feroxyhite is nearly antiferromagnetic, in accordance with the finding of Pernet et al. (1984). The small deviation from an area ratio of 50:50 suggests that the unbalance was due to defects in the structure.

The spectra of the heated samples showed that the heating had a significant effect on the magnetic properties, as was also observed by Okamoto (1968). For the smallest particles, heating resulted in an increase in the relative area of the magnetically split components at high temperatures and a better resolved magnetic hyperfine splitting at lower temperatures. Similar effects of heating have been observed for microcrystals of goethite (α-FeOOH) (Koch et al., 1986) and spinels (Tronc and Bonnin, 1985). In goethite, the effect was explained by an enhancement of the magnetic coupling among the crystallites due to loss of water molecules between the crystallites. The present results are probably due to a similar mechanism.

After samples A and B were heated, the magnetic spins were only slightly aligned by an external field. Thus, heating substantially reduced the magnetic moment of these particles. The surface disorder may have affected more atoms after the heating; however, the relatively sharp lines in the spectra of samples Av and Bv also indicated that the magnetic structure was more perfect than that of the samples Cv and Dv. Inasmuch as the ferrimagnetic structure presumably was due to defects, the decrease in magnetization was probably due to an annealing of these defects to yield a more perfect antiferromagnetic order.

Previous magnetization measurements have led to an estimated Curie temperature of about 450 K (Okamoto, 1968). The magnetic transition temperature may, however, depend on the type of magnetic order. For example, the transition temperature of the small crystallites with speromagnetic order may be different from that found for the larger crystallites with ferri- or antiferromagnetic order. The present Mössbauer results do in fact indicate such a difference. For example, the room-temperature spectra of samples C and D showed no magnetic splitting, even when an external magnetic field was applied, indicating that the transition temperature was less than 300 K for these samples. Even in speromagnetic particles one might expect a small magnetic moment which will interact with the external field and lead to a substantial change in the Mössbauer spectra (Madsen et al., 1986). At 80 K the spectra of samples C and D were in fact affected by the external field in a way that indicates a small magnetic moment in the particles (Figure 8). Thus, the magnetic transition temperature of the speromagnetic particles was between 300 and 80 K, whereas it was about 450 K for the ferri- or antiferromagnetic crystallites.

ACKNOWLEDGMENTS

Thanks are due to B. Bloch for preparing the electron micrographs. The project was supported by the Danish Natural Science Research Council.

REFERENCES

Carlson, L. and Schwertmann, U. (1980) Natural occurrence of feroxyhite (δ'-FeOOH): Clays & Clay Minerals 28, 272–280.

Chukhrov, F. V., Zvyagin, B. B., Gorshkov, A. J., Yermilova, L. P., Korovushkin, V. V., Rudnitszkaya, Ye. S., and Yakubovskaya, N. Ya. (1977) Feroxyhyte, a new modification of FeOOH: Intern. Geol. Rev. 19, 873–890.

Coey, J. M. D. (1971) Noncollinear spin arrangement in ultrafine ferrimagnetic crystallites: Phys. Rev. Letters 27, 1140–1142.

Klug, H. P. and Alexander, L. E. (1974) X-ray Diffraction Procedures for Polycrystalline and Amorphous Materials, Wiley, New York, 966 pp.

Koch, C. J. W., Madsen, M. B., Mørup, S., Christiansen, G., Gerward, L., and Villadsen, J. (1986) Effect of heating on microcrystalline goethite: Clays & Clay Minerals 34, 17–24.

Krakow, W., Colijn, H., and Muller, O. (1980) δ-FeO(OH) and its solid solutions, Part 2. High resolution transmission electron microscopy of pure δ-FeO(OH): J. Materials Science 15, 119–126.

Madsen, M. B., Mørup, S., and Koch, C. J. W. (1986) Magnetic properties of ferrihydrite: Hyperfine Interactions 27, 329–332.

Madsen, M. B., Mørup S., Koch, C. J. W., and Borggaard, O. K. (1985) A study of microcrystals of synthetic feroxyhite (δ'-FeOOH): Surface Science 156, 328–334.

Morrish, A. H., Haneda, K., and Schurer, P. J. (1976) Surface magnetic structure of small γ-Fe_2O_3 particles: J. Physique Colloq. 37, C6-301–C6-305.

Mørup, S., Dumesic, J. A., and Topsøe, H. (1980) Magnetic microcrystals: in Applications of Mössbauer Spectroscopy, Vol. 2, R. L. Cohen, ed., Academic Press, New York, 1–53.

Mørup, S., Topsøe, H., and Clausen, B. S. (1982) Magnetic properties of microcrystals studied by Mössbauer spectroscopy: Physica Scripta 25, 713–719.

Okamoto, S. (1968) Structure of δ-FeOOH: J. Amer. Ceram. Soc. 51, 594–599.

Patrat, G., de Bergevin, F., Pernet, M., and Joubert, J. C. (1983) Structure locale de δ-FeOOH: Acta Crystallogr. B39, 165–170.

Pernet, M., Obradors, X., Fontcuberta, J., Joubert, J. C., and Tejada, J. (1984) Magnetic structure and supermagnetic properties of δ-FeOOH: IEEE Transactions Magnetics MAG-20, 1524–1525.

Pyman, M. A. F. and Posner, A. M. (1978) The surface areas of amorphous mixed oxides and their relation to potentiometric titration: J. Coll. Interface Sci. 66, 85–93.

Tronc, E. and Bonnin, D. (1985) Magnetic coupling among spinel iron oxide microparticles by Mössbauer spectroscopy: J. Physique Lett. 46, L437–L443.

Proceedings of the International Clay Conference, Denver, 1985, L. G. Schultz, H. van Olphen, and F. A. Mumpton, eds.,
The Clay Minerals Society, Bloomington, Indiana, 221–226 (1987).

EFFECT OF HUMIC AND FULVIC ACIDS ON THE FORMATION OF ALLOPHANE

K. Inoue[1] and P. M. Huang

Department of Soil Science, University of Saskatchewan
Saskatoon, Saskatchewan S7N 0W0, Canada

Abstract—The effects of humic substances on the interaction of hydroxy-Al ions and orthosilicic acid at a Si concentration of 1.6×10^{-3} M, a Si/Al molar ratio of 1.0, an OH/Al molar ratio of 3.0, and humic or fulvic acid concentrations of 1 to 1000 mg/liter were studied. Humic substances at concentrations >100–300 mg/liter significantly perturbed this interaction and thus inhibited the formation of allophane. The Na-pyrophosphate- and Na_2CO_3-soluble fractions of the precipitates (>0.01 μm), formed in the presence of the humic substances at a concentration of 300 mg/liter, were, respectively, 71% and 22% in the humic acid system and 17–24% and 36–40% in the fulvic acid systems. Selective dissolution data and differential infrared spectroscopy showed that "proto-imogolite" allophane was a major component in the pyrophosphate- and Na_2CO_3-insoluble fraction of the precipitates formed in the presence of fulvic acid. The pyrophosphate- and Na_2CO_3-insoluble fraction of the precipitates formed in the presence of humic acid showed a very weak infrared spectrum. The present results show that humic substances are pedogenically significant in impeding the formation of allophane.

Key Words—Allophane, Fulvic acid, Humic acid, Infrared spectroscopy, Selective dissolution, Soils.

INTRODUCTION

Current understanding of the effects of low-molecular-weight organic acids on the formation of allophane and imogolite is very limited (Farmer, 1981). Non-humified low-molecular-weight organic acids, such as citric acid, which has a strong affinity for Al, perturb the interaction of hydroxy-Al ions and orthosilicic acid and thus inhibit the subsequent formation of allophane and imogolite, leading to the formation of hydroxy-Al-organic complexes, ill-defined aluminosilicate complexes, and pseudoboehmite (Inoue and Huang, 1984a, 1985). The formation of imogolite from the solutions containing hydroxy-Al ions and orthosilicic acid is also strongly perturbed by humic substances extracted from soils (Inoue and Huang, 1984b); however, the formation of allophane as influenced by humic substances still remains to be investigated.

The objective of the present study was to investigate the effects of humic and fulvic acids on the formation of allophane and to interpret the pedogenic significance of humic substances in the formation of poorly crystalline to noncrystalline aluminosilicates.

MATERIALS AND METHODS

Humic substances

A soil sample from the A_1 horizon of a Dystrandept (Ando soil) in Morioka, Japan, was extracted with 0.1

M NaOH (Kumada, 1981) to obtain fulvic acid I (FA-I) and humic acid (HA). The extracted HA was purified with a large excess of 0.5% (v/v) HCl-HF (Schnitzer, 1977). The extracted FA was passed through a H+-Dowex 50X-X8 cation-exchange resin and partially purified according to the method described by Schnitzer and Skinner (1968), except that, instead of Sephadex G-10, Bio-gel P-2 (about 200 g of gel, in a 4 × 45 cm column) was used to prepare the desalted FA.

The FA-II purified from the B_h horizon of the Armadale soil, a Spodosol (Podzol) in Prince Edward Island, Canada (Schnitzer and Skinner, 1968; Hansen and Schnitzer, 1969), was also used in the present study.

The HA and FA products were neutralized to pH 7.0 with a 1 M NaOH solution immediately prior to the experiments.

Reactions in the presence of humic substances

Orthosilicic acid, prepared by the method described by Inoue and Huang (1984a, 1985), and Na-humate or Na-fulvate solution were simultaneously mixed with an $AlCl_3$, solution to give a Si/Al molar ratio of 1.0. About 400 ml of the resultant solution was titrated with 0.1 M NaOH (with continuous stirring) at a rate of about 0.5 ml NaOH/min to an OH/Al molar ratio of 3.0. Solutions containing humic substances were adjusted with small amounts of 0.1 M NaOH or 0.1 M HCl to the same pH as the solution without humic substances and then diluted to 500 ml. The Si concentration of the resulting solution was 1.6×10^{-3} M; the final concentrations of humic substances in the parent

---

[1] Permanent address: Department of Agricultural Chemistry, Faculty of Agriculture, Iwate University, Morioka 020, Japan.

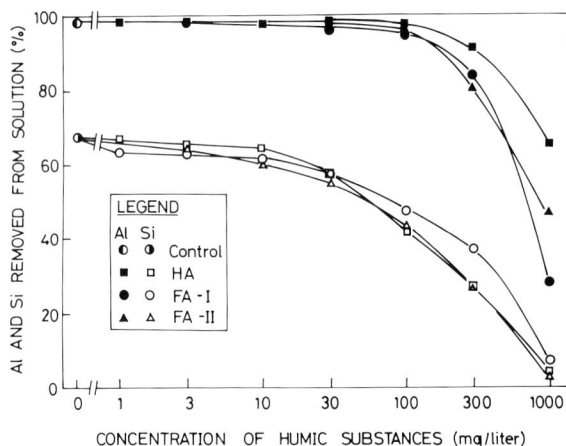

Figure 1. Influence of humic substances on interaction of hydroxy-Al ions with orthosilicic acid at a Si/Al ratio of 1.0 and an OH/Al ratio of 3.0. HA = humic acid; FA = fulvic acid. Initial Si concentration = 1.6×10^{-3} M.

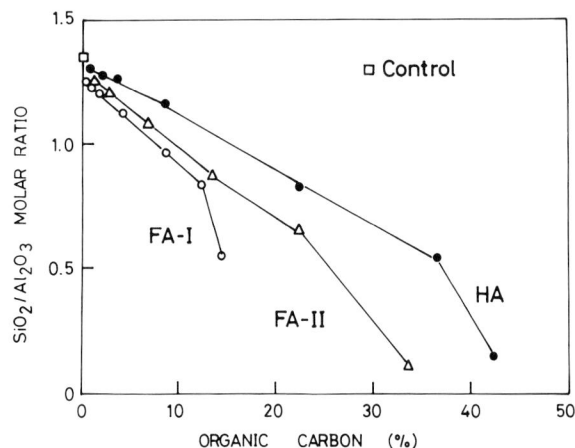

Figure 2. Relationship between SiO_2/Al_2O_3 molar ratios and organic carbon contents of precipitates formed from solutions containing hydroxy-Al ions, orthosilicic acid, and humic substances.

solutions were 0, 1, 3, 10, 30, 100, 300, and 1000 mg/liter.

About 400 ml of each parent solution was heated at 96°–100°C for 110 hr. After cooling to room temperature, the precipitates were collected by ultrafiltration using a Sartorius cellulose nitrate membrane filter (0.01-μm pore size) by the procedure described by Inoue and Huang (1984a, 1985).

Analyses of solutions and precipitates

Si and Al concentrations in the filtrates were determined by the methods described previously (Inoue and Huang, 1984a, 1985). The nature of the precipitates was examined by infrared (IR) spectroscopy. IR spectra were obtained using a Perkin Elmer 983 spectrometer (1.0 mg oven-dried sample/170 mg KBr). Transmission electron micrographs of the precipitates were taken at 60 kV with a Philips EM 400 instrument. The organic C content of the precipitates was determined by dry combustion.

Dissolution analysis and differential IR spectroscopy

Twenty milligrams of precipitate formed in the presence of HA and FA-I was subjected to successive extraction by 0.1 M $Na_4P_2O_7$ and 2% Na_2CO_3; differential IR spectra of the materials extracted by these treatments were obtained using the procedure outlined by Wada and Tokashiki (1972). The Al and organic C contents of the pyrophosphate extract (Al_{pyr} and C_{pyr}) and the Na_2CO_3 extract (Al_{car} and C_{car}) were determined by the procedure described by Wada and Higashi (1976).

RESULTS AND DISCUSSION

Interaction of hydroxy-Al ions with orthosilicic acid as influenced by humic substances

The percentage of Al and Si precipitated from the parent solutions decreased with increasing initial concentration of humic substances in the system (Figure

Table 1. Amounts of Al and organic carbon in pyrophosphate- and Na_2CO_3-soluble and -insoluble fractions of the precipitates.[1]

Humic substance	0.1 M Na-pyrophosphate-soluble fraction			2% Na_2CO_3-soluble fraction			Pyrophosphate-Na_2CO_3-insoluble fraction	Al_{pyr}/Al_{car}	Al_{pyr}/C_{pyr}	Al_{car}/C_{car}
	Total weight (%)	Al_{pyr}[2] (%)	C_{pyr}[2] (%)	Total weight (%)	Al_{car}[2] (%)	C_{car}[2] (%)	Total weight (%)	Molar ratio		
Humic acid	71.2	7.5	19.3	22.3	2.8	4.7	6.5	2.68	0.17	0.26
Fulvic acid-I	16.7	1.8	1.7	35.5	5.5	3.3	47.8	0.33	0.47	0.74
Fulvic acid-II	24.1	2.4	4.0	40.1	5.3	4.5	35.8	0.45	0.27	0.52

[1] 0.1 M Na-pyrophosphate, 2% Na_2CO_3, 300 mg humic substance/liter; oven-dry basis.

[2] Al_{pyr}, Al_{car}, C_{pyr}, and C_{car} denote, respectively, the amounts of Al and organic carbon extracted with pyrophospate and Na_2CO_3.

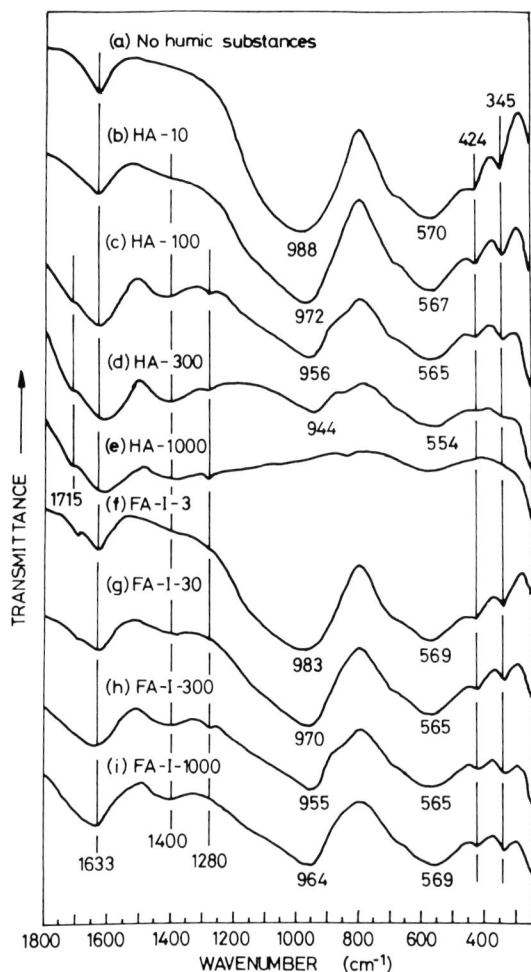

Figure 3.   Infrared spectra of precipitation products (>0.01 μm) formed by reaction of hydroxy-Al ions with orthosilicic acid both in absence and presence of humic substances: (a) no humic substances, (b) humic acid, 10 mg/liter, (c) humic acid, 100 mg/liter, (d) humic acid, 300 mg/liter, (e) humic acid, 1000 mg/liter, (f) fulvic acid-I, 3 mg/liter, (g) fulvic acid-I, 30 mg/liter, (h) fulvic acid-I, 300 mg/liter, and (i) fulvic acid-I, 1000 mg/liter.

Figure 4.   Selective dissolution-differential infrared spectra of precipitates formed in the presence of (A) humic acid and (B) fulvic acid-I at the concentration of 300 mg/liter. (a) original, (b) 0.1 M Na-pyrophosphate-soluble fraction, (c) 2% Na_2CO_3-soluble fraction, and (d) pyrophosphate- and Na_2CO_3-insoluble fraction.

1). In precipitates, the SiO_2/Al_2O_3 ratio decreased with increasing organic carbon content (Figure 2). Humic substances must have a large and sufficiently flexible surface in order to have many points of contact with the surfaces of mineral colloids (Greenland, 1965). The strong competition of such humic substances with orthosilicic acid for the coordination sites of Al is thought to account for the marked impedence of the humic substances (even at such low concentration) on the formation of allophane.

Nature of the precipitates and soluble products

IR spectra of the precipitates. Figure 3 shows the IR spectra of the precipitates formed from the solutions.

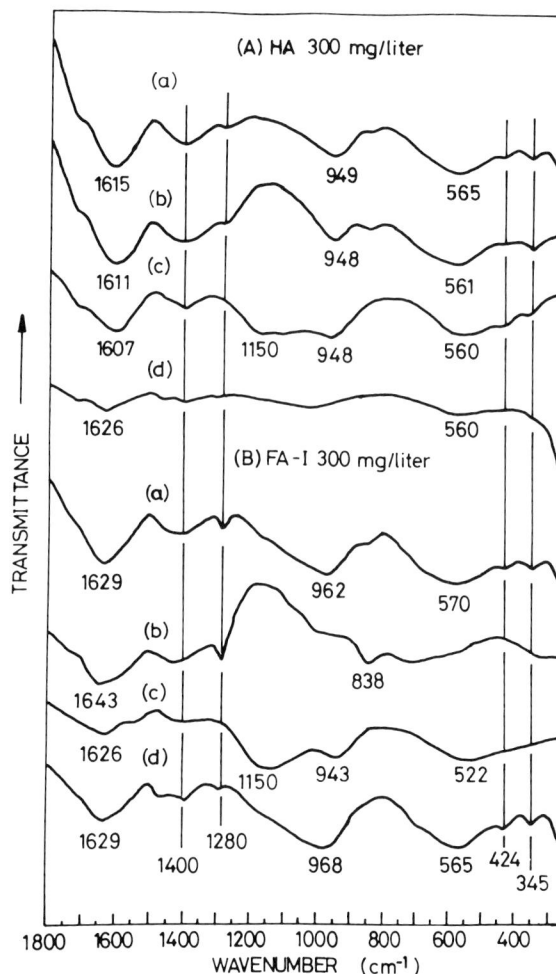

The IR spectrum of the precipitates formed in the absence of humic substances (Figure 3a) gave strong absorption maxima at 1633, 988, and 570 cm^{-1} and weak absorption bands at 424 and 345 cm^{-1}. These bands are common to soil allophane (Wada and Harward, 1974; Wada, 1977, 1980). At increasing concentrations of humic substances, however, the absorption maxima of the Si–O stretching band at 900–1000 cm^{-1} shifted gradually from 988 to 944 cm^{-1} (Figures 3b–3d) and from 983 to 964 cm^{-1} (Figures 3f–3i); their intensities all weakened (Figure 3). The IR bands (972–944, 569–554, 424, and 345 cm^{-1}) of the precipitates formed in the presence of humic substances (Figure 3) were similar to those of the "allophane-like constituents" of Wada and Harward (1974) and Wada (1977, 1980) and to the "proto-imogolite" allophane of Parfitt and Hen-

Table 2. Mineralogical composition of the precipitates formed both in the absence and presence of humic substances.[1]

Concentration of humic substance (mg/liter)	Precipitates (>0.01 μm)		
	Humic acid	Fulvic acid-I	Fulvic acid-II
0	A	A	A
1	A	A	—
3	A	A	A
10	PIA	PIA	PIA
30	PIA	PIA	PIA
100	PIA, Al-H	PIA	PIA, Al-F
300	PIA, Al-H	PIA, Al-F	PIA, Al-F
1000	Al-H	PIA, Al-F	Al-F

[1] A = allophane; Al-F = hydroxy-Al-fulvate complexes; Al-H = hydroxy-Al-humate complexes; PIA = "proto-imogolite" allophane. Mineralogical composition was established by chemical analysis, infrared spectroscopy, selective dissolution-differential IR spectroscopy, and electron microscopy.

mi (1980) and Farmer et al. (1980). HA at a concentration of 1000 mg/liter almost completely inhibited the formation of allophanes or even of "allophane-like" materials or "proto-imogolite" allophane and resulted in the formation of hydroxy-Al-humate complexes (Figure 3e). FA-I at a concentration of 1000 mg/liter did not completely perturb the interaction of hydroxy-Al ions with orthosilicic acid, however, but led to the formation of "proto-imogolite" allophane (Figure 3i) complexed with FA as described below. FA-II at the same concentration inhibited the formation of even "proto-imogolite" allophane (not shown).

Selective dissolution-differential IR spectroscopy. In the presence of HA, the weight percentage of each fraction was in the following order: pyrophosphate-extractable fraction > Na_2CO_3-extractable fraction > pyrophosphate-Na_2CO_3-insoluble fraction. Almost the reverse order was noted in the presence of FA-I and FA-II. The total weight percentage of the pyrophosphate-soluble fraction and the high Al_{pyr}/Al_{car} ratio (Table 1) indicate that most of the precipitates formed in the presence of HA were composed chiefly of hydroxy-Al-humate complexes.

The IR spectrum of the original precipitates (Figure 4A-a) shows that they were mainly composed of "proto-imogolite" allophane-humate complexes (bands at 1615, 1400, 1280, 949, 565, 424, and 345 cm^{-1}). The IR spectrum of the pyrophosphate-soluble fraction of the precipitates formed in the presence of HA (Figure 4A-b) is similar to that of the original precipitates, indicating the predominance of the pyrophosphate-soluble fraction in the sample. The IR spectrum of the Na_2CO_3-soluble fraction of the precipitates formed in the presence of HA (Figure 4A-c) shows that it also contained "proto-imogolite" allophane-humate complexes. The pyrophosphate- and Na_2CO_3-insoluble fraction shows a weak IR absorption spectrum (Figure

4A-d). On the other hand, the IR spectrum of the original precipitates formed in the presence of FA-I (Figure 4B-a) is similar to that of the pyrophosphate- and Na_2CO_3-insoluble fraction (Figure 4B-d). The pyrophosphate- and Na_2CO_3-insoluble fraction was a major component of the precipitates formed in the presence of FA-I (Table 1). The data indicate that this fraction consisted predominantly of "proto-imogolite" allophane-fulvate complexes.

The IR spectrum of the Na_2CO_3-soluble fraction of the precipitates formed in the presence of FA-I (Figure 4B-c) is similar to that of the Na_2CO_3-soluble fraction of the precipitates formed in the presence of HA (Figure 4A-c). These fractions contained "proto-imogolite" allophane-humic substance complexes (948 or 943 cm^{-1}). In addition to these complexes, the Na_2CO_3 treatment after the pyrophosphate extraction resulted in the removal of adsorbed pyrophosphate as shown in the broad absorption band at about 1150 cm^{-1} (Figures 4A-c and 4B-c). Only a small amount of "proto-imogolite" allophane-humic substance complexes appeared to be present in the pyrophosphate-soluble fraction formed in the presence of FA-I (Figure 4B-b).

Electron micrographs. In the absence of humic substances, the precipitates consisted of irregular aggregates (Figure 5a), in which allophane was dominant (Figure 3a). Electron micrographs of the precipitates formed in the presence of humic substances (Figures 5d–5i), however, show a different morphology from microaggregates of allophane (Figure 5a). The precipitates formed at the 100 mg/liter level of humic substances (Figures 5d–5f) had a cross-linked morphology of small particles and/or aggregates. The precipitates formed at the concentration of 1000 mg/liter HA (Figure 5g) show an irregular morphology and are characterized by the predominance of hydroxy-Al-humate complexes as suggested by IR (Figure 3e). Electron micrographs of the precipitates formed at a concentration of 1000 mg/liter FA (Figures 5h and 5i) show cross-linked, rigid, and spheroidal particles having diameters of 0.02–0.145 μm. Ghosh and Schnitzer (1980) reported that the macromolecular configurations of HA and FA molecules vary from rigid spherocolloidal to flexible linear, depending upon their concentration, the pH of the system, and the ionic strength of the medium. Hydroxy-Al-humic substance complexes and "proto-imogolite" allophane-humic substance complexes formed show no flexible linear morphology, but rather a cross-linked, rigid, spheroidal morphology or a cross-linked, irregular morphology, different from those of allophane (Figure 5a) and humic substances (Figures 5b and 5c).

Pedogenic implications

Shoji and Masui (1972), Tokashiki and Wada (1975), Wada and Higashi (1976), and Higashi and Wada (1977)

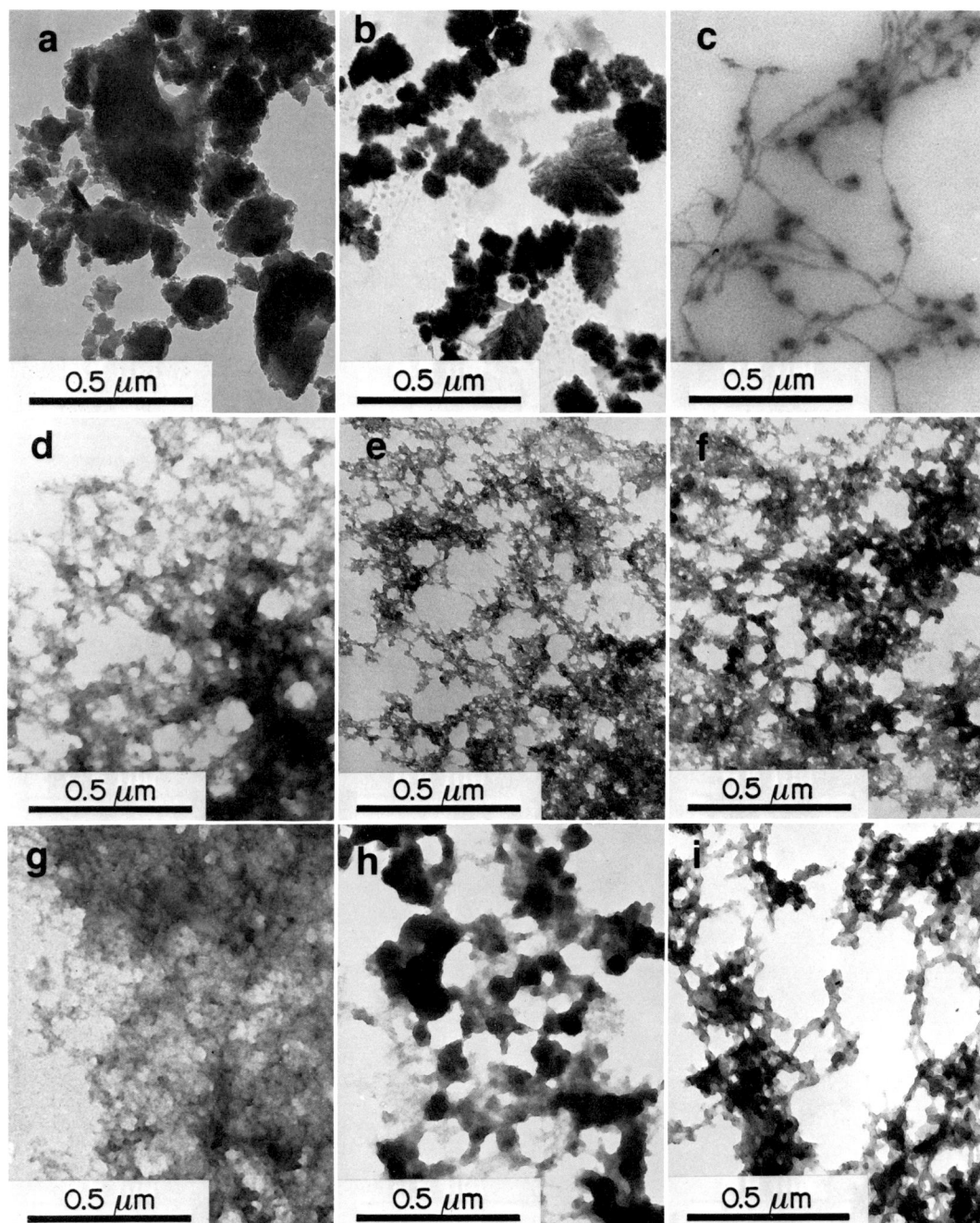

Figure 5. Transmission electron micrographs of humic substances and precipitates formed by reaction of hydroxy-Al ions with orthosilicic acid both in the absence and presence of humic substances. (a) no humic substances; (b) Na-humate (pH 9.5); (c) fulvic acid-I (pH 4.0); (d) humic acid, 100 mg/liter; (e) fulvic acid-I, 100 mg/liter; (f) fulvic acid-II, 100 mg/liter; (g) humic acid, 1000 mg/liter; (h) fulvic acid-I, 1000 mg/liter; and (i) fulvic acid-II, 1000 mg/liter.

suggested that the accumulation of humus in the Dystrandept A₁ horizon may favor the formation of opaline silica, if the supply of silica is plentiful, and may retard the formation of "allophane-like" constituents, allophane, and imogolite, particularly the last two. The

formation of imogolite from the solutions containing hydroxy-Al ions and orthosilicic acid was strongly perturbed by humic substances (Inoue and Huang, 1984b).

The present study, as summarized in Table 2, reveals that humic substances inhibited the formation of allophane, leading to the formation of hydroxy-Al-humic substance complexes and/or "proto-imogolite" allophane. Al released from parent volcanic ash by weathering is strongly bound to humic substances, which limits the coprecipitation of Si and Al (Wada and Higiashi, 1976; Higashi and Wada, 1977). Previous results (Inoue and Huang, 1984a, 1984b, 1985) and the present findings substantiate the hypothesis that the accumulation of humus in the Dystrandepts A_1 and Spodosol A_2 and B_h horizons favors the formation of hydroxy-Al-humus complexes, "allophane-like" constituents, and/or opaline silica and may retard the formation of allophane and imogolite, which are formed in the Dystrandept B and (B) and Spodosol B_s horizons (Wada and Higashi, 1976; Higashi and Wada, 1977; Farmer et al., 1980, 1983; Farmer, 1981, 1984; Anderson et al., 1982).

ACKNOWLEDGMENTS

We are indebted to M. Schnitzer for kindly providing the fulvic acid (FA-II) sample purified from a Podzol B_h horizon in Prince Edward Island, Canada. The research was supported by the Natural Sciences and Engineering Research Council of Canada Grants A2348- and G1296—Huang.

REFERENCES

Anderson, H. A., Berrow, M. L., Farmer, V. C., Hepburn, A., Russell, J. D., and Walker, A. D. (1982) A reassessment of podzol formation processes: J. Soil Sci. 33, 125–136.

Farmer, V. C. (1981) Possible roles of a mobile hydroxy-aluminium orthosilicate complex (proto-imogolite) and other hydroxyaluminium and hydroxy-iron species in podzolization: in Migrations Organominerales dans les Sols Tempérés, Colloques Internationaux du C.N.R.S. No. 303, 275–279.

Farmer, V. C. (1984) Distribution of allophane and organic matter in podzol B horizons: reply to Buurman and Reeuwijk: J. Soil Sci. 35, 453–458.

Farmer, V. C., Russell, J. D., and Berrow, M. L. (1980) Imogolite and proto-imogolite allophane in spodic horizons: evidence for a mobile aluminium silicate complex in podzol formation: J. Soil Sci. 31, 673–684.

Farmer, V. C., Russell, J. D., and Smith, B. F. L. (1983) Extraction of inorganic forms of translocated Al, Fe and Si from a podzol Bs horizon: J. Soil Sci. 34, 571–576.

Ghosh, K. and Schnitzer, M. (1980) Macromolecular structures of humic substances: Soil Sci. 129, 266–276.

Greenland, D. J. (1965) Interaction between clays and organic compounds in soils. Pt. 1. Mechanisms of interaction between clays and defined organic compounds: Soils & Fertilizers 28, 415–425.

Hansen, E. H. and Schnitzer, M. (1969) Molecular weight measurements of polycarboxylic acids in waters by vapour pressure osmometry: Anal. Chem. Acta 46, 247–254.

Higashi, T. and Wada, K. (1977) Size fractionation, dissolution analysis, and infrared spectroscopy of humus complexes in Ando soils: J. Soil Sci. 28, 653–663.

Inoue, K. and Huang, P. M. (1984a) Influence of citric acid on the natural formation of imogolite: Nature 308, 58–60.

Inoue, K. and Huang, P. M. (1984b) Effect of humic and fulvic acids on the formation of imogolite: Agronomy Abstracts, 76th Ann. Meet. ASA, CSSA, SSSA, Las Vegas, Nevada, 1984, p. 273.

Inoue, K. and Huang, P. M. (1985) Influence of citric acid on the formation of short-range ordered aluminosilicates: Clays & Clay Minerals 33, 312–322.

Kumada, K. (1981) Chemistry of Soil Organic Matter: 2nd ed., Japan Scientific Societies Press, Tokyo, 304 pp. (in Japanese).

Parfitt, R. L. and Henmi, T. (1980) Structure of some allophanes from New Zealand: Clays & Clay Minerals 28, 285–294.

Schnitzer, M. (1977) Recent findings on the characterization of humic substances extracted from soils from widely differing climatic zones: in Soil Organic Matter Studies, Vol. II, International Atomic Energy Agency, Vienna, 117–132.

Schnitzer, M. and Skinner, S. I. M. (1968) Alkali versus acid extraction of soil organic matter: Soil Sci. 105, 392–396.

Shoji, S. and Masui, J. (1972) Amorphous clay minerals of recent volcanic ash soils. Pt. 3. Mineral composition of fine clay fractions: J. Sci. Soil Manure Japan 43, 187–193.

Tokashiki, Y. and Wada, K. (1975) Weathering implications of the mineralogy of clay fractions of two Ando soils, Kyushu: Geoderma 14, 47–62.

Wada, K. (1977) Allophane and imogolite: in Minerals in Soil Environments, J. B. Dixon and S. B. Weed, eds., Soil Sci. Soc. Amer., Madison, Wisconsin, 603–638.

Wada, K. (1980) Mineralogical characteristics of Andisols: in Soils with Variable Charge, B. K. G. Theng, ed., New Zealand Soc. Soil Sci., Palmerston North, 87–107.

Wada, K. and Harward, M. E. (1974) Amorphous clay constituents of soils: Adv. Agron. 26, 211–260.

Wada, K. and Higashi, T. (1976) The categories of aluminium- and iron-humus complexes in Ando soils determined by selective dissolution: J. Soil Sci. 27, 357–368.

Wada, K. and Tokashiki, Y. (1972) Selective dissolution and difference infrared spectroscopy in quantitative mineralogical analysis of volcanic-ash soil clays: Geoderma 7, 199–213.

Proceedings of the International Clay Conference, Denver, 1985, L. G. Schultz, H. van Olphen, and F. A. Mumpton, eds.,
The Clay Minerals Society, Bloomington, Indiana, 227–230 (1987).

MINERALOGY AND GEOCHEMISTRY OF LIMONITE FROM THE LONE STAR IRON ORES, EAST TEXAS

Annabelle M. Foos

Department of Geology, University of Akron
Akron, Ohio 44325

Abstract—The mineralogy and geochemistry of limonite from the Middle Eocene Lone Star iron ores of east Texas were investigated. X-ray powder diffraction analysis indicates that the major minerals are goethite, kaolinite, and quartz, with minor amounts of lepidocrocite. Electron microprobe data show an excellent correlation (r = .98) between Al_2O_3 and SiO_2, with both oxides inversely proportional to Fe_2O_3 content. These relationships indicate a mixture of goethite and kaolinite. All the Al in the limonite ore can be accounted for by the kaolinite, suggesting that goethite does not contain appreciable amounts of Al. The degree of Al substitution in goethite as determined by the position of the 111 goethite reflections indicates that Al substitution is <5%, in agreement with the conclusion based on the microprobe data.

The meager Al substitution in goethite and the presence of lepidocrocite indicates that the Lone Star limonite formed in a hydromorphic environment. Two processes are responsible for the precipitation of goethite; *in situ* oxidation of primary iron-rich minerals (berthierine and siderite) and oxidation of Fe^{+2} in solution.

Key Words—Al-substitution, Goethite, Hydromorphic environment, Kaolinite, Lepidocrocite, Limonite.

INTRODUCTION

The Lone Star iron ores of east Texas occur in the Middle Eocene Weeches Formation of the Gulf coastal plain. Three facies have been recognized in the Weeches Formation at Lone Star, Texas: greensand, sideritic, and limonite facies (Foos, 1984). The greensands are composed of varying amounts of fecal pellets, berthierine ooids, kaolinite plus berthierine matrix, siderite cement, and detrital quartz (Figure 1). The sideritic facies is composed of rocks containing >50% siderite and which range from sideritic greensands and mudstones (50–80% siderite) to dense siderite nodules (>80% siderite). Other minerals in the sideritic facies include berthierine, kaolinite, quartz, and pyrite (Foos, 1984). In outcrop the limonite facies always occurs above the greensand and sideritic facies. Previous workers (Eckel, 1938; Roe, 1961) have suggested that the water table controls the distribution of these facies, with the greensand and sideritic facies occurring below and limonite facies occurring above the water table.

The focus of the present paper is the limonite facies, which apparently formed by alteration of the greensand and sideritic facies in the near-surface environment. The mineralogy and geochemistry of the limonite and the processes involved in alteration of primary iron-rich minerals are discussed.

SAMPLING AND ANALYTICAL TECHNIQUES

Samples of the Lone Star iron ores were collected from mining faces on the property of Lone Star Steel Company in Morris and Cass Counties, northeast Texas. Fresh samples were collected in which oxidation of the primary iron-rich minerals had occurred prior to exposure by the mining operations. Standard petrographic techniques were used to study the texture of the ores. The mineralogy of the ores was determined by X-ray powder diffraction (XRD) on a Philips APD-3600 instrument with Ni-filtered CuKα radiation. The bulk chemistry of the samples was determined by X-ray fluorescence (XRF) on a Philips PW1400 X-ray spectrometer. Modal estimates were calculated from the XRF data. The chemistry of the individual phases in the limonite was determined with wave-length dispersive spectrometry on an ARL-EMX-SEM electron microprobe, operated at 15 kV and 0.15 mA beam current. A Bence-Albee correction was applied to the data to correct for fluorescence, absorption, and atomic number effects (Bence and Albee, 1968).

RESULTS

Thin section petrography indicates that texturally, two distinct types of limonite are present—massive-to-nodular limonite and goethite-replaced greensands. The major component of the massive-to-nodular limonite was found to be laminated, finely crystalline goethite (80%), accompanied by pellets replaced by goethite (14%), and detrital quartz (6%) (Figure 2). The samples were porous to vuggy, and the interior of vugs and fractures were commonly lined with fibrous goethite. The goethite-replaced greensands were texturally similar to the greensands (compare Figures 1 and 3) and were composed of pellets replaced by goethite (50%) and detrital quartz (4%) in a matrix of goethite + kaolinite (46%). Pseudomorphs of goethite after siderite

227

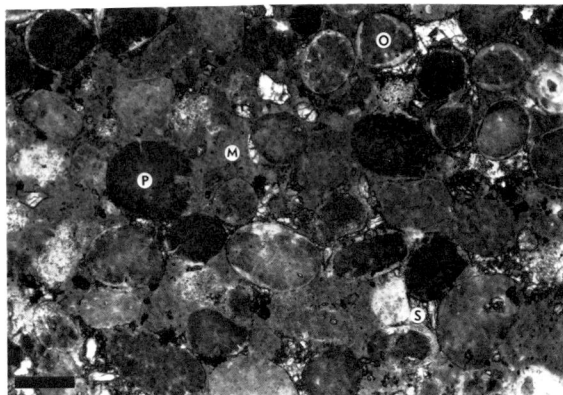

Figure 1. Photomicrograph of greensand from the Weeches Formation, east Texas. P = berthierine pellet; O = berthierine ooid; M = berthierine + kaolinite matrix; S = siderite; plane light; scale bar = 0.3 mm.

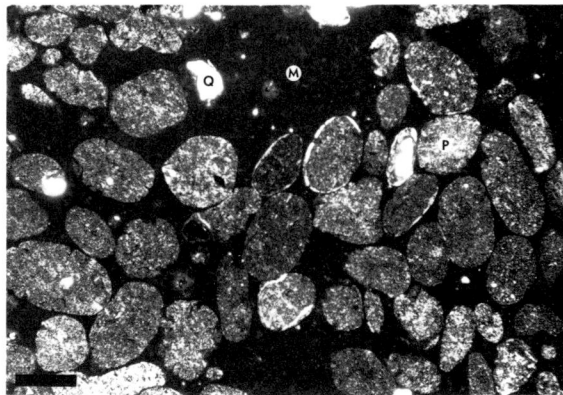

Figure 3. Photomicrograph of goethite-replaced greensand. P = goethite pellet; M = goethite + kaolinite matrix; Q = detrital quartz; crossed nicols; scale bar = 0.3 mm.

were observed in some of the goethite-replaced greensands.

Qualitative XRD analysis indicated that the major minerals in the limonite are goethite, kaolinite, and quartz, with minor amounts of lepidocrocite.

XRF data are summarized in Table 1. Relative to the greensands (column III), the limonites (columns I and II) are depleted in MgO, Al_2O_3, SiO_2, K_2O, P_2O_5, CaO, TiO_2, and MnO and enriched in Fe_2O_3. The difference between the whole-rock chemical composition of massive-to-nodular limonites and goethite-replaced greensands is not significant. Modal analysis indicated that the goethite content ranged from 38 to 88% (mean = 72%); kaolinite ranged from 8 to 27% (mean = 16%); and quartz ranged from 4 to 35% (mean = 13%). No relationship was found between the limonite textural type and the relative concentrations of goethite, clay, and quartz.

One objective of the microprobe analysis was to de-

termine the distribution of Al, Si, and Fe in the limonite ores. The data show an excellent correlation (r = .98) between the Al and Si contents; moreover both oxides are inversely proportional to the Fe content (Figure 4). This relationship indicates that the microprobe analyses represent a mixture of primarily two phases, kaolinite and goethite. Four texturally distinct types of limonite were analyzed: goethite-replaced pellets, finely crystalline goethite matrix, laminated goethite, and fibrous goethite (Table 2). The laminated and fibrous goethite in the massive-to-nodular limonite contains the least Al and Si and the most Fe. The goethite-replaced pellets and finely crystalline goethite matrix that are most common in goethite-replaced greensands are rich in Al and Si. This relationship suggests that kaolinite and goethite are more highly segregated in the massive-to-nodular limonites.

The average Al/Si atomic ratio of the goethite-replaced pellets and finely crystalline goethite matrix are 0.83 and 0.90 respectively. These values are signifi-

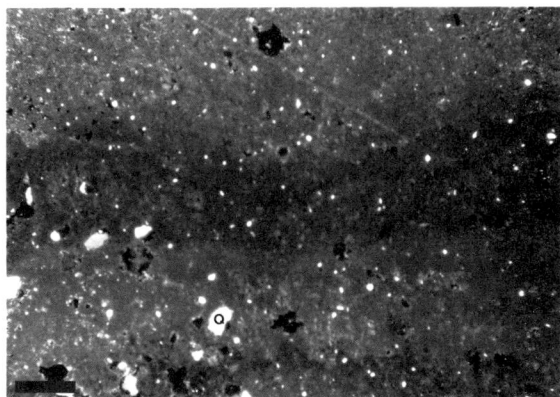

Figure 2. Photomicrograph of massive-to-nodular limonite showing laminated, finely crystalline goethite. Q = detrital quartz; crossed nicols; scale bar = 0.3 mm.

Table 1. Average X-ray fluorescence analyses of limonites and greensands from the Lone Star iron ores.

Weight percent	I	II	III
MgO	0.04	0.08	1.46
Al_2O_3	6.43	6.59	12.67
SiO_2	22.09	18.68	35.34
Fe_2O_3[1]	63.87	68.76	32.86
K_2O	0.17	0.25	0.63
P_2O_5	0.16	0.19	0.23
CaO	0.08	0.09	0.42
TiO_2	0.33	0.28	0.42
MnO	0.05	0.05	0.11
Total[2]	93.22	94.97	82.72
No. of samples	8	8	9

I = massive-to-nodular limonite; II = goethite-replaced greensand; III = greensand.
[1] Total iron expressed as Fe_2O_3.
[2] Difference from 100 is water.

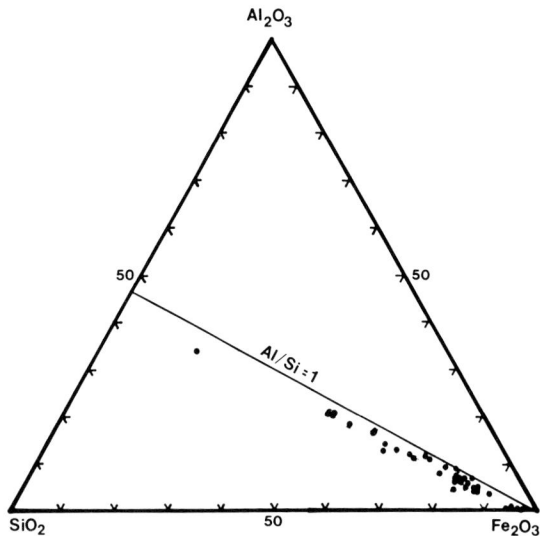

Figure 4. Triangular diagram showing the relationship between the weight percentages of Al_2O_3, SiO_2, and Fe_2O_3 in limonite from the Lone Star iron ores.

Table 2. Electron microprobe analyses of limonite from the Lone Star iron ores.

Weight percent	I	II	III	IV
MgO	0.30	0.23	0.09	—
Al_2O_3	6.63	11.81	1.34	0.71
SiO_2	9.44	15.61	3.07	1.84
K_2O	0.38	0.24	0.05	0.02
TiO_2	0.19	0.18	0.08	0.05
MnO	0.09	0.08	0.15	0.11
Fe_2O_3[1]	69.92	57.10	81.16	84.39
Total[2]	86.95	85.25	85.94	87.16
Formula weight[3]				
Al	1.48	2.45	0.33	0.17
Si	1.79	2.74	0.64	0.38
Fe	10.14	7.89	12.81	13.31
Al/Si	0.83	0.90	0.52	0.45
No. of analyses	20	10	8	6

I = goethite replaced pellets; II = finely crystalline goethite matrix; III = laminated goethite; IV = fibrous goethite.
[1] Total iron expressed as Fe_2O_3.
[2] Difference from 100 is water.
[3] Calculated to an anhydrous base of 21 oxygens.

cantly lower than the theoretical value for kaolinite (1.00). The low Al/Si ratio in these samples could result from the presence of a silica phase or the substitution of Fe^{3+} for Al in the kaolinite structure (Fysh et al., 1983). The average Al/Si ratio of the laminated and fibrous goethite are 0.52 and 0.45, respectively. These values are half that of the theoretical value for kaolinite suggesting that, along with goethite and kaolinite, a silica phase is present. The reason the Al/Si ratios plotted in Figure 4 appear to follow so closely the ideal ratio for kaolinite is that samples which deviate most from the 1.0 ideal value are all laminated or fibrous goethite (columns III and IV in Table 2). These samples contain small amounts of kaolinite and plot near the Fe_2O_3 corner of the triangle; here, even a few percent of a silica mineral distorts the Al/Si ratio considerably, but the plotted points on Figure 4 are still nearly parallel to and all below the ideal ratio.

The silica phase present is most likely either finely crystalline quartz or opal-CT. The amount of quartz estimated from the XRF data is higher than the amount determined petrographically. This excess quartz could be present as finely crystalline quartz, which might not be seen petrographically, rather than detrital quartz. The other option is that the silica phase is opal-CT. Opal-CT was not detected on the XRD patterns; however, the XRD pattern of opal-CT is characteristically weak, so the detection limit is high—probably several percent. Opal-CT was observed lining a vug in one of the samples.

An interpretation of the microprobe data is that all of the Al in the Lone Star limonite occurs in the kaolinite phase. This interpretation was tested with X-ray powder diffraction. The amount of Al substitution in the goethite was estimated using the Vegard's rule for the 111 reflection (Fitzpatrick and Schwertmann, 1982). The results indicate that the Al substitution ranged from 0 to 4.5 mole %. Considering the accuracy of this method (±4%, Schulze, 1984), these values support the microprobe data.

DISCUSSION

The degree of Al substitution in goethite reflects the geochemical environment of formation. Fitzpatrick and Schwertmann (1982) observed that moderately acid, hydromorphic soils and calcareous environments are characterized by small amounts of Al substitution, but that nonhydromorphic environments are characterized by goethite with large amounts of Al substitution. The small amount of Al substitution in goethite from the Lone Star iron ores suggest that the limonite formed in a hydromorphic environment. This interpretation is supported by the presence of lepidocrocite in these samples which is an indicator of hydromorphic conditions (Schwertmann and Taylor, 1977). The hydromorphic environment is characterized by horizons that are water-saturated either permanently or periodically. In this environment, reducing conditions bring about mobilization of iron as Fe^{2+} and reprecipitation upon reoxidation (Van Breemen and Brinkman, 1976).

The source of iron in the Lone Star limonite is berthierine, with a minor contribution from siderite in the greensands. Both of these minerals are stable under reducing conditions, but unstable in an oxidizing environment (Maynard, 1983). Fluctuating oxidizing and reducing conditions are necessary for the mobilization of iron in the greensands. Oxidation of berthierine would break it down into goethite + kaolinite. This reaction, followed by reduction of Fe^{3+}, would mobilize the iron until it was reprecipitated as goethite in areas of higher oxidation potential.

Two processes, thus, appear to be responsible for the precipitation of goethite in the Lone Star iron ores: the *in situ* oxidation of primary iron-rich minerals, and the oxidation of Fe^{2+} in solution. Both processes were apparently active for the massive-to-nodular limonite and the goethite-replaced greensands; however, the relative importance of these processes differed for the two textural types. *In situ* oxidation of berthierine and siderite was more important in the goethite-replaced greensands relative to the massive-to-nodular limonite.

REFERENCES

Bence, A. E. and Albee, A. L. (1968) Empirical correction factors for the electron microanalysis of silicates and oxides: *J. Geol.* **76**, 382–403.

Eckel, E. B. (1938) The brown iron ores of eastern Texas: *U.S. Geol. Surv. Bull.* **902**, 157 pp.

Fitzpatrick, R. W. and Schwertmann, U. (1982) Al-substituted goethite—an indicator of pedogenic and other weathering environments in South Africa: *Geoderma* **27**, 335–347.

Foos, A. M. (1984) The mineralogy, petrography, and geochemistry of the Eocene Lone Star iron ores, east Texas and the Ordovician Hooker ironstone, northwest Georgia: Ph.D. dissertation, Univ. Texas at Dallas, Dallas, Texas, 260 pp.

Fysh, S. A., Cashion, J. D., and Clark, P. E. (1983) Mössbauer effect studies of iron in kaolin. I Structural iron: *Clays & Clay Minerals* **31**, 285–292.

Maynard, J. B. (1983) *Geochemistry of Sedimentary Ore Deposits:* Springer Verlag, New York, 309 pp.

Roe, D. G. (1961) Mineralogy and some trace element studies of east Texas iron ores: M.A. Thesis, Texas Christian Univ., Fort Worth, Texas, 62 pp.

Schulze, D. G. (1984) The influence of aluminum on iron oxides. VIII. Unit-cell dimensions of Al-substituted goethites and estimation of Al from them: *Clays & Clay Minerals* **32**, 36–44.

Schwertmann, U. and Taylor, R. M. (1977) Iron oxides: in *Minerals in Soil Environments,* J. B. Dixon and S. B. Weed, eds., Amer. Soc. Agron., Madison, Wisconsin, 145–180.

Van Breemen, N. and Brinkman, R. (1976) Chemical equilibria and soil formation: in *Soil Chemistry. A. Basic Elements,* G. H. Bolt and M. G. M. Buggenwert, eds., Elsevier, New York, 213 pp.

Proceedings of the International Clay Conference, Denver, 1985, L. G. Schultz, H. van Olphen, and F. A. Mumpton, eds.,
The Clay Minerals Society, Bloomington, Indiana, 231–236 (1987).

EFFECT OF PHOSPHATE ANION ON THE
FORMATION OF IMOGOLITE

T. Henmi[1] and P. M. Huang

Department of Soil Science, University of Saskatchewan
Saskatoon, Saskatchewan S7N 0W0, Canada

Abstract—The influence of phosphate anion on the interaction of hydroxy-Al ions and monomeric silicic acid was studied in systems having an initial Si concentration of 1.54×10^{-3} M, a Si/Al molar ratio of 0.5, an OH/Al molar ratio of 2.0, and P/Al molar ratios of 0–0.04. Parent solutions were heated to 95°–100°C for 110 hr (1 atm), and the precipitation (>0.01 μm) and soluble products (<0.01 μm) formed were examined by X-ray powder diffraction, infrared spectroscopic, electron optical, and chemical analyses. The amount of imogolite in the precipitates decreased with increasing the P/Al molar ratio of the parent solution from 0 to 0.005. The tube length and the degree of order of imogolite formed decreased as the P/Al molar ratio of the parent solution increased. Imogolite was not found in the precipitates formed in the solution at a P/Al molar ratio of 0.01. As the P/Al molar ratio of the solution rose, the amount of "proto-imogolite" (soluble complexes of hydroxy-Al ions and orthosilicic acid) in freeze-dried soluble products increased; bayerite and/or boehmite also were present in the precipitates. No material precipitated from the solution having a P/Al molar ratio of 0.04. The phosphate anion apparently strongly inhibits the growth of "proto-imogolite" nuclei and the subsequent formation of imogolite tube structure. The present study indicates that phosphate anion, common in soil solutions and natural waters, affects the genesis of imogolite in nature.

Key Words—Hydroxy-Al, Imogolite, Infrared spectroscopy, Phosphate, "Proto-imogolite," Transmission electron microscopy, X-ray powder diffraction.

INTRODUCTION

Poorly ordered hydrous aluminosilicates, such as allophanes and imogolite, are commonly the principal mineral components of the clay fraction of Andisols and Podzols (Wada, 1980; Farmer, 1982; Parfitt and Henmi, 1982; Ross and Kodama, 1979). These hydrous aluminosilicates affect the physical and chemical properties of the soils: namely, cation- and anion-exchange capacities, ion-retention, water holding capacities, consistency, and the degree of cementation of subsoils (Wada, 1977; McKeague and Kodama, 1981; Warkentin and Maeda, 1980; Farmer et al., 1984). Therefore, the genesis and transformation of these hydrous aluminosilicates in soils are important subjects in soil and environmental sciences.

Imogolite occurs as long tubular fibers and is formed by the growth of "proto-imogolite" nuclei which itself forms by the interaction of hydroxy-Al ions and orthosilicic acid at pH of 3.5–5.0 (Farmer and Fraser, 1979). The formation of imogolite is hindered by the presence of low-molecular-weight organic ligands, such as citric acid (Inoue and Huang, 1984). In addition to organic ligands, inorganic ligands, such as phosphate, may also inhibit the formation of imogolite. Phosphate

in soil solutions and natural waters originates from parent rocks (Lindsay and Vlek, 1977), applied fertilizers (Cooke and Williams, 1973), and domestic sewage (Devey and Harkness, 1973). The objective of the present study was to investigate the effect of phosphate on the formation of imogolite under well controlled laboratory conditions.

MATERIALS AND METHODS

Monomeric orthosilicic acid solution was prepared by passing Na-metasilicate through a H^+-saturated Dowex 50W-X8 cation-exchange resin (Luciuk and Huang, 1974). The initial concentration of silicic acid solution was kept below 0.002 M Si to avoid polymerization of orthosilicic acid. Aliquots of 0.2 M $AlCl_3$ solution and 0.0005–0.01 M NaH_2PO_4 solutions were simultaneously mixed with the orthosilicic acid solution to yield a Si/Al molar ratio of 0.5 and P/Al molar ratios that ranged from 0 to 0.04. The P concentrations of the reaction systems were selected to simulate those of soil solution, which are of the order of 0.1–1 ppm (10^{-5}–10^{-6} M) in nature (Bohn et al., 1979). The solution mixtures (about 900 ml) were titrated with 0.1 M NaOH (with continuous stirring) at a rate of about 0.5 ml NaOH/min to an OH/Al molar ratio of 2.0, and then diluted to 1000 ml. The Si concentration of the resulting solution (parent solution) was 0.00154 mole Si/liter. All parent solutions had pHs of 4.02–4.04. About 900 ml of each parent solution was heated,

[1] Permanent address: Department of Agricultural Chemistry, Faculty of Agriculture, Ehime University, Matsuyama 790, Japan.

Table 1. Influence of phosphate anion on the interaction of hydroxy-Al ions with orthosilicic acid.

P/Al	pH(i)[1]	pH(f)[2]	Si and Al removed from solution (%)		Si/Al molar ratio of precipitates	P/Al molar ratio in precipitates	Imogolite (%)	Mineral composition of precipitates[3]
			Si	Al				
0	4.04	3.01	62.8	78.3	0.42	0	88	Im, By, Bm
0.0025	4.02	3.30	47.1	61.4	0.40	1.6×10^{-3}	79	Im, By, Bm
0.005	4.03	3.32	35.3	50.6	0.36	4.4×10^{-3}	52	Im, By, Bm
0.01	4.02	3.41	7.6	21.7	0.18	21.4×10^{-3}	0	By, Bm, X
0.04	4.02	3.58	0	0	NP[4]	NP[4]	0	NP[4]

[1] Initial pH of the parent solution before heating.
[2] Final pH of the parent solution after heating at 95°–100°C for 110 hr.
[3] Im = imogolite, By = bayerite, Bm = boehmite, X = unknown, poorly ordered aluminosilicates. Mineral composition of precipitates was determined by X-ray powder diffraction analysis and infrared spectroscopy.
[4] No precipitate (>0.01 μm).

using reflux condensers, at 95°–100°C on a hotplate for 110 hr.

After cooling to room temperature, the pH of the suspensions was measured, and the precipitates formed were separated from the filtrate by ultrafiltration using a Sartorius cellulose nitrate membrane filter having a pore size of 0.01 μm. The collected precipitates were dialyzed using cellulose tubes having a molecular weight cut-off of 3500 against deionized water until they were free of Cl ion.

A portion of each precipitate was examined by X-ray powder diffraction (XRD) methods using glass slide mounts and a Philips X-ray diffractometer and Ni-filtered CuKα radiation generated at 35 kV and 16 mA. For electron optical observations, a drop of a diluted suspension of the precipitate was spread on a copper grid covered with Formvar film reinforced by a coating of carbon. Transmission electron micrographs were taken at 100 kV with a JEM 7A microscope. The rest of the precipitate was freeze-dried and analyzed by infrared (IR) spectroscopy. IR spectra of 13-mm KBr disks (170 mg) that contained 1.0 mg of the precipitate were recorded on a Perkin-Elmer 983 IR spectrometer. The total amounts of Al, Si, and P in the parent solutions and the filtrates were determined by colorimetry. Al was measured by the aluminon method (Hsu, 1963), Si by the silicomolybdate method using an amino sulfonic acid reducing agent (Weaver et al., 1968), and P by the molybdate blue method, following the procedure of Jackson (1958). The percentage of Al and Si removed from parent solutions was estimated from the difference between the Al and Si concentrations in the parent solutions and those in the filtrates. The amount of P incorporated into the precipitates was calculated by subtracting the P concentration of the filtrates from that of the parent solutions.

The soluble products in filtrates which passed through the membrane filter (pore size: 0.01 μm) were freeze-dried, dialyzed in the same way as the precipitates, and then freeze-dried again to obtain powder samples for examination by IR and XRD.

RESULTS AND DISCUSSION

Table 1 shows the pH of the parent solutions in the absence and in the presence of phosphate anion both before and after the solutions were heated. In the absence of phosphate anions and on heating, the pH of the parent solutions decreased markedly from 4.04 to 3.01, and a transparent, colloidal precipitate formed. The decrease of pH is attributed chiefly to the reactions involved in the hydrolysis and polymerization of $Al(H_2O)_6^{3+}$ and the interaction of the hydroxy-Al ions with orthosilicic acid. The decrease in pH of the parent solutions was less as the P/Al molar ratio increased, suggesting that the phosphate anion interfered with these chemical reactions in the solution upon heating. The phosphate ion may also have served as a buffer to accept protons in the systems.

The percentage of Si and Al removed from the solutions by the ultrafiltration procedure decreased with increasing P/Al molar ratio of the solution (Table 1). No precipitate was collected by ultrafiltration from the solutions having a P/Al molar ratio of 0.04 (Table 1), suggesting that precipitation was hindered by the presence of phosphate anion. The Si/Al molar ratio of the products decreased as the P/Al molar ratio of the parent solutions increased. The amount of P incorporated into the precipitates increased with an increase of the P/Al molar ratio of the parent solutions (Table 1), suggesting that part of the added phosphate anion was complexed with and/or adsorbed on the precipitates as they formed.

The IR spectrum of the precipitate that formed from the parent solution without phosphate anion (Figure 1a) closely resembles that of natural (Wada, 1980) and synthetic (Farmer and Fraser, 1979) imogolite. The absorption maxima at 990 and 935 cm^{-1}, assigned to Si–O-stretching vibrations, are attributed to the tubular morphology and the orthosilicate anion in the structure of imogolite, respectively (Cradwick et al., 1972). The bands at 590 and 693 cm^{-1} are probably due to an octahedral sheet resembling a gibbsitic sheet which makes up the framework of the wall of imogolite

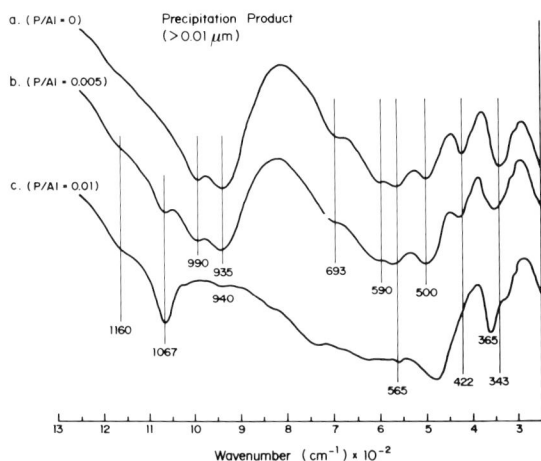

Figure 1. Infrared spectra of precipitates of the reaction of hydroxy-Al ions and orthosilicic acid as influenced by phosphate.

tube. The bands at 500, 422, and 343 cm^{-1} are prominent in the spectrum of natural imogolite.

The XRD pattern (Figure 2a) shows that the main component of the precipitate formed in the parent solutions in the absence of phosphate was imogolite (strong peaks at 8.67 and 6.43 Å). Small amounts of bayerite (4.71 Å) and boehmite (6.12 Å) (Hsu, 1977) are also indicated (Table 1). The difference in the Si/Al molar ratio of the precipitate (0.42) from that of imogolite (0.5) can be explained by the admixture of such small amounts of bayerite and boehmite in the precipitates.

The characteristic IR absorption bands of imogolite formed in the parent solutions with P/Al molar ratios of 0.005 and 0.01 are distorted and/or featureless (Figure 1), and the imogolite content, as estimated from the absorbance near 343 cm^{-1} (Farmer et al., 1977), decreased (Table 1) as the P/Al ratio of the solutions rose. At a P/Al molar ratio of 0.01, features of imogolite were absent in the IR spectrum (Figure 1c) and the XRD pattern (Figure 2c) of the precipitate. These results indicate that the formation of imogolite was strongly inhibited by phosphate anions.

Compared with the organic ligand, citric acid (Inoue and Huang, 1984), phosphate more strongly inhibited the formation of imogolite. The IR spectrum (Figure 1c) is similar to that of boehmite (strong absorption maxima at 1067, 750–480, and 365 cm^{-1}, and a shoulder at 1160 cm^{-1}) (Wada et al., 1979), but the XRD pattern (Figure 2c) shows only the prominent reflection of bayerite (4.71 Å). Thus, the boehmite in the precipitate at a P/Al molar ratio of 0.01 was poorly ordered pseudoboehmite. The intensity of the IR absorption near 3138 cm^{-1} also increased with an increase in the P/Al molar ratio (not shown). This observation coupled with the increase in the intensity of the absorbance

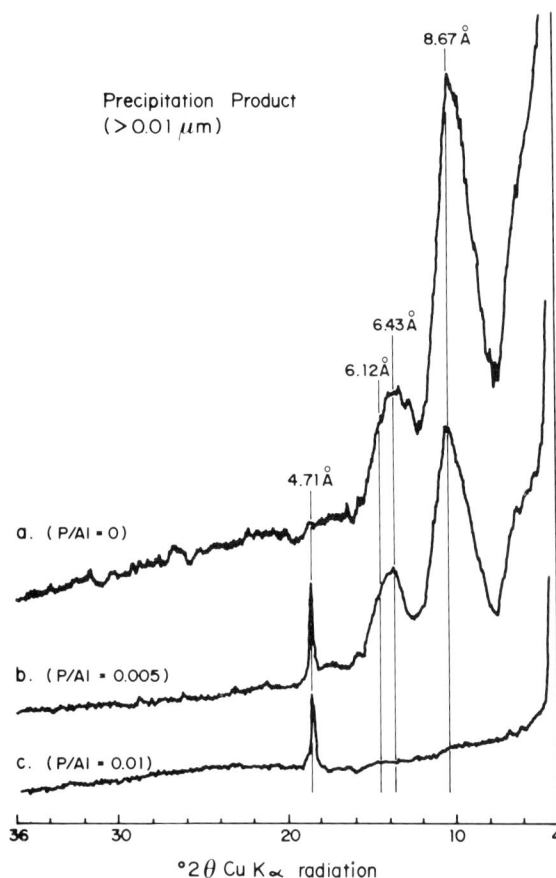

Figure 2. X-ray powder diffraction patterns of precipitates of the reaction of hydroxy-Al ions and orthosilicic acid as influenced by phosphate.

near 1067 cm^{-1} as the P/Al ratio rose indicates that the formation of poorly ordered pseudoboehmite was promoted by increasing the P/Al ratio of the system. In addition, the spectrum (Figure 1c) contains weak but significant absorption bands at about 940 and 565 cm^{-1}, indicating that small amounts of noncrystalline or poorly ordered aluminosilicates may have been present in the precipitate formed from the solution having a P/Al molar ratio of 0.01. The Si/Al molar ratio of the product (0.18) (Table 1) supports this possibility.

The precipitate from the parent solution in the absence of phosphate anion consisted mostly of long threads several micrometers in length (Figure 3A). The magnified view of the threads (Figure 4) shows that they are bundles of fine tubes 25 Å in diameter, characteristic of imogolite (Henmi and Wada, 1976). The electron diffraction patterns of the product were those of imogolite (not shown) (Cradwick et al., 1972). The amount and length of the threads of imogolite tubes in the precipitate decreased as the P/Al molar ratio of the parent solutions increased from 0.0025 to 0.005

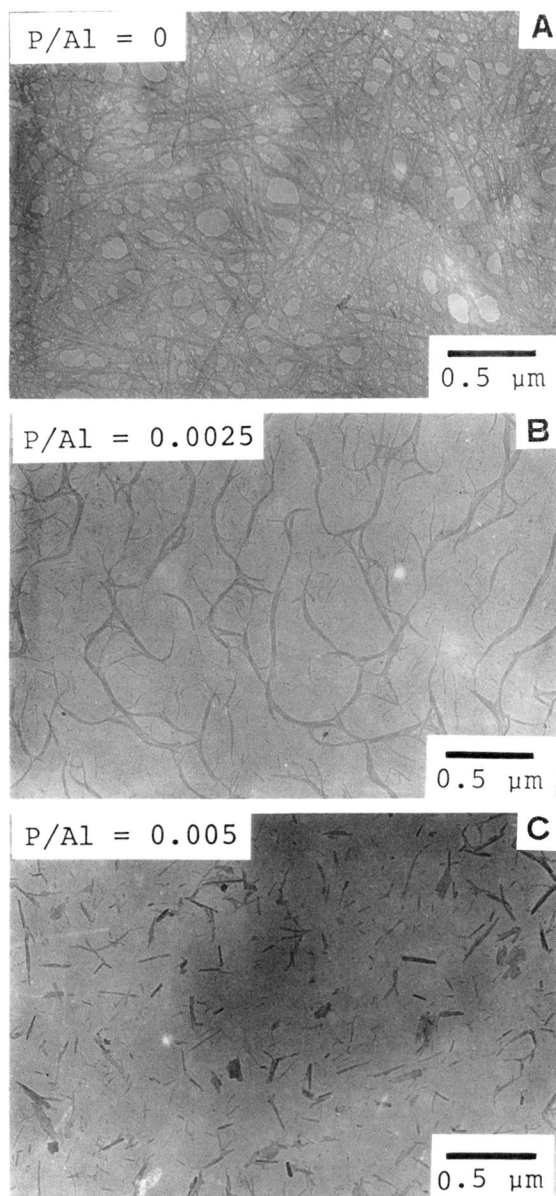

Figure 3. Transmission electron micrographs of precipitates of the reaction of hydroxy-Al ions and orthosilicic acid as influenced by phosphate.

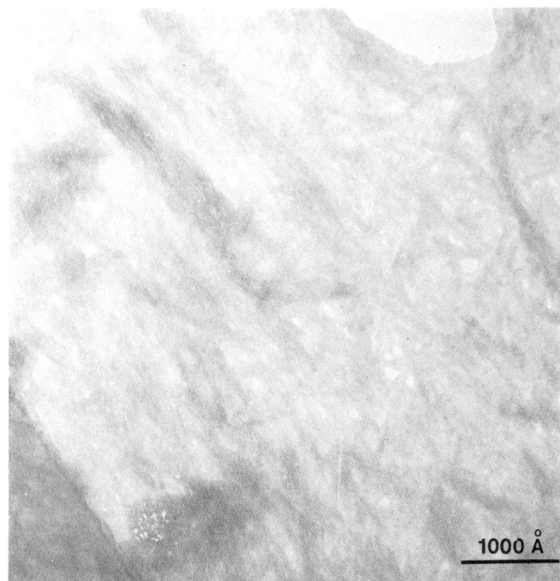

Figure 4. High-resolution transmission electron micrograph of precipitate of the reaction of hydroxy-Al ions and orthosilicic acid. Magnified version of Figure 3A.

(Figures 3B and 3C), indicating that the growth of imogolite tubes was impeded by the presence of phosphate anions. Phosphate can be adsorbed on the broken edges of imogolite tubes (Henmi and Huang, 1985). Therefore, the phosphate anions were probably adsorbed on the edge of newly formed imogolite tubes, the adsorbed anion covering the edges and inhibiting the growth of the tubes.

The IR spectra of the soluble products are shown in Figure 5. The spectrum of the product formed in the absence of phosphate contains absorption maxima or shoulders at 1200, 1080, 800, and 465 cm^{-1} (Figure 5a) which are close to those of pure silica gel (Farmer et al., 1979). The freeze-dried products of the filtrate, however, may have been a slightly impure silica gel (containing Al), inasmuch as only 78.3% of Al was removed from solution by the filtration (Table 1). Wada and Nagasato (1983) also found that freeze-drying a solution of orthosilicic acid yielded silica gel. Part of the orthosilicic acid that was not involved in the formation of precipitates apparently changed to the silica gel by polymerization when the soluble products were being freeze-dried. The IR spectra of the soluble products show fewer features characteristic of silica gel as the P/Al molar ratio of the parent solutions increased (Figure 5). At a ratio of 0.04, no precipitate formed, and the soluble product yielded the IR spectrum having absorption maxima at 973, 565, 422, and 343 cm^{-1} (Figure 5c) and a noncrystalline XRD pattern (not shown). These data indicate that the major component of the soluble product was "proto-imogolite" (973, 565, 422, and 343 cm^{-1}), which is considered to be a hydroxyaluminum–orthosilicate complex and a small fragment of the structure of imogolite (Farmer, 1981). The IR spectrum of the soluble product formed at a P/Al molar ratio of 0 shows no absorption band at about 343 cm^{-1} (Figure 5a), indicating that "proto-imogolite" was absent in the soluble product. An increase in the absorbance at 343 cm^{-1} with increasing P/Al molar ratio (Figure 5) indicates an increase in the content of "proto-imogolite" in the soluble products with increasing P/Al ratio of the parent solutions. These findings suggest that the co-existence of phosphate an-

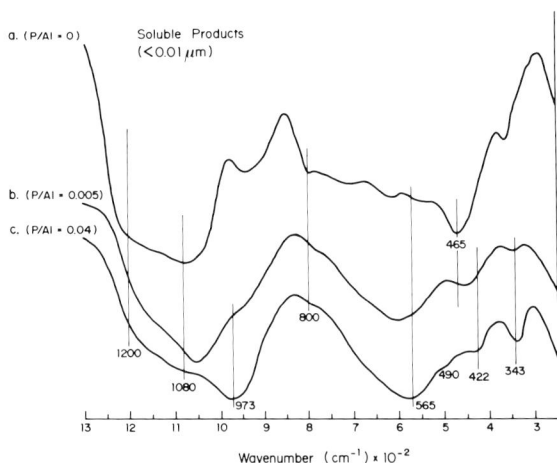

Figure 5. Infrared spectra of soluble products formed in systems containing hydroxy-Al ions, orthosilicic acid, and phosphate ions.

ions with hydroxy-Al ions and orthosilicic acid in the reaction systems contributed to the formation of "proto-imogolite" in the soluble products. The transformation of "proto-imogolite" to imogolite in the precipitates was inhibited by the complexing of the former with phosphate. The shoulders at 1200, 1080, 800, and 490 cm^{-1} (Figure 5c) indicate the presence of unreacted silicic acid in the soluble products formed from a parent solution having a P/Al molar ratio of 0.04. The IR spectra of all soluble products contained no indication of the presence of phosphate, probably because of the very low P content compared with those of Si and Al.

CONCLUSIONS

The data obtained here reveal that phosphate anions in concentration ranges as low as those of P levels of natural soil solutions significantly interfered with the interaction of hydroxy-Al ions and orthosilicic acid and thus perturbed the nucleation, growth, and formation of imogolite by retarding the polymerization of "proto-imogolite." The positively charged "proto-imogolite" apparently complexed with phosphate, the resulting complex becoming stabilized in solution where the P/Al molar ratios were sufficiently high. Thus, the amount of phosphate present is important in the genesis of imogolite in soils and sediments.

ACKNOWLEDGMENTS

This study was supported by the Natural Sciences and Engineering Research Council of Canada Grants A2348 and G1296—Huang. The Travel Fellowship awarded to the first author by the Ministry of Education of Japan is greatly appreciated.

REFERENCES

Bohn, H. L., McNeal, B. L., and O'Connor, G. A. (1979) *Soil Chemistry:* Wiley, New York, 290 pp.

Cooke, G. W. and Williams, R. J. B. (1973) Significance of man-made sources of phosphorus: fertilizer and farming. The phosphorus involved in agricultural systems and possibilities of its movement into natural water: in *Phosphorus in Freshwater and the Marine Environment,* S. H. Jenkins and K. J. Ives, eds., Pergamon Press, Oxford, 111–128.

Cradwick, P. D. G., Farmer, V. C., Russell, J. D., Masson, C. R., Wada, K., and Yoshinaga, N. (1972) Imogolite, a hydrated aluminium silicate of tubular structure: *Nature (London) Phys. Sci.* **240,** 187–189.

Devey, D. G. and Harkness, N. (1973) The significance of man-made sources of phosphorus: detergents and sewage: in *Phosphorus in Freshwater and the Marine Environment,* S. H. Jenkins and K. J. Ives, eds., Pergamon Press, Oxford, 35–54.

Farmer, V. C. (1981) Possible roles of a mobile hydroxy-aluminium orthosilicate complex (proto-imogolite) and other hydroxyaluminium and hydroxy-iron species in podzolization: in *Migrations Organominérales dans les Sols Tempérés, Colloques Internationaux du C.N.R.S. No. 303,* 275–279.

Farmer, V. C. (1982) Significance of the presence of allophane and imogolite in Podzol Bs horizons for podzolization mechanisms: a review: *Soil Sci. Plant Nutr.* **28,** 571–578.

Farmer, V. C. and Fraser, A. R. (1979) Synthetic imogolite, a tubular hydroxyaluminium silicate: in *Proc. Int. Clay Conf., Oxford, 1978,* M. M. Mortland and V. C. Farmer, eds., Elsevier, Amsterdam, 547–553.

Farmer, V. C., Fraser, A. R., Robertson, L., and Sleeman, J. R. (1984) Proto-imogolite allophane in podzol concretions in Australia: possible relationship to aluminous ferrallitic (lateritic) cementation: *J. Soil Sci.* **35,** 333–340.

Farmer, V. C., Fraser, A. R., Russell, J. D., and Yoshinaga, N. (1977) Recognition of imogolite structure in allophanic clays by infrared spectroscopy: *Clay Miner.* **12,** 55–57.

Farmer, V. C., Fraser, A. R., and Tait, J. M. (1979) Characterization of the chemical structures of natural and synthetic aluminosilicate gels and sols by infrared spectroscopy: *Geochim. Cosmochim. Acta* **43,** 1417–1420.

Henmi, T. and Huang, P. M. (1985) Removal of phosphorus by poorly ordered clays as influenced by heating and grinding: *Applied Clay Sci.* **1,** 133–144.

Henmi, T. and Wada, K. (1976) Morphology and composition of allophane: *Amer. Mineral.* **61,** 379–390.

Hsu, P. H. (1963) Effect of initial pH, phosphate, and silicate on the determination of aluminium with aluminon: *Soil Sci.* **96,** 230–238.

Hsu, P. H. (1977) Aluminum hydroxides and oxyhydroxides: in *Minerals in Soil Environments,* J. B. Dixon and S. B. Weed, eds., Soil Sci. Soc. Amer., Madison, Wisconsin, 99–143.

Inoue, K. and Huang, P. M. (1984) Influence of citric acid on the natural formation of imogolite: *Nature (London)* **308,** 58–60.

Jackson, M. L. (1958) *Soil Chemical Analysis:* Prentice-Hall, Englewood Cliffs, New Jersey, 134–182.

Lindsay, W. L. and Vlek, P. L. G. (1977) Phosphate minerals: in *Minerals in Soil Environments,* J. B. Dixon and S. B. Weed, eds., Soil Sci. Soc. Amer., Madison, Wisconsin, 639–672.

Luciuk, G. M. and Huang, P. M. (1974) Effect of monosilicic acid on hydrolytic reactions of aluminum: *Soil Sci. Soc. Amer. Proc.* **38,** 235–244.

McKeague, J. A. and Kodama, H. (1981) Imogolite in cemented horizons of British Columbia soils: *Geoderma* **25,** 189–197.

Parfitt, R. L. and Henmi, T. (1982) Comparison of an oxalate-extraction method and an infrared spectroscopic method for determining allophane in soil clays: *Soil Sci. Plant Nutr.* **28,** 183–190.

Ross, G. J. and Kodama, H. (1979) Evidence for imogolite in Canadian soils: *Clays & Clay Minerals* **27,** 297–300.

Wada, K. (1977) Allophane and imogolite: in *Minerals in Soil Environments,* J. B. Dixon and S. B. Weed, eds., Soil Sci. Soc. Amer., Madison, Wisconsin, 603–638.

Wada, K. (1980) Mineralogical characteristics of Andisols: in *Soils with Variable Charge,* B. K. G. Theng, ed., New Zealand Soc. Soil Sci., Palmerston North, New Zealand, 87–107.

Wada, S.-I., Eto, A., and Wada, K. (1979) Synthetic allophane and imogolite: *J. Soil Sci.* **30,** 347–355.

Wada, S.-I. and Nagasato, A. (1983) Formation of silica microplates by freezing dilute silicic acid solutions: *Soil Sci. Plant Nutr.* **29,** 93–95.

Warkentin, B. P. and Maeda, T. (1980) Physical and mechanical characteristics of Andisols: in *Soils with Variable Charge,* B. K. G. Theng, ed., New Zealand Soc. Soil Sci., Palmerston North, New Zealand, 281–301.

Weaver, R. M., Syer, J. K., and Jackson, M. L. (1968) Determination of silica in citrate-bicarbonate-dithionite extracts of soils: *Soil Sci. Soc. Amer. Proc.* **32,** 497–501.

Proceedings of the International Clay Conference, Denver, 1985, L. G. Schultz, H. van Olphen, and F. A. Mumpton, eds.,
The Clay Minerals Society, Bloomington, Indiana, 237–243 (1987).

AMMONIUM OXALATE REACTIVITY OF
SYNTHETIC HYDROXIDES AND SILICA-ALUMINA GELS

J. M. Hernández Moreno,[1] V. A. Cubas,[1] J. Hernández Brito,[1]
E. Fernández Caldas,[1] and A. Herbillon[2]

[1] Departamento de Edafología y Química Agrícola, Universidad de La Laguna, Tenerife, Spain

[2] Laboratoire de Physico-Chimie Minérale et de Catalyse, Université Catholique de Louvain
Louvain-la-Neuve, Belgium

Abstract—Hydroxyl reactivity of synthetic goethite, gibbsite, and two series of silica-alumina gels having a wide range of Si/Al composition (one series having characteristics of allophane) were studied by phosphate adsorption and "ammonium oxalate reactivity" (Ro), where Ro = amount of hydroxyls released in the reaction of amonium oxalate with hydroxylated surfaces at pH 6.3 over 25 min. Ro ranged from 14 mmole/100 g for goethite to 61 and 106 mmole/100 g for gibbsite prepared at pH 4.5 and 6.5, respectively. In the silica-alumina gels, Ro ranged from 90 to 442 mmole/100 g. In both series of gels, Ro reached a maximum at a SiO_2/Al_2O_3 ratio near 1. At this composition the Al and Si that dissolved during the reaction had the highest Al/Si ratio.

Ro was also determined on samples heated to 105°C. In all but one sample of the allophanic series, Ro decreased on heating, the largest decreases being observed for the more reactive samples. In the most Si-rich member of the series ($SiO_2/Al_2O_3 = 1.5$), however, Ro increased on heating by 24 mmole/100 g. This result may explain the Ro values observed for Andisols containing Si-rich allophanes. Here, Ro increased on heating to 105°C. Active OH groups on goethite, as determined by Ro, were found to be in agreement with those determined by phosphate adsorption. The maximum phosphate adsorption for the allophanic gels was displayed by samples having the highest Ro values.

Key Words—Allophane, Andisols, Gibbsite, Goethite, Hydroxyl ion, Phosphate.

INTRODUCTION

The difficulties in characterizing short-range ordered materials commonly found in soils are related to their particular surface chemical properties and the problems incurred in the measurement of surface area. The concept of surface area in short-range ordered materials loses its conventional meaning if their microporous character is considered (Paterson, 1977; Rousseaux and Warkentin, 1976); important changes take place on the exposed surface with drying, sometimes irreversibly (Egashira and Aomine, 1974; Uehara and Gillman, 1981), and a high proportion of crystal defects or broken bonds may give rise to differences in reactivity between materials having similar surface areas. On the other hand, the surface chemistry of these short-range ordered materials is a function of the reactivity of surface-hydroxyl groups, which determine surface charge, acidity, and anion reactions by ligand exchange.

Various methods have been proposed to estimate the hydroxyl reactivity of such surfaces, partly to overcome the difficulties incurred and the length of time involved in the determination of surface-charge characteristics. Generally, these methods are based on determining the amount of hydroxyls released by ligand exchange with anions such as fluoride (Bracewell *et al.*, 1970; Perrott *et al.*, 1976) and phosphate (Rajan and Perrott, 1975). Depending on the conditions employed,

the ligand-exchange reaction may be strictly a surface reaction, or it may cause substantial disruption of the surface.

González Batista *et al.* (1982a) and Hernández Moreno *et al.* (1985) described a method to estimate the hydroxyl reactivity of allophanic materials using ammonium oxalate at pH 6.3. They defined the amount of hydroxyl release during 25 min at 20°C as oxalate reactivity (Ro). The pH and reaction time was selected on the basis of the kinetics observed when this technique was applied to Andisols (Hernández Moreno *et al.*, 1985). Their results indicated that Ro represents a means of characterizing surfaces that is intermediate between selective dissolution and the determination of strictly surface properties. The major advantage of this method over fluoride- and phosphate-exchange reactions is that the amount of Al and Fe involved in the reaction can be analyzed because these ions form soluble chelated complexes with oxalate.

From Ro data on Andisols and other soils having variable charge, siliceous allophanic materials appear to be less reactive than alumina-rich allophanic materials (Hernández Moreno *et al.*, 1985). On the other hand, the moisture content of the samples and thermal pretreatment appear to have influenced Ro values considerably. Drying the samples always led to an increase in Ro, whereas, the effect of heating was variable. The

Ro of alumina-rich samples decreased with heating, with the largest decreases being observed for Hydrandepts and related soils. Their study of Ro for different types of Andisols demonstrated that this method may be used to define the andic properties of and to differentiate among some of the Great Soil Groups (ICOMAND, 1984; Fernández Caldas *et al.*, 1985).

In the present work, an attempt was made to explain the significance of Ro by using synthetic reference materials: goethite, gibbsite, allophanes, and other silica-alumina gels. For comparison, phosphate-adsorption characteristics of these materials were also determined.

MATERIALS AND METHODS

Samples were synthesized as described below and were characterized by X-ray powder diffraction (XRD), differential thermal analysis (DTA), thermogravimetric analysis (TGA), and transmission electron microscopy (TEM). Chemical analyses of allophane samples was made by dissolving them in boiling 0.5 N NaOH (Hasimoto and Jackson, 1960). Dissolution with ammonium oxalate at pH 3 was carried out by the method of Blakemore *et al.* (1981).

Synthesis

Goethite was prepared by the method of Atkinson *et al.* (1968). Two samples of gibbsite, g-4 and g-6, were prepared by the method of Gastuche and Herbillon (1962), wherein the mineral was precipitated at pH 4.5 and 6.5, respectively. A series of allophanes was prepared by the method of Wada *et al.* (1979). Throughout the text these samples will be referred to as allophanes or allophanic gels. All samples were freeze-dried and stored in a P_2O_5 atmosphere.

Silica-alumina gels were provided by the Laboratoire de Physico-Chimie Minérale et de Catalyse, Louvain-la-Neuve, Belgium, and belonged to the same series as those prepared by Rouxhet and Semples (1974). The samples were prepared by coprecipitation through hydrolysis of ethyl orthosilicate and aluminum isopropoxide. The samples were pretreated for 16 hr at 500°C in air and 2 hr at 500°–600°C under vacuum.

Ammonium oxalate reactivity

Ten milliliters of saturated ammonium oxalate solution was placed in a centrifuge tube containing 200 mg of sample and 50 mg of solid ammonium oxalate. The pH was maintained at 6.3 for 25 min by means an automatic titrator using 0.1 or 1.0 M HCl with constant stirring. The temperature was controlled with an accuracy of ±1°C at 20°C. The amount of OH displaced during the reaction was estimated from the amount of HCl consumed during the 25-min treatment. This amount is herein referred to as oxalate reactivity (Ro) (mmole/100 g). Ro values refer to samples dried in a P_2O_5 atmosphere. The difference be-

Table 1. Oxalate reactivity and Al(Fe) dissolution for goethite and gibbsite.

Sample	Moisture[1] condition	Ro[2] (mmole/ 100 g)	ΔRo[3] (mmole/ 100 g)	Al(Fe)[4] (mmole/ 100 g)	Al(Fe)[5] (%)
Goethite	P_2O_5	14 ⎱	−2	1.3	0.7
	105°C	16 ⎰		1.5	—
Gibbsite (g-4)	P_2O_5	61 ⎱	5	13.8	4.7
	105°C	56 ⎰		—	—
Gibbsite (g-6)	P_2O_5	106 ⎱	3	26.3	9.5
	105°C	75 ⎰		—	—

[1] Dried over P_2O_5 or heated to 105°C.
[2] Ro = amount of OH released during reaction with ammonium oxalate at pH 6.3 during 25 min.
[3] ΔRo = Ro variation on heating to 105°C (Ro = Ro − Ro_{105}).
[4] Al(Fe) dissolved after 25-min reaction with ammonium sulfate at pH 6.3.
[5] Wt. % Al(Fe) dissolved by ammonium oxalate at pH 3 (Blakemore *et al.*, 1981).

tween the reactivity values determined for the P_2O_5-dried sample (Ro) and the heated sample (Ro_{105}) is referred to as ΔRo, where ΔRo = Ro − Ro_{105}. After a reaction time of 25 min, the suspensions were immediately centrifuged and the Al, Fe, and Si content of the supernatant determined by atomic absorption methods.

Phosphate adsorption

The adsorption of phosphate was carried out by the method of Parfitt and Henmi (1980), in which phosphate was measured by the standard phosphomolybdate colorimetric method. In the allophanic gels, the amount of phosphate-desorbed silicon was analyzed by the method of Weaver *et al.* (1968) or by atomic absorption.

RESULTS

Characterization of samples

Goethite. By TEM goethite particles were lath-shaped and had average dimensions of 0.15 × 0.02 μm. Ammonium oxalate at pH 3 dissolved 0.7% Fe (Table 1).

Gibbsite. By TEM, sample g-4 was generally well crystallized, occurring as pseudohexagonal crystals having average dimensions of 0.1 × 0.05 μm. Some monoclinic prismatic crystals (0.08 × 0.03 μm) were also present. Sample g-6 was similar in shape and size, but the crystals were thinner and showed less developed crystal shapes; monoclinic crystals were absent in this sample. Ammonium oxalate at pH 3 dissolved 4.7% Al and 9.5% Al in samples g-4 and g-6, respectively (Table 1), which indicates the presence of poorly crystalline phases.

Synthetic allophanes. The chemical compositions of members of the synthetic allophane series are listed in Table 2. The molar SiO_2/Al_2O_3 ratios ranged from 0.15

Table 2. Oxalate reactivity and Al-Si dissolution in the allophanic series.

Sample[1]	Moisture[2] conditions	Ro[3] (mmole/ 100 g)	ΔRo[3] (mmole/ 100 g)	Al[4] (mmole/ 100 g)	Si[4] (mmole/ 100 g)	Ro/ Al	Al/ Si[5]
SA 0.15	P_2O_5	162	7	34	16.4	4.8	2.1
	105°C	155		32	15.8	4.8	2.0
SA 0.70	P_2O_5	286	44	86	23.2	3.3	3.7
	105°C	242		85	20.5	2.8	4.1
SA 0.85	P_2O_5	355	49	119	30.3	3.0	3.9
	105°C	306		114	28.0	2.7	4.1
SA 0.95	P_2O_5	390	52	113	36.2	3.4	3.1
	105°C	338		135	28.0	2.5	4.8
SA 1.00	P_2O_5	400	77	108	31.4	3.7	3.4
	105°C	323		124	26.3	2.6	4.7
SA 1.10	P_2O_5	442	142	125	24.5	3.5	5.1
	105°C	300		110	21.4	2.7	5.1
SA 1.35	P_2O_5	270	66	90	32.0	3.0	2.8
	105°C	204		75	20.6	2.8	3.6
SA 1.40	P_2O_5	248	6	92	27.7	2.7	3.3
	105°C	242		108	27.0	2.2	4.0
SA 1.50	P_2O_5	200	−24	68	21.0	2.9	3.2
	105°C	224		87	21.2	2.6	4.1

[1] SiO_2/Al_2O_3 molar ratio.
[2] Dried over P_2O_5 or heated to 105°C.
[3] Ro and ΔRo are defined as in Table 1.
[4] Al, Si dissolved after 25-min oxalate reaction at pH 6.3.
[5] Molar ratio.

to 1.5. No significant XRD peaks were observed. The thermal properties of the series were characteristic of allophanes, in that the exothermic DTA effect shifted from 990° to 1000°C as the ratio SiO_2/Al_2O_3 increased, a tendency that was also observed by Henmi (1980) for natural allophanes. TEM showed spheroidal or chain-like compound aggregates made up of simple spheroidal aggregates having average diameters between 0.04 and 0.02 μm. The simple aggregates had a very diffuse appearance at high magnifications.

The properties of sample SA 0.15 were not typical of allophane. This sample gave a XRD reflection at a d-value corresponding to pseudoboehmite; by TGA, dehydroxylation was observed near 400°C. The electron micrographs showed a fibrous gel, which at high

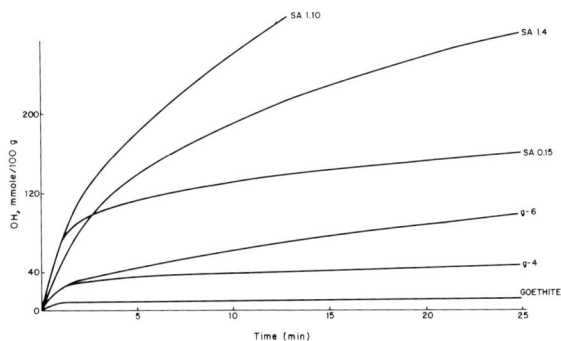

Figure 2. Ammonium oxalate reactivity at pH 6.3 (Ro) vs. composition (SiO_2/Al_2O_3 molar) of (a) allophanic gels and (b) silica-alumina gels. (c) Variation of Ro on heating at 105°C (ΔRo = Ro − Ro_{105}) vs. composition in allophanes.

magnifications appeared to be a network of short, unoriented fibers (0.03 × 0.01 μm).

Complete dissolution in ammonium oxalate at pH 3 was not always obtained in the allophane series.

Ammonium oxalate reactivity

The reaction kinetics of ammonium oxalate with some of the hydroxides and allophanes are presented in Figure 1. For the crystalline samples and allophane sample SA 0.15, the rate of release of OHs stabilized within only a few minutes. The reactivity of both P_2O_5-dried and heated samples and the amounts of Al(Fe) dissolved are presented in Tables 1 and 2. The Ro values for the allophanes ranged from 162 to 442 mmole/100 g; those for the heated samples (Ro_{105}) ranged from 155 and 338 mmole/100 g. The ΔRo values ranged between 142 and −24 mmole/100 g. Ro is plotted against the composition of the allophanes in Figure 2a. Note that the maximum Ro was found for

Figure 1. Reaction kinetics of ammonium oxalate with goethite, gibbsites g-4 and g-6, and some allophanes (SA).

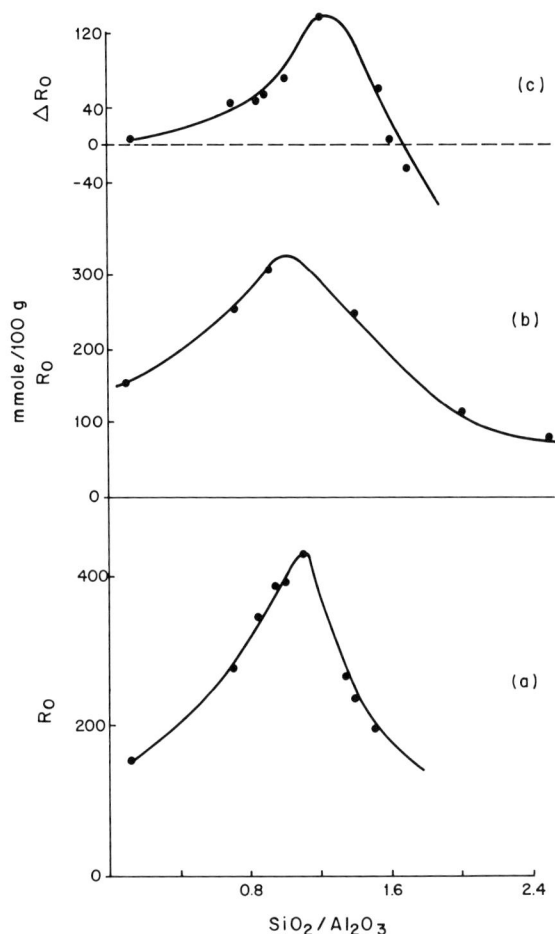

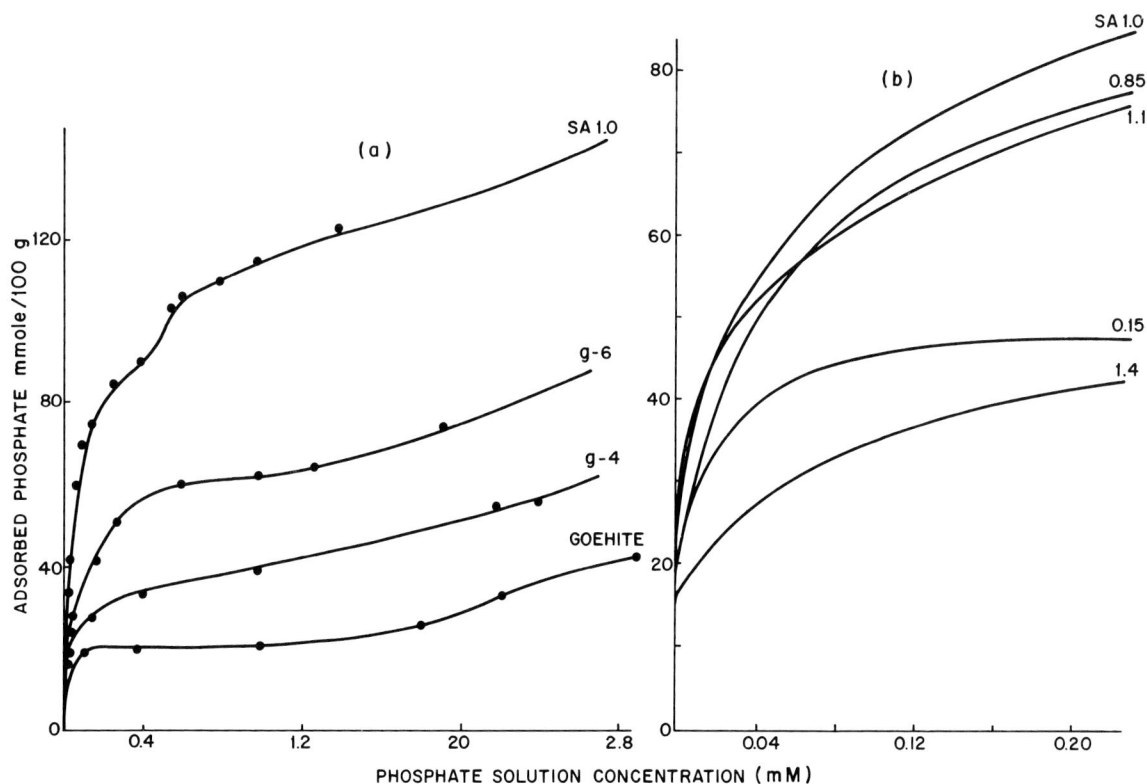

Figure 3.   Phosphate adsorption isotherms for: (a) goethite, gibbsites g-4 and g-6, and allophane sample SA 1.0; (b) some allophanes at low concentration.

allophane sample SA 1.1. The oxalate reactivity of the silica-alumina gels is shown in Figure 2b. Similar trends can be seen in Figures 2a and 2b, wherein a maximum is at a composition corresponding to a SiO_2/Al_2O_3 ratio near 1. The variation of ΔRo vs. composition for the allophane series is also shown in Figure 2c. Here also, samples having a SiO_2/Al_2O_3 ratio near 1 have the greatest value of ΔRo. The ratios of Al and Si dissolved during the 25 min of reaction ranged from 2.1 for sample SA 0.15 to 5.1 for sample SA 1.1. The Al/Si ratios for the heated samples tended to increase on heating (Table 2).

Phosphate-adsorption isotherms

Phosphate-adsorption isotherms for some of the samples are presented in Figures 3 and 4. The adsorption isotherm of goethite reached a plateau at 18 mmole/ 100 g of adsorbed phosphate; the plateau continued to a phosphate concentration of the order of 1.6 mM. Gibbsite samples g-4 and g-6 reached plateaus at 35 and 60 mmole/100 g, respectively, and then showed an increase in adsorption rate near 2 mM phosphate concentration. At low concentrations ($<100\ \mu M$ phosphate), the maximum phosphate adsorption for the allophane series was displayed by samples having SiO_2/ Al_2O_3 ratios near 1, a trend similar to that observed

for Ro. In this concentration range the adsorption values were similar to those obtained by Clark and McBride (1984) for a series of synthetic allophanes having similar characteristics as those studied here. At higher concentrations of phosphate, the isotherms contained adsorption steps, in agreement with the results obtained by others for natural and synthetic allophanes (Rajan, 1975; Rajan and Perrott, 1975; Kawai, 1980).

DISCUSSION

The time vs. hydroxyl-release curves (Figure 1) show that the rate of OH release stabilized in the first minutes of the reaction for the crystalline samples and allophane sample SA 0.15; however, the other allophanes were very reactive over longer periods. The time of 25 min is a compromise which seems sufficient to differentiate members of the series in the range of SiO_2/ Al_2O_3 composition studied. The Ro-composition relationship in allophanes (Figure 2a) is similar to that observed in the other silica-alumina gel series studied (Figure 2b). This similarity is remarkable, considering the great difference in conditions of preparation and treatment between both series of samples. The interpretation of the relationships between the surface properties of the gels and their composition is difficult, because the composition, structure, and surface-chemical

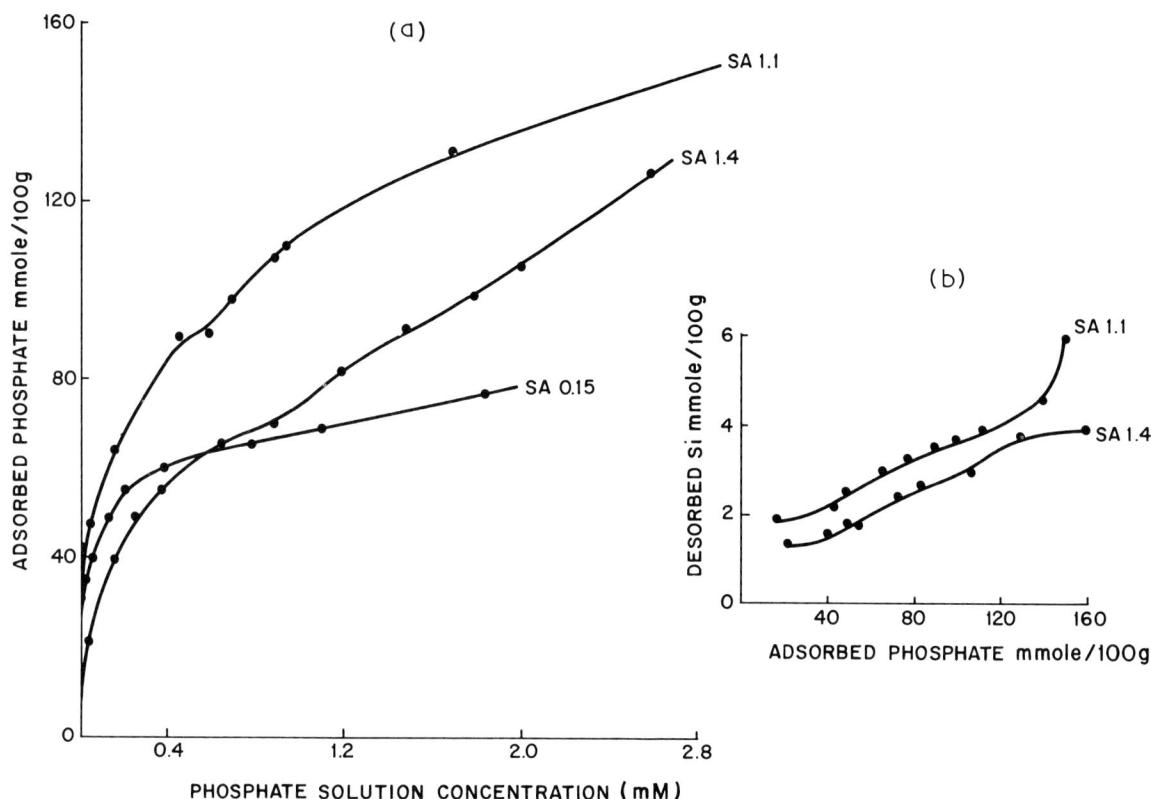

Figure 4. (a) Phosphate adsorption isotherms in allophane samples SA 0.15, 1.1, and 1.4. (b) Si desorption in allophane samples SA 1.1 and 1.4.

characteristics may be interdependent. Other authors have pointed out that both natural and synthetic allophanes display a maximum development of surface properties at a SiO_2/Al_2O_3 ratio near 1. This fact has been attributed to a considerable surface area because of microporosity and to the existence of pores and defects in the walls of the elemental units of allophane (Bracewell *et al.*, 1970; Rajan and Perrott, 1975; Parfitt and Henmi, 1980; Parfitt *et al.*, 1980; Clark and McBride, 1984). Also, the amounts of Al and Si dissolved during the reaction with oxalate led to the highest Al/Si ratio in the extracts.

Influence of the moisture content

Synthetic allophanes provided valuable information on the variation of ΔRo with composition. The most reactive members of the series had the greatest values of ΔRo. It must be remembered that thermal instability was used as an index of reactivity and was related to the presence of "imogolite units" in the structure of allophane. In accord with the above considerations and the hypothesis that surface acidity (and therefore Ro) probably increases on drying and heating (Hernández Moreno *et al.*, 1985), the decrease in reactivity on heating can be attributed to an irreversible reduction of exposed surface (Egashira and Aomine, 1974; Maeda

et al., 1977). The greater sensitivity to heating found in members of the series having SiO_2/Al_2O_3 near 1 is similar to the results obtained by Henmi *et al.* (1981) and indicates that the effect of heating was already acting at temperatures of about 100°C. In the sample SA 1.5 of the series, the increase in Ro on heating can be attributed to the fact that microporosity was scarcely affected or that the change was reversible. This result helps to explain the negative values of ΔRo found in the more siliceous Andisols (Hernández Moreno *et al.*, 1985). The most alumina-rich member of the series (sample SA 0.15) also gave a small value of ΔRo; however, this material was more ordered and therefore less sensitive to heating. The goethite sample gave a ΔRo value near zero, which agrees with its highly crystalline nature; however, gibbsite samples gave relatively large values of ΔRo, confirming the presence of a poorly crystalline phase which was readily affected by heating.

The increase of the Al/Si ratios of the extracts on passing from P_2O_5 to 105°C drying conditions can be related to the results of Henmi *et al.* (1981) who found that heating affects the degree of condensation of silica. This condensation probably was associated with restructuring of the surface, exposing more Al. On the other hand, heating led to a decrease in the ratios Ro/Al in the allophanic gels; i.e., the number of OHs per

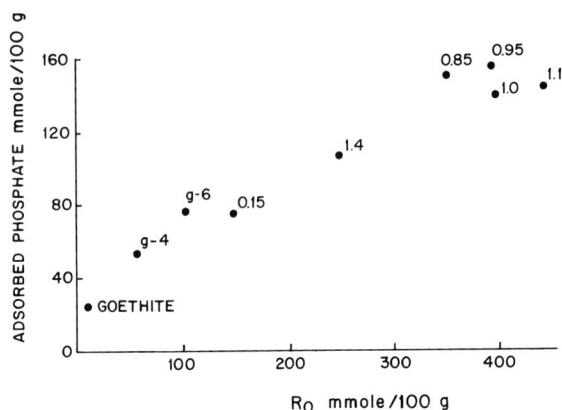

Figure 5. Ammonium oxalate reactivity at pH 6.3 (Ro) vs. phosphate adsorption at 2 mM concentration in allophanes (SA 0.15, 0.85, 0.95, 1.0, and 1.1) and hydroxides (goethite and gibbsites g-4 and g-6).

Al dissolved decreased (Table 2). In sample SA 0.15 the ratio Ro/Al was not affected by heating, probably due to its highly crystalline character.

Phosphate adsorption and Ro

The results obtained on the goethite sample allow a comparison to be made between Ro and phosphate adsorption. This comparison is important in order to establish an equivalence between the reactions involved. The amount of Fe dissolved with oxalate reaction at pH 6.3 was very small (0.02%), despite the high oxalate concentration (~ 1 M). On the other hand, a high affinity adsorption value was found near 18 mmole/100 g, which is of the same order as that found for a similar preparation having 40 mmole/100 g of $OH(OH_2)$ active groups (Parfitt et al., 1977a and 1977b). The balance between phosphate adsorbed and the number of OH groups released is as follows (Parfitt, 1978): at pH 5.1, OH/ΔP = 0.3; at pH 8.1, OH/ΔP = 1. In the present study, OH/ΔP = Ro/ΔP = 0.7 at pH 6.3. This value is reasonable and indicates that Ro is very close to the actual number of hydroxyl active groups on goethite surface at pH 6.3. In the gibbsite samples, the equivalence between ΔP and Ro is not as clear due to a continuous range of high- to low-affinity adsorption sites, which probably were due to the existence of phases having a range of crystallinities.

Phosphate adsorption at low concentrations has recently been suggested as an index of active Al–OH groups in allophanes and imogolite (Parfitt and Henmi, 1980). This index, however, cannot predict the capacity of these materials to adsorb phosphate at large concentrations. Thus, for the series of allophanes studied here, the sequence of phosphate adsorption values depended on the position on the adsorption isotherm. Except for the most reactive samples, Ro values correlated well with ΔP values at concentrations greater

than about 1 mM phosphate (Figure 5). As mentioned above, the adsorption isotherms of allophane show various steps of phosphate adsorption that generally correspond to Si desorption inflections (Figure 4). Taking into account the "proto-imogolite" model of allophane (Farmer et al., 1979; Parfitt et al., 1980), these different types of desorption can be attributed to the position of Si tetrahedra with respect to the distorted gibbsite-like layer and their degree of condensation. This could explain the fact that allophane sample SA 1.4 desorbed less Si than the alumina-rich allophanes, i.e., it required larger phosphate potential for Si to be displaced. The same reasoning may be applied to explain the decrease in oxalate reactivity from allophanes having SiO_2/Al_2O_3 ratios near 1 to the more siliceous members. In natural allophanic clays, González Batista et al. (1982a, 1982b) observed an inverse relationship between Ro and Si content, despite an increase in surface area (determined with ethylene glycol monoethyl ether) in siliceous samples.

ACKNOWLEDGMENT

The authors thank D. L. Norton who kindly revised an early draft of the present paper.

REFERENCES

Atkinson, R. J., Posner, A. M., and Quirk, J. P. (1968) Crystal nucleation in Fe(III) solution and hydroxide gels: *J. Inorg. Nucl. Chem.* **30**, 2371–2381.

Blakemore, L. C., Searle, P. L., and Daly, B. K. (1981) Methods for chemical analysis of soils: *New Zealand Soil Bureau Sci. Rept.* **10A**, 90 pp.

Bracewell, J. M., Campbell, A. S., and Mitchell, D. B. (1970) An assessment of some thermal and chemical techniques in the study of the poorly ordered alumino-silicates in soil clays: *Clay Miner.* **8**, 325–335.

Clark, C. J. and McBride, M. B. (1984) Cation and anion retention by natural and synthetic allophanes and imogolite: *Clays & Clay Minerals* **32**, 291–299.

Egashira, K. and Aomine, S. (1974) Effects of drying and heating on the surface area of allophane and imogolite: *Clay Sci.* **4**, 231–242.

Farmer, V. C., Fraser, A. R., and Tait, J. M. (1979) Characterization of the chemical structure of natural and synthetic alumino-silicate gels and sols by infrared spectroscopy: *Geochim. Cosmochim. Acta* **43**, 1417–1420.

Fernández Caldas, E., Hernández Moreno, J. M., Tejedor Salguero, M. L., González Batista, A., and Cubas, V. A. (1985) Behaviour of oxalate reactivity (Ro) in different types of Andisols. II: in *Volcanic Soils. Catena Supplement 7,* E. Fernández Caldas and D. H. Yaalon, eds., Catena Verlag, Cremlingen-Destedt, Germany, 25–34.

Gastuche, M. C. and Herbillon, A. (1962) Etude des gels d'alumine: cristallization en milieu déionisé: *Bull. Soc. Chim. Fr.*, 1404–1412.

González Batista, A., García Hernández, J. E., Hernández Moreno, J. M., and Fernández Caldas, E. (1982a) Estudio de la cinética de la reacción de oxalato amónico con arcillas alofánicas: *Anal. Edafol. Agrobiol.* **12**, 915–926.

González Batista, A., Hernández Moreno, J. M., Fernández Caldas, E., and Herbillon, A. (1982b) Influence of silica content on the surface charge characteristics of allophanic clays: *Clays & Clay Minerals* **30**, 103–110.

Hasimoto, I. and Jackson, M. L. (1960) Rapid dissolution of allophane and kaolin-halloysite after dehydration: in *Clays and Clay Minerals, Proc. 7th Natl. Conf. Washington, D.C., 1958,* Ada Swineford, ed., Pergamon Press, New York, 102–113.

Henmi, T. (1980) Effect of SiO_2/Al_2O_3 ratio on the thermal reactions of allophane: *Clays & Clay Minerals* **28,** 92–96.

Henmi, T., Tangue, K., Minawa, T., and Yoshinaga, N. (1981) Effect of SiO_2/Al_2O_3 ratio on the thermal reactions of allophane. II. Infrared and X-ray powder diffraction data: *Clays & Clay Minerals* **29,** 124–128.

Hernández Moreno, J. M., Cubas, V. A., González Batista, A., and Fernández Caldas, E. (1985) Study of ammonium oxalate reactivity at pH 6.3 (Ro) in different types of soils with variable charge. I: in *Volcanic Soils. Catena Supplement 7,* E. Fernández Caldas and D. H. Yaalon, eds., Catena Verlag, Cremlingen-Destedt, Germany, 9–23.

ICOMAND (1984) International committee on the classification of Andisols: *New Zealand Soil Bureau Circ.* **7,** 12 pp.

Kawai, K. (1980) The relationships of phosphorus adsorption to amorphous aluminium for characterizing Andosols: *Soil Sci.* **129,** 186–190.

Maeda, T., Takenaka, T. H., and Warkentin, B. P. (1977) Physical properties of allophane soils: *Adv. Agron.* **29,** 229–264.

Parfitt, R. L. (1978) Anion adsorption by soil materials: *Adv. Agron.* **30,** 1–50.

Parfitt, R. L., Farmer, V. C., and Russell, J. D. (1977a) Adsorption on hydrous oxides. I. Oxalate and benzoate on goethite: *J. Soil Sci.* **28,** 29–39.

Parfitt, R. L., Fraser, A. R., Russell, J. D., and Farmer, V. C. (1977b) Adsorption on hydrous oxides. II. Oxalate, benzoate and phosphate on gibbsite: *J. Soil Sci.* **28,** 40–47.

Parfitt, R. L., Furkert, R. J., and Henmi, T. (1980) Identi-
fication and structure of two types of allophane from volcanic ash soils and tephra: *Clays & Clay Minerals* **28,** 328–334.

Parfitt, R. L. and Henmi, T. (1980) Structure of some allophanes from New Zealand: *Clays & Clay Minerals* **28,** 258–294.

Paterson, E. (1977) Specific surface area and pore structure of allophanic soil clays: *Clay Miner.* **12,** 1–9.

Perrott, K. W., Smith, B. F. L., and Inkson, R. H. E. (1976) The reaction of fluoride with soils and soil minerals: *J. Soil Sci.* **27,** 58–67.

Rajan, S. S. S. (1975) Phosphate adsorption and the displacement of structural silicon on allophane clay: *J. Soil Sci.* **26,** 250–255.

Rajan, S. S. S. and Perrott, K. W. (1975) Phosphate adsorption by synthetic amorphous aluminosilicates: *J. Soil Sci.* **26,** 257–266.

Rousseaux, J. M. and Warkentin, B. P. (1976) Surface properties and forces holding water in allophane soils: *Soil Sci. Amer. J.* **40,** 446–451.

Rouxhet, P. G. and Semples, R. C. (1974) Hydrogen bond strengths and acidities of hydroxyl groups on silica alumina surfaces and in molecules in solution: *J. Chem. Soc. Faraday I* **70,** 2021–2032.

Uehara, G. and Gillman, G. (1981) *The Mineralogy, Chemistry, and Physics of Tropical Soils with Variable Charge Clays:* D. L. Plucknett, ed., Westview Press, Boulder, Colorado, 170 pp.

Wada, S. I., Eto, A., and Wada, K. (1979) Synthetic allophane and imogolite: *J. Soil Sci.* **30,** 347–355.

Weaver, R. M., Syers, J. K., and Jackson, M. L. (1968) Determination of silica in citrate-bicarbonate-dithionite extracts of soils: *Soil Sci. Soc. Amer. Proc.* **32,** 497–501.

PHYSICAL
AND
CHEMICAL PROPERTIES

Proceedings of the International Clay Conference, Denver, 1985, L. G. Schultz, H. van Olphen, and F. A. Mumpton, eds.,
The Clay Minerals Society, Bloomington, Indiana, 247–256 (1987).

THE CLAY-WATER INTERFACE[1]

Philip F. Low

Department of Agronomy, Purdue University
West Lafayette, Indiana 47907

Abstract—Pertinent equations of electric double-layer theory have been derived and used with experimental data on ζ, the zeta potential, and V_{ex}, the anion exclusion volume, to determine values of ψ_δ, the electric potential at the outer Helmholtz plane; σ_δ, the charge density in this plane; and τ, the distance between this plane and the plane of shear for different clay minerals. At all electrolyte concentrations, ψ_δ has a constant value of 50–60 mV and $\tau = 0$. Also, σ_δ increases with electrolyte concentration but remains small relative to the charge density at the clay-water interface. Therefore, the diffuse layer of these minerals is poorly developed, and the plane of shear coincides with the outer Helmholtz plane. J_i/J_i^0, the ratio of the value of any water property, i, in the clay-water system to that in pure bulk water, has been shown to be exponentially related to $1/t$, the reciprocal of the *average* thickness of the water films on the particle surfaces. Inasmuch as this relation does not depend on the nature of the clay, clay-water interaction is believed to be nonspecific in character.

An equation was derived for the osmotic component of Π, the swelling pressure of the clay, and has been used to calculate this component at several values of λ, the interlayer distance, for samples in which ψ_δ has a value close to that observed experimentally. A comparison of the calculated and observed results indicates that the osmotic contribution to Π is relatively insignificant. The relation between Π and λ suggests that Π is exponentially related to $1/\lambda$. A similar relation was observed between Π and $1/t$; hence, Π appears to be related to J_i/J_i^0. The latter relation is substantiated experimentally by using data on Π and ϵ, the molar absorptivity, for the same systems. The development of Π therefore depends on the same factor that causes J_i to differ from J_i^0, i.e., the non-specific interaction of water with the particle surfaces.

Key Words—Anion exclusion, clay-water interface, Double-layer theory, Swelling pressure, Water, Zeta potential.

INTRODUCTION

The colloidal properties of clays exert a profound influence on many physical and chemical processes that occur in the Earth's crust. They also play an important role in many industrial processes. Most of these properties depend directly or indirectly on the clay-water interface. Although much is known about this interface, it is still a controversial subject. Among the issues that remain unresolved are the characteristics of the electrical double layer and interfacial water and the relative effects that they have on such processes as swelling. In the present paper, these issues are addressed with the intent of developing fundamental concepts.

RESULTS AND DISCUSSION

Characteristics of the electrical double layer of clays

Electric double-layer theory is treated in several textbooks (e.g., Verwey and Overbeek, 1948; van Olphen, 1963); however, to facilitate understanding, a brief derivation of the basic differential equation of the theory, i.e., the Poisson-Boltzmann equation, is given here.

One of the equations involved in this derivation is the Poisson equation, which describes the effect of a space charge in an electric field on the rate of change of the electric potential gradient; another is the Boltzmann equation, which describes the distribution of particles in a force field. To derive the Poisson equation in the form that applies to a clay particle, the particle will be regarded as a negatively charged condenser plate and the neighboring ions as point charges. Also, in keeping with the conventions of elementary electrostatic theory: (1) 4π lines of force are assumed to emanate from each positive electrostatic unit of charge (esu) and terminate on each negative esu; and (2) the electric field intensity, i.e., the force acting on a positive esu, is assumed to equal the number of lines of force per square centimeter in a vacuum. Consider now an infinitesimal, rectangular parallelepiped in the electric field of the particle as illustrated in Figure 1. The parallelepiped has faces of unit area that are dx units apart in the direction of the field. Because the parallelepiped includes an excess of positive ions, more lines of force leave its left-hand face than enter its right-hand face, and the difference amounts to $4\pi\rho\,dx$, where ρ is the *net* space charge density. This difference must equal the change in electric field intensity, ξ, across the parallelepiped. Therefore,

---
[1] Journal paper 10,421, Purdue University Agricultural Experiment Station, West Lafayette, Indiana 47907.

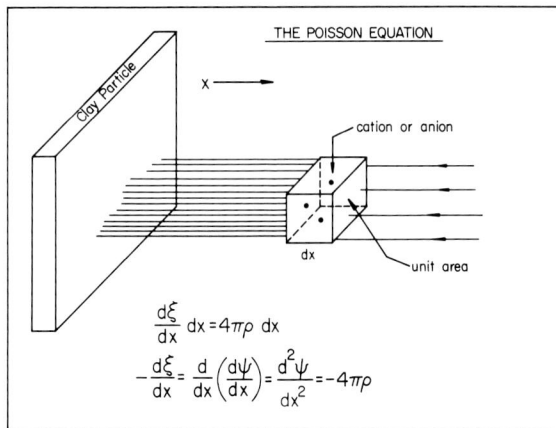

Figure 1. Illustration of concepts underlying the Poisson equation.

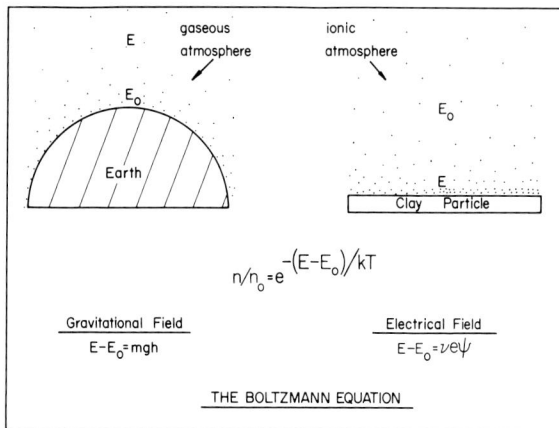

Figure 2. Illustration of concepts underlying the Boltzmann equation.

$$(d\xi/dx)\,dx = 4\pi\rho\,dx;$$

however, $\xi = -d\psi/dx$, where ψ is the electric pontential. Consequently,

$$-(d\xi/dx) = d(d\psi/dx)/dx$$
$$= (d^2\psi/dx^2) = -4\pi\rho. \qquad (1)$$

If, instead of being in a vacuum, the particle and ions are in a medium having a dielectric constant, D, the preceding equation becomes

$$(d^2\psi/dx^2) = -4\pi\rho/D. \qquad (2)$$

This is the Poisson equation.

In its general form, the Boltzmann equation may be written

$$n/n_0 = \exp[-(E - E_0)/kT], \qquad (3)$$

where n is the concentration of particles at a point in a force field where their potential energy is E, n_0 is the concentration of the same particles at a reference point where their potential energy is E_0, k is the Boltzmann constant, and T is the absolute temperature. By convention, E_0 is usually assigned a value of zero. As indicated in Figure 2, if the Boltzmann equation is applied to the distribution of gas molecules in the gravitational field of the Earth, $E - E_0 = mgh$, where m is the molecular mass, g is the acceleration of gravity, and h is the height above the Earth's surface. When it is applied to the distribution of ions in the electric field of a clay particle, $E - E_0 = ve\psi$, where v is the ionic valence and e is the electronic charge. Hence, in the latter case, Eq. (3) becomes

$$n/n_0 = \exp(-ve\psi/kT). \qquad (4)$$

It is evident that

$$\rho = e \Sigma\, v_i n_i, \qquad (5)$$

where the subscript i designates any ionic species. Combining this equation with Eqs. (2) and (4) yields:

$$(d^2\psi/dx^2) = -\frac{4\pi e}{D} \Sigma\, v_i n_{i0} \exp(-v_i e\psi/kT), \qquad (6)$$

which is the Poisson-Boltzmann equation. If a single symmetrical electrolyte is present, the summation in Eq. (6) includes only two terms and the equation can be written in the following hyperbolic form:

$$(d^2\psi/dx^2) = \frac{8\pi e v n_0}{D} \sinh \frac{v e\psi}{kT}. \qquad (7)$$

Nearly all the equations of electrical double-layer theory are derived by the integration of Eq. (6) or Eq. (7) between the appropriate limits.

In electrical double-layer theory, three planes are particularly significant (see Figure 3). One is the plane of the clay-water interface; a second is the outer Helmholtz plane (O.H.P.); and a third is the plane of shear. The O.H.P. is the plane that defines the outer limit of the Stern layer, i.e., the layer of counterions that are condensed on the particle surface. The plane of shear is the plane in which shear occurs between the envelope of water that moves with the particle and the water in the surrounding solution. We shall let δ be the distance between the clay-water interface and the O.H.P. and τ be the distance between the O.H.P. and the plane of shear. The electric potentials in the three planes are herein designated ψ_0, ψ_δ, and ζ, respectively, and the surface charge densities in the first two of them are designated σ_0 and σ_δ, respectively. Note that the value of σ_0 can be determined directly, without recourse to theory, by dividing the cation-exchange capacity by the specific surface area of the clay. In the Stern layer, the ions are assumed to oscillate about fixed adsorption sites, whereas, in the diffuse layer, they are assumed to undergo Brownian motion, like the ions in a solution. In applying double-layer theory to clays, it is often assumed that the Stern layer is absent (i.e., that δ is zero) and, hence, that the diffuse layer begins at the clay-water interface. Inasmuch as the Poisson-Boltz-

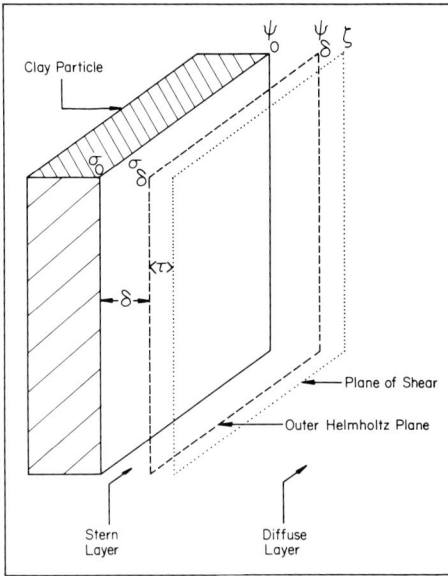

Figure 3. Model of the electrical double layer of clays.

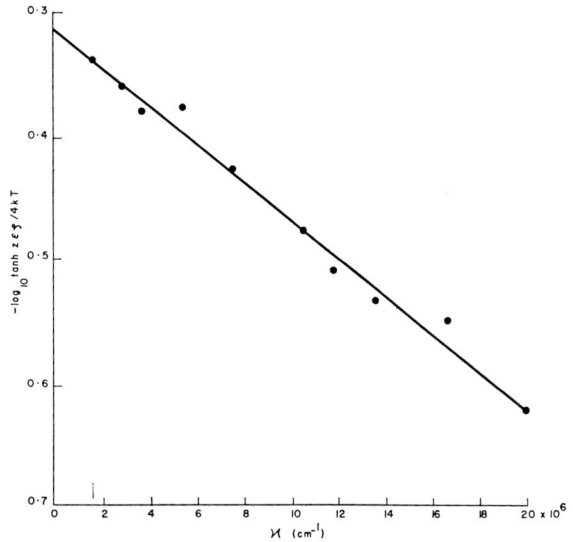

Figure 4. Relation between $-\ln\tanh(ve\zeta/4kT)$ and κ for Na-kaolinite (after Hunter and Alexander, 1963).

mann equation applies *only within the diffuse layer*, this assumption simplifies the theory; however, as will be shown below, the diffuse layer does not begin at the clay-water interface.

If Eq. (7) is applied to the diffuse layer of a single clay particle and is integrated between the limits $d\psi/dx = 0$ at $x = \infty$ and $\psi = \psi_\delta$ at $x = 0$, i.e., at the O.H.P.,

$$\tanh(ve\psi/4kT) = \tanh(ve\psi_\delta/4kT)e^{-\kappa x}, \qquad (8)$$

where

$$\kappa^2 = \frac{8\pi v^2 e^2 n_0}{DkT}. \qquad (9)$$

By applying the logarithmic form of Eq. (8) to the plane of shear (where $\psi = \zeta$ and $x = \tau$), Eversole and Boardman (1941) obtained:

$$\ln\tanh(ve\zeta/4kT) = \ln\tanh(ve\psi_\delta/4kT) - \kappa\tau. \qquad (10)$$

Utilizing experimentally determined values of ζ for glass in different electrolyte solutions, they then calculated corresponding values of $\ln\tanh(ve\zeta/4kT)$ and κ and plotted the former against the latter. The result was a straight line, indicating that both τ and ψ_δ were constant with changing electrolyte concentration. From the slope of this line they determined τ, and from its intercept they determined ψ_δ.

Following the procedure of Eversole and Boardman (1941), Hunter and Alexander (1963) determined, by micro-electrophoresis, the values of ζ for kaolinite in NaCl solutions of different concentration at a pH of 7.4 and constructed a plot of $\ln\tanh(ve\zeta/4kT)$ vs. κ (Figure 4). From it they determined that $\tau = 3.6$ A and $\psi_\delta = -55$ mV. Later, Miller (1984) determined, by both micro-electrophoresis and the moving-boundary

method, values of ζ for homoionic montmorillonites in chloride solutions of the respective cations. He also constructed plots of $\ln\tanh(ve\zeta/4kT)$ vs. κ. Representative of his plots are the two for Upton, Wyoming, Na-montmorillonite shown in Figure 5. Note that both plots have zero slope, indicating that $\tau = 0$ and $\psi_\delta = \zeta$ and that the O.H.P. is a plane of constant potential. From these plots, the values of ψ_δ are -59.5 and -58.4 mV, respectively.

The negative adsorption of anions, Γ_-, by a clay particle is the difference between the number of anions that would exist per square centimeter of surface if the particle were uncharged and the number that actually

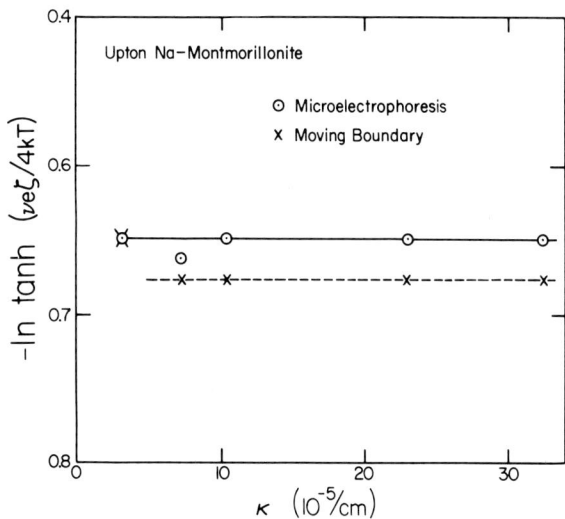

Figure 5. Relation between $-\ln\tanh(ve\zeta/4kT)$ and κ for Upton, Wyoming, Na-montmorillonite.

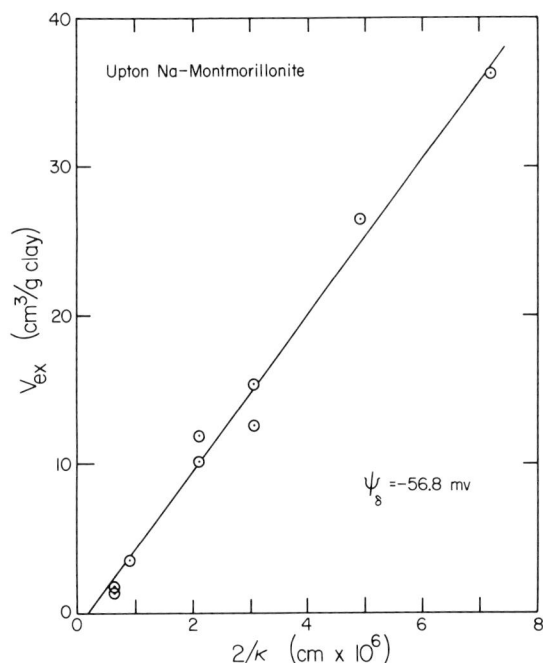

Figure 6. Relation between V_{ex} and $2/\kappa$ for Upton, Wyoming, Na-montmorillonite.

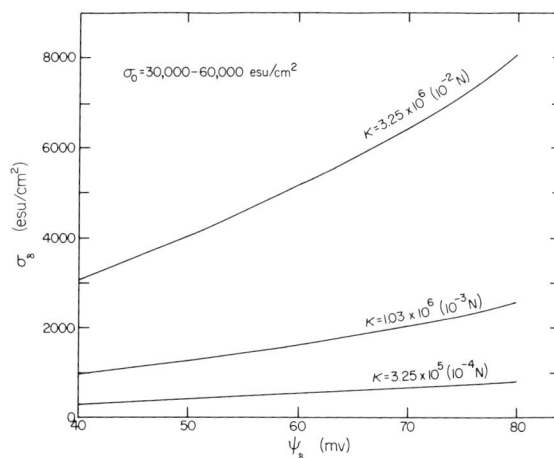

Figure 7. Relation between σ_δ and ψ_δ for clays at three electrolyte concentrations.

exists. It is given, when a single symmetrical electrolyte is present, by

$$\Gamma_- = \int_0^\infty (n_0 - n_-)\, dx, \qquad (11)$$

where n_- is the concentration of anions at any value of x. By employing Eq. (4) to express n_- in terms of n_0, replacing dx by $d\psi/(d\psi/dx)$, and substituting the right-hand side of the first integral of Eq. (7) for $d\psi/dx$, Eq. (11) can be cast in an integrable form and then integrated to give:

$$V_{ex} = \Gamma_- S/n_0 = \frac{2S}{\kappa}[1 - \exp(ve\psi_\delta/2kT)], \qquad (12)$$

where V_{ex} is the anion exclusion volume and S is the specific surface area of the clay. Eq. (12) was derived by Chan et al. (1984). It shows that, provided the value of S is known, the value of ψ_δ can be determined from the slope of the line obtained by plotting experimental values of V_{ex} against the corresponding values of $2/\kappa$. A similar equation for V_{ex} was derived earlier by Scho-

field (1947), but the assumptions on which it was based now appear to be untenable.

Presented in Figure 6 is the relation between V_{ex} and $2/\kappa$ for Upton Na-montmorillonite in NaCl solutions, obtained using the data of Miller (1984). Because this relation is linear, ψ_δ must be independent of electrolyte concentration. Further, because $S = 8 \times 10^6$ cm²/g (Low, 1980), ψ_δ for the Upton Na-montmorillonite, determined as just described, is -56.8 mV, which is close to the values obtained by means of Eq. (10). It should be mentioned here that two other independent methods have yielded essentially the same value of ψ_δ (Miller, 1984).

Electrical neutrality requires that the charge at the O.H.P. be balanced by the space charge in the diffuse layer. It follows, therefore, that

$$\sigma_\delta = -\int_0^\infty \rho\, dx,$$

in which x is measured from the O.H.P. If Eq. (2) is used to substitute for ρ in this equation and the result is integrated,

$$\sigma_\delta = \frac{DkT\kappa}{2\pi ve} \sinh(ve\psi_\delta/2kT). \qquad (13)$$

Thus, σ_δ is a function of κ, i.e., of electrolyte concentration, and of ψ_δ as illustrated in Figure 7. Note from the illustration that, because ψ_δ is constant for a given clay, σ_δ increases with electrolyte concentration.

Table 1. Representative values of ψ_δ, σ_δ, τ, and σ_δ/σ_0 for Na-saturated kaolinite, montmorillonite, and illite.

Clay mineral	ψ_δ (mV)	σ_δ (esu/cm²)	$\bar{\tau}$ (Å)	σ_δ/σ_0	Reference
Kaolinite	-55	456[1]	3.6	~ 0.013	Hunter and Alexander (1963)
Montmorillonite	-56.8	474	0.0	0.015	Miller (1984)
Illite	-51.0	412	—	0.005	Chan et al. (1984)

[1] Data in this column are for an electrolyte concentration of 10^{-4} N.

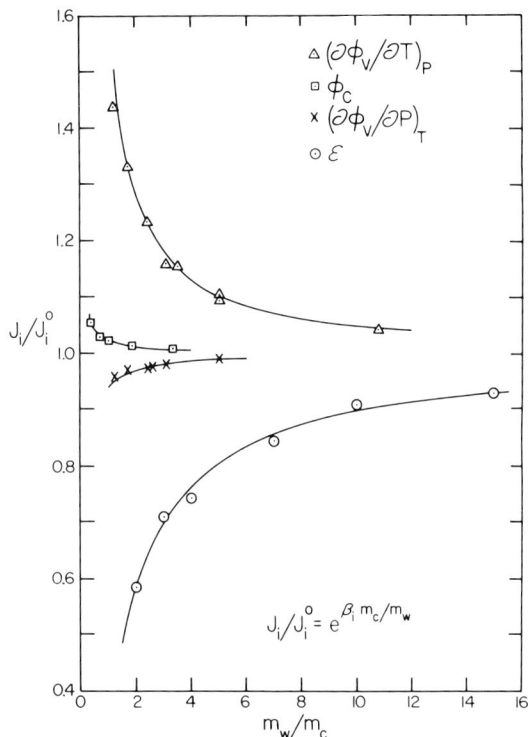

Figure 8. Relation between J_i/J_i^0 and m_w/m_c for four different properties of the water mixed with Upton, Wyoming, Na-montmorillonite.

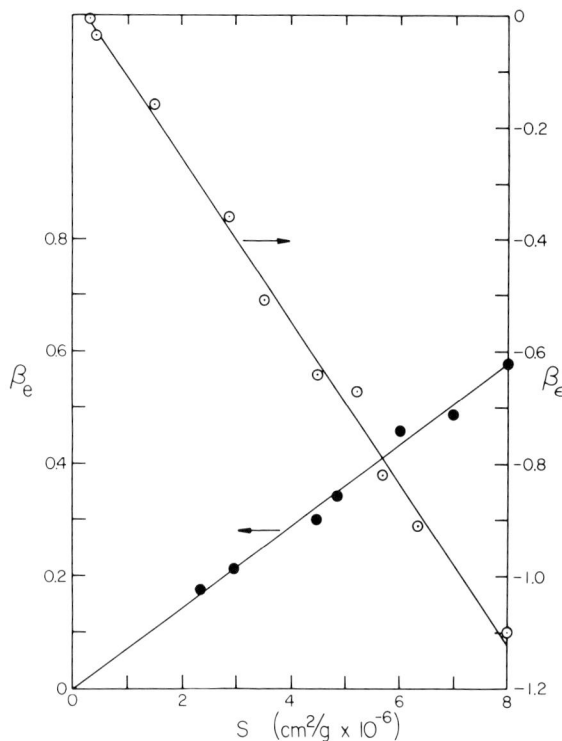

Figure 9. The effect of S on β_e, the parameter in Eq. (14) for the apparent specific expandability, and on β_ϵ, the parameter in Eq. (14) for the molar absorptivity.

Representative values of ψ_δ, τ, σ_δ, and σ_δ/σ_0 for three Na-saturated clay minerals are listed in Table 1. The values of ψ_δ and τ for the Na-kaolinite and Na-montmorillonite were reported above, and the value of ψ_δ for the Na-illite was reported by Chan et al. (1984). All values of σ_δ were calculated for an electrolyte concentration of 10^{-4} N by means of Eq. (13). It is important to note that σ_δ/σ_0 gives the fraction of the counterions that are in the diffuse layer. Hence, for the given clay minerals, very few counterions occupy this layer; almost all of them are in the Stern layer. This means that the diffuse layer is very weak.

Based on the information that has been presented here on the electrical double layer of clays, we can draw the following conclusions: (1) $\tau \simeq 0$, i.e., the plane of shear is coincident or nearly coincident with the O.H.P. and $\psi_\delta \simeq \zeta$; (2) ψ_δ is essentially constant—clays can therefore be regarded as colloids having a surface of constant potential; (3) σ_δ increases with electrolyte concentration; and (4) $\sigma_\delta/\sigma_0 \ll 1.0$—therefore relatively few counterions are in the diffuse layer. These conclusions are reinforced by additional data presented by Low (1981), Chan et al. (1984), and Miller (1984).

Nature of water near surfaces of clay particles

Low (1979), Mulla and Low (1983), and Sun et al. (1985) showed that all of the properties of water in clay-water systems can be described by a common equation, viz.,

$$J_i/J_i^0 = \exp(\beta_i m_c/m_w), \qquad (14)$$

where J is the value of any property i of the water in the clay-water system, J_i^0 is the value of the same property for pure bulk water, β is a parameter that depends on the property and the nature of the clay, m_c is the mass of clay, and m_w is the mass of water. The validity of this equation is illustrated for four different water properties in Figure 8. These properties are: (1) the apparent specific expandability, $(\partial\phi_v/\partial T)_P$; (2) the apparent specific heat capacity, ϕ_c; (3) the apparent specific compressibility, $(\partial\phi_v/\partial P)_T$; and (4) the molar absorptivity at the frequency of O–D stretching, ϵ. In determining ϵ, which is sometimes called the molar absorption coefficient or molar extinction coefficient, a small amount of D_2O was dissolved in the water to produce a dilute solution of HOD. The data in Figure 8 were extracted from the work of Low (1979), Mulla and Low (1983), and Sun et al. (1985). Note from the illustration that for each property the data points fall on the curve (solid line) described by Eq. (14) with the appropriate value of β. Note also that J/J^0 differs from unity for every property and that the distribution of J/J^0 depends on the specific property. These facts suggest that a common factor, e.g., the surface of the clay, disturbs the structure of the adjacent water to an appreciable depth and that this disturbance affects the structure-sensitive properties of the water to different degrees. In other words, the properties of the water are differentially sensitive to the rearrangement of the water molecules induced by the surface of the clay.

Recent work (Mulla and Low, 1983; Sun *et al.,* 1985) indicates that, in general,

$$\beta_i = k_i S, \tag{15}$$

where k is a proportionality constant that depends *only* on the given property. The proportionality between β and S is illustrated in Figure 9, which was constructed using data taken from the works cited above. In the illustration, β_e is the appropriate parameter if the property under consideration is the apparent specific expandibility, $(\partial\phi_v/\partial T)_P$, and β_ϵ is the appropriate parameter if the property under consideration is the molar absorptivity, ϵ.

If the right-hand-side of Eq. (15) is substituted for β in Eq. (14),

$$J_i/J_i^0 = \exp(k_i S m_c/m_w). \tag{16}$$

Further, by eliminating $S m_c/m_w$ between the analogues of Eq. (16) for any two water properties, designated by the subscripts 1 and 2,

$$(J_1/J_1^0) = (J_2/J_2^0)^{k_1/k_2}. \tag{17}$$

Eqs. (16) and (17) are potentially very useful. By means of the former, the value of S for the clay in a clay-water system can be determined at a known value of m_c/m_w from the measured value of any property of the water in the system once the value of k for that property has been determined. By means of the latter, the value of any property of the water in a clay-water system can be determined from the measured value of any other property of the water in the system provided the respective values of k_1 and k_2 have once been determined. Eq. (16) was the basis for a new method of measuring S (Mulla *et al.,* 1985).

Reason dictates that the *average* thickness, t, of the water films on the surfaces of the clay layers or particles is given by

$$t = m_w/m_c\rho_w S, \tag{18}$$

where ρ_w is the density of the water. If Eq. (18) is solved for m_c/m_w and the result combined with Eq. (16),

$$J_i/J_i^0 = \exp(k_i/\rho_w t), \tag{19}$$

which indicates that J_i is a single-valued function of t. In other words, J_i does not depend on the geometry, charge density, or any other characteristic of the particle surfaces, but only on the thickness of the films of water adsorbed on them. The validity of this conclusion is demonstrated by Figure 10 which shows the data of Mulla and Low (1983) for the relation between ϵ and t and the data of Sun *et al.* (1985) for the relation

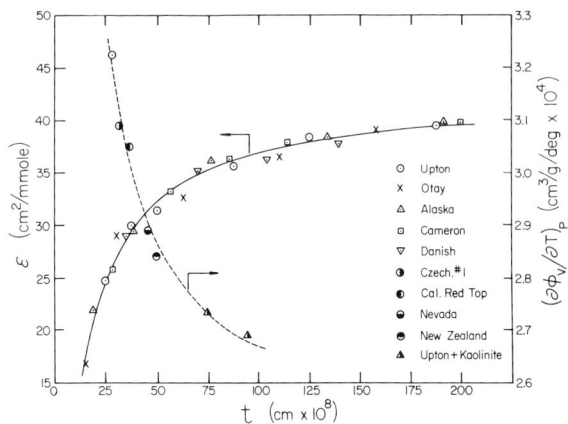

Figure 10.   The dependence of ϵ and $(\partial\phi_v/\partial T)_P$ on t.

between $(\partial\phi_v/\partial T)_P$ and t. Note that the smectites represented therein have a wide range of ionic substitution and, hence, of surface area, surface charge density, and degree of ditrigonality. Yet, all the data points fall close to the respective curves (solid and dashed lines). These curves obey Eq. (19) with the appropriate values of k_i and with ρ_w assigned a value of 1.0 g/cm^3.

The non-specific interaction of water with the surfaces of clay particles cannot be fully explained at the present time. Nevertheless, three different explanations appear to be plausible. The first is that there is insufficient space in the interstices between the clay particles (or layers) for the characteristic structure of bulk water to develop (Clifford and Pethica, 1968); the second is that the mere presence of the solid surface of the clay alters the motion and arrangement of the neighboring water molecules (Mulla *et al.,* 1984); the third is that the dispersion forces emanating from the clay particles (or layers) alter the motion and arrangement of the neighboring water molecules, but the intensity of these forces is not affected significantly by ionic substitution, at least to the degree that it has occurred. Regardless of which of these explanations is correct, the following conclusions are still warranted: (1) the surfaces of clay particles have an in-depth effect on the structure-sensitive properties of the interparticle (or interlayer) water; (2) the magnitude of this effect changes exponentially with 1/t, the reciprocal of the thickness of the water films on the particle surfaces; and (3) the rate of change, as governed by k_i, depends on the given property.

Clay swelling. The swelling of clay is one of the most important of all natural phenomena because it affects the structure, permeability, and erodibility of soils; the hydrology of geological formations; the stability of roads, runways, and buildings; the drilling of oil wells, etc. The most popular concept of swelling is that it has its origin in the excess osmotic pressure of the solution located between the surfaces of adjacent clay particles.

To test this concept, the relevant equations must be

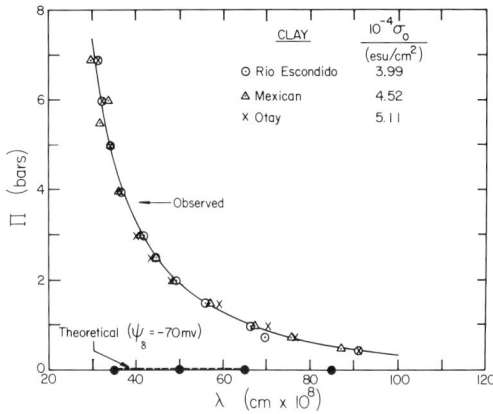

Figure 11. Theoretical and observed relations between Π and λ when $N = 10^{-4}$.

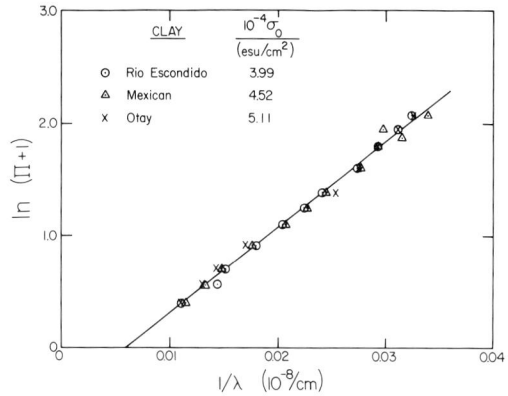

Figure 12. Relation between $\ln(\Pi + 1)$ and $1/\lambda$ for three Na-smectites.

derived. Following the traditional procedure (Langmuir, 1938), assume that the repulsive pressure, p, that tends to force the particles apart equals the osmotic pressure midway between their parallel surfaces less the osmotic pressure in the external solution. The osmotic pressure is then represented by the product of kT and the sum of the ionic concentrations in accordance with van't Hoff's law, giving

$$p = kT[(n_+ + n_-)_h - (n_+ + n_-)_0],$$

where the $+$ and $-$ subscripts denote the cation and anion, respectively, the subscript h signifies the midplane between the surfaces of the particles, and the zero subscript signifies the external solution beyond the reach of the electric double layers. By utilizing Eq. (4), this equation can be transformed to

$$p = 2n_0kT\left(\cosh\frac{ve\psi_h}{kT} - 1\right), \qquad (20)$$

where ψ_h is the electric potential at the mid-plane. The van der Waals attractive force, f, opposes p and is given by

$$f = A/6\pi\lambda^3, \qquad (21)$$

where A is the Hamaker constant and λ is the distance between the clay layers. According to Israelachvili and Adams (1978) $A = 2.2 \times 10^{-20}$ joules. Because Π, the swelling pressure of the clay, is the net repulsive force,

$$\Pi = p - f. \qquad (22)$$

Therefore, Π can be determined as a function of λ, if ψ_h is known as a function of λ.

Integrating Eq. (7) for overlapping double layers yields

$$\int_z^u (2\cosh y - 2\cosh u)^{-1/2} \, dy = -\kappa h, \qquad (23)$$

where $y = ve\psi/kT$, $z = ve\psi_\delta/kT$, $u = ve\psi_h/kT$, and h is the distance from the O.H.P. to the mid-plane between the clay layers. Note that $\lambda = 2(\delta + h)$. A numerical solution of Eq. (23) is obtainable for any combination of the limits, z and u. For a clay having a constant σ_δ, permissable combinations of these limits are given by

$$\delta_\sigma = (Dn_0kT/2\pi)^{1/2}(2\cosh z - 2\cosh u)^{1/2}; \qquad (24)$$

whereas, for a clay having a constant ψ_δ, the value of z is fixed and u is independently variable. Thus, regardless of whether σ_δ or ψ_δ is constant, ψ_h can be determined as a function of λ. It follows, therefore, that the same is true of Π.

In the present study, δ was assumed to be 5.5 A and ψ_δ was held constant at -70 mV, which is slightly more negative than the experimental value reported for montmorillonite in Table 1. Then, the distribution of Π with λ was determined as described above. The results are presented as the theoretical curve in Figure 11. Also presented in this figure is the distribution of Π with λ that was observed for three Na-smectites by Viani et al. (1983). From Figure 11, the osmotic contribution to Π (represented by the theoretical curve) appears to be relatively insignificant. Evidently, the excess of ions in the diffuse layer is too small to produce an appreciable value of p.

In Figure 11, the data points for the three smectites fall on a common curve. The equation for this curve was obtained by plotting $\ln(\Pi + 1)$ against $1/\lambda$ as shown in Figure 12 and determining the slope and intercept of the resulting straight line by linear regression analysis. It is

$$(\Pi + 1) = \exp k_\alpha\left(\frac{1}{\lambda} - \frac{1}{\lambda^0}\right) = b\exp(k_\alpha/\lambda), \qquad (25)$$

where λ^0 is the value of λ when $\Pi = 0$, $b = \exp(-k_\alpha/\lambda^0) = 0.642$, and $k_\alpha = 75.73 \times 10^{-8}$ cm. Therefore, Π is apparently a single-valued function of $1/\lambda$ and the relation between Π and λ apparently does not depend on the specific characteristics of the clay surface. Previously, Viani et al. (1983, 1985) reached the same conclusion from their studies of the swelling of eight

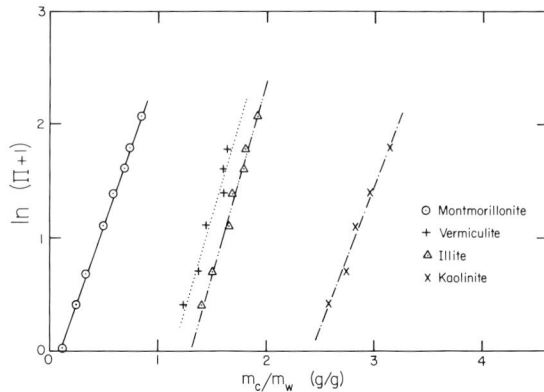

Figure 13. Relation between $\ln(\Pi + 1)$ and m_c/m_w for four Na-saturated clay minerals.

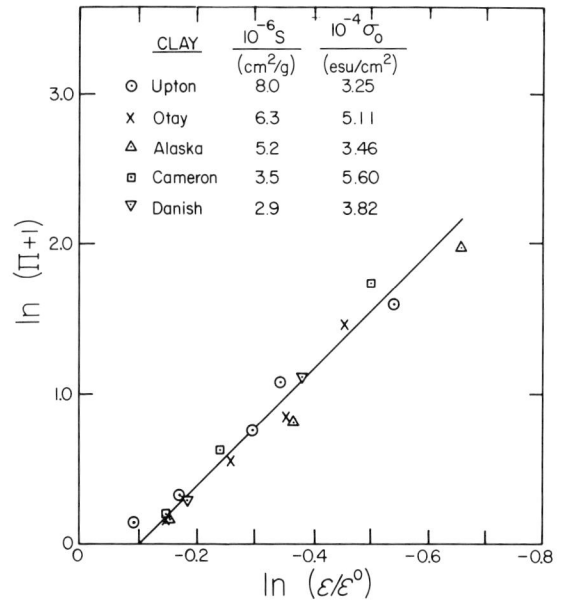

Figure 14. Relation between $\ln(\Pi + 1)$ and $\ln(\epsilon/\epsilon^0)$ for five different Na-smectites.

smectites and a vermiculite. Moreover, they showed that Eq. (25) does not have the form predicted by double-layer theory.

The validity of Eq. (25) is reinforced by the results presented in Figure 13, where $\ln(\Pi + 1)$ is shown to be linearly related to m_c/m_w for montmorillonite, vermiculite, illite, and kaolinite. The same kind of relation was found for 35 different smectites by Low (1980). This means that, in general,

$$(\Pi + 1) = \exp \alpha[(m_c/m_w) - (m_c/m_w)^0]$$
$$= B \exp \alpha(m_c/m_w), \qquad (26)$$

where $(m_c/m_w)^0$ is the value of (m_c/m_w) when Π is zero, and α and B are parameters that are characteristic of the clay. Note that $B = \exp[-\alpha(m_c/m_w)^0]$. If Eq. (18) is solved for m_c/m_w and combined with Eq. (26),

$$(\Pi + 1) = \exp \frac{\alpha}{\rho_w S}\left(\frac{1}{t} - \frac{1}{t^0}\right)$$
$$= B \exp (\alpha/\rho_w St), \qquad (27)$$

where t^0 is the *average* thickness of the water films when $\Pi = 0$ and $B = \exp(-\alpha/\rho_w St^0)$. For smectites,

$$\lambda = 2\gamma m_w/m_c\rho_w S, \qquad (28)$$

in which γ is the fraction of the total water that occupies interlayer regions. Comparison of Eq. (18) with Eq. (28) shows that $t = \lambda/2\gamma$. By using this relation between t and λ in Eq. (27),

$$(\Pi + 1) = \exp \frac{2\gamma\alpha}{\rho_w S}\left(\frac{1}{\lambda} - \frac{1}{\lambda^0}\right) \qquad (29)$$

which is identical to Eq. (25), if $k_\alpha = 2\gamma\alpha/\rho_w S$. It should be noted that, because α is proportional to S (Low, 1980), $k_\alpha = 2\gamma k'/\rho_w$, where k' is the proportionality constant.

Although clay swelling has been shown not to be caused primarily by excess osmotic pressure in the interlayer solution, no alternative cause has been proposed. If Eq. (19) in its logarithmic form is combined with Eq. (27) in the same form and α is replaced by $k's$,

$$\ln(\Pi + 1) = \frac{k'}{k_i} \ln(J_i/J_i^0) + \ln B. \qquad (30)$$

The applicability of Eq. (30) is demonstrated in Figure 14 for the specific case in which $J_i/J_i^0 = \epsilon/\epsilon^0$ and $k'/k_i = k'/k_\epsilon$. This case is particularly apropos because clays are transparent to infrared radiation at the frequency used in measuring ϵ, and thus only the nature of the water affects the results. Therefore, the development of Π must be affected by the same factor that causes J_i to differ from J_i^0. As noted above, three possible explanations can be proposed for the difference between J_i and J_i^0, but all of them rely on the idea that interaction of the water with the surfaces of the clay particles modifies the motion and arrangement of the water molecules and, hence, the structure of the water. This interaction may be regarded as an hydration of the surface and the resulting component of Π may be called the hydration force or the structural component of the swelling (disjoining) pressure. Therefore, we propose that hydration of the clay surfaces is the primary cause of swelling. Evidently, the hydration lowers $\bar{G}_w - G_w^\circ$, the relative partial molar free energy of the water and, thereby, enhances Π in keeping with the thermodynamic relation

$$\bar{G}_w - G_w^\circ = -\bar{v}_w\Pi, \qquad (31)$$

where $\bar{v}_w$ is the partial molar volume of the water.

Many investigators, including Schofield (1946), Norrish (1954), Bolt and Miller (1955), Warkentin *et al.*

(1957), Norrish and Rausell-Colom (1963), van Olphen (1963), Quirk (1968), Barclay and Ottewill (1970), and Callaghan and Ottewill (1974) have claimed that clay swelling is osmotic in character and can be described by double-layer theory. All of these investigators found satisfactory agreement between the observed and theoretical curves of II vs. λ. The agreement is questionable, however, because in locating the individual curves at least one of the following two assumptions was used: (1) all of the water in a swollen clay is in interlayer regions, i.e., $\gamma = 1.0$, and (2) there is no Stern layer on the surfaces of clay particles, i.e., the diffuse layer begins at the clay-water interface. The first assumption was used to convert measured values of m_w/m_c to the equivalent values of λ by means of Eq. (28); the second assumption was used to set the limits of integration in Eq. (23). By utilizing the combined data of Low (1980) and Viani et al. (1983) in Eq. (28), the value of γ is ~0.7 in a clay-water phase that is subjected to an applied pressure in order to equilibrate it with an external solution at atmospheric pressure. Although there is reason to believe that γ may have higher values under other conditions (e.g., Fink and Nakayama, 1972), the above conditions are the ones that obtain during experiments on swelling. Also, the evidence presented in Table 1 and elsewhere (Low, 1981; Chan et al., 1984; Miller, 1984) indicates that clay particles have well-developed Stern layers and that, as a result, $\sigma_0 \gg \sigma_\delta$ and $\psi_0 \gg \psi_\delta$. Therefore, neither one of the above assumptions appears to be tenable, and serious errors may be incurred if they are employed.

An obvious weakness with the concept that swelling results from surface hydration is that it lacks a sound theoretical basis. In other words, it is not supported by equations that have been derived from fundamental principles. All of the relevant equations are empirical and, hence, are not intellectually satisfying. Nevertheless, the empirical nature of these equations does not preclude their importance. Many essentially empirical equations have served science well. Good examples are the equations expressing Darcy's law, Ohm's law, the ideal gas laws, and van't Hoff's law of osmotic pressure.

In the present paper, substantial evidence has been provided to support the conclusion that hydration of clay surfaces is largely responsible for the swelling or disjoining pressure that tends to separate them. Other investigators (e.g., Derjaguin and Churaev, 1974, 1981; Israelachvili and Adams, 1978; Cowley et al., 1978; Pashley, 1981; Peschel et al., 1982; Pashley and Quirk, 1984; Pashley and Israelachvili, 1984; Churaev and Derjaguin, 1985) have reached essentially the same conclusion on the basis of their studies of the repulsive forces between various kinds of surfaces. These investigators, however, do not disregard the double-layer contribution to II, especially if the surfaces are far

enough apart that their influence on the intervening water no longer extends to the mid-plane. Also, Pashley and his co-workers have claimed that it is the adsorbed counterions that hydrate rather than the surfaces per se. Recent computer experiments have also shown that closely spaced solid surfaces modify the water between then (e.g., Gruen et al., 1981; Christou et al., 1981; Jonsson, 1981; Lee et al., 1984; Mulla et al., 1984) and that this modification generates a repulsive force (Magda et al., 1985). It is reasonable to assert, therefore, that the hydration force is a major component of the swelling pressure of clays and other colloids.

REFERENCES

Barclay, L. M. and Ottewill, R. H. (1970) The measurement of forces between colloidal particles: *Spec. Disc. Faraday Soc.* **1**, 138–147.

Bolt, G. H. and Miller, R. D. (1955) Compression studies of illite suspensions: *Soil Sci. Soc. Amer. Proc.* **19**, 285–288.

Callaghan, J. C. and Ottewill, R. H. (1974) Interparticle forces in montmorillonite gels: *Faraday Disc. Chem. Soc.* **57**, 110–118.

Chan, D. Y. C., Pashley, R. M., and Quirk, J. P. (1984) Surface potentials derived from co-ion exclusion measurements on homoionic montmorillonite and illite: *Clays & Clay Minerals* **32**, 131–138.

Christou, N. I., Whitehouse, J. S., Nicholson, D., and Parsonage, N. G. (1981) A Monte Carlo study of fluid water in contact with structureless walls: *Symp. Faraday Soc.* **16**, 139–149.

Churaev, N. V. and Derjaguin, B. V. (1985) Inclusion of structural forces in the theory of stability of colloids and films: *J. Colloid Interface Sci.* **103**, 542–553.

Clifford, J. and Pethica, B. A. (1968) Hydrogen bonding in aqueous colloid systems: in *Hydrogen-bonded Solvent Systems*, A. K. Covington and P. Jones, eds., Taylor and Francis, Ltd., London.

Cowley, A. C., Fuller, N. L., Rand, R. P., and Parsegian, V. A. (1978) Measurement of repulsive forces between charged phospholipid bilayers: *Biochemistry* **17**, 3163–3168.

Derjaguin, B. V. and Churaev, N. V. (1974) Structural component of disjoining pressure: *J. Colloid Interface Sci.* **49**, 249–255.

Derjaguin, B. V. and Churaev, N. V. (1981) Structure of the boundary layers of liquids and its influence on the mass transfer in fine pores: *Prog. Surface Membrane Sci.* **14**, 69–130.

Eversole, W. G. and Boardman, W. W. (1941) The effect of electrostatic forces on electrokinetic potentials: *J. Chem. Phys.* **9**, 798–801.

Fink, D. H. and Nakayama, F. S. (1972) Equation for describing the free-swelling of montmorillonite in water: *Soil Sci.* **114**, 355–358.

Gruen, D. W. R., Marcelja, S., and Pailthorpe, B. A. (1981) Theory of polarization profiles and the "hydration force": *Chem. Phys. Lett.* **82**, 315–320.

Hunter, R. J. and Alexander, A. E. (1963) Surface properties and flow behavior of kaolinite. Part II. Electrophoretic studies of anion adsorption: *J. Colloid Sci.* **18**, 833–845.

Israelachvili, J. N. and Adams, G. E. (1978) Measurement of forces between two mica surfaces in aqueous electrolyte solutions in the range 0–100 nm: *J. Chem. Soc. Faraday Trans. I* **74**, 975–1001.

Jonsson, B. (1981) Monte Carlo simulations of liquid water between two rigid walls: *Chem. Phys. Lett.* **82**, 520–525.

Langmuir, I. (1938) The role of attractive and repulsive forces in the formation of tactoids, thixotropic gels, protein crystals and coacervates: *J. Chem. Phys.* **6,** 873–900.

Lee, C. Y., McCammon, J. A., and Rossky, P. J. (1984) The structure of liquid water at an extended hydrophobic surface: *J. Chem. Phys.* **80,** 4448–4455.

Low, P. F. (1979) Nature and properties of water in montmorillonite-water systems: *Soil Sci. Soc. Amer. J.* **43,** 651–658.

Low, P. F. (1980) The swelling of clay: II. Montmorillonites: *Soil Sci. Soc. Amer. J.* **44,** 667–676.

Low, P. F. (1981) The swelling of clay. III. Dissociation of exchangeable cations: *Soil Sci. Soc. Amer. J.* **45,** 1074–1078.

Magda, J. J., Tirrell, M., and Davis, H. T. (1985) Molecular dynamics of narrow, liquid-filled pores: *J. Chem. Phys.* **83,** 1888–1901.

Miller, S. E. (1984) Characterization of the electrical double layer of montmorillonite: Ph.D. thesis, Purdue Univ., West Lafayette, Indiana, 93 pp.

Mulla, D. J. and Low, P. F. (1983) The molar absorptivity of interparticle water in clay-water systems: *J. Colloid Interface Sci.* **95,** 51–60.

Mulla, D. J., Low, P. F., Cushman, J. H., and Diestler, D. J. (1984) A molecular dynamics study of water near silicate surfaces: *J. Colloid Interface Sci.* **100,** 576–580.

Mulla, D. J., Low, P. F., and Roth, C. B. (1985) Measurement of the specific surface area of clays by internal reflectance spectroscopy: *Clays & Clay Minerals* **33,** 391–396.

Norrish, K. (1954) The swelling of montmorillonite: *Faraday Soc. Disc.* **18,** 120–134.

Norrish, K. and Rausell-Colom, J. A. (1963) Low-angle X-ray diffraction studies of the swelling of montmorillonite and vermiculite: in *Clays and Clay Minerals, Proc. 10th Natl. Conf., Austin, Texas, 1961,* Ada Swineford and P. F. Franks, eds., Pergamon Press, New York, 123–149.

Pashley, R. M. (1981) DLVO and hydration forces between mica surfaces in Li⁺, Na⁺, K⁺, and Cs⁺ electrolyte solutions: A correlation of double-layer and hydration forces with surface cation exchange properties: *J. Colloid Interface Sci.* **83,** 531–546.

Pashley, R. M. and Israelachvili, J. N. (1984) Molecular layering of water in thin films between mica surfaces and its relation to hydration forces: *J. Colloid Interface Sci.* **101,** 511–523.

Pashley, R. M. and Quirk, J. P. (1984) The effect of cation valency on DLVO and hydration forces between macroscopic sheets of muscovite mica in relation to clay swelling: *Colloids and Surfaces* **9,** 1–17.

Peschel, G., Belouschek, P., Müller, M. M., Müller, M. R., and König, R. (1982) The interaction of solid surfaces in aqueous systems: *Colloid Polymer Sci.* **260,** 444–451.

Quirk, J. P. (1968) Particle interaction and soil swelling: *Israel J. Chem.* **6,** 213–234.

Schofield, R. K. (1946) Ionic forces in thick films of liquid between charged surfaces: *Trans. Faraday Soc.* **42B,** 219–225.

Schofield, R. K. (1947) Calculation of surface areas from measurements of negative adsorption: *Nature* **147,** 408–410.

Sun, Y., Lin, H., and Low, P. F. (1985) The non-specific interaction of water with the surfaces of clay minerals: *J. Colloid Interface Sci.* **112,** 556–564.

van Olphen, H. (1963) *An Introduction to Clay Colloid Chemistry:* Interscience, London, 318 pp.

Verwey, E. J. W. and Overbeek, J. Th. G. (1948) *Theory of the Stability of Lyophobic Colloids:* Elsevier, New York, 205 pp.

Viani, B. E., Low, P. F., and Roth, C. B. (1983) Direct measurement of the relation between interlayer force and interlayer distance in the swelling of montmorillonite: *J. Colloid Interface Sci.* **96,** 229–244.

Viani, B. E., Roth, C. B., and Low, P. F. (1985) Direct measurement of the relation between swelling pressure and interlayer distance in Li-vermiculite: *Clays & Clay Minerals* **33,** 244–250.

Warkentin, B. P., Bolt, G. H., and Miller, R. D. (1957) Swelling pressure of montmorillonite: *Soil Sci. Soc. Amer. Proc.* **21,** 495–497.

Proceedings of the International Clay Conference, Denver, 1985, L. G. Schultz, H. van Olphen, and F. A. Mumpton, eds.,
The Clay Minerals Society, Bloomington, Indiana, 257–260 (1987).

EFFECT OF COMMINUTION ON THE CATION-EXCHANGE CAPACITY OF TALC AND CHLORITE FROM TRIMOUNS, FRANCE

J. Yvon, J. M. Cases, R. Mercier, and J. F. Delon

Centre de Recherche sur la Valorisation des Minerais
U.A. 235 du C.N.R.S., B.P. 40
54501 Vandoeuvre Cédex, France

Abstract—From an analysis of cation-exchange capacity (CEC), surface area, and structural data of talc-chlorite mixtures from Trimouns, France, ground in various manners, the talc component appears to have surface properties that can be predicted from its ideal structure. The CEC of 0.15 to 0.35 meq/100 g results only from the dissociation of silanol groups on lateral surfaces. The experimental equivalent size of an exchangeable site is 15 ± 8 Å compared with the theoretical value of 19.5 Å.

The CEC of the chlorite component, ranging from 7 to 10 meq/100 g, has three different origins: (1) structural CEC of about 5.8 meq/100 g; (2) CEC resulting from Si–OH and ^{IV}Al–OH dissociation on lateral surfaces, where the equivalent size of an exchange site is 29.5 Å²; and CEC resulting from hydroxide-sheet leaching. Estimates show that half of the Mg is leached from a basal face if it consists of a hydroxide sheet and that the contribution of Mg leaching from lateral surfaces leads to a CEC three times that of the tetrahedral border CEC. These results suggest that the surface properties of chlorite cannot be understood if the mineral is assumed to be stable and in electrostatic equilibrium.

Key Words—Cation-exchange capacity, Chlorite, Comminution, Surface area, Talc.

INTRODUCTION

The surface charge of sheet silicates, neutralized by exchangeable cations, may have two different origins: one origin is the amphoteric dissociation of lateral surface OH groups and which depends on pH (Parks and De Bruyn, 1962; Parks, 1965; Cases, 1969; Prédali and Cases, 1973); the other arises from a structural deficiency of cationic charges in the layer. Thus (1) the density of amphoteric sites or their average equivalent surface is a lateral surface constant, and (2) the structural cation-exchange capacity (CEC) is a material constant. These parameters are independent of particle size.

According to Mukherjee and Roy (1973), the CEC of talc results only from the dissociation of silanol groups on the lateral surface. The pKa corresponding to the dissociation of lateral –Mg–OH is 15.5 (Hair and Hertl, 1970), whereas the pKa corresponding to the dissociation of lateral $\equiv$Si–OH is <5 (Cases *et al.*, 1985). Therefore, only tetrahedral silanol groups bordering the crystals are dissociated at pH 7.

On the lateral surface of a magnesian chlorite, the $\equiv$Si–OH and –Mg–OH groups play the same role as those on the lateral surface of talc. In addition =Al–OH groups also must be taken into consideration, because the $\equiv^{IV}$Al–OH sites bordering the tetrahedral sheet are more acid than silanol groups (Rouxhet, 1974) and are completely dissociated at pH 7. The $\equiv^{VI}$Al–OH groups bordering the octahedral sheet, however,

are most likely undissociated because the point of zero charge of aluminum hydroxides ranges from 5 to 9.5 (Parks, 1965), which places the dissociation pKa at >10.

This study concerns the effect of dry grinding on the CEC of natural mixtures of talc and chlorite from Trimouns, Ariège, France. Its purpose is to describe CEC by models that allow the application of additivity rules.

EXPERIMENTAL

Materials

Talc-rich and chlorite-rich samples of ore were hand sorted, crushed, homogenized, and quartered into subsamples. The subsamples were dry ground in different ways, without classification, according to the flow-sheet in Figure 1. Samples were coded as follows: To and Co are, respectively, crude talc-rich and chlorite-rich samples. TMF is the talc-rich sample ground n times in the FORPLEX mill; CnF is the chlorite-rich sample ground n times in the FORPLEX mill; TmS is the talc-rich sample ground during m seconds in the AUREC mill; and CmS is the chlorite-rich sample ground during m seconds in the AUREC mill.

Analysis of the pure constituents shows that talc in these samples is essentially pure mineral talc. Infrared spectra and electron microprobe analysis suggest a minor substitution of Fe for Mg of about 1%. The chlorite in these samples is a magnesian chlorite, the structural formula of which is close to

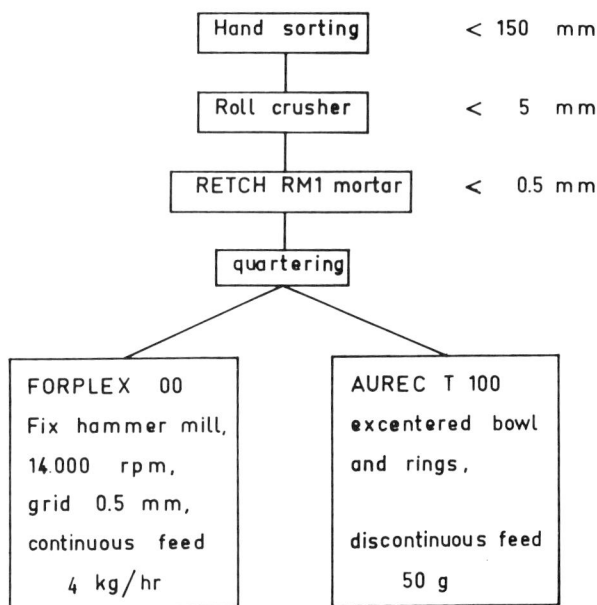

Figure 1. Flow-sheet of sample preparations.

Figure 2. 001 projection of tetrahedral lattices of a 2:1 trioctahedral sheet silicate.

$$(Mg_{4.78}Fe_{0.12}Al_{1.1})(Si_{2.9}Al_{1.1})O_{10}(OH)_8.$$

The talc-rich sample contains about 6% chlorite as an impurity, and the chlorite-rich sample contains about 20% talc, as estimated by thermal gravimetric analysis. Neither of these ores contains detectable amounts of smectite by X-ray powder diffraction.

Methods

The CEC was measured by titration of surface acidities to pH 7, after decationization by resins, according to the method of Rouiller-Doirisse (Thomas, 1982). The specific surface area was measured using the BET method for nitrogen adsorption isotherms at 77 K and the apparatus described by Delon (1970). The specific lateral surfaces were calculated from particle-size distribution curves (Sedigraph 5000 D), according to the method developed by Liétard *et al.* (1980). For some samples, the results were verified by transmission electron microscopy of shadowed preparations (Yvon, 1984).

The quality of the ore did not allow sufficient quantities of pure mineral to be obtained for the grinding tests. All grinding tests were therefore carried out on impure talc-rich and chlorite-rich samples. The properties of the separate pure phases were calculated using Eq. (1), where M(A) and M(B) are extensive properties measured on talc-rich and chlorite-rich samples, respectively, ground under the same conditions:

$$\begin{pmatrix} M(A) \\ M(B) \end{pmatrix} = \begin{pmatrix} T(A) & C(A) \\ T(B) & C(B) \end{pmatrix} \times \begin{pmatrix} P(T) \\ P(C) \end{pmatrix}. \quad (1)$$

Here, P(T) and P(C) are the values of the property "M" for pure talc and pure chlorite, T(A) and C(A) are the

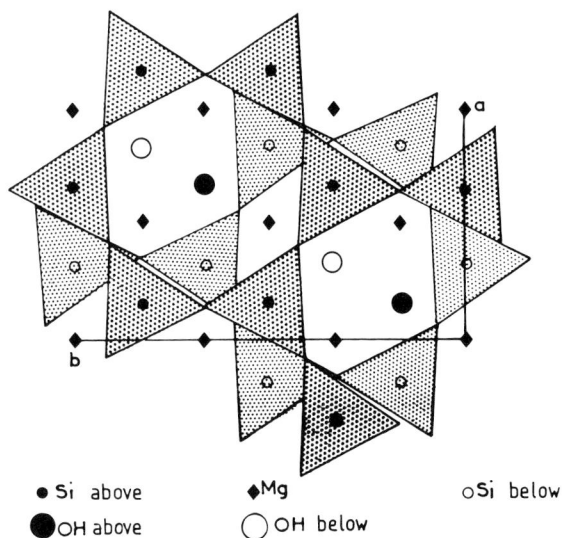

amounts of talc and chlorite in the talc-rich product, and T(B) and C(B) are the amount of talc and chlorite in the chlorite-rich product.

The theoretical values of site areas were calculated as follows: the orthogonal projection of silicate layers on the (001) plane of a 2:1 sheet silicate (Figure 2) constitutes a plane lattice of $\equiv Si-O$ bonds. Moving through this lattice along a random path of length "L", N number of $\equiv Si-O$ bonds are encountered. The value of $En \equiv L/N$ (Figure 3) converges toward a limit which is the "linear equivalent area" of a tetrahedral site on the lateral surface (along any random break parallel to the c^* axis). The average value of En is 2.11 Å (Figure 3). Multiplying En by the 001 spacing c yields the theoretical area of one exchange site on the lateral surface of 19.5 Å² for talc and 29.5 Å² for chlorite.

RESULTS AND DISCUSSION

The data calculated for the pure phases, after grinding the natural mixtures under a variety of conditions, are shown in Table 1. The specific surface, SS, has two components: a specific lateral surface SS_l and a specific basal surface SS_b. Considering a mineral the CEC of which results only from dissociation of lateral tetrahedral groups (CEC_t),

$$CEC_t = \frac{SS_l}{En \cdot c} \cdot \frac{1}{N_A} \quad (2)$$

where N_A is Avogadro's number and c the 001 spacing. For talc, application of Eq. (2) gives an average equivalent size of exchangeable sites of about 15 ± 8.5 Å² (arithmetic average) or 16 ± 7.8 Å² (weighted average). Despite the poor accuracy of this determination, resulting from very low CEC values of talc (Table 1), the

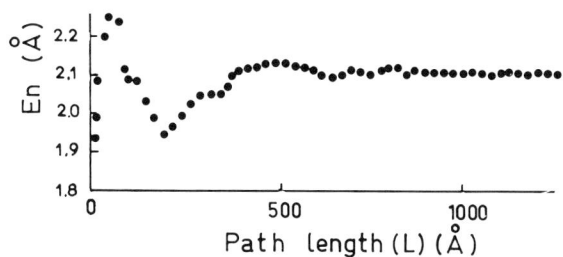

Figure 3. Average distance between two ≡Si–O bonds along a random path of length (L).

CEC of talc appears to be due to the dissociation of silanol groups, the theoretical size of which is 19.5 Å².

Applying Eq. (2) to chlorite yields a very small equivalent surface which has no physical significance, even if a structural CEC equal to the total CEC of unground product is substracted from the experimental CEC to maximize En. Yet, wet grinding data, especially in saline media, show that trioctahedral hydroxide sheets release significant amounts of Mg into the solution (Bonneau, 1982; Yvon, 1984). The Mg of the hydroxide sheet of chlorite is probably partially leached from the structure upon decationization, which creates a cationic defect on both the lateral and the basal surfaces (Figure 4).

The CEC resulting from hydroxide sheet dissolution on lateral sites can be considered to be proportional to

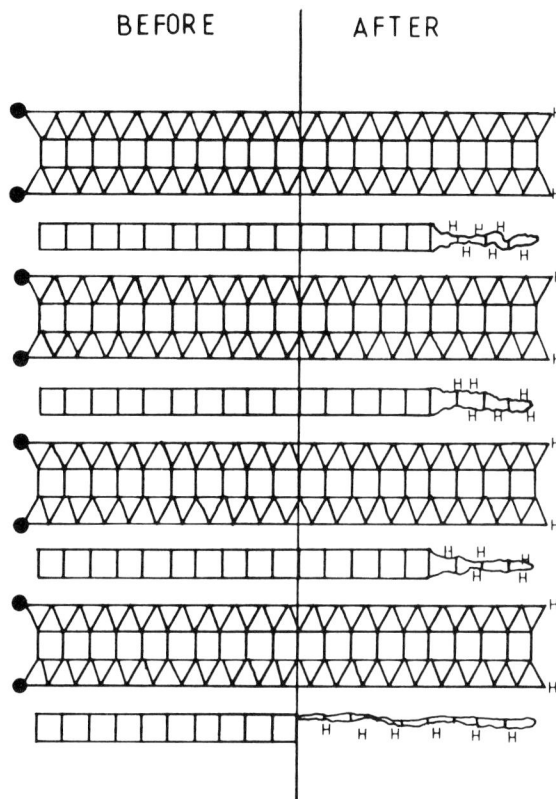

Figure 4. Schematic diagram of the effect of decationization on brucite-like sheet outcroppings. ● = exchangeable cation; H = proton.

Table 1. Specific surface (SS) and cation-exchange capacity (CEC) of pure talc and pure chlorite ground in different ways.

Sample code[1]	SS (m²/g)	[2]SS$_l$ (m²/g)	[3]SS$_b$ (m²/g)	CEC (meq/100 g)
T0	1.6	0.11	1.45	0.15
T1F	1.8	0.17	1.60	0.35
T2F	1.8	0.16	1.65	0.25
T3F	2.5	0.16	2.35	0.25
T4F	2.4	0.16	2.35	0.25
T5F	2.4	0.17	2.20	0.30
T5S	2.6	0.22	2.35	0.15
T10S	3.2	0.21	3.00	0.15
T20S	4.0	0.25	3.75	0.30
T40S	7.3	0.16	7.15	0.35
T120S	13.0	0.66	12.35	0.35
C0	2.5	0.30	2.15	7.35
C1F	2.7	0.32	2.35	7.15
C2F	2.6	0.28	2.35	8.95
C3F	2.8	0.27	2.55	9.30
C4F	3.0	0.27	2.75	9.30
C5F	3.1	0.27	2.80	10.00
C5S	3.2	0.25	2.95	9.00
C10S	4.6	0.26	4.30	9.20
C20S	4.2	0.34	3.85	9.30
C40S	5.2	0.41	4.80	9.40
C80S	7.9	0.53	7.35	10.25
C120S	9.8	0.47	9.35	10.25

[1] See text for explanations.
[2] Specific lateral surface area.
[3] Specific basal surface area.

the extension of the lateral specific surface and consequently to the tetrahedral CEC (CEC$_t$). The basal hydroxide faces consist of units having a surface area of $a_0 \cdot b_0$, where a_0 and b_0 are the cell parameters. The cationic deficiency resulting from the Mg removal from a basal hydroxide face can be written as:

Table 2. Calculated and experimentally determined CEC of the pure chlorite (meq/100 g).

Grinding sequence	CEC$_t$ tetrahedral from Eq. (2) in text	CEC$_b$ basal from Eq. (3) in text	CEC calculated from Eq. (7) in text	CEC experimental
C0	0.15	2.75	7.90	7.35
C1F	0.20	3.05	8.05	7.15
C2F	0.15	3.00	7.95	8.95
C3F	0.15	3.30	8.10	9.30
C4F	0.15	3.50	8.20	9.30
C5F	0.15	3.60	8.25	10.00
C5S	0.15	3.80	8.25	9.00
C10S	0.15	5.50	9.15	9.20
C20S	0.20	4.90	9.10	9.30
C40S	0.25	6.15	9.80	9.40
C80S	0.30	9.45	11.80	10.25
C120S	0.30	12.00	12.95	10.25

$$CEC_b = 2(3 - x)\frac{SS_b}{a_0 \cdot b_0 N_A},\qquad(3)$$

where x is the stoichiometric coefficient of Al in the chlorite formula. The tetrahedral defect of charge is assumed to be equilibrated in the brucite-like sheet (Caillère and Hénin, 1960). Thus, the CEC of chlorite can be written as:

$$CEC = \alpha CEC_t + \beta CEC_b + CEC_m,\qquad(4)$$

where α and β are proportionality coefficients and CEC_m is a possible "material property" CEC. Grouping by three (x, y, z), the twelve observations of Table 1 allow α, β, and CEC_m to be determined by solving systems of the type:

$$\begin{bmatrix} CEC(x) \\ CEC(y) \\ CEC(z) \end{bmatrix} = \begin{bmatrix} CEC_t(x) & CEC_b(x) & 1 \\ CEC_t(y) & CEC_b(y) & 1 \\ CEC_t(z) & CEC_b(z) & 1 \end{bmatrix} \times \begin{bmatrix} \alpha \\ \beta \\ CEC_m \end{bmatrix}.$$

CEC vector   shape matrix   coefficients vector

$$(5)$$

The shape matrix only depends on morphological characteristics as shown by Eqs. (2) and (3).

Resolving such a system into all possible configurations shows that the most probable solution gives the following coefficient values: $\alpha = 4$; $\beta = 0.5$; $CEC_m = 5.85$. Thus, the CEC of chlorite in agreement with this model is:

$$CEC = 4CEC_t + 0.5CEC_b + 5.85\qquad(6)$$

and can be written using Eqs. (2) and (3) as:

$$CEC = \frac{1}{N_A}\left(4 \cdot \frac{SS_l}{En \cdot c} + (3 - x)\frac{SS_b}{a_0 \cdot b_0}\right) + 5.85.\qquad(7)$$

The experimental values of CEC_t, CEC_b, the experimental value of CEC, and the CEC calculated from Eq. (7) are given in Table 2.

According to the proposed description, the CEC resulting from Mg leaching from lateral faces is three times that of the tetrahedral CEC. Only half of the Mg ions are leached from a hydroxide face. For finely ground chlorites, the CEC is experimentally underestimated which may result from the neutralization of exchangeable sites by Mg^{2+} migrating from a structure disturbed by grinding.

CONCLUSIONS

The CEC of talc only results from the dissociation of silanol groups on the lateral faces. The apparent very high CEC of chlorite with regard to its morphology and expected pKa of OH surface groups can be explained by release of Mg from the hydroxide sheet which creates negative charges, and by the possible presence of struc-

turally disequilibrated layers. These aspects must be considered when evaluating the surface reactivity of this mineral.

ACKNOWLEDGMENTS

This study was supported by the "Délégation Générale à la Recherche Scientifique et Technique" A.C. VRSS Grant no. 78-07-0749 and the company S.A. Talcs de Luzenac.

REFERENCES

Bonneau, L. (1982) Etude des sites actifs en surface des amiantes. Détermination et activité: Thèse 3ème cycle, Univ. Paris, 127 pp.

Caillère, S. and Hénin, S. (1960) Relation entre la constitution cristallochimique des phyllites et leur température de déshydratation. Application au cas des chlorites: *Bull. Soc. Fr. Céram.* **68**, 63–67.

Cases, J. M. (1969) Point de charge nulle et structure des silicates: *J. Chim. Phys.* **66**, 1602–1611.

Cases, J. M., Doerler, N., and François, M. (1985) Influence de différents types de broyage fin sur les minéraux. I. Cas du quartz: in *Proc. 15th Int. Cong. Miner. Process., Vol. 1,* Gédim ed. Paris, 169–179.

Delon, J. F. (1970) Contribution à l'étude de la surface spécifique et de la microporosité des roches: Thèse doc., Univ. Nancy, Nancy, France, 197 pp.

Hair, M. L. and Hertl, W. (1970) Activity of surface hydroxyl groups: *J. Phys. Chem.* **74**, 91–94.

Liétard, O., Yvon, J., Delon, J. F., Mercier, R., and Cases, J. M. (1980) Determination of the basal and lateral surfaces of kaolin: variations with types of crystalline defects: in *Fine Particle Processing, Vol. 1,* P. Somasundaran, ed., Society of Mining Engineers of AIME, New York, 558–582.

Mukherjee, D. K. and Roy, S. (1973) Effect of dry grinding on the BEC of some Indian talc: *Indian Ceramics* **16**, 215–219.

Parks, G. A. (1965) The isoelectric point of solid oxides, solid hydroxides and aqueous hydroxo-complex systems: *Chem. Rev.* **65**, 177–197.

Parks, G. A. and De Bruyn, P. L. (1962) The zero point of charge of oxides: *J. Phys. Chem.* **66**, 967–973.

Prédali, J. J. and Cases, J. M. (1973) Zeta potential of magnesian carbonates in inorganic electrolytes: *J. Coll. Int. Sci.* **45**, 449–458.

Rouxhet, P. (1974) Hydrogen bond strengths and acidity of hydroxyl groups on silica alumina surface: *J. Chem. Soc. Faraday Trans.* **70**, 2021–2052.

Thomas, F. (1982) Etude de l'adsorption des molécules azotées à la surface du kaolin. Thèse 3ème cycle, Univ. Nancy, Nancy, France, 87 pp.

Yvon, J. (1984) Eléments sur les propriétés cristallochimiques, morphologiques et superficielles des minéraux constitutifs de minerais de talc: Thèse doc., Inst. Nat. Polytech. Lorraine, Nancy, France, 303 pp.

Proceedings of the International Clay Conference, Denver, 1985, L. G. Schultz, H. van Olphen, and F. A. Mumpton, eds.,
The Clay Minerals Society, Bloomington, Indiana, 261–266 (1987).

MAGNETIC ORDER IN TRIOCTAHEDRAL SHEET SILICATES: A REVIEW

J. M. D. COEY

Department of Pure and Applied Physics, Trinity College
Dublin 2, Ireland

Abstract—Data on the magnetic susceptibility, magnetization, Mössbauer spectra, magnetic neutron scattering, and computer simulation of the magnetic ground state of representative Fe-rich trioctahedral phyllosilicates are reviewed. The principal exchange interaction, between Fe^{2+} in edge-sharing octahedra in the octahedral sheets, is ferromagnetic with $J = 1$–2 K; intersheet coupling is about 20 times weaker. Anisotropy due to the Fe^{2+} may be represented by a trigonal crystal field which stabilizes the $^5A_{1g}$ orbital singlet as the ferrous ground state, the trigonal, hard axis being normal to the sheets.

Ferrous end members of the trioctahedral series order as planar antiferromagnets at 20–30 K. The planar antiferromagnetic ground state can tolerate the small amounts of Fe^{3+} found in greenalite or minnesotaite; however, it is destabilized in biotite and thuringite, in which octahedral sheets contain more than a small percentage of ferric pairs, in favor of a spin glass state having only short-range ferromagnetic correlations (~ 10 Å) in the planes. Except for a reduction by a factor of 4 in the magnetic ordering temperature, the bulk magnetic properties of the spin glass and planar antiferromagnetic states are remarkably similar.

Key Words—Antiferromagnetism, Ferromagnetism, Iron, Magnetic order, Mössbauer spectroscopy, Spinglass, Trioctahedral.

INTRODUCTION

The magnetic properties of minerals depend essentially on the Fe they contain, because Fe is 40 times more abundant than the sum of all other magnetic elements in the crust. Only a handful of natural iron oxides, hydroxides, and sulfides exhibit collective magnetic order at room temperature and above, but Fe-rich silicates should order magnetically, albeit at much lower temperatures. Two common types of magnetic order in ionic compounds are *antiferromagnetism* and *ferrimagnetism*. In an antiferromagnet (e.g., goethite), the ionic moments are aligned on two equal antiparallel sublattices. No *net* magnetization exists, and the order can only be detected by special techniques, such as neutron diffraction. In a ferrimagnet (e.g., magnetite) the moments are also aligned on two antiparallel sublattices, but these are unequal here, so that the material has a net spontaneous magnetization. On heating a magnetically ordered material above a critical temperature, T_c, the ionic moments become completely disordered and fluctuate about zero—such material is *paramagnetic*. Minerals containing Fe in insufficient quantities to order magnetically are paramagnetic at all temperatures.

In recent years, condensed-matter physicists have lavished attention on noncrystalline and disordered solids. A group of these materials known as *spin glasses* exhibit freezing of the atomic moments in more or less random orientations in the vicinity of a spin-freezing temperature, T_f. Several noncrystalline varieties have been distinguished on the basis of their short-range magnetic order (Moorjani and Coey, 1984). Spin glasses also revert to a paramagnetic state above T_f.

The present paper is concerned with magnetic order in trioctahedral, largely ferrous phyllosilicates. A significant factor in the study of these materials is the magnetic anisotropy of Fe^{2+} which arises from the crystal field interaction and spin-orbit coupling. To understand the anisotropy, it is useful to consider first the paramagnetic susceptibility of phyllosilicate crystals that contain only a small amount of Fe, in the ferrous form. Anisotropy of Fe^{3+} is negligible by comparison. Susceptibility measurements also provide information about the exchange interactions which couple the Fe moments together. The interaction between ions on neighboring cation sites having spins $\vec{S}_i$ and $\vec{S}_j$ (the spin of the ion being proportional to its magnetic moment) is usually represented by the expression $-2J\vec{S}_i \cdot \vec{S}_j$, where J is the exchange constant. Positive J indicates a ferromagnetic interaction, whereas negative J indicates an antiferromagnetic interaction. Exchange is really important only if Fe ions on neighboring sites share a common oxygen ligand (Fe–O–Fe superexchange bonds). Thus, silicates whose Fe content is less than the percolation concentration should not order magnetically. Some of the relevant theory of magnetism

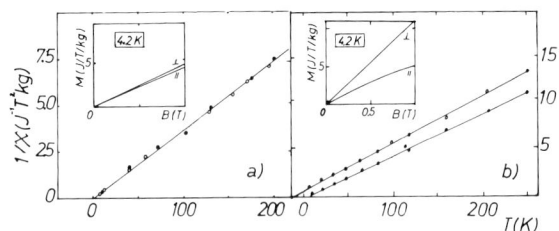

Figure 1. Inverse susceptibility and low-temperature magnetization curves of (a) vermiculite dilute in Fe^{3+} and (b) clinochlore dilute in Fe^{2+}. $\parallel$ (●) and $\perp$ (○) denote the orientation of the applied field relative to c′, the normal to the flakes.

was recently summarized by Coey (1986) with specific reference to Fe^{2+} and Fe^{3+}. Details of the magnetic properties of 1:1, 2:1, and 2:1:1 phyllosilicates, including descriptions of the samples, were reported by Coey et al. (1981, 1984), Ballet and Coey (1982), and Ballet et al. (1985a, 1985b). The present review deals with results for trioctahedral minerals in the paramagnetic and magnetically ordered states and shows how sensitively magnetic order depends on cation distribution.

PARAMAGNETISM

Crystal field

The susceptibility of two crystals which contain a small proportion of Fe^{2+} or Fe^{3+} were measured with the field parallel or perpendicular to c′, the normal to the sheets. The data are presented in Figure 1 as inverse susceptibility as a function of temperature because one anticipates a Curie-Weiss law of the form:

$$\chi = C/(T - \theta). \tag{1}$$

For vermiculite, the dilute ferric mineral, the intercept θ of the χ^{-1} vs. T plot is zero for both directions of applied field; the susceptibility is isotropic. The susceptibility, however, is anisotropic for the dilute ferrous mineral, clinochlore. The anisotropy is clearly associated with the Fe^{2+}. Susceptibility is greatest if the

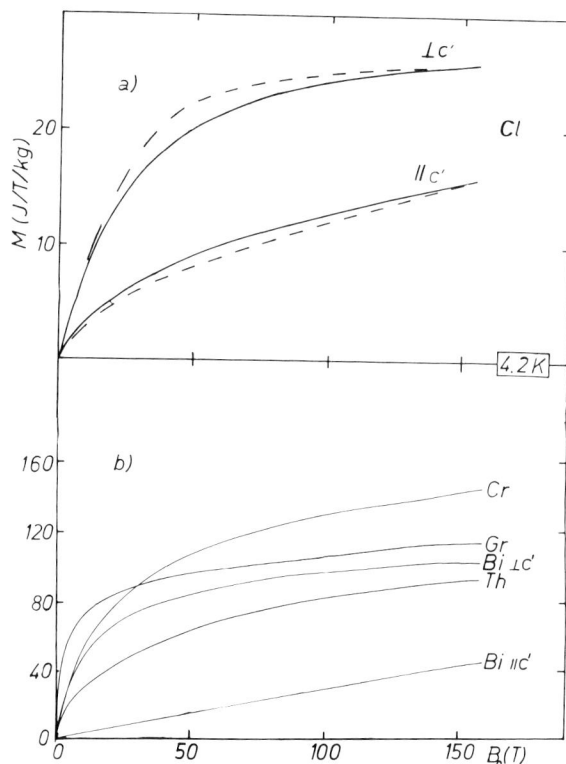

Figure 2. Magnetization curves at 4.2 K in large magnetic fields (a) clinochlore, dashed lines are calculated from the trigonal model with D = 20 K; (b) Fe-rich phyllosilicates, greenalite (Gr), cronstedtite (Cr), biotite (Bi), and thuringite (Th). All are powders except biotite.

field is in an "easy" direction, i.e., in the plane of the sheets, and least if it is parallel to c′. The difference in the intercepts ($\theta_\perp - \theta_\parallel$), after correction for the presence of a small fraction p_3 of Fe^{3+}, reflects the crystal field acting on Fe^{2+} in the octahedral sheet. A simple crystal field model which accounts well for a range of experimental information on ferrous sheet silicates, including the sign, magnitude, and direction of the nuclear

Table 1. Magnetic properties of trioctahedral phyllosilicates.[1]

	Layer type	Octahedral sheet spacing (Å)	D (K)	δ_t (K)	Z	p_3 (%)	$p_{3,3}$ (%)	θ (K)	J_{av} (K)	T_m (K)	Magnetic order
Ferrobrucite	1	4.6			6.0	0	0	30	1.3	34	planar antiferromagnet
Greenalite	1:1	7.2	22	1000	5.2	12	3	24	1.2	17	planar antiferromagnet
Cronstedtite	1:1				tetrahedral	ferric iron		18		6	antiferromagnet
Minnesotaite	2:1	9.6			4.5	8	1	38	2.1	19	planar antiferromagnet
Biotite	2:1	10.0	19	1200	5.3	13	4	43	2.0	7	spin glass
Thuringite	2:1:1	7.1	20	1100	3.9	25	8	17	1.1	5	spin glass

[1] D = anisotropy parameter; δ_t = trigonal crystal-field splitting; Z = magnetic coordination number; p_3 = percent Fe^{3+}; $p_{3,3}$ = percent Fe^{3+} having an Fe^{3+} nearest neighbor; θ = paramagnetic Curie temperature; J_{av} = average exchange interaction; T_m = temperature of susceptibility maximum.

electric field gradient, is a trigonally distorted octahedral field which splits the T_{2g} orbital triplet by an amount δ_t, so as to stabilize the $^5A_{1g}$ orbital singlet as the ground state. The spin Hamiltonian in the presence of spin-orbit coupling is then DS_z^2 (Varret, 1976) and $D \approx +15$ K. The pseudotrigonal c′ axis is thus a "hard" direction, and the ferrous moments tend to lie in the plane of the sheets. Values of δ_t and D for various minerals are listed in Table 1. The trigonal model is only an approximation; no distinction is made between M1 and M2 sites, nonaxial terms are neglected, and the distribution in crystal field parameters due to different cation environments is ignored. Nevertheless, the model provides a basis for understanding the magnetic properties of ferrous sheet silicates. Figure 2a shows magnetization curves measured $\parallel$ and $\perp$ to c′ for the clinochlore, which contains 8% Fe^{2+} in the octahedral sheets, and fits based on the trigonal model.

Exchange

A study of exchange interactions requires an examination of Fe-rich silicates. The magnetic coordination number Z is the number of Fe nearest-neighbors of an Fe ion in the octahedral sheet. The maximum value of Z is 6 in a trioctahedral sheet. By using powders, the effects of crystal field on the intercept θ of the χ^{-1} vs. T plot almost average to zero (Ballet and Coey, 1982). The value of θ is then proportional to Z and J as follows:

$$\theta = 2ZJS(S + 1)/3k, \qquad (2)$$

where S is the spin quantum number of the ion, and k is Boltzmann's constant (1.38×10^{-23} J/K). For Fe^{2+}, S = 2 and Eq. (2) reduces to $J = k\theta/4Z$.

Values of θ and Z for a range of predominantly ferrous phyllosilicates are listed in the table. Data of Miyamoto (1976) for synthetic ferrobrucite, $Fe(OH)_2$, are included for comparison. The average value of $J^{(2,2)}$, the exchange between ferrous ions, is 1.0 K for 1:1 phyllosilicates and thuringite, which has roughly equal Fe populations in its talc-like and brucitic sheets (Ballet et al., 1985a). For the 2:1 phyllosilicates, $J \approx 2.0$ K. In both types of phyllosilicates, J is positive, hence, the main exchange coupling between ferrous ions in adjacent edge-sharing octahedra of the octahedral sheet is rather weak, but ferromagnetic.

Inasmuch as ferrous phyllosilicates generally contain some Fe^{3+}, the sign and magnitude of exchange bonds involving Fe^{3+} is also relevant. The Fe^{3+}–O–Fe^{3+} exchange interaction ($J^{(3,3)}$ can be derived from the θ-values of the dioctahedral ferric minerals ferripyrophyllite and nontronite to be $J^{(3,3)} \approx -1.2$ K. The interaction is weak and antiferromagnetic. The sign of the Fe^{2+}–O–Fe^{3+} interaction may be inferred from the θ-value of glauconite. Two samples having ferric fractions $p_3 =$ 8% and 9% have $\theta = 10$ K and 8 K, respectively,

suggesting that $J^{(2,3)}$ is sufficiently strong and positive (~ 7 K) to overcome the negative $J^{(3,3)}$ interactions.

This survey of the paramagnetic properties of phyllosilicates establishes the main parameters that determine the magnetic order in the ground state, namely easy-plane anisotropy and ferromagnetic $J^{(2,2)}$ and $J^{(2,3)}$ coupling, but antiferromagnetic $J^{(3,3)}$ coupling. The exchange coupling between sheets, although much weaker than the intrasheet interactions, is also an important parameter.

COLLECTIVE MAGNETIC ORDER

Evidence that some form of magnetic ordering occurs at low temperatures in phyllosilicates comes from their magnetic properties and Mössbauer spectra. A selection of data is shown in Figures 3–5. A susceptibility maximum exists at a temperature, T_m, somewhere in the range 5–20 K. The maximum only appears if low fields, <0.1 T, are used for the measurement, because the magnetization <T_m is nonlinear in field. At temperatures much less than T_m, the magnetization curves of powder samples show that about half of the collinear-saturation magnetization is achieved in an applied field of about 2 T, but fields >20 T are required to approach saturation (Figure 2b). One reason why such large fields are needed is that the hard anisotropy axes of the crystallites in powder samples are randomly oriented with respect to the applied field. The magnetocrystalline anisotropy is represented by an effective anisotropy field in each crystallite $B^{\parallel} = 2DS^2/g\mu_B S$, which is about 50 T, where g is the Lande factor for Fe^{2+} (~ 2), μ_B is the Bohr magneton, and D is the anisotropy constant. Note, however, that the magnetization of a biotite crystal fails to saturate even if the field is applied perpendicular to c′. Hysteresis loops appear if the magnetization is measured in increasing and decreasing field, as shown in Figure 3. Substantial coercivity and remanence are present, but the latter decays slowly in time, proportional to ln t (Beausoleil et al., 1983).

Further evidence for magnetic order is provided by magnetic splitting in Mössbauer spectra, some of which at 4.2 K are shown in Figure 5. Fits of the spectra show that the quadrupole interaction $eqV_{zz}\sqrt{1 + \eta^2/3} \approx +3$ mm/s and $\eta \approx 0$, which is consistent with the orbital singlet ground state in a trigonally distorted octahedral crystal field. Hyperfine fields are in the range 13–16 T. The angle between the hyperfine field direction and the field gradient axis is close to 90°, as expected for moments lying in the plane of the sheets.

The nature of the magnetic order cannot be clearly inferred from the above results. Exchange interactions are predominantly ferromagnetic and ferromagnetic hysteresis is seen, but the susceptibility peak in low fields is typical of antiferromagnetic order. Magnetic neutron diffraction has therefore been used to establish the magnetic structures of five iron-rich trioctahedral

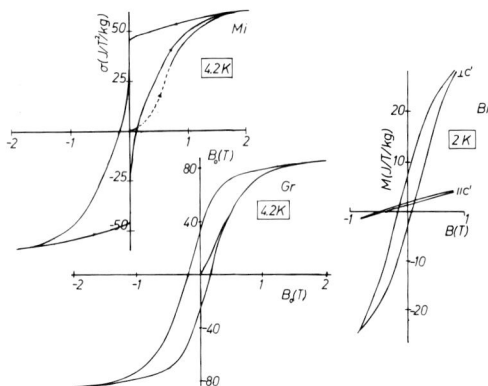

Figure 3. Hysteresis loops at low temperatures for greenalite (Gr), minnesotaite (Mi), and biotite (Bi). Biotite was cooled in a field of 0.85 T.

minerals; greenalite, cronstedtite, minnesotaite, biotite, and thuringite. Data were obtained at the Institute Laue Langevin, Grenoble, on the powder diffractometer D1B. Magnetic scattering in each mineral was obtained as the difference between the diffraction pattern at a temperature of 1.5 or 4.2 K and at a temperature much greater than T_m.

For greenalite and minnesotaite (ideal formulae: $Fe^{2+}_3Si_2O_5(OH)_4$ and $Fe^{2+}_3Si_4O_{10}(OH)_2$), a sequence of magnetic Bragg peaks can be indexed as $0\ 0\ (2n + 1)/2$ or $1\ 1\ (2n + 1)/2$, indicating a doubling of the magnetic periodicity in the c-direction. From an analysis of the peak intensities, it follows that the minerals order as planar antiferromagnets in which the moments lie in the planes of the sheets, as shown in the insert to Figure 6 (Coey et al., 1981; Ballet et al., 1985b; Townsend et al., 1985). This magnetic structure provides a natural explanation of the observed properties. Weak interplane exchange coupling is antiferromagnetic, leading to the low-field susceptibility peak. Fields in excess of 0.1 T destroy the antiferromagnetic order via a spin flop or metamagnetic transition, causing the moments in successive planes to align parallel. An analysis of the spin flop leads to a value of the total interplane interaction which is about 5% of the total intraplane interaction.

Cronstedtite (ideal formula: $Fe^{3+}Fe^{2+}_2(SiFe^{3+})O_5 \cdot (OH)_4$) is a special case because of its tetrahedral Fe^{3+}. It shows a different kind of antiferromagnetic order. Preliminary analysis of the data suggests that charge ordering and antiferromagnetic order both occur within the octahedral sheets.

The results on biotite and thuringite are unusual. No coherent magnetic Bragg scattering has been detected at temperatures as low as 1.5 K; only a minor amount of small-angle scattering of magnetic origin has been found which is indicative of short-range magnetic order. The minerals appear to be spin glasses. Apart from the smaller values of T_m and the coexistence of para-

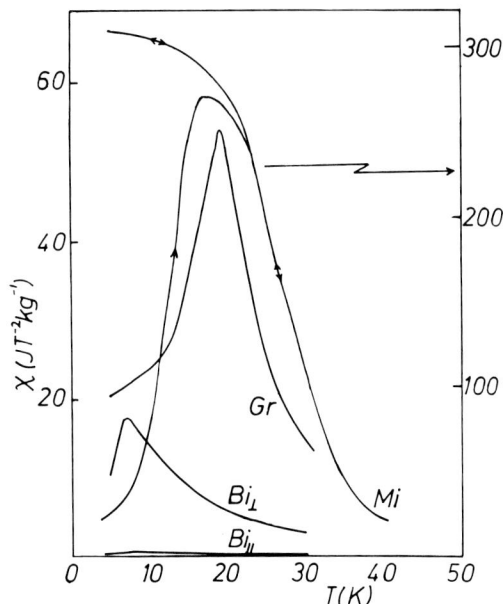

Figure 4. Susceptibility for Fe-rich trioctahedral minerals, measured in fields of order 0.05 T. Gr = greenalite, Mi = minnesotaite, Bi = biotite.

magnetic ferrous peaks in the magnetically split Mössbauer spectrum, the magnetic properties do not distinguish these spin glass minerals from their planar antiferromagnetic counterparts. Possible reasons for the absence of magnetic long-range order in biotite and thuringite are considered below.

One reason for the absence of magnetic long-range order might be that the interplane interactions are so weak that these minerals are true, two-dimensional magnetic systems (the theoretical 2d xy model which would correspond to the easy-plane anisotropy of Fe^{2+} does not exhibit long-range order at any temperature). Unfortunately, this explanation is difficult to sustain. The difference between the two minerals that exhibit planar antiferromagnetism and the two that only show spin freezing with no long-range magnetic order is unrelated to the distance between octahedral sheets: greenalite and thuringite are both 7-Å minerals from a magnetic viewpoint because Fe^{2+} in the latter occurs

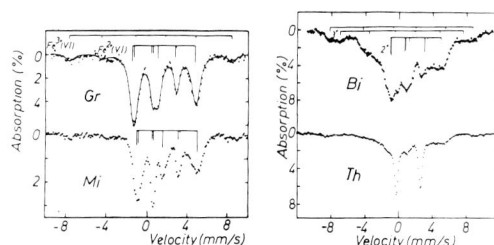

Figure 5. Mössbauer spectra at 4.2 K for Fe-rich trioctahedral minerals. Abbreviations are the same as in Figure 2. Mi = minnesotaite.

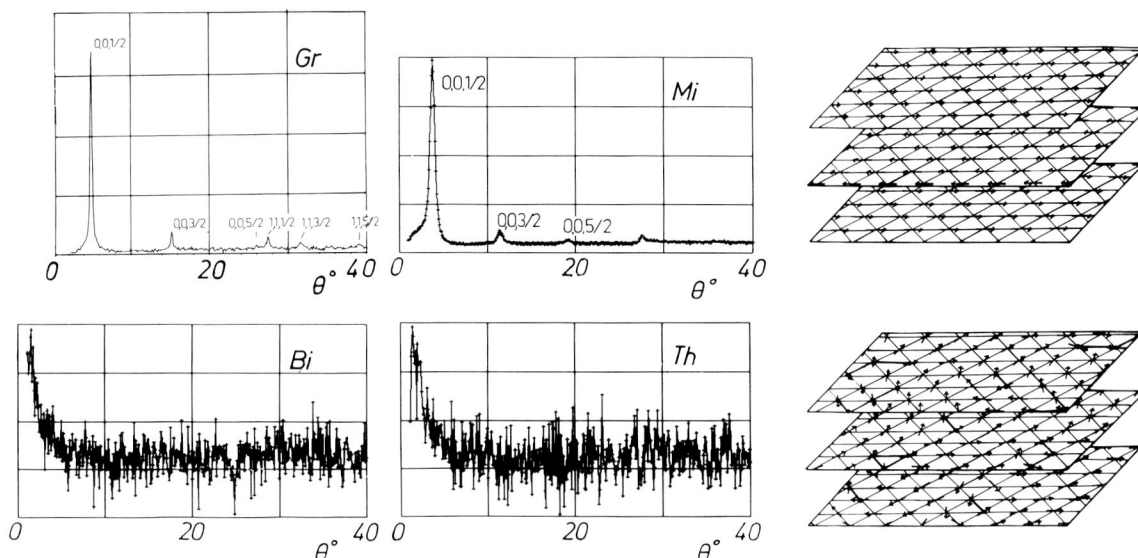

Figure 6. Magnetic neutron scattering for Fe-rich trioctahedral minerals. The proposed magnetic structures are illustrated.

in both the talc-like and brucitic sheets; minnesotaite and biotite are both 10-Å minerals. A more likely reason is the greater proportion of Fe^{3+} in the minerals that show no long-range order. Of critical importance is the proportion of Fe^{3+}-Fe^{3+} pairs on adjacent octahedral sites, because these pairs try to couple antiparallel and therefore tend to destroy the ferromagnetic order within each plane. Assuming random cation distributions, the proportions of these Fe^{3+}-Fe^{3+} pairs in thuringite and biotite are 8% and 4%, respectively, whereas in minnesotaite and greenalite, they are 1 and 3%, respectively.

The influence of antiferromagnetic pairs is best seen by computer simulation (Ballet *et al.,* 1985a; Coey, 1986). A few strongly coupled Fe^{3+}-Fe^{3+} pairs introduce magnetic vortices which reduce the magnetization of a plane of vector moments on a triangular lattice. A larger concentration of Fe^{3+}, particularly if nonmagnetic sites are also present (corresponding to Mg, Al . . . in the octahedral sheet), will destroy the ferromagnetism completely for quite reasonable values of the ratio of exchange interactions. Figure 7b shows only short-range order, with a correlation length of about 3 interatomic spacings (~ 10 Å).

SUMMARY

In summary, magnetocrystalline anisotropy of Fe^{2+} in phyllosilicates causes the spins to be in or near the planes of the sheets. Magnetic order in end-member ferrous trioctahedral phyllosilicates is of the planar antiferromagnetic variety. This ordering is not modified by modest substitution of nonmagnetic cations (e.g., Mg for Fe^{2+}); however, Fe^{3+} in quantities sufficient to introduce a few ferric pairs on a few percent of neighboring sites in the octahedral sheet will destroy the magnetic long-range order, planar antiferromagnetism giving way to random spin freezing having only short-range ferromagnetic correlations within the plane. Neutron diffraction is needed to distinguish the two possibilities, as the bulk magnetic properties are similar.

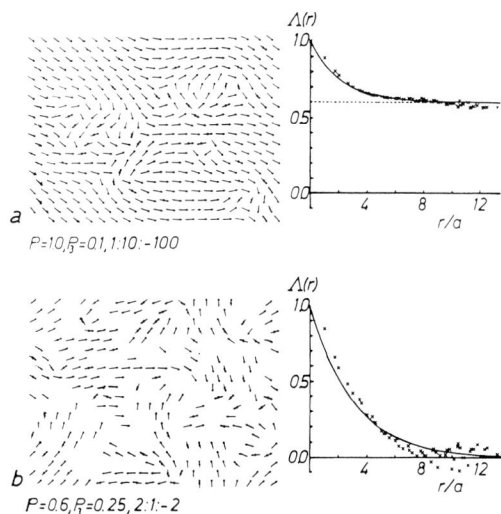

Figure 7. Monte Carlo simulations of magnetic ground state of triangular lattice and spin-spin correlation function $\lambda(r)$ plotted as function of distance (in units of interaction spacing). (a) p = 1; p_3 = 0.1; $J^{(2,2)}$:$J^{(2,3)}$:$J^{(3,3)}$ = 1:10:−100. (b) p = 0.6, p_3 = 0.25; $J^{(2,2)}$:$J^{(2,3)}$:$J^{(3,3)}$ = 2:1:−2.

REFERENCES

Ballet, O. and Coey, J. M. D. (1982) Magnetic properties of sheet silicates; 2:1 layer minerals: *Physics Chem. Minerals* **8**, 218–229.

Ballet, O., Coey, J. M. D., and Burke, K. J. (1985a) Magnetic properties of sheet silicates; 2:1:1 layer minerals: *Physics Chem. Minerals* **12**, 370–378.

Ballet, O., Coey, J. M. D., Mangin, P., and Townsend, M. G. (1985b) Ferrous talc—a planar antiferromagnet: *Solid State Commun.* **55**, 787–790.

Beausoleil, N., Levallée, P., Yelon, A., Ballet, O., and Coey, J. M. D. (1983) Magnetic properties of biotite micas: *Appl. Physics* **54**, 906–915.

Coey, J. M. D. (1986) Magnetic properties of iron in soil oxides and clay minerals: in *Iron in Soils and Clay Minerals,* J. W. Stucki and B. A. Goodman, eds., Reidel, Dordrecht, The Netherlands (in press).

Coey, J. M. D., Ballet, O., Moukarika, A., and Soubeyroux, J. L. (1981) Magnetic properties of sheet silicates; 1:1 layer minerals: *Physics Chem. Minerals* **7**, 141–148.

Coey, J. M. D., Chukhrov, F. V., and Zvyagin, B. B. (1984) Cation distribution, Mössbauer spectra, and magnetic properties of ferripyrophyllite: *Clays & Clay Minerals* **32**, 198–204.

Miyamoto, H. (1976) The magnetic properties of Fe(OH)$_2$: *Materials Research Bull.* **22**, 329–336.

Moorjani, K. and Coey, J. M. D. (1984) *Magnetic Glasses:* Elsevier, Amsterdam, 525 pp.

Townsend, M. G., Longworth, G., and Roudaut, E. (1985) Field-induced ferromagnetism in minnesotaite: *Physics Chem. Minerals* **12**, 9–12.

Varret, F. (1976) Crystal-field effects on high spin ferrous iron: *J. Physique* **37**, 437–456.

Proceedings of the International Clay Conference, Denver, 1985, L. G. Schultz, H. van Olphen, and F. A. Mumpton, eds.,
The Clay Minerals Society, Bloomington, Indiana, 267–272 (1987).

RAPID DETERMINATION OF SURFACE AREAS OF MINERAL POWDERS USING ADSORPTION CALORIMETRY[1]

P. P. S. Saluja,[2] D. W. Oscarson, H. G. Miller, and J. C. LeBlanc

Atomic Energy of Canada Limited, Whiteshell Nuclear Research Establishment
Pinawa, Manitoba R0E 1L0, Canada

Abstract—Adsorption-desorption calorimetry (ADCAL) was used to determine surface areas of 10 mineral powders, including 4 clay minerals, and carbon black. The method involved flowing a carrier liquid (pure heptane) through a small amount (<0.4 g) of sample in the sorption cell of a commercial microcalorimeter. When thermal equilibrium was established, pure heptane was replaced by a heptane solution containing n-butanol (n-BuOH) as an adsorbate. The maximum integral enthalpy of adsorption of n-BuOH on the surface was assumed to be proportional to the surface area of the sample. The empirical proportionality constant was determined using a reference sample (SiO_2 or kaolinite (KGa-1)) of known BET surface area.

Well-characterized powders of TiO_2 (anatase) and very poorly crystalline SiO_2, ZnO, and carbon black of known BET specific surface areas ranging from 24.3 to 0.64 m^2/g were used as standards to determine the internal consistency of the method. A precision of $\pm 4\%$ was obtained; the specific surface areas determined by ADCAL were within $\pm 5\%$ of the BET values. The method was applied to three nonexpanding clay minerals (a well crystallized (KGa-1) and a poorly crystallized (KGa-2) kaolinite and palygorskite (PFl-1)), one expanding clay mineral (Wyoming montmorillonite (SWy-1)), and two minerals with low specific surface areas (<0.5 m^2/g), olivine and microcline. Except for the expanding clay and the very low surface area microcline, the specific surface areas determined by ADCAL were within $\pm 4\%$ of the BET values. The advantages of this method are that it is fast (<2 hr/sample), only a small amount of sample is required (<0.4 g), and it can be conducted at room temperature.

Key Words—Adsorption calorimetry, n-Butanol, Kaolinite, Montmorillonite, Palygorskite, Surface area.

INTRODUCTION

A knowledge of the surface area of powdered material is of considerable importance in geological and engineering studies, for example, in evaluating the extent of mineral or nuclear wasteform dissolution, for quality assurance of powders in many industries, and as an aid in selecting effective sorbents for air pollutants. The most common methods of measuring surface area used in clay and soil research involve the determination of the amount of adsorbed inert gas, such as nitrogen or krypton (BET method), or of polar molecules, such as glycerol, ethylene glycol, or water, required to form a monolayer on the solid surface (van Olphen, 1970; van Olphen and Fripiat, 1979). The surface area is then calculated from the cross sectional area of the adsorbed molecules at monolayer coverage.

These methods, however, have a number of limitations, such as the need for low temperatures ($-196°C$) and extensive outgassing (as in the BET method) and the long time and the relatively large amount of sample required per run. We have explored an alternative method in which the surface area is derived from the enthalpy of saturation adsorption of n-butanol (n-BuOH) from a heptane solution, using adsorption-desorption calorimetry (ADCAL). The surface area of the sample is assumed to be proportional to the enthalpy of adsorption. The proportionality constant is determined using a reference powder of known BET area.

EXPERIMENTAL

Materials

Four, well-characterized materials, SiO_2, TiO_2, ZnO, and carbon black powder (CBP), obtained from Duke Scientific Corporation, Palo Alto, California, were used as standards to determine the internal consistency of the method. The TiO_2 was anatase (β-TiO_2); the SiO_2, ZnO, and CBP were very poorly crystalline materials as indicated by the presence of only one or two broad peaks on X-ray powder diffraction (XRD) patterns. No crystalline phases were identified. The specific surface areas and particle size distributions of these materials are given in Table 1.

Minerals. Two kaolinites (KGa-1 and KGa-2), palygorskite (PFl-1), and montmorillonite (SWy-1) were obtained from the Source Clays Repository of The Clay Minerals Society. The mineralogical composition and physicochemical data on these clays were listed by van

[1] Issued as Atomic Energy of Canada Limited publication AECL-8886.

[2] To whom correspondence should be addressed.

267

Table 1. Specific surface area and particle-size distribution of four reference materials.

Material	Specific surface area (m^2/g)[1]	Particle-size distribution[2] (%)						
		$<44\ \mu m$	$44–74\ \mu m$	$74–149\ \mu m$	$149–177\ \mu m$	$177–250\ \mu m$	$250–420\ \mu m$	$>420\ \mu m$
SiO_2 (209)[3]	24.3	14.5	23.7	33.8	3.6	8.2	9.1	7.1
TiO_2 (203)	10.3	0	0.8	23.6	6.5	17.6	28.1	23.4
ZnO	0.64	17.9	19.1	50.6	2.7	7.1	1.5	1.1
Carbon black powder (199)	6.6	1.1	6.8	26.9	5.7	14.9	24.3	20.3

[1] BET surface areas obtained from Duke Scientific Corporation, Palo Alto, California.

[2] Determined by dry sieving.

[3] Numbers in parentheses are the catalogue numbers used by Duke Scientific Corporation; no catalog number is available for ZnO.

Olphen and Fripiat (1979). Samples of microcline and olivine were obtained from the Geological Survey of Canada; they were fractionated by dry sieving, and the 74–149-μm particle-size fraction was used in this study. The specific surface areas of this particle-size fraction of the microcline and olivine were determined by the BET method by Particle Data Laboratories, Elmhurst, Illinois.

The mineralogical composition of the minerals was confirmed by XRD. Two reference materials (SiO_2 and ZnO) and the microcline and olivine were outgassed at ~200°C under vacuum (0.1 Pa) for 16 hr prior to the surface-area measurement. The surface areas of the TiO_2, CBP, and clay mineral samples were determined without pretreatment.

Apparatus and procedure

ADCAL experiments. ADCAL experiments were performed in a commercial heat-conduction-type sorption microcalorimeter (Model LKB 2107-122 from LKB-Produkter AB, Bromma, Sweden) capable of operating in both flow and batch modes (Figure 1). The microcalorimeter was contained in an air thermostat (Model LKB 2107-210) maintained at 25° ± 0.005°C, using a refrigerated circulating bath. A peristaltic pump (Microperpex Model LKB 2132) was used to flow solutions, either carrier fluid (pure heptane in this work) or a dilute solution of probe adsorbate (n-BuOH in heptane in this work) from the reservoir bottles through the sorption cell (Figure 1). Additional heat-exchanger tubing inside the air bath provided a pre-thermostatting of the solutions to the calorimeter temperature before solutions came in contact with the powder in the flow cell.

The flow-sorption cell had an inside diameter of 0.6 cm, and the adsorbent bed in the cell was 1.8 cm long (Figure 1). This cell, containing a weighed amount of sample (typically 0.05–0.4 g), was inserted in the sample compartment of the twin-cell (differential) calorimeter, and the tube at the top was connected to the waste container downstream. The reference compartment was unoccupied, or an empty batch sorption cell was inserted into it. An upward flow of solution through the sample in the cell eliminated the formation of air bubbles and the possibility of clogging the filters in the cell; thus, efficient and uniform contact was achieved between the liquid and the solid phases. Thermal equilibrium between the sample and carrier liquid was reached in <2 hr at a typical flow rate of 0.003 cm³/s. The method can be made even faster by pre-thermostatting the sample preloaded in the cell with heptane. The calorimetric response arose from heat conduction across thermopiles located between the flow-sorption vessel and the heat-sink assembly. When the temperatures of the flow-sorption vessel and the heat-sink assembly were different, a voltage was produced at the output of the thermopiles. This output was amplified and fed either to an integrating recorder or a digital readout system.

At equilibrium, heptane molecules adsorbed on the powder surface were continuously exchanged with heptane molecules in solution; no net heat was evolved or absorbed. This condition yielded a stable (center zero) base line. A three-way chromatographic valve was then switched to replace pure heptane with a heptane solution containing n-BuOH. The heptane molecules on the powder surface were continuously replaced by n-BuOH molecules (the rate of replacement being proportional to the concentration of n-BuOH in solution) until a dynamic equilibrium was reestablished.

The voltage output from the thermopiles in ADCAL experiments was proportional to the amount of heat evolved or absorbed during preferential adsorption or desorption of n-BuOH, which in turn, was proportional to the surface area of the sample powder. Therefore, the voltage-time curve had to be integrated to find the peak area. Peak areas were quantified by an electronic integrating recorder or calculated from a cubic spline integration routine using the digital data collected by a Tektronix 4051 computer (Noll and Burchfield, 1982). To convert peak area to heat of adsorption ($\Delta_a H$) or desorption ($\Delta_d H$), an electrical calibration was performed by passing current through a built-in heater coil embedded in the flow-cell compartment of the calorimeter.

Electrical calibration experiment. The duration and magnitude of the calibration current were selected to

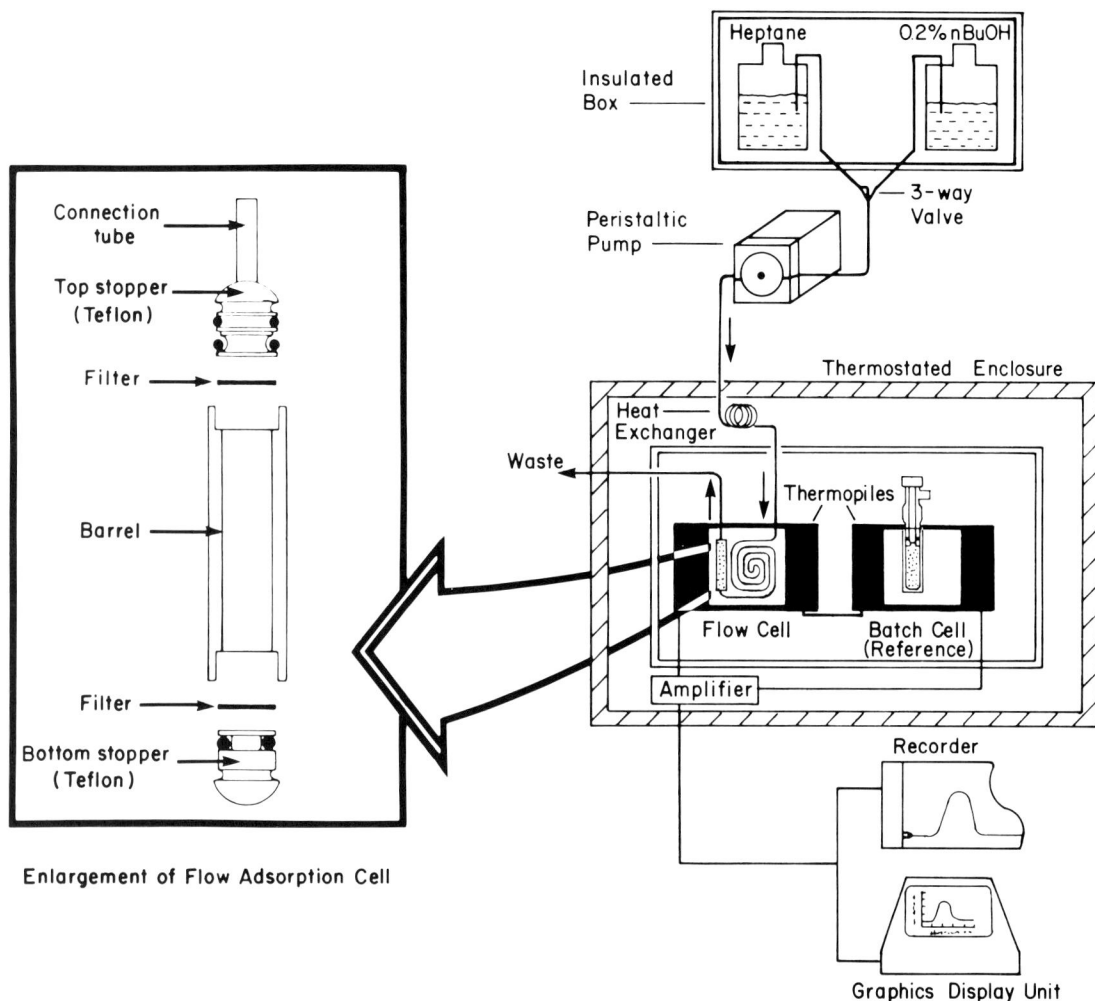

Figure 1. Schematic diagram of the adsorption microcalorimeter.

produce a response curve, similar to the peaks observed during actual ADCAL experiments on a sample. The flow rate of the liquid and the temperature of the bath were held constant during this calibration. The heater response to the calibration was evaluated as:

$$\Delta H_{ec} = I^2 Rt = K_{ec} \int_0^t V \, dt = K_{ec} A_{ec} \, ,$$

where ΔH_{ec} is the electrical energy in joules, I is the calibration current in amperes, R is the resistance of the flow-cell heater in ohms (49.76 Ω for this calorimeter block), t is the time in seconds, V is the amplified voltage output of the thermopiles, A_{ec} is the area under the calibration peak in arbitrary units, and K_{ec} (J/area) is the calibration constant relating these area units to the calibration energy in joules. Inasmuch as the heat flow from the flow cell across the thermopiles was symmetrical, exothermic calibration was applicable to both exothermic and endothermic experiments.

Figure 2 shows a typical adsorption-desorption thermogram. From a series of runs with increasing n-BuOH concentrations ranging from 0.01 to 2.0 vol. %, the concentration was determined beyond which no incremental change in the enthalpy of adsorption was detected (0.2 vol. % n-BuOH in these experiments). At this concentration a complete monolayer was adsorbed. Because no data were available, however, on the amount adsorbed or on the orientation of the adsorbed molecules on the surface, it is not clear whether or not additional n-BuOH molecules were associated with those in a monolayer, for example, by dipole–dipole association, which could also have contributed to the observed enthalpy. Therefore, the coverage at maximum enthalpy production was defined as "saturation coverage" rather than "monolayer coverage." In routine work, a higher concentration (0.2 vol. % in these experiments) was used ("saturation experiments"). The saturation coverage was assumed to be

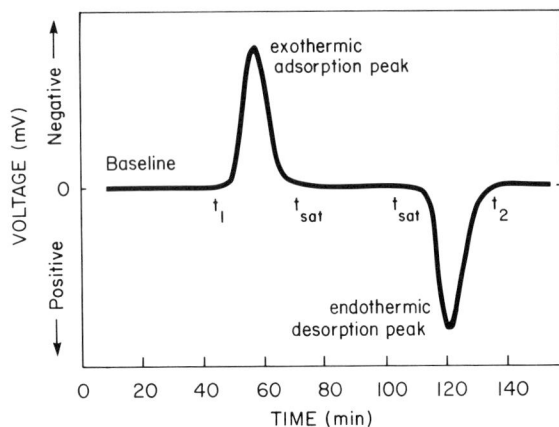

Figure 2. Typical adsorption-desorption thermogram.

a measure of the number of adsorption sites on the surface and therefore of the surface area; the composition of the adsorbed layer(s) was assumed to be identical for solids of a given class, e.g., oxides or clay minerals. Many experiments with a variety of adsorbents need to be carried out to establish an appropriate classification of solid surfaces. The first phase of our surface area investigations on selected classes of materials (oxides, clay minerals, and carbon black powder) is reported in this paper.

The reversibility of adsorption and desorption of n-BuOH was tested by alternately flowing heptane and 0.2 vol. % n-BuOH solution through the sample several times. Essentially complete reversibility was obtained for powders with specific surface areas >5 m²/g. Thus, several runs were made on the same starting sample. For powders with specific surface areas <1 m²/g, reversibility decreased with the number of adsorption-desorption cycles, probably because of a significant contribution from impurities on the surface of the samples and in the flowing organic solutions. Thus, several independent surface-area measurements on freshly outgassed samples of minerals with relatively low surface areas (e.g., ZnO, olivine, and microcline) were conducted.

RESULTS AND DISCUSSION

Enthalpy of adsorption at saturation (ΔH_{sat})

Once calibrations were complete, the net heat of adsorption ($\Delta_a H_{sat}$) or desorption ($\Delta_d H_{sat}$) at saturation was calculated by multiplying the area of the ADCAL peaks (A_a or A_d) by K_{ec}. For adsorption, $\Delta_a H_{sat} = K_{ec} A_a$; for desorption, $\Delta_d H_{sat} = K_{ec} A_d$, where

$$A_a = \int_{t_1}^{t_{sat}} V_a \, dt \quad \text{and} \quad A_d$$

$$A_d = \int_{t_{sat}}^{t_2} V_d \, dt \qquad \text{(see Figure 2).}$$

(In Figure 2, $t_{sat} - t_1$ and $t_2 - t_{sat}$ are the duration of the "saturation" adsorption and desorption experiments, respectively.)

For complete reversibility,

$$A_a = -A_d \text{ and } \Delta_a H_{sat} = -\Delta_d H_{sat} = \Delta H_{sat}.$$

Calculation of specific surface area

The specific surface area of the materials examined ranged from <1 to ~130 m²/g. The surface area was deduced from the measured ΔH_{sat} during a saturation adsorption experiment. Inasmuch as both the specific surface area (S) and the integral enthalpy of adsorption (ΔH_{sat}) of n-BuOH per unit weight (w) of sample at saturation were proportional to the number of adsorption sites,

$$S(m^2/g) = k_{SA}(m^2/J)\Delta H_{sat}(J)/w(g) \qquad (1)$$

where k_{SA} is a proportionality constant representing an adsorbent-adsorbate surface calibration factor. This proportionality constant was determined by conducting a saturation adsorption experiment using a well-characterized reference material—a surface area standard—for the class of minerals under investigation. Using the n-BuOH/heptane system and the SiO₂ as the surface area standard (specific surface area $= 24.3$ m²/g) for oxides, the k_{SA} value was determined to be 9.7 m²/J. This value is in fair agreement with a value of 11.5 m²/J obtained for nonporous oxide powders, reported by Groszek (1966) and Steinberg (1981).

The k_{SA} factor was assumed to be the same for a given class of adsorbents, e.g., oxides, and the same adsorbate and carrier liquid. The factor may be different for different classes of adsorbents.

Using the k_{SA} factor, determined for the SiO₂ reference material, the applicability of ADCAL was first examined for three other reference materials, TiO₂, ZnO, and CBP.

Reference materials

The surface areas of the three reference materials determined by ADCAL were reasonably reproducible and in fair agreement with those determined by the BET method as given by the suppliers of these materials (Table 2). The good agreement between the ADCAL and BET values suggests that the surfaces accessible to N₂ in the BET method and n-BuOH in the ADCAL method were very similar for these materials.

Clay minerals

The specific surface areas of kaolinite KGa-2 and the palygorskite determined by ADCAL, using kaolinite KGa-1 as a standard, are 23.4 ± 2.5 and 126 ± 11 m²/g, respectively. The k_{SA} factor for kaolinite KGa-1 was 9.7 m²/J. These specific surface area values agree with the values of 23.5 and 136 m²/g, respectively, obtained by the BET method (Table 2). Because AD-CAL gave reasonable values for the surface areas of nonexpanding clays, such as kaolinite and palygorskite, the method shows promise and may be a very useful tool for exploring the surface chemistry of clays, such as structural changes on heating and chemical changes resulting from various surface pretreatments (e.g., lime treatment of coal powders for decreasing SO_x emissions).

In addition to the minerals listed in Table 2, a specific surface area of ~ 200 m^2/g for montmorillonite SWy-1 was determined using ADCAL. This value is higher than the N_2 BET area of 31 m^2/g (van Olphen and Fripiat, 1979) and much lower than the total theoretical specific surface area of montmorillonite of ~ 750 m^2/g (van Olphen, 1977) and the glycerol area of ~ 660 m^2/g (van Olphen and Fripiat, 1979). Apparently the interlayer region was not fully occupied by n-BuOH, possibly because of competition from residual interlayer water in the sample which was not outgassed. In addition, the k_{SA} factor determined for kaolinite KGa-1 may not have been applicable for montmorillonite. Thus, an appropriate internal standard (e.g., another swelling clay, rather than kaolinite) should be used to obtain k_{SA} and the surface area for smectite clays. In principle, the total surface area (or just the external surface area) of swelling clays may possibly be measured by an educated choice of a carrier liquid and/or probe adsorbate. For montmorillonite, for example, it may be necessary to use an adsorbate molecule (e.g., glycerol or ethylene glycol) with a greater bonding affinity or more favorable stereochemistry for the interlayer region of montmorillonite than n-BuOH. The applicability of the ADCAL method for determining the surface areas of different groups of minerals is under investigation.

Minerals with low surface area

Microcline and olivine, minerals with relatively low surface areas, were also examined using ADCAL. In preliminary experiments, surface area determinations were conducted using untreated sample powders as received. Inasmuch as minerals with low surface areas must be free of impurities, experiments were repeated after outgassing the samples for 16 hr at $\sim 200°C$ under vacuum (0.1 Pa). The results obtained from ADCAL for outgassed and untreated microcline and olivine and the values determined by the BET method are compared in Table 2. For olivine, the surface area of the untreated and outgassed samples determined by AD-CAL agrees well with the BET results (Table 2). For outgassed microcline, the surface area measured by ADCAL is about twice the BET value (Table 2). For minerals with specific surface areas <0.2 m^2/g, however, the BET method is unreliable (Lowell and Karp, 1972). Further work to refine the optimum outgassing

Table 2. Comparison of specific surface areas determined by n-BuOH adsorption using a flow microcalorimeter and by N_2 adsorption using the BET method.

Material	Specific surface area (m^2/g)	
	Adsorption calorimetry[1]	BET
SiO_2	24.3[2]	24.3[3]
TiO_2	11.6, 10.0	
Average	10.8 ± 1.1[4]	10.3[3]
Carbon black powder	6.29, 6.11	
Average	6.20 ± 0.13	6.6[3]
ZnO (outgassed)	0.61, 0.53	
Average	0.57 ± 0.06	0.64[3]
Olivine (outgassed)	0.49	
	0.26	
	0.33	
	0.34	
Average	0.36 ± 0.10	0.34[5]
Olivine (untreated)	0.15, 0.13	
Average	0.14 ± 0.01	0.13[5]
Microcline (outgassed)	0.29	
	0.23	
	0.23	
	0.13	
	0.18	
	0.13	
Average	0.20 ± 0.06	0.08[5]
Microcline (untreated)	0.07, 0.05	
Average	0.06 ± 0.01	0.11[5]
Kaolinite (KGa-1)	10.1[2]	10.1[6]
Kaolinite (KGa-2)	24.7	
	19.5	
	22.4	
	24.5	
	25.9	
Average	23.4 ± 2.5	23.5[6]
Palygorskite	135	
	114	
	129	
Average	126 ± 11	136[6]

[1] Specific surface areas for all materials determined using ADCAL were calculated from Eq. (1) in text using a value of 9.7 m^2/J for the proportionality factor, k_{SA}. This value was determined from heat of adsorption experiments on two reference materials, SiO_2 and kaolinite (KGa-1), having BET specific surface areas of 24.3 and 10.1 m^2/g, respectively. The k_{SA} factor was assumed to be the same for each class of material reported in this table.
[2] Used as internal standards in this study.
[3] BET surface areas were obtained from the supplier, Duke Scientific Corporation, Palo Alto, California.
[4] Average ± one standard deviation.
[5] Values determined by BET method by Particle Data Laboratories, Elmhurst, Illinois.
[6] Values from van Olphen and Fripiat (1979).

temperature, time, and vacuum for minerals with low surface areas is continuing.

CONCLUSIONS

A technique using adsorption-desorption calorimetry was developed for the rapid determination of the surface area of a wide variety of mineral powders. The method is fast (<2 hr/sample) and can be adapted for almost routine operation. Other advantages of this method are that only a small amount of sample is required (<0.4 g) and the determination can be conducted at room temperature, rather than at a very low temperature ($-196°C$), as required for the BET method. Furthermore, subtle changes in the surface structure and chemistry of minerals can probably be detected which are difficult to detect by other methods of analysis. Thus, further exploitation of sorption microcalorimetry for the determination of surface area and as a tool for examining the surfaces of minerals should provide valuable information.

ACKNOWLEDGMENTS

We thank A. G. Wikjord and T. T. Vandergraaf for providing several of the samples and K. V. Ticknor for conducting experiments on untreated microcline and olivine.

REFERENCES

Groszek, A. J. (1966) Determination of surface areas of powders by flow microcalorimetry: *Chem. and Ind.,* 1754–1756.

Lowell, S. and Karp, S. (1972) Determination of low surface areas by the continuous flow method: *Anal. Chem.* **44,** 1706–1707.

Noll, L. A. and Burchfield, T. E. (1982) Silica gel as a model surface for adsorption calorimetry of enhanced oil recovery systems: *Bartlesville Energy Tech. Center Rept.* **DOE/BETC/R1-82/7,** 1–45.

Steinberg, G. (1981) What you can do with surface calorimetry: *CHEMTECH,* 730–737.

van Olphen, H. (1970) Determination of surface areas of clays—evaluation of methods: in *IUPAC Supplement to Pure and Applied Chemistry, Surface Area Determination,* Butterworths, London, 255–271.

van Olphen, H. (1977) *An Introduction to Clay Colloid Chemistry:* 2nd ed., Interscience Publishers, New York, 301 pp.

van Olphen, H. and Fripiat, J. J. (1979) Surface area: in *Data Handbook for Clay Materials and other Non-Metallic Minerals,* H. van Olphen and J. J. Fripiat, eds., Pergamon Press, Oxford, 203–216.

Proceedings of the International Clay Conference, Denver, 1985, L. G. Schultz, H. van Olphen, and F. A. Mumpton, eds.,
The Clay Minerals Society, Bloomington, Indiana, 273–276 (1987).

HYDROXY-COPPER INTERLAYERING IN MONTMORILLONITE BY THE TITRATION METHOD

SHOJI YAMANAKA, KOICHI NUMATA, AND MAKOTO HATTORI

Department of Applied Chemistry, Faculty of Engineering, Hiroshima University
Higashi-Hiroshima 724, Japan

Abstract—Hydroxy-copper montmorillonite was prepared by titrating copper solutions with sodium hydroxide in the presence of Na-montmorillonite. If a copper nitrate solution was used, basic copper nitrate $Cu_2(OH)_3NO_3$ was precipitated without forming hydroxy-copper interlayers. If a copper acetate solution, however, was titrated similarly in the presence of montmorillonite, the basal spacing of the montmorillonite increased to 20.2 Å, and no basic copper acetate $Cu_2(OH)_3(OCOCH_3)\cdot H_2O$ was detected until hydroxy-copper layers had fully developed in the interlayer spaces. Chemical analyses showed that the resulting hydroxy interlayers were basic copper acetate, although the amount of acetate ions was small compared with that in $Cu_2(OH)_3(OCOCH_3)\cdot H_2O$. A similar hydroxy-copper interlayering occurred in montmorillonite even when Na-montmorillonite was stirred together with freshly precipitated $Cu_2(OH)_3(OCOCH_3)\cdot H_2O$ in water. The basic copper acetate interlayers decomposed on heating above 300°C, and CuO deposited on the outer surfaces of the montmorillonite.

Key Words—Copper, Hydroxy-copper interlayering, Montmorillonite, Pillared interlayer complex, Titration.

INTRODUCTION

The properties of Cu^{2+}-exchanged smectites have been extensively studied. Pinnavaia and Mortland (1971) showed that the Cu^{2+} in the interlayer spaces forms novel arene complexes which have not been found in homogeneous solutions. Cu^{2+} in the interlayer spaces is also a useful electron spin resonance (ESR) probe for the study of hydration and dehydration mechanisms of clays, chiefly because Cu^{2+} has a single unpaired electron with a spin of $S = \frac{1}{2}$ and gives an anisotropic ESR signal. ESR studies of such materials were reviewed by Pinnavaia (1980). If adsorbed hydrocarbons having an olefinic double bond are heated in Cu^{2+}-montmorillonite interlayers, the respective dimers and trimers are obtained (Thomas, 1982). Other unusual chemical conversions of organic molecules using Cu^{2+}-montmorillonite were reviewed by Thomas *et al.* (1977).

Recently, Cu^{2+}-exchanged fluor-tetrasilicic mica ($Cu_{0.5}Mg_{2.5}Si_4O_{10}F_2$) has been found to catalyze the dehydrogenation of methanol to methyl formate selectively (Morikawa *et al.*, 1982). The high selectivity of the tetrasilicic mica in the reaction was attributed to its weak solid acidity and to high resistivity of the exchanged Cu^{2+} toward the reduction to Cu metal.

The above studies, however, were all performed on simply ion-exchanged Cu^{2+}-montmorillonite. If Cu^{2+} ions of more than the cation-exchange capacity (CEC) can be introduced into montmorillonite, the montmorillonite should behave differently. Yamanaka and Brindley (1978) showed that Ni^{2+} ions in excess of the

CEC precipitated in the form of hydroxy-nickel ions into the interlayer region of montmorillonite, when a nickel nitrate solution was titrated in the presence of montmorillonite. In the present study, an attempt has been made to introduce Cu^{2+} ions in excess of the CEC into montmorillonite by a similar titration method.

EXPERIMENTAL

Materials

Sodium montmorillonite used in this study was from the Tsukinuno district, Yamagata Prefecture, Japan, the same sample as used in a previous study (Yamanaka *et al.*, 1984). Its structural formula is $(Na_{0.35}K_{0.01}Ca_{0.02})(Si_{3.89}Al_{0.11})(Al_{1.60}Fe_{0.09}Mg_{0.32})O_{10}(OH)_2\cdot n\,H_2O$; its CEC is 100 meq/100 g.

Copper basic acetate $Cu_2(OH)_3(OCOCH_3)\cdot H_2O$ was prepared by hydrolysis of a copper acetate solution with sodium hydroxide (Gauthier, 1956). The precipitate was separated by filtration and washed with water. Part of the filtrate was dried in air and submitted for chemical analysis (Found: C, 9.28; H, 3.13; Cu, 49.5%. Calculated for $Cu_2(OH)_3(OCOCH_3)\cdot H_2O$: C, 9.41; H, 3.14; Cu, 49.8%). The other part of the washed filtrate was dispersed again in distilled water and used for the reaction with clay.

Titrations

Potentiometric titrations were carried out by the batch method according to the following procedure. Separate 400-mg samples of Na-montmorillonite were dispersed in 30 ml of 0.1 M $Cu(NO_3)_2$ or 0.1 M

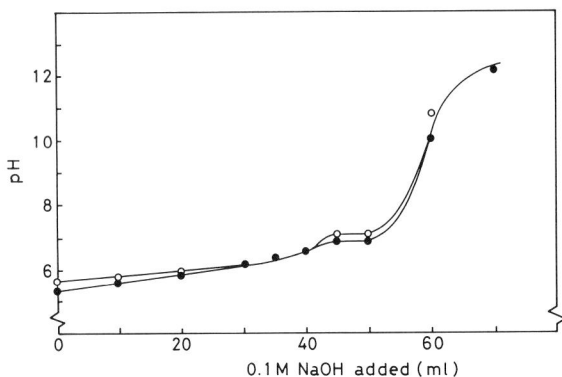

Figure 1. Potentiometric titration curves of copper acetate solution (30 ml, 0.1 M solution) with 0.1 M NaOH solution. ● = in presence of Na-montmorillonite (400 mg); ○ = without clay.

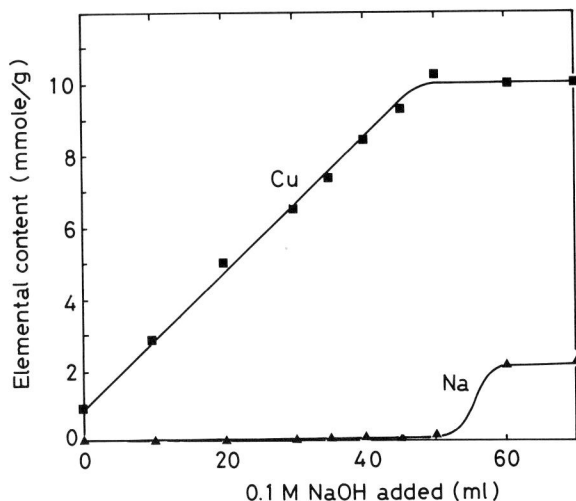

Figure 2. Copper (■) and sodium (▲) contents (mmole/g of silicate layer) in samples separated at different titration stages.

$Cu(OCOCH_3)_2$ solutions, and different amounts of 0.1 M NaOH were added to the dispersions. After constant stirring in an air-tight bottle for 6 days, the pH of each dispersion was recorded, and the solid was separated and washed with water repeatedly by centrifugation. Part of the wet sample was spread on glass slides and dried in air for X-ray powder diffraction (XRD) measurements (Ni-filtered CuKα radiation). The rest of the wet sample was dried by evacuation at room temperature and submitted for Cu and Na analyses. For comparison, similar titrations of copper nitrate and acetate solutions were performed without clay.

RESULTS

As shown in a previous study (Yamanaka and Brindley, 1978), nickel hydroxide is not precipitated during the titration of a nickel nitrate solution with NaOH in the presence of montmorillonite until the brucite-like hydroxy-nickel interlayers form in the clay. In the present study, however, during the titration of a copper nitrate solution, copper basic nitrate crystals, $Cu_2(OH)_3NO_3$, precipitated irrespective of whether montmorillonite was present or not. This finding suggests that copper hydroxide was introduced into the interlayer spaces of montmorillonite by the titration of the nitrate solution.

On the contrary, during the titration of a copper acetate solution with NaOH in the presence of montmorillonite, copper basic acetate did not precipitate in the early stages of titration, although in the absence of the clay, copper basic acetate precipitated with the addition of NaOH. The potentiometric titration curves of copper acetate solutions with and without the presence of clay are shown in Figure 1. The pHs measured in the presence of the clay were slightly lower than those measured in the absence of the clay. The two titration curves meet at about 40 ml of NaOH added. If additional NaOH was added to the copper acetate

solution in the absence of clay, the precipitated basic salt was copper hydroxide, and the titration curve showed slight increases in the pHs. It is well known that copper hydroxide is subject to dehydration even in a solution (Cotton and Wilkinson, 1972). If the copper hydroxide precipitate was kept in contact with the solution for several days, it gradually converted into brown CuO by dehydration. Similar changes were also observed even when clay was present in the acetate solution. In the presence of clay, however, the copper hydroxide dehydrated to CuO so rapidly that the dispersion changed to deep brown within several hours. The clay surfaces seem to have catalyzed the dehydration of the hydroxide to CuO.

The solid obtained at each titration stage shown in Figure 1 was analyzed for Cu and Na (Figure 2). The copper content increased with titration until all the copper ions in the solution had precipitated by the addition of 50 ml of NaOH solution. The clay in equilibrium with the acetate solution contained 1.0 mmole of Cu^{2+}/g of clay, that is, twice as much Cu^{2+} ions as the CEC of the clay before titration. During the final stages of the titration, most of the Cu^{2+} precipitated; therefore, the concentration of Cu^{2+} in the solution decreased markedly, whereas that of Na^+ increased linearly as the titration proceeded. The increase in the Na^+ content of the solid separated at the final stages of titration can be interpreted in terms of the reverse ion exchange between the interlayer hydroxy-copper ions and Na^+ in the solution.

Figure 3a shows the XRD pattern of the solid separated after titrating 30 ml of 0.1 M copper acetate solutions with 40 ml of 0.1 M NaOH in the presence of 400 mg of Na-montmorillonite. The basal spacing is regular, although the 001 reflection corresponding

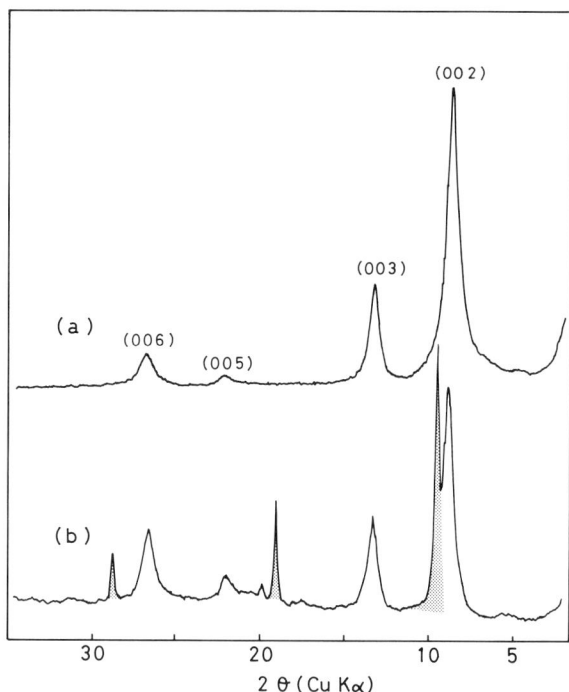

Figure 3. X-ray powder diffraction pattern of the montmorillonite (400 mg) titrated with 0.1 M NaOH (40 ml) in 30 ml of copper acetate solution (a); that of 300 mg of montmorillonite similarly titrated (b); shaded diffraction peaks are of $Cu_2(OH)_3(OCOCH_3) \cdot H_2O$.

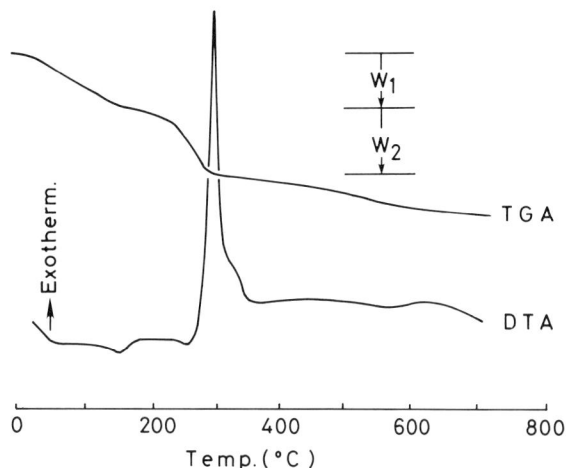

Figure 4. Differential thermal and thermogravimetric analysis curves of the 20.2-Å phase measured for a heating rate of 10°C/min.

to a spacing of 20.2 Å is missing. Similar titrations carried out in the presence of smaller amount of clay caused excess Cu^{2+} to be deposited as $Cu_2(OH)_3(OCOCH_3) \cdot H_2O$, which gave an additional sharp XRD pattern, as shown in Figure 3b.

An elemental analysis of the 20.2-Å phase shown in the XRD pattern in Figure 3a gave Cu and C contents of the air-dried sample of 25.02 and 1.89%, respectively. The weight loss on heating to 900°C was 20.3%. The considerable carbon content suggests that acetate ions in the solution were incorporated into the hydroxy-copper interlayers. From these analytical data, the composition of the interlayer material was calculated to be $[Cu_{2.8}(OH)_{4.6}(OCOCH_3)_{0.56} \cdot n\,H_2O]^{+0.4}$, on the basis of $O_{10}(OH)_2$. In the calculation, the composition and charge of the silicate layer were assumed to be unchanged by the titration. The analytical results indicate that the solid phase having a basal spacing of 20.2 Å contained almost fully developed hydroxy-copper interlayers, part of the hydroxy groups being replaced with acetate groups.

Freshly precipitated $Cu_2(OH)_3(OCOCH_3) \cdot H_2O$ was separated, washed with water, and mixed with Na-montmorillonite in water. After constant stirring for 1 day, the montmorillonite changed into a hydroxy-copper intercalated phase having a basal spacing of 20.2 Å.

Differential thermal (DTA) and thermogravimetric (TGA) analysis curves of the hydroxy-copper intercalated phase are shown in Figure 4. The first weight loss, W_1 (7.1%), is attributed to dehydration of adsorbed water; the second loss, W_2 (5.8%), below 280°C can be interpreted in terms of decomposition of the hydroxy interlayers into CuO. The intense exothermic peak at the completion of the decomposition seems to be due to the crystallization of the CuO. A similar exothermic peak was also observed at 250°C in the DTA curve of $Cu_2(OH)_3(OCOCH_3) \cdot H_2O$. On heating above 300°C, the hydroxy-copper intercalated montmorillonite collapsed to 9.6 Å, and CuO deposited on the outer surfaces of the montmorillonite.

DISCUSSION

The titration method has been used to introduce hydroxy-cation interlayers into montmorillonite, the cations of which are hydrolyzed only slightly before precipitation occurs, such as magnesium and nickel (Slaughter and Milne, 1960; Gupta and Malik, 1969). The hydroxides can preferentially precipitate in the interlayer region of montmorillonite as a result of the acidic character of montmorillonite interlayers, as discussed by Yamanaka and Brindley (1978). Water molecules associated with exchanged cations in the interlayer spaces are polarized and tend to be dissociated to a greater extent than in solutions; the dissociated protons can be ion exchanged and further metal ions in the solutions taken up by the clay as the titration proceeds. This process results in fully developed brucite-like interlayers. Even if the hydroxide is precipitated on the outer surfaces of the clay due to the unavoidable local concentration of NaOH during the titration, the precipitate is dissolved by the acidity of clay.

This mechanism is evidently applicable to the formation of hydroxy-copper interlayers from a copper acetate solution; however, such interlayers can not be formed from a copper nitrate solution. This difference seems to be due to the difference in pH at which the two types of copper basic salts are precipitated. The basic acetate precipitates at pH = 5.7, whereas the basic nitrate precipitates at a pH as low as 4.2. Hydroxides of magnesium and nickel which can be intercalated by the titration method are precipitated at pHs as high as 7. Although the montmorillonite interlayer has an acidic character, the acidity is most likely not strong enough to dissolve the copper basic nitrate which precipitates from such a low pH solution.

The basal spacing of 20.2 Å is much larger than the range of spacings 14.4–14.8 Å of typical chlorite-like phases. This large spacing must reflect the bulky acetate ions replacing part of the interlayer hydroxy ions, and the water molecules accompanying the hydroxy interlayers.

After the copper hydroxy interlayers decomposed to CuO on heating, the oxide was unable to remain between the silicate layers, but precipitated on the edge or outer surfaces. Chromium oxide pillars are also released from the interlayer spaces of montmorillonite on heating above 300°C (Brindley and Yamanaka, 1979), in contrast to the pillar-forming oxides, Al_2O_3, ZrO_2, and Fe_2O_3, and nickel and magnesium hydroxy interlayers which remain between the silicate layers in the form of the respective oxides, even after dehydration. The contrast in the behavior of the oxides between the silicate layers presents an interesting problem for future study.

ACKNOWLEDGMENT

This work was supported in part by a Grant-in-Aid for Scientific Research (No. 58850169) from the Ministry of Education, Science, and Culture, Japan.

REFERENCES

Brindley, G. W. and Yamanaka, S. (1979) A study of hydroxy-chromium montmorillonites and the form of the hydroxy-chromium polymers: *Amer. Mineral.* **64**, 830–835.

Cotton, F. A. and Wilkinson, G. (1972) *Advanced Inorganic Chemistry:* 3rd ed., Wiley, New York, p. 916.

Gauthier, J. (1956) On novel copper basic acetate: *Compt. Rend.* **242**, 644–647.

Gupta, G. C. and Malik, W. U. (1969) Transformation of montmorillonite to nickel-chlorite: *Clays & Clay Minerals* **17**, 233–239.

Morikawa, Y., Takagi, K., Moro-oka, Y., and Ikawa, T. (1982) Cu-fluor tetrasilicic mica: a novel effective catalyst for the dehydrogenation of methanol to form methyl formate: *Chem. Letters,* 1805–1808.

Pinnavaia, T. J. (1980) Application of ESR spectroscopy to inorganic-clay systems: in *Advanced Chemical Methods for Soil and Clay Minerals Research,* J. W. Stucki and W. L. Banwarf, eds., D. Reidel, Dordrecht, Holland, 391–421.

Pinnavaia, T. J. and Mortland, M. M. (1971) Interlamellar metal complexes on layer silicates. I. Copper(II)-arene complexes on montmorillonite: *J. Phys. Chem.* **75**, 3957–3962.

Slaughter, M. and Milne, I. H. (1960) The formation of chlorite-like structures from montmorillonite: in *Clays and Clay Minerals, Proc. 7th Natl. Conf., Washington, D.C., 1958,* Ada Swineford, ed., Pergamon Press, New York, 114–124.

Thomas, J. M. (1982) Sheet silicate intercalates: new agents for unusual chemical conversions: in *Intercalation Chemistry,* M. S. Whittingham and A. J. Jacobson, eds., Academic Press, New York, 55–99.

Thomas, J. M., Adams, J. M., Graham, S. H., and Tennakoon, D. T. B. (1977) Chemical conversion using sheet-silicate intercalates: *Adv. Chem. Ser.* **163**, 298–315.

Yamanaka, S. and Brindley, G. W. (1978) Hydroxy-nickel interlayering in montmorillonite by titration method: *Clays & Clay Minerals* **26**, 21–24.

Yamanaka, S., Doi, T., Sako, S., and Hattori, M. (1984) High surface area solids obtained by intercalation of iron oxide pillars in montmorillonite: *Mat. Res. Bull.* **19**, 161–168.

Proceedings of the International Clay Conference, Denver, 1985, L. G. Schultz, H. van Olphen, and F. A. Mumpton, eds.,
The Clay Minerals Society, Bloomington, Indiana, 277–283 (1987).

IDENTIFICATION OF EXPANDING LAYER SILICATES:
LAYER CHARGE VS. EXPANSION PROPERTIES[1]

Prakash B. Malla and Lowell A. Douglas

Department of Soils and Crops, New Jersey Agricultural Experiment Station
Rutgers University, New Brunswick, New Jersey 08903

Abstract—Layer charges of several reference clays and several soil clays were determined by an alkylammonium ion-exchange technique. The diffentiation of smectite from vermiculite by glycerol or ethylene glycol solvation did not coincide with the layer charges assigned by the nomenclature committee of the Association Internationale pour l'Etude des Argiles (AIPEA). Two-layer ethylene glycol complexes were obtained for clays with layer charge as high as $0.72/(Si,Al)_4O_{10}$. Two-layer glycerol complexes were obtained for clays with layer charge as high as $0.65/(Si,Al)_4O_{10}$. Glycerol solvation failed to distinguish between mixtures of minerals and minerals with heterogeneous charge, whereas n-alkylammonium ion exchange distinguished such components.

Saturation of clays with K^+ caused the structure to collapse to 11.2–12.8 Å if the layer charge was less than $0.57/(Si,Al)_4O_{10}$, and to 10–10.6 Å if the layer charge was equal to or greater than $0.63/(Si,Al)_4O_{10}$. Clays with a double layer of glycerol (18 Å) after K^+-saturation and heat treatment at 300°C re-expanded if the layer charge was 0.36 or less. Minerals having layer charges of 0.39 to 0.43 re-expanded with a single layer of glycerol (14 Å). Minerals having layer charges of 0.63 or more failed to re-expand beyond 10 Å, whereas minerals having layer charges between 0.46 and 0.57 expanded and showed two X-ray powder diffraction peaks between 10.0 and 14.0 Å.

Key Words—Expansion, Layer charge, Smectite, Solvation, Vermiculite, X-ray powder diffraction.

INTRODUCTION

The 2:1 clay minerals, particularly smectites and vermiculites, are widely distributed in soils and sediments. The precise identification and characterization of smectites and vermiculites is important because of their role in physical and chemical reactions in soils and sediments and their wide industrial applications. MacEwan (1944) was the first to differentiate smectite from vermiculite and chlorite by treating the clay with glycerol and evaporating the mixture to apparent dryness. Montmorillonite gave a basal spacing of 17.7 Å, whereas the basal spacings of chlorite and vermiculite remained at 14.0 Å. Bradley (1945) used ethylene glycol rather than glycerol and suggested that the 17.1-Å basal spacing of a clay saturated with Mg^{2+} and solvated with ethylene glycol was diagnostic of smectite. Even after the formation of ethylene glycol-glycerol complexes in a controlled environment, certain vermiculites expand to form double-layer interlayer complexes with ethylene glycol, but others fail to do so. Some smectites, however, form only single-layer complexes with glycerol (Bradley, 1945; Walker, 1958; Brindley, 1966; Harward *et al.,* 1969). In general, smectites form a double-layer complex and vermiculites a single-layer complex after Mg^{2+}-saturation and glycerol solvation.

Walker (1958) concluded, therefore, that such treatment is a valid means of distinguishing smectite from vermiculite.

The collapse of the d(001) spacing of Mg- or Ca-saturated materials from 14 to 11–12 or 10.0–10.6 Å with K-saturation at relative humidity greater than 20% has also been employed as a means of differentiating smectite from vermiculite (Sayegh *et al.,* 1965; Harward *et al.,* 1969). Barshad (1960) differentiated high-charge smectite from low-charge smectite and vermiculite in soils by saturating the clays with K^+ and solvating them with a glycerol-ethanol solution. Schultz (1969) reported a proportional relationship between the re-expansion and total net layer charge of smectite that had been saturated with K^+, heat treated, and solvated with ethylene glycol. Ross and Kodama (1984) used glycerol instead of ethylene glycol to solvate clays after K-saturation and heat treatment in order to differentiate typical smectites from vermiculites; however, they reported that smectites separated from podzolic Ae horizons failed to re-expand upon glycerol solvation when they were first saturated by K^+ and heat treated.

The nomenclature committee of the Association Internationale pour l'Etude des Argiles (AIPEA) proposed to distinguish smectites and vermiculites by setting the ranges of layer charge per formula unit as 0.2–0.6 for smectite and 0.6–0.9 for vemiculite (Bailey, 1980). Many soil smectites or soil vermiculites, however, do not respond to ethylene glycol or glycerol in

[1] Journal Series, New Jersey Agricultural Experiment Station, Cook College, Rutgers University, New Brunswick, New Jersey; NJAES Publication No. K-15107-1-85.

Table 1. Location and sample characteristics.

Sample symbol[1] (particle size) (μm)	Location	Soil/clay classification	Soil series	pH (water)
SWy-1	Crook County, Wyoming	Na-montmorillonite		
STx-1	Gonzales, Texas	Ca-montmorillonite		
SAr-1	Apache County, Arizona	Ca-montmorillonite		
Camargo	Oklahoma	High-charge montmorillonite		
BJM (<0.2)	Black Jack Mine, Idaho	Beidellite		
Libby (<0.5)	Montana	Vermiculite		
Transvaal (<0.5)	Republic of South Africa	Vermiculite		
6082 (<0.2)	Black Marl, New Jersey			3.6
6211 (<0.2)	B2, Hauran basin, Syria	Xerochrept		6.7
6201 (<0.2)	B2, Hauran basin, Syria	Xerochrept		6.9
15[2] (<0.2)	2Bt/ES, Ohio	Typic Glossaqualf	Clermont	4.7
19[2] (<0.2)	2Bt1, Ohio	Typic Glossaqualf	Clermont	6.8
21[2] (<0.2)	2Bt2, Ohio	Typic Glossaqualf	Clermont	7.1
25[2] (<0.2)	2Bt5, Ohio	Typic Glossaqualf	Clermont	7.5
7150 (<0.2)	B2, Maryland	Ultic Hapludalf	Faquier	—
2303 (<2.0)	B2, Guatemala	Alfisol	—	—
320 (<2.0)	A1, New Jersey	Spodic Quartizipsamment	Lakewood	3.9

[1] Sample obtained from the Source Clay Repository of The Clay Minerals Society.

[2] Samples were provided by J. Bigham, Department of Agronomy, Ohio State University, Columbus, Ohio.

the manner of the clays studied by MacEwan (1944) or Bradley (1945). Some soil clay minerals behave like smectites when treated with glycerol, but have the charge of and collapse to 10 Å on K-saturation, like vermiculite (Egashira et al., 1982; Malla and Douglas, 1984). Therefore, Douglas (1982b) emphasized the use of layer charge rather than expansion with ethylene glycol or glycerol as a distinguishing criterion for smectites and vermiculites. Środoń and Eberl (1980) indicated that the ethylene glycol test tends to overestimate smectites and the glycerol test tends to underestimate them, compared with the AIPEA definition. The purpose of the present paper was to see whether a better correlation between layer charge as estimated by n-alkylammonium ion-exchange technique and swelling behavior of 2:1 clay minerals could be achieved by the modification of the swelling tests. Such a study is important for better identification and characterization of smectites and vermiculites in order to understand their exact reaction behaviors. The use of layer charge and/or glyceral solvation of K-saturated, heat-treated materials rather than expansion of Mg-saturated and glycerol or ethylene glycol solvation is proposed to distinguish between smectites and vermiculites.

The layer charge of 2:1 clay minerals can be determined by chemical analysis and cation-exchange capacity (CEC) determinations. Both of these techniques, however, involve questionable assumptions when applied to mixtures of several clay minerals; consequently a meaningful estimate of layer charge of soil clay minerals is questionable. Moreover, both chemical analyses and CEC provide only mean values and no information about charge distribution (Lagaly and Weiss, 1970). We have therefore used the n-alkylammonium ion-exchange technique developed by Lagaly and Weiss

(1969) to estimate the layer charge. This technique is more direct compared with chemical analysis and CEC determination and enables both magnitude and heterogeneity of layer charge of 2:1 swelling clay to be estimated even in a complex mixture.

MATERIALS AND METHODS

A wide range of reference and soil clays from different locations were characterized, as described below (Table 1).

Pretreatments, saturation, and solvation

Air-dry samples (<2 μm) were treated with sodium hypochlorite (Clorox) to destroy organic matter (Anderson, 1961). The clay fraction was extracted by centrifuge methods (Jackson, 1969). The sodium citrate-bicarbonate (CBD) method was employed to remove the easily reducible iron (Mehra and Jackson, 1960). Sodium citrate was used to remove the hydroxy-aluminum interlayers (Tamura, 1958). Oriented samples for X-ray powder diffraction (XRD) were prepared by transferring clays from membrane filters to glass slides (Drever, 1973).

Re-expansion properties of clays after K+-saturation were studied by the method proposed by Ross and Kodama (1984) and Ross (G. J. Ross, Chemistry and Biology Research Institute, Agriculture Canada, Ottawa, Ontario K1A 0C6, personal communication, 1984) as follows: Clay samples were saturated with 1 N KCl solution, washed free of chlorides with distilled water, and then heated at 300°C for about 1 hr. Next, the samples were suspended in 1–2 ml of 3% glycerol solution (3 ml of glycerol in 97 ml of water) and dispersed thoroughly with a Fisher 150 ultrasonic disperser fitted with microtip. The clay-glycerol suspen-

Table 2. Layer charge per $(Si,Al)_4O_{10}$ estimated from various arrangements of cations in the interlayer spaces.

Samples[1]	Layer charge			Estimated from	
	Upper	Lower	Mean	Transition layers	Paraffin-type structures
SWy-1	0.37	0.21	0.28	mono-bi layer	
STx-1	0.40	0.24	0.31	mono-bi layer	
SAr-1	0.49	0.31	0.39	mono-bi layer	
Camargo	0.47	0.40	0.43	bi-pseudotri[2]	
BJM[3]	0.53	0.40	0.46	bi-pseudotri	
Libby	0.82	0.59	0.70	bi-pseudotri	
Transvaal	0.82	0.64	0.73	bi-pseudotri	0.75
6082	0.53	0.42	0.47	bi-pseudotri	
6211[4]	0.60	0.44	0.52	bi-pseudotri	
6201[4]	0.65	0.49	0.57	bi-pseudotri	
2303	0.43	0.30	0.36	mono-bi layer	
2303	0.78	0.60	0.70	bi-pseudotri	
15	0.71	0.56	0.63	bi-pseudotri	
19	0.78	0.56	0.66	bi-pseudotri	
21	0.78	0.60	0.69	bi-pseudotri	
25	0.78	0.65	0.72	bi-pseudotri	
7150[4]	0.78	0.56	0.66	bi-pseudotri	
320[4]	0.78	0.60	0.70	bi-pseudotri	0.75

[1] See Table 1 for sources of samples.
[2] Pseudotri = pseudotrimolecular layer.
[3] BJM = Black Jack Mine beidellite.
[4] 1 N Na-citrate treated to remove hydroxy interlayers.

sions were then transferred to a glass slide and examined by XRD after air drying.

Estimation of layer charge with n-alkylammonium ion exchange

Alkylammonium chlorides were prepared from the respective amines by the Lagaly and Weiss (1969) procedure as modified by Ruhlicke and Kohler (1981). The terminology of Lagaly and Weiss (1969, 1976) is used here to describe the interlayer packing of alkylammonium ions. Layer charges were estimated from the relation for layer transitions (Lagaly, 1982):

$$(1.27)(4.5)nC + 14 = \lambda[d(010)d(100)]$$
$$\div 2(x + y + z), \quad (1)$$

where $\lambda = 1$ or 2 for monolayer to bilayer and bilayer to pseudotrimolecular transition, respectively; $(x + y + z)$ = interlayer cation density (=average layer charge in $eq/(Si,Al)_4O_{10}$; and $d(010)d(100) = 49$ and 46.5 $Å^2$ for trioctahedral and dioctahedral minerals, respectively. The layer charge in high-charged minerals in which basal spacings increased linearly with nC (number of carbon atoms in the chain) owing to paraffin-type chain packing was estimated from the relation (Lagaly and Weiss, 1969):

$$\sin \alpha = \Delta d/1.26, \quad (2)$$

where Δd = change of basal spacings with change in nC, and α = tilting angle.

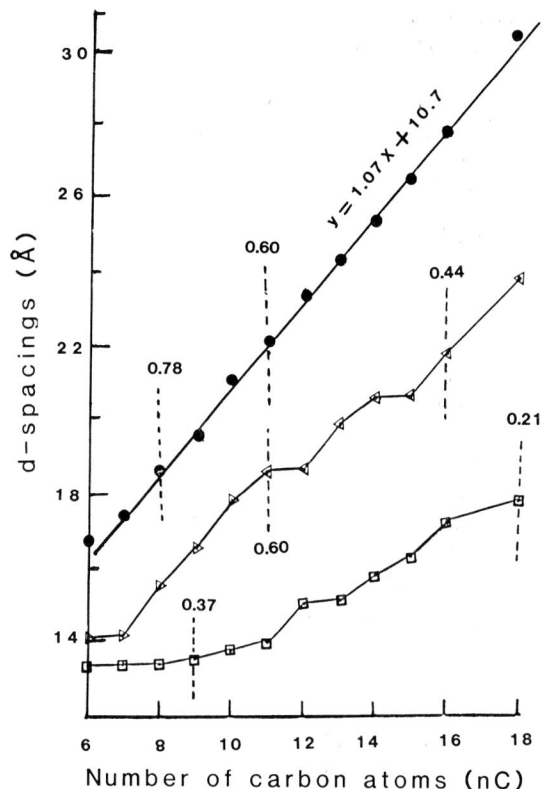

Figure 1. Variations in basal spacings of samples SWy-1 (□), 6211 (△), and 320 (●) with chain length (nC) of alkylammonium ions. Dashed vertical lines indicate upper and lower limits of charge densities.

X-ray powder diffraction

Samples were X-rayed using a Siemens diffractometer with Ni-filtered CuK_α radiation and a Johanson's graphite crystal monochromator. The diffractometer was controlled by, and data stored on, a Thetaplus software system (Dapple Systems, Inc., Sunnyvale, California), operating on an Apple IIe microcomputer.

RESULTS AND DISCUSSION

Charge density

Layer charges of the clays estimated from various arrangements of alkylammonium ions in the interlayer spaces are given in Table 2. Lagaly and Weiss (1969) described the relations among basal spacing, chain length of the organic cations, and the layer charge (Figure 1). In estimating layer charges, soil clays were assumed to be dioctahedral. The layer arrangements were recognized from the following spacings: 13.6 ± 0.2 Å for monomolecular, 17.8 ± 0.3 Å for bimolecular, and 22.0 ± 0.4 Å for pseudotrimolecular layers. Some integral series of reflections of soil clays having relatively small layer charges were difficult to recognize because of the presence of other clays in the mixture or the

Table 3. Comparison of layer charge per $(Si,Al)_4O_{10}$ and basal spacings of clays with glycerol (Mg + G), ethylene glycol (Mg + EG) and K-saturation (K[+]).

Samples[1]	Mean layer charge density	Mg + EG[2] (Å)	Mg + G[2] (Å)	K[+] (RT)[2] (Å)
SWy-1	0.28	17.04	18.12	12.10
STx-1	0.31	16.90	17.72	12.31
SAr-1	0.39	17.09	17.81	12.56
Camargo	0.43	17.38	18.30	12.73
BMJ	0.46	16.89	17.93, 14.58	11.93
Libby	0.70	14.05	14.15	10.22
Transvaal	0.73, 0.75	14.15	14.18	10.03
6082	0.47	17.36	18.25	11.20
6211[3]	0.52	18.18	18.74	12.75
6201[3]	0.57	18.47	19.60	11.53
⌠2303	0.36	17.34	18.15	12.12 ⌉
⌡2303	0.70	14.22	14.23	10.04 ⌋
15	0.63	16.96	18.56, 14.61	10.65
19	0.66	16.94	18.38, 14.53	10.64
21	0.69	16.83	17.86, 14.55	10.58
25	0.72	16.53	18.31, 14.33	10.4
7150[3]	0.66	17.03, 14.27	14.16	10.29
320[3]	0.70, 0.75	16.90, 14.14	18.22, 14.09	10.15

[1] See Table 1 for sources of samples.
[2] EG = ethylene glycol; G = glycerol; RT = room temperature.
[3] 1 N Na-citrate treated to remove hydroxy interlayers.

absence of the integral series of reflections. These samples expanded with a broad transition range. The transition range consists of a non-integral series of 001 reflections between monomolecular and bimolecular or bimolecular and pseudotrimolecular layers. The magnitude of transition zones has been equated to the extent of layer charge heterogeneity in smectites (Lagaly and Weiss, 1970; Stul and Mortier, 1974; Lagaly, 1981) and vermiculites (Lagaly, 1982), that is, a small transition range indicates a more homogeneous charge distribution and vice versa.

In clays having layer charges greater than $0.63/(Si,Al)_4O_{10}$, a paraffin-type chain packing was indicated by the larger number of integral series of reflections and spacings greater than those of a pseudotrimolecular layer. This type packing indicates a slight heterogeneity of charge (Lagaly, 1982). Our observation is somewhat contrary to Lagaly (1982), who observed a paraffin-type interlayer only when the layer charge was greater than $0.75/(Si,Al)_4O_{10}$. The relationship between d(001) and nC for soil clays was not as linear (except for sample 320) as determined for vermiculites of comparable layer charges. The difference is probably due to interference from hydroxy interlayer components, particle size effects, the location of charge (octahedral vs. tetrahedral), and the di- or trioctahedral nature of octahedral sheet. The effect of hydroxy interlayers on estimating layer charges and the proportion of octahedral and tetrahedral charge by the alkylammonium ion-exchange technique will be discussed elsewhere (manuscript in preparation). Estimating the layer charge of soil clays from a paraffin-type structure may give er-

roneous high values because of less-than-perfect, paraffin-type, chain packing.

Sample 2303 contains a mixture of minerals having layer charges of $0.70/(Si,Al)_4O_{10}$ (vermiculite) and $0.36/(Si,Al)_4O_{10}$ (smectite). This nature of this mixture was ascertained only after the sample was treated with alkylammonium, which produced a series of peaks for every nC belonging to high- and low-charge minerals. The XRD patterns for even nCs are given in Figure 3. A similar mixture of two minerals with different layer charges was also reported by Lagaly and Weiss (1970) for montmorillonite.

Layer charge vs. expansion and collapse

In Table 3, the mean layer charges of the clays are compared with their expansion and/or collapse behavior with ethylene glycol, glycerol, and K[+]-saturation at room temperature. The data indicate that expansion after Mg^{2+}-saturation and ethylene glycol solvation was achieved for minerals having a mean layer charge as high as $0.72/(Si,Al)_4O_{10}$. Suquet et al. (1977) reported that Mg^{2+}-saturated, synthetic trioctahedral smectites accommodate two layers of ethylene glycol if their layer charge is ≤0.70 per half unit cell. Both 17-Å and 14-Å peaks, corresponding to two-layer and one-layer glycol interlayers, respectively, were observed for samples 2303, 7150, and 320. The 14-Å peaks were due to the presence of hydroxy interlayers that resisted expansion.

The results of solvation of the clays with glycerol following Mg^{2+}-saturation indicate that the natural dioctahedral minerals with mean layer charges of 0.63 or more accommodated both single (14 Å) and double

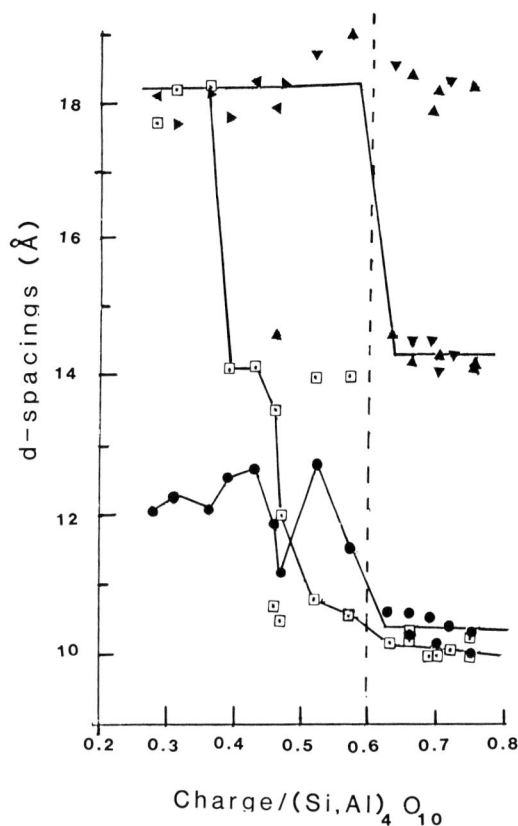

Figure 2. Charge density vs. basal spacings (Å) of Mg^{2+}-saturated and glycerol solvated (◄), K$^+$-saturated clay followed by heat treatment at 300°C and glycerol solvation (▣), and K$^+$-(room temperature) saturated clays (●).

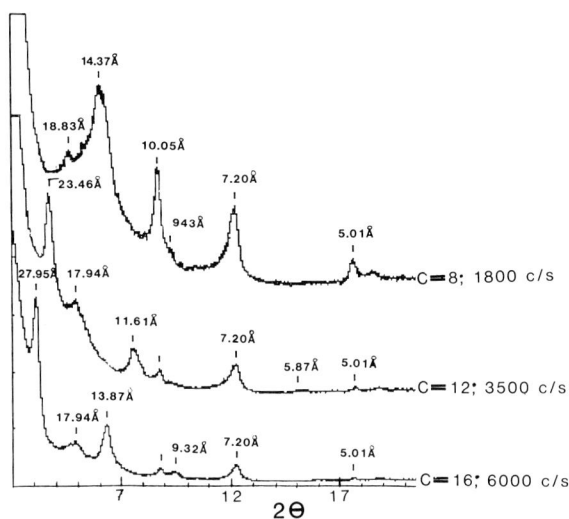

Figure 3. X-ray powder diffraction patterns of the sample 2303 for three even nCs showing two d(100) values for each nC. Similar patterns were observed for all nCs used. (CuKα radiation.)

(18 Å) layers of glycerol (Figure 2). Trioctahedral minerals, however (Transvaal and Libby vermiculites), did not expand beyond 14 Å with either glycerol or ethylene glycol. Suquet *et al.* (1977) reported that trioctahedral minerals with a layer charge <0.6/(Si,Al)$_4$O$_{10}$ expand with glycerol solvation.

Structural collapse was achieved by K$^+$-satuation at room temperature after air drying. Minerals having layer charges <0.57 collapsed to 11.2–12.8 Å. Minerals with layer charges >0.63 collapsed to less than 10.65 Å. Samples 6211 and 6201 collapsed and gave a broad XRD peak. Sample 2303 collapsed to 12.12 and to 10.04 Å, indicating the presence of two minerals with large and small layer charges, as confirmed by alkyl-ammonium ion-exchange data (Figure 2).

Re-expansions vs. layer charges are plotted in Figure 2. On the basis of re-expansion, the minerals can be classified into four groups, three groups in the smectite charge range and one in the vermiculite charge range. Minerals having a layer charge of 0.36 or less expanded and accommodated two layers of glycerol in the interlayer. Minerals having a layer charge between 0.39 and

0.43 re-expanded to 14 Å. Minerals having a layer charge of 0.63 or more failed to re-expand beyond 10 Å, whereas minerals having layer charges between 0.46 and 0.57 expanded and showed two XRD peaks between 10 and 14 Å. Although they employed slightly different procedures, similar trends were reported for smectites by Schultz (1969) and smectites and vermiculites by Barshad (1960).

Evaluation of glycerol test

Distinguishing soil vermiculite from soil smectite is complicated by: (1) the layer charge heterogeneity, as well as the presence of mixtures of minerals having different layer charges and the presence of nonexpandable interlayers; (2) the solvation technique; (3) the sensitivity of the technique used; (4) the nature of the octahedral sheet and distribution of charge in the octahedral and tetrahedral sheets; and (5) the lack of standard nomenclature usage.

Charge heterogeneity in smectites and vermiculites is well established (Stul and Mortier, 1974; Lagaly and Weiss, 1969, 1970; Lagaly, 1982). Lagaly and Weiss (1976) and Lagaly (1981) reported that 50% of 300 smectites examined exhibited mixed layer-charge distribution (i.e., the presence of groups of low-charge and high-charge interlayers). Double-layer and single-layer glycerol complexes in samples of mean layer charges of 0.63 or more may be visualized as follows: the low-charge interlayers accommodate double layers of glycerol, whereas high-charge interlayers accommodate only single layers greater than a certain value. Based on the data from the present study, this value is about 0.65/(Si,Al)$_4$O$_{10}$. As in heterogeneous vermiculite, sample

Table 4. Basal spacings (Å) of glycerol-solvated beidellites.

Experimental condition	Spacing (Å)	Layer charge per $(Si,Al)_4O_{10}$	Source
BJM[1] + Mg + liquid glycerol	17.93, 14.58	0.46	This paper
BJM + Mg + liquid glycerol	18.00	0.46	Suquet et al. (1977)
BJM + Mg + vapor glycerol	14.40		Harward et al. (1969)
Natural BJM (Ca,Mg) + liquid glycerol	17.60	0.46	Weir and Greene-Kelley (1962)
Rupsroth beidellite + Mg + liquid glycerol	17.90	0.54	Suquet et al. (1975, 1977)
Synthetic beidellite + Mg + liquid glycerol	17.80–17.90	0.48	Harward and Brindley (1965)
Syn. beidellite + vapor glycerol	14.20–14.50	0.48	Harward and Brindley (1965)

[1] BJM = Black Jack Mine beidellite.

2303 appeared to be a mixture of minerals having two different layer charges and, thus, gave glycerol XRD peaks at both 14 and 18 Å. Here, the 18-Å peak may have been contributed by a low-charge mineral (e.g., smectite) or by a low-charge interlayer within a high-charge mineral (e.g., vermiculite). Thus, the glycerol test failed to differentiate the mineral having a heterogenous charge distribution (Table 2) from a simple mixture of minerals.

The glycerol test used to identify smectites and vermiculites is strongly influenced by (Table 4) the glycerol solvation techniques used (Ross and Kodama, 1984). In the present study, samples 24621 and 24625 did not expand to 18 Å on glycerol solvation, if the solvation was carried out in water-glycerol-clay suspension. Expansion to 18 and 14 Å occurred, however, if the Mg^{2+}-saturated and glycerol-solvated clay suspension was dispersed ultrasonically and incubated overnight at 65°C before preparing an oriented sample for XRD. Thus, the interpretation of spacings of clay minerals for identification purposes by the glycerol test is misleading because of its dependence on the way the solvation is carried out.

As far as the sensitivity of the glycerol test is concerned, only the presence of smectite or vermiculite or a mixture of these minerals can be estimated. A quantitative estimation of charge is not possible. The presence of both high- and low-charge smectites (Nash, 1979; Senkayi et al., 1983), high-charge smectites (Egashira et al., 1982), and the abundance of smectite with tetrahedral substitution (Douglas, 1982a) in soils have been reported on the basis of the glycerol test. A careful examination of the results indicates that those workers were probably dealing with minerals having the charge of vermiculite but called them high-charge smectite.

The fourth difficulty in differentiating smectites and vermiculites with the glycerol test lies in the dioctahedral nature of some soil vermiculites as compared to the trioctahedral nature of reference vermiculites and the presence of both octahedral and tetrahedral charge in soil clays. The dioctahedral vermiculite usually found in soils expanded with glycerol, but trioctahedral vermiculite failed to do so (Table 3). The use of Mg^{2+}-saturation plus the glycerol solvation test to

differentiate vermiculite from smectite, advocated by Walker (1958), was based on the inability of trioctahedral vermiculite to expand with glycerol. If this test is used on soil clays, it must be realized that the reaction of glycerol with trioctahedral minerals differs from this reaction with dioctahedral minerals. The distribution of charge in both octahedral and tetrahedral sheets, which is expected in soil clays due to weathering, probably plays an important role in the expansion behavior of these materials with glycerol. Suquet et al. (1977) observed less expansion of synthetic saponite with both glycerol and ethylene glycol because of the charge excess in the octahedral sheet.

The fifth difficulty in distinguishing smectite from vermiculite stems from the inconsistent use of a standard system of the nomenclature as defined by the AIPEA. The glycerol test was used before the AIPEA system of nomenclature was put forward. The test has been widely used despite the fact that anomalous identifications are frequently made. Furthermore, the terms smectite and vermiculite encompass minerals having wide ranges of layer charges and that may vary widely in their reaction behaviors. This simple fact necessitates the use of layer-charge criteria to distinguish smectite from vermiculite for precise characterization of their reaction behaviors. The layer-charge parameter to distinguish smectite from vermiculite at $0.6/(Si,Al)_4O_{10}$, as defined by AIPEA, is only arbitrary inasmuch as smectites and vermiculites form a continuum in terms of their layer charge (Walker, 1958; Barshad and Kishk, 1970; Malla and Douglas, 1984).

CONCLUSIONS

Both the ethylene glycol and glycerol tests tend to overestimate soil smectites and underestimate soil vermiculites, as defined by AIPEA. K^+-saturation of clays (relative humidity >20%), as well as re-expansion of heat-treated clays with glycerol (Figure 2), is a more reliable test to differentiate smectite from vermiculite, because of the favorable geometry of the K^+-unit layer complexes in the latter mineral (van Olphen, 1966). The use of re-expansion technique as suggested by Ross and Kodama (1984) is recommended to distinguish low-charge smectite from high-charge smectite. Ross and Kodama (1984), however, did not verify the technique with layer charge, as has been done in the present study. The n-alkylammonium ion-exchange technique is, however, indispensable for an estimate of the exact magnitude of the layer charge.

ACKNOWLEDGMENTS

This research was supported partly by state (New Jersey) and partly by federal (Hatch) funds.

REFERENCES

Anderson, J. U. (1961) An improved pretreatment for mineralogical analysis of samples containing organic matter: in *Clays and Clay Minerals, Proc. 10th Natl. Conf., Austin, Texas, 1961,* Ada Swineford, ed., Pergamon Press, New York, 380–388.

Bailey, S. W. (1980) Summary of recommendations of AIPEA nomenclature committee: *Clays & Clay Minerals* **28,** 73–78.

Barshad, I. (1960) X-ray analysis of soil colloids by modified salted paste method: in *Clays and Clay Minerals, Proc. 7th Natl. Conf., Washington, D.C., 1958,* Ada Swineford, ed., Pergamon Press, New York, 350–364.

Barshad, I. and Kishk, F. M. (1970) Factors affecting potassium fixation and cation exchange capacities of soil vermiculite clays: *Clays & Clay Minerals* **18,** 127–137.

Bradley, W. F. (1945) Diagnostic criteria for clay minerals: *Amer. Miner.* **30,** 704–713.

Brindley, G. W. (1966) Ethylene glycol and glycerol complexes of smectites and vermiculites: *Clay Miner.* **6,** 237–260.

Douglas, L. A. (1982a) Smectites in acidic soils: in *Proc. Inter. Clay Conf., Bologna, Pavia, Italy, 1981,* H. van Olphen and F. Veniale, eds., Elsevier, Amsterdam, 635–640.

Douglas, L. A. (1982b) Smectites in acidic soils—the problem: *Agron. Abst.,* 1982 Annual Meetings of ASA, CSSA, SSSA, Anaheim, California, p. 281.

Drever, I. (1973) The preparation of oriented clay mineral specimens for X-ray diffraction analysis by a filter-membrane peel technique: *Amer. Miner.* **58,** 553–554.

Egashira, K., Dixon, J. B., and Hossner, L. R. (1982) High charge smectite from lignite overburden of east Texas: in *Proc. Inter. Clay Conf., Bologna, Pavia, Italy, 1981,* H. van Olphen and F. Veniale, eds., Elsevier, Amsterdam, 335–345.

Harward, M. E. and Brindley, G. W. (1965) Swelling properties of synthetic smectites in relation to lattice substitutions: in *Clays and Clay Minerals, Proc. 13th Natl. Conf., Madison, Wisconsin, 1964,* W. F. Bradley and S. W. Bailey, eds., Pergamon Press, New York, 209–222.

Harward, M. E., Carstea, D. D., and Sayegh, A. H. (1969) Properties of vermiculites and smectites: expansion and collapse: *Clays & Clay Minerals* **16,** 437–447.

Jackson, M. L. (1969) *Soil Chemical Analysis—Advanced Course:* 2nd ed., publ. by author, Dept. Soil Sci., Univ. Wisconsin, Madison, Wisconsin, 895 pp.

Lagaly, G. (1981) Characterization of clays by organic compounds: *Clay Miner.* **16,** 1–21.

Lagaly, G. (1982) Layer charge heterogeneity in vermiculites: *Clays & Clay Minerals* **30,** 215–222.

Lagaly, G. and Weiss, A. (1969) Determination of the layer charge in mica-type layer silicates: in *Proc. Intern. Clay Conf., Tokyo, 1969, Vol. 1,* L. Heller, ed., Israel Univ. Press, Jerusalem, 61–80.

Lagaly, G. and Weiss, A. (1970) Inhomogeneous charge distribution in mica-type layer silicates: *Reunion Hispano-Belga de Minerals de la Arcilla,* J. M. Serratosa, ed., Consejo Superior de Investigaciones Cientificas, Madrid, 179–187.

Lagaly, G. and Weiss, A. (1976) The layer charge of smectitic layer silicates: in *Proc. Inter. Clay Conf., Mexico City, 1975,* S. W. Bailey, ed., Applied Publishing, Wilmette, Illinois, 157–172.

MacEwan, D. M. C. (1944) Identification of the montmorillonite group of minerals by X-rays: *Nature* **154,** 577–588.

Malla, P. B. and Douglas, L. A. (1984) Charge density of some smectite-like soil clays: *Agr. Abst.,* 1984 Annual Meetings of ASA, CSSA, SSSA, Las Vegas, Nevada, p. 274.

Mehra, O. P. and Jackson, M. L. (1960) Iron oxide removal by a dithionate-citrate system buffered with sodium bicarbonate: in *Clays and Clay Minerals, Proc. 7th Natl. Conf., Washington, D.C., 1958,* Ada Swineford, ed., Pergamon Press, New York, 317–327.

Nash, V. E. (1979) Mineralogy of soils developed on Pliocene-Pleistocene terraces of the Tombigbee River in Mississippi: *Soil Sci. Soc. Amer. J.* **43,** 616–623.

Ross, G. J. and Kodama, H. (1984) Problems in differentiating soil vermiculites and soil smectites: *Agron. Abst.,* 1984 Annual Meetings of ASA, CSSA, SSSA, Las Vegas, Nevada, p. 275.

Ruhlicke, G. U. and Kohler, E. E. (1981) A simplified procedure for determining layer charge by n-alkylammonium method: *Clay Miner.* **10,** 305–307.

Sayegh, A. H., Harward, M. E., and Knox, E. G. (1965) Humidity and temperature interaction with respect to K-saturated expanding minerals: *Amer. Mineral.* **50,** 490–495.

Schultz, L. G. (1969) Lithium and potassium absorption, dehydroxylation temperature, and structural water content of aluminous smectites: *Clays & Clay Minerals* **17,** 115–149.

Senkayi, A. L., Dixon, J. B., and Viani, B. E. (1983) Mineralogical transformations during weathering of lignite overburden in east Texas: *Clays & Clay Minerals* **31,** 49–56.

Środoń, J. and Eberl, D. D. (1980) The presentation of X-ray data for clay minerals: *Clay Miner.* **15,** 317–320.

Stul, M. S. and Mortier, W. J. (1974) The heterogeneity of the charge density in montmorillonites: *Clays & Clay Minerals* **22,** 392–396.

Suquet, H., de la Calle, C., and Pezerat, H. (1975) Swelling and structural organization of saponite: *Clays & Clay Minerals* **23,** 1–9.

Suquet, H., Iiyama, J. T., Kodama, H., and Pezerat, H. (1977) Synthesis and swelling properties of saponites with increasing layer charge: *Clays & Clay Minerals* **25,** 231–242.

Tamura, T. (1958) Identification of clay minerals from acid soils: *J. Soil Sci.* **9,** 141–147.

van Olphen, H. (1966) Collapse of potassium montmorillonite clays upon heating—"potassium fixation": in *Clays and Clay Minerals, Proc. 14th Natl. Conf., Berkeley, California, 1965,* S. W. Bailey, ed., Pergamon Press, New York, 393–405.

Walker, G. F. (1958) Reactions of expanding lattice minerals with glycerol and ethylene glycol: *Clay Miner. Bull.* **3,** 302–313.

Weir, A. H. and Greene-Kelly, R. (1962) Beidellite: *Amer. Mineral.* **47,** 137–146.

Proceedings of the International Clay Conference, Denver, 1985, L. G. Schultz, H. van Olphen, and F. A. Mumpton, eds.,
The Clay Minerals Society, Bloomington, Indiana, 284–291 (1987).

WATER DYNAMICS IN THE CLAY-WATER SYSTEM: A QUASIELASTIC NEUTRON SCATTERING STUDY

C. POINSIGNON,[1,2] J. ESTRADE-SZWARCKOPF,[3] J. CONARD,[3] AND A. J. DIANOUX[1]

[1] Institut Laue-Langevin, 38042 Grenoble Cedex, France

[2] Centre National de la Recherche Scientifique, Centre de Recherche sur la Valorisation des Minerais
Ecole Nationale Supérieure de Géologie et de Prospection Minière, 54501 Vandoeuvre Cedex, France

[3] Centre National de la Recherche Scientifique
Centre de Recherche sur les Solides à Organisation Cristalline Imparfaite
45045 Orléans Cedex, France

Abstract — Water dynamics was studied in three smectites—hectorite, montmorillonite (bentonite), and vermiculite having water contents of 0.5, 1, and 2 water layers, respectively—by quasielastic neutron scattering. The study began with the lowest stable hydration state, i.e., the hydration shell of the exchangeable cation. The water content related to the planar hydrate $Li^+ \cdot 3 H_2O$ adsorbed on hectorite was found to be equivalent to the presence of one half of a water monolayer. The water molecules were found to be involved in two uniaxial motions: (1) a slow motion of the entire hydrate around an axis parallel to the c-axis of the clay layer, and (2) a fast rotation of water molecules around their c_2 axes. The slow motion occurring in the interlayer space was anisotropic, the component along the c axis was found to be nil. Neither motion was thermally activated between 300 and 210 K, and both stopped simultaneously at 190 K. The fast correlation time, τ_{1f} was 2.7×10^{-12} s; the slow one was 21.4×10^{-12} s.

Similar measurements on $Li^+ \cdot 3 H_2O$ adsorbed on the bentonite and vermiculite suggest that the correlation times of both motions were sensitive to the electrical charge of the clay structure; i.e., $\tau_{1f} = 4.4 \times 10^{-12}$ and 6.9×10^{-12} s and $\tau_{1s} = 23.4 \times 10^{-12}$ and 29×10^{-12} s, respectively, for Li-bentonite and Li-vermiculite. Both motions persisted with increasing water content, but slowed and could be thermally activated. For Li-hectorite, τ_{1f} and τ_{1s} changed to 4.1×10^{-12} and 23.9×10^{-12} s for one adsorbed water layer and 4.4×10^{-12} and 29×10^{-12} s for two adsorbed water layers.

The long-range motion of water appeared to be a jump diffusion in a bidimensional space as inferred from the Q dependence of half width at half maximum of the corresponding energy distribution having a residence time of 2×10^{-10} s and a jump length of 3 Å. For water contents greater than that necessary to form two layers, porosity water became predominant; i.e., its dynamics was found to be similar to that of bulk water. The molecule was involved in an isotropic rotational motion with a corresponding time τ_{1f} of 3.1×10^{-12} s. The clay-water interfacial region can thus be described as follows: The water molecules are distributed over a superlattice (3a.b) and remain an average of 10^{-10} s on a site. During this time, they spin around their c_2 axis and have a correlation time τ_{1f} of $4–5 \times 10^{-12}$ s. If they are in the hydration shell of a charge-balancing cation, they are also involved in the slow rotation of the entire hydrate that has a correlation time of $\sim 3 \times 10^{-11}$ s.

Key Words — Hectorite, Lithium, Montmorillonite, Quasielastic neutron scattering, Vermiculite, Water.

INTRODUCTION

Swelling clays (or smectites) are phyllosilicates whose particular behavior in water makes them appropriate candidates for the study of water adsorbed at the interface and in the pore spaces of materials. In the clay-water system one can distinguish adsorbed water on external and interlayer surfaces, water associated with cations or hydrogen bonded, and pore water. A study of the water dynamics in the clay-water interfacial region assumes a knowledge of the number of water molecules having a structure different from that of liquid water. Woessner (1979), in a ^1H and ^2H nuclear magnetic resonance (NMR) study of Li-exchanged hectorite gels showed that the surface influence was limited to the first two molecular water layers. On the basis of wide-angle X-ray scattering data for hectorite and

montmorillonite gels, Pons et al. (1980) reported that short-range, but nonetheless strong interactions existed only between water molecules of the first molecular layer and the clay surface. The molecules are distributed over a superlattice (3a.b) as shown in Figure 1. These observations were confirmed by Fripiat et al. (1982) from heats of immersion and NMR data in the same system. The good agreement between the results obtained by such different methods justifies limiting the present water dynamics study to the first two adsorbed water layers. Moreover, at greater water content, pore water becomes predominant, and the individual contributions of adsorbed and pore water become difficult to distinguish.

Because water in porous materials and clays "freezes" at temperatures <273 K (Anderson and Tice, 1971;

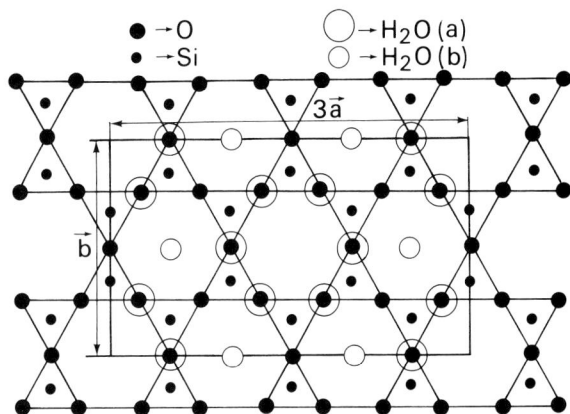

Figure 1. Water monolayer at the clay surface. H_2O (a) = water molecules above surface oxygens (large open circles); H_2O (b) = water molecules or cations above hexagonal holes (small open circles). The water molecules are distributed over the superlattice (3a,b). The large solid circle represents surface oxygens of the network, the small solid circle, the silicon atoms.

Homshaw, 1980), the reference state chosen in the present study for the bulk water was that defined for supercooled water by Teixeira *et al.* (1984). In the liquid state, neutron diffraction studies (Dore, 1984) show that the water molecules form a tetrahedral structure. In this state, water dynamics can be described by two components: a fast component related to isotropic rotational motion and a slow component related to diffusional translational motion (Dore, 1984). In the present study, quasielastic neutron scattering was used to study water dynamics in the first two layers of water molecules adsorbed at the clay surface. This technique allows a direct means of obtaining proton dynamics in time and space at the microscopic scale. Specifically, the study was aimed at confirming the geometry of the motions of water molecules, as deduced from NMR data, in order to determine their correlation time and to evaluate the influence of the crystal chemical properties of the clays on the water dynamics at the interface, i.e., the influence of the electrical charge on the clay and its location in the structure.

EXPERIMENTAL

Preparation of homoionic clay

The smectites used in this study were (1) hectorite from Hector, California, having the composition $M^+_{0.29}(Mg_{2.71}Li_{0.29})^{VI}(Si_4)^{IV}O_{10}(OH,F)_2$; (2) bentonite from Upton, Wyoming, having the composition $M^+_{0.312}(Al_{1.576}Fe^{3+}_{0.194}Fe^{2+}_{0.005})^{VI}(Si_{3.393}Al_{0.07})^{IV}O_{10}(OH)_2$; and (3) vermiculite from Libby, Montana, having the composition $M^+_{0.5}(Mg_{2.27}Al_{0.73})^{VI}(Si_{2.72}Al_{1.24})^{IV}O_{10}(OH)_2$; where M^+ is an exchangeable cation. Homoionic smectites were prepared by exchange with salts of Li^+, Cu^{2+}, and Ca^{2+} by the method of Glaeser and Mering (1968).

Self-supporting films were prepared by evaporating dilute suspensions on a flat surface. Samples were equilibrated at constant water vapor pressure, P/P_0, using sulfuric acid or salt solutions of appropriate concentrations to produce one or two interlamellar water layers.

Table 1 summarizes the principal crystal chemical properties of the samples.

Quasielastic neutron scattering experiments

Quasielastic neutron scattering experiments were carried out at the Institut Laue-Langevin in Grenoble, France, on a multichopper time-of-flight (T.O.F.) spectrometer IN5, a thermal backscattering spectrometer IN13, and a high-resolution backscattering spectrometer IN10 (Maier, 1983). The T.O.F. spectra were recorded for two incident wavelengths, $\lambda = 10.05$ Å (20 μeV resolution) and $\lambda = 5.14$ Å (138 μeV resolution), where $1\ \mu eV = 1.519 \times 10^9$ rd/s. The scattered neutrons were detected at 12 scattering angles having transferred moment values between 0.136 and 1.14 Å^{-1} for low incident energy and between 0.27 and 2.28 Å^{-1} for high incident energy. The experiment on the IN13 spectrometer was conducted using a wavelength of 2.52 Å and a resolution of 10 μeV. Spectra were recorded at 11 scattering angles from 0.287 to 4.878 Å^{-1} and at three temperatures, 100, 260, and 300 K. For $\lambda = 6.28$ Å, 7 spectra were recorded on the IN10 spectrometer at 0.21, 0.28, 0.67, 0.73, 1.28, 1.65, and 1.88 Å^{-1} at a resolution varying from 0.7 to 1.2 μeV according to

Table 1. Properties, structural charge, exchangeable cation, and hydration state of bentonite, hectorite and vermiculite.

	Bentonite[1]	Hectorite[2]	Vermiculite[3]
Structural charge	0.32	0.29	0.50
Cation-exchange capacity (meq/100 g)	99	74	200
Charge localization	100% octahedral	100% octahedral	32% oct. 45% tet.
Exchangeable cations	Li, Na, Ca	Li, Cu	Li, Ca
Number of water layers	1, 2	0.5, 1, 2	1, 2

[1] Upton, Wyoming.
[2] Hector, California.
[3] Libby, Montana.

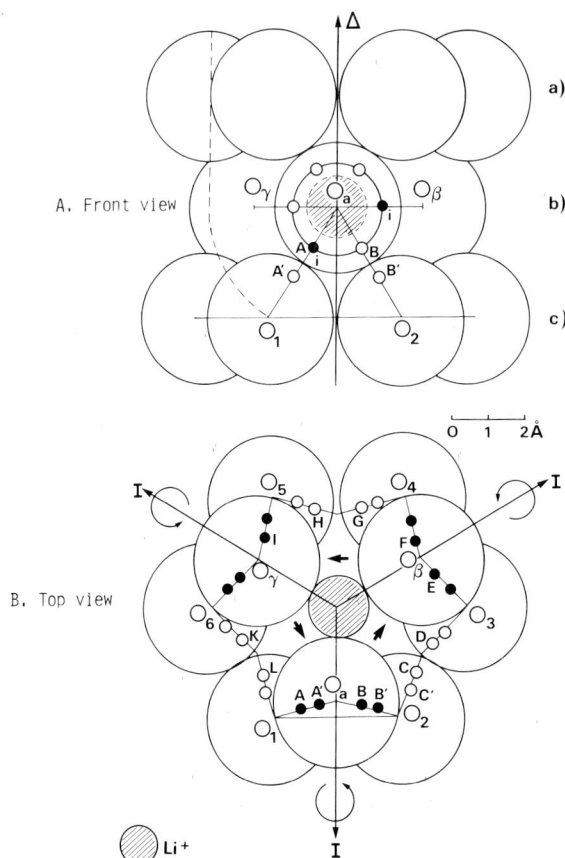

Figure 2. Model of Li$^+$·3 H$_2$O-hectorite. O$_1$, O$_2$, ... O$_6$ = oxygens of one clay layer; O$_\alpha$, O$_\beta$, ... O$_\gamma$ = water oxygens of the hydrate; A$_i$ A, and the small open circles (Figure 2B) are the six possible sites (sp^3 orbitals of lattice network) for the protons of the water molecule rotating around I, the molecular C$_2$ axis; A, B, ... K, L = the six possible sites for the protons during the slow rotation of the hydrate around Δ, the layer c axis. During this slow motion, oxygens of water molecules of the hydrate must move. The Li oscillation between (a) and (c) induce the splitting of site A in A', B in B'. The Li atom (shadowed circle) is drawn in Figure 2B in an average position.

the position of the detectors. The experiment was conducted at temperatures of 300, 278, 268, and 233 K.

BACKGROUND

Water structure in the interlameller space of smectite

Many spectroscopic studies of interlamellar water by X-ray (de la Calle et al., 1977; Fornés et al., 1980; Alcover and Gatineau, 1980) and infrared (IR) spectroscopy (Prost, 1975; Poinsignon et al., 1978) confirm the structuring power of exchangeable cations for water. The cation and the water form the following coordination polyhedra in the interlamellar space of the smectites: a triangle for Li·3 H$_2$O; a square for Cu·4 H$_2$O; and an octahedron for Ca·6 H$_2$O and Cu·6 H$_2$O whose structure is not rigid. For Li-hectorite, it is useful to

recall the main experimental features obtained by Conard et al. (1984) for the lower, stable hydration state, because those features allow a model to be constructed that can be checked by QNS and that can be applied to other homoionic smectites.

IR and thermogravimetric (TGA) data (Poinsignon et al., 1978) obtained on Li-hectorite showed that the Li$^+$·3 H$_2$O hydrate is the lower stable hydration state, corresponding to one half of a monolayer. ^{1}H- and ^{7}Li-NMR data (Conard, 1976) led to the following dynamical model of Li·3 H$_2$O. Lithium is in an axial symmetry site, and the water protons are involved in two rotational motions around two perpendicular axes (Figure 2); namely, rotation of the entire hydrate around the c-axis of the clay platelet, and rotation of the water molecule around its C$_2$ axis. The different moments of inertia of the motions suggest different correlation times for these two motions. The trihydrate is centered above the hexagonal hole of the sheet. The hydrate rotation involves 12 possible sites for the proton on the structural oxygens, distributed on a 2.49-Å radius circle (Figure 2A). The water rotation needs six sites on a 1.2-Å radius circle.

Theory

QNS by the clay-water system is dominated by the incoherent cross section of hydrogen. The coherent contribution can then be neglected, and the double differential cross section, describing the neutron intensity scattered per solid angle and energy unit, may be written as follows:

$$\frac{d^2\sigma}{d\Omega\,d\omega} = \frac{k}{k_0} \cdot \frac{1}{4\pi} \sigma_{\text{inc}} S_S(Q,\omega) \qquad (1)$$

where k and k$_0$ are the incident and scattered wave vectors; k − k$_0$ = Q, the momentum transfer; E − E$_0$ = $\hbar\omega$ = the energy transfer; σ_{inc} is the incoherent cross section; and S$_S$(Q,ω) is the incoherent scattering law.

QNS data exhibit two main features: an elastic incoherent structure factor (EISF), and a characteristic time of the motion related to the broadening Γ of the quasi-elastic part. The EISF and Γ variations with scattering angle are related to the geometry and the correlation time of the motion expressed in the scattering law S$_S$(Q,ω).

S$_S$(Q,ω) must take account of all the motions involving water protons and must consider the complete proton content of the sample (protons are the only incoherent scatterers in this sample). Elastic incoherent scattering is due only to the structural hydroxyl groups. This elastic contribution is given by Cδ(ω) in the Eq. (2) below. The scattering law may then be written (Lechner and Leadbetter, 1979) as the folding product of the individual scattering law related to any motion in which protons are involved.

For Li$^+$·3 H$_2$O, protons are involved in the water molecule rotation, R$_1$, in the hydrate rotation, R$_2$, and

Table 2. Proton correlation time and diffusion constant of water in the Li-exchanged hectorite[1] at 298 K.

Number of water layer	Type of proton motion	$1/\tau_1$ (μeV)	τ_1 (s)	τ (s)
0.5	Slow rotation	31	2.14×10^{-11}	2.8×10^{-12}
	Fast rotation	240	2.7×10^{-12}	1.3×10^{-12}
	Translation	$D_t = 4.5 \times 10^{-7}$ cm^2/s		
0.8	Slow rotation	29.5	2.23×10^{-11}	2.9×10^{-12}
	Fast rotation	210	3.1×10^{-12}	1.6×10^{-12}
	Translation	$D_t = 6 \times 10^{-7}$ cm^2/s		
1	Slow rotation	27.5	2.39×10^{-11}	3.1×10^{-12}
	Fast rotation	160	4.11×10^{-12}	2.1×10^{-12}
	Translation	6.6×10^{-7} cm^2/s		
2	Slow rotation	22.75	2.9×10^{-11}	3.8×10^{-12}
	Fast rotation	150	4.4×10^{-12}	2.2×10^{-12}
	Translation	2.6×10^{-6} cm^2/s		

[1] τ = the residence time; motional correlation time τ_1 is given by $\tau_1 = \tau/(1 - \cos \pi/3)$ for the rotation over 6 positions and $\tau_1 = \tau/(1 - \cos \pi/6)$ for the slow rotation over 12 positions.

in a long-range motion. These motions are assumed to be non-correlated. The Debye-Waller factor $e^{-\langle u^2 \rangle Q}$, where $\langle u^2 \rangle$ is the vibrational amplitude of the proton, is the contribution of vibration of higher energy to the spectral intensity.

$$S_S(Q,\omega) = e^{-\langle u^2 \rangle Q^2}\{S_S^{R_1} \times S_S^{R_2} \times S_S^{trans}$$
$$\times [(1 - C\delta(\omega)] + C\delta(\omega)\}, \quad (2)$$

where C is the ratio of fixed protons to the total proton number of the sample and $\times$ stands for the convolution product. $S_S^R(Q,\omega)$ are the double Fourier transform in space and time of the self correlation function $G_s(r,t)$ (Van Hove, 1954) calculated from the model determined by NMR and IR experiments. Using the formalism established by Dianoux et al. (1975) for a random jump motion among N equidistant sites on a circle of radius a, the scattering function $S_S^R(Q,\omega)$ is given by:

$$S_S^R(Q,\omega) = A_0(Qa \sin \theta)\delta(\omega)$$
$$+ \frac{1}{\pi} \sum_{n=1}^{N-1} A_n(Qa \sin \theta)\frac{\tau_n}{1 + (\omega\tau_n)^2}, \quad (3)$$

where A_0 is the EISF and A_n are the quasielastic structure factors. They are given by:

$$A_n(Qa \sin \theta) = \frac{1}{N} \sum_{p=1}^{N} j_0\left(2Qa \sin \theta \sin \frac{\pi P}{N}\right)$$
$$\cdot \cos(n2\pi P/N), \quad (4)$$

where j_0 is the Bessel function of zero order, i.e., $j_0(x) = \sin(x)/x$; θ is the angle between the momentum transfer Q and the rotation axis; and τ_n, the reciprocal half width at half maximum (hwhm) of the nth Lorentzian, is related to τ, the residence time, and τ_1 the reciprocal hwhm of the first Lorentzian, as follows:

$$\tau_n = \tau_1 \frac{\sin^2\pi/N}{\sin^2 n\pi/N}$$
$$\tau_1 = \tau/[1 - (\cos 2\pi/N)]. \quad (5)$$

For large N values (N > 6), τ/τ_1 tends towards $1/D_R$, and the uniaxial jump motion is similar to a continuous uniaxial motion.

S_S^{trans}, the scattering law for a translational motion, is a Lorentzian curve with a hwhm of:

$$S_S^{trans} = (1/\pi)[\Gamma/(\Gamma^2 + \omega^2)]. \quad (6)$$

For a jump diffusion process, the variation of Γ with Q was given by Chudley-Elliott (1961) as:

$$\Gamma = (1/\tau_0)(1 - [\sin(Ql \sin \theta)/(Ql \sin \theta)]), \quad (7)$$

where τ_0 is the residence time and l is the jump length. For small Q values ($Ql \ll 1$), Γ approaches $DQ^2 \sin\theta$, where $D = l^2/6\tau_0$. D is the diffusion constant of the mobile species: the plot of Γ vs. Q^2 is a straight line whose slope is equal to $D \sin\theta$.

RESULTS AND DISCUSSION

Rotational motions

$Li^+ \cdot 3 H_2O$-hectorites. Because of their different moments of inertia, rotations R_1 and R_2 (Eq. (1)) were analyzed with two different resolutions on the IN5 spectrometer (18 and 138 μeV). The fit of the spectra with the theoretical scattering law (Eq. (2)) was performed with two floating parameters: a, the gyration radius and τ_1, the first correlation time. Table 2 summarizes the parameters giving the best fit, using $a_1 = 1.23$ Å and $a_2 = 2.18$ Å. To study the anisotropy of water motion, measurements were performed on samples oriented by $\alpha = 45°$ and $135°$ to the beam. The scattering function depends on the angle θ between the displacement vector and Q.

For spectra recorded at 90°, Q is parallel or perpendicular to the rotation plane depending on the sample orientation in the beam and the rotation examined (inasmuch as they have perpendicular rotation axes). The axis of rotation of the water molecule is in the ab plane, and the rotation axis of the hydrate is parallel to the c direction (Conard et al., 1984). Motional anisotropy manifests itself through the amplitude variation of the spectra ($A_\parallel$ or $A_\perp$). $A_\parallel$ = amplitude of the broadened part of the 90° spectra obtained for Q perpendicular to the rotation axis; $A_\perp$ = amplitude of the 90° spectra for Q parallel to the rotational axis (Figure 3).

Despite angular distribution of the platelets in the sample at 24°, the anisotropy is clear: for R_2, $A_\parallel/A_\perp = 2.9$ and is very weak; for R_1 (fast rotation of the water molecule), $A_\parallel/A_\perp \sim 1$. The lack of anisotropy for the

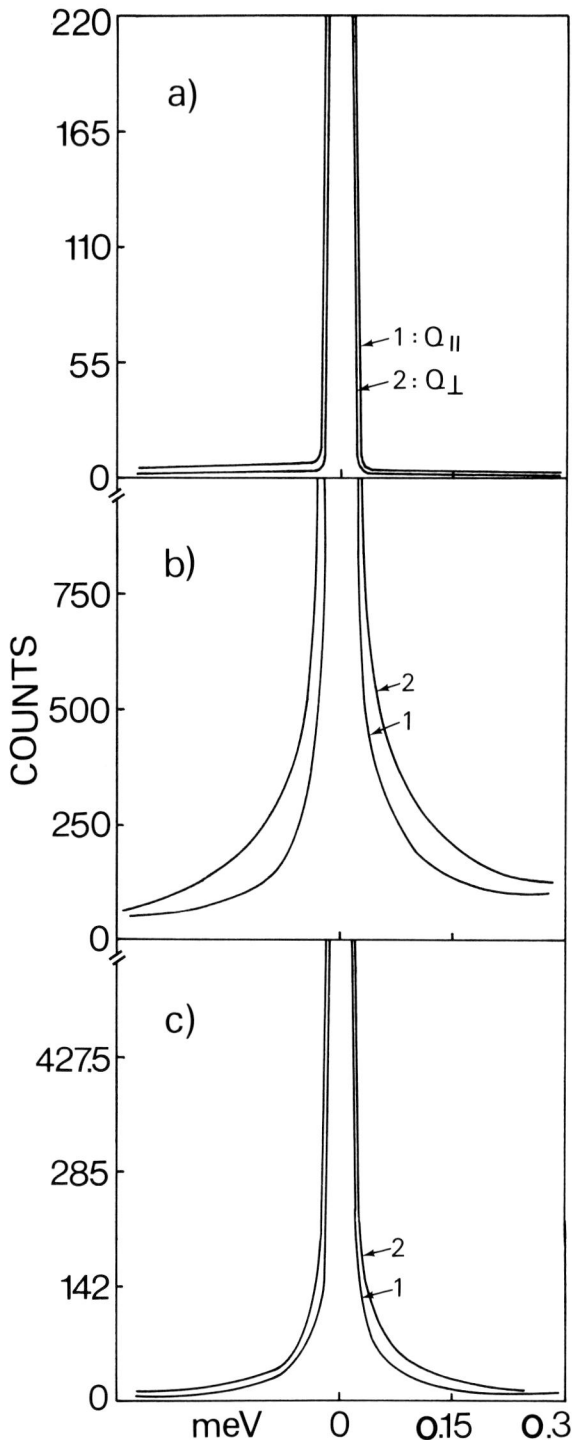

Figure 3. Anisotropy of the slow rotation for (a) vanadium, (b) Li-vermiculite, and (c) Ca-bentonite.

fast motion (R_1) is understandable if the results of Prost (1975) from partial deuteration experiments of this sample are considered. The rotation axis of H_2O, colinear with the transition momentum of the antisym-

metric bending vibrations of water, is at 50° to the clay sheet.

The thermal variation of the correlation times was studied between 300 and 190 K. Between 300 and 210 K, the correlation times of both rotations did not vary. They increased markedly at temperatures less than 200 K, and the motions stopped at 190 K.

Li-hectorite containing one water layer. Similar measurements on Li-hectorite containing one water layer showed a slowing down of the fast motion R_1 correlation time, which may be attributed to the formation of hydrogen bonds with the "filling water" of the first layer. R_2 was not sensitive to such bonding. Anisotropy was always observed for the slow motion R_2.

Li-bentonite and Li-vermiculite containing one water layer. The water superlattice (3a,b) established by Pons *et al.* (1980) for Li- and Na-bentonite (Figure 1) and the similar observations reported by de la Calle *et al.* (1977) allow an extension of the trihydrate model to Li-bentonite and Li-vermiculite. The fit of several spectra to Eq. (4) led to a determination of the correlation times of both motions and to a confirmation of the uniqueness of the gyration radius, a, and thus the validity of the model. The slow motion anisotropy was always present, as shown by the value of the ratio $A_\parallel/A_\perp$. At 260 K, the correlation time increased, but was always well resolved. The slow motion was then thermally activated to give correlation times of 24×10^{-12} s at 300 K and 34×10^{-12} s at 260 K. Spectra obtained from these samples containing two water layers always exhibited the same properties, as summarized in Table 3.

Cu-hectorite, Ca-bentonite, and Ca-vermiculite. From an electron paramagnetic resonance (EPR) study of a series of Cu^{2+}-exchanged smectites and vermiculites at different levels of hydration, Clementz *et al.* (1973) determined the stereochemistry of the hydrated Cu^{2+}. In the one-water layer state, Cu^{2+} is in the center of a square, with a cation-water molecule distance of ~ 2 Å. In the two-water layer state, Cu^{2+} is octahedrally coordinated, similar to the coordination in the structure shown by de la Calle *et al.* (1977) for Ca^{2+} in a two-water layer vermiculite (Figure 4).

The fit of QNS spectra of these samples was performed using the scattering law defined by Eq. (4), modified in light of the literature data. The best simultaneous fits of several spectra are listed in Table 4. The same properties as those previously observed were noted. The anisotropy of the slow motion was always observed and was more pronounced in the two-water layer state than in the one-water layer state for Cu-hectorite. For $Cu^{2+} \cdot 4 H_2O$, $A_\parallel/A_\perp = 1.66$; for $Cu^{2+} \cdot 8 H_2O$, $A_\parallel/A_\perp = 2.7$. The anisotropy was not measured for Ca-vermiculite, but for Ca-bentonite containing two water layers it was found to be weak. Structural ob-

Table 3. Proton correlation times and diffusion constant of water in the two first layers of the Li-exchange smectites at 295 K.[1]

Number of water layer	Type of proton motion	Li-bentonite			Li-vermiculite		
		$1/\tau_1$ μeV	τ_1 (s)	τ (s)	$1/\tau_1$ μeV	τ_1 (s)	τ (s)
1	Slow rotation	28	2.34×10^{-11}	3.1×10^{-12}	22.75	2.9×10^{-11}	3.75×10^{-12}
	Fast rotation	150	4.38×10^{-12}	2.2×10^{-12}	95	6.9×10^{-12}	3.4×10^{-12}
	Translation[2]		5×10^{-7} cm^2/s				
2	Slow rotation						
	Fast rotation				80	8.2×10^{-12}	4.1×10^{-12}
	Translation[2]		6.7×10^{-6} cm^2/s			3.4×10^{-6} cm^2/s	

[1] Symbols as in Table 2.
[2] Values from Hall and Ross (1978).

servations reported by de la Calle *et al.* (1977) and IR measurements of the present study on this type of sample showed C_1 and C_6-H_2O of the cube to be embedded into the hexagonal hole (Figure 4). The rotation axis, thus, is along the C_1-C_6 line, instead of being along the lattice c axis, as in hectorite. These data explain the low anisotropy.

Long-range motion

Spectra were obtained on the high-resolution backscattering spectrometer IN10 for Li-hectorite at two water contents corresponding to 0.5 and 2.5 layers, and at three temperatures, 298, 278, and 255 K. The spectra allowed a plot of Γ vs. Q to be drawn (Figure 5). Curve 1 is related to Li·3 H_2O-hectorite at 298 K; it is characteristic of a jump diffusion having a jump length of ~3 Å and a residence time of 2×10^{-10} s. A diffusion constant of 4.5×10^{-7} cm^2/s was derived from the initial slope of the plot of Γ vs. Q^2. Values of D at water contents of 0.5, 1.0, and 2.5 layers for Li-hectorite at 298 K are listed in Table 2. The macroscopic diffusion constant obtained by Laï and Mortland *et al.* (1968) using a radioactive tracer technique is of the same order of magnitude (6×10^{-7} cm^2/s). The diffusing species

could not have been free acidic H^+ inasmuch as their concentration (1 or 2%) was too low; neither could they have been free water molecules, because the bond energy for cation-water is too high, i.e., 98 kJ/mole (Poinsignon *et al.*, 1982). This long-range motion may be explained as a jump motion of the hydrated cation in the interlayer space.

The plots of hwhm of spectra obtained by high-resolution measurements at 298, 273, and 255 K for a water content of 2.5 layers (curves 2, 3, and 4, respectively, in Figure 5) show an unusual shape. At small Q values, the slope of curve 4 (298 K) is very steep. For Q values >0.8 Å^{-1}, the slope first is negative, then zero, and finally positive. Curve 3 (273 K) displays a similar shape, whereas curve 2 (255 K) has a shape similar to that of the curve for Li·3 H_2O (curve 1). The shapes of curves 2–4 in Figure 5 may be due to differences in the diffusion behavior of the several types of water in the sample. Indeed, for the two-water layer hydrate of Na-hectorite, Prost (1975) showed from IR spectra that $>50\%$ of the hydration water is not between the layers. At small Q values, curves 3 and 4 manifest this extra-layer water, i.e., the pore-water diffusion. For Q > 0.8 Å^{-1}, the width of this motion is greater than the energy window of the instrument. For

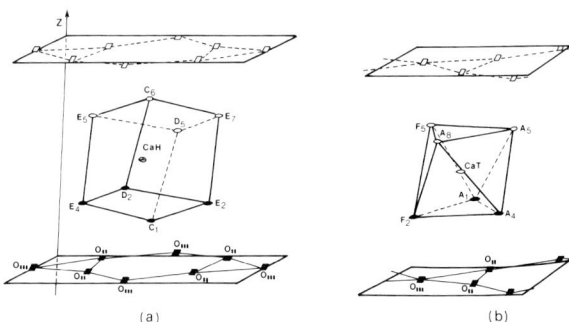

Figure 4. Coordination polyhedron of water around a Ca atom in a 2-water-layer Ca-vermiculite. Ca$_T$ is in front of a silicate tetrahedron; Ca$_H$ is in front of a ditrigonal hole (taken from de la Calle *et al.*, 1977).

Table 4. Correlation times of water in the two first layers of Ca^{2+}/Cu^{2+}-exchanged smectites at 295 K.[1]

Number of water layer	Type of proton motion	Cu-hectorite		Ca-bentonite		Ca-vermiculite	
		τ_{1ps}	A[2]	τ_1	A[2]	τ_1	A[2]
1	Slow rotation	Q$_\perp$ 34		36			
		Q$_\parallel$ 24	2.7	23	1.4	31	2.77
	Fast rotation	4.9	1				
2	Slow rotation	Q$_\perp$ 42		29			
		Q$_\parallel$ 18	2.33				
	Fast rotation						

[1] Symbols as in Table 2.
[2] Anisotropy of the motion (A) is characterized by $A_\perp/A_\parallel$, ratio of the amplitudes of the spectra recorded at a 90° scattering angle for the sample oriented at 135° and 45° to the beam.

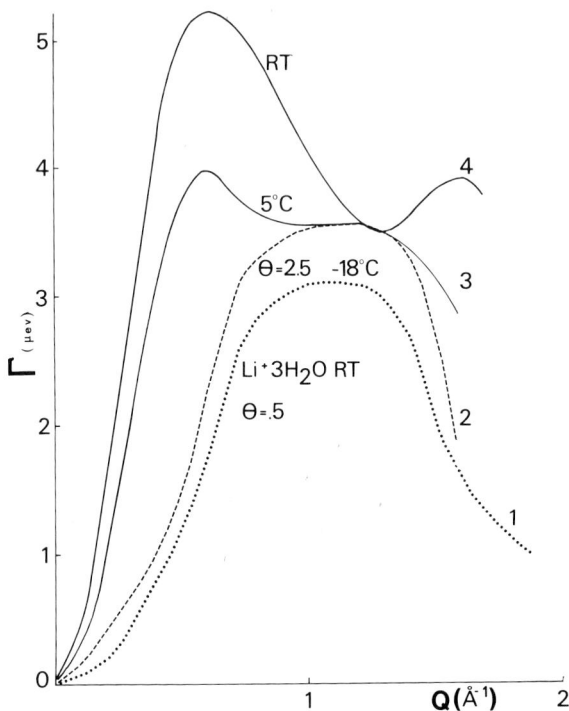

Figure 5.   Variation of Γ (half width at half maximum of the high-resolution spectra) corresponding to long-range motion of water protons vs. Q for the Li-exchanged hectorite.

$Q > 1$ Å^{-1}, the jump motion of the hydrated cation on the clay surface can be seen.

This interpretation is confirmed by curve 2, recorded at 255 K. At this temperature, the long-range motion of pore water was frozen. Curve 2 manifests the jump motion of the cation hydration shell, having a shape similar to that of curve 1. Also, at 298 K, the two motions occurred simultaneously and contributed to the broadening of the spectra. The higher sensitivity of pore water than that of interlayer water to temperature allowed their separation.

DISCUSSION AND CONCLUSIONS

For the smectites studied, having one or two water layers and electrical charges that differed in magnitude and in location in the structure, the exchangeable cation was found to be in the center of a water polyhedron, that shape of which was determined by the cation charge and by the water content of the clay. The hydration shell was involved in an uniaxial anisotropic rotation about the lattice c axis whose gyration radius varied from 2.18 Å (Li$^+\cdot 3$ H$_2$O-hectorite) to 2.4 Å (Cu$^{2+}\cdot$ 4 H$_2$O-hectorite), depending on the ionic radius and the shape of the polyhedron. Its correlation time was not sensitive to the clay charge (Table 2; one- and two-water layers). This motion was also not sensitive to temperature ($\tau_c = 24 \times 10^{-12}$ s at 300 K and 34 $\times$

10^{-12} s at 260 K for the monolayer Li-vermiculite), the activation energy being of the order of 3 kJ/mole.

The self rotation of the water molecule about its c_2 axis was more sensitive that the slow motion to (1) the water content (2.7 $\times$ 10^{-12} s for $\theta_w = 0.5$, 4.1 and 4.4 $\times$ 10^{-12} s for $\theta_w = 1$ and 2 in Li-hectorite, and faster in Li$\cdot 3$ H$_2$O than in supercooled water at the same temperature (295 K), 3.2 $\times$ 10^{-12} s (Texeira et al., 1984)); and (2) the clay structural charge, i.e., 4.1, 4.4, and 7 $\times$ 10^{-12} s for one-water layer Li-hectorite, Li-bentonite, and Li-vermiculite, respectively. Unfortunately, the influence of decreasing temperature on this motion in the monolayer hydrate was not checked at a temperature <260 K.

At the clay-water interface, the adsorbed water had its own structure in the first two layers. Proton dynamics were described by essentially three motions: two uniaxial reorientational motions and one jump-diffusional motion. The other types of water (interlayer water not belonging to the first layer or the pore water) were involved in an isotropic rotational diffusional motion. The correlation time remained similar to the broad component of supercooled water, but the sharp component was sensitive to the geometrical environment (Hall and Ross, 1978, 1981). The lifetime of protons in the adsorbed water was given by the residence time τ_0 of the jump diffusion process (10^{-10} s). During this time, the water molecules spin around their c_2 axis with a correlation time τ_{1f} of about 10^{-12} s, whereas they rotate with the cation hydrate with a correlation time τ_{1s} of about 10^{-11} s. The residence time τ of the proton between two jumps in one or the other direction is of the same order of magnitude (10^{-12} s).

REFERENCES

Alcover, J. F. and Gatineau, L. (1980) Interlayer space of vermiculites: Clay Miner. 15, 239–248.

Anderson, D. M. and Tice, A. R. (1971) Low temperature phase of interfacial water in clay water systems: Soil Sci. Soc. Amer. Proc. 35, 47–54.

Chudley, C. T. and Eliott, T. S. (1961) Neutron scattering from a liquid on a jump diffusion model: Proc. Phys. Soc. 77, 353–361.

Clementz, D. M., Pinnavaia, T. J., and Mortland, M. M. (1973) Stereochemistry of hydrated copper(II) ions on interlamellar surface of layer silicate: an electron spin resonance study: J. Phys. Chem. 77, 196–200.

Conard, J. (1976) Structure of water and hydrogen bonding on clays studied by ⁷Li and ¹H NMR: Magn. Res. in Coll. Interf. Science: Amer. Chem. Soc. Symposium Series T 34, 85–93.

Conard, J., Estrade-Szwarckopf, H., Dianoux, A. J., and Poinsignon, C. (1984) Water dynamics in a planar lithium hydrate in the interlayer space of a swelling clay. A neutron scattering study: J. Physique 45, 1361–1371.

de la Calle, C., Pezerat, H., and Gasperin, M. (1977) Problème d'ordre-désordre dans les vermiculites. Structure du minéral calcique hydraté à 2 couches: J. Physique 38, C7–128.

Dianoux, A. J., Volino, F., and Hervet, H. (1975) Incoherent scattering law for neutron quasielastic scattering in liquid crystals: *Mol. Phys.* **30**, 1181–1194.

Dore, J. C. (1984) Neutron diffraction studies of water in the normal and supercooled liquid phase: *J. Physique C7,* **9**, 49–64.

Fornès, V., de la Calle, C., Suquet, H., and Pezerat, H. (1980) Etude de la couche interfoliaire des hydrates à 2 couches des vermiculites calcique et magnésienne: *Clay Miner.* **15**, 399–412.

Fripiat, J. J., Cases, J. M., François, M., and Letellier, M. (1982) Thermodynamic and microdynamic behaviour of water in clay suspension and gels: *J. Colloid Interface Sci.* **267**, 463–466.

Glaeser, R. and Mering, J. (1968) Domaines et hydration homogène des smectites: *Compt. Rend. Acad. Sci.* **267**, 463–466.

Hall, P. and Ross, D. K. (1978) Incoherent neutron scattering function for molecular diffusion in lamellar systems: *Mol. Phys.* **36**, 1549–1554.

Hall, P. and Ross, D. K. (1981) Incoherent neutron scattering function for random jump diffusion in bounded and infinite media: *Mol. Phys.* **42**, 673–682.

Homshaw, L. G. (1980) Freezing and melting temperature hysteresis of water in porous materials: application to the study of pore form: *J. Soil Science* **31**, 398–414.

Laï, T. M. and Mortland, M. M. (1968) Cationic diffusion in clay minerals: *Soil Sci. Soc. Amer. Proc.* **32**, 56–61.

Lechner, R. and Leadbetter, A. (1979) Neutron scattering studies: in *Plastically Crystallite State,* J. N. Sherwood, ed., Wiley, New York, 285–320.

Maier, B. (1983) Neutron Research Facilities at the ILL High Flux Reactor: Institut Laue-Langevin, 156X, 38042 Grenoble Cedex, France, 120 pp.

Poinsignon, C., Cases, J., and Fripiat, J. J. (1978) Electrical polarisation of water molecules adsorbed on smectite. An infrared study: *J. Chem. Phys.* **82**, 1855–1860.

Poinsignon, C., Yvon, J., and Mercier, R. (1982) Dehydration energy of exchangeable cations in montmorillonite — a DTA study: *Israel J. Chem.* **22**, 253–255.

Pons, C. H., Tchoubar, C., and Tchoubar, D. (1980) Organisation des molécules d'eau à la surface des feuillets dans un gel de montmorillonite-Na: *Bull. Mineral.* **103**, 452–456.

Prost, R. (1975) Etude de l'hydratation des argiles: *Ann. Agro.* **26**, 463–597.

Teixeira, J., Bellissent-Funel, M. C., Chen, S. H., and Dianoux, A. J. (1984) Dynamics of supercooled water studied by neutron scattering: *J. Physique C7* **45**, 65–71.

Van Hove, L. (1954) Correlations in space and time and Born approximation. Scattering in systems of interacting particles: *Phys. Rev.* **95**, 249–262.

Woessner, D. E. (1979) An NMR investigation into the range of the surface effect on the rotation of water: *J. Mag. Res.* **39**, 297–303.

Proceedings of the International Clay Conference, Denver, 1985, L. G. Schultz, H. van Olphen, and F. A. Mumpton, eds.,
The Clay Minerals Society, Bloomington, Indiana, 292–297 (1987).

FACTORS AFFECTING THE MICROSTRUCTURE OF SMECTITES: ROLE OF CATION AND HISTORY OF APPLIED STRESSES

H. Ben Rhaïem,[1] C. H. Pons,[1] and D. Tessier[2]

[1] Université d'Orléans, B.P. 6759, Rue de Chartres, 45067 Orléans Cédex 2

[2] Station de Science du Sol, I.N.R.A., 78000 Versailles, France

Abstract—Small-angle X-ray scattering and transmission electron microscopy were used to determine the microstructure of Ca- and Na-smectite pastes. Samples were examined during drying and during rewetting at room temperature in an ultrafiltration cell by varying the suction pressure. The microstructure of Ca-smectite in (10^{-3} M CaCl$_2$) at suction pressures corresponding to high water contents (≤ 1 bar) appears to consist of a network of quasi-crystals containing pores about 1 μm in size. The network walls result from the face-to-face bonding of 50 to about 400 layers, depending on the suction pressure and the previous degree of drying. For Na-smectites in a 10^{-3} M NaCl solution at low suction pressure (≤ 0.1 bar), the tactoids were found to consist of a few layers (≤ 10), and the water was found to be located essentially between the layers inside the tactoids.

Key Words—Microstructure, Small-angle X-ray scattering, Smectite, Suction pressure, Tactoid.

INTRODUCTION

As part of a study on the effects of wetting and drying of soils, Pons *et al.* (1982) previously reported on the structure of pastes of a Greek montmorillonite in both the sodium and the calcium form containing different amounts of electrolyte solutions. Wetting and drying of the pastes was carried out at room temperature in a specially designed ultrafiltration cell (Tessier and Berrier, 1979). Structure determination on the pastes was made using both transmission electron microscopy (TEM), and small-angle X-ray scattering (SAXS). The structure description comprised a description of texture (particle size and particle arrangement) as well as a description of the crystalline structure from the Ångstrom to the micrometer range to provide a better definition of particle and pore structure. In the present study, analogous experiments were carried out on various smectite pastes as a function of wetting and drying history in the ultrafiltration cell.

METHODS AND MATERIALS

The ultrafiltration cell in which isothermal drying and wetting cycles were carried out consists of a membrane supported by a metal grid at the bottom and an inlet for compressed air at the top, the pressure of which can be varied. The bottom of the cell was placed in a beaker containing water or an electrolyte solution. The clay paste was placed on the membrane so that the liquid in the beaker could be imbibed by the clay paste. Water take-up by the clay was governed by capillary action and by osmotic swelling. Together, these forces constitute the "suction" pressure of the system. The suction pressure at any stage of liquid uptake in the wetting and drying process is equal to the air pres-

sure applied above the paste. In the following suction pressure is expressed in bars. Curves of first drying of the pastes were obtained by applying air pressure ≤ 25 bar and then a relative vapor pressure to achieve equilibrium. Rehydration effects were studied by applying a maximum pressure to the clays as a drying step followed by rewetting at decreasing air pressures. At each equilibrium reached at a given pressure, the water content was measured by weighing the clay sample before and after heating to 150°C.

The following smectite clays were studied: Wyoming bentonite, a montmorillonite from Greece (layer charge 0.4 per half unit cell), and hectorite from Hector, California. From these clays in the sodium form, <2-μm fractions were prepared by suspending the sodium clays in water, followed by repeated sedimentation. Both Na- and Ca-saturated suspensions were prepared in 10^{-3} M NaCl and 10^{-3} M CaCl$_2$ solutions, respectively, and homogenized by mechanical stirring.

Transmission electron microscopy (TEM) was carried out on ultrathin sections of about 500 Å thickness. These sections were prepared by successive replacements of the liquid phase by acetone, propylene oxide, and, finally an epoxy resin (Spurr, 1969; Tessier, 1984).

Small-angle X-ray scattering (SAXS) experiments were carried out with the X-rays emitted by the synchrotron storage ring at L.U.R.E..[3] The experimental conditions were described by Pons *et al.* (1981). Using such an X-ray source, the diagrams can be directly compared with theoretical SAXS curves. Computation models were used which assumed the particles to be

---

[3] L.U.R.E. = Laboratoire pour l'Utilisation du Rayonnement Electromagnétique.

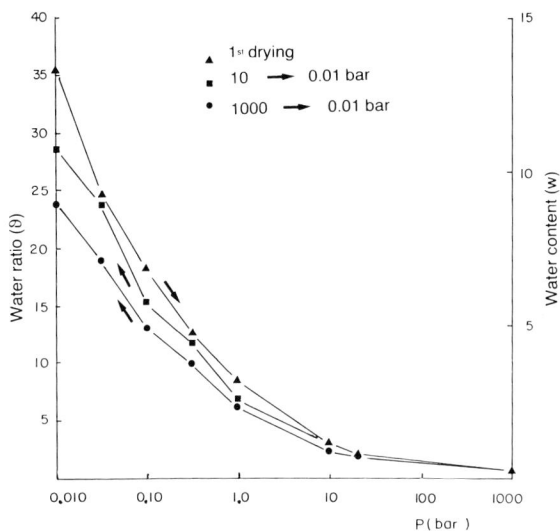

Figure 1. Hydration-dehydration behavior of Greek montmorillonite in 10^{-3} M NaCl (w = wt. %).

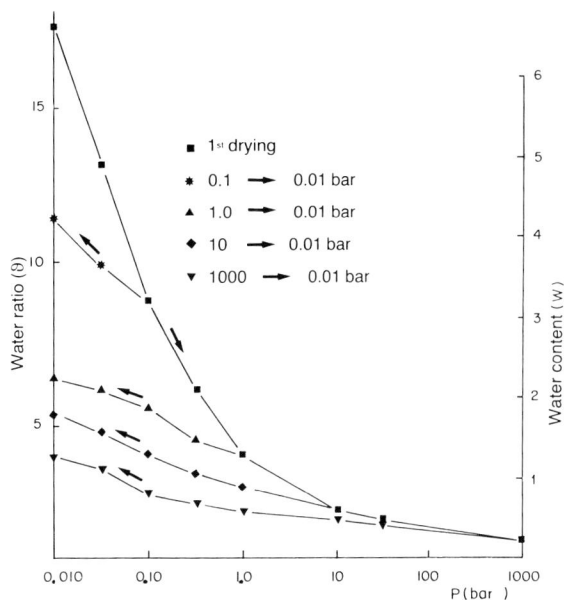

Figure 2. Hydration-dehydration behavior of Wyoming Ca-montmorillonite in 10^{-3} CaCl$_2$ (w = wt. %).

stacks of M parallel silicate layers. The internal structure of such a particle is given by the degree of order in the succession of layers which in turn is determined by the interlayer distance distribution (a set of i translations d_1, d_2 ... d_i and their respective probabilities P_1, P_2 ... P_i (Pons et al., 1981; Pons et al., 1982).

RESULTS AND DISCUSSION

Hydration-dehydration behavior

Results summarized in Figure 1 show that after drying at suction pressures of 10 or 1000 bar, the Greek Na-smectite when rewetted with a solution of 10^{-3} M NaCl recovered about 80 and 70%, respectively, of its original water content at 0.010 bar. On the other hand, results summarized in Figure 2 show that after drying at suction pressures of 0.1, 1, 10, or 1000 bar, the Wyoming Ca-smectite when rewetted with a solution of NaCl 10^{-3} M solution recovered about 64%, 34%, 27%, and 19%, respectively, of its original water content at 0.010 bar. Ca- and Na-smectites prepared using NaCl or CaCl$_2$ dilute solution thus behaved differently. Rehydration of Ca-smectites apparently depends on the pressures applied previously to the material. This result, which is in accord with the data of Croney and Coleman (1954), shows that any study of the hydration behavior of Ca-smectites and of clays in general in the presence of divalent cations should take into account the pristine state of the clay which may have been affected not only by such factors as drought or frost, but also by previously applied pressure. Consequently, this result should be kept in mind whenever water adsorption isotherms are made, especially in the range of very high water activities, and when the microstructure is described.

Changes in interlayer spacing and particle-size distribution

Ca-montmorillonites. The simultaneous variation of interlayer spacing and particle-size distribution throughout the initial drying and rewetting steps, as obtained from SAXS studies, are described below. Clay particles in the gel or paste state are herein defined by the following parameters (Saez-Aunon et al., 1983): (1) the average distance, $\bar{d}$, characterizing the average interlayer spacing; (2) the ratio $\hat{\delta}^2/\bar{d}^2$, where $\bar{\delta}^2$ is the variance of the interlayer distance distribution (the higher this ratio, the more disordered the system); and (3) the ratio $\bar{d}/d_{max}$ which characterizes the lack of symmetry (d_{max} is the interlayer distance with highest probability).

The data on first drying summarized in Table 1 show that the interlayer spacing was not affected by suction pressures ≤ 10 bar. Only at p > 50 bar did the interlayer spacing decrease (Figure 3a). Moreover, at p ≤ 1 bar, the number of layers (M) in each particle (55 layers) remained unchanged (Figure 3b). Hence, at p < 1 bar, interlayer hydration and particle size were constant. Thus, for Ca-montmorillonites at suction pressures ≤ 1 bar, only interparticle water was partly withdrawn and the spacing between particles became smaller. At pressures >10 bar, the number of layers in each particle increased markedly. Thus, drying led to an association of the initial particles and to the formation of particles having more layers (Figure 3b). As also shown by the data in Table 1 and Figure 3c, drying at p = 1000 bar, caused the internal structure of the particle to become more organized ($\bar{\delta}^2/\bar{d}^2$ decreases at p = 1000 bar).

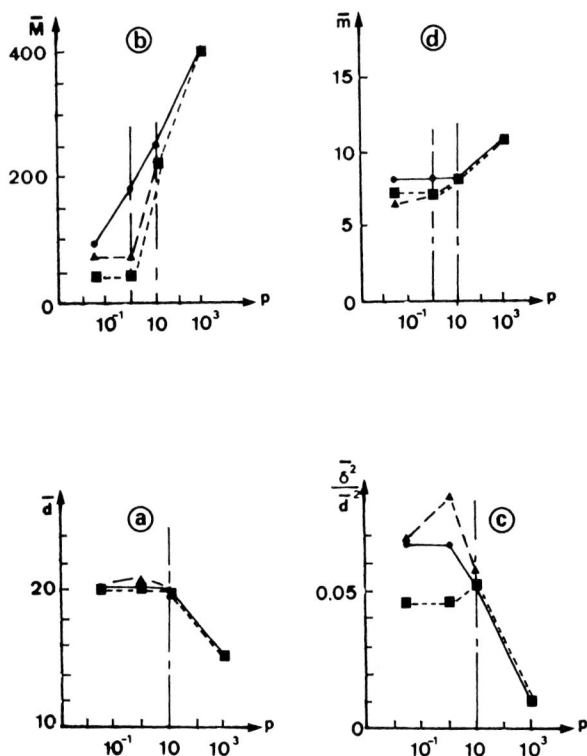

Figure 3.   Small-angle X-ray scattering parameters for Wyoming Ca-montmorillonite in 10^{-3} M CaCl$_2$. ■ = drying; ▲ = 10 → 0.032 bar; ● = 10^3 → 0.032 bar rewetting; $\bar{M}$ = average number of parallel silicate layers in particles; $\bar{m}$ = average number of parallel silicate layers in sub-stackings; P = precision; $\bar{d}$ = ; average interlayer distance; $\bar{\delta}^2$ = variance of the interlayer distance distribution.

Figure 4.   Small-angle X-ray scattering diagram during first drying of Wyoming Na-montmorillonite in 10^{-3} M NaCl at 1 bar. Intensity (arbitrary units) vs. S, where S = 2 sin θ/λ.

This process was accompanied by an increase in the number of layers in uniform sub-stackings (m) (Figure 3d) (Pons *et al.*, 1982; Ben Rhaïem, 1983). Such particle structure as evidenced by SAXS analysis is characteristic of Ca- and Mg-montmorillonites.

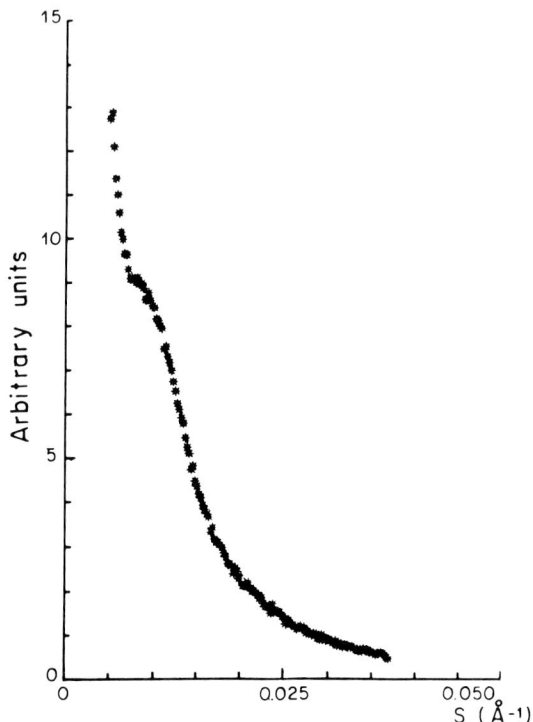

During rewetting, data on interlayer hydration show that from 1000 bar, $\bar{d}$ increased to 20.15 Å at 10 bar (Table 1). The resulting interlayer spacing was close to that observed after initial drying to 10 bar (Figure 3a). Therefore interlayer hydration was perfectly reversible at this stage. By comparing the data from initial drying (Table 1) with those from rewetting (Table 1), however, a change in the layer number (M) occurs at p = 0.032 bar (Figure 3b). This hysteresis in particle size was

Table 1.   Small-angle X-ray scattering results obtained during the first drying and rewetting for Wyoming Ca-montmorillonite.[1]

Stress (bar)	M[2]	m[2]	$\bar{d}^3$ (Å)	$d_{max}{}^3$ (Å)	$\bar{d}/d_{max}$	$\bar{\delta}^2/\bar{d}^{2}$ [4]	Water content (wt. %)
			First drying				
0.032	55	7.7	20.09	18.6	1.080	0.046	4.90
1.0	55	7.7	20.09	18.6	1.080	0.046	1.28
10	225	8.5	19.97	18.6	1.074	0.058	0.60
1000	400	11	15.36	15.6	0.985	0.010	0.26
			Rewetting[5]				
10 → 0.032	65	6.85	19.15	18.6	1.03	0.066	1.6
10 → 1.0	65	7.5	20.67	18.6	1.11	0.084	0.92
10^3 → 0.032	90	8.20	20.15	18.6	1.08	0.066	1.10
10^3 → 10	170	8.40	20.15	18.6	1.08	0.066	0.60

[1] For 1 hr in 10^{-3} M CaCl$_2$.
[2] M and m = number of layers in the particles and in the sub-stackings, respectively.
[3] $\bar{d}$ = interlayer average distance; d_{max} = interlayer distance with the highest probability.
[4] $\bar{\delta}^2$ = variance of the interlayer distance distribution.
[5] 10 → 0.032 = 10 bar to 0.032 bar rewetting, etc.

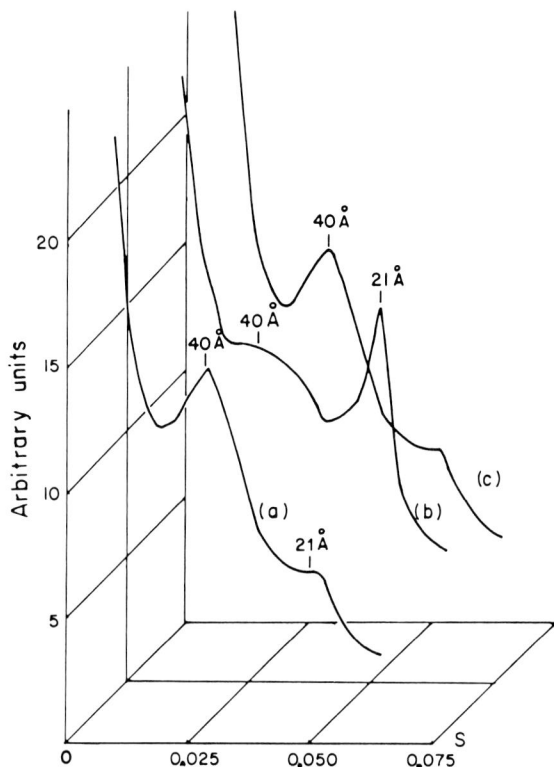

Figure 5. Small-angle X-ray scattering diagrams obtained for Wyoming Na-montmorillonite in 10^{-3} M NaCl. (a) 10 bar, first drying; (b) 25 bar, first drying; (c) 10 bar, after rewetting from 1000 bar. Intensity (arbitrary units) vs. S, where $S = 2 \sin \theta / \lambda$.

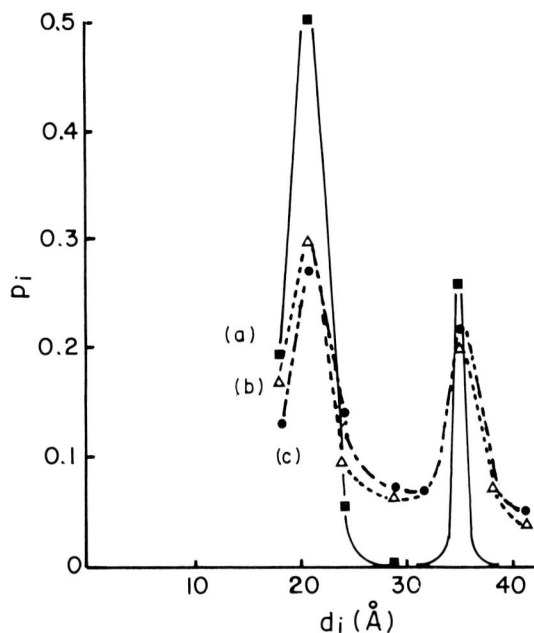

Figure 6. Interlayer distance (d_i) distribution for Wyoming Na-montmorillonite in 10^{-3} M NaCl. (a) 10 bar, first drying; (b) 25 bar, first drying; (c) 10 bar, after rewetting from 1000 bar. p_i = probability.

more noticeable when the initial drying was more intense. This hysteresis in particle size was accompanied by hysteresis in water content (Figure 2).

The hydration behavior of Ca-clays should therefore have been related essentially to a change in particle size and arrangement, i.e., to a textural change. A similar behavior for Mg- and Na-smectites in the presence of concentrated solutions was noted by Tessier and Pédro (1982) and Tessier (1984).

Na-montmorillonites. The range of hydration corresponding to low suction pressures was examined, i.e., on conditions that allowed an optimum characterization of strong interlayer hydration states (≤25 bar).

The SAXS diagram obtained at 1 bar from a Na-montmorillonite (10^{-3} M NaCl) is presented in Figure 4. The scattering curve shows a diffuse peak at very small angles (~100 Å). This diagram is characteristic of Na-montmorillonite in the gel state (Pons, 1980; Pons *et al.*, 1982).

For the same clay examined at 10 and 25 bar (Figures 5a and 5b), two peaks were observed at 40 and 20 Å; at 25 bar, the second peak was more intense than the first. By fitting the theoretical curves to SAXS diagrams, interstratified phases (a gel phase (40 Å) and a hydrated solid phase (~20 Å)) appear to coexist in the particles. The interlayer distance distribution (Figures 6a and 6b) suggests that the transition of the gel to the hydrated solid is accompanied by an abrupt change of spacing from 35 Å (gel phase) to 21.6 Å (hydrated solid). The latter spacing corresponds to four water layers in the interlayer position.

The pattern and the interlayer distance distribution obtained at 10 bar after rewetting from 1000 bar (Fig-

Table 2. Small-angle X-ray scattering results obtained during the first drying and rewetting for the Wyoming Na-montmorillonite.[1]

	Stress (bar)	M	d̄	d_{max} (Å)	$d̄/d_{max}$ (Å)	$\bar{\delta}^2/\bar{d}^2$	Water content (g/g)
Drying	1	8	84.73	70.00	1.21	0.101	3.70
Rewetting	10 → 1	15	57.93	50.00	1.16	0.120	2.60
Rewetting	10^3 → 1	20	46.55	40.00	1.16	0.085	2.34

[1] See Table 1 for explanations of symbols.

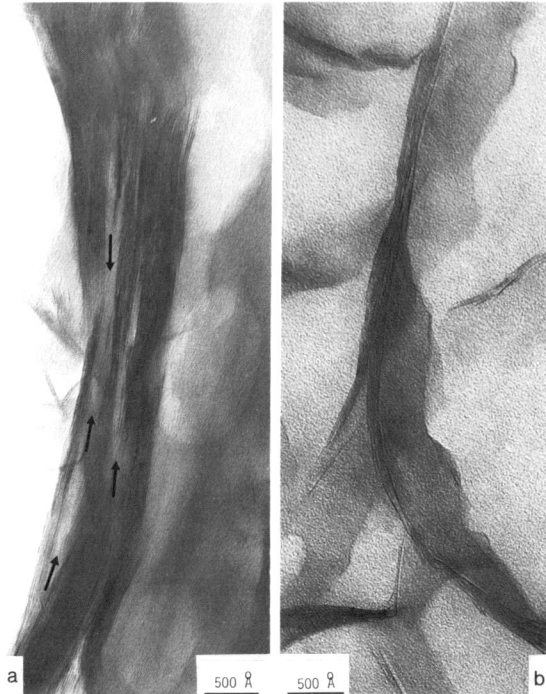

Figure 7. Transmission electron micrographs of samples prepared at 0.032 bar. (a) Greek montmorillonite in 10^{-3} M $CaCl_2$; (b) Na-hectorite from Hector, California, in 10^{-3} M NaCl. Micrographs by C. Clinard.

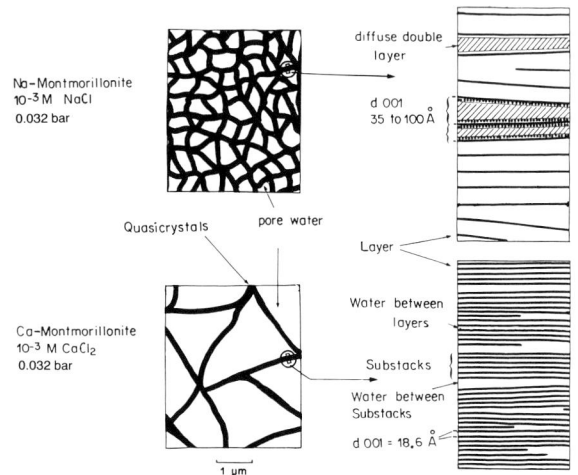

Figure 9. Schematic representation of the microstructure of Ca- and Na-smectites prepared with diluted solutions (10^{-3} M).

ures 5c and 6c) were the same as those obtained at 10 bar after the initial drying. Thus, the changes in interlayer spacing in water-Na-smectite systems prepared with dilute solution were reversible at suction pressures of 10–25 bar.

For lower suction pressures, data in Table 2 obtained from Na-montmorillonites (10^{-3} M), during initial drying at 1 bar and during rewetting from 10 and 1000 bar to 1 bar indicate that (1) the particle layer number increased with increasing intensity of initial drying, and (2) the interlayer spacing decreased with increasing intensity of initial drying.

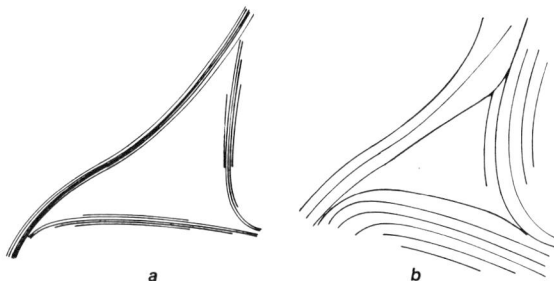

Figure 8. Schematic representation of microstructure of Wyoming Na-montmorillonite. (a) as observed by transmission electron microscopy; (b) as predicted from small-angle X-ray scattering results.

For both the Ca- and Na-smectites, the changes in the system during drying and wetting involved a change in particle size and, for Na-smectites a change of the particle's internal structure.

Microstructure of clays

SAXS data provided quantitative information on interlayer spacings. The water status was thus established by distinguishing the water within the interlayer spacing (i.e., within particles) from the water in larger pores (i.e., between particles). Actually, TEM data give a more complete picture of water-clay systems by providing a link between internal particle structure and system geometry. For example, for a Ca-montmorillonite prepared at 0.032 bar (Figure 7a), sections analyzed by TEM showed the pore-wall structure to be made up of several packed layers (~ 50) arranged face-to-face. Within the 50-layer aggregate, internal discontinuities as large as 50 Å in size were observed. Layers were substantially extended in the ab plane (Tessier, 1984). The type of packing in Ca-smectites corresponded to the definition of quasi-crystals as given by Aylmore and Quirk (1971). Thus, considering the SAXS data and the morphology of the quasi-crystals, the water distribution can be obtained and used as the basis of a model of Ca-smectite organization (Figure 9). At high water contents, i.e., under low suction pressures, the pore water amounted to 92% of the sample water. During drying, the pore size decreased and pore walls became thicker as the system lost water.

For low-charge Na-smectites (layer charge <0.45 negative charge per half-unit cell) TEM data showed that particles at low suction pressures (0.010 and 0.032 bar) had fewer layers than a Ca-smectite. Particles shown in the micrographs consist of 4–10 layers (Figures 7b). In this respect, micrographs obtained after embedding samples in epoxy resin are not accurate representations of the system in the wet state. For example, the layer spacing was always 15.6 Å. Under such conditions, TEM picture showed the same number of layers as found for the tactoids (Figure 8a) by SAXS analysis, and the visible pores were much larger than those that actually occurred at high water contents. An arrangement such as that shown in Figures 8b and 9 in which layers not strictly parallel (Pons, 1980) may be the actual arrangement.

CONCLUSION

For the water-Ca-smectite system, a suction pressure >50 bar must be applied to change interlayer spacing. Thus, over a wide range of suction pressures, these smectites are highly stable, hydrated, quasi-crystalline structures. The quasi-crystals, however, vary in arrangement and size. The transformation of the system therefore involves both a change in pore size (gathering or splitting of quasi-crystals) and a change in number of pores. At high water contents, the water is essentially located in the pores between the quasi-crystals. These pores reach their maximum size when the following conditions are met: (1) large lateral extension of the particles in the *ab* plane, (2) hydrated interlayer space having organized water layers, and (3) high flexibility of the quasi-crystals.

In water-Na-smectite systems prepared with dilute solutions ($<10^{-3}$ M), the system was different in arrangements, sizes, and internal structures of the tactoids, even at the lowest suction pressures. SAXS data showed that: (1) for $p > 25$, the sub-stackings (number of layers ≤ 4) were separated by the 35-Å interlayer spacings; (2) for $p < 10$ bar, the interlayer spacings (≥ 35 Å) were compatible with the existence of a diffuse double layer. Interlayer water content can therefore increase significantly and account for most of the soil water content, especially for a low-charge smectite having exchangeable cations.

REFERENCES

Aylmore, L. A. G. and Quirk, J. P. (1971) Domains and quasi-crystalline regions in clay systems: *Soil Sci. Soc. Amer. Proc.* **35,** 652–654.

Ben Rhaïem, H. (1983) Etude du comportement hydrique des montmorillonites calciques et sodiques par analyse de la diffusion de rayons X aux petits angles: Thesis Univ. Orléans, Orléans, France, 136 pp.

Croney, D. and Coleman, J. D. (1954) Soil structure in relation to soil suction (pF): *J. Soil Sci.* **5,** 75–84.

Pons, C. H. (1980) Mise en évidence des relations entre la texture et la structure dans les systèmes eau-smectites par diffusion aux petits angles du rayonnement synchrotron: Thesis Univ. Orléans, Orléans, France, 115 pp.

Pons, C. H., Rousseaux, F., and Tchoubar, D. (1981) Utilisation du rayonnement synchrotron en diffusion aux petits angles pour l'étude du gonflement des smectites. I: Etude du système eau-montmorillonite-Na en fonction de la température: *Clay Miner.* **16,** 23–42.

Pons, C. H., Tessier, D., Ben Rhaïem, H., and Tchoubar, D. (1982) A comparison between X-ray studies and electron microscopy observations of smectite fabric: in *Proc. Int. Clay. Conf., Bologna, Pavia, 1981,* H. van Olphen and F. Veniale, eds., Elsevier, Amsterdam, 165–186.

Saez-Aunon, J., Pons, C. H., Iglesias, J. E., and Raussell-Colom, J. A. (1983) Etude du gonflement des vermiculites ornithine en solution saline par analyse de la diffusion des rayons X aux petits angles. Méthode d'interprétation et recherche des paramètres d'ordre: *J. Appl. Crystallogr.* **16,** 439–448.

Spurr, A. R. (1969) A low-viscosity epoxy resin embedding medium for electron microscopy: *Ultrastructure Research* **26,** 31–43.

Tessier, D. (1984) Etude expérimentale de l'organisation des matériaux argileux: Thesis. Univ. Paris, Paris, France, I.N.R.A. Versailles Publ., 360 pp.

Tessier, D. and Berrier, J. (1979) Utilisation de la microscopie électronique à balayage dans l'étude des sols. Observation de sols humides soumis à différents pF: *Science du Sol* **1,** 67–82.

Tessier, D. and Pédro, G. (1982) Electron concentration and suction parameters: in *Proc. Int. Clay Conf., Bologna, Pavia, 1981,* H. van Olphen and F. Veniale, eds., Elsevier, Amsterdam, 165–176.

CATALYSIS
AND
SURFACE CHEMISTRY

Proceedings of the International Clay Conference, Denver, 1985, L. G. Schultz, H. van Olphen, and F. A. Mumpton, eds.,
The Clay Minerals Society, Bloomington, Indiana, 301–304 (1987).

CATION ADSORPTION IN AQUEOUS CLAY SYSTEMS: AN INTRODUCTORY REVIEW

G. H. BOLT

Department of Soils & Plant Nutrition, Agricultural University Wageningen
6703 BC Wageningen, The Netherlands

Abstract—A major development in the last four years in the field of ion adsorption by clays is the convergence of the "surface complexation" model and the Gouy-Stern description in terms of the electric double-layer structure. A common ground was reached in the form of so-called "triple" layer models which lend themselves well to generalized modeling efforts based on experimental data. An excessive number of system parameters, however, then becomes available for fitting purposes. Moreover, the apparent need to refine descriptions of lateral surface heterogeneity leads to sobering thoughts as to the ability of such theories to predict, for example, heavy metal adsorption in natural formations. A recent examination of ion-exchange equations in ternary systems has led to the reassuring suggestion that many such systems can be described by means of the information available from constituent binary systems.

Key Words—Adsorption, Cation exchange, Double layer structure, Surface complexation model, Triple layer model.

INTRODUCTION

Paraphrasing Heller-Kallai (1982) as she introduced the subject of clay–salt interactions at the 1981 International Clay Conference, it is impossible to review the entire field of cation adsorption in aqueous clay systems in the space alloted for this paper in the present Proceedings. For those interested in gaining insight into present-day thinking on the surface chemistry of clays, Sposito's (1984) book entitled *The Surface Chemistry of Soils* is recommended. It gives a comprehensive and up-to-date account of developments in this area, and, as pointed out by Bolt and van Olphen (1985), its review of the relevant literature allows the reader to obtain a solid overview of the subject, especially adsorption phenomena and current model theories.

Some of the most important advances in the field of ion adsorption that have emerged in the last few years are reviewed below.

IMPORTANT ADVANCES

Surface complexation vs. ionic double layers

Developments in the field of ion adsorption have been dominated in the last several years by the convergence of theories that use "surface complexation" as the chief principle for describing ion binding by solid (soil) particles (perhaps beginning with Stumm *et al.,* 1970) and those that employ "colloid chemical" models based on extended double layer theory. As is often the case, this "coming together" took place from both sides. The original "constant capacity" surface-complexation model, in which the counterion swarm plays only a subordinate role, was extended by the recognition of a

second layer that contains so-called "outer sphere complexes" (cf. James and Park, 1982; several chapters in Anderson and Rubin, 1981). At the same time, the Gouy-Stern model of the diffuse double layer (first outlined clearly in Grahame's classical paper in 1947) was brought to bear on ion-adsorption phenomena when the Stern adsorption energy was specified in terms of pair-formation constants. The meeting point was thus reached when workers decided to fill the "inner Helmholtz plane" of the Stern layer with specifically adsorbed "pair formers" which coincided with the "outer-sphere" surface complexes mentioned above. Examples of, and references to the application of such "triple" layer models to cation adsorption are given in Tewari (1981). The comparative merits of these multilayer models were discussed at length by Bolt and Van Riemsdijk (1982) and especially Sposito (1984).

Although numerous reports in which such models were described were published in the late 1970s, this information was largely ignored at the 1981 International Clay Conference, a possible exception being Madrid *et al.* (1982), probably because the literature did not focus on specific clay mineral species. Instead, these multilayer models were used chiefly to describe variable-charge minerals, in particular oxy-hydroxides. Lately this situation has changed because workers concerned with ion exchange on clay minerals have revived interest in specifying the Stern layer in terms of pair-formation constants. This trend, in turn, may be traced to increased interest in the adsorption of trace levels of transition metal ions in soil, the cation selectivity of which differs markedly for various cations.

A first attempt in this direction was made by Heald *et al.* (1964), but their efforts remained largely ignored, although Shainberg and Kemper (1967) and the ensuing discussion by Bolt et al. (1967) used essentially the same approach. In practice, these efforts were seriously hampered by the then limited access to computer modeling. In the past three years, numerical attempts to apply diffuse-layer extensions to the modeling of exchange behavior of "constant charge" clays appeared; see, for example, Neal and Cooper (1983). As could be expected, "constant-charge" surfaces allow a fair check on the applicability of such models. In addition, such checks serve to solidify trust in the principles behind the application of multilayer computer models that have been used to describe variable charge surfaces. Nevertheless, some suspicion is aroused when following the increasingly complex computer-simulation models as applied to the interpretation of ion adsorption by clay-size materials from natural deposits. One is soon faced with a minimum of 4–5 system parameters that are assumed to be characteristic of the system. A reasonable fit of medium-quality adsorption data to such 4–5-parameter models often proves little about the trustworthiness of the model (see, e.g., Westall and Hohl, 1980; Bolt and Van Riemsdijk, 1982).

Trace-metal adsorption by soil clays

Reliable experimental data are still indispensable for understanding, e.g., heavy (trace) metal adsorption in natural environments such as soil. Societal interest in the fate of these (often industrially derived) contaminants has grown rapidly. Relevant references (to 1980) to such situations are found in Kinniburgh and Jackson (1981) and in several chapters of Tewari (1981). More recent studies of the "high-level" adsorption of Cu on hectorite, gibbsite, and hydroxy-Al-coated hectorite and montmorillonite were published by McBride (1982a, 1982b), McBride et al. (1984), and Harsh et al. (1984) and Harsh and Doner (1984), respectively. Sposito et al. (1981) reported complete exchange isotherms for Cu on Na-montmorillonite. In addition to numerous reports of trace level adsorption of various transition metals on "local" soils, the papers by Inskeep and Baham (1983a, 1983b) on the adsorption of trace amounts of Cd and Cu in the absence and presence of certain organic ligands on Na-montmorillonite are noteworthy, as are the studies of Al- and Fe-polymers on clays as possible seats of high preference for heavy metal ions in soil (Oades, 1984; M. G. M. Bruggenwert, Dept. Soil Science and Plant Nutrition, Agricultural University, Wageningen, The Netherlands, personal communication).

Lateral heterogeneity of the adsorbent surface

Further on the modeling of adsorption phenomena, the handling of the increasing complexity of the adsorption zone in the direction normal to the clay surface may well be in vain if it does not keep pace with possible lateral heterogeneity. Several types of such heterogeneity have been pointed out recently. In this context, the term "demixing" has been used to describe the sorting out of exchangeable ions between clay-plate condensates (domains) and those on exterior surfaces. Evidence for this effect in mixed Na/Ca-montmorillonite suspensions was reported by Shainberg et al. (1982) from conductance measurements, thereby confirming earlier indications to this effect in Bar-on et al. (1970) and relevant double-layer calculations of Bolt (1979). The above "plate-condensation-induced-demixing" approach is similar to the earlier use of highly preferential "interlattice" exchange sites to explain, e.g., K-fixation in clays and intercalation phenomena in general. In addition, evidence has been reported of crystallographically anchored heterogeneity of stacked layers in several smectites of different origin (Talibudeen and Goulding, 1983a, 1983b). One might also assume that intra-face heterogeneity is not unlikely in view of lattice imperfections in many crystals. Such presumed site-heterogeneity has been applied to constant-capacity adsorption models (Sposito, 1980, following the works of Sips, 1948). This approach may be used to show that a Langmuir distribution law applied to a "family" of sites exhibiting a symmetric distribution of the corresponding ln K values leads to a Freundlich adsorption isotherm. In addition, Van Riemsdijk et al. (1986) recently showed that for a multi-ionic system the presence of a distributed ln K function for site-ion pairs could lead to a combined Langmuir-Freundlich adsorption isotherm for each of the adsorbates in the presence of a constant concentration of the others, thereby giving the often-observed finite slope of the adsorption isotherm at vanishing concentration in addition to the Freundlich curve at higher concentrations.

As should be clear from the above discussion, such refinements, when added to the heterogeneity of the adsorbed layer itself in the direction "normal" to the adsorbing surface, will rapidly surpass unambiguous verification by experimentation. Multi-parameter models (comprising 5 to 7 parameters) can be computer fitted without proving much about the credibility of the model. Apparently, a best-accepted simplification (e.g., dismissing either site- or layer-heterogeneity) must be selected before fitting a model containing a moderate number of parameters to the data at hand. This selection will be even more important when examining the adsorption properties of "mixed" systems, such as a natural soil. A comprehensive survey of the effect of site-energy distribution on adsorption isotherms was reported by Jaroniec (1983).

Ion adsorption in ternary systems

A final comment concerns the available information on ternary exchange/adsorption systems. It is significant that in the last 50 years, cation-exchange studies were for all practical purposes limited to binary systems, despite the fact that in nature binary systems are rare. In the past few years, this situation has changed; Elprince *et al.* (1980) and Chu and Sposito (1981) have revived the still available precise experimental data on ternary systems collected by Vanselow in the 1960s. They wrote out the thermodynamics of ternary systems on the basis of the subregular liquid-mixture model. For the system NH_4-Ba-La-montmorillonite, they showed that the third order terms that covered deviations of the ternary systems from predictions based on knowledge of the underlying three binary systems were negligible in magnitude. Thus, the chances are good that a guess based on the relevant binary systems will suffice for the prediction of the behavior of the ternary system, thereby saving a large amount of laboratory work. More systems would have to be investigated, of course, to substantiate this approach. Of equal interest is the series of papers by Fletcher and Townsend (1981a, 1981b, 1981c, 1981d, 1983). Here, the well-known Gaines and Thomas method of data analysis was extended to ternary systems and enlightening illustrations were presented suggesting how to plot ternary exchange data. Additional information on Na-Ca-Mg systems were also reported in Fletcher *et al.* (1984a, 1984b).

REFERENCES

Anderson, M. A. and Rubin, A. J., eds. (1981) *Adsorption of Inorganics at Solid-Liquid Interfaces:* Ann Arbor Science, Ann Arbor, Michigan, 357 pp.

Bar-on, P., Shainberg, I., and Michaeli, I. (1970) Electrophoretic mobility of montmorillonite particles saturated with Na/Ca ion: *J. Colloid Interface Sci.* 33, 471–472.

Bolt, G. H. (1979) Theories of cation adsorption by soil constituents: distribution equilibrium in electrostatic fields: in *Soil Chemistry B. Physico-chemical Models,* G. H. Bolt, ed., 2nd ed., Elsevier, Amsterdam, 47–76.

Bolt, G. H., Shainberg, I., and Kemper, W. D. (1967) Discussion on the paper by I. Shainberg and W. D. Kemper entitled: "Ion exchange equilibria on montmorillonite": *Soil Sci.* 104, 444–453.

Bolt, G. H. and van Olphen, H. (1985) Book Review of *G. Sposito, The Surface Chemistry of Soils. Clays & Clay Minerals* 33, p. 367.

Bolt, G. H. and Van Riemsdijk, W. H. (1982) Ion adsorption on inorganic variable charge constituents: in *Soil Chemistry B. Physico-chemical Models,* G. H. Bolt, ed., 2nd ed., Elsevier, Amsterdam, 459–504.

Chu, S.-Y. and Sposito, G. (1981) The thermodynamics of ternary cation exchange systems and the subregular model: *Soil Sci. Soc. Amer. J.* 45, 1084–1089.

Elprince, A. M., Vanselow, A. P., and Sposito, G. (1980) Heterovalent ternary cation exchange equilibria: NH_4-Ba-La exchange on montmorillonite: *Soil Sci. Soc. Amer. J.* 44, 964–969.

Fletcher, I., Sposito, G., and Levesque, C. S. (1984a) Sodium-calcium-magnesium exchange reactions on a montmorillonitic soil: I. Binary exchange reactions: *Soil Sci. Soc. Amer. J.* 48, 1016–1021.

Fletcher, I., Holtzclaw, K. M., Jouany, C., Sposito, G., and LeVesque, C. S. (1984b) Sodium-calcium-magnesium ex-

change reactions on a montmorillonitic soil: II. Ternary exchange reactions. *Soil Sci. Soc. Amer. J.* 48, 1022–1025.

Fletcher, P. and Townsend, R. P. (1981a) Ternary ion exchange in zeolites, part 1.—Problem of predicting equilibrium compositions: *J. Chem. Soc. Faraday Trans.* 77, 955–963.

Fletcher, P. and Townsend, R. P. (1981b) Ternary ion exchange in zeolites, part 2.—A thermodynamic formulation: *J. Chem. Soc. Faraday Trans.* 77, 965–980.

Fletcher, P. and Townsend, R. P. (1981c) Ternary ion exchange in zeolites, part 3.—Activity coefficients in multicomponent electrolyte solutions: *J. Chem. Soc. Faraday Trans.* 77, 2077–2089.

Fletcher, P. and Townsend, R. P. (1981d) Ternary ion exchange in zeolites, part 4.—Activity correction for the solution phase: *J. Chem. Soc. Faraday Trans.* 79, 419–432.

Grahame, D. C. (1947) The electrical double layer and the theory of electrocapillarity: *Chem. Rev.* 41, 441–501.

Harsh, J. B. and Doner, H. E. (1984) Specific adsorption of copper on an hydroxy-aluminum-montmorillonite complex: *Soil Sci. Soc. Amer. J.* 48, 1034–1039.

Harsh, J. B., Doner, H. E., and McBride, M. B. (1984) Chemisorption of copper on hydroxy-aluminum-hectorite: an electron spin resonance study: *Clays & Clay Minerals* 32, 407–413.

Heald, W. R., Frere, M. H., and De Wit, C. T. (1964) Ion adsorption on charged surfaces: *Soil Sci. Amer. Proc.* 28, 622–647.

Heller-Kallai, L. (1982) Clay-salt interactions: in *Proc. Int. Clay Conf., Bologna, Pavia, 1981,* H. van Olphen and F. Veniale, eds., Elsevier, Amsterdam, 127–132.

Inskeep, W. P. and Baham, J. (1983a) Adsorption of Cd(II) and Cu(II) by Na-montmorillonite at low surface coverage: *Soil Sci. Soc. Amer. J.* 47, 660–665.

Inskeep, W. P. and Baham, J. (1983b) Competitive complexation of Cd(II) and Cu(II) by water-soluble organic ligands and Na-montmorillonite: *Soil Sci. Soc. Amer. J.* 47, 1109–1115.

James, R. O. and Parks, G. A. (1982) Characterization of aqueous colloids by their electrical double layer and intrinsic chemical properties: *Surface Colloid Sci.* 12, 119–216.

Jaroniec, M. (1983) Physical adsorption on heterogeneous solids: *Adv. Coll. Interface Sci.* 18, 149–225.

Kinniburgh, D. G. and Jackson, M. L. (1981) Cation adsorption by hydrous metal oxides and clay: in *Adsorption of Inorganics at Solid-Liquid Interfaces,* M. A. Anderson and A. J. Rubin, eds., Ann Arbor Science, Ann Arbor, Michigan, 91–160.

Madrid, L., Diaz, E., Cabrera, F., DeArambarri, P., and Balbontin, J. (1982) Charge properties of some clays from soil with variable charge: in *Proc. Int. Clay Conf.,* Bologna, Pavia, 1981, H. Van Olphen and F. Veniale, eds., Elsevier, Amsterdam, 157–163.

McBride, M. B. (1982a) Cu^{2+}-adsorption characteristics of aluminum hydroxide and oxyhydroxides: *Clays & Clay Minerals* 30, 21–28.

McBride, M. B. (1982b) Hydrolysis and dehydration reactions of exchangeable Cu^{2+} on hectorite: *Clays & Clay Minerals* 30, 200–206.

McBride, M. B., Fraser, A. R., and McHardy, W. J. (1984) Cu^{2+} interaction with microcrystalline gibbsite. Evidence for oriented chemisorbed copper ions: *Clays & Clay Minerals* 32, 12–18.

Neal, C. and Cooper, D. M. (1983) Extended version of Gouy-Chapman electrostatic theory as applied to the exchange behavior of clay in natural waters: *Clays & Clay Minerals* 31, 367–376.

Oades, J. M. (1984) Interactions of polycations of aluminum and iron with clays: *Clays & Clay Minerals* 32, 49–57.

Shainberg, I. and Kemper, W. D. (1967) Ion exchange equilibria on montmorillonite: *Soil Sci.* **103,** 4–9.

Shainberg, I., Oster, J. D., and Wood, J. D. (1982) Electrical conductivity of Na/Ca-montmorillonite gels: *Clays & Clay Minerals* **30,** 55–62.

Sips, R. (1948) On the structure of a catalyst surface: *J. Chem. Phys.* **16,** 490–495.

Sposito, G. (1980) Derivation of the Freundlich equation for ion exchange reactions in soils: *Soil Sci. Soc. Amer. J.* **44,** 652–654.

Sposito, G. (1984) *The Surface Chemistry of Soils:* Oxford University Press, New York, 234 pp.

Sposito, G., Holtzclaw, K. M., Johnston, C. T., and LeVesque-Madore, C. S. (1981) Thermodynamics of sodium-copper exchange on Wyoming bentonite at 298° K. *Soil Sci. Soc. Amer. J.* **45,** 1079–1084.

Stumm, W., Huang, C. P., and Jenkins, S. R. (1970) Specific chemical interaction affecting the stability of dispersed systems: *Croat. Chim. Acta* **42,** 223–245.

Talibudeen, O. and Goulding, K. W. T. (1983a) Charge heterogeneity in smectites: *Clays & Clay Minerals* **31,** 37–42.

Talibudeen, O. and Goulding, K. W. T. (1983b) Apparent charge heterogeneity in kaolins in relation to their 2:1 phyllosilicate content: *Clays & Clay Minerals* **31,** 137–142.

Tewari, P. H., ed. (1981) *Adsorption from Aqueous Solutions:* Plenum Press, New York, 248 pp.

Van Riemsdijk, W. H., Bolt, G. H., Koopal, L. K., and Blaakmeer, J. (1986) Electrolyte adsorption on heterogeneous surfaces: adsorption models: *J. Coll. Interface Sci.* **109,** 219–228.

Westall, J. and Hohl, H. (1980) A comparison of electrostatic models for the oxide/solution interface: *Adv. Coll. Interface Sci.* **12,** 265–294.

Proceedings of the International Clay Conference, Denver, 1985, L. G. Schultz, H. van Olphen, and F. A. Mumpton, eds.,
The Clay Minerals Society, Bloomington, Indiana, 305–310 (1987).

ACIDITY AND CATALYTIC PROPERTIES OF PILLARED MONTMORILLONITE AND BEIDELLITE

A. Schutz,[1] D. Plee,[2] F. Borg,[3] P. Jacobs,[4]
G. Poncelet,[1] and J. J. Fripiat[2]

[1] Groupe de Physico-Chimie Minérale et de Catalyse, Place Croix du Sud 1
B-1348 Louvain-la-Neuve, Belgium

[2] CNRS-CRSOCI, rue de la Férollerie, F-45045 Orléans, France

[3] Total-Compagnie Française de Raffinage, Centre de Recherches
B.P. 27, F-76700 Harfleur, France

[4] Laboratorium voor Colloidale en Oppervlaktescheikunde, Kardinaal Mercierlaan 92
B-3030 Heverlee, Belgium

Abstract—Montmorillonite and synthetic beidellite were pillared with hydroxy-Al-polymers, and the physicochemical properties and catalytic activities of the pillared products were characterized by several techniques. The basal 001 reflection of the pillared clays was found to be about 18 Å and to decrease slightly after calcination. Infrared spectra of pillared beidellite exhibited an OH-stretching band at 3440 cm^{-1}, similar to that observed for H^+-exchanged beidellite or for the calcined form of NH_4-beidellite. These OH groups, which interact with pyridine to give pyridinium ions (Brönsted acidity), were formed by the proton attack of Si–O–Al linkages of the tetrahedral layer to form Si–OH . . . Al entities. Such a band was not evident in the pillared montomorillonite. In both pillared clays, Lewis acidity was also detected by pyridine adsorption and was generated by the aluminum pillars.

Hydroisomerization-hydrocracking of n-decane was performed in a continuous flow reactor at atomspheric pressure. The catalytic activity and the distribution of the products were determined for the pillared beidellite and montmorillonite and compared with the same data obtained for noncrystalline silica-alumina, ultrastable zeolite Y, and zeolite H-ZSM-5. Pillared montmorillonites and beidellites showed substantially different catalytic activities. Pillared beidellites had an overall activity only slightly less than that of the zeolites. The clays behaved similarly to zeolite-based bifunctional catalysts; i.e., they displayed high selectivity for isomerization at high conversion and symmetrical distribution of the individual products. They appear to be promising dewaxing catalysts. The pillared montmorillonites were much less active than the pillared beidellites. The high activities and selectivities of the pillared beidellites are likely due to the presence of Si–OH acid sites exposed in the interlamellar space.

Key Words—Acidity, Beidellite, Catalysis, Montmorillonite, n-Decane, Pillared interlayer clay, Pyridine.

INTRODUCTION

Brindley and Sempels (1977) and Lahav et al. (1978) showed that montmorillonite and, more generally, dioctahedral phyllosilicates may be expanded to form thermally stable structures by pillaring the bidimensional structure with hydroxy-aluminum polymers. The 001 reflection of the air-dried solid is slightly greater than 18 Å and decreases by about 0.5 Å after the sample is calcined between 300° and 500°C. Therefore, the interlamellar space has a thickness of about 7.5 Å. The BET surface area of these pillared clays is in the range 250–300 m^2/g.

Acid sites in pillared clays may be located either on the surface of the clay or on the surface of the pillar. The nature of the acid sites of the pillar is dependent on the structure of the pillar. Vaughan and Lussier (1980) and Pinnavaia et al. (1984) suggested that the pillaring cation is most likely an Al_{13} polyhydroxypolymer related to the known cation $[Al_{13}O_4(OH)_{24}$-$(H_2O)_{12}]^{7+}$ described by Johansson et al. (1960). This Al_{13} polymer is characterized by a tetrahedral aluminum in the center of three layers of aluminum octahedra. The structure contains the four layers of oxygen or hydroxyls required to produce an 18-Å 001 spacing. The calculated surface area of the pillaring cation is ≈ 110 Å^2. This arrangement provides enough void space in the interlayer region to account for the rather large surface area of the pillared clay. Until recently, nothing was known about the nature of the transformation of the Al_{13} species after thermal activation above 300°C; however, a spinel-like structure may be formed, similar to that obtained by calcining bayerite or gibbsite.

The pillaring of a beidellite, i.e., a dioctahedral smectite wherein Si is substituted in part by Al, may create another source of acidity. Indeed, the Si–O–Al^{IV} linkage in beidellite is easily attacked by protons, and Chourabi and Fripiat (1981) showed that by decomposing an ammonium beidellite, Si–OH infrared

Table 1. Surface areas, pore volumes, and residual cation-exchange capacity (CEC) of pillared montmorillonite (CPM) and pillared beidellite (CPB) calcined at 400°C.

Smectite sample	Surface area (m^2/g)	Total porous volume (cm^3/g)	Microporous volume (cm^3/g)	Residual CEC (meq/g)
CPM	250	0.21	0.10	~0.1
CPB	320	0.25	0.12	~0.2

stretching bands are produced at 3500 cm^{-1} and 3420 cm^{-1}. They assigned these bands to silanol groups formed by a deamination reaction, as suggested by Uytterhoeven et al. (1965) for zeolites X and Y. Inasmuch as the Al_{13} polymer is acidic, the thermal activation of the pillared beidellite will probably create similar Brönsted acid sites and enhance its catalytic properties related to acidity.

The aim of this paper is to correlate the acidity and catalytic properties of pillared beidellites and montmorillonites and to compare these properties with those of synthetic zeolite catalysts.

EXPERIMENTAL

The procedure used to prepare 18-Å pillared smectites was described by Jacobs et al. (1981). The average structural formula obtained for 6 different samples of pillared montmorillonite, after calcination at 900°C, was:

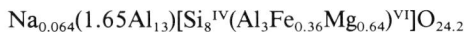

$$Na_{0.064}(1.65Al_{13})[Si_8^{IV}(Al_3Fe_{0.36}Mg_{0.64})^{VI}]O_{24.2}$$

assuming the pillars to consist of 13 Al atoms. Al_{13} in this formula stands for one aluminum pillar.

Beidellite was synthesized and pillared according to the procedure described by Plee et al. (1984) and Poncelet and Schutz (1986). The average formula of the beidellite, from 5 different samples calcined at 900°C, was

$$Na_{0.9}[(Si_{7.1}Al_{0.9})^{IV}Al_4^{VI}O_{22}].$$

When pillared, the average formula was:

$$Na_{0.06}(1.51Al_{13})[(Si_{7.1}Al_{0.9})^{IV}Al_4^{VI}]O_{23.8},$$

again on the basis of the solid calcined at 900°C.

X-ray powder diffraction (XRD) patterns were recorded with a Philips diffractometer using Ni-filtered CuKα radiation. Infrared (IR) spectra of self-supporting wafers (10–20 mg) were scanned using a 180 Perkin-Elmer grating instrument. The samples were outgassed and heated in an IR cell. Pyridine was used as a probe molecule for the investigation of acidity. Nitrogen adsorption-desorption isotherms (BET specific surface areas and porosities) were measured at liquid N_2 temperature in a conventional glass volumetric apparatus equipped with Bell & Howell pressure gauges. ^{29}Si and ^{27}Al magic angle spinning–nuclear magnetic resonance (MAS-NMR) spectra were recorded using

two spectrometers operating at 8.45 and 11.7 Tesla and spinning frequencies of about 2.6 and 3.5 kHz, respectively. Additional details were given by Plee et al. (1985).

The n-decane hydroconversion was performed in a continuous flow reactor operated at atmospheric pressure, with a hydrogen/hydrocarbon molar ratio of 100 and a weight-hourly-space velocity (WHSV) of 0.5 g of n-decane per g of catalyst per hour. The acid catalysts were impregnated with Pt(II) tetramine complex to give a loading of 1% by weight of platinum metal. The catalyst powder was compressed, crushed, and sieved, and the 0.3–0.6-mm fraction was loaded into the reactor. The catalyst was activated by in situ calcination at 400°C, followed by hydrogen reduction at the same temperature. The catalysts used were: (1) a pillared beidellite, prepared using a solution having an OH/Al molar ratio of 1.2 and containing 30 meq Al/g of clay; (2) a pillared montmorillonite prepared in the same way; (3) a commercial silica-alumina catalyst containing 25% by weight of Al_2O_3 and having a specific surface area of 275 m^2/g; (4) two reference catalysts, an ultrastable Y zeolite (USY) prepared by steaming a NH_4-Y zeolite at 750°C, and a H-ZSM-5 zeolite having a Si/Al ratio of 30.

RESULTS AND DISCUSSION

Structure of the pillared smectites

Because of the small number of 001 reflections, no quantitative information was obtained from the XRD patterns except that both types of solids had a turbostratic structure poorly ordered along the c axis. The most intense reflections were undoubtedly due to the 001 reflections of layers with stacking defects and perhaps to some mixed-layer contribution. As far as the long range order is concerned, the pillared montmorillonite could not be distinguished from the pillared beidellite.

The BET specific surface area, the microporosity, and the total pore volume observed from N_2 adsorption at −190°C of the two pillared smectites are listed in Table 1. The specific surface area of the calcined pillared montmorillonite (CPM) is lower than that of calcined pillared beidellite (CPB).

The results in Table 1 do not clearly distinguish between pillared montmorillonite and pillared beidellite. For this reason, short-range ordering was studied using high-resolution solid state ^{27}Al and ^{29}Si NMR under the conditions of MAS-NMR. Results from this study (reported in more detail by Plee et al., 1985) can be summarized as follows:

(1) The Al_{13} polymer is indeed the pillaring agent for all the investigated smectites.
(2) The calcination of the pillared clays does not transform the pillar into a pseudo-spinel structure.
(3) The calcination of pillared smectites without tet-

rahedral substitutions does not lead to a modification of the tetrahedral layer of the sheet silicate.

(4) The calcination of pillared smectites with tetrahedral substitutions (beidellites) modifies the tetrahedral layer and leads to a structural modification of the pillar.

Moreover, it was suggested by Plee *et al.* (1985) that the reaction of the hydroxylated pillars with the tetrahedral layers is due to coupling of the OH apex of an inverted aluminum tetrahedron to an OH of an aluminum octahedron belonging to the Al_{13} polymer. This arrangement leads to a $Si-O-Al^{IV}$ linkage in which the negative charge of the inverted tetrahedron is no longer buried in a continuous tetrahedral network but is located in the open interlamellar space. The pillaring of beidellite with Al_{13} followed by thermal activation thus results in seeding the growth of a three-dimensional network grafted on the two-dimensional network of the clay. The resulting high surface area solid can be considered as a "two-dimensional zeolite." In fact, Vaughan and Lussier (1980) reported that pillared clays are indeed two-dimensional molecular sieves because they sorb 1, 3, 5 trimethylbenzene (D = 6.7 Å) but not 1, 2, 3, 5 tetramethylbenzene (D = 8.0 Å).

Acidity

The origin of the Brönsted acidity of pillared clays may be protons from different sources. The water molecules belonging to the hydration shell of charge-balancing cations are subjected to a strong electrical polarizing field and therefore may have a degree of dissociation several orders of magnitude larger than that of liquid water (Mortland and Raman, 1968). Water molecules hydrating the aluminum pillars are thus a potential source of acidity.

In beidellites, protons can be captured by tetrahedral $Si^{IV}-O-Al^{IV}$ linkages yielding $Si^{IV}-OH \ldots Al^{IV}$ groups (Poncelet and Schutz, 1986) similar to those found in zeolite Y (Uytterhoeven *et al.*, 1965). For instance, when Na-beidellite was treated by the present authors with a diluted acid solution (e.g., 0.05 M HCl), Na^+ was exchanged by H_3O^+ and the IR spectrum obtained after this treatment showed a strong band at 3440 cm^{-1} which was attributed to bridging OH ($Al^{IV} \ldots OH-Si$). The band at 3640 cm^{-1} was apparently due to the OH-stretching vibration of the octahedral cage.

The band at 3440 cm^{-1} increased in intensity to 300°C and remained to as high as 500°C. This bridging OH reacted with pyridine to give the characteristic band of the pyridinium ion at 1540 cm^{-1}, as shown in Figure 1, indicating that the band at 1454 cm^{-1}, indicative of pyridine coordinated to Lewis acid sites, was very weak. A similar band at 3440 cm^{-1} was also observed for pillared beidellite outgassed at 80°C. The intensity of this band increased as the temperature was increased to 300°C and decreased when the samples

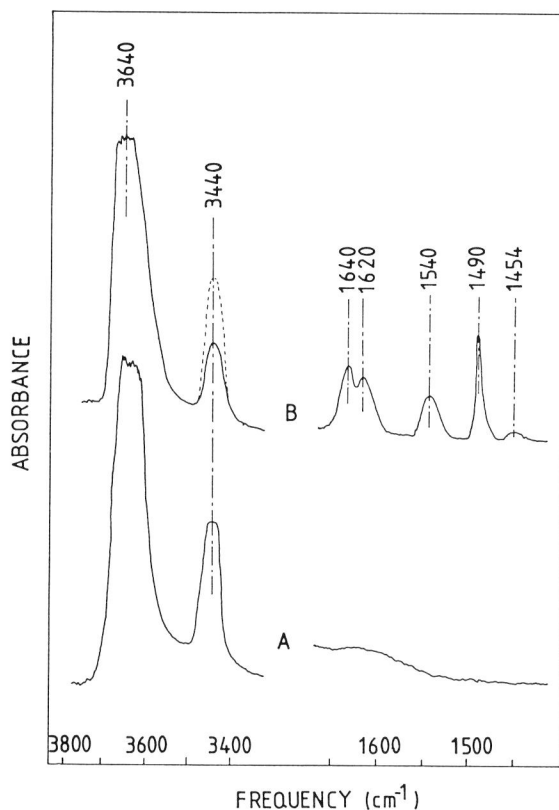

Figure 1. Infrared spectra of H_3O^+-beidellite after outgassing at 300°C (A) and adsorption of pyridine (B).

were heated to higher temperatures. This OH bridging band was not observed in pillared montmorillonite.

Figure 2 shows the IR spectra of pyridine adsorbed on pillared-calcined (300°C) beidellite at different temperatures. Here also, the 1540-cm^{-1} band characteristic of pyridinium ions (Brönsted acidity) was observed. In addition, a strong band at 1454 cm^{-1} indicative of pyridine coordinated to Lewis sites was observed, in contrast with the data shown in Figure 1. In pillared montmorillonite, the band at 1454 cm^{-1} was essentially the only one observed, the amount of pyridinium being much smaller than in pillared beidellite. Some Al-for-Si substitutions, which are never absent in montmorillonite, may have been responsible for the very small amount of $Al^{IV} \ldots OH-Si$ entities present. Occelli and Tindwa (1983) reported the presence of both Brönsted and Lewis acid sites in Na-bentonite pillared with alumina clusters and used the solid acidity and microporosity to explain the high cracking activity of the clay catalyst for gas-oil conversion.

The Lewis acidity, therefore, appears to have been due to the pillars in both types of clay; but the Lewis sites did not appear to have been the same as those on the surface of the spinel-like structure of transition alumina obtained by calcining aluminum hydrates.

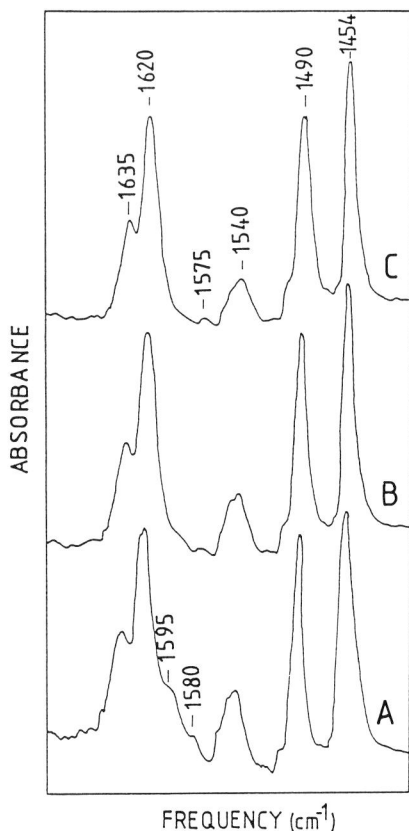

Figure 2. Infrared spectra of pyridine adsorbed on pillared beidellite outgassed at 300°C after heating under vacuum at 150°C (A), 250°C (B), and 310°C (C).

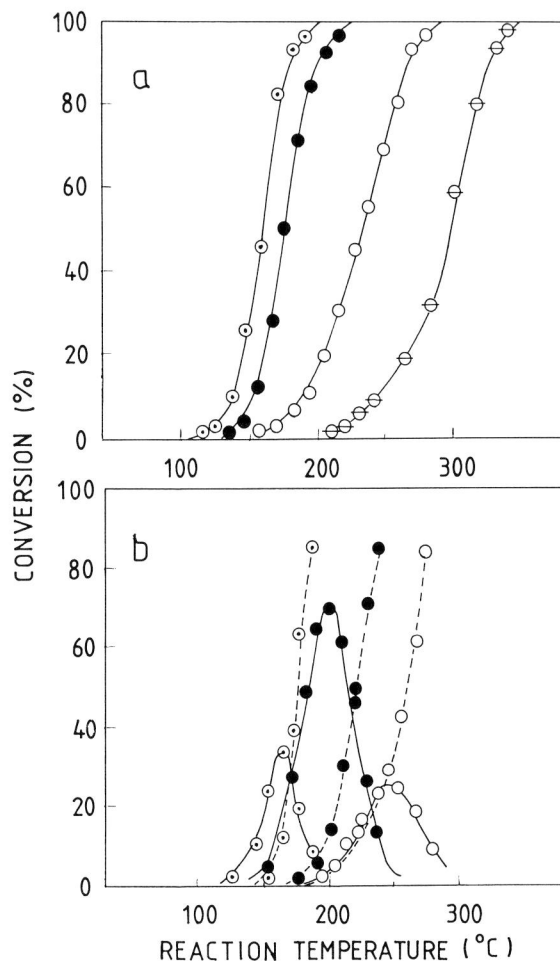

Figure 3. (a) Evolution of the overall conversion of n-decane vs. reaction temperature for: ⊙ = USY (ultrastable zeolite Y); ● = pillared beidellite; ○ = pillared montmorillonite; ⊖ = silica alumina. (b) Evolution of the percentages of hydroisomerization (solid lines) and hydrocracking (dotted lines) vs. reaction temperature for: ⊙ = USY (ultrastable zeolite Y); ● = pillared beidellite; ○ = pillared montmorillonite.

The intensity ratio of the 3440-cm^{-1} band observed for the pillared beidellite and the same band relative to the acid-treated beidellite calcined at 300°C was 0.45, corresponding to 0.3 OH per unit cell for the pillared clay. This value is higher than the residual cation-exchange capacity. Either proton hydrolysis on the surface and/or the inversion of alumina tetrahedra of the tetrahedral layer after calcination was likely responsible for this appreciable discrepancy.

In the range of temperatures used in the catalytic reactions, both Brönsted and Lewis acid sites appear to have coexisted. In pillared montmorillonite, there were merely Lewis sites; the Brönsted sites were mainly due to hydration water. The difference in acidity between the two pillared clays (strength and number of sites) is clearly illustrated by thermodesorption of pyridine preadsorbed at 150°C in the vapor phase on precalcined samples (Plee *et al.*, 1985). From pillared beidellite, 2.45 mmole of pyridine per gram of clay were desorbed; from pillared montmorillonite, only 1.05 mmole/g was desorbed. Furthermore, at 500°C, desorption was not complete for the pillared beidellite.

Catalytic properties:
isomerization of n-decane

Hydroconversion of long-chain n-paraffins is performed industrially to decrease the pour points of the hydrocarbon fractions in which they are present. As a test reaction to determine the dewaxing properties of the catalysts concerned, n-decane was used as feedstock. This molecule, in fact, represents the largest hydrocarbon chain for which most of the individual isomers can still be separated by high-resolution capillary gas-liquid chromatography.

The overall activity of the USY zeolite, the noncrystalline silica-alumina, and the pillared beidellite and montmorillonite is shown in Figure 3a. The activity of the pillared montmorillonite was between that

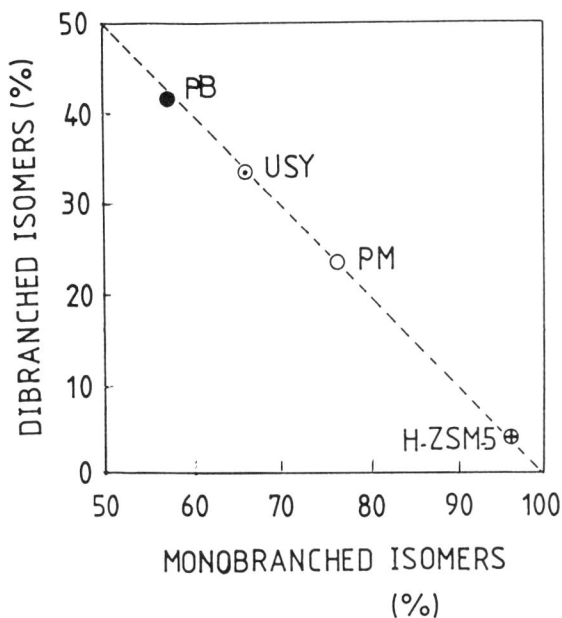

Figure 4. Percentages of dibranched isomers vs. percentages of monobranched isomers obtained at maximum isomerization for pillared beidellite (PB), ultrastable zeolite Y (USY), pillared montmorillonite (PM), and zeolite H-ZSM-5.

of the zeolites and the noncrystalline silica-alumina, whereas, that of the pillared beidellite was close to that of the zeolites. This activity sequence was predictable from the information obtained from pyridine sorption measurements reported above and was therefore in agreement with the IR data.

More information on the practical use of these materials as dewaxing catalysts can be obtained from Figure 3b which represents the overall isomerization-hydrocracking selectivity of the aforementioned catalysts. For the pillared beidellite, an extremely high isomerization selectivity was obtained; as much as 70% of feed isomerization was found at 200°C, whereas only 15% of the feed was hydrocracked. Such high isomerization selectivities, as far as we are aware, have never been reported for any other isomerization-hydrocracking catalyst. This behavior can be explained by a combination of porosity and acid strength. For materials missing the high acid-strength sites found in USY zeolites, a large degree of feed isomerization is possible, provided these branched molecules can easily diffuse out of the system, as is to a large extent true for the pillared beidellite, but much less so for the pillared montmorillonite.

Figure 4 shows the composition of the feed isomers in terms of mono- and dibranched feed hydrocarbons obtained for the catalysts investigated. At the maximum isomerization conversion, more dibranched isomers should be expected for the more open materials.

The observed sequence

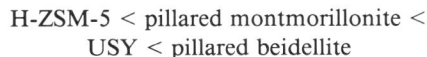

$$\text{H-ZSM-5} < \text{pillared montmorillonite} < \text{USY} < \text{pillared beidellite}$$

is therefore one in which the intracrystalline space is increasing. In other words, the interlamellar space available in the beidellite exceeded considerably that available in the pillared montmorillonite, making the former material with its higher acidity a selective isomerization or dewaxing catalyst.

As far as the hydrocracking properties of the pillared beidellite are concerned, it behaved similarly to the USY. Such behavior was characterized by the following product distribution characteristics:

(1) absence of secondary hydrocracking; C_3 and C_7, as well as C_4 and C_6 fragments from n-decane were present in equimolar amounts;
(2) low absolute quantities of C_3 and C_7 fragments;
(3) high yields of isomers in each carbon number fraction; and
(4) complete absence of C_1 and C_2 (and the corresponding C_8 and C_9) fragments.

These specific features were much less pronounced for the pillared montmorillonite because of its lower acid strength and its lower volume of interlamellar space.

The catalytic stability of the pillared beidellite was also tested. As much as 100 kg of n-decane was fed to 1 kg of catalyst and in this period (200 hr), only a minor decline in catalytic activity was observed.

Although the mechanistic aspects and detailed product distributions will be published elsewhere, the present data suggest that pillared beidellite is an extremely selective isomerization catalyst which can be used for dewaxing purposes. As a hydrocracking catalyst, this pillared beidellite produces high yields of isomers in the hydrocracked products, making it an attractive material for producing high octane gasoline from, for example, a naphtha fraction.

ACKNOWLEDGMENTS

The authors thank the "Compagnie Française de Raffinage" for financial support of this research. P. Jacobs acknowledges FNRS-NFWO (Belgium) for a position as "Senior Research Fellow."

REFERENCES

Brindley, G. W. and Sempels, R. E. (1977) Preparation and properties of some hydroxy-aluminum beidellites: Clay Miner. 12, 229–237.

Chourabi, B. and Fripiat, J. J. (1981) Determination of tetrahedral substitutions and interlayer surface heterogeneity from vibrational spectra of ammonium in smectites: Clays & Clay Minerals 29, 260–268.

Jacobs, P., Poncelet, G., and Schutz, A. (1981) Procédé de préparation d'argiles pontées, argiles préparées par ce procédé et application desdites argiles: Fr. Pat. 2.512.043, Aug. 27, 1981, 24 pp.

Johansson, G., Lundgren, G., Sillèn, L. G., and Söderquist, R. (1960) On the crystal structure of a basic aluminum sulfate and the corresponding selenate: *Acta Chim. Scand.* **14,** 769–773.

Lahav, N., Shani, U., and Shabtai, J. (1978) Cross-linked smectites. I. Synthesis and properties of hydroxy-aluminum montmorillonite: *Clays & Clay Minerals* **26,** 107–115.

Mortland, M. M. and Raman, K. U. (1968) Surface acidity of smectites in relation to hydration, exchangeable cation, and structure: *Clays & Clay Minerals* **16,** 393–398.

Occelli, M. L. and Tindwa, R. M. (1983) Physico-chemical properties of montmorillonite interlayered with cationic oxyaluminum pillars: *Clays & Clay Minerals* **31,** 22–28.

Pinnavaia, T. J., Tzou, Ming-Shin, Landau, S. D., and Raythatha, R. (1984) On the pillaring and delamination of smectite clay catalysts by polyoxo cations of aluminum: *J. Mol. Catal.* **27,** 195–212.

Plee, D., Borg, F., Gatineau, L., and Fripiat, J. J. (1985) High-resolution solid-state ^{27}Al and ^{29}Si nuclear magnetic resonance study of pillared clays: *J. Amer. Chem. Soc.* **107,** 2362–2369.

Plee, D., Schutz, A., Borg, F., Poncelet, G., Jacobs, P., Gatineau, L., and Fripiat, J. J. (1984) Nouvelle zéolite à structure bidimensionnelle et application de ladite zéolite: *Fr. Pat.* **2.563.446,** Apr. 27, 1984, 37 pp.

Poncelet, G. and Schutz, A. (1986) Pillared montmorillonite and beidellite: Acidity and catalytic properties: *NATO Workshop on Chemical Reactions in Organic and Inorganic Systems,* R. Setton, ed., D. Reidel Publ. Co., Doordrecht, The Netherlands, 165–178.

Uytterhoeven, J. B., Christner, L. G., and Hall, W. K. (1965) Studies on the hydrogen held by solids. VIII. The decationated zeolites: *J. Phys. Chem.* **29,** 2117–2125.

Vaughan, D. E. W. and Lussier, R. J. (1980) Preparation of molecular sieves based on pillared interlayered clays (PILC): in *Proc. 5th Int. Conf. Zeolites, Naples, 1980,* L. V. C. Rees, ed., Heyden and Sons, London, 94–101.

Proceedings of the International Clay Conference, Denver, 1985, L. G. Schultz, H. van Olphen, and F. A. Mumpton, eds.,
The Clay Minerals Society, Bloomington, Indiana, 311–318 (1987).

NEW DIFFERENTIAL THERMOGRAVIMETRIC METHOD USING CYCLOHEXYLAMINE FOR MEASURING THE CONCENTRATION OF INTERLAMELLAR PROTONS IN CLAY CATALYSTS

JAMES A. BALLANTINE, PETER GRAHAM, ILA PATEL, J. HOWARD PURNELL,
AND KEVIN J. WILLIAMS

Department of Chemistry, University College of Swansea, Singleton Park
Swansea, Wales SA2 8PP, United Kingdom

JOHN M. THOMAS

Department of Physical Chemistry, University of Cambridge, Lensfield Road
Cambridge, England CB2 1EP, United Kingdom

Abstract—A new method for the estimation of proton concentrations in montmorillonite catalysts was developed that makes use of the well-resolved thermal desorption of intercalated cyclohexylamine. The clay sample was emersed in liquid cyclohexylamine, dried, and its thermal desorption plotted by means of a thermal gravimetric balance fitted with differential thermal gravimetric analyzer (TGA-DTG). The differential thermogravimetric (DTG) desorption curves for this cyclohexylamine-adsorbed material had four distinct desorptive zones, one of which was assigned to the thermal desorption of intercalated cyclohexylammonium ions. The percentage weight loss corresponding to this region of the curve gave a measure of the concentration of cyclohexylammonium ions in the montmorillonite sample. This value was then expressed in terms of the concentration of protons in the clay available to cyclohexylamine. A comparison, for a series of Al_{13}-exchanged and heat-treated montmorillonites, of these proton concentrations with the catalytic activities of the clays indicated that high proton concentration correlated with strong catalytic activity.

Key Words—Catalyst, Cyclohexylamine, Montmorillonite, Proton, Thermogravimetric analysis.

INTRODUCTION

Air-dried montmorillonites have been shown to have unusually high acidities, particularly if most of the water is removed to leave only one layer of intercalated water between the charged sheets (Mortland *et al.,* 1963; Mortland, 1968). Mortland (1968) also showed that ion exchange with polyvalent cations increases the acidity of the intercalated water in montmorillonites due to the hydrolysis of the solvated water molecules according to the equation:

$$[Al(H_2O)_6]^{3+} \rightarrow [Al(OH)(H_2O)_5]^{2+} + H^+.$$

This release of protons into the interlamellar region is thought to be the major source of the high Brönsted acidity in these materials (Mortland, 1968).

The variety of methods that have been devised to study the acidities in solids such as montmorillonites are of four general types.

Thermal desorption of adsorbed nitrogenous bases. Most workers have chosen ammonia as the solvate molecule and have examined the desorption of ammonia on heating. According to Barth and Ballou (1961), strong Brönsted acid sites bind ammonia as the ammonium ion which is desorbed at a higher tem-

perature than physically adsorbed or hydrogen-bonded molecules. The quantity of ammonia evolved between 175° and 515°C was found to be in excess of the proton content of the solid and suggested that Lewis acid sites were also involved (Hall *et al.,* 1964; Peri, 1965).

Infrared spectroscopy of adsorbed nitrogenous bases. In this technique nitrogenous bases are intercalated in montmorillonite films and their infrared (IR) spectra examined for the presence of N^+–H bands in the 2500- and 1400–1500-cm^{-1} regions (Fripiat *et al.,* 1962; Russell, 1965; Swoboda and Kunze, 1966; Farmer, 1971; Laura and Cloos, 1975). Pyridine is a widely used probe molecule and binds to both Brönsted and Lewis sites (Parry, 1963), but its N^+–H absorption band at 1550 cm^{-1} is often suppressed in the presence of strong hydrogen bonds. This fact casts doubts on conclusions based on the appearance of this band (Farmer and Mortland, 1966). By using a variety of bases of different base strengths, acid strengths can be determined rather than concentrations by this IR technique. For example, Swoboda and Kunze (1968) showed that although pyridine and aniline are protonated in Ca^{2+}- and Mg^{2+}-exchanged montmorillonites, o-chloroaniline is not.

Nonaqueous titrations using adsorbed Hammett indi-

cators. This method involves the adsorption of colored organic base indicators on montmorillonite suspended in a nonpolar solvent and titrated with butan-1-amine until a color change indicates the end point (Walling, 1950; Benesi, 1957; Atkinson and Curthoys, 1979). Different indicator bases of different H_0 end points can be used to define the acidity of the various different acidic sites present in different montmorillonites (Frenkel, 1974).

Carbocation formation in solid–surface arylmethanol complexes. Triphenylcarbinol derivatives can be used as a probe for acidity in clay materials because they give a colored carbocation on protonation and the end point can be estimated in terms of an acidity function H_R (Helsen, 1970).

These four methods vary as to their utility and ease of application. Obtaining information on both the total acid concentration and the range of different acid strengths present in a montmorillonite catalyst is very tedious, inasmuch as the equilibria involved with Hammett H_0 indicators are only very slowly established, and numerous different solid samples are required. In this publication a relatively simple new method for the determination of the proton concentration in montmorillonites is described that involves thermal desorption of an adsorbed base. The method makes use of a unique four-stage desorption of cyclohexylamine which is observed during differential thermogravimetry (DTG) measurements.

MATERIALS

Al^{3+}- and Zr^{4+}-exchanged montmorillonites

The montmorillonite used in this study was a Wyoming bentonite (British Drug Houses, cation-exchange capacity of 86 meq/100 g), which was exchanged under standard conditions (Ballantine *et al.,* 1985) with Al^{3+} or Zr^{4+} solutions, dried at 50°C in a vacuum oven, ground, and sieved to 120–140 mesh (A.S.T.M.) before use.

Cyclohexylammonium-exchanged montmorillonite

Cyclohexylamine hydrochloride was prepared by bubbling dry hydrogen chloride through a solution of cyclohexylamine in dry ether. The hydrochloride separated as a hygroscopic white solid which was subsequently crystallized from ethanol. The Wyoming bentonite was cation exchanged twice as described by Ballantine *et al.* (1985) with excess aqueous cyclohexylamine hydrochloride (0.5 M) to yield cyclohexylammonium-exchanged montmorillonite, which was dried at 50°C in a vacuum oven.

Al_{13}-pillared montmorillonite

An aqueous solution of $AlCl_3 \cdot 6H_2O$ was treated with sodium carbonate solution at 95°C; the resultant Al_{13} solution (Brindley and Sempels, 1977) was diluted and used to cation exchange the Wyoming bentonite. After total exchange of the sodium ions, the clay was dried in a vacuum oven at 50°C. Samples of the Al_{13}-exchanged clay were calcined for 4 hr at 165°, 200°, 300°, 400°, and 500°C to produce different Al_{13}-exchanged and pillared montmorillonites. These are herein designated as $Al_{13}(165°)$, $Al_{13}(200°)$, etc.

METHODS

Vapor intercalation of cyclohexylamine: Time study

Al^{3+}-exchanged montmorillonite was spread thinly on the surface of an open glass dish in a closed container in contact with excess cyclohexylamine vapor at room temperature. Aliquots of the clay were sampled after different vapor-contact times of as much as 500 hr, and their thermal desorption measured using a Stanton-Redcroft Model TG-750 Thermobalance which was programmed to provide simultaneous plots of the weight change (TGA) and the differential thermal gravimetric data (DTG) as a function of temperature. An inert atmosphere of nitrogen was maintained throughout the heating period, and the temperature was increased at a rate of 30°C/min. Cyclohexylamine intercalation at different contact times was easily determined from the different thermal desorption curves.

Cyclohexylamine-DTG method for measuring interlamellar proton concentrations

A small sample (~0.5 g) of clay was placed in a sample tube, covered with carbonate-free liquid cyclohexylamine, shaken thoroughly, and allowed to stand for 4 hr at room temperature. The excess cyclohexylamine was removed by decantation, and the clay paste was placed on a small piece of filter paper and dried with a gentle current of warm air. A 5–10-mg sample of the dried clay was transferred to the TGA-DTG apparatus, and the thermal desorption of cyclohexylamine was measured under a constant stream of nitrogen at a heating rate of 30°C/min. The weight loss corresponding to the DTA peak centered at ~318°C was measured, converted to millimoles of cyclohexylammonium ions in the original sample of clay and hence to a measurement of $[H^+]$ in mmole/g of clay.

Monitoring catalytic activity

Two organic reactions were used to monitor the catalytic activity of the clay samples.

Intermolecular elimination of ammonia from cyclohexylamine. Clay samples (0.5 g) were heated with cyclohexylamine (5.0 ml) in a sealed reactor at 210°C for 4 hr and the yield of dicyclohexylamine monitored by gas chromatographic (GLC) analysis (Ballantine *et al.,* 1985).

Dehydraton of pentan-1-ol. Clay samples (0.5 g) were heated with pentan-1-ol (5.0 ml) in a sealed reactor at

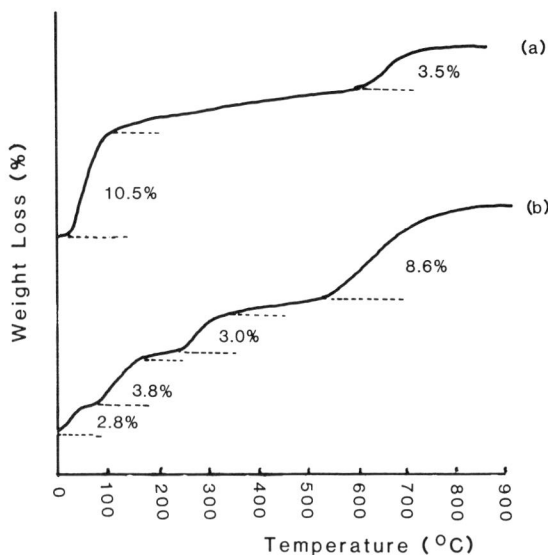

Figure 1. (a) Thermogravimetric analysis plot for Al^{3+}-exchanged montmorillonite; (b) after 5-hr exposure to cyclohexylamine vapor.

200°C for 1.5 hr and the yields of di-pent-1-yl ether and pent-1-ene monitored by GLC (Ballantine *et al.*, 1984). Under these conditions the yields of 1,2-ether were insignificantly small.

RESULTS AND DISCUSSION

Vapor intercalation of cyclohexylamine

The intercalation of cyclohexylamine vapor in Al^{3+}-exchanged montmorillonite was studied over a prolonged period at room temperature. A weighed sample of Al^{3+}-exchanged montmorillonite was placed in an enclosed container and exposed to excess cyclohexylamine vapor; the gain in weight was monitored each day. The increases in weight for a 0.5-g sample of montmorillonite were as follows: 24 hr, 17.1%; 48 hr, 20.2%; 72 hr, 21.5%; 168 hr, 24.2%.

Vapor intercalation of cyclohexylamine: DTA-DTG

More distinctive changes were observed if the adsorbed cyclohexylamine was thermally desorbed in a stream of nitrogen using the TGA-DTG apparatus. Figures 1 and 2 illustrate the type of curves obtained. Figure 1a represents the weight loss characteristics (TGA) of the original Al^{3+}-exchanged montmorillonite before treatment with cyclohexylamine vapor. Three temperature zones at which desorption of water took place can clearly be distinguished. The first is from about 40° to 160°C and corresponding to the expulsion of physically adsorbed water. A second gradual loss of water from about 160° to 600°C and a third loss from about 600° to 800°C correspond to the dehydroxylation of the Al–OH bonds in the clay crystal (Al-Owais *et al.*, 1986).

The DTG curve (Figure 2a) for water loss from the

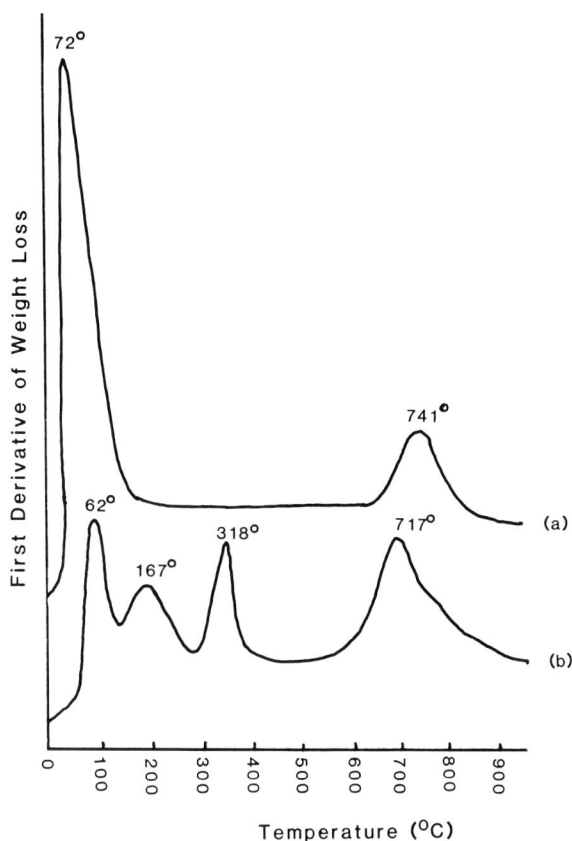

Figure 2. (a) Differential thermogravimetric plot for Al^{3+}-exchanged montmorillonite; (b) after 5-hr exposure to cyclohexylamine vapor.

original Al^{3+}-exchanged montmorillonite before cyclohexylamine treatment shows two distinct maxima, one centered at about 72°C, corresponding to the loss of physical water, and the one centered at about 741°C, corresponding to the dehydroxylation of the clay. The central region, 160°–600°C, produced no DTG maxima because the thermal desorption was gradual without a rapid change in slope.

The adsorption of cyclohexylamine from the vapor phase caused dramatic changes in the TGA-DTG curves, and four distinct desorptive zones are apparent in Figures 1b and 2b for Al^{3+}-exchanged montmorillonite which had been exposed to cyclohexylamine vapor at room temperature. The weight losses in these four zones can be measured from the TGA curve (Figure 1b). For samples exposed to cyclohexylamine vapor for 5 hr, these are: peak I (55°–100°C) 2.8% loss; peak II (150°–180°C) 3.8% loss; peak III (300°–330°C) 3.0% loss; and peak IV (650°–800°C) 8.6% loss. In addition, a gradual desorption took place between 330° and 650°C which did not give rise to a DTG peak (Figure 2b), but which contributed to the total desorption weight loss.

A study of the adsorption of cyclohexylamine by

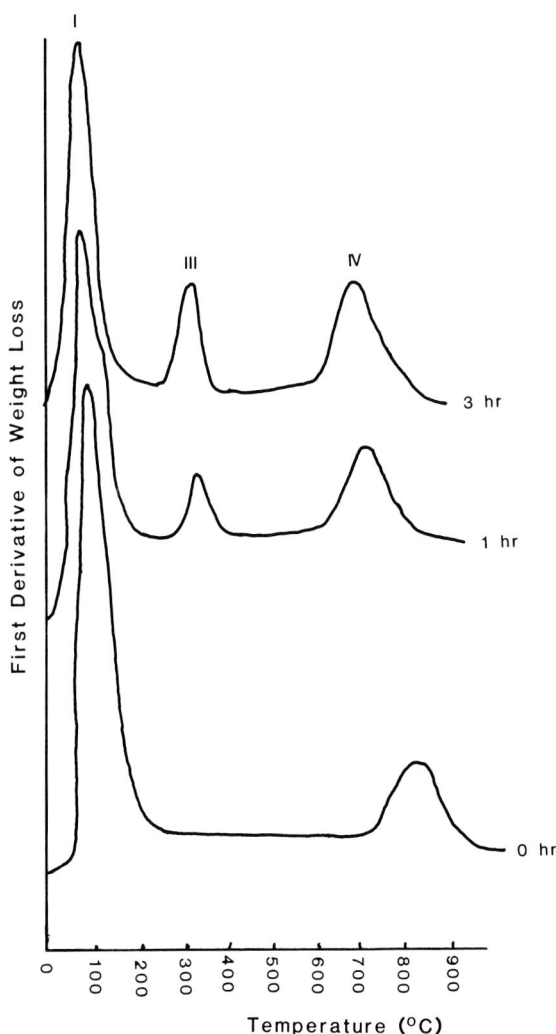

Figure 3. Differential thermogravimetric plots of Al³⁺-exchanged montmorillonite after contact times of 0, 1, and 3 hr with cyclohexylamine vapor.

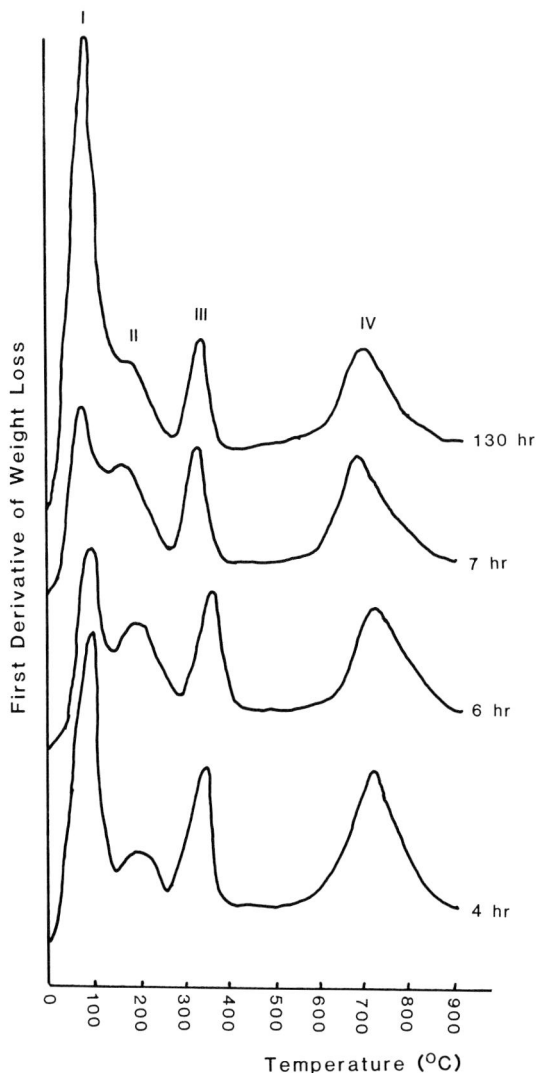

Figure 4. Differential thermogravimetric plots of Al³⁺-exchanged montmorillonite after contact times of 4, 6, 7, and 130 hr with cyclohexylamine vapor.

Al^{3+}-exchanged montmorillonite shows that the size of the DTG peaks are dependent on the length of exposure to the vapor. Figures 3 and 4 and Table 1 show the dependence of the DTG curve on the duration of exposure to cyclohexylamine vapor.

The TGA-DTG weight-loss curves for Al^{3+}-exchanged montmorillonite immersed in liquid cyclohexylamine for about 4 hr and dried are almost identical to those corresponding to 500 hr of absorption from the vapor phase. The total weight of cyclohexylamine desorbed (Table 1) increased steadily from 16.8% to 27% over a period of about 100 hr, with very little increase thereafter. These data are not very helpful for our study inasmuch as the total amount desorbed was the sum of at least four different processes which took place at different rates and came to a stop at different times.

The data in Table 1 suggest the following trends: (1) Peak I gradually decreased to a minimum at 2.8% after 5–6 hr, then increased gradually to 12.0% after 500 hr. (2) Peak II did not appear until after 4 hr, then rapidly rose to a maximum of 4.8% at 6 hr, after which it remained essentially constant. (3) Peak III increased rapidly to 2.5% after 2 hr then remained essentially constant. (4) Peak IV increased rapidly in the first hour, gradually increased to a maximum after about 4 hr, and then remained essentially constant.

Assignment of TGA-DTG zones

Peak I. In the zone represented by peak I, the adsorbates were clearly the most easily removed by heat treatment, and desorption was complete at 100°C under these experimental conditions. In this thermal region water and cyclohexylamine were obviously physically adsorbed. In the initial stages of vapor contact the water was displaced; hence this peak decreased, but after about 7 hr of vapor contact, it increased when cyclohexylamine replaced the adsorbed water and the clay swelled.

Peak III. The zone represented by peak III is one in which the adsorption of cyclohexylamine came to completion and the clay became "filled" with cyclohexylamine after only 2 hr of vapor contact. In this region, protonated cyclohexylamine was chemisorbed, as indicated by the following observation: A TGA-DTG examination of water-washed alkylammonium montmorillonite prepared by exchanging the Na$^+$ in the clay with either cyclohexylammonium or diazobicyclo[2,2,2]octane dihydrochloride (DABCO-2H$^+$) ions showed peaks characteristic of these alkylammonium protonated species in the zone represented by peak III, but not in the zone represented by peak II.

Peak II. The time study shows that in the zone represented by peak II, the clay adsorbent started to "fill" further only after it had become "full" in the peak III zone, and became completely "full" before peak I began to increase. The desorption temperature in the peak II zone (150°–180°C) was greater than that for physically absorbed water or cyclohexylamine, and less than that for protonated cyclohexylamine. This sequence suggests that adsorption in the peak II zone was by another chemisorbed species, such as coordinated cyclohexylamine in which the nitrogen atom is coordinated either directly to an interlamellar cation, as amine of solvation, via a water bridge on a hydrated cation (Farmer and Mortland, 1966), or to the proton of a cyclohexylammonium species via a hydrogen bond (Farmer and Mortland, 1965).

Peak IV. The large size of the zone represented by peak IV was not anticipated. This region of the TGA-DTG curve is normally associated with water loss caused by dehydroxylation of Al–OH bonds in the clay structure at these high temperatures. The adsorption of cyclohexylamine vapor, however, indicates that cyclohexylamine corresponding to about 5% of the weight of the clay was bound to the clay so strongly that it did not desorb at a temperature of less than 650°C. The adsorption data also indicates that this cyclohexylamine was adsorbed in the first hour and that adsorption was complete after about 4 hr. Peak IV was also enhanced in the TGA-DTG measurement of cyclohexylamine hydrochloride-exchanged montmorillonite if zone IV was measured at 7.4% cyclohexylamine.

Very strong chemical bonds must have formed to cause desorption in this thermal zone; how such strong bonds were established with cyclohexylamine is not clear. Alternatively this loss may have been due to the

Table 1. Thermogravimetric analysis of Al^{3+}-montmorillonite treated with cyclohexylamine vapor.

Exposure time (hr)	Weight losses				
	Peak I (%)	Peak II (%)	Peak III (%)	Peak IV (%)	Total (%)
0	10.5	—	—	3.5	16.8
1	8.0	—	1.8	6.0	18.0
2	7.5	—	2.5	6.0	18.2
3	7.5	—	2.3	6.6	19.3
4	5.5	1.9sh	2.6	8.7	20.1
5	2.8	3.8sh	3.0	8.6	20.5
6	2.8	4.8sh	2.8	8.2	20.8
7	3.5	4.6sh	2.8	7.4	21.0
130	10.0	4.6sh	2.9	7.9	27.0
505	12.0	4.4sh	2.9	7.6	28.0

sh = shoulder.

desorption of products derived from cyclohexylamine through pyrolytic or other processes.

TGA-DTG measurement of proton concentration

The assignment of peak III to the desorption of protonated cyclohexylamine provides a basis for a new method for the estimation of the proton concentration in a clay sample. In applying this method, a small sample of the clay was not merely exposed to the vapor, but actually immersed in liquid cyclohexylamine and dried before TGA-DTG measurements. The size of peak III provided a measure of the concentration of protonated cyclohexylamine within each clay sample; hence the concentration of protons in mmole/g of dry clay could be calculated.

Obviously this technique measured only those protonic sites which had sufficient acidic strength to react with cyclohexylamine. Other bases such as ammonia and pyridine were unsuitable for this TGA-DTG method of measuring proton concentration because they showed no sharp zones of desorption. On the contrary, they exhibited merged zones having only poor resolution.

The proton concentrations of several montmorillonites as measured by this cyclohexylamine DTG technique are listed in Table 2. The [H$^+$] values for Na- and Al^{3+}-exchanged montmorillonite reported by Frenkel (1974), who used bromocresol green, a colored indicator base, are 0.10 mmole/g and 0.38 mmole/g respectively. Therefore, the two methods agree closely.

As the Al$_{13}$-exchanged montmorillonite was heated to higher temperatures to induce pillar formation by the dehydroxylation of the interlamellar polyoxycations, the concentrations of protons available to cyclohexylamine decreased markedly. The TGA-DTG curves of pillared montmorillonites show that these materials have cyclohexylamine desorption characteristics similar to those of layered clays, with almost identical zones of desorption, which suggests similar desorption processes. Thus, a protonated cyclohexyl-

Table 2. Correlation of catalyst acidity with reaction yields.

Montmorillonite[1]	[H+] (mmole/g)	Dipent-1-yl ether (wt. %, 1.5 hr)	Pent-1-ene (wt. %, 1.5 hr)	Dicyclo-hexylamine (wt. %, 4 hr)
Na+	0.09	—	—	>1.5
Al³+	0.42	42.1	6.0	24.0
Zr⁴+	0.40	29.7	4.8	16.0
Al₁₃ (165°C)	0.53	24.1	1.3	12.0
Al₁₃ (200°C)	0.53	23.7	1.4	12.4
Al₁₃ (300°C)	0.40	26.1	1.2	10.8
Al₁₃ (400°C)	0.35	19.5	1.1	10.0
Al₁₃ (500°C)	0.18	4.5	0.3	2.7

[1] See text for preparation details.

amine which was produced by interaction of the liquid cyclohexylamine with either the polyoxycation or the pillars should have contributed to the proton concentration as measured by peak III.

The weight-loss curves of the parent Al_{13}-pillared materials, in the absence of cyclohexylamine, showed negligible desorption of water at 300°C, indicating that although dehydroxylation of the pillars or interlayer hydroxides probably took place at this temperature, it was a negligible factor in the measurements of zone III peaks during the cyclohexylamine desorption of pillared materials under the rapid heating in the DTG experiment.

Comparison of catalytic activity with proton concentrations

Intermolecular elimination of ammonia from cyclohexylamine. Table 2 gives details of the yields of dicyclohexylamine obtained in a 4-hr reaction at 210°C and proton concentrations for several montmorillonites. This reaction time was selected to ensure that the reaction had not reached equilibrium but that the product yield was still large enough to measure. The results show that for Na+-, Al³+-, and Zr⁴+-exchanged montmorillonites, the yields of dicyclohexylamine correlates with the proton concentration of the clays, although not in a strictly linear fashion. Within the series of Al_{13}-pillared montmorillonites a similar situation was observed. The high proton concentration of the $Al_{13}(165°)$ and $Al_{13}(200°)$ materials, however, was not reflected in correspondingly high yields of dicyclohexylamine compared with those of the layered clays. Furthermore, this reaction required the presence of both protonated and non-protonated cyclohexylamine species at the active site (Ballantine *et al.,* 1985) (see scheme I below), so that a combination of a high concentration of protons and a restricted internal volume as found in samples $Al_{13}(165°)$ and $Al_{13}(200°)$ was not the best combination of conditions to produce high yields of dicyclohexylamine.

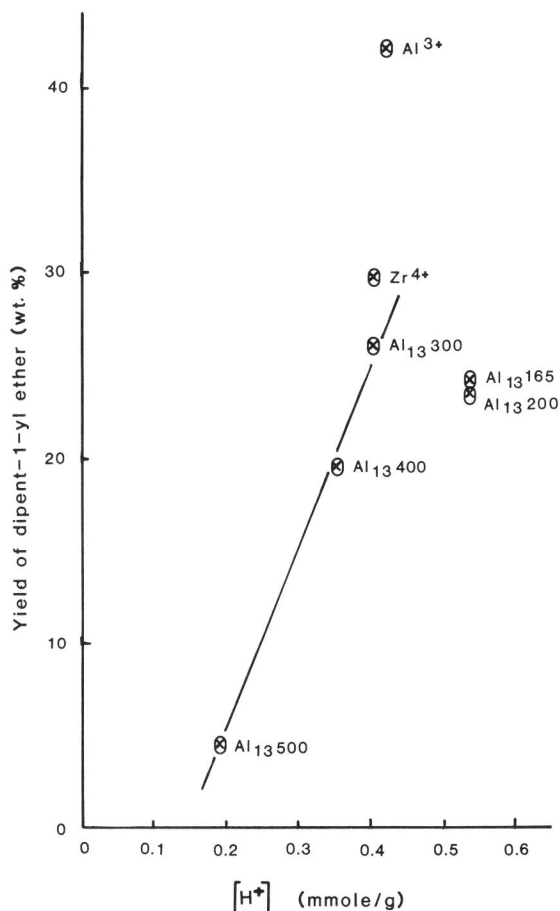

Figure 5. Yield of dipent-1-yl ether vs. proton concentration for several montmorillonite catalysts. ($Al_{13}165$, e.g., refers to a montmorillonite which has been exchanged with the Al_{13} ion and heated to 165°C for 4 hr.)

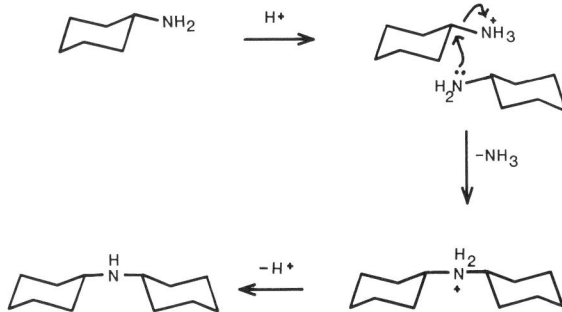

Dehydration of pentan-1-ol. Table 2 lists details of the yields of dipent-1-yl ether and pent-1-ene obtained from the dehydration of pentan-1-ol at 200°C after 1.5 hr for several montmorillonites as well as their proton concentrations. This reaction time also was selected to ensure that the reaction had not reached equilibrium but that the product yields were large enough to mea-

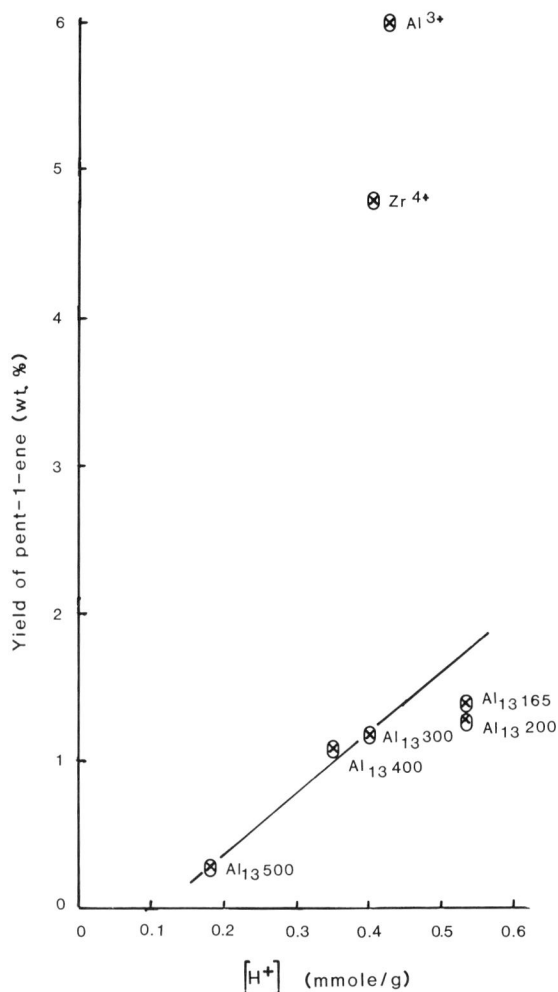

Figure 6. Yield of pent-1-ene vs. proton concentration for several montmorillonite catalysts. (Symbols as in Figure 5.)

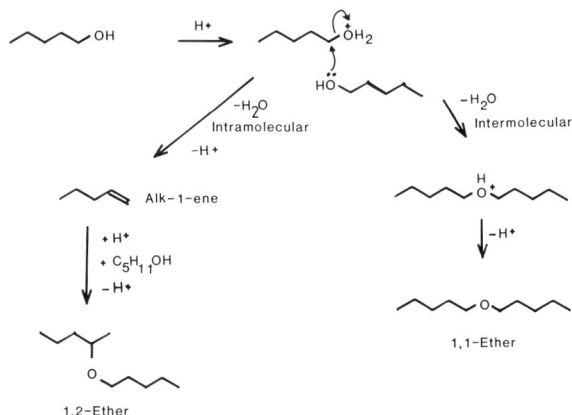

Figure 7. Simultaneous yields of dipent-1-yl ether and pent-1-ene for several montmorillonites. (Symbols as in Figure 5.)

sure. Scheme II shows that the dipent-1-yl ether arises from intermolecular dehydration, whereas pent-1-ene arises from intramolecular dehydration (Ballantine *et al.*, 1984).

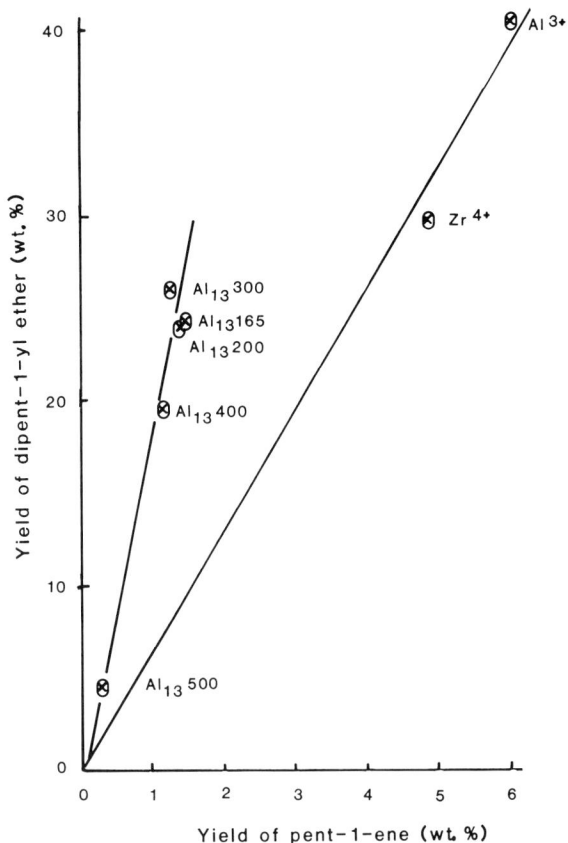

The swelling clays having high proton concentrations gave higher yields of both products (Table 3). Figures 5 and 6 show that once the Al_{13}-exchanged montmorillonites became pillared (>300°C), their catalytic activity correlated well with their proton concentrations as measured by the cyclohexylamine DTG technique. The $Al_{13}(165°)$ and $Al_{13}(200°)$ montmorillonites, however, had not yet dehydroxylated; presumably, they contained little physically adsorbed water and were of high acidity (Mortland, 1968). They contained large Al_{13} polyoxy cations which could effectively reduce diffusion to the catalytic sites such that their reactivity was not reflected by their high acidity.

The Al_{13} materials showed a distinctly different selectivity for dipent-1-yl ether and pent-1-ene compared with that of the swelling clays. The simultaneous yields of dipent-1-yl ether vs. pent-1-ene yields for each montmorillonite sample are shown in Figure 7. Clearly, all of the Al_{13} materials are on a line which favors the production of the 1,1-ether by the intermolecular process, whereas the swelling clays are on a line which favors alkene production by the intramolecular process. Obviously the different structural makeup of these catalysts strongly influences the selectivity of the reaction.

ACKNOWLEDGMENTS

We acknowledge with thanks the interest and financial support of the B.P. Research Centre, Sunbury-on-Thames, for studentships to P.G., I.P., and K.J.W. and for the purchase of the thermoanalyzer.

REFERENCES

Al-Owais, A. A., Ballantine, J. A., Purnell, J. H., and Thomas, J. M. (1986) Thermogravimetric study of the intercalation of acetic acid and of water by Al^{3+}-exchanged montmorillonite: *J. Mol. Catal.* **35**, 201–212.

Atkinson, D. and Curthoys, G. (1979) The acidity of solid surfaces and its determination by amine titration and adsorption of coloured indicators: *Chem. Soc. Reviews* **8**, 475–497.

Ballantine, J. A., Davies, M., Patel, I., Purnell, J. H., Rayanakorn, M., Williams, K. J., and Thomas, J. M. (1984) Organic reactions catalysed by sheet silicates: ester production by the direct addition of carboxylic acids to alkenes: *J. Mol. Catal.* **26**, 37–56.

Ballantine, J. A., Purnell, J. H., Rayanakorn, M., Williams, K. J., and Thomas, J. M. (1985) Organic reactions catalysed by sheet silicates: intermolecular elimination of ammonia from primary amines: *J. Mol. Catal.* **30**, 373–388.

Barth, R. T. and Ballou, E. V. (1961) Determination of acid sites on solid catalysts by ammonia gas adsorption: *Anal. Chem.* **33**, 1080–1084.

Benesi, H. A. (1957) Acidity of catalyst surfaces. II. Amine titration using Hammett indicators: *J. Phys. Chem.* **61**, 970–973.

Brindley, G. W. and Sempels, R. E. (1977) Preparation and properties of some hydroxy-aluminium beidellites: *Clay Miner.* **12**, 229–237.

Farmer, V. C. (1971) Characterisation of adsorption bonds in clays by infrared spectroscopy: *Soil Sci.* **112**, 62–68.

Farmer, V. C. and Mortland, M. M. (1965) An infrared study of complexes of ethylamine with ethylammonium and copper ions in montmorillonite: *J. Phys. Chem.* **69**, 683–686.

Farmer, V. C. and Mortland, M. M. (1966) An infrared study of the co-ordination of pyridine and water to exchangeable cations in montmorillonite and saponite: *J. Chem. Soc. A,* 344–351.

Frenkel, M. (1974) Surface acidity of montmorillonites: *Clays & Clay Minerals* **22**, 435–441.

Fripiat, J. J., Servais, A., and Leonard, A. (1962) Adsorption of amines by montmorillonites. II. Nature of the amine-montmorillonite bond: *Bull. Soc. Chim. Fr.*, 635–644.

Hall, K., Lutinkski, F. E., and Gerberich, H. R. (1964) Hydrogen held by solids. IV. The hydroxyl groups of alumina and silica-alumina as catalytic sites: *J. Catal.* **3**, 512–527.

Helsen, J. J. (1970) Determination of the type of acidity and spectrometric measurement of the dissociation of water from montmorillonites through adsorption of triphenylcarbinol: *Bull. Groupe Fr. Argiles* **22**, 139–155.

Laura, R. D. and Cloos, P. (1975) Adsorption of ethylenediamine (EDA) on montmorillonite saturated with different cations. IV. Aluminum-, calcium-, and magnesium-montmorillonite: protonation, ion-exchange, co-ordination, and hydrogen bonding: *Clays & Clay Minerals* **23**, 343–348.

Mortland, M. M. (1968) Protonation of compounds at clay mineral surfaces: in *Trans. Int. Cong. Soil Sci. 9th, Vol. 1.,* J. W. Holmes, ed., Elsevier, New York, 691–699.

Mortland, M. M., Fripiat, J. J., Chaussidon, J., and Uytterhoeven, J. (1963) Interaction between ammonia and the expanding lattices of montmorillonite and vermiculite: *J. Phys. Chem.* **67**, 248–258.

Parry, E. P. (1963) An infrared study of pyridine adsorbed on acidic solids. Characterisation of surface activity: *J. Catal.* **2**, 371–379.

Peri, J. B. (1965) Infrared study of adsorption of ammonia on dry gamma-alumina: *J. Phys. Chem.* **69**, 231–239.

Russell, J. D. (1965) Infrared study of the reactions of ammonia with montmorillonite and saponite: *Trans. Faraday Soc.* **61**, 2284–2294.

Swoboda, A. R. and Kunze, G. W. (1966) Infrared techniques for studying the adsorption of volatile vapors by clays: *Soil Sci.* **101**, 373–377.

Swoboda, A. R. and Kunze, G. W. (1968) Reactivity of montmorillonite surfaces with weak organic bases: *Soil Sci. Soc. Amer. Proc.* **32**, 806–811.

Walling, C. (1950) The acid strength of surfaces: *J. Amer. Chem. Soc.* **72**, 1164–1168.

Proceedings of the International Clay Conference, Denver, 1985, L. G. Schultz, H. van Olphen, and F. A. Mumpton, eds.,
The Clay Minerals Society, Bloomington, Indiana, 319–323 (1987).

SURFACE AND CATALYTIC PROPERTIES OF SOME PILLARED CLAYS

MARIO L. OCCELLI[1]

Gulf Research & Development Company, P.O. Drawer 2038
Pittsburgh, Pennsylvania 15230

Abstract—Exchange reactions of Ca-bentonite with zirconyl chloride, zirconyl-alumino-halohydroxy complexes, colloidal aluminosilicate solutions, and colloidal alumina particles generated a family of microporous materials which, after heating in air at 400°C for 4 hr, have Langmuir surface areas of 300–500 m^2/g, pore volumes of 0.16–0.21 cm^3/g, and basal spacings in the range 17.6–19.0 Å.

The most stable structures were obtained by pillaring the bentonite with either hydroxyaluminum oligomers or colloidal alumina particles. After heating in air for 4 hr at 750°C, these products retained ~70% of their original surface area and most of their catalytic activity for gas-oil conversion. They were also found to be hydrothermally stable (10 hr with 95% steam at 1 atm) to 700°C. These pillared bentonites were thermally more stable (to ~700°C vs. ~600°C) and exhibited greater cracking activity for gas-oil than similarly prepared hectorites; however, published data indicate that hectorite catalysts minimize secondary cracking reactions and produce greater gasoline yields.

Key Words—Catalysis, Hydroxy-aluminum complex, Pillared interlayer clays, Surface area, Thermal stability, Zirconium.

INTRODUCTION

In 1982, catalyst sales in the petroleum industry amounted to almost 800 million dollars, representing 33.8% of the total worldwide catalyst sales. Catalyst requirements in the petrochemical industry are projected to increase 5% annually during 1982–1990 (Oil and Gas Journal, Nov. 1, 1982, p. 28), and pillared clays have the potential to play an important role in this expanding market. In fact, bentonites pillared with heat-stable inorganic oxide clusters have gas-oil cracking activity similar to commercial fluid cracking catalysts (FCC) containing zeolites (Vaughan *et al.*, 1979; Lussier *et al.*, 1980; Vaughan *et al.*, 1981); in addition, they have unique selectivity for light-cycle gas-oil (LCGO, used as furnace oil) production without affecting gasoline yields (Occelli, 1983). Similar results have been observed for pillared hectorites (Occelli and Finseth, 1986).

In addition to gas-oil cracking, pillared clays are potentially useful in (1) hydrocracking (Vaughan *et al.*, 1979), (2) upgrading heavy oils (Occelli and Rennard, 1984), (3) converting methanol to hydrocarbons; (4) alkylation (Occelli *et al.*, 1985a), and (5) oligomerizing propane-propylene mixtures to transportation fluids (Occelli *et al.*, 1985b). More details on the properties of pillared clays can be found in Pinnavaia (1983).

The present paper reports the preparation of several novel pillared bentonites and compares their surface properties and cracking activity for gas-oil to that of similarly prepared pillared hectorites and delaminated clays (Pinnavaia, 1984).

EXPERIMENTAL

Physical properties

A DIGISORB 2600 instrument from Micromeritics Instrument Corporation was used to measure N_2 sorption, Langmuir or BET surface areas, and pore-size distributions. Mercury-penetration-porosimetry measurements were made using a Quantachrome porosimeter. X-ray powder diffraction (XRD) patterns were obtained with a standard Philips diffractometer equipped with a pulse height analyzer, using $CuK\alpha$ radiation.

Preparation of pillared bentonites

Aluminum chlorhydroxide bentonite. The natural (Wyoming) bentonite used was supplied by the American Colloid Company. It had a BET surface area of ~50 m^2/g; its chemical analysis is shown in Table 1. An excess (150 g) of aluminum chlorhydroxide (ACH) solution (chlorhydrol from the Reheis Chemical Company) was added dropwise to a 10-liter slurry containing 100 g bentonite; vigorous stirring was continued for 1 hr at 50°C. The slurry was filtered (under vacuum) and washed with 6 liters of hot distilled water at 50°C. The wet cake was then reslurried in 10 liters of distilled water, and the slurry was stirred for 1 hr and filtered again. The pillared clay product (ACH-bentonite) was washed, air dried, crushed, and sized into 100 × 325 mesh granules for evaluation and catalytic testing.

---

[1] Present address: Union Oil Company of California, P.O. Box 76, Brea, California 92621.

Table 1. Chemical analysis of bentonites before and after pillaring.

	Oxide composition (wt. %)[1]					
	Ca-bentonite	ACH-bentonite	CAL-bentonite	ZACH-bentonite	ZCH-bentonite	ALSI-bentonite
SiO_2	62.9	55.9	62.83	55.3	52.0	71.2
Al_2O_3	20.0	31.6	27.36	25.8	15.9	18.1
Fe_2O_3	4.38	3.60	4.24	3.60	2.73	3.01
CaO	2.58	0.43	0.15	0.24	0.16	0.28
MgO	2.16	1.55	2.11	1.58	1.96	1.61
K_2O	2.32	0.37	0.87	0.35	0.62	0.76
Na_2O	0.53	0.67	0.54	0.50	0.39	0.52
ZrO_2	—	—	—	6.12	17.10	—
Total	94.87	94.12	98.10	93.49	90.86	95.48

[1] The difference from 100% is due to chemically bound water. ACH = aluminum chlorhydroxide, CAL = colloidal alumina, ZACH = zirconium-aluminum chlorhydrated, ZCH = zirconyl chlorhydroxide, and ALSI = alumina-coated silica.

Zirconyl chlorhydroxide bentonite. Aqueous zirconyl chloride (ZCH) was prepared by diluting 483 g of 20% $ZrOCl_2$ solution (from Magnesium Elektron, Inc.) to 3 liters with distilled water and then aging the solution at 60°C for 48 hr. This solution was added dropwise to a slurry containing 100 g of bentonite in 10 liters of distilled water; the final pH of the slurry was 2.2. After stirring for 0.5 hr at 20°C, the pillared clay (ZCH-bentonite) was washed by repeating twice the washing steps described above for the preparation of ACH-bentonite. The wet clay was then air dried and crushed to the desired size.

Zirconium-aluminum chlorhydrated bentonite. A zirconium-aluminum-chlorhydrated pillared clay (ZACH-bentonite) was prepared using the procedure employed for the preparation of ACH-bentonite. The interlayering cation was prepared by diluting 73 g of REZAL 67 (from the Reheis Chemical Company) to 4 liters with distilled water and aging the solution for 48 hr at 60°C. REZAL-67 is a solution containing randomly polymerized polynuclear cations with an approximate Al/Zr ratio of about 6.7. This solution is stable to pHs as high as 4.0, whereas zirconium chlorhydrate salts precipitate at pHs between 2.0 and 2.5. The synthesis of a Zr-Al chlorhydrated complex with the empirical formula $[ZrOCl_2 \cdot Al_8(OH)_{20}]^{+4}$ was described by Beckman (1959).

Alumina-coated silica bentonite. An aqueous dispersion of micrometer-size, alumina-coated silica (ALSI) was used to generate pillared materials by a procedure similar to that employed for the preparation of the ACH-bentonite. A 30% solution of ALSI is commercially available from Nalco Chemical Company (NALCO ISJ-612). This solution is stable to pHs from 2 to 6 and contains less than 1.5% chloride ions as counter ions of the positively charged colloidal particles. The

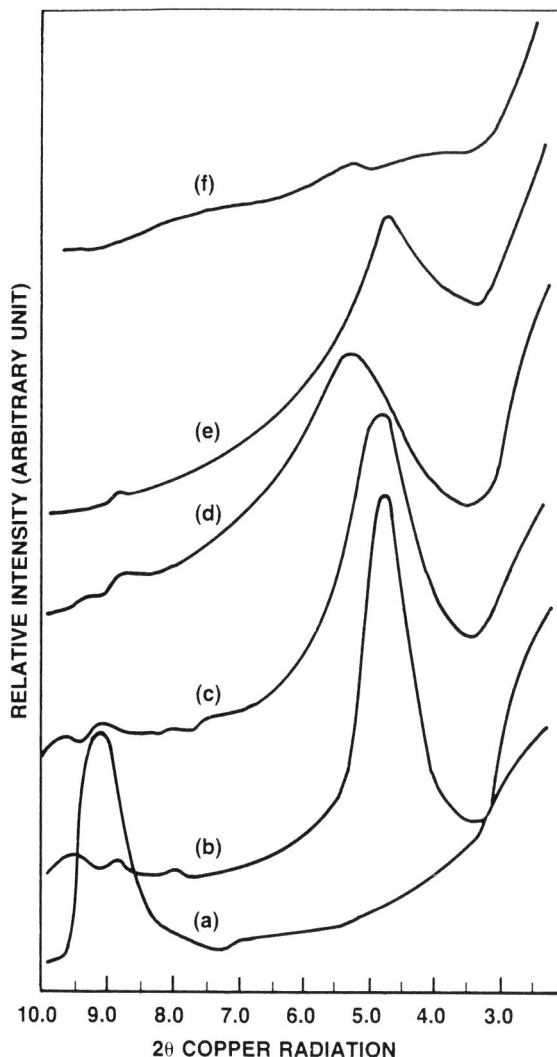

Figure 1. X-ray powder diffractograms of Ca-bentonite (a) before and after pillaring with: (b) aluminum chlorhydroxide, (c) colloidal alumina, (d) zirconium-aluminum chlorhydrated, (e) alumina-coated silica, (f) zirconyl chlorhydroxide solutions. Samples were dried in air at 300°C for 4 hr.

low chloride concentration greatly facilitates the removal of excess reactants after pillaring.

Colloidal alumina bentonite. A 10% aqueous colloidal dispersion of alumina (CAL) particles (NALCO ISJ-614) (800 g) was used to pillar bentonite (100 g) according to the procedure described above. The absence of chloride impurities simplifies the washing step, and reslurrying of the clay after pillaring was not required.

Cracking activity

Catalyst cracking activity was evaluated with a flow system similar to that described by Ciapetta and Anderson (1969). Feedstock was a gas-oil (260°–426°C

Table 2. Surface area[1] variation of pillared bentonites[2] with calcination temperature.[3]

Calcination temperature (°C)	CAL-bentonite		ZACH-bentonite		ZCH-bentonite		ALSI-bentonite		ACH-bentonite[4]	
	BET S.A.	LANG. S.A.	BET S.A.	LANG. S.A.	BET S.A.	LANG. S.A.	BET S.A.	LANG. S.A.	BET S.A.	LANG. S.A.
300	291	372	256	329	243	314	247	320	398	494
400	288	370	246	315	243	314	165	212	400	496
500	273	350	238	306	225	290	148	191	368	455
600	239	306	197	252	180	243	137	174	336	417
650	218	279	176	226	192	247	—	—	—	—
700	195	261	147	190	135	173	148	191	282	349
750	100	134	96	129	76	—	—	—	169	—
800	30	—	60	—	29	—	—	—	60	—

[1] In m²/g, ±5%.
[2] CAL = colloidal alumina; ZACH = zirconium-aluminum chlorhydrated; ZCH = zirconyl chlorhydroxide; ALSI = alumina-coated silica; ACH = aluminum chlorhydroxide.
[3] A 4-hr heating period was used.
[4] Sample prepared from (flash dried) extrudates, 0.16 cm in diameter.

boiling range) containing 23.0 wt. % furnace oil and 74.3 wt. % slurry oil. The charge consisted of 2.5 g of 100 × 325 mesh granules calcined in air at 300°–800°C. Typical test conditions were: 515°C reactor temperature, 80-s contact time, and 15 weight-hourly space velocity (WHSV). Percentage conversion is defined as: $(V_f - V_p) \times 100/V_f$, where V_f is the volume of cracked fresh feed (ff) and V_p is the volume of product boiling above 204°C.

RESULTS AND DISCUSSION

Reaction of bentonites with Al, Al-Zr, and Zr polyoxycations and with colloidal alumina and colloidal silica particles produced clays characterized, after dehydroxylation, by basal spacings of 18.0–19.0 Å, N_2 pore volumes of 0.16–0.21 cm³/g, and Langmuir surface areas of 300–500 m²/g. XRD data of these products are shown in Figure 1.

The pillared bentonite products retained most of their initial surface area after heating in air to ~600°C; however, if air was replaced by a 95% steam–5% nitrogen mixture, the stability was reduced (see also, Occelli, 1983); between 590° and 650°C, the structure of these clay catalysts collapsed with loss of microporosity, surface properties, and catalytic activity. Thermal stability data (expressed as retention of surface area with calcination temperature) are given in Table 2.

The surface and catalytic properties of pillared bentonites depended on the method used to prepare the final product. If, after pillaring, the clays were formed into 0.16-cm extrudates and flash dried, crushing to the desired size (i.e., 100 × 325 mesh granules) took place with relative ease. Shear stresses otherwise necessary to break the hard, air-dried cake (from the filtration step) into granules for catalytic evaluation were not required. After air drying at 400°C for 4 hr, an extruded and flash-dried sample of ACH-bentonite had a d(001) = 18.2 Å, and a Langmuir surface area of 494 m²/g. At 700°C, the surface area decreased to 349 m²/g; even after heating in air at 800°C (4 hr), this pillared clay retained its structure (Figure 2). Over the 400°–700°C temperature range, gas-oil cracking conversion decreased slightly from 87 to 84%. Steaming at 700°C (for 10 hr using a 95% steam–5% N_2 mixture) reduced the conversion to 76%; however, steaming at 730°C resulted in a near total loss of surface properties and catalytic activity.

ZCH-bentonite is amorphous to X-rays (Figure 1f), indicating the absence of long-range face-to-face layer stacking. Its 243 m²/g BET surface area (after heating for 4 hr at 400°C) probably is due to face-to-edge and edge-to-edge assemblages like those found for delaminated clays (Pinnavaia et al., 1984). ZACH- and ZCH-bentonites (that is, pillared and delaminated bentonites) with similar BET surface areas showed comparable thermal stability and cracking activity (Table 3). Cracking activity and product selectivities for gasoil conversion depended mainly on the surface area generated by pillaring and on the nature of the expanded layer silicates, not on the pillaring agent used (Occelli and Finseth, 1986; Occelli, 1985). Because of their greater surface area, these catalysts were more active than the corresponding pillared hectorites (see Figure 3).

In the 50–70% conversion range, the less-active pillared hectorites produce greater yields of gasoline and LCGO (Figure 3) by cracking more of the heavier fractions (slurry oil, SO); they also minimize secondary

Table 3. Surface and cracking properties of ZACH- and ZCH-bentonite.[1]

Proppant[1]	Area (m²/g)	Conversion (V%)	Gasoline (V%)	LCGO[2] (V%)	SO[3] (V%)	Carbon (wt. %)
ZACH	256	79.1	56.6	15.3	5.5	12.1
ZCH	252	80.6	55.2	14.8	4.8	12.2

[1] ZACH = zirconium-aluminum chlorhydrated; ZCH = zirconyl chlorhydroxide.
[2] LCGO = light-cycle gas-oil.
[3] SO = slurry oil.

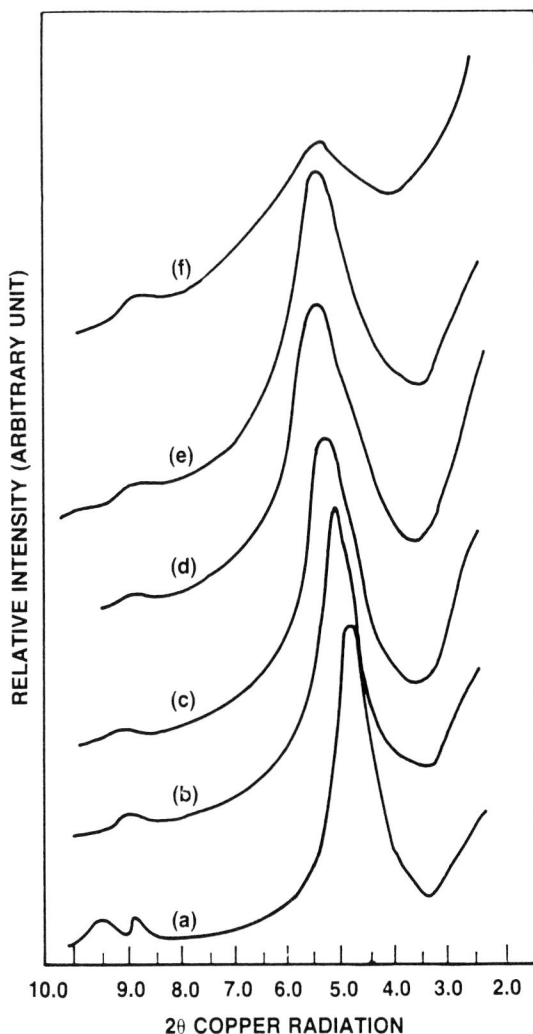

Figure 2.   X-ray powder diffractograms of ACH-bentonite
after heating for 4 hr in air at: (a) 300°, (b) 500°, (c) 600°, (d)
700°, (e) 750°, and (f) 800°C. Samples prepared from extrudates 0.16 cm in diameter.

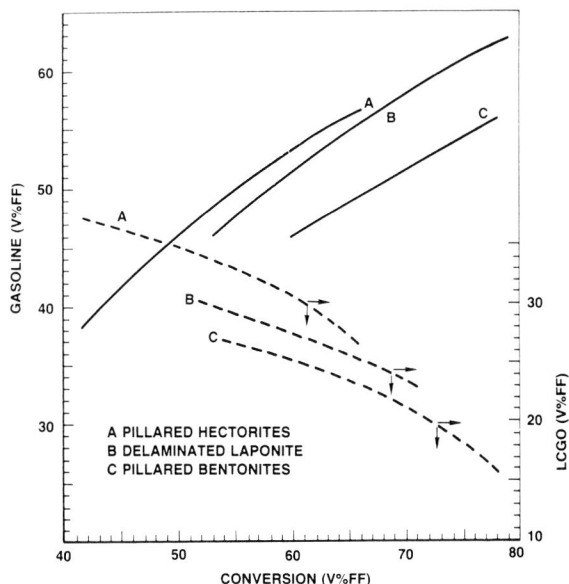

Figure 3.   Liquid product selectivities.

cracking reactions responsible for light gas (C_2–C_4) generation at the expense of gasoline yields (Occelli, 1985).

In the aforementioned conversion range, pillared
bentonites with 3–4% Fe_2O_3 impurities generate as
much carbon as a pillared hectorite that contains <0.3%
Fe_2O_3 (Occelli and Finseth, 1986). Delaminated clays
(Figure 3) had better carbon selectivity because their
macroporosity favored desorption of high molecular
weight hydrocarbons which otherwise would have been
retained (occluded) as coke (Occelli et al., 1984).

The ease of deactivation of pillared clays from formation of high carbonaceous deposits has been attributed to their Lewis acidity and microporosity (Occelli
and Lester, 1985). Polymerization and condensation
reactions on Lewis centers are believed to take place
and to aid the retention of polyalkyl aromatics which

generate highly carbonaceous deposits. As reported by
Occelli et al. (1985b), coke formed from the oligomerization of propylene using bentonite pillared with
Al_2O_3 and Al_2O_3-ZrO_2 clusters contains 93.2 and 68.1%
aromatic and olefinic carbon, respectively.

CONCLUSIONS

Reaction of Ca-bentonite with Zr, Zr-Al, Al polyoxycations, colloidal alumina, or alumina-coated colloidal
silica particles generates (on heating) microporous materials stable (in air) to ~700°C. Thermal and hydrothermal stability depends on the method used to prepare the finished catalyst. Pillared bentonites are more
active and more stable than the corresponding pillared
hectorites; liquid product selectivities from gas oil
cracking (i.e., gasoline, light-cycle gas oil, and slurry
oil) are controlled by the nature of the silicate layer.
The various pillars used do not have distinguishable
selectivity effects. Pillared and delaminated hectorites
are more selective cracking catalysts compared with
delaminated laponite which minimizes coke formation
(Occelli et al., 1984). All the aforementioned clay catalysts, however, lack the hydrothermal stability and
low coke selectivity of commercially available cracking
catalysts containing zeolites.

ACKNOWLEDGMENTS

Special thanks are due to S. S. Pollack (Department
of Energy, Pittsburgh, Pennsylvania) for many useful
suggestions and support during the completion of this
work.

REFERENCES

Beckman, S. M. (1959) Zirconyl and aluminum halohy-
droxy complex: U.S. Patent 2,906,668, 3 pp.

Ciapetta, F. G. and Anderson, D. (1969) Microactivity test for cracking: *Oil Gas J.* **65,** 88–93.

Lussier, R. J., Magee, J. S., and Vaughan, D. E. W. (1980) Pillared interlayered clay cracking catalysts—preparation and properties: *7th Canadian Symp. Catal., Edmonton, Alberta,* 88–95 (preprint).

Occelli, M. L. (1983) Catalytic cracking with an interlayered clay, a two dimensional molecular sieve: *Ind. Eng. Chem. Prod. Res. Dev.* **22,** 553–559.

Occelli, M. L. (1985) New routes to the preparation of pillared montmorillonite catalysts: *J. Mol. Catal.* (in press).

Occelli, M. L. and Finseth, D. H. (1986) Surface and catalytic properties of some pillared hectorites: *J. Catal.* **99,** 316–326.

Occelli, M. L., Hsu, J. T., and Galya, L. G. (1985b) Propylene oligomerization with pillared clays: *J. Mol. Catal.* **33,** 371–389.

Occelli, M. L., Innes, R. I., Hwu, F. S. S., and Hightower, J. W. (1985a) Sorption and catalysis on sodium-montmorillonite interlayered with aluminum clusters: *Ind. Eng. Chem. Prod. Res. Dev.* **24,** 69–82.

Occelli, M. L., Landau, S. D., and Pinnavaia, T. J. (1984) Cracking selectivity of a delaminated clay catalyst: *J. Catal.* **90,** 256–262.

Occelli, M. L. and Lester, J. E. (1985) The nature of active sites and coking reactions in a pillared clay mineral: *Ind. Eng. Chem. Prod. Res. Dev.* **24,** 27–32.

Occelli, M. L. and Rennard, R. J. (1984) Hydrocracking with pillared clays: in *Amer. Chem. Soc., Div. Fuel Chem. Preprints,* 30–39.

Pinnavaia, T. J. (1983) Intercalated clay catalysts: *Science* **220,** 365–371.

Pinnavaia, T. J. (1984) Preparation and properties of pillared and delaminated clay catalysts: in *Heterogeneous Catalysis,* B. L. Shapiro, ed., Texas A&M Univ. Press, College Station, Texas, 142–164.

Pinnavaia, T. J., Tzou, M. S., Landau, S. D., and Raythatha, R. H. (1984) On the pillaring and delamination of smectite clay catalysts by polyoxycations of aluminum: *J. Mol. Cat.* **27,** 195–212.

Vaughan, D. E. W., Lussier, R., and Magee, J. S. (1979) Pillared interlayered clay materials useful as catalysts and sorbents: *U.S. Patent* **4,176,090,** 7 pp.

Vaughan, D. E. W., Lussier, R., and Magee, J. S. (1981) Stabilized pillared interlayered clays: *U.S. Patent* **4,248,739,** 10 pp.

Proceedings of the International Clay Conference, Denver, 1985, L. G. Schultz, H. van Olphen, and F. A. Mumpton, eds.,
The Clay Minerals Society, Bloomington, Indiana, 324–328 (1987).

CATALYSIS OF DIELS-ALDER CYCLOADDITION REACTIONS BY ION-EXCHANGED MONTMORILLONITES

J. M. ADAMS, K. MARTIN, AND R. W. McCABE

Edward Davies Chemical Laboratories, University College of Wales
Aberystwyth, Dyfed, SY23 1NE, United Kingdom

Abstract—A study was made of the ability of cation-exchanged montmorillonites to catalyze some typical Diels-Alder cyclization reactions in organic solvents. Although Cr^{3+}- and Fe^{3+}-clays were effective catalysts, samples exchanged with nontransition metals, such as Na^+, Mg^{2+}, and Al^{3+}, were not. The differences between the activities, for example, of Fe^{3+}- and Al^{3+}-montmorillonites suggest that the clays did not act as Lewis acid Diels-Alder catalysts nor as proton donors, but rather that a one-electron transfer was involved, leading to a radical cation-catalyzed process. The Cr^{3+}- and Fe^{3+}-clay samples were of comparable activity to the Fe^{3+}-exchanged, acid-treated montmorillonite (K10, Süd-Chemie) suggested recently as a useful Diels-Alder catalyst. Under comparable conditions, for example, Fe^{3+}-K10, Fe^{3+}-, and Cr^{3+}-montmorillonites all produced >90% reaction between cyclopentadiene and methyl vinyl ketone in 20 min at 22°C in dichloromethane solvent. The clay-based catalysts gave rapid and stereoselective reactions and were easy to use. The type of activity displayed in these reactions emphasizes that ion-exchanged montmorillonites are not merely solid sources of protons, but can with suitable pretreatment be effective catalysts in a range of organic syntheses.

Key Words—Catalysis, Cation exchange, Diels-Alder reaction, Montmorillonite, Transition metal.

INTRODUCTION

Diels-Alder reactions are useful in organic synthesis for forming 6-ring systems. The concerted $4\pi + 2\pi$ cycloaddition reaction involves combination of a diene with a dienophile to form a cyclohexene ring.

Scheme 1

The relative electronic energies for the orbitals in these reactions are such that usually the highest occupied molecular orbital (HOMO) of the diene overlaps with the lowest unoccupied molecular orbital (LUMO) of the dienophile. Although many of the reactions can be carried out at reasonable rates without a catalyst, a large energy difference between the HOMO and LUMO may require rather high temperatures for the reaction to proceed.

It is well known that if α,β-unsaturated carbonyl compounds are used as dienophiles, Lewis acid complexation of the dienophile leads to a significant enhancement of the reaction rate—factors of 10^6 are not uncommon—and the product distribution tends towards the more thermodynamically favored form (see, e.g., Bednarski and Danishevsky, 1983; Houk and Strozier, 1973; Oppolzer and Chapuis, 1983). Polar solvents are most efficient, and Breslow *et al.* (1983) showed that use of water as solvent increases both the rate and stereoselectivity. In addition, Bellville and

Bauld (1981, 1982) showed that the rate of the Diels-Alder reaction between unactivated hydrocarbons can be increased markedly when a catalyst is used that is designed to produce radical cation intermediates from the dienophile, which is thus made highly electron deficient.

Laszlo and Lucchetti recently extended the range of catalyst types to include Fe^{3+}-exchanged K10 (an acid-activated bentonite supplied by Süd-Chemie AG, Munich). This material can be employed successfully alone, when using oxygen-containing dienophiles (Laszlo and Lucchetti, 1984a, 1984b, 1984c, 1984d), or, more efficiently, when hydrocarbon dienophiles are involved, in conjunction with 4-t-butylphenol (Laszlo and Lucchetti, 1984d), which is known to form radical cations readily. This work is said to "take advantage of the joint presence in a clay of pools of water internal to the layered structure . . . and Lewis acid sites" (Laszlo and Lucchetti, 1984b). The advantages of this type of catalyst over others are low cost, the simple reaction set-up, and the ease of purification of products, inasmuch as the clay catalyst can easily be filtered off. Laszlo and Lucchetti (1984d) reported that for reactions employing 4-t-butylphenol, Fe^{3+}-K10 is more efficient than K10 itself or even Zn^{2+}-, Co^{2+}-, or Al^{3+}-exchanged K10 materials.

The above work can be related to other Diels-Alder reactions that are carried out over ion-exchanged montmorillonites. Downing *et al.* (1978) showed that Cu^+-montmorillonite is a good catalyst for butadiene cyclodimerization; however, although the product (4-

vinyl cyclohexane) was that expected from a Diels-Alder reaction, it was not certain that the mechanism was truly of the Diels-Alder type. Maxwell (1982) showed that Cu^+ cations are catalysts for this process even when not supported on a clay. He suggested a stepwise (non-concerted) mechanism which involves direct bonding of the diene to the Cu^+ cation:

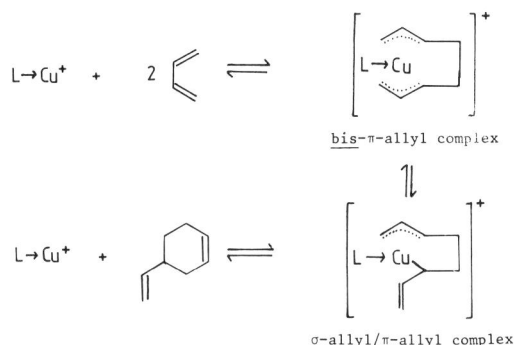

bis-π-allyl complex

σ-allyl/π-allyl complex

Scheme 2

The remaining reaction of Diels-Alder type catalyzed by montmorillonite is that of dimerization of oleic acid (Newton, 1984)—one of the few large-scale processes that uses a clay catalyst at the present time. Here, the overall reaction involves many steps before the Diels-Alder cyclization takes place, namely, *cis/trans* isomerization, skeletal isomerization, and hydrogen transfer (den Otter, 1970a, 1970b, 1970c). Although the clay catalyzes these other processes, the final step could simply be a thermal reaction inasmuch as the process is carried out at 230°C.

Recent work by Adams and Clapp (1986) showed that montmorillonites exchanged with a variety of transition metal cations will catalyze the cyclodimerization of butadiene and isoprene. Whereas all of the clays they tested (Cr^{3+}, Co^{2+}, Ni^{2+}, Ag^+) were efficacious to some degree, one test carried out with an Al^{3+}-clay showed it to be ineffective, suggesting that these reactions could involve one-electron transfers, as suggested by Laszlo and Lucchetti (1984d).

In this present study the reaction rates of some standard Diels-Alder reactions have been examined using a range of ion-exchanged montmorillonite catalysts. Because reactions catalyzed by ion-exchanged or acid-treated clays can depend very strongly on the solvent (see, e.g., Adams *et al.,* 1982; Choudary *et al.,* 1985; McCabe *et al.,* 1985) the reactions were carried out in a variety of solvents.

EXPERIMENTAL

Catalyst characterization

The bentonite used for most of this study came from Moosburg, Federal Republic of Germany (Tonsil 13, Süd-Chemie AG). The <2-μm clay fraction was se-lected by sedimentation. An X-ray powder diffraction (XRD) examination showed that this fraction was ~98% montmorillonite with ~2% quartz impurity. An X-ray fluorescence (XRF) analysis gave 64.0% SiO_2, 20.8% Al_2O_3, 3.86% Fe_2O_3, 0.07% TiO_2, 1.5% CaO, 1.42% MgO, 0.04% K_2O, and 1.20% Na_2O. The loss on ignition was 7.1%. The cation-exchange capacity (CEC) totaled 103 meq/100 g of air-dry material and was made up of 68, 33, and 2 meq/100 g of Na^+, Ca^{2+}, and Mg^{2+}, respectively.

The K10 catalyst (Süd-Chemie AG) is derived from the Tonsil clay by acid treatment. It has nitrogen surface area of ~250 m²/g.

Catalyst preparation

To aid later cation-exchange processes, a large quantity of the Tonsil 13 clay was initially exchanged with Na^+. The clay was added to an 0.1 mole/liter solution of NaCl and heated to 70°C with vigorous stirring for 1 hr. Upon sedimentation it was immediately apparent that two fractions were present, the finer material of which was subjected to ultrasonic treatment for 10 min, and the <2-μm fraction was collected by sedimentation.

Cation exchange was carried out by dissolving the metal chloride (10-fold molar excess, based on the CEC of the clay) in deionized water (40 ml/g clay), after which the Na^+-clay was added and the temperature raised to 70°C for 1 hr. Repetitive centrifugation and resuspension was then carried out until the solution was free of chloride. The clay was then dried at 80°C for ~15 hr and ground before use. Fe^{3+}-K10 samples were produced in a similar manner, but without the preliminary Na^+-exchange.

Large-scale reactions

In large-scale experiments the reaction conditions were either: (1) 0.5 g clay, 50 ml solvent, 10 ml diene, and an equimolar amount of dienophile; or (2) 0.5 g clay, 10 ml solvent, and ~0.2 g of reactants. In all experiments the dienophile was added to the suspension of the clay in the solvent before the diene.

Reactions carried out in nuclear magnetic resonance spectrometer

The rates of the reactions were conveniently studied by nuclear magnetic resonance spectroscopy (NMR). The following method for studying the processes in situ was developed.

1. The NMR tube was charged with 0.025 g clay and 0.5 ml solvent and was then sealed.
2. After 5 min shaking (for homogenization) 20 μl of the dienophile was injected, and the tube was shaken again for 5 min.
3. 20 μl of diene was injected, and the NMR tube was shaken manually.

4. The reaction was followed with time by NMR spectroscopy. Before each NMR spectrum was recorded, the sample tube was centrifuged to sediment the clay. Between spectra the tube was shaken using a mechanical shaker.

5. The rate of formation of products was calculated from integrals of methyl protons on the dienophile and product. For cyclopentadiene dimerization, the rate of signal loss of the cyclopentadiene CH_2 group was monitored.

RESULTS AND DISCUSSION

A wide range of Diels-Alder reactions were carried out, some of which are detailed below. Some of the reactions involved activated, α,β-unsaturated ketones or esters as dienophiles. Generally clay-based catalysts exchanged with nontransition metal cations (such as, Na^+, Mg^{2+}, or Al^{3+}) gave no significant degree of catalysis. Because Lewis acids catalyze Diels-Alder reactions, experiments were also carried out with Al^{3+}-clays which had been dehydrated by azeotropic removal of the water initially present with toluene. Even with this pretreatment, the Al^{3+}-clays were not active. Apparently only non-coordinated interlayer water had been removed, leaving the Al^{3+} cations coordinatively saturated so that they could not function as true Lewis acids.

Effects of catalyst type

The reaction of cyclopentadiene and methyl vinyl ketone in dichloromethane solvent was studied as a direct comparison with the work of Laszlo and Lucchetti (1984b).

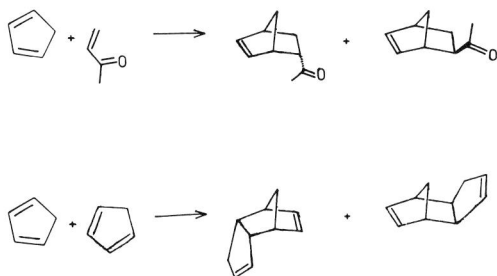

Scheme 3

Their work was extended to the dimerization of cyclopentadiene (in dichloromethane) and by comparing the Fe^{3+}-K10 catalyst with various cation-exchanged montmorillonites. As shown in Table 1, Fe^{3+}- and Cr^{3+}-clays have activities for both reactions which are comparable to that of Fe^{3+}-K10, even though this material has a much larger nitrogen surface area (250 vs. 50 m²/g). Thus, the reactions were not catalyzed solely by surface Fe^{3+} or Cr^{3+} cations, but must also have involved these cations in the interlayer region of the K10 or montmorillonite (it should be noted that the CECs

Table 1. Comparison of the activity of clay catalysts for the reactions of cyclopentadiene with methyl vinyl ketone for cyclopentadiene dimerization.[1]

| Catalyst | Cyclopentadiene + methyl vinyl ketone | | Cyclopentadiene dimerization[2] |
	Yield after 20 min (%)	Endo : exo ratio of product	Yield after 15 min (%)
Fe^{3+}-K10	97	9:1	53
Fe^{3+}-montmorillonite	93	7:1	39
Cr^{3+}-montmorillonite	91	9:1	30
Al^{3+}-montmorillonite	20	8:1	—
No catalyst (control)	17	2:1	0

[1] Reactions were carried out at 22°C with dichloromethane solvent.
[2] Endo : exo ratios for the cyclopentadiene dimerization could not be obtained from the ¹H nuclear magnetic resonance data measured.

of these materials are similar). An involvement of interlayer cations implies that diffusion between the clay layers was not likely to have been the rate-determining process. The fact that the transition metal cations are active, but Al^{3+} cations are not, suggests that one-electron transfer processes were as important here as they are in radical cation catalyzed Diels-Alder reactions (Bellville and Bauld, 1981, 1982; Laszlo and Lucchetti, 1984b, 1984d). In addition, the NMR analysis techniques showed that the reactions catalyzed by Fe^{3+}-K10 and Fe^{3+}-clay were not as "clean" as those carried out over Cr^{3+}-montmorillonite, in that numerous unidentified by-products were observed.

Solvent effects

The reaction of cyclopentadiene and methyl vinyl ketone was also investigated here. The solvents studied were chosen because: (1) Laszlo and Lucchetti (1984b) had found dichloromethane to be effective; (2) halogenated solvents are moderately polar; and (3) the NMR peaks from the solvents did not obscure those from the reactants or products. As shown in Figure 1, apart from chloroform, all of the solvents tested gave broadly comparable activity despite the differences observed in the basal spacings of the clays—16.4 Å for dichloromethane, 17.1 Å for carbon tetrachloride, 17.4 Å for tetrachloroethane, and 18.3 Å for benzene. It is noteworthy that addition of chloroform to the clay caused some coagulation of the particles, leading to the possibility of phase transfer problems, and thereby contributed to the slow reaction rate with this solvent.

Reactions with dienophiles activated to differing degrees

The highly conjugating, and thus highly electron withdrawing, properties of various carbonyl (and other) functional groups lowers the relative energy of the LUMO of the dienophile, thereby allowing more effi-

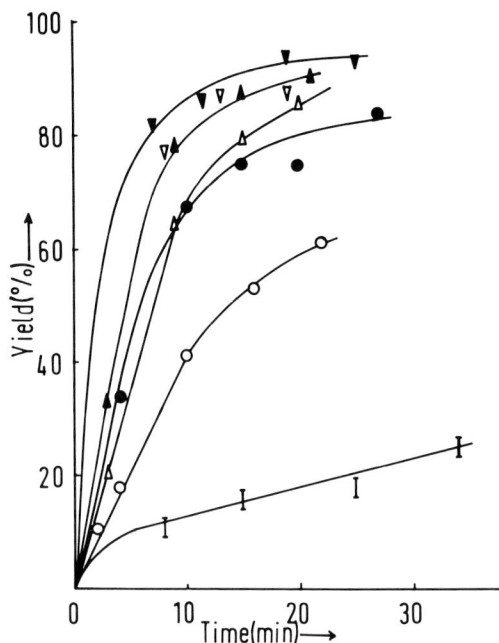

Figure 1.   Reaction of cyclopentadiene and methyl vinyl ketone at 22°C using Cr^{3+}-montmorillonite as catalyst. Solvents employed were: chloroform (○), benzene (●), carbon tetrachloride (△), dichloromethane (▲), chlorobenzene (▽) and tetrachloroethane (▼). (The reaction profiles for dichloromethane and chlorobenzene coincide—only one line has been drawn.) For comparison a control reaction is shown which was carried out in dichloromethane solvent with no catalyst (I).

Table 2.   Effect of the variation of the dienophile on the reaction rate.[1]

Dienophile	Yield (%)	Endo : exo ratio in product
Methyl acrylate	100	9:1
Methyl vinyl ketone	~60	9:1
Cyclopentadiene	10–15	—
Allyl chloride	0[2]	—

[1] Reactions were between cyclopentadiene and the dienophile specified and carried out with Cr^{3+}-montmorillonite catalysts at 22°C in dichloromethane solvent. Yields quoted are after 5-min reaction.

[2] No product detected; cyclopentadiene dimers only.

cient overlap with the diene HOMO (Houk and Strozier, 1973). In such circumstances, reactions proceeded at enhanced rates compared with those observed for dienophiles such as allyl chloride, which has little activation through the inductive effect of the chlorine atom.

The rates of reaction with cyclopentadiene of methyl acrylate (highly activated), methyl vinyl ketone (moderately activated), cyclopentadiene (poorly activated), and allyl chloride (very poorly activated) were compared. In each reaction the same solvent (dichloromethane) and catalyst (Cr^{3+}-montmorillonite) were used so that direct comparisons could be made. As expected, the order of reactivity of methyl vinyl ketone and methyl acrylate followed that predicted by consideration of the degree of activation of the double bond by the carbonyl group (see Table 2). Although no product was detected with allyl chloride, an appreciable rate increase was observed for the dimerization of cyclopentadiene.

Two mechanisms can be suggested for the catalytic process. Both involve activation of the dienophile either by protonation of, or electron transfer from, the carbonyl function or second alkene residue (i.e., for cyclopentadiene or isoprene) of the dienophile. Neither

process was able to increase the reactivity of the allyl chloride because no second π system was available for such reaction. The protonation process can be dismissed because no rate increase was observed with the highly acidic Al^{3+}-exchanged montmorillonite (Table 1). Therefore, the mechanism appears to have involved one-electron transfer from the dienophile to the transition metal ion, reaction of the activated dienophile with the diene, and return of the electron to the radical from the metal.

The acidity of the clay also appeared to affect the amount of by-products in the reaction. Although the rate of dimerization of cyclopentadiene on its own was roughly half that for reaction of methyl vinyl ketone with cyclopentadiene (see Table 1), little or no cyclopentadiene dimer was observed in the latter reaction. These results may be rationalized by the fact that the carbonyl group of the methyl vinyl ketone (or methyl acrylate) is more polar and more readily protonated than the double bond of cyclopentadiene and would therefore be held more firmly in the outer coordination sphere of the transition metal (i.e., the active site for the reaction), thereby preventing cyclopentadiene from reacting with itself.

CONCLUSIONS

From the present work, and that of Adams and Clapp (1986), it is apparent that:

1. Transition-metal-exchanged montmorillonites (Cr^{3+}, Fe^{3+}, Co^{2+}, Ni^{2+}, Ag^+) are good catalysts for Diels-Alder processes, whereas nontransition-metal-exchanged clays (Al^{3+}, Mg^{2+}, Na^+) are not.
2. Cr^{3+}- and Fe^{3+}-clays are the most active, but of these the Fe^{3+}-clays give reactions which have more by-products.
3. Transition-metal-exchange montmorillonites will catalyze reactions involving activated dienophiles and non-activated hydrocarbon dienophiles which have conjugated double bonds, but not, apparently, dienophiles which are unable to donate electrons easily.
4. As Laszlo and Lucchetti (1984b) first pointed out

for Fe^{3+}-K10, it is not necessary to use inconvenient water solution or suspension to promote rapid, stereoselective Diels-Alder reactions; they can be carried out in the presence of clay catalysts. Moreover, for organic synthesis, these catalysts lead to a remarkable ease of "work-up" after reaction, the catalysts can be easily removed by filtration.

In a more general sense, the confirmation of Diels-Alder activity for transition-metal-exchanged montmorillonites demonstrates that clay catalysts are more versatile than they are often considered to be—they do not merely act as solid acid.

ACKNOWLEDGMENTS

We are indebted to P. Laszlo's group at Liege, especially A. Cornelis and S. Chalais, for most helpful discussions. We are also grateful to English Clays Lovering Pochin & Co. plc for financial and analytical assistance and Süd-Chemie AG for gifts of K10 and the Tonsil 13 clay.

REFERENCES

Adams, J. M. and Clapp, T. V. (1986) Reactions of the conjugated dienes butadiene and isoprene alone and with methanol over ion-exchanged montmorillonites: *Clays & Clay Minerals* **34**, 287–296.

Adams, J. M., Clement, D. E., and Graham, S. H. (1982) Synthesis of methyl t-butyl ether from methanol and isobutene using a clay catalyst: *Clays & Clay Minerals* **30**, 129–134.

Bednarski, M. and Danishefsky, S. (1983) Mild Lewis acid catalysis: tris (6,6,7,7,8,8,8-heptafluoro-2,2-dimethyl-3,5-octanedionato) europium-mediated hetero-Diels-Alder reaction: *J. Amer. Chem. Soc.* **105**, 3716–3717.

Bellville, D. J. and Bauld, N. L. (1981) The cation-radical catalyzed Diels-Alder reaction: *J. Amer. Chem. Soc.* **103**, 718–720.

Bellville, D. J. and Bauld, N. L. (1982) Selectivity profile of the cation radical Diels-Alder reaction: *J. Amer. Chem. Soc.* **104**, 2665–2667.

Breslow, R., Maitra, U., and Rideout, D. (1983) Selective

Diels-Alder reactions in aqueous solutions and suspensions: *Tetrahedron Letters* **24**, 1901–1904.

Choudary, B. M., Kumar, K. R., Jamil, Z., and Thyagarajan, G. (1985) A novel 'anchored' palladium(II) phosphinated montmorillonite: the first example in the interlamellars of smectite clay: *J. Chem. Soc. Chem. Comm.* 931–932.

den Otter, M. J. A. M. (1970a) The dimerization of oleic acid with a montmorillonite catalyst I. Important process parameters; some main reactions: *Fette Siefen Anstrichmittel* **72**, 667–673.

den Otter, M. J. A. M. (1970b) The dimerization of oleic acid with a montmorillonite catalyst II. Glc analysis of the monomer; the structure of the dimer; a reaction model: *Fette Siefen Anstrichmittel* **72**, 875–883.

den Otter, M. J. A. M. (1970c) The dimerization of oleic acid with a montmorillonite catalyst III. Test of the reaction model: *Fette Siefen Anstrichmittel* **72**, 1056–1066.

Downing, R. S., von Amstel, J., and Joustra, A. H. (1978) Dimerization process catalyst: *U.S. Patent* **4,125,483**, Nov. 14, 3 pp.

Houk, K. N. and Strozier, R. W. (1973) On Lewis acid catalysis of Diels-Alder reactions: *J. Amer. Chem. Soc.* **95**, 4094–4096.

Laszlo, P. and Lucchetti, J. (1984a) Catalyse de la réaction de Diels et Alder: *L'actualité chimique,* October, 42–44.

Laszlo, P. and Lucchetti, J. (1984b) Acceleration of the Diels-Alder reaction by clays suspended in organic solvents: *Tetrahedron Letters* **25**, 2147–2150.

Laszlo, P. and Lucchetti, J. (1984c) Easy formation of Diels-Alder cycloadducts between furans and α,β-unsaturated aldehydes and ketones at normal pressure: *Tetrahedron Letters* **25**, 4387–4388.

Laszlo, P. and Lucchetti, J. (1984d) Catalysis of the Diels-Alder reaction in the presence of clays: *Tetrahedron Letters* **25**, 1567–1570.

Maxwell, I. E. (1982) Non-acid catalysis with zeolites: in *Advances in Catalysis,* D. D. Eley, H. Pines, and P. B. Weisz, eds., Academic Press, New York, 2–73.

McCabe, R. W., Adams, J. M., and Martin, K. (1985) Clay and zeolite catalyzed cyclic anhydride formation: *J. Chem. Res.,* S357–S358.

Newton, L. S. (1984) Dimer acids and their derivatives—potential applications: *Speciality Chemicals,* May, 17–21.

Oppolzer, W. and Chapuis, C. (1983) Asymmetric Diels-Alder reaction of a chiral allenic ester: enantioselective synthesis of $(-)-\beta$-santalene: *Tetrahedron Letters* **24**, 4665–4668.

Proceedings of the International Clay Conference, Denver, 1985, L. G. Schultz, H. van Olphen, and F. A. Mumpton, eds.,
The Clay Minerals Society, Bloomington, Indiana, 329–334 (1987).

USE OF AN ORGANIC-CLAY MINERAL COMPLEX FOR CHROMATOGRAPHIC SEPARATION OF OPTICAL ISOMERS

AKIHIKO YAMAGISHI[1]

Department of Chemistry, Faculty of Science
Hokkaido University, Sapporo 060, Japan

Abstract—A silica gel coated with a film of Λ-Ru(phen)$_3{}^{2+}$-montmorillonite (phen = 1,10-phenanthroline) was used as a packing material in high-performance liquid-column chromatography. The resolution of Co(acac)$_3$ (acac = acetyl-acetonato) was studied by varying the flow-rate, the component of an eluting solvent, and the temperature of the column. The column showed remarkable resolution efficiency compared with known column materials, such as natural and synthetic organic polymers. The mechanisms of chirality recognition were interpreted in terms of the intermolecular stacking of Co(acac)$_3$ with preadsorbed Λ-Ru(phen)$_3{}^{2+}$ either on the external surface or in the interlayer space of the clay film. The column also resolved a wide variety of organic compounds, some of which are key molecules in asymmetric syntheses, such as 2,2'-binaphthol and phenyl alkyl sulfoxides.

Key Words—Adsorption, Chromatography, Chirality, Montmorillonite, Optical isomers, Separation.

INTRODUCTION

Yamagishi (1982) recently investigated the adsorption of an optically active metal complex by a colloidal clay. Sodium ions in Na-montmorillonite, for example, were exchanged quantitatively by optically active tris (1,10-phenanthroline)metal(II) (M(phen)$_3{}^{2+}$), but no stereoselectivity was observed when enantiomeric M(phen)$_3{}^{2+}$ was adsorbed. Thus, the complex was thought to be bound to the clay in units of racemic pairs rather than as single enantiomers. When M(phen)$_3{}^{2+}$ was added as a pure enantiomer, the complex was adsorbed in an amount equivalent to the cation-exchange capacity (schematic diagram 1 below). In contrast, when the same complex was added as a racemic mixture, it was adsorbed in an amount twice that of the cation-exchange capacity (diagram 2).

$$\underline{\quad{}_{|}\; \Delta\; {}_{|}\quad{}_{|}\; \Delta\; {}_{|}\quad{}_{|}\; \Delta\; {}_{|}\quad{}_{|}\; \Delta\; {}_{|}\quad{}_{|}\; \Delta\; {}_{|}\quad} \qquad (1)$$

$$\underline{\quad{}_{|}\; \Delta\; {}_{|}\; \Lambda\; {}_{|}\; \Delta\; {}_{|}\; \Lambda\; {}_{|}\; \Delta\; {}_{|}\; \Lambda\; {}_{|}\; \Delta\; {}_{|}\; \Lambda\; {}_{|}\; \Delta\; {}_{|}\quad} \qquad (2)$$

A clay containing optically active M(phen)$_3{}^{2+}$ in state (1) recognized the absolute configuration of a new adsorbate very distinctly. Δ-Ni(phen)$_3{}^{2+}$-montmorillonite, for example, adsorbed the Λ-enantiomer of Fe(phen)$_3{}^{2+}$ but not the Δ-enantiomer of Fe(phen)$_3{}^{2+}$. These facts suggest that a clay containing an optically active metal chelate could be used as a packing material for chromatographic resolution (Yamagishi, 1983a, 1983b).

The present work is concerned with the application of these materials in high-performance, liquid-column chromatography. A silica gel coated with the film of Λ-Ru(phen)$_3{}^{2+}$-montmorillonite was found to exhibit remarkable resolution efficiency toward a wide variety of inorganic and organic compounds.

EXPERIMENTAL

Materials

The packing material was prepared as follows: 2 g of silica gel (230–400 mesh, Wako Pure Chemical Ind., Japan), 50 mg of polyvinylalcohol (Nakarai Chemical Ind., Japan), and 40 mg of sodium montmorillonite (Kunimine Ind. Co., Japan) were mixed in 50 ml of water. The mixture was heated at 80°C until the water was completely evaporated. The solid was ground to <200 mesh. Then, 2×10^{-5} mole of Λ-Ru(phen)$_3$(ClO$_4$)$_2$ was added to this solid dispersed in 50 ml of water. The Ru-chelate was completely adsorbed, and a brown material was formed. The slurry was poured into a 30×0.4 cm (i.d.), 2-ml capacity, stainless steel column.

The racemic mixtures used in the present study were obtained as follows: M(acac)$_3$ was synthesized following the procedure of Fay *et al.* (1970). Binaphthol and its analogues were donated by Prof. Noyori (Nagoya University, Japan). Phenyl alkyl sulfoxides were synthesized according to conventional procedures. Pyrazine derivatives were prepared as described previously (Yamagishi, 1983b). Bridged p-benzoquinone were obtained from Prof. S. Nishida (Hokkaido University, Japan). Other reagents were commercially available reagents which were used without purification.

---

[1] Present address: Department of Chemistry, Faculty of Arts and Sciences, The University of Tokyo, Komaba, Meguro, Tokyo 153, Japan.

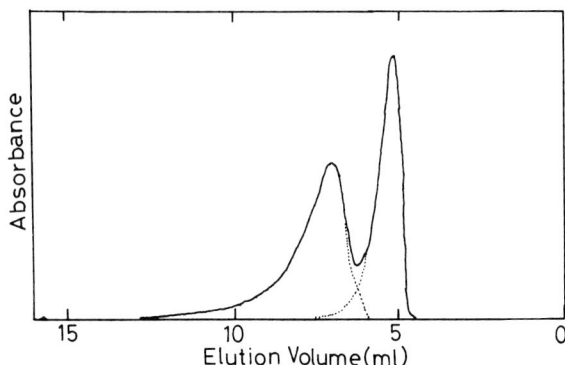

Figure 1. Chromatogram of racemic Co(acac)₃ on a
Λ-Ru(phen)₃²⁺-montmorillonite column. Eluate = water; flow
rate = 0.1 ml/min.

RESULTS AND DISCUSSION

Resolution of tris-acetylacetonato complexes

The M(acac)₃ (I) complex shown below has a spherical shape and no electric charge. It contains no functional groups that can interact with resolving reagents. Thus, M(acac)₃ resists conventional resolution by means of diastereomer formation. In 1980, Co(acac)₃ and Cr(acac)₃ were resolved for the first time to pure enantiomers on a chromatographic column of optically active poly(triphenylmethyl metacrylate) polymer (Yuki et al., 1980).

Instrumental

High-performance, liquid-column chromatography was carried out using a JASCO BIP-II chromatograph equipped with a JASCO UV-100 detector. The ultraviolet (UV) and circular dichroism (CD) spectra of the fractions were recorded on a Hitachi EPS-3T spectrophotometer and a JASCO J-500A spectropolarimeter, respectively.

In a typical run, 50 μl of solution was injected into the chromatograph. A solvent was passed through the column thermostated at a constant temperature. Eluted fractions were collected at every 1 ml. The concentrations were determined by absorbance. The optical purity was estimated from the CD amplitude by knowing the molecular circular dichroism of the species.

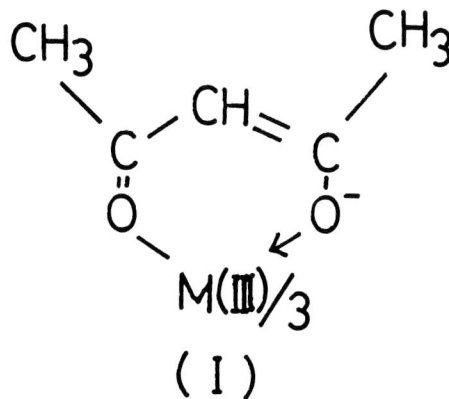

Figure 1 shows the chromatogram produced when racemic Co(acac)₃ was placed in the column at 20°C and eluted with water at a flow rate of 0.1 ml/min. Two peaks were observed at the elution volumes (V) of 5.1 and 7.0 ml. From the CD spectra, the first and

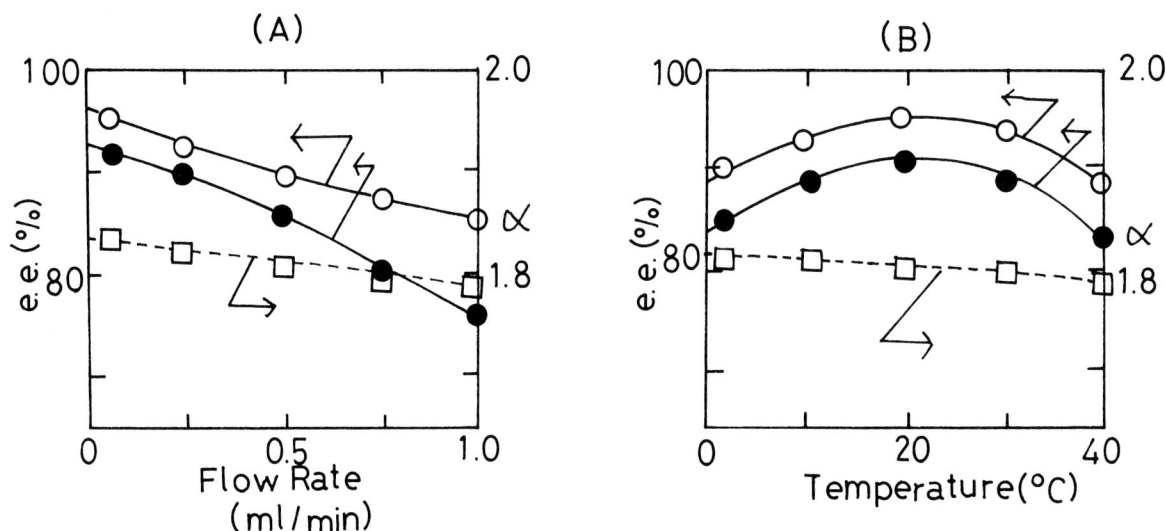

Figure 2.   A. Dependence of the optical purity (e.e.) of Δ-Co(acac)₃ (–O–) and Λ-Co(acac)₃ (–●–) and the separation factor (α) on flow rate. B. Dependence of optical purity (e.e.) and α on temperature of the column.

(A)

Flow Rate →

(B)

Temperature →

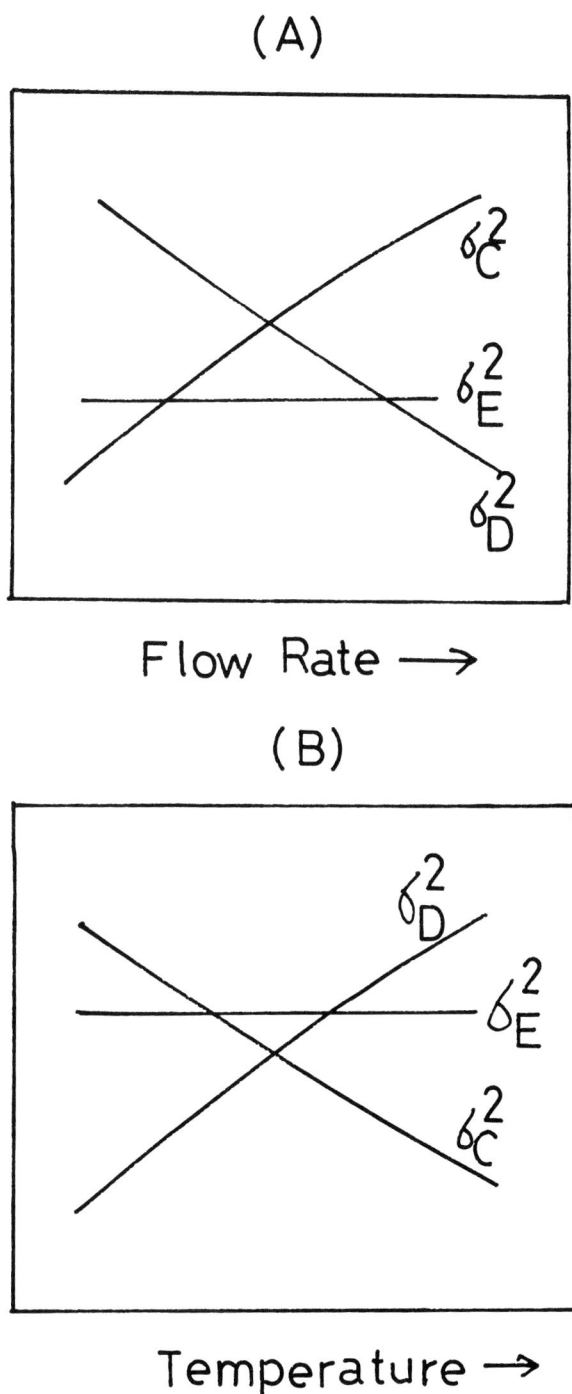

Figure 3.   A. Dependence of δ_C^2, δ_E^2, and δ_D^2 on flow rate. B. Dependence of δ_C^2, δ_E^2, and δ_D^2 on temperature of the column.

(A)

(B)

(C)

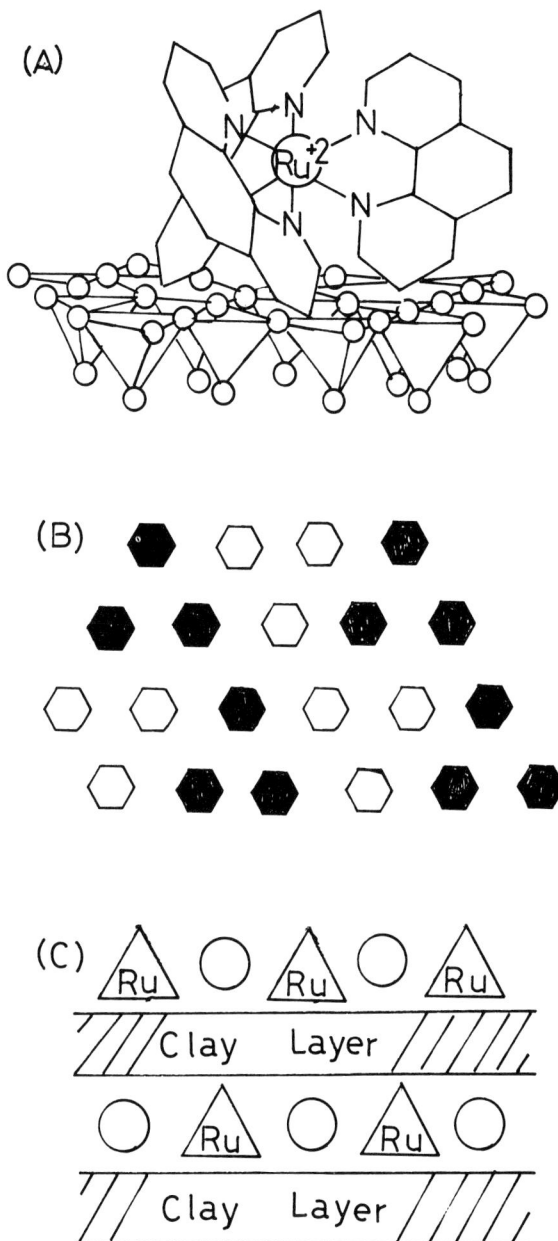

Figure 4.   A. Adsorbed state of Ru(phen)$_3^{2+}$ on the clay mineral surface. B. The distribution of Ru(phen)$_3^{2+}$ on the silicate layer. The hexagons denote hexagonal holes surrounded by six [SiO$_4^{4-}$] tetrahedra. The three solid hexagons are occupied by one Ru(phen)$_3^{2+}$. C. Schematic drawing of the interlayer structure. Ru-chelates are located both on an external surface and in the interlayer space. Open circle denotes solvent molecule.

second peaks can be assigned to the Δ- and Λ-enantiomers of Co(acac)$_3$, respectively. The elution curves of both enantiomers were constructed from the measured optical purity of a collected fraction. If the whole eluent was divided into two parts at V = 6.2 ml (minimum in the chromatogram), both fractions contained the enantiomers with an optical purity (e.e.) of 96 and 93%, respectively. Although the separation was not complete, the present column resolved (Co(acac)$_3$ at a

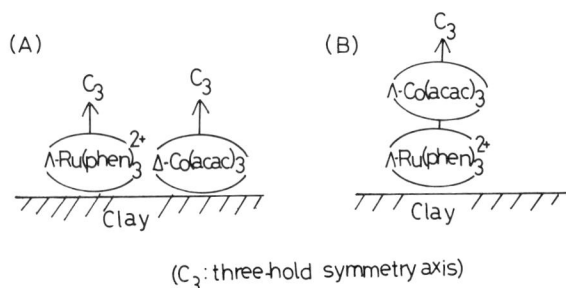

(C$_3$: three-fold symmetry axis)

Figure 5. A. Stacking between Λ-Ru(phen)$_3$$^{2+}$ and Δ-Co(acac)$_3$. B. Stacking between Λ-Ru(phen)$_3$$^{2+}$ and Λ-Co(acac)$_3$.

high efficiency compared with the typical column materials, such as lactose, potato starch, and ion-exchange resins (Fay et al., 1970).

Figure 2A shows the effects of the flow rate on the separation factor (α) defined by $\alpha = (V_2 - V_d)/(V_1 - V_d)$, elution volumes at the first and second peaks. The illustration also shows the dependence of the optical purity (e.e.) of the first and second fractions divided at the minimum point of a chromatogram on the flow rate of an eluting solvent. As seen in Figure 2, α was little affected by the variation of the flow rate from 0.05 to 1.0 ml/min, but the optical purity (e.e.) decreased appreciably. Figure 2B shows similar plots of α and e.e. as a function of the temperaturae of the column. α changed little between 0° and 40°C, but e.e. reached a maximum at about 22°C.

Apparently, the decrease of optical purity (Figure 2) was caused by the broadening of the peak but not by a decrease of the separation factor. Generally, the width of a peak in a chromatogram is a function of: (1) the disturbed flow of the eluent in the column ($\delta_E{}^2$); (2) the molecular diffusion of the solute in the mobile phase ($\delta_D{}^2$); and (3) the nonequilibrium distribution between the stationary and mobile phases ($\delta_C{}^2$). Figure 3 shows typical dependences of these factors on the flow rate and temperature. Comparing Figures 2A and 2B with

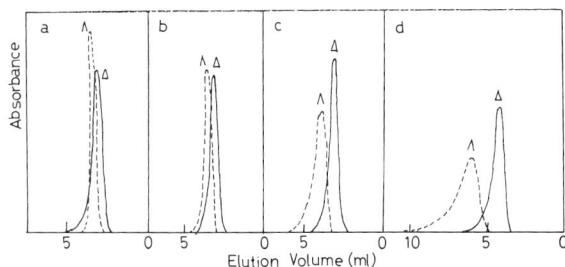

Figure 6. Chromatograms of Δ-Co(acac)$_3$ (——) and Λ-Co(acac)$_3$ (– – –) wherein the enantiomers were eluted with methanol-water mixtures; (a) water, (b) 20% (v/v) methanol-water, (c) 25% (v/v) methanol-water, and (d) 33% (v/v) methanol-water.

Table 1. Chromatographic resolution of racemic mixtures on a Λ-Ru(phen)$^{32+}$-montmorillonite at 20°C.[1]

Racemic sample	Composition of eluate[2]	Configuration of more strongly retained isomer	Separation factor	Optical yield (%)
Cr(acac)$_3$	1:2	(Λ)	1.8	90
Ru(acac)$_3$	1:3	(Λ)	1.6	85
Rh(acac)$_3$	1:3	(Λ)	1.8	90
Binaphthyl derivatives				
(II)	1:1	(S)	1.8	95
(III)	1:1	(S)	1.9	95
(IV)	1:1	(S)	1.5	78
(V)	1:1		1.0	0
Phenyl alkyl sulfoxides				
(VI)	0:1	(R)	1.7	80
(VII)	0:1	(R)	2.0	89
(VIII)	1:2	(R)	1.7	80
(IX)	1:2	(R)	1.8	86
(X)	1:1	(R)	1.7	91
1,2-diphenylethane derivatives				
(XI)	1:2	(R)	1.4	76
(XII)	1:2	(R)	1.3	70
(XIII)	1:2	(R)	1.2	64
(XIV)	0:1	(S)	1.8	91
(XV)	1:4	(RR)	2.0	95
(XVI)	0:1		1.0	0
2,3-dihydropyrazine derivatives				
(XVII)	1:2	(R)	3.5	95
(XVIII)	1:2	(SS)	2.4	84
(XIX)	1:3		1.0	0
Bridged quinone and related compounds				
(XX)	1:2	(+)$_D$	1.0	0
(XXI)	1:2	(+)$_D$	2.0	98
(XXII)	1:2	(+)$_D$	2.0	96
Others				
(XXIII)	1:3	(+)$_D$	1.5	80
(XXIV)	1:2	(+)$_D$	1.3	78

[1] Elution was performed after about 1×10^{-7} mole of the racemic mixture (Figure 7) was added to the column. The flow rate of the solvent was 0.1 ml/min.
[2] Ratio of methanol to water.

Figure 3, for high flow rates (>0.1 ml/min) and low temperature (<20°C), $\delta_C{}^2$ appears to be the principal cause of the peak width and the resolution efficiency. At low flow rates (<0.1 ml/min) and high temperature (>30°C), $\delta_E{}^2$ determined the peak width.

On the basis of this analysis, the following means of improving resolution efficiency are suggested: (1) The particle size of the column material (silica gel, in the present experiment) should be as uniform as possible, minimizing the contribution of $\delta_E{}^2$. (2) The clay film should be as thin as possible, thereby increasing the equilibration rate of adsorption between the clay and the mobile phase, deducing $\delta_C{}^2$.

The mechanism of chirality recognition is understood in a straightforward way based on the molecular staking models. As shown earlier (Yamagishi, 1982),

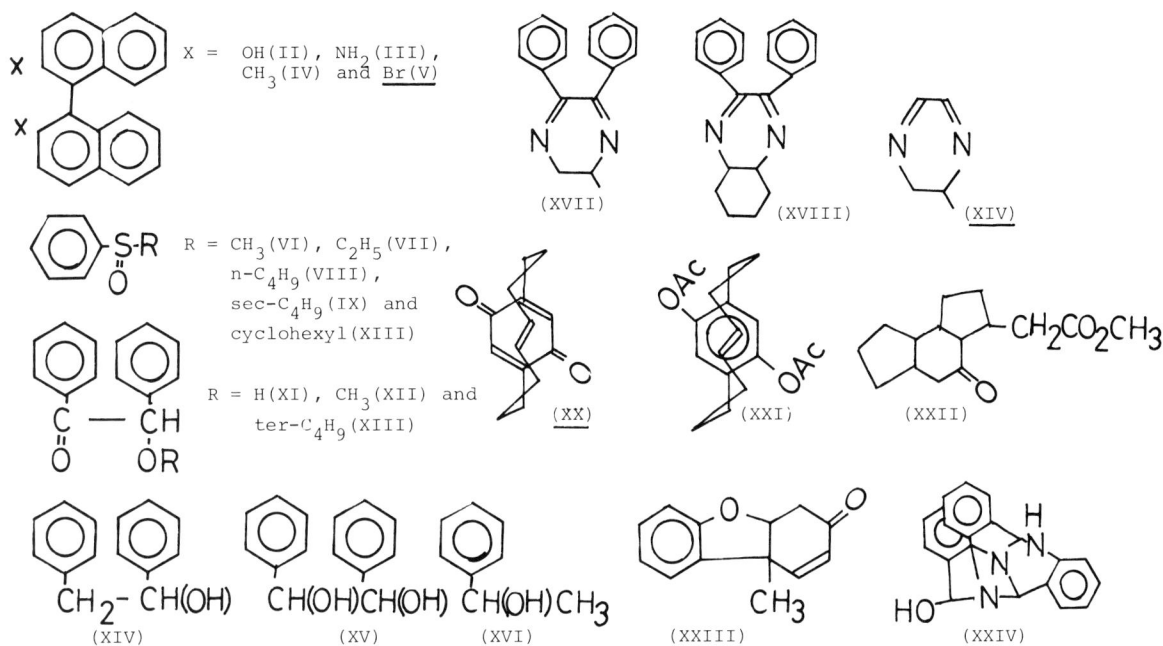

Figure 7.   Structures of several compounds used in this study.

Ru(phen)$_3$$^{2+}$ ions were adsorbed with their three-fold symmetry axes normal to the clay surface (Figure 4A). The Ru-complexes adsorbed within the cation-exchange capacity of the clay were distributed on the silicate surface as shown schematically in Figure 4B. When racemic Co(acac)$_3$ comes into contact with the adsorption layers, the absolute configuration of Co(acac)$_3$ can be recognized by the intermolecular stacking of Co(acac)$_3$ with the pre-adsorbed Ru(phen)$_3$$^{2+}$ complexes.

Two possible ways of stacking exist for Co(acac)$_3$ and Ru(phen)$_3$$^{2+}$, side-by-side (Figure 5A) one above the other (Figure 5B). In the former stacking mode, Δ-Co(acac)$_3$ fits preferably to Λ-Ru(phen)$_3$$^{2+}$, whereas Λ-Co(acac)$_3$ fits better to Λ-Ru(phen)$_3$$^{2+}$ in the latter stacking mode, and the phenanthroline ligands of Λ-Ru(phen)$_3$$^{2+}$ are in close contact with the acetylacetonato ligands in Co(acac)$_3$. Stacking as in Figure 5A is possible both on the external surfaces and in the interlayer space (Figure 4C), whereas stacking as in Figure 5B will occur on the external surface only.

To obtain information about the type of stacking, it was determined which of the enantiomers of Co(acac)$_3$ was bound more strongly. Figure 6 shows the elution curves of Δ- and Λ-Co(acac)$_3$ molecules when the complexes were eluted with methanol-water mixtures. In pure water (Figure 6A), Λ-Co(acac)$_3$ was more strongly retained on the column than Δ-Co(acac)$_3$. Using 20% methanol (Figure 6B), the elution curves of the enantiomers did not change significantly except for a decrease of the separation factor. With increasing amounts

of methanol (Figures 6C and 6D), the elution curve of Δ-Co(acac)$_3$ tailed more than did that of Λ-Co(acac)$_3$. For example at 33% methanol, the Δ-Co(acac)$_3$ peak was broader than the Λ-Co(acac)$_3$ peak. The results are interpreted in terms of the above two stacking modes. If water is used as an eluate, Δ-Co(acac)$_3$ cannot be bound in the interlayer space as in Figures 5A and 5B, because water molecules are strongly adsorbed and occupy the sites around Λ-Ru(phen)$_3$$^{2+}$. If a methanol-rich solvent is used, methanol molecules are partially replaced by water molecules in the interlayer space, which in turn are displaced by Δ-Co(acac)$_3$. This penetration of Δ-Co(acac)$_3$ into the interlayer space might be responsible for the tailing of the Δ-Co(acac)$_3$ peak in methanol-rich solvents. Based on this interpretation, reduced tailing and improved resolution efficiency should make the clay film thinner, because the number of available sites is smaller.

Other acetylacetonato complexes were also resolved on the column with reasonable efficiency (Table 1).

Resolution of organic compounds

If the intermolecular forces between Λ-Ru(phen)$_3$$^{2+}$ and a resolved molecule were responsible for the resolution, the present column should have resolved other organic molecules, too. Some part of the molecule might stack with the phenanthroline ligands in Λ-Ru(phen)$_3$$^{2+}$ just as the ligands of M(acac)$_3$. Suitable should be molecules with bulky planar groups such as aliphatic and aromatic rings.

Several organic molecules were studied (Figure 7).

The chromatographic results for these compounds are summarized in Table 1. The following characteristic features were noted for the present column:

1. 1,1'-binaphthol (II) and its analogues (III) and (IV) were resolved efficiently. These compounds are used as asymmetric ligands in the hydrogenation of ketones (Noyori and Tomino, 1979). Comparing the results for (V) with those of (II)–(IV), the substituents at the 2 and 2' positions of binaphthyl should be electron-donating in order to achieve high-resolution efficiency. The stacking interaction between Λ-Ru(phen)$_3^{2+}$ and the molecules to be resolved seems to be dominated by the charge-transfer interaction. The Ru-complex acts as the electron acceptor.

2. Phenyl alkyl sulfoxides were resolved (VI–X). An optically active phenyl alkyl sulfoxide is now currently attracting industrial attention as an intermediate in the enantioselective syntheses of cycloalkanones (Posner et al., 1981). High resolution in methanol-water solvent was attained when a bulky alkyl group, such as cyclohexyl (X) or sec-butyl (IX) was attached.

3. 1,2-Diphenylethane (XVI) was not resolved at all. Comparing three benzoin derivatives, the $\rangle$C=O groups enhanced the resolution efficiency. The results may indicate the importance of hydrogen bonds or ion-dipole interactions between these molecules and the clay mineral.

4. 2,3-Dihydropyrazine derivatives were resolved if they contained phenyl groups (XVII and XVIII), indicating also the importance of the stacking of the planar part of the molecule with the Ru-complex.

5. A bridged p-benzoquinone (XX) was not resolved, but its precursor, a bridged p-hydroquinone diester (XXI), was resolved almost completely. More interestingly, a photo-induced intramolecular adduct of (XX) was also efficiently resolved. This adduct compound (XXII) has three condensed aliphatic rings, similar to alkaloids. Accordingly, the present column should resolve valuable natural alkaloid groups.

In conclusion, the present column material, an adduct of montmorillonite with an optically active metal complex, has been shown to be a promising packing material for resolving a variety of important organic compounds.

REFERENCES

Fay, R. C., Girgis, A. Y., and Klabunde, U. (1970) Stereochemical arrangement of metal tris-diketonato: *J. Amer. Chem. Soc.* **92,** 7056–7060.

Noyori, R. and Tomino, I. (1979) Virtually complete enantioface differentiation in carbonyl group reduction by a complex aluminum hydride reagent: *J. Amer. Chem. Soc.* **101,** 3129–3131.

Posner, G. H., Mallamo, J. P., and Miuva, K. (1981) High asymmetric induction during organometallic β addition to α,β-ethylenic sulfoxides: *J. Amer. Chem. Soc.* **103,** 2886–2887.

Yamagishi, A. (1982) Racemic adsorption of Fe(phen)$_3^{2+}$ on a colloidal clay: *J. Phys. Chem.* **86,** 2472–2479.

Yamagishi, A. (1983a) Clay column chromatography for optical resolution: *J. Chromatography* **262,** 342–348.

Yamagishi, A. (1983b) Clay as a medium for optical resolution: *J. Chem. Soc. Chem. Comm.* 9–10.

Yuki, H., Okamoto, Y., and Okamoto, I. (1980) Resolution of racemic compounds by optically active poly(triphenylmethyl methacrylate): *J. Amer. Chem. Soc.* **102,** 6356–6358.

Proceedings of the International Clay Conference, Denver, 1985, L. G. Schultz, H. van Olphen, and F. A. Mumpton, eds.,
The Clay Minerals Society, Bloomington, Indiana, 335–339 (1987).

THERMOGRAVIMETRIC ANALYSIS STUDY OF
THE ADSORPTION OF ETHANOIC ACID VAPOR BY
ION-EXCHANGED MONTMORILLONITE

J. Howard Purnell, Ahmad Al-Owais, and James A. Ballantine

Department of Chemistry, University College of Swansea
Swansea, SA2 8PP, Wales, United Kingdom

John M. Thomas

Department of Physical Chemistry, Lensfield Road
Cambridge, CB2 1EP, England, United Kingdom

Abstract—Thermogravimetric analyses (TGA) of ion-exchanged montmorillonites that were exposed to organic vapors or liquids showed multiple desorptive weight losses. Thermograms of the montmorillonite showed a low-temperature (~100°C) weight loss corresponding to the removal of physically sorbed material and further losses at higher temperatures (180°–600°C) corresponding to the removal of species chemisorbed via different mechanisms. The corresponding peaks appearing in the differential form of the thermograms (DTG) were frequently sufficiently well resolved to allow direct and independent measurement. Thus, separate isotherms relating to the individual modes of sorption could be determined simultaneously. Ethanoic acid sorbed on Al^{3+}-exchanged bentonite was characterized by a single higher temperature DTG peak centered at about 230°C. Isotherms for vapor uptake by the clay at 35° and 60°C by both physical and chemical processes suggested that the chemisorption was rapid and provided the first layer of intercalated material. The physical sorption was slow and proceeded via the formation of a second acid layer and, at the highest relative pressures, a third. Each layer appears to have contained the same amount of acid.

The physical sorption occurred predominantly internally, and the properties of the intercalated acid were similar to those of the liquid acid. Chemisorption occurred via the formation of basic aluminum acetates, protons having been liberated in the process. This observation is of relevance to the mechanism of catalytic esterification of acid by alkenes because it establishes both the origin of the protons needed to form the alkene carbocations and that these then yield ester by attack on the basic acetate rather than acid.

Key Words—Adsorption, Ethanoic acid, Intercalation, Montmorillonite, Thermogravimetric analysis.

INTRODUCTION

Al-Owais *et al.* (1986) recently reported that thermogravimetric analysis (TGA) provides valuable information on the intercalation of organic vapors and liquids by montmorillonite clays. They noted that differential thermogravimetric (DTG) peaks of such intercalates can be used to characterize the organic material because the temperatures at which these weight losses occur are little affected by the nature of the exchanged ion in the clay, although the number of peaks and their mid-temperatures vary from intercalate to intercalate. Cyclohexylamine, for example (Ballantine *et al.,* 1987) gives four peaks; ethanoic acid, in contrast, only two (Al-Owais *et al.,* 1986). Each peak appears to correspond to desorption of material originally intercalated via a particular sorption mechanism. For ethanoic acid, the first DTG peak, centered at about 100°C, represents thermal desorption of physically intercalated acid; the second peak, at about 230°C, represents desorption of chemisorbed acid, as Al-Owais

et al. (1986) have established. Classical adsorption techniques cannot distinguish between these processes. Thus, some of the curious features of published sorption isotherms may result from the fact that the physical and chemical sorptions occur on vastly different time scales; the chemisorption reflected by the 230°C weight loss of ethanoic acid, for example, was complete in only a few minutes, whereas, the physical sorption giving rise to the lower temperature weight loss required, in some experiments, as much as 36 hr for equilibrium to be attained.

TGA-DTG results reported by Al-Owais *et al.* (1986) offer a direct route to simultaneous quantitative evaluation of different modes of sorption. To illustrate this new technique we have studied ethanoic acid sorption by Al^{3+}-exchanged bentonite at two temperatures over a wide range of relative pressures. This particular system is of interest because a knowledge of the physical and chemical sorption processes involved should extend our understanding of the mechanism of the very

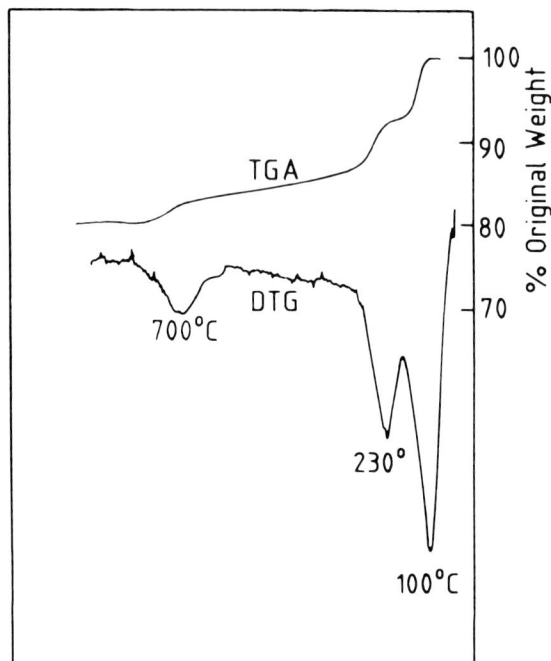

Figure 1.　Integral thermal gravimetric analysis (TGA) and differential thermal gravimetric analysis (DTG) traces of Al^{3+}-exchanged bentonite after exposure to ethanoic acid vapor. Peak in the DTG at 700°C corresponds to structural collapse.

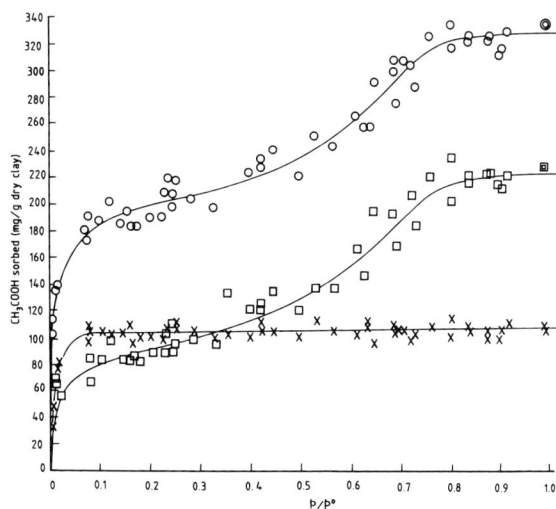

Figure 2.　Sorption isotherms for chemisorbed ($\times$), physically sorbed ($\square$) and total ($\bigcirc$) ethanoic acid vapor sorbed at 35°C by Al^{3+}-exchanged bentonite. Curves drawn are best fit.

efficient catalysis, by ion-exchanged montmorillonites, of alkene-acid esterification, a process of commercial potential.

EXPERIMENTAL

The clay material used in this study was an Al^{3+}-exchanged (Wyoming) bentonite (British Drug Houses). The exchange was carried out in an aqueous solution of $Al_2(SO_4)_3$, the clay then being centrifuged and washed repeatedly until Al^{3+} could not be detected in the washings. The clay was then dried, ground, and sieved to 120–140 mesh A.S.T.M.

Several 250-mg portions of this material were placed in an 80-liter thermostatically enclosed vessel. An atmosphere of dry air and ethanoic acid vapor at some relative pressure (p/p^0) of the acid in the range .01–1.00 was produced by introducing a weighed amount of acid. Inasmuch as the container volume and the initial weight of acid in the gas phase were known, the equilibrium partial pressure of acid in the experiment was calculable by making allowance for the acid sorbed by the clay as determined by the TGA measurements. The experiment was then repeated at a different value of p/p^0.

A sample was taken from each weighed portion of clay for thermal analysis at discrete time intervals, increasing progressively from 5 min initially to 4 hr at long exposure times, at each p/p^0. The results were averaged to allow for clay inhomogeneity. The pro-

cedure was continued until there was no ambiguity about the attainment of equilibrium. Depending on conditions, the equilibration required 10–36 hr for the acid uptake characterized by the weight loss peak at 100°C, but no more than a few minutes for the acid uptake leading to the weight loss peak at 230°C. The experiments were repeated at 50 relative pressures of acid vapor at 35°C and, subsequently, at about 30 relative pressures at 60°C.

The data presented refer only to sorption, and it is possible that somewhat different data would be obtained in desorption experiments.

RESULTS

Inasmuch as the experiments provided data relating the weight of ethanoic acid vapor sorbed per unit weight of clay to the relative pressure of acid in the gas phase, sorption isotherms were constructed for both the 100°C and the 230°C peaks of the DTG which are illustrated in the typical TGA/DTG curves given in Figure 1. Figure 2 illustrates the results for sorption at 35°C. The data for the 230°C peak, indicated by crosses, show that above a relative pressure of only about .05 the isotherm reaches a plateau (saturation). The data for the 100°C peak, represented by squares, show in contrast a continuing dependence on p/p^0 over most of its range and, in addition, evidence of a double plateau. The uppermost curve is that of the sum of the others, that is, the conventional isotherm. The plateaus in these three curves lie close to 100, 200, and 300 mg acid/g clay, respectively.

Figure 3 shows similar data for sorption at 60°C. Here also, the 230°C peak data lie on a plateau at almost all values of p/p^0 employed. The 100°C peak data also produce a curve showing two separate pla-

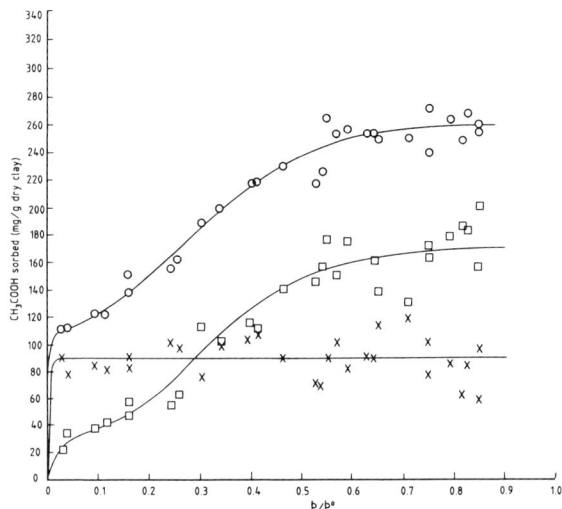

Figure 3. Sorption isotherms for chemisorbed ($\times$), physically sorbed ($\square$) and total ($\bigcirc$) ethanoic acid vapor sorbed at 60°C by Al^{3+}-exchanged bentonite. Curves drawn are best fit.

teaus. Although the scatter in the data for this temperature is greater than is evident in Figure 2 the plots are similar in shape.

Figure 4 is a BET plot of the 35°C data for the 100°C peak. Type IV systems are not well represented by the general equation, and linearity over the range $0 < p/p^0 < .35$, is often taken as satisfactory confirmation of the physical nature of the sorptive process in question. Thus, the BET plot in Figure 4 is consistent with physical sorption. Figure 5 shows a Langmuir plot of the few 230°C peak data for the experiment at 35°C which lie below the saturation plateau. Despite the sparseness of the data, the linearity of this plot is consistent with the proposed (Al Owais et al., 1986) chemisorptive nature of the process leading to this weight-loss peak. After heating the acid-saturated clay to about 270°C,

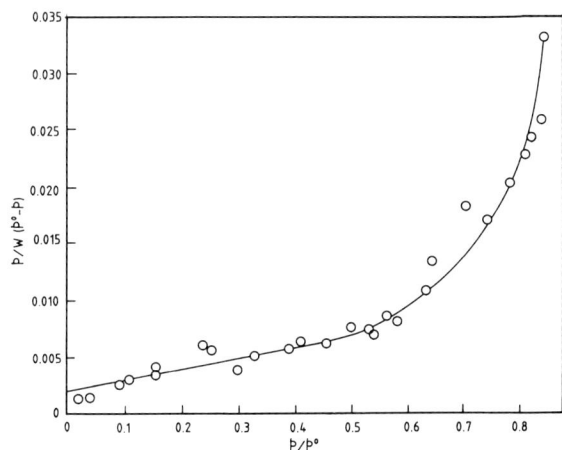

Figure 4. BET plot for sorption at 60°C of physically sorbed (100°C peak of Figure 1) ethanoic acid by Al^{3+}-exchanged bentonite.

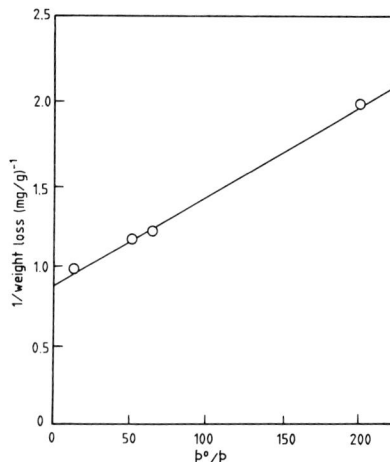

Figure 5. Langmuir plot of data for chemisorbed (230°C peak of Figure 1) acid at low relative pressure (p/p^0) at 35°C.

where all acid was removed, re-exposure to acid vapor led to fresh intercalation to the original levels (see Figure 1). Thus, the removal of the acid in no way modified the clay.

DISCUSSION

The results presented strongly suggest that the intercalation of ethanoic acid by Al^{3+}-exchanged montmorillonite proceeded via two distinct mechanisms. Considering first the rapid, early uptake of acid characterized by the 230°C weight-loss peak in the DTG of Figure 1, Al Owais et al. (1986) showed that the size of this peak is totally unaffected if acid-exposed clay was subjected to high vacuum at the initial sorption temperature for long periods of time. Further, diminution of this peak at atmospheric pressure only became perceptible after a period of many hours at temperatures above about 150°C. Indeed, little change was seen after 1 hr at 200°C. Therefore, at the sorption temperature, the process involved appears to have been irreversible. In sharp contrast, the sorption-desorption processes responsible for the 100°C peak of the DTG of Figure 1 were reversible, even though, occasionally, these processes took a long time. A significant consequence of the strength of the acid sorption that gave rise to the 230°C peak is that, except at very low p/p^0, the extent of sorption was constant. Compliance with the Langmuir equation indicates monolayer formation.

Little confirmatory evidence of these findings can be found in the literature, inasmuch as direct study of concurrent sorptions by alternative mechanisms do not appear to have been carried out previously. Some workers have attempted to separate "internal" and "external" sorption isotherms from the composite results; e.g., Annabi-Bergaya et al. (1979) tried this approach for the system isopropanol/Li^+-Ca^{2+} clay. The result of such a separation is strikingly similar to our results, because the "internal" isotherm (1) reached a plateau at low p/p^0, much earlier than did the "external" isotherm and, (2) the "internal" isotherm showed Langmuir behavior. From unpublished studies in this laboratory alkanol-exposed clays show a DTG peak analogous to the 230°C peak of Figure 1. The "internal" isotherm of Annabi-Bergaya *et al.* (1979) possibly reflects this feature.

Validity of the TGA technique

As was noted above, the three plateaus shown in Figure 2 are at about 100, 200, and 300 mg acid/g clay. These results indicate the formation of three acid layers, all of about the same volume. The first of these layers is clearly chemisorptive; the others arose via physical sorption. Moreover, ammonia-clay systems are well known to be highly resistant to desorption under vacuum at moderate temperatures (see e.g., James and Harward, 1963; Bissada and Johns, 1969; Russell, 1965). James and Harward (1963) attributed this resistance to ammoniation of the interlamellar cation; Russell (1965) subsequently produced confirmatory infrared evidence. Much the same has been shown for amines, as the work of Fripiat and his colleagues, reviewed by Laura and Cloos (1975) and Cloos *et al.* (1975) has demonstrated. Rapid initial formation of a single, chemisorbed amine layer preceding other intercalative processes has clearly been indicated; for example, Ballantine *et al.* (1987) showed that for simple amines as many as four distinct processes may be involved, the first chemisorbed layer resulting from protonation, a view entirely consistent with that of Fripiat and others already referred to. Adams *et al.* (1977) reported too that X-ray powder diffraction analyses show that ethylene glycol and lactones form as many as three intercalated layers. For the intercalate examined in the present study, the first layer appears certainly to be chemisorbed, later layers arising via predominantly physical processes. Al Owais *et al.* (1986) have found that other organic species of varying functional type also commonly exhibit a higher temperature weight loss in the TGA. Thus, first layer chemisorption may well be a general characteristic of intercalation.

The present findings are seen in no way to conflict with previous observation, and the TGA method is obviously more direct and less open to ambiguity, a decided advantage.

Nature of sorption

Chemisorption of basic species involves protonation, as has been confirmed by Ballantine *et al.* (1987). Ethanoic acid, however, being a weak acid, should itself ionize strongly in the electrostatic field existing in the interlamellar space. Thus, some alternative explanation of its chemisorption must be sought, the obvious probability being the sequence:

$$CH_3COOH \rightleftharpoons CH_3COO^- + H^+$$

$$X^{n+} + CH_3COO^- \rightarrow [X \cdot COOCH_3]^{(n-1)+},$$

where X represents an interlamellar cation.

In the present study, X is nominally Al, and, if the cation present is actually Al^{3+}, a large amount of acetate should have formed because, although the relevant stability constants are unknown, those for indium acetates are large (Sunden, 1953). Those for Al should therefore be even larger because of the smaller ionic radius of Al^{3+}. In fact, in aqueous environments Al^{3+} is both hydrated and hydrolyzed. Illustratively, therefore,

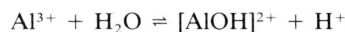

$$Al^{3+} + H_2O \rightleftharpoons [AlOH]^{2+} + H^+$$

$$[AlOH]^{2+} + H_2O \rightleftharpoons [Al(OH)_2]^+ + H^+;$$

hence, basic acetates, e.g., $[AlOHCOOCH_3]^+$, rather than acetates should form, which should be even more stable. This proposal has been confirmed by X. X. Tennakoon and J. M. Thomas (Department of Physical Chemistry, Cambridge University, Cambridge, United Kingdom, 1985) by nuclear magnetic resonance that demonstrated directly the presence of such basic aluminum acetates in acid exposed montmorillonite.

That basic acetate formation and equivalent proton liberation is the origin of the chemisorption seems to be established. The expected stability of these acetates provides an immediate explanation of the strength and rapidity of the chemisorption, and limited availability of ionic species accounts for monolayer formation in the process.

With reference to the 100°C peak of the DTG, the underlying sorption process is clearly physical in origin. First, it is significantly reduced in size (desorption) by exposure of the clay to high vacuum at the sorption temperature. Second, it can be eliminated by heating to about 100°C. Third, as has been shown elsewhere (Al Owais *et al.*, 1986), the enthalpy of sorption associated with this peak corresponds very closely to that of condensation of ethanoic acid vapor. Fourth, the BET isotherm provides a semi-quantitative account of the results.

Sorption sites

The only significant issue remaining is that of the site or sites of the sorption and the relationship with the shape of the resulting isotherm. It is well known that polar molecules are not only intercalated by montmorillonites but that they are also adsorbed on the external surfaces of the particles. In montmorillonites the external surface area is about 5% of that available in the interlamellar space so that the maximum external uptake cannot exceed about 10 mg acid/g dry clay. Figures 2 and 3 suggest that such a contribution can not be separated from the total uptake. That external adsorption is negligible with our material is further suggested by (1) the remarkable consistency of plateau height ratio in the isotherms, which is indicative of the formation of one, two, and, subsequently, three internal layers; and (2) the actual plateau uptake value of ~100 mg/g dry clay (35°C) that implies an intercalated "volume" of each layer of slightly greater than 0.1 ml/g dry clay. Because the d(001) value of a one-layer clay is ~12.5 Å and that of the collapsed clay is ~9.5 Å (corresponding to a specific volume of 0.45 ml/g), the 3-Å layer must have volume of about 0.14 ml/g of dry clay. It is thus highly probable that the physical sorption process in our material was, essentially, intercalation and that adsorption on the external surface was negligible.

The isotherms therefore indicate progressive formation of two physically sorbed acid layers of equal size. As noted above, the enthalpy associated with the physical process corresponds to that for condensation of the acid vapor. In addition, the Van't Hoff plot yielding this result was linear over the whole range of formation of the two physical layers. The total absence of any break in this plot is thus evidence that the layers are physically identical.

Successive layers are often assumed to form abruptly, as shown by abrupt changes in 001 spacings (Mooney et al., 1952). Study of the isotherms of Figures 2 and 3 indicates that this process is by no means true in detail. In Figure 2, in the region $0 < p/p^0 < 0.1$, the second (first physical) layer began to form before the first (chemisorbed) layer was complete. Further, inasmuch as sorption of this second layer never reached a clearly defined plateau on the isotherm, the third layer must have begun to form before the second layer was complete. This observation is consistent with earlier work. In 1952, Mooney et al. reported that the change of the 001 spacings was not exactly coincident with the change of slope on their isotherms. Messina (1963) reported variations of 001 spacings as a function of the relative pressure of water that showed 001 spacings to change progressively rather than abruptly. More recently, Annabi-Bergaya et al. (1979) reported similar results for alkanol uptake. They suggested that layer completion does not occur simultaneously for all particles because the particles are inhomogeneous; hence, the simultaneous presence of different 001 spacings indicating mixed layers. Obviously, as they pointed out, differences in structure and composition of particles adds to the effect. The foregoing explanation, however, retains the concept of abrupt expansion of the clay crystal on layer completion.

Like the chemisorption data, those that we have adduced for the alternative physical intercalation are consistent with results derived from other techniques. The TGA approach to isotherm derivation thus is competitive with and, to a large extent, less ambiguous and more informative than other methods currently employed. Perhaps its greatest value lies in the fact that

it can reveal directly a variety of competitive chemisorptive processes; Ballantine et al. (1987) showed for simple amines, as many as three such processes. Such information will be of value in guiding experiment and in elucidating spectroscopic studies, in particular. The present work has shown that in the clay-catalyzed esterification of aliphatic acid by alkene, the reaction probably involves the exceptionally favorable nucleophilic reaction of an alkene carbo-cation with acetate anion.

REFERENCES

Adams, J. M., Lukawski, K. S., Reid, P. I., Thomas, J. M., and Walters, M. J. (1977) Surface and intercalate chemistry of the layered silicates. Part VII. Sequential solid-state transformations of the intercalates formed between cation-exchanged montmorillonites and γ-butyrolactone and γ-valerolactone: *J. Chem. Res.(S)* 20–21.

Al-Owais, A. A., Ballantine, J. A., Purnell, J. H., and Thomas, J. M. (1986) Thermogravimetric study of the intercalation of acetic acid and of water by Al³⁺-exchanged montmorillonite: *J. Mol. Cat.* **35**, 201–212.

Annabi-Bergaya, F., Gruz, M. I., Gatineau, L., and Fripiat, J. J. (1979) Adsorption of alcohol by smectites. I. Distinction between internal and external surfaces: *Clay Miner.* **14**, 249–258.

Ballantine, J. A., Graham, P., Patel, I., Purnell, J. H., Williams, K. J., and Thomas, J. M. (1986) New differential thermogravimetric method using cyclohexylamine for measuring the concentration of interlamellar protons in clay catalysts: in *Proc. Int. Clay Conf., Denver, 1985*, L. G. Schultz, H. van Olphen, and F. A. Mumpton, eds., The Clay Minerals Society, Bloomington, Indiana, 311–318.

Bissada, K. K. and Johns, W. D. (1969) Montmorillonite-organic complexes: gas chromatographic determination of energies of interactions: *Clays & Clay Minerals* **17**, 197–204.

Cloos, P., Laura, R. D., and Badot, C. (1975) Adsorption of ethylenediamine (EDA) on montmorillonite saturated with different cations. V. Ammonium and trimethylammonium montmorillonite: ion-exchange, protonation and hydrogen bonding: *Clays & Clay Minerals* **23**, 417–423.

James, D. W. and Harward, M. E. (1963) Mechanism of ammonia adsorption by montmorillonite and kaolinite: in *Clays and Clay Minerals, Proc. 11th Natl. Conf., Ottawa, Ontario, 1962*, Ada Swineford, ed., Pergamon Press, New York, 301–320.

Laura, R. D. and Cloos, P. (1975) Adsorption of ethylenediamine (EDA) on montmorillonite saturated with different cations. IV. Aluminum, calcium and magnesium montmorillonite: protonation, ion-exchange, co-ordination, and hydrogen bonding: *Clays & Clay Minerals* **23**, 343–348.

Messina, M. L. (1963) Expansion of fractionated montmorillonites under various relative humidities: *Clays & Clay Minerals* **21**, 617–632.

Mooney, R. W., Keenan, A. G., and Wood, L. A. (1952) Adsorption of water vapor by montmorillonite. II. Effect of exchangeable cations and lattice swelling as measured by X-ray diffraction: *J. Amer. Chem. Soc.* **74**, 1371–1374.

Russell, J. D. (1965) Infra-red study of the reactions of ammonia with montmorillonite and saponite: *Trans. Faraday Soc.* **61**, 2284–2294.

Sunden, N. (1953) The complex chemistry of the indium ion. I. The acetate, formate, propionate and glycolate systems: *Svensk. Kem. Tidskr.* **65**, 257–274.

ORGANICS ON CLAYS

Proceedings of the International Clay Conference, Denver, 1985, L. G. Schultz, H. van Olphen, and F. A. Mumpton, eds.,
The Clay Minerals Society, Bloomington, Indiana, 343–351 (1987).

CLAY-ORGANIC INTERACTIONS:
PROBLEMS AND RECENT RESULTS

GERHARD LAGALY

Institut für Anorganische Chemie der Universität Kiel, Olshausenstraße 40
D-2300 Kiel, Federal Republic of Germany

Abstract—The adsorption of organic molecules by hydrophobic and hydrophilic clay minerals is more strongly governed by structural aspects than previously assumed. In this review this structural dependence is illustrated (1) by methanol molecules on Li- and Ca-smectite, (2) the clustering of polar molecules (ethanol, formamide, dimethyl sulfoxide) in alkylammonium smectites and vermiculites, and (3) the adsorption of methanol-benzene mixtures by alkylpyridinium smectites. The reactions of clay minerals with small, but complicated organic compounds and with monomeric and polymeric anions have also been reported recently, such as interactions with paraquat, diquat, thionine, parathione, purines, pyrimidines, nuclein bases, adenosine mono- and triphosphate, and chelated Co-complexes.

Interesting new results on polymer clay interactions have been reported for polyanion adsorption, e.g., polyacrylate and anionic polyacrylamides, and behavior of smectites towards polycations is explained by the reactions with ionenes. The protein lysozyme can be intercalated by a disaggregation-reaggregation mechanism. An example of enzyme immobilization is the binding of glucose oxidase by sodium and hexadecyl trimethylammonium montmorillonite. Organic molecules of low and high molecular weight exert a large influence on rheological properties. This influence is illustrated by the effect of cationic and anionic surfactants on the viscosity of kaolinite dispersions and of polyvinyl alcohol on dispersions of sodium montmorillonite.

Key Words—Alkylammonium, Clay-organic, Enzyme, Kaolinite, Polymer, Rheological properties, Smectite.

INTRODUCTION

Clay minerals bind numerous organic compounds by different types of interactions. The interactions have been classified as adsorption, ion exchange, and intercalation. They may take place on both external and internal surfaces, or they may be restricted to the external surface. This simplified classification may give the impression that clay-organic reactions are easy to understand. The following examples, reported in the last four years, illustrate the complexity and the interplay of the different forces that govern the binding of organic materials on clay surfaces.

STRUCTURAL ASPECTS OF
THE ADSORPTION OF
ORGANIC MOLECULES

In dealing with clay-alcohol adsorption, Annabi-Bergaya *et al.* (1981) wrote, "The intermolecular forces between adsorbed species have been generally underestimated in studies of adsorption processes by smectites. This type of interaction is important not only for molecules within the interlamellar space but also for their occlusion in micropores." This statement was based on the behavior of methanol molecules on smectite surfaces which carried different types of gegen ions (Figures 1a and 1b). The methanol molecules tended to maintain orientations similar to those in the solid and, also likely, the liquid forms. Intermolecular interactions had to compete with the solvation power of the gegen ions. A liquid-like association was maintained in the presence of lithium ions. In contrast, solvation shells of methanol molecules formed around the calcium ions as a result of the stronger electrical forces arising from the divalent cations.

Lateral interactions between adsorbed species should be of decisive importance for all strongly associated liquids. The interactions between liquid molecules must compete with the surface forces that tend to break the association. Systems are conceivable in which the self-association of liquid molecules is promoted by distinct surface structures to an extent where it can resist the dissociation. On surfaces primed with alkyl chains, polar molecules orient themselves around the alkyl chains in a way that optimizes the screening of the charges. This effect favors the clustering of polar molecules near the hydrocarbon moieties (Figure 1d). Thus, polar molecules such as ethanol, formamide, and dimethyl sulfoxide sorbed by alkylammonium smectites and vermiculites must cluster in the interlayer space between the alkyl chains and on the external surfaces (Lagaly and Witter, 1982; Lagaly *et al.,* 1983).

ADSORPTION FROM BINARY
LIQUID MIXTURES

The foregoing concept was clearly demonstrated by Dékány *et al.* (1985a). Smectites were made hydro-

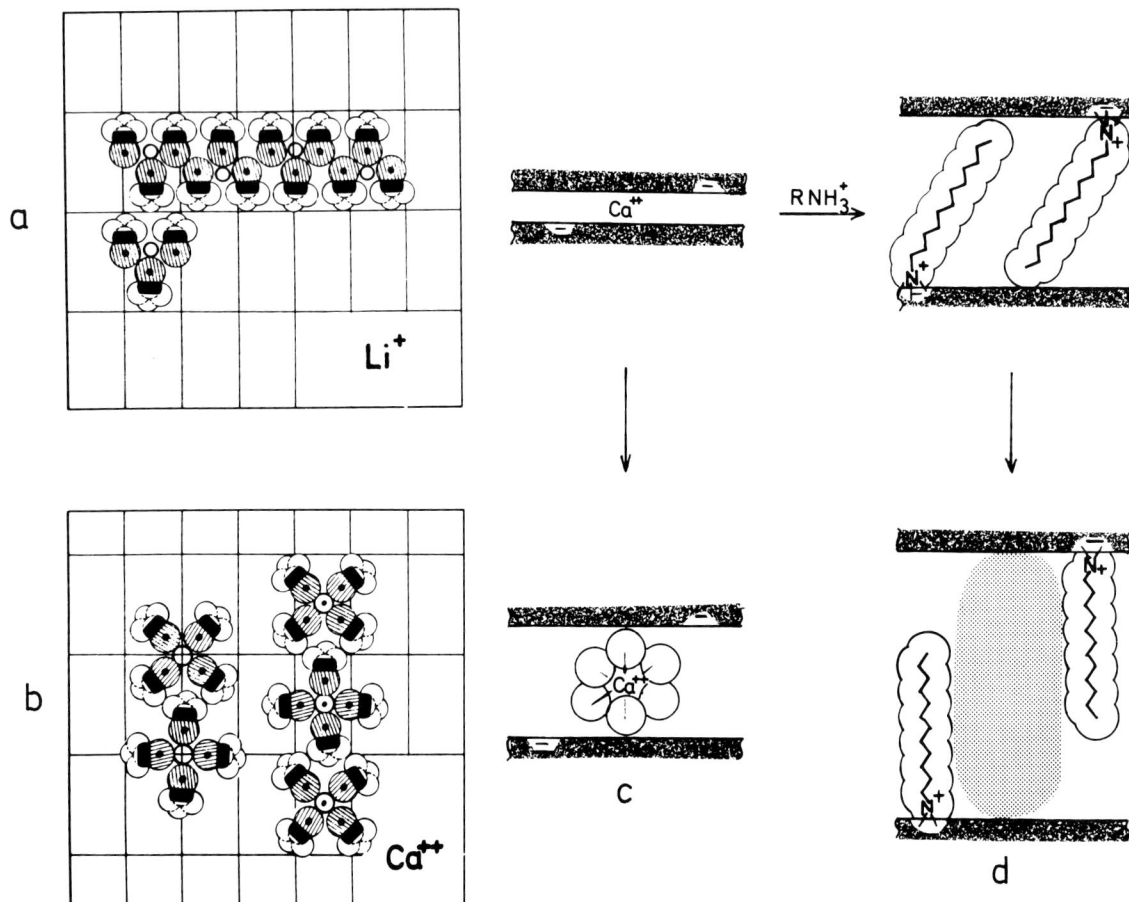

Figure 1. Polar molecules on clay mineral surfaces: (a–c) Methanol molecules on the surface of smectites. In the presence of Li$^+$ ions (a), they are associated as in the solid and (probably) liquid states but they form hydration shells (b, c) around Ca^{2+} ions (Annabi-Bergaya et al., 1981). (d) On hydrophobic surfaces, polar molecules can aggregate to clusters between the alkyl chains.

phobic by coating them with different types of alkyl-ammonium and alkylpyridinium ions. The hydrophobicity increased with the degree of coverage and the alkyl chain length. The related changes of the adsorption capacity were evaluated from the surface excess isotherms of binary mixtures.

Smectites covered with alkylpyridinium ions were contacted with binary mixtures of methanol and benzene in molar ratios ranging from $x_1 = 0$ (pure benzene) to $x_1 = 1$ (pure methanol). Because both compounds were adsorbed, the changes of the methanol concentration did not measure the amount of methanol adsorbed but determined the surface excess, $n_1^{\sigma(n)}$. The function $n_1^{\sigma(n)} = f(x_1)$ described the surface excess isotherm (Figure 2). The real composition of the adsorbed phase (n_1^s, n_2^s) was derived from this isotherm by methods of Schay and Nagy (cf. Lagaly and Witter, 1982). Figure 2b shows the surface excess isotherms $n_1^{\sigma(n)} = f(x_1)$ and the real adsorption isotherms (individual isotherms) $n_1^s = f(x_1)$ and $n_2^s = f(x_1)$ for a mont-

morillonite with different degrees of coverage by hexadecyl pyridinium ions. The molar ratio benzene : methanol of the plateaux increased from $n_2^s/n_1^s = 0.25$ (sample I) to 2.33 (sample V) and expressed the gain in hydrophobicity with increasing coverage.

The volume of the adsorption phase was easily obtained from n_1^s and n_2^s and the molecular volumes of methanol and benzene. The "free interlayer volume" was derived from the basal spacings and the volume of the interlayer alkylpyridinium ions. Even when the error in estimating the molecular volumes from the liquid densities and other possible errors (interlayer composition, purity of smectite, etc.) were taken into consideration, the free interlayer volume exceeded considerably the volume of the adsorption phase. This intriguing result suggests that only a part of the interlayer liquid molecules belonged to the adsorption phase. The remaining space in the interlayer must have been filled with methanol and benzene molecules in a ratio similar to that in the equilibrium liquid mixture. Such

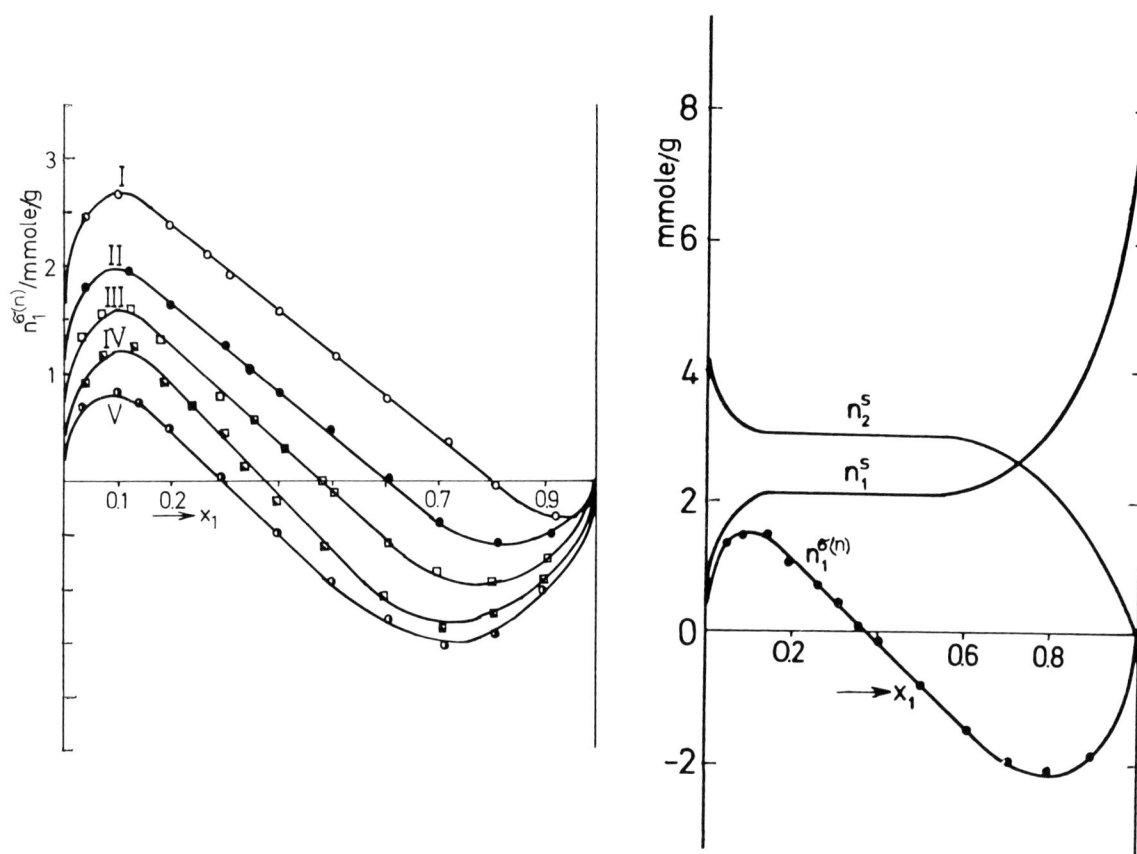

Figure 2. Adsorption of methanol and benzene on hexadecyl pyridinium-sodium montmorillonite, from Mád, Hungary: (a) Surface excess $n_1^{\sigma(n)}$ as a function of the equilibrium mole fraction x_1 of methanol. Mole fractions of hexadecyl pyridinium ions on the surface: I = 0.22, II = 0.44, III = 0.62, IV = 0.90, V = 1.00. (b) Surface excess isotherm $n_1^{\sigma(n)}$ and the individual isotherms showing the amount of adsorbed methanol (n_1^s) and benzene (n_2^s); hexadecyl pyridinium-sodium montmorillonite, mole fraction of hexadecyl pyridinium ions = 0.81 (Dékány et al., 1985a).

a process contributes essentially to a positive entropy change which, at longer alkyl chains, becomes imperative for swelling (cf. Dékány et al., 1985b).

These recent studies increase our understanding of the role of structure during the swelling of three-layer clay minerals in organic liquids.

INTERACTION WITH SMALL 'COMPLICATED' ORGANIC MOLECULES

The binding of small 'complicated' organic molecules or ions has recently received increasing attention for different reasons. Studies on biocide-clay interactions have been encouraged by the present concern about pollution control. The possible adsorption of dangerous compounds by clays also has stimulated many researchers to study clay systems. Furthermore, facts and speculations about the role of clay minerals in the earlier states of life have provoked investigations of the reaction of clays and biologically important molecules.

Usually, the factors governing the binding of com-

pounds on clays are concentration (or vapor pressure), pH, temperature, type of interlayer cation, and type of clay mineral. Additional factors which may be important are the true nature of the mineral within its subgroup, the magnitude and distribution of layer charges, the degree of dispersion, the specific procedure used during the pretreatment of the clay, and the presence of salts in and the ionic strength of the medium (e.g., paraquat, diquat, and thionine on smectites (Narine and Guy, 1981; nuclein bases, Samii and Lagaly, 1987). Enhanced levels of adsorption may be attained by surface-induced associations. Organic cations bound at the surface enhance the adsorption of other molecules ("surface condensation") or act as nuclei for stacking arrangements (Narine and Guy, 1981; Mingelgrin and Tsvetkov, 1985, for organophosphate esters). Synergistic effects may also be observed; e.g., increased adsorption of adenosine monophosphate in the presence of adenosine triphosphate was reported by Graf and Lagaly (1980). A second well-documented illustration is the adsorption of purines, pyrimidines,

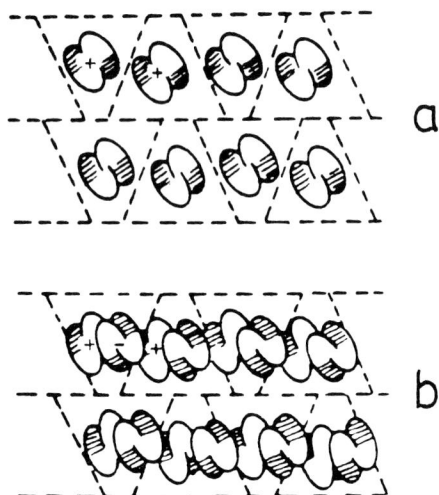

Figure 3. Schematic representation of racemic adsorption of an optically active chelated Co(III) complex (bis 1-(2-pyridylazo)-2-naphthol-cobalt(III) chloride) on sodium montmorillonite. (a) Close-packed monolayer formed by adsorption from solutions of the (+) or (−) isomers; (b) close-packed layer of (+) and (−) isomers formed by adsorption from the racemic solution.

and nuclein bases by smectites and their pronounced co-adsorption. Thymine that is adsorbed in only trace amounts from pure solutions can be adsorbed in greater amounts in the presence of the complementary base adenine, similar to base-pairing in nucleic acids. The co-adsorption of the non-complementary pair thymine-cytosine is less pronounced (Lailach and Brindley, 1969; Samii and Lagaly, 1987).

An interesting co-adsorption phenomenon was observed by Yamagishi and Fujita (1984) and Yamagishi (1987). Optically active chelated Co^{3+} complexes were adsorbed on montmorillonite from racemic solution and solutions of the enantiomeric forms. The amount adsorbed from the racemic solution was twice that from solutions of the pure enantiomers. If, as with the latter material, all surface sites were occupied by the same enantiomer, e.g., (+) forms, further (+) forms were unable to fit between the adsorbed complexes. In "racemic" adsorption, the (−) isomers easily fit between the (+) isomers (Figure 3).

For divalent organic cations, the question arises whether the fit or misfit between the charge distances in the adsorbate and in the silicate surface contributes to the adsorption affinity. For "simple" compounds, such as short-chain alkyldiammonium ions and their methyl derivatives, the selectivity increases with increasing substitution, e.g., from the ethylene diammonium ion to the hexamethyl derivative. The main cause of the increased free enthalpy of exchange is the increased charge delocalization in the ion (Maes et al., 1980). The concept of charge pattern interactions (Lagaly, 1986) substantiates these results. The charge de-

localization in the end groups of the ions improves the fit to the surface charge pattern.

Presently, our knowledge about the desorption of organic compounds (Narine and Guy, 1981), especially biocide molecules, is still very limited. The desorption of residues bound by surface soils pose possible threats to the environment. Few studies have been made of the stability of adsorbed organic molecules or ions. The increased acidity of interlayer water can accelerate hydrolysis of organic compounds, the rate of which is dependent on the moisture content. El-Amamy and Mill (1984) observed maximal rates for montmorillonite with 20–50% water.

The non-biological degradation of compounds in soils has two aspects: chemical degradation or photodegradation (Margulies and Rozen, 1985) can limit the application of certain agrochemicals to soils, but at the same time, the possibility of accumulation of dangerous residues in soils is reduced.

INTERACTIONS WITH POLYMERS

Studies on the binding of organic anions by clay minerals have generally been neglected, because anion adsorption is thought to be impeded by the high negative charge of the layers. Positive edge charges are often considered as the sole possible binding sites for anions. The binding is then expected to be highly pH-dependent and to cease at pH above the p.z.c. (pH ~5) (cf. Welzen et al., 1981). Recent papers clearly prove that the pH-dependence is not as expected, as demonstrated in Figure 4 for the adsorption of oxalic acid, polyacrylate, polyacrylamide, and sulfonated polyacrylamide (Siffert and Espinasse, 1980; Hollander et al., 1981). A straightforward explanation of these reactions, however, is often difficult. For example, if the polymer charge is sensitive to pH, the pH-dependence of polyanion adsorption can also be influenced by conformational changes. As seen in Figure 4c, the adsorption of polyanions on an adsorbent with high negative charge increases with the ionic strength. This increase is explained by the compression of the diffuse ionic layers at higher salt concentrations which reduces the electrostatic repulsion between the surface and the polyanions. Figure 4d shows that the adsorption increases with increasing negative charge density of hydrolyzed polyacrylamide. The high affinity of clay minerals towards neutral polyacrylamide is, at least partially, an ionic phenomenon. Due to the increased acidity of the surface water around the cations, the amide group is protonated and is held by negative surface charges (Siffert and Espinasse, 1980).

The results presented in Figures 4a–4d clearly show that binding mechanisms other than gegen-ion binding must be taken into consideration, such as the formation of hydrogen bonds, the exchange of structural OH groups, and the complexing of surface metal ions. Formation of chelate complexes between surface Al-ions

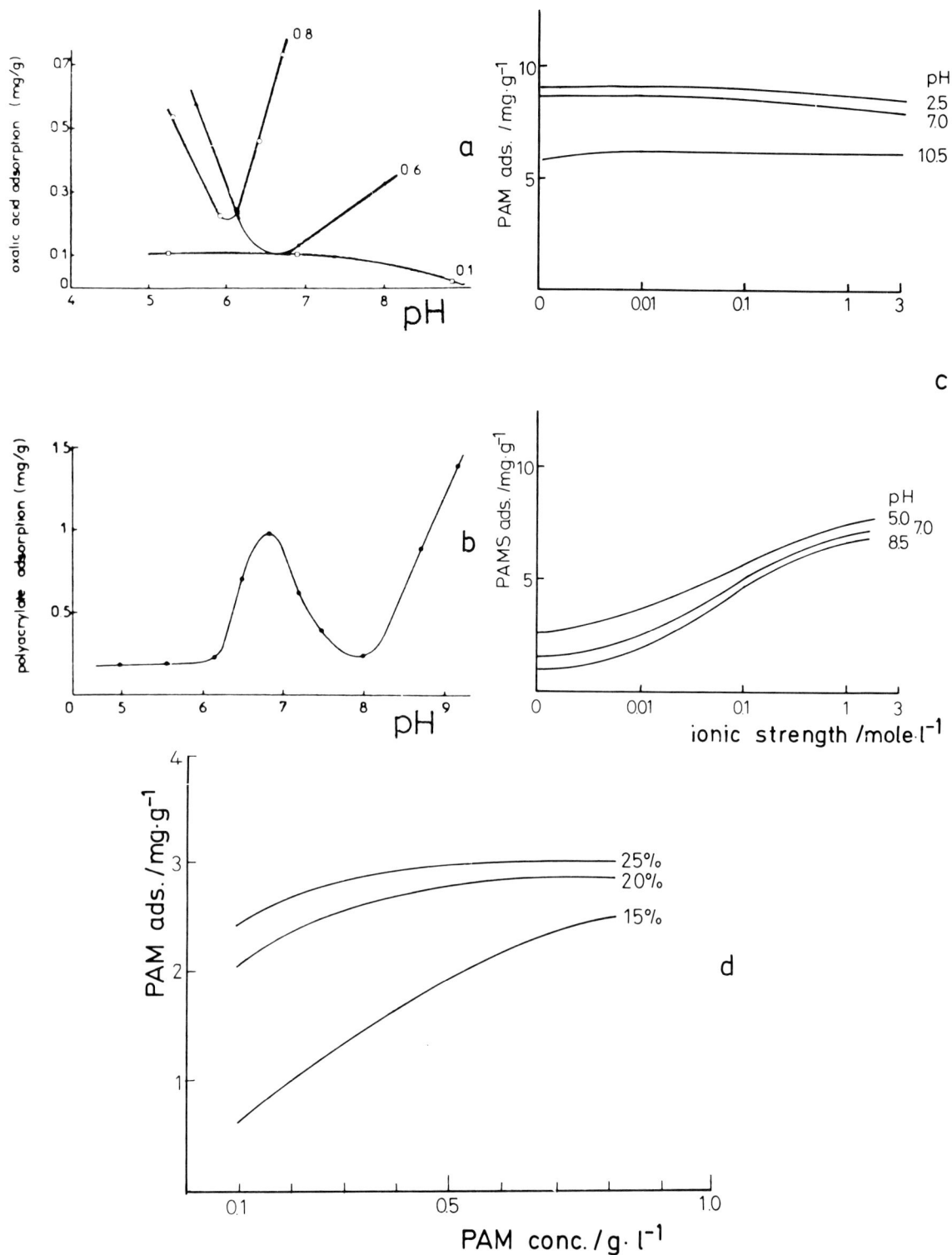

Figure 4. Influence of pH on the adsorption of anionic compounds by smectites: (a) oxalic acid adsorption onto sodium montmorillonite for different initial acid concentrations (g/liter) (Siffert and Espinasse, 1980); (b) polyacrylate adsorption onto sodium montmorillonite (Siffert and Espinasse, 1980); (c) adsorption of polyacrylamide (PAM) and sulfonated polyacrylamide (PAMS) on sodium kaolinite at different pHs and ionic strength (NaCl) (Hollander *et al.,* 1981); (d) influence of the degree of hydrolysis (increasing anionic charge) on the adsorption of polyacrylamide (PAM) onto sodium montmorillonite, pH ~7.5 (Stutzman and Siffert, 1977).

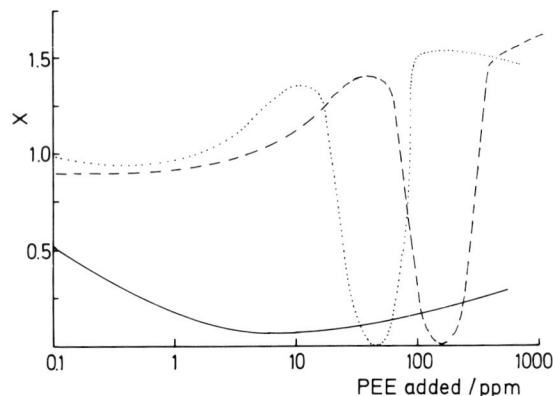

Figure 5. Flocculation of montmorillonite by the cationic polymer PEE (ethanolamine-epichlorhydrine polycondensate); X = amount of montmorillonite + PEE (ads.) in solution/total amount of montmorillonite; natural water (~200 μs/cm, ——) or deionized water (<1 μs/cm, -----, ·····), initial concentration of montmorillonite: 100 mg/liter (——, ·····) and 300 mg/liter (-----) (Kim *et al.*, 1983).

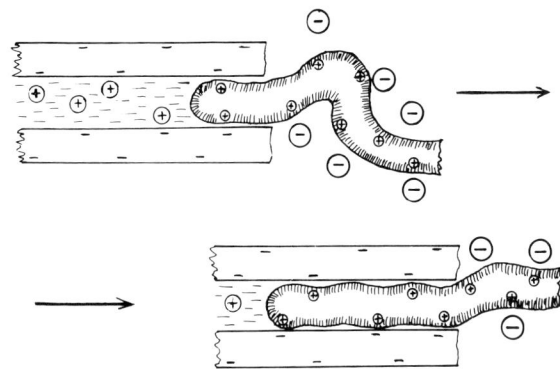

Figure 6. Penetration of polycations into interlayer spaces. When penetration proceeds, increasing number of contacts reduce the mobility of the polycations and impede complete surface coverage. In the extreme, polycations remain accumulated near crystal edges.

and carboxylate groups is probably the chief mechanism for binding organic acid anions. Chelation explains why the adsorption continues into the alkaline region. In the presence of divalent cations, anions can be bound by cation bridges. The adsorption behavior becomes particularly complex when anionic surfactants are also present (Bocquenet and Siffert, 1984).

The sites for anion adsorption (except for cation bridges) are predominantly on the surface of the crystal edges. Increasing adsorption with decreasing particle size is expected, but it is not generally observed (Herrmann and Lagaly, 1985). The adsorption can also be sensitive to the degree of dispersion. In the presence of polymers the levels of adsorption can be affected by the solid-to-liquid ratio (Hollander *et al.*, 1981). This effect is probably due to major interactions between the particles themselves and to nonequilibrium states for the adsorbed macromolecules—a phenomenon commonly observed for polymer adsorption.

Despite numerous practical applications, the number of papers on clay polymer interactions is limited. Relevant data were reported by Burchill *et al.* (1983). The addition of polymers to colloidal clay dispersions causes flocculation at low polymer concentration and stabilization at higher polymer concentration. After the addition of small amounts of polyacrylamide to kaolinite, Nabzar *et al.* (1984) noted a region of stabilization, followed by regions of flocculation and stabilization at higher polymer concentrations. The first stabilization resulted from the disrupture of edge(+)/face(−) contacts by small amounts of polymer adsorbed. Flocculation was caused by bridging, and at still higher concentration, sterical stabilization occurred.

Polymers are used extensively as flocculating agents.

In practical applications polyanions are more effective flocculating agents than polycations. Polyanions are attached to the clay particles at a few sites, and larger parts of the macromolecules remain free in solution and bridge between neighboring particles, thereby promoting flocculation. Polycations are strongly adsorbed on and almost entirely attached to the negative surface and do not bridge between the particles as easily as polyanions. Flocculation then occurs as a consequence of the elimination of the diffuse character of the double layer when organic cations are adsorbed. This type of flocculation requires very distinct conditions of salt and clay concentrations (Howard and Leung, 1981; Kim *et al.*, 1983; Figure 5).

Polyanions are adsorbed exclusively on external surfaces. Polycations can penetrate between the layers, but complete coverage of the internal surfaces is rarely achieved. The increasing number of contacts to the surface retards and, eventually, impedes the penetration of the polycation (Figure 6). Even ionenes $\{-N^+(CH_3)_2-(CH_2)_x\}_n$, where x = 4–12, that easily penetrate between the layers, did not completely cover the internal surfaces; the coverage seldom exceeded 50–60% (Lagaly, 1986).

A complete surface coverage of smectite can be attained by disaggregation-reaggregation in the presence of sodium ions. An instructive example is the adsorption of the protein lysozyme (Larsson and Siffert, 1983; Figure 7). As shown schematically in Figure 7, crystals of sodium smectite in dilute suspensions disintegrated into very thin lamellae and isolated silicate layers (a). Lysozyme molecules were then adsorbed on the surface of these isolated layers or on the external surface of the lamellae (b). The positively charged lysozyme molecules reduced the negative surface charge and promoted the reaggregation of the lamellae or silicate layers to interstratified crystals (c). The silicate layers were held together by lysozyme molecules more strongly

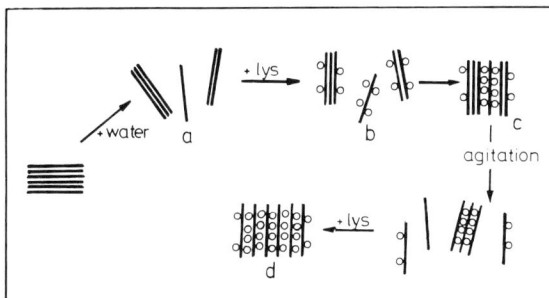

Figure 7. Lysozyme (lys) adsorption by a disaggregation-reaggregation mechanism (Larsson and Siffert, 1983).

Figure 8. Interaction of a cationic surface active agent, cetyltrimethylammonium bromide (CTA$^+$Br$^-$), with sodium kaolinite: (a) adsorption; (b) Bingham yield stress as a function of equilibrium CTA$^+$ concentration at pH = 3.3 and 10 (Welzen et al., 1981).

than by sodium ions. Further agitation then broke the interlamellar bonds of the sodium interlayers and created additional surfaces for lysozyme adsorption. Finally, complete saturation with lysozyme was attained (d).

Polymer adsorption commonly appears to be irreversible in nature. Macromolecules become attached at many separate sites on the surface and cannot be removed simultaneously from all sites before detached segments readsorb. If vacated sites could be blocked by competing agents immediately after desorption of the segments, some displacement of the polymer might be attained. Dodson and Somasundaran (1984) reported that polyacrylamide and its 9% hydrolyzed derivative could be desorbed to an extent of 20–50% by sodium poly- and metaphosphates.

The interaction of clay minerals with enzymes was extensively studied in the 1950s. The number of papers diminished in the following years, however, improved analytical procedures and availability of many enzymes have recently stimulated such studies. An example is the immobilization of glucose oxidase by montmorillonite (Garwood et al., 1983). Sodium montmorillonite binds this enzyme between the layers below the i.e.p. by an ionic mechanism and denaturates the enzyme. If, however, the enzyme was bound by non-ionic forces on hexadecyl trimethylammonium montmorillonite, it retained its activity. Another example, reported by Mortland (1984), was the deamination of glutamic acid by pyridoxal phosphate, which was promoted by Cu^{2+}-montmorillonite. The Cu^{2+}-smectite-pyridoxal phosphate complex acted like a pseudoenzyme, and the silicate framework substituted for the apo-enzyme. These studies demonstrated the inhibiting or promoting effect of clays on enzyme activity.

INFLUENCE OF ORGANIC MATERIALS ON RHEOLOGICAL PROPERTIES

The influence of organic molecules of low and high molecular weight on rheological properties is sometimes underestimated. Despite many bewildering ex-

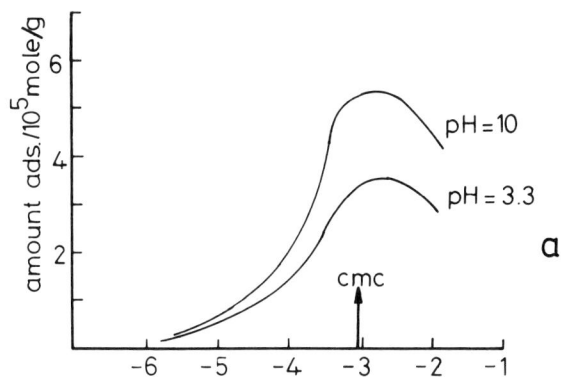

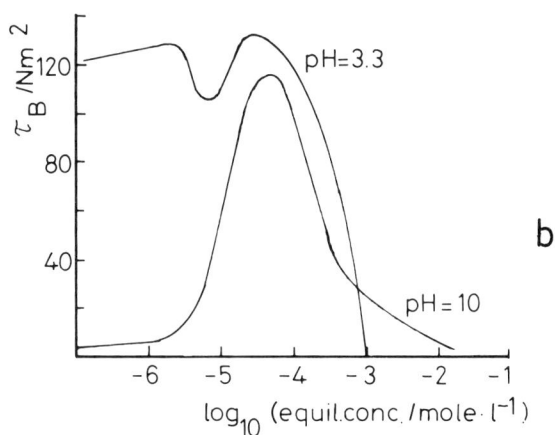

perimental results primarily due to incomplete or inadequate rheological measurements, some basic principles are discernible. For example, the viscosity of dispersions of hydrophobic smectites in polar organic solvents (tetradecylammonium montmorillonite in dimethyl sulfoxide) could be increased considerably by certain salts, e.g., CsI and LiCl (Sander and Lagaly, 1983). This effect was explained by the influence of the salts on the liquid structure.

The many effects of cationic and anionic surfactants on the viscosity of kaolinite dispersions are illustrated by the work of Welzen et al. (1981). As shown in Figure 8, the cationic surface active agent, cetyl trimethylammonium bromide (CTABr) influenced the Bingham yield stress of a sodium kaolinite dispersion at pH 3.3. The curious minimum at 10^{-5} mole CTABr/liter is apparently due to the attraction between the alkyl chains of the surfactant in the following way (Figure 9). At low CTABr concentrations and low levels of adsorp-

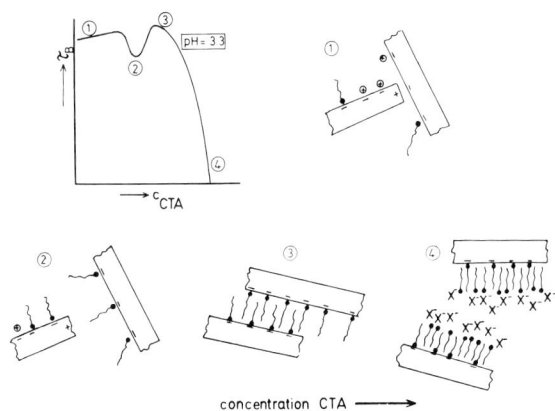

Figure 9. Different types of particle-particle interactions (in acidic medium) with increasing concentration of cetyltri-methylammonium ions (CTA).

tion, edge-to-face contacts between the particles (1) are the cause of a high yield stress; with increasing CTABr concentration the negative charges of the layers are compensated by surfactant cations and the number of edge-to-face contacts is strongly reduced (2). The yield stress decreases. At a slightly higher coverage by CTA^+ cations, new contacts form by interacting chains (3), and the yield stress increases again. Finally, the mono-layer is completed by adsorption of CTA^+ cations to-gether with their gegen ions (4). All particles become negatively charged, and the Bingham yield stress falls to zero. In the same way, the behavior at higher pHs and also with anionic surfactants is explained by the influence of the surface active agents on the contacts between the particles.

The effect of small amounts of organic materials is further illustrated by the addition of polyvinyl alcohol (PVA) to sodium montmorillonite dispersions (Heath and Tadros, 1983). The addition of PVA to a 4% (wt/v) dispersion (at pH 7 and in the presence of 0.1 mole NaCl/liter) gradually increased the pseudoplastic flow with no significant thixotropy. At ~0.5% PVA (wt/v), a maximum was reached after which the yield was reduced and became zero at 1% PVA. One conclusion is that relatively small changes in the amounts of or-ganic materials adhering to clay minerals can strongly influence the flow behavior of clay water systems and the gel strength.

PRACTICAL APPLICATIONS

Various practical applications of organic bentonites were summarized by Jones (1983). In the quest for additional applications, novel adsorbent materials and catalysts have been prepared from smectites, e.g., tri-ethylene diamine ("DABCO") derivatives (van Leem-put *et al.,* 1983; Mortland and Berkheiser, 1976) and pillared clays (Manos *et al.,* 1984; Pinnavaia, 1983; Pinnavaia *et al.,* 1985).

Some success has been attained concerning the res-olution of optically active metal complexes (Yama-gishi, 1981, 1983, 1986). A recent example of stabili-zation of emulsions by solid particles is the stabilization of O/W emulsions by sodium montmorillonite in the presence of monoglycerides without hydrophilic sur-factants (Tsugita *et al.,* 1983).

REFERENCES

Annabi-Bergaya, F., Cruz, I. M., Gatineau, L., and Fripiat, J. J. (1981) Adsorption of alcohols by smectites. IV. Models: *Clay Miner.* **16,** 115–122.

Bocquenet, Y. and Siffert, B. (1984) Polyacrylamide ad-sorption onto kaolinite and illite in the presence of sodium dodecylbenzene sulfonate: *Colloids and Surfaces* **9,** 147–161.

Burchill, S., Hall, P. L., Harrison, R., Hayes, M. H. B., Lang-ford, J. I., Livingston, W. R., Smedley, R. J., Ross, D. K., and Tuck, J. J. (1983) Smectite-polymer interactions in aqueous systems: *Clay Miner.* **18,** 373–397.

Dékány, I., Szántó, F., Weiss, A., and Lagaly, G. (1985a) Interlamellar liquid sorption on hydrophobic silicates: *Ber. Bunsenges. Physik. Chem.* **89,** 62–67.

Dékány, I., Szántó, F., and Nagy, L. G. (1985b) Sorption and immersional wetting on clay minerals having modified surface: *J. Colloid Interf. Sci.* **103,** 321–331.

Dodson, P. J. and Somasundaran, P. (1984) Desorption of polyacrylamide and hydrolyzed polyacrylamide from ka-olinite surface: *J. Colloid Interf. Sci.* **97,** 481–487.

El-Amamy, M. M. and Mill, Th. (1984) Hydrolysis kinetics of organic chemicals on montmorillonite and kaolinite sur-faces as related to moisture content: *Clays & Clay Minerals* **32,** 67–73.

Garwood, G. A., Mortland, M. M., and Pinnavaia, T. J. (1983) Immobilization of glucose oxidase on montmorillonite clay: hydrophobic and ionic modes of binding: *J. Molecular Ca-talysis* **22,** 153–163.

Graf, G. and Lagaly, G. (1980) Interaction of clay minerals with adenosine-5-phosphates: *Clays & Clay Minerals* **28,** 12–18.

Heath, D. and Tadros, Th. F. (1983) Influence of pH, elec-trolyte and polyvinyl alcohol addition on the rheological characteristics of aqueous dispersions of sodium mont-morillonite: *J. Colloid Interf. Sci.* **93,** 307–319.

Herrman, H. and Lagaly, G. (1985) ATP-clay interactions. *Proc. 6th European Clay Conf., Prague, 1983,* 269–277.

Hollander, A. F., Somasundaran, P., and Gryte, C. C. (1981) Adsorption of polyacrylamide and sulfonated polyacryl-amide on Na-kaolinite: in *Adsorption from Aqueous Solu-tions,* P. H. Tewari, ed., Plenum Press, New York, 143–162.

Howard, G. J. and Leung, W. M. (1981) Polyvinylpyridine as flocculating agents: *Colloid Polymer Sci.* **259,** 1031–1039.

Jones, T. R. (1983) The properties and uses of clays which swell in organic solvents: *Clay Miner.* **18,** 399–410.

Kim, H. S., Lamarche, C., and Verdier, A. (1983) Étude des interactions entre un polyélectrolyte cationique de type am-monium tertiaire et une suspension de bentonite: *Colloid Polym. Sci.* **261,** 64–69.

Lagaly, G. (1986) Smectitic clays as ionic macromolecules: in *Developments in Ionic Polymers, vol. 2,* A. D. Wilson, ed., Elsevier, London 77–140.

Lagaly, G. and Witter, R. (1982) Clustering of liquid mol-ecules on solid surfaces: *Ber. Bunsenges. Physik. Chem.* **86,** 74–80.

Lagaly, G., Witter, R., and Sander, H. (1983) Water on hydrophobic surfaces: in *Adsorption from Solution,* R. H.

Ottewill, C. H. Rochester, and A. L. Smith, eds., Acad. Press, London, 65–77.

Lailach, G. and Brindley, G. W. (1969) Specific co-adsorption of purines and pyrimidines by montmorillonite. *Clays & Clay Minerals* **17**, 95–100.

Larsson, N. and Siffert, B. (1983) Formation of lysozyme-containing crystals of montmorillonite: *J. Colloid Interf. Sci.* **93**, 424–431.

van Leemput, L., Stul, M. S., Maes, A., Uytterhoeven, J. B., and Cremers, A. (1983) Surface properties of smectites exchanged with mono- and biprotonated 1.4-diazabicyclo(2,2,2)-octane: *Clays & Clay Minerals* **31**, 261–268.

Maes, A., van Leemput, L., Cremers, A., and Uytterhoeven, J. B. (1980) Electron density distribution as a parameter in understanding organic cation exchange in montmorillonite: *J. Colloid Interf. Sci.* **77**, 14–20.

Manos, Ch. G., Mortland, M. M., and Pinnavaia, T. J. (1984) Tris(acetylacetonato)silicon(IV) binding to montmorillonite and hydrolysis to interlayer silicic acid: *Clays & Clay Minerals* **32**, 93–98.

Margulies, L. and Rozen, H. (1985) Energy transfer at the surface of clays. A method of preventing the photodegradation of agrochemicals: *Abstracts, Internat. Clay Conference, Denver, Colorado, 1985*, p. 153.

Mingelgrin, U. and Tsvetkov, F. (1985) Surface condensation of organo-phosphate esters on smectites: *Clays & Clay Minerals* **33**, 62–70.

Mortland, M. M. (1984) Deamination of glutamic acid by pyridoxal phosphate-Cu^{2+}-smectite catalysts: *J. Molecul. Catal.* **27**, 143–155.

Mortland, M. M. and Berkheiser, V. (1976) Triethylenediamine-clay complexes as matrices for adsorption and catalytic reactions: *Clays & Clay Minerals* **24**, 61–63.

Nabzar, L., Pefferkorn, E., and Varoqui, R. (1984) Polyacrylamide sodium kaolinite interactions: flocculation behavior of polymer clay suspensions: *J. Colloid Interf. Sci.* **102**, 380–388.

Narine, D. R. and Guy, R. D. (1981) Interaction of some large organic cations with bentonite in dilute aqueous systems: *Clays & Clay Minerals* **29**, 205–212.

Pinnavaia, T. J. (1983) Intercalated clay catalysts: *Science* **220**, 365–371.

Pinnavaia, T. J., Tzou, M.-Sh., and Landau, St. D. (1985) New chromia pillared clay catalysts: *J. Amer. Chem. Soc.* **107**, 4783–4785.

Samii, A. M. and Lagaly, G. (1987) Adsorption of nuclein bases on smectites: in *Proc. Int. Clay Conf., Denver, 1985,* L. G. Schultz, H. van Olphen, and F. A. Mumpton, eds., The Clay Minerals Society, Bloomington, Indiana, 363–369.

Sander, H. and Lagaly, G. (1983) Viskositätssteuerung organischer Bentonitdispersionen durch Salze: *Keram. Z.* **35**, 584–587.

Siffert, B. and Espinasse, P. (1980) Adsorption of organic diacids and sodium polyacrylate onto montomorillonite. *Clays & Clay Minerals* **28**, 381–387.

Stutzmann, Th. and Siffert, B. (1977) Contribution to the adsorption mechanism of acetamide and polyacrylamide onto clays: *Clays & Clay Min.* **25**, 392–406.

Tsugita, A., Takemoto, S., Mori, K., Yoneya, T., and Otani, Y. (1983) Studies on O/W emulsions stabilized with insoluble montmorillonite-organic complexes: *J. Colloid Interf. Sci.* **95**, 551–560.

Welzen, I. T. A. M., Stein, H. N., Stevels, J. M., and Siskens, C. A. M. (1981) The influence of surface-active agents on kaolinite: *J. Colloid Interf. Sci.* **81**, 455–467.

Yamagishi, A. (1981) Optical resolution of metal chelates by use of adsorption on a colloidal clay: *J. Amer. Chem. Soc.* **103**, 4640–4642.

Yamagishi, A. (1983) Clay column chromatography for optical resolution of tris(chelated) and bis(chelated) complexes on a Ru(1.10-phenanthroline) montmorillonite column: *J. Chromatography* **262**, 41–60.

Yamagishi, A. (1987) Clay column chromatography for optical resolution: resolution of aromatic molecules on a lambda-Ru(phen)$_3^{2+}$ montmorillonite column: in *Proc. Int. Clay Conf., Denver, 1985,* L. G. Schultz, H. van Olphen, and F. A. Mumpton, ed., The Clay Minerals Society, Bloomington, Indiana, 329–334.

Yamagishi, A. and Fujita, N. (1984) Racemic adsorption of a bis(chelated) cobalt(III) complex by colloidally dispersed sodium montmorillonite: *J. Colloid Interf. Sci.* **100**, 136–142.

Proceedings of the International Clay Conference, Denver, 1985, L. G. Schultz, H. van Olphen, and F. A. Mumpton, eds.,
The Clay Minerals Society, Bloomington, Indiana, 352–358 (1987).

SPECTROSCOPIC STUDY OF THE SURFACE CHEMISTRY
OF PROFLAVINE ON CLAY MINERALS

J. Cenens,[1] D. P. Vliers,[1] R. A. Schoonheydt,[1] and F. C. De Schryver[2]

[1] Laboratorium voor Oppervlaktechemie, Katholieke Universiteit Leuven
Kardinaal Mercierlaan 92, B-3030 Leuven (Heverlee), Belgium

[2] Departement Scheikunde, Katholieke Universiteit Leuven
Celestijnenlaan 200 F, B-3030 Leuven (Heverlee), Belgium

Abstract—Proflavine, ion-exchanged on synthetic mica-type montmorillonite Barasym SSM-100, hectorite, Laponite and sepiolite in aqueous suspensions, gives in addition to a monoprotonated molecule, a dimer and a diprotonated molecule. The first two molecules are found only on hectorite, Laponite, and sepiolite, whereas all three species are found on Barasym. The first 1–2 μmole of proflavine protonate in contact with 1 g of Barasym. The reaction occurs on the external surface and is similar to the highly selective Cs^+ uptake by illite. Dimerization is the result of a simple concentration effect on the surface and gives rise to metachromatic behavior. No π-electron interaction with surface oxygens need be invoked. On the surface the monomeric, monoprotonated molecules have an extinction coefficient of 58,000 at 453 nm, and their dimerization constant is 1700. The quantum yield of dimer emission is 0.017. The emission of the monomer is quenched by the dimer both on the external surface and on the interlamellar surface by Perrin's mechanism of quenching, in absence of diffusion, with an effective quenching radius of 2.7 nm. At loadings < 1% of the cation-exchange capacity, adsorption takes place mainly on the external surface.

Key Words—Absorption spectroscopy, Adsorption, Cation-exchange capacity, Emission spectroscopy, Hectorite, Montmorillonite, Proflavine, Sepiolite.

INTRODUCTION

The study of dye molecules on clay surfaces has accelerated in recent years because: (1) The interfacial interactions between dyes and surface atoms are reflected in the spectroscopic properties of the adsorbed molecules. Thus, surface properties can be studied. (2) Adsorption is a valuable technique to direct photochemical or photo-catalytic reactions in the desired direction (de Mayo, 1982; Van Damme *et al.*, 1984). (3) Supramolecular assemblies may be synthesized on the surface and may have valuable photochemical properties (Levitz *et al.*, 1984; DellaGuardia and Thomas, 1983).

Cationic dyes have been especially investigated because of the ease of their adsorption by ion exchange. Bergmann and O'Konski (1963) found that methylene blue showed metachromatic behavior in aqueous montmorillonite suspensions even when the exchange solution contained only monomers. This reaction was explained by dye aggregation on the surface, leading to dimers and higher aggregates. These observations were confirmed by Yariv and Lurie (1971) and extended to acridine orange by Cohen and Yariv (1984). The latter authors explained the metachromatic behavior in terms of a strong electronic interaction between the π-electron system of the dye and the valence electrons of the surface oxygens. The reasons for this interpretation are (1) the observation of metachromasy far below mono-

layer coverage; and (2) the absence of metachromasy with rhodamine 6G, in which the $C_6H_5COOC_2H_5$ substituent prevents efficient π-interaction (Grauer *et al.*, 1984) (the formulas of the dyes are given in Figure 1).

Interpretations of metachromasy of dyes on clays in terms of dye aggregation or electronic interactions with surface oxygens are hampered by (1) the site heterogeneity of clay surfaces; (2) side reactions such as protonation; and (3) the absence of emission from dimers or higher aggregates. We have shown the importance of these three factors for proflavine on several clays after freeze-drying (Schoonheydt *et al.*, 1986). In this paper, the study is extended to aqueous suspensions. Small loadings of proflavine were used to avoid flocculation of the clay and residual proflavine in solution.

EXPERIMENTAL

Preparation of the clays

Hectorite (HEC) and Barasym SSM-100 (BS) from the Source Repository of The Clay Minerals Society, Laponite (LAP) from Laporte Industries, and sepiolite (SEP) from Tolsa, Spain, were investigated. BS is a 2:1 layered aluminosilicate consisting of randomly interstratified, expandable and non-expandable dioctahedral layers (Wright *et al.*, 1972). Laponite is a synthetic hectorite. The clays were chosen for their low Fe content and widely different external surface areas, as shown in Table 1 (van Olphen and Fripiat, 1979). The clays

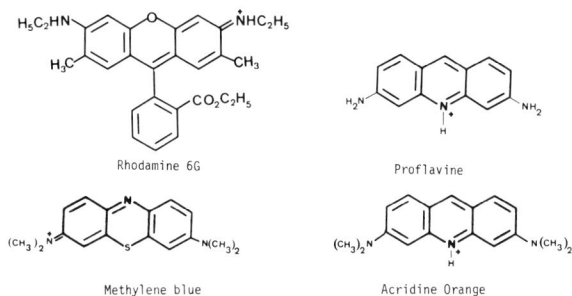

Figure 1.   Formulae of the dye molecules.

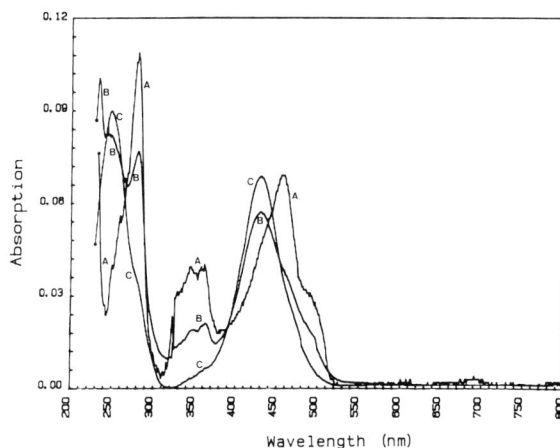

Figure 2.   Absorption spectra of proflavine-Barasym suspensions at loadings of (A) 0.2%, (B) 1.8%, and (C) 9.1% of the cation-exchange capacity.

were saturated with Na$^+$ by repeated (three times) exchange with 1 molar NaCl solutions. The <2-μm fraction was separated by centrifugation, concentrated, washed free of Cl$^-$, and freeze-dried. Stable suspensions suitable for spectroscopic studies were prepared by suspending the freeze-dried clay in doubly distilled H$_2$O and by centrifugation to obtain the <0.3-μm fraction. These suspensions contained ~4 g clay/dm^3 and were kept in the dark at 277 K to prevent the growth of microorganisms. The cation-exchange capacities (CEC) were determined by the ^{22}Na-method (Peigneur et al., 1975) (Table 1). Clays loaded with proflavine were prepared by shaking 10 cm^3 of an aqueous 10^{-5} molar solution of monoprotonated proflavine (PFH$^+$) with various amounts of clay suspensions to obtain loadings in the range 0–25% of the CEC and simultaneous dilution to 40 cm^3. In this range, proflavine is quantitatively taken up by HEC, LAP, and BS. The uptake by SEP is quantitative only for the first 4% of the CEC. The absorbance of the systems at 450 nm is then ~0.1 absorbance units, whatever the loading. Proflavine hemisulfate (99.9%) from Fluka A.C. was used as received.

Techniques

Absorption spectroscopy. Absorption spectra of proflavine-clay suspensions and identical Na$^+$-clay suspensions were taken in the range 800–230 nm on a Cary 17 instrument connected to a HP 9825B desktop computer. The spectra of the Na$^+$-clay suspensions were subtracted from those of the corresponding proflavine-clay suspensions to obtain the spectra of adsorbed pro-

flavine. The latter were plotted with a HP 7470A plotter.

Fluorescence spectroscopy. The emission spectra were taken in the range 460–700 nm with a Spex fluorolog spectrofluorimeter at an excitation wavelength of 458 nm. The emitted signal was corrected with a Spex Emission Corrector. The spectral measurements were controlled with a Tectronix desktop computer and transmitted to a PDP 11/23 computer for storage on tape, integration, and plotting. For the quantum yield determinations, freshly prepared 2.5 × 10^{-6} molar solutions of proflavine at pH = 4 (acetate buffer) having an absorbance of ~0.1 and a quantum yield of 0.34 were used (Haugen and Melhuish, 1964). The quantum yield was calculated by the formula:

$$\emptyset = \emptyset_0 \frac{S_{clay}}{S_0} \cdot \frac{1 - 10^{-A_0}}{1 - 10^{-A_{clay}}} ,$$

where $\emptyset_0$, S_0, and A_0 are, respectively, the quantum yield, emission intensity, and absorbance of the aqueous

Table 1.   N$_2$-surface areas and ^{22}Na cation-exchange capacity (CEC) of the clays.

Clay	Surface area (m^2/g)	CEC (μeq/g)
Hectorite	63	526
Barasym[2]	133	464
Sepiolite[2]	276	200[1]
Laponite	360	733

[1] Determined from the adsorption isotherm of PFH$^+$.

[2] Surface area provided by supplier.

Table 2.   Band positions and band intensities of adsorbed proflavine.

Species	Absorption band maximum (nm)	Absorption coefficient (mole^{-1} cm^{-1} ($\pm$2000))	Emission maximum (nm)
(PF)$_2$	430	28,000	580–600
	250	37,000	
PFH$_2$$^{2+}$	490	shoulder	
	455–460	26,400	530
	360	15,440	
	345	not determined	
	280	54,500	
	245	not determined	
PFH$^+$	453	58,000	497
	250	not determined	

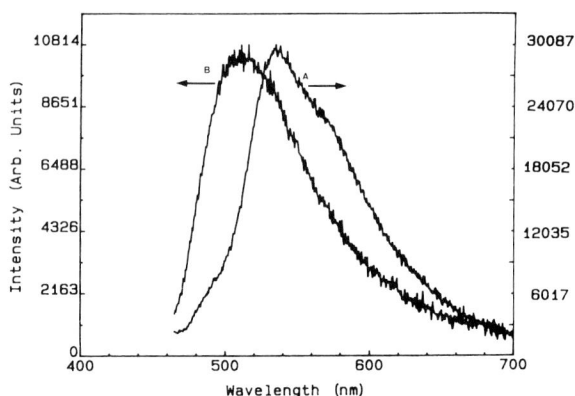

Figure 3. Emission spectra of proflavine-Barasym suspensions at loadings of (A) 2.0% and (B) 18.5% of the cation-exchange capacity.

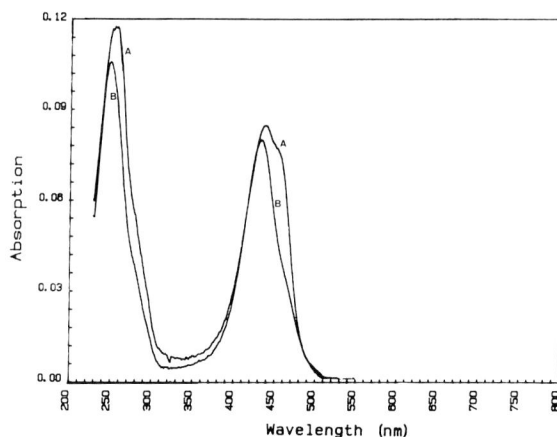

Figure 4. Absorption spectra of proflavine-hectorite suspensions at loadings of (A) 0.3% and (B) 20.3% of the cation-exchange capacity.

proflavine solutions having $\emptyset_0 = 0.34$ (Haugen and Melhuish, 1964). $\emptyset$, S_{clay}, and A_{clay} are the same quantities of the clay suspensions.

RESULTS

Synthetic mica/montmorillonite Barasym SMM-100

The absorption spectra of proflavine on BS are shown in Figure 2. The band maxima and the absorption coefficients are summarized in Table 2. At the lowest loading the spectrum of adsorbed proflavine is identical to that of an aqueous solution at pH < 1, i.e., diprotonated proflavine (PFH_2^{2+}). Two additional features should be noted: (1) The intensity ratio of the 460-nm band to the 360-nm band is somewhat larger on the clay than in solution, suggesting the presence of some monoprotonated proflavine (PFH^+). (2) A weak shoulder is present at 430 nm, which is indicative for dimers of proflavine (($PF)_2$).

As the loading increased the relative amount of adsorbed PFH_2^{2+} decreased, and that of the dimers increased. The latter was the only detectable species at >9% of the CEC. In the visible region the absorption coefficients of adsorbed proflavine agreed well with those in aqueous solution (Schoonheydt et al., 1986). That of PFH_2^{2+} (Table 2) was somewhat too high because the presence of small amounts of $(PF)_2$ and PFH^+ was not taken into account. In the UV range, especially at <280 nm, the accuracy on the absorption coefficients was smaller, because the subtraction of the spectra of Na^+-clay suspensions was never perfect and light scattering became important.

The emission spectra of proflavine adsorbed on Barasym are shown in Figure 3. At small loadings (<3% of the CEC) the emission maximum was at 530 nm with a shoulder at 580 nm, independent of the excitation wavelength (490, 460, 430, 360 nm). This spectrum was due to PFH_2^{2+}. At a loading of >3% of the

CEC an emission band at 502 nm grew, gradually intensifying with increasing loading to become the only emission present at the highest loading investigated (25% of the CEC). This emission was due to PFH^+. The corresponding excitation spectrum had a maximum at about 455 nm. If, however, the excitation spectra were recorded with the emission fixed at 530 nm or 580 nm, two excitation bands were oberved at 462 nm and 360 nm, corresponding to PFH_2^{2+}. Thus, at the smallest loadings PFH_2^{2+} was dominant. As the loading increased $(PF)_2$ became the main species on the clay, but its emission was undetectable, except for the 580-nm shoulder. PFH^+, which was not clearly visible in the absorption spectra, was clearly present in the emission spectra, especially at higher loadings. This does not mean that PFH^+ was not present at the small loadings; indeed, the PFH_2^{2+} absorption extended beyond 500 nm, and the eventual emission of excited PFH^+ in that region was reabsorbed. Only if there

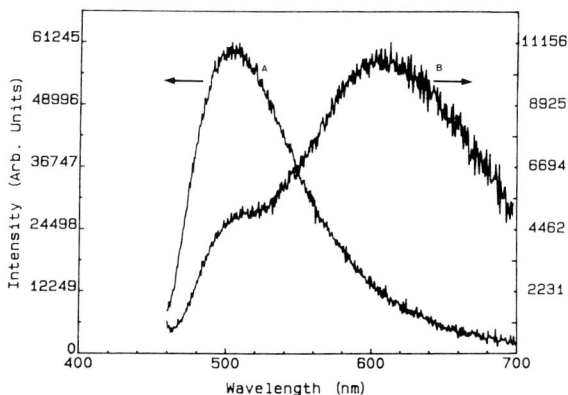

Figure 5. Emission spectra of proflavine-hectorite suspensions at loadings of (A) 0.3% and (B) 20.3% of the cation-exchange capacity.

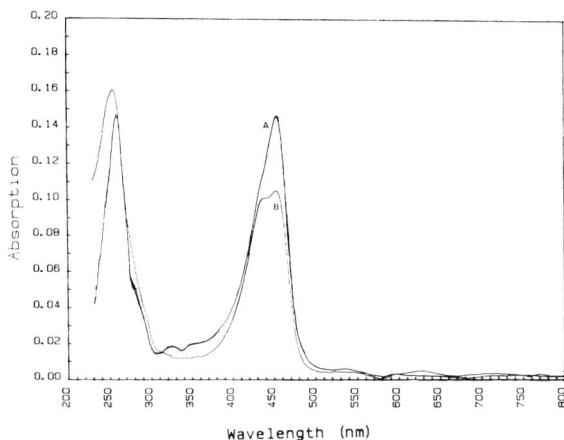

Figure 6. Absorption spectra of proflavine-Laponite suspensions at loadings of (A) 0.03% and (B) 0.3% of the cation-exchange capacity.

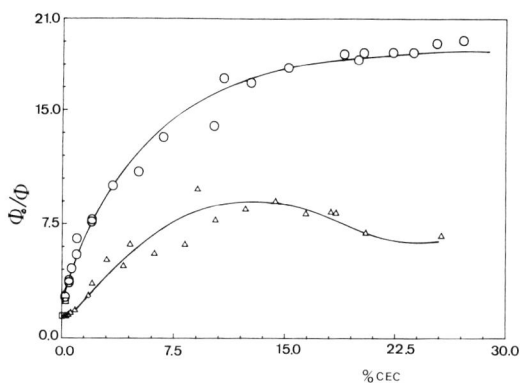

Figure 7. Variation of the quantum yield Ø of proflavine on hectorite (O) and Barasym (△) at different loadings (relative to the quantum yield $Ø_0$ of a 2.5×10^{-6} molar aqueous solution of proflavine).

was no significant absorption at about 500 nm, i.e., in the region where dimers are predominant in the absorption spectra, was the PFH$^+$ emission visible.

Hectorite. The absorption and emission spectra of the hectorite suspensions are shown in Figures 4 and 5. The absorption spectra were dominated by (PF)$_2$, with a small amount of PFH$^+$ at the smallest loading, as evidenced by the shoulder at about 453 nm and the absence of PFH$_2{}^{2+}$ bands at about 360 and 280 nm. The emission of PFH$^+$ at 502 nm was predominant. A second, weak emission maximum at about 600 nm grew with increasing loading and became the most important emission at the highest loading. This was the dimer emission.

Laponite. LAP has a small average particle diameter. Therefore, concentrated suspensions can be prepared without appreciable turbidity, and very small loadings can be investigated. A loading of 0.03% of the CEC gave an absorption spectrum having a maximum at 453 nm (PFH$^+$) and a weak shoulder at 430 nm, due to (PF)$_2$ (Figure 6). The extinction coefficient of the 453 nm band of PFH$^+$ was 58,000, significantly larger than its value in aqueous solution (40,000) and in alcoholic solution (52,000) (Hida and Sanuki, 1970). At higher loadings the dimer band increased relative to the monomer band. The emission spectra were dominated by the PFH$^+$ emission at 497 nm, which shifted to 503 nm with increasing loading.

Sepiolite. For SEP only loadings of <4% of the CEC could be investigated because of residual PFH$^+$ in solution at higher loadings. The absorption spectra were dominated by the 453-nm band of PFH$^+$ which had an extinction coefficient of 46,000. With increasing loadings the spectrum broadened on its high frequency side due to the formation of dimers. The emission was that of PFH$^+$.

Quantum yields. Figure 7 shows that the quantum yields, Ø, were smaller than that of an aqueous solution of PFH$^+$ at all loadings. They decreased with increasing loading in a characteristic way for each clay, and at a given loading, Ø(HEC, LAP) < Ø(BS) < Ø(SEP). The decrease of Ø or increase of $Ø_0/Ø$ was monotonous for HEC. For BS, $Ø_0/Ø$ reached a shallow maximum at a loading of about 15% of the CEC. Such a minimum in the quantum yield was also observed for Ru(2,2'-bipy)$_3{}^{2+}$ on BS (Schoonheydt *et al.*, 1984), although here the quantum yield was about the same as in aqueous solution.

DISCUSSION

In addition to classical absorption spectroscopy, fluorescence spectroscopy and quantum yield determinations are valuable tools to study surface reactions. Whereas in absorption spectroscopy small amounts of PFH$^+$ cannot be detected in excess PFH$_2{}^{2+}$ or (PF)$_2$, they can be detected in emission because the quantum yield of PFH$^+$ (0.34) is twice that of PFH$_2{}^{2+}$ (0.18) and orders of magnitude larger than that of (PF)$_2$. The reverse is also true; PFH$_2{}^{2+}$ and (PF)$_2$, not detectable in the emission spectra in the presence of PFH$^+$, are easily seen in the absorption spectra.

The weak emission of (PF)$_2$, which cannot be observed in aqueous solution, was observed on clays. Although the molecules have a certain mobility on the surface (Yamagishi and Soma, 1981), this mobility is smaller than in solution. Such a reduced mobility is probably the reason for the observation of a dimer emission.

In general, the positions of the absorption and emission bands on clay surfaces are the same as in aqueous solution. Only the pronounced 495-nm shoulder of PFH$_2{}^{2+}$, which was only a weak shoulder in solution, and the blue shift of the emission maxima of PFH$^+$ and PFH$_2{}^{2+}$ by 5–15 nm were noted. This means that

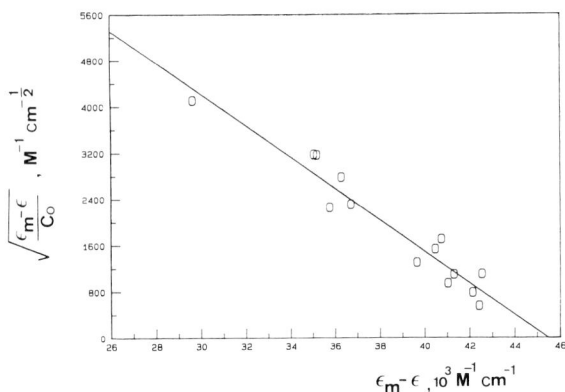

Figure 8.   Plot to determine the equilibrum constant K_D for dimerization of proflavine on the surface of hectorite.

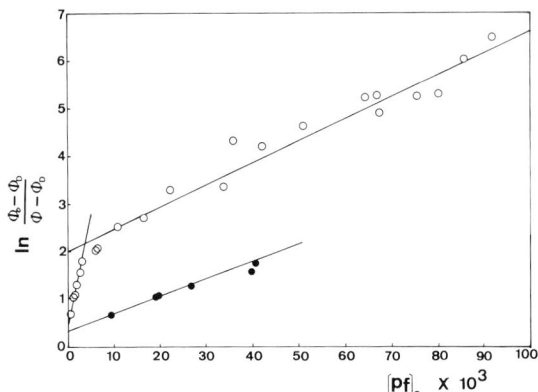

Figure 9.   Variation of the quantum yield of PFH$^+$ on hectorite as a function of $(PF)_2$ concentration at the surface of the clay, plotted according to the Perrin formulation. Calculation was made with volumes of 750×10^{-9} m^3/g ($\bigcirc$) and 63×10^{-9} m^3/g ($\bullet$).

the excited state interacted less strongly with the clay surroundings than with H_2O (Nicholls and Leermakers, 1971; Gerisher, 1974). The pure monomeric, nonprotonated form (PFH$^+$) was only found on LAP at the smallest loadings. Its extinction coefficient at 453 nm is 58,000, and the band width at half maximum is 2390 cm^{-1}. These values are to be compared to 40,000 and 2900 cm^{-1} respectively in aqueous solution. The differences can be explained by the combination of two factors: the change of environment from pure H_2O to a less polar water-surface oxygen environment, and the presence of dimers in aqueous solution even at concentrations as low as 2.5×10^{-6} mole/dm^3, causing a band broadening and a decrease of the extinction coefficient.

Surface reactions

Three reactions occur when aqueous solutions of PFH$^+$ are brought in contact with clays: ion-exchange, protonation to PFH$_2^{2+}$, and dimerization. The extent of these reactions depends on the type of clay and the extent of loading. PFH$_2^{2+}$ was detected only on the surface of BS at small loadings (<9% of the CEC). From the intensity of the 280-nm band of PFH$_2^{2+}$ and its absorption coefficient (Table 2), the amount of PFH$_2^{2+}$ was estimated to be between 1 and 2 μmole/g for loadings in the range 0.30–3% of the CEC. Schoonheydt *et al.* (1986) showed that at these loadings, the adsorbed proflavine molecules are mainly located on the external surface. Therefore, protonation of PFH$^+$ to PFH$_2^{2+}$ occurs on the frayed edges or edge-interlattice sites that contain strongly acidic OH groups or H_2O molecules. The number of these sites is of the same order as that of the highly selective Cs$^+$ ion-exchange sites of illites (Brouwer *et al.,* 1983). The similarity between these two clays is evident. Illites are dioctahedral hydromicas, whereas BS is a synthetic, mica-type, dioctahedral clay having NH$_4^+$ in the interlamellar space. K$^+$ is the main exchange cation of illites, but K$^+$ and NH$_4^+$ have about the same ionic radius. Both materials may con-

tain noncrystalline alumina (Wright *et al.,* 1972; van Olphen and Fripiat, 1979; Nemecz, 1981). Whether the Cs$^+$ sites of illites are identical to the PFH$_2^{2+}$ sites of BS is not clear. The high negative charge density of BS may also play a role in the protonation reaction on BS. It is well known that the protonation constant of ethylenediamine on montmorillonites exceeds its aqueous solution value and that the increase is proportional to the negative charge density of the minerals (Maes and Cremers, 1981). At >9% of the CEC, PFH$_2^{2+}$ was not observed with the absorption spectra, because its relative amount was too small and it dimerized.

Dimerization was the only side reaction for the other clays. From the true extinction coefficient of PFH$^+$ at 453 nm, determined on LAP, the dimerization constant of PFH$^+$ on the surface of HEC can be calculated by the method of Inaoka *et al.* (1978). The data are summarized in Figure 8 from which K_D, the dimerization constant was derived to be 1700 dm^3/mole. This value contrasts with K_D = 2300 dm^3/mole of Hida and Sanuki (1970), and K_D = 500 dm^3/mole of Haugen and Melhuish (1964). The latter authors used a monomer extinction coefficient of 40,000. From the previous discussion, this is not a pure monomer extinction coefficient.

This dimerization reaction is expected on the basis of the known solution chemistry of proflavine (Hida and Sanuki, 1970). If the molecules are adsorbed at random on a surface of 750 m^2/g in a monolayer of 1-nm thickness, 1 μmole/g corresponds to an equivalent solution concentration of 1.3 mmole/g. For K_D = 1700, this concentration gives 0.37 μmole/g monomer and 0.32 μmole/g dimer. Therefore, special electronic π-interaction with surface oxygens need not be invoked to explain metachromatic behavior.

Explaining qualitatively the decrease of the quantum yield with increasing loading is not a problem, because the quantum yield of PFH$_2^{2+}$ is only half that of PFH$^+$

and because dimers do not emit in solution and emit weakly on the clay surface. The minimum in the quantum yield of BS at about 15% of the CEC must therefore be ascribed to an increase in the PFH^+ emission. Probably, at this loading, interlamellar adsorption of PFH^+ sets in without dimer formation. A similar hypothesis was advanced to explain the evolution of the quantum yield of $Ru(2,2'\text{-bipy})_3^{2+}$ with loading (Schoonheydt et al., 1984). For HEC and LAP the asymptotic behavior of $\emptyset_0/\emptyset$ can be explained in detail in the following way. At $\emptyset_0/\emptyset = 20$, only dimer emission is present (Figure 7), and its quantum yield is 0.017. This dimer quantum yield can be subtracted from the experimental quantum yields to give the PFH^+ quantum yields, because PFH^+ and $(PF)_2$ are the only two species present on the surface. Monomer and dimer concentrations on the surface can be calculated as explained above. If, as shown in Figure 9, $\ln(\emptyset_0 - \emptyset_D/\emptyset - \emptyset_D)$ is plotted against the dimer concentration on the surface, two straight lines are obtained: one at loadings <1% of the CEC with a slope of 423/mole, the other, at loadings >1% of the CEC with a slope of 48/mole. These data indicate that the monomer emission is quenched by the dimer as in solution (Haugen and Melhuish, 1964); the quenching follows Perrin's mechanism of quenching without diffusion of monomer and dimer (within the lifetime of the excited state of the monomer).

The presence of two straight lines is indicative of two types of surfaces: external and interlamellar. Thus, if at a loading of <1% of the CEC, adsorption occurs exclusively on the external surface having an area of 63 m²/g or an effective volume of 63×10^{-9} m³/g, the first straight line in Figure 9 transforms to a straight line parallel to that for interlamellar adsorption. From the slope the effective quenching radius can be calculated to be 2.7 nm (Turro, 1978).

The physical picture which emerges from these data is that, at <1% of the CEC, adsorption occurs exclusively on the external surface. At >1% of the CEC the total surface of HEC is available for adsorption. The proflavine molecules dimerize, mainly because of the concentration effect around the clay particles. They do not diffuse within the lifetime of the excited state. As a consequence, the emission of PFH^+ is quenched by $(PF)_2$ following Perrin's mechanism with an effective quenching radius of 2.7 nm.

ACKNOWLEDGMENTS

Acknowledgment is made to the National Fund of Scientific Research (Belgium) for support of this research and for the research tenure to R.A.S. as Senior Research Associate. J.C. thanks the "Instituut voor Aanmoediging van het Wetenschappelijk Onderzoek in Nijverheid en Landbouw (Belgium)" for a Ph.D. grant.

REFERENCES

Bergmann, K. and O'Konski, C. T. (1963) A spectroscopic study of methylene blue monomer, dimer and complexes with montmorillonite: J. Phys. Chem. **67**, 2169–2177.

Brouwer, E., Baeyens, B., Maes, A., and Cremers, A. (1983) Cesium and rubidium ion equilibria in illite clay: J. Phys. Chem. **87**, 1213–1219.

Cohen, R. and Yariv, S. (1984) Metachromasy in clay minerals. Sorption of acridine orange by montmorillonite: J. Chem. Soc., Faraday Trans. I **80**, 1705–1715.

DellaGuardia, R. A. and Thomas, J. K. (1983) Photoprocesses on colloidal clay systems. 2. Quenching studies and the effect of surfactants on the luminescent properties of pyrene and pyrene derivatives adsorbed on clay colloids: J. Phys. Chem. **87**, 3550–3557.

de Mayo, P. (1982) Superficial photochemistry: Pure Applied Chem. **54**, 1623–1632.

Gerisher, H. (1974) Photochemistry of adsorbed species: Faraday Disc. Chem. Soc. **58**, 219–236.

Grauer, Z., Avnir, D., and Yariv, S. (1984) Adsorption characteristics of rhodamine 6G on montmorillonite and laponite, elucidated from electronic absorption and emission spectra: Can. J. Chem. **62**, 1889–1894.

Haugen, G. R. and Melhuish, W. H. (1964) Association and self-quenching of proflavine in water: Trans. Faraday Soc. **60**, 386–394.

Hida, M. and Sanuki, T. (1970) Studies of the aggregation of dyes. The scope of application of the maximum slope method: Bull. Chem. Soc. Japan **43**, 2291–2296.

Inaoka, W., Harada, S., and Yasunaga, T. (1978) Kinetic studies of the dimerization reaction of thionine by the laser Raman temperature-jump method: Bull. Chem. Soc. Japan **51**, 1701–1703.

Levitz, P., Van Damme, H., and Keravis, D. (1984) Fluorescence decay study of the adsorption of nonionic surfactants at the solid-liquid interface. 1. Structure of the adsorption layer on a hydrophilic solid: J. Phys. Chem. **28**, 2228–2235.

Maes, A. and Cremers, A. (1981) Influence of charge density on the acid-base equilibrium of ethylenediamine in montmorillonites: J. Chem. Soc., Faraday Trans. I **77**, 1553–1559.

Nemecz, E. (1981) Clay Minerals: Akademiai Kiado, Budapest, Hungary, 547 pp.

Nicholls, C. H. and Leermakers, P. A. (1971) Photochemical and spectroscopic properties of organic molecules in adsorbed and other perturbing polar environments: Adv. Photochem. **8**, 315–335.

Peigneur, P., Maes, A., and Cremers, A. (1975) Heterogeneity of charge density in montmorillonite as inferred from cobalt adsorption: Clays & Clay Minerals **23**, 71–75.

Schoonheydt, R. A., Cenens, J., and De Schryver, F. C. (1986) Spectroscopy of proflavine on clays: J. Chem. Soc. Faraday Trans. I **82**, 281–283.

Schoonheydt, R. A., De Pauw, P., Vliers, D., and De Schryver, F. C. (1984) Luminescence of tris(2,2'-bipyridine)ruthenium(II) in aqueous clay mineral suspensions: J. Phys. Chem. **88**, 5113–5118.

Turro, N. J. (1978) Modern Molecular Photochemistry: Benjamin/Cummings, Menlo Park, California, 628 pp.

Van Damme, H., Nijs, H., and Fripiat, J. J. (1984) Photocatalysed reactions on clay surfaces: J. Molecular Catal. **27**, 123–142.

van Olphen, H. and Fripiat, J. J. (1979) Data Handbook for Clay Minerals and Other Non-metallic Minerals: Pergamon Press, New York, 346 pp.

Wright, A. C., Granquist, W. T., and Kennedy, J. V. (1972) Catalysis by layer lattice silicates. I. The structure and thermal modification of a synthetic ammonium dioctahedral clay: *J. Catalysis* **25**, 65–80.

Yamagishi, A. and Soma, M. (1981) Aliphatic tail effects on adsorption of acridine orange cation on a colloidal surface of montmorillonite: *J. Phys. Chem.* **85**, 3090–3092.

Yariv, S. and Lurie, D. (1971) Metachromasy in clay minerals. Part 1. Sorption of methylene blue by montmorillonite: *Israel J. Chem.* **9**, 537–552.

Proceedings of the International Clay Conference, Denver, 1985, L. G. Schultz, H. van Olphen, and F. A. Mumpton, eds.,
The Clay Minerals Society, Bloomington, Indiana, 359–362 (1987).

ADSORPTION OF COLLAGEN BY SEPIOLITE

R. Pérez-Castells,[1] A. Alvarez,[1] J. Gavilanes,[2] M. A. Lizarbe,[2]
A. Martínez Del Pozo,[2] N. Olmo,[2] and J. Santarén[1]

[1] Departamento de Investigación y Desarrollo, TOLSA, S.A., Madrid, Spain

[2] Departamento de Bioquímica, Facultad de Ciencias Químicas, Universidad Complutense, Madrid, Spain

Abstract—Interaction between sepiolite and collagen has been studied using two different micronizing procedures for the sepiolite. Micronizing in the dry state yields shorter fibers than micronizing in the wet state. The adsorption of collagen was followed by ultraviolet-absorbance and measurement of the amino acid content of the protein-clay complexes. The interaction primarily involved the high molecular weight aggregates of the protein. A collagen-clay weight ratio of 0.4 was reached at pH ~4. The adsorption decreased with increasing ionic strength. A model is proposed wherein rod-shaped protein molecules are bound on the stepped surface of the sepiolite.

Key Words—Adsorption, Collagen, Micronization, Protein, Sepiolite, Ultraviolet absorbance spectroscopy.

INTRODUCTION

Natural sepiolite occurs in compact form, even though its crystalline structure is fibrous. These fibers are 0.2–2 μm in length and 100–300 Å in width and have a thickness of 50–100 Å (Martin-Vivaldi and Robertson, 1971). The structure of sepiolite was established by Longchambon (1936), Nagy and Bradley (1955), and Brauner and Preisinger (1956). This structure, which is widely accepted at present, contains channels 3.6 × 10.6 Å in cross section, oriented along the fiber axis. The cation-exchange capacity of sepiolite is about 10 meq/100 g. The surface area of sepiolite by N_2 adsorption (BET method) is 340 m^2/g. Micropores account for 60–70% of this surface area. The channels are not accessible to nitrogen. Sepiolite has high sorption capacity and can adsorb as much as 200–250% of its own weight in water. Three types of active centers of adsorption have been distinguished on sepiolite surfaces (Serratosa, 1979): oxygen atoms in the tetrahedral sheet of silica, water molecules coordinated with magnesium ions at the edges, and Si–OH groups located at intervals of 5 Å along the axis of the fiber.

Collagen is the main polypeptide component of the connective tissue. The many varieties of this tissue are due to the presence of different forms of collagen as well as to its interaction with other molecules. Tendons from some birds and bones have specific mechanical properties due to interactions between collagen and inorganic structures (White *et al.*, 1977). Collagen in these forms of connective tissue corresponds to the type I collagen (Bornstein and Sage, 1980) and is composed of three polypeptide chains arranged in a specific helical structure (triple helix of collagen) (Linsenmayer, 1981).

Recently, Olmo *et al.* (1985) described interactions between type I collagen and sepiolite. Aqueous solutions of the protein contain a mixture of collagen forms ranging from monomers to high molecular weight aggregates; sepiolite mainly interacts with such aggregates (Olmo *et al.*, 1985). Considering that many forms of connective tissue exhibit specific properties depending on collagen-inorganic compound interactions, sepiolite-collagen complexes may be of potential usefulness in the prosthetics field; e.g., as cell culture substratum and bone replacement. The biological role of collagen is dependent on the specific features of the molecule arising from its rod-like shape. Thus, the importance of the sepiolite-collagen complexes would increase if the overall structure of the protein is preserved after its reaction with the mineral. Bearing these biological implications in mind, we have studied the mode of interaction between sepiolite clay and the protein.

MATERIALS AND METHODS

The type I collagen employed for this study was purified from fetal calf skin as described by Trelstad *et al.* (1976). Adsorption of collagen was studied using 10 mg of sepiolite, dry (DM) or wet micronized (WM), and different amounts of collagen in a constant volume of 3 ml. The wet and dry micronized samples used had essentially the same chemical and physical properties, but their suspensions showed different rheological behavior. WM-sepiolite was prepared by wet micronization as described by Alvarez *et al.* (1984); DM-sepiolite was prepared by dry micronization in a fluid energy mill. The WM-sepiolite contains a higher percentage of and longer fibers than the DM-sepiolite (Figure 1), probably accounting for their different rheological properties. DM-sepiolite did not yield stable gels

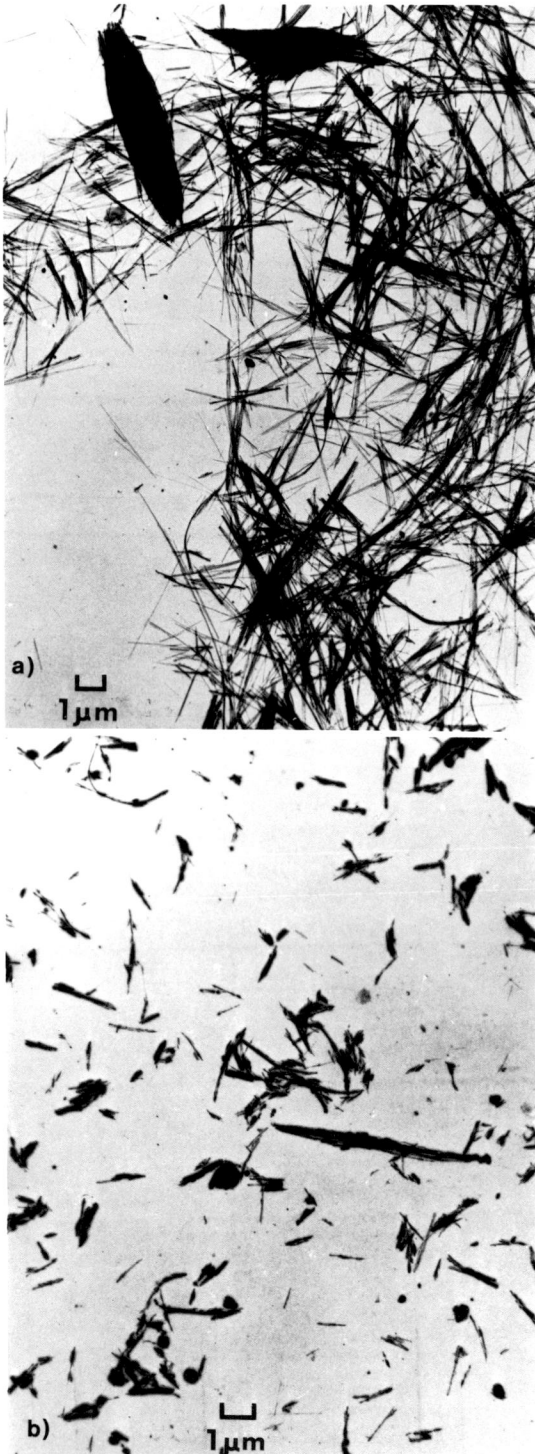

a)

1μm

b)

1μm

Figure 1.   Electron micrographs of (a) wet micronized sepiolite and (b) dry micronized sepiolite.

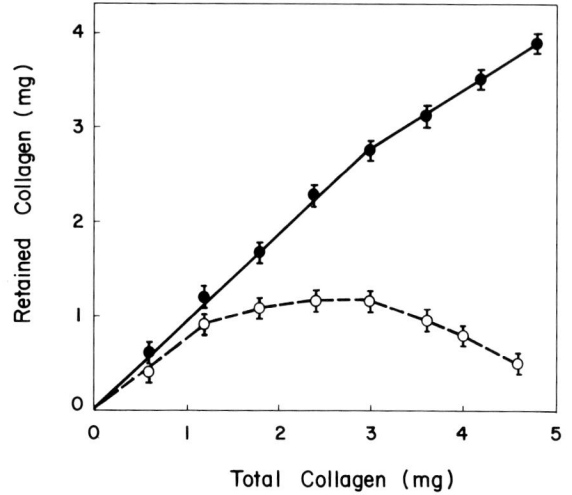

Figure 2.   Retained collagen vs. total collagen added in mg of protein ± standard deviation; ● = wet-micronized sepiolite, ○ = dry-micronized sepiolite (10 mg of clay, 3 ml solution, pH adjusted to 3.6).

in solvents, whereas WM-sepiolite formed viscous, pseudoplastic gels in water and could be used together with surfactants as gelling agents for organic solvents.

Adsorption of collagen was studied at 20°C as a function of pH and ionic composition of the solution. The collagen-clay dispersion was mixed with a magnetic stirrer for 30 s and maintained for 20 min at 20°C. The dispersion was then centrifuged at 1000 g for 15 min, and the supernatant liquid was analyzed to determine

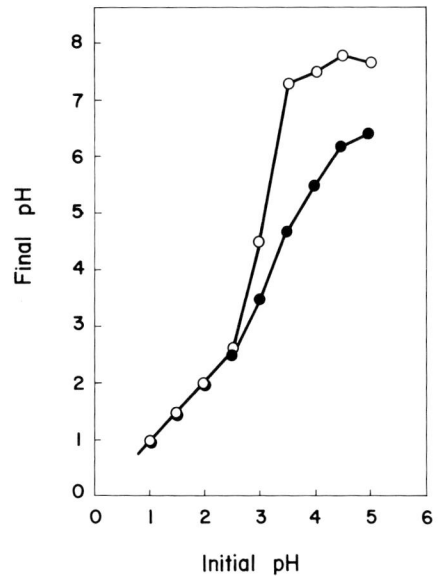

Figure 3.   Final pH vs. initial solution pH after addition of (○) wet-micronized sepiolite (10 mg) and (●) sepiolite-collagen complex (10 mg of clay, 4 mg of protein in 3 ml water).

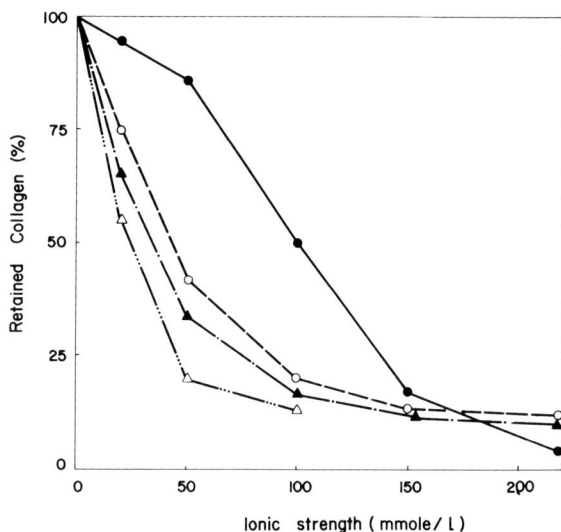

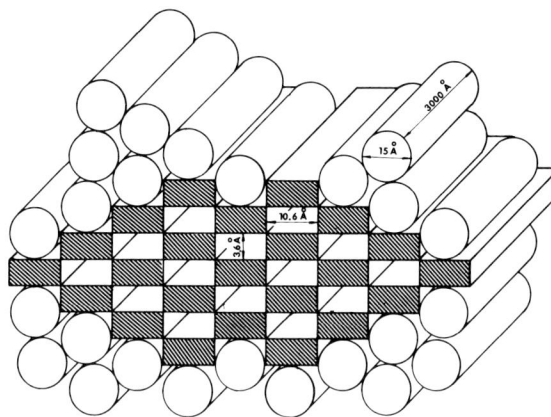

Figure 5.   Model proposed for sepiolite-type I collagen complex. Cylinders correspond to collagen: aggregates of collagen are represented by associated cylinders. Prisms are fibers of sepiolite.

Figure 4.   Percentage of collagen retained vs. ionic strength in solution. Experiments were performed using 10 mg of clay, 4 mg of protein in 3 ml solution of (●), NaCl; (○), MgCl₂; (▲), CaCl₂; and (△), sodium phosphate. pH = 3.6.

the amount of protein present. This analysis was carried out by determining the amino acid content and by UV-absorbance measurements (Olmo *et al.*, 1985).

RESULTS AND DISCUSSION

Adsorption of collagen on sepiolite in acid solution is shown in Figure 2. WM-sepiolite exhibited a higher capacity for protein retention than DM-sepiolite. Inasmuch as the differences between these two forms of the sepiolite are related to the micronization process, the interaction must have been a function of the length of and the degree of aggregation of individual fibers.

A collagen-clay weight ratio of 0.4 (WM-sepiolite) can be attained for the complex, which corresponds to 83% of total collagen added. The shape of the adsorption curve suggests that the sepiolite is not saturated. It was not possible to obtain the saturation rate because of the difficulty of working with more concentrated collagen suspensions. The results in Figure 2 were obtained for 20 min of reaction time, and they did not change by using reaction times as long as 1 hr. The high capacity of WM-sepiolite for the adsorption of collagen may be explained as follows: Collagen is positively charged in acid solution (i.e., the isoelectric point is in the alkaline range, Davison and Drake, 1966). Adsorption of collagen by sepiolite will therefore initially occur by direct electrostatic interaction between negative clay and positive protein. A further increase of the protein retention is undoubtedly due to subsequent collagen-collagen interactions.

The addition of sepiolite to aqueous non-buffered solutions modified the pH, as indicated in Figure 3. Apparently, H⁺ ions associated with the negative clay

surface. By addition of a collagen-WM-sepiolite complex, the pH variation was different and the increase of pH was somewhat reduced (Figure 3). This difference should be due to masking of the negative charge sites of the sepiolite by adsorbed protein.

Figure 4 shows the effect of ionic strength of the solution on collagen adsorption. The observed decrease of collagen retention with increasing strength is also indicative of electrostatic bonding between sepiolite and type I collagen. The collagen structure does not change in the studied pH and ionic strength ranges (Stewart and Mandelkern, 1964).

Type I collagen has a rod-like shape (15 × 3000 Å, Veis *et al.*, 1967). The charge on the protein is located on the surface of the structure. Collagen aggregates to linear structures (Piez and Trus, 1981). Considering these facts and the structure of sepiolite, we propose that protein-clay complexes have the structure schematically indicated in Figure 5. The protein molecules would be located along the stepped surface of the clay. The complex would be maintained by ionic interactions between the positive charges of the protein molecules (cylinders in Figure 5) and the negative charges located at the steps of the surface of the clay (prisms in Figure 5). Protein-protein interaction is shown by accumulated cylinders.

"In vitro" assays are being carried on to study the biocompatibility of sepiolite-collagen complexes.

CONCLUSIONS

Type I collagen interacts with sepiolite. The amounts adsorbed depend on the pre-treatment of the sepiolite: wet-micronized sepiolite adsorbs more collagen than dry-micronized sepiolite. The adsorption is sensitive to pH and ionic strength, suggesting an electrostatic bonding between clay and protein.

REFERENCES

Alvarez, A., Aragón, J. J., and Pérez Castells, R. (1984) Procedimiento de fabricación de sepiolita de grado reológico: *Spanish Patent* **534891**, 23 pp.

Bornstein, P. and Sage, H. (1980) Structurally distinct collagen types: *Ann. Rev. Biochem.* **49**, 957–1003.

Brauner, K. and Preisinger, A. (1956) Struktur und Entstehung des Sepiolites: *Tcher. Min. Petro. Mitteilungen* **6**, 120–140.

Davison, P. F. and Drake, M. P. (1966) The physical characterization of monomeric tropocollagen: *Biochemistry* **5**, 313–321.

Linsenmayer, T. F. (1981) Collagen: in *Cell Biology of the Extracellular Matrix,* E. D. Hay, ed., Academic Press, New York, 5–32.

Longchambon, H. (1936) Palygorskites: *C.R. Acad. Sci., Paris* **203**, 672–674.

Martin Vivaldi, J. L. and Robertson, R. H. S. (1971) Palygorskite and sepiolite (the hormites): in *Electron Optical Investigation of Clays,* J. A. Gard, ed., Mineralogical Society, London, 255–276.

Nagy, B. and Bradley, W. F. (1955) Structural scheme of sepiolite: *Amer. Mineral.* **40**, 885–892.

Olmo, N., Martínez Del Pozo, A., Lizarbe, A., and Gavilanes, J. G. (1985) Sepiolite-collagen interaction: *Collagen Rel. Res.* **5**, 9–16.

Piez, K. A. and Trus, B. L. (1981) A new model for packing of type-I collagen molecules in the native fibril: *Bioscience Reports* **1**, 801–810.

Serratosa, J. M. (1979) Surface properties of fibrous clay minerals (palygorskite and sepiolite): in *Proc. Int. Clay Conf., Oxford, 1978,* M. M. Mortland and V. C. Farmer, eds., Elsevier, Amsterdam, 99–109.

Stewart, W. E. and Mandelkern, L. (1964) The effect of neutral salt on the melting temperature and regeneration kinetics of the ordered collagen structure: *Biochemistry* **3**, 1135–1137.

Trelstad, R. L., Catanese, V. M., and Rubin, D. F. (1976) Collagen fractionation: separation of native types I, II and III by differential precipitation: *Anal. Biochem.* **71**, 114–118.

Veis, A., Anesey, J., and Mussel, S. (1967) A limiting microfibril model for the three-dimensional arrangement within collagen fibers: *Nature* **215**, 931–934.

White, S. W., Hulmes, D. J. S., Miller, A., and Timmins, P. A. (1977) Collagen mineral axial relationship in calcified turkey leg tendon by X-ray and neutron diffraction: *Nature* **266**, 421–425.

Proceedings of the International Clay Conference, Denver, 1985, L. G. Schultz, H. van Olphen, and F. A. Mumpton, eds.,
The Clay Minerals Society, Bloomington, Indiana, 363–369 (1987).

ADSORPTION OF NUCLEIN BASES ON SMECTITES

Abbas M. Samii and G. Lagaly

Department of Chemistry, The University of Texas at Austin, Austin, Texas 78712 and
Institut für Anorganische Chemie, Universität Kiel, Olshausenstraße 40
D-2300 Kiel, Federal Republic of Germany

Abstract—The adsorption of the nuclein bases adenine, cytosine, thymine, and uracil from aqueous solutions exhibits several unusual features, especially if montmorillonites are used as the adsorbents. Adenine and cytosine are more strongly adsorbed than uracil and thymine. The amount adsorbed is sensitive to the layer charge and is at a maximum at layer charges of about 0.29 eq/(Si,Al)$_4$O$_{10}$. In the presence of adenine, the adsorption of thymine and uracil increases, probably due to base-pairing as in nucleic acids. The increased adsorption provides an instructive example of co-adsorption phenomena; however, the molar ratio of adsorbed adenine/thymine or adenine/uracil does not reach 1. Only some of the molecules are bound as base-pairs. Thymine and uracil are more strongly adsorbed if the smectite is previously loaded with adenine.

The adsorption of adenine and cytosine are highly dependent on the type and concentration of added salts. Adenine adsorption is strongly increased by salts in the following order: NaCl < NaNO$_3$ < NaI; for cytosine adsorption, the order is: NaNO$_3$ < NaI < NaCl. The influence of the salts is explained by their effect on the water structure.

Key Words—Adsorption, Layer charge, Nuclein base, Salt, Smectite.

INTRODUCTION

Adsorption studies of pyrimidines, purines, nucleic bases, and nucleosides by clay minerals have been prompted by suggestions that clay organic interactions may have played an important part in the initial development of living forms (Bernal, 1951; Oparin, 1957; Hofmann, 1961; Cairns-Smith, 1966; Weiss, 1981). Accordingly, Brindley and co-workers (Lailach *et al.,* 1968a, 1968b; Lailach and Brindley, 1969; Thompson and Brindley, 1969) measured the adsorption of various pyrimidines, purines, and nucleosides on montmorillonite (bentonite from Texas) and illite (from Skytop, Pennsylvania) in the presence of different types of gegen ions (Li, Na, Mg, Ca, Co, Ni, Cu, Fe(III)) and at different pHs. A spacing of 12.5 Å for the Na- and Ca-forms of the montmorillonite indicated that the bases lie flat on the silicate layers (Lailach *et al.,* 1968a).

Lailach and Brindley (1969) also described pronounced co-adsorption from mixtures of two-bases on the Texas montmorillonite. Co-adsorption phenomena were neglected for a long time in colloid chemistry. The results of Lailach and Brindley (1969) thus offered a means of studying this effect in more detail. The outstanding advantage of smectites as adsorbents arises from the variability of their layer charge. The surface charge density and, thus, the concentration of gegen ions can be notably different from montmorillonite to montmorillonite, whereas the geometric arrangement of the surface atoms remains almost constant.

Our first experiments indicated that not only the extent of co-adsorption but also the amounts of bases adsorbed from one component solutions were highly dependent on the layer charge density and were also very sensitive to the type and concentration of added salts. The present paper examines further some of the co-adsorption phenomena reported earlier by Brindley and co-workers.

EXPERIMENTAL

Samples

The samples examined in this study are listed in Table 1. The "greenbond" montmorillonite from Wyoming and the montmorillonite from Niederschönbuch, Bavaria, were supplied by Südchemie, Moosburg, Federal Republic of Germany. The other smectites were obtained from the Source Clays Repository of The Clay Minerals Society. Charge distributions (Table 1) were determined by alkylammonium ion exchange (Lagaly, 1981).

Prior to the adsorption measurements, iron oxides and organic materials were removed from the clays. The carbonates in hectorite were decomposed by dropwise addition of 0.01 M HCl so that the pH never decreased below 5. Iron oxides and organic matter were removed following the method of Stul and van Leemput (1982).

Treatments

Samples of bentonite (50 g) were dispersed in 500 ml of water, to which 600 ml of citrate buffer was added (citrate buffer: 0.3 mole sodium citrate, 1 mole NaHCO$_3$, and 1.2 mole NaCl in 1 liter H$_2$O, pH ad-

Table 1. Source and layer-charge data of the smectites studied.

Origin	Layer charge ($\bar{\xi}$)[1]	Charge distribution[2]	Equivalent area, A_e[3] (Å^2)	Interlayer-exchange capacity[4] (meq/100 g)
Hectorite				
A. California, San Bernardino County[3] (SHCa-1)	0.23	0.20–0.29	107	62
Montmorillonites				
B. Wyoming, Crook County[3] (SWy-1)	0.27	0.23–0.33	86	73
C. Niederschönbuch, Bavaria	0.29	0.20–0.39	80	78
D. Wyoming "greenbond"	0.30	0.25–0.39	77	81
E. Texas, Gonzales County[3] (STx-1)	0.31	0.24–0.42	75	84
F. Arizona, Apache County[3] (SAz-1)	0.35	0.26–0.45	66	95

[1] $A_e = a_0 b_0 / 2\bar{\xi} = 23.25/\bar{\xi}$ (Å^2).
[2] $eq/(Si,Al)_4 O_{10}$.
[3] Obtained from the Source Clays Repository of The Clay Minerals Society (van Olphen and Fripiat, 1975).
[4] Interlayer-exchange capacity = $1000\bar{\xi}/3.70$ (meq/100 g).

justed to 7.3). After heating to 75°C, 10 g of sodium dithionite was added and the suspension was reacted for 48 hr at 70°C. The clay was then separated and washed with 500 ml of acidified NaCl solution (0.05 mole HCl and 1 mole NaCl in 1 liter H_2O). After repeating the dithionite reduction, the acidified suspension was reacted (30 hr, 90°C) with a mixture of 250 ml of H_2O_2 (30% in water) and 750 ml of 0.5 M sodium acetate. The clay was washed first with 500 ml of 1 M NaCl and then with water until the smectite began to disperse. The dispersion was diluted with 5–6 liter of water, and all particles >2 μm in size were removed by sedimentation. The very dilute dispersions of the <2-μm fraction was coagulated by the addition of $CaCl_2$. The precipitate was treated with 250 ml of 0.5 M $CaCl_2$ solution to ensure its complete transformation into the Ca^{2+}-form. The sample was freeze-dried and dried in vacuo (p ≤ 10 Pa) at 120°C.

Adsorption experiments

Weighed amounts of the nuclein bases adenine, thymine, cytosine, and uracil (Figure 1) were dissolved in water to give 0.5×10^{-3} M, 1×10^{-3} M, and 2×10^{-3} M solutions. The pH was adjusted to 5 by a few drops of dilute HCl.

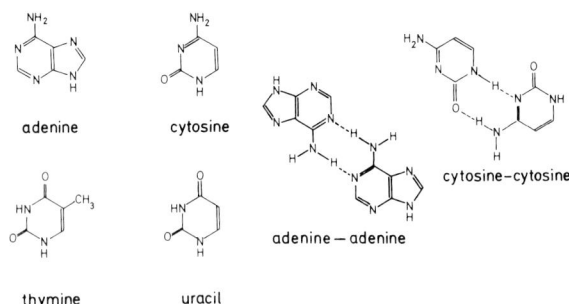

Figure 1. Nuclein bases and examples of possible configurations of adenine-adenine and cytosine-cytosine pairs.

Adsorption on the smectites was measured by the concentration changes of ^{14}C-labeled bases. About 100 mg of the ^{14}C-bases having an activity of 0.1 mCi was dissolved in 100 ml water, and 2 ml of this solution was added to 100 ml of the solutions of the unlabeled bases.

Typically, 10–80-mg samples of the dried and finely powdered Ca-smectite were weighed into 20-ml glass vessels, and 10 or 20 ml of the base solution was added. This dispersion was agitated in an ultrasonic bath for several minutes and then allowed to stand 24 hr at 20°C under occasional shaking. One milliliter of the clear supernatant was transferred into 15 ml of scintillation liquid (Unisolve 100, W. Zinsser, Frankfurt, Germany) in a 20-ml scintillation vessel. The activity was measured by a β-scintillation counter.

The adsorption data refer to the dried samples (120°C, p ≤ 10 Pa). The molecular weight of the Ca-smectite was taken to be 370 g/mole $(Si,Al)_4 O_{10}$.

RESULTS AND DISCUSSION

Influence of layer charge

The amounts of bases adsorbed from 10^{-3} M solutions varied with the mean layer charge (Figure 2). The adsorption reached a maximum for layer charges of 0.27 eq/$(Si,Al)_4 O_{10}$ (adenine) or 0.29 eq/$(Si,Al)_4 O_{10}$ (cytosine, thymine) and decreased sharply beyond 0.29 eq/$(Si,Al)_4 O_{10}$. More adenine and cytosine were adsorbed than thymine and uracil.

The adsorption isotherms of adenine and cytosine, up to the highest possible equilibrium concentration (2×10^{-3} M), are shown in Figure 3. The large adsorption capacity of the Bavarian montmorillonite (sample C) is apparent.

The solution pH (pH = 5) slightly exceeded the pK_a of adenine and cytosine (4.2 and 4.6, respectively). In the interlayer space of montmorillonite the ratio of protonated to unprotonated species is enhanced by the

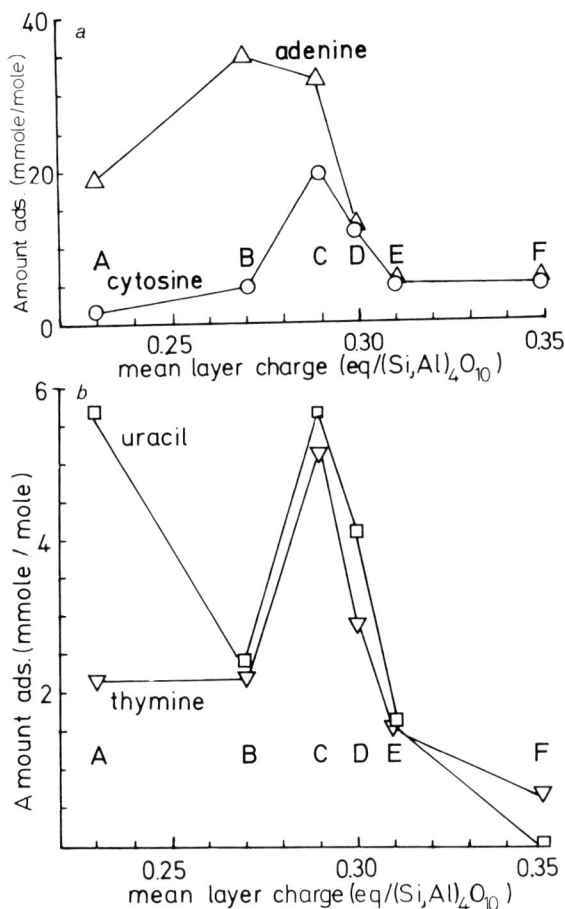

Figure 2. Adsorption of bases from 10^{-3} M aqueous solution as a function of the mean layer charge $\bar{\xi}$; 40 mg Ca-smectite, 10 ml solution, pH = 5. A = hectorite; B–F = montmorillonites: B = Wyoming, Crook County; C = Bavaria; D = Wyoming, "greenbond"; E = Texas; F = Arizona.

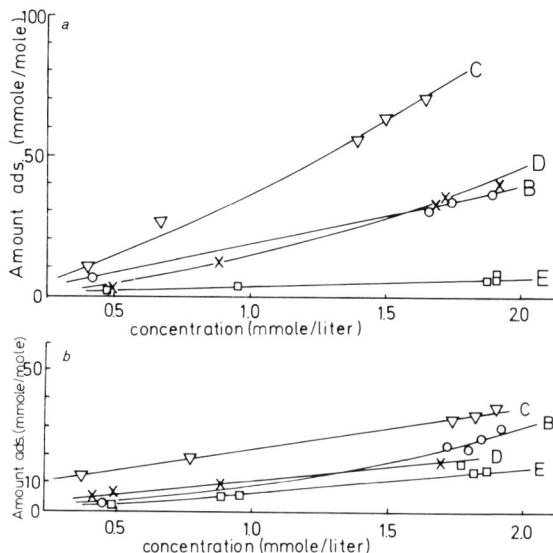

Figure 3. Adsorption isotherms of adenine (a) and cytosine (b) on Ca-montmorillonites. B = Wyoming, Crook County; C = Bavaria; D = Wyoming, "greenbond"; E = Texas; pH = 5, 20°C.

increased acidity of interlayer water (Mortland and Raman, 1968; Touillaux et al., 1968; El-Amamy and Mill, 1984). Thus, both compounds were probably adsorbed as protonated species and were bound as interlayer cations. In contrast, the pK_a of thymine and uracil (0 and 0.5, respectively) is considerably different from the solution pH; hence protonation is unlikely, even in the interlayer space. Thymine and uracil were therefore adsorbed as neutral molecules.

The dependence of adsorption on layer charge suggests that lateral interactions between adsorbed molecules were important and that they were promoted by the pronounced tendency of the bases to associate to base-pairs. Formation of pairs in which two organic bases share the same proton (Farmer and Mortland, 1966; Mortland, 1966) may have been of fundamental importance. Formation of such pairs explains the adsorption maxima that have been observed as a function of pH (e.g., Yariv and Heller-Kallai, 1975). Adenine-adenine, cytosine-cytosine, and thymine-thymine pairs

require areas of 76–88 Å^2 (molecular areas: adenine, 44 Å^2; cytosine, 38 Å^2; thymine, 41 Å^2; uracil, 36 Å^2). In fact, montmorillonites C and B have mean equivalent areas of 80 and 86 Å^2, respectively (Table 1).

In addition to the compatibility between the mean equivalent area and the molecular area of the base-pairs, the influence of the real distribution of layer charges was also important. The layer-charge distribution curve of montmorillonite C that displayed the maximum adsorption capacity of the samples investigated passes through a maximum at $\xi = 0.25$–0.29 eq/(Si,Al)$_4$O$_{10}$. In contrast, other smectites, such as montmorillonite F, which has a low adsorption capacity, showed minima in this range of layer charge. This influence of the layer-charge distribution on adsorption can be explained in terms of the model discussed below, but other factors cannot be completely excluded. The extent of tetrahedral substitution in samples B–E is minor (Vogt and Köster, 1978); hence, the adsorption maximum cannot be related to a distinct distribution between tetrahedral and octahedral sites of layer charge.

A preferential adsorption of base-pairs is obvious, but special factors seem to operate for nucleins (vide infra).

Effect of salts

The adsorption of the bases was sensitive to the presence of salts. Sodium salts increased considerably the adsorption of adenine on Wyoming montmorillonite (sample B) if a threshold equilibrium concentration of 0.5–1 × 10^{-3} mole/liter was exceeded. The

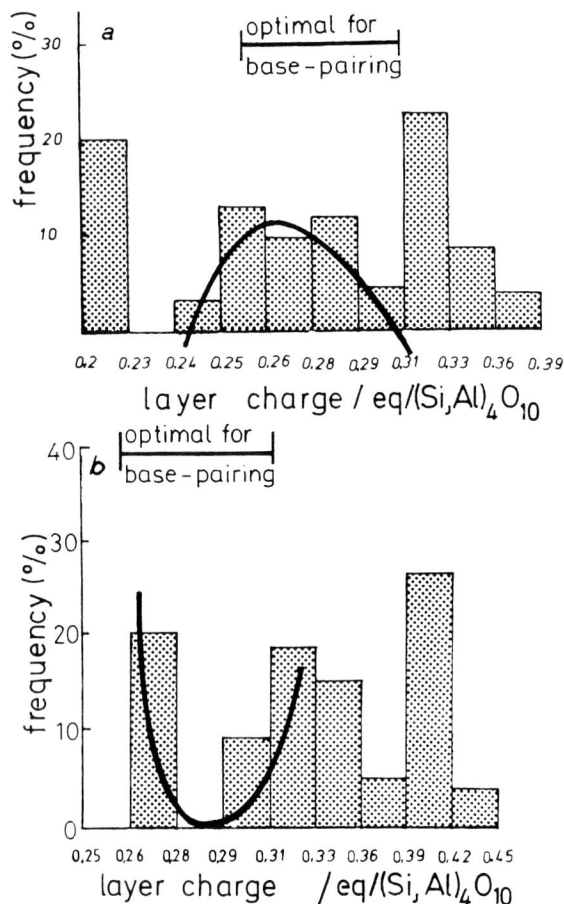

Figure 4. Charge distribution diagrams of (a) montmorillonite C (from Bavaria) and (b) montmorillonite F (from Arizona).

efficacy of the salts increased in the order: NaCl < NaNO₃ < NaI (Figure 5a). The influence of the salts was modest for Texas montmorillonite (sample E); only NaI enhanced the adsorption (Figure 5b).

The adsorption of cytosine on Wyoming montmorillonite (sample B) was also augmented above distinct cytosine concentrations (Figure 5c). The threshold concentrations were higher than for adenine, and plateaus were not reached. The reversed order of the salts was striking: NaNO₃ < NaI < NaCl. For montmorillonite E (Figure 5d), the influence of the salts was in the same order.

A possible explanation of the salt effect is based on the influence of the salts on the water structure. The anions Cl⁻, NO₃⁻, and I⁻ exert an increasing "breaking power" towards the water structure (Hofmeister series, see, e.g., Luck, 1978). They increase the proportion of OH groups not taking part in hydrogen bonds (i.e., free OH groups; Luck, 1978, 1984), which promotes hydration of the bases. The stronger hydration of the bases reduces their adsorption. A similar behavior was

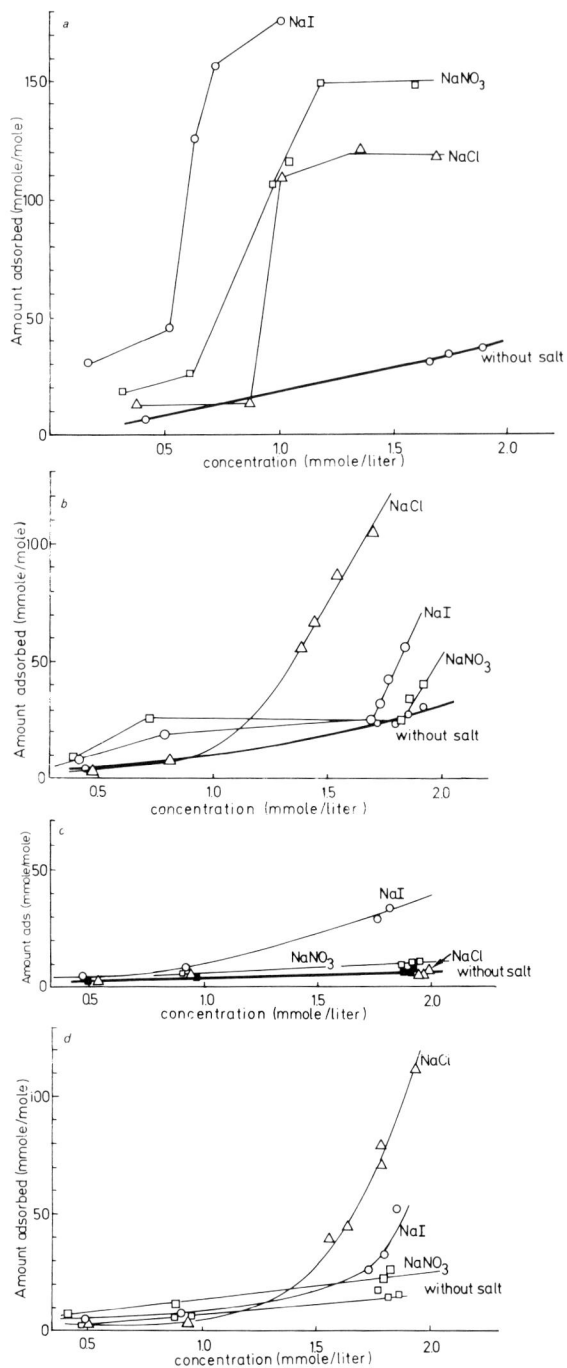

Figure 5. Effect of sodium salts on the adsorption of (a) adenine and (b) cytosine on montmorillonite B (from Wyoming, Crook County) and (c) adenine and (d) cytosine on montmorillonite E (from Texas). x = amounts of base adsorbed; c = equilibrium concentration of the base; salt concentration = 0.2 M, pH = 5, 20°C.

observed for polyethylene glycols. Agents that break the water structure promote hydration of the polyether chain, and the turbidity points are shifted to higher temperatures (Luck, 1978, 1984).

Table 2. Co-adsorption of nuclein bases.[1]

Montmorillonite	Base A	Base B	From mixture[3]			From "pure" solutions[4]		
			Base A	Base B	B/A	Base A	Base B	B/A
C. Niederschönbuch	Adenine	Thymine	21.2	10.3	0.49	32.2	5.5	0.17
	Adenine	Uracil	23.9	5.8	0.24	32.2	6.2	0.19
	Cytosine	Thymine	23.5	6.7	0.29	20.0	5.5	0.28
	Cytosine	Uracil	24.1	4.1	0.17	20.9	6.2	0.31
B. Wyoming SWy-1	Adenine	Thymine	27.7	13.2	0.48	34.9	2.3	0.07
	Adenine	Uracil	23.9	9.1	0.38	34.9	2.7	0.08
	Cytosine	Thymine	13.8	8.8	0.64	5.0	2.3	0.46
	Cytosine	Uracil	16.5	5.8	0.35	5.0	2.7	0.54

Amounts adsorbed[2]

[1] Initial concentration of each base 1×10^{-3} M, pH = 5, 20°C.
[2] mmole/mole silicate.
[3] 40 mg Ca-smectite in 5 ml aqueous solution of base A (2×10^{-3} M) + 5 ml aqueous solution of base B (2×10^{-3} M).
[4] 40 mg Ca-smectite in 10 ml of aqueous solution of base A or B (1×10^{-3} M).

Adenine and cytosine molecules show pronounced self-association in solution (Ts'o et al., 1963; Chan et al., 1964; Pullman et al., 1966). Pairing of cytosine molecules is energetically more favored than the formation of adenine pairs (Pullman et al., 1966). The addition of salts also promotes the self-association of the bases. This effect was more pronounced for adenine, because in pure water it was associated to a lesser extent than cytosine. Increased self-association enhanced adsorption, inasmuch as association of the bases on the surface played an important role. Thus, the addition of salt appears to exert two opposite effects: (1) it decreases adsorption if hydration effects are dominant, and (2) it increases adsorption if base-base associations are intensified.

The experimental observations with adenine and cytosine are in agreement with this model. The addition of salts to adenine solutions favored base-base association and increased the adsorption in the order of increasing structure breaking (NaCl < NaNO₃ < NaI).

Because the self-association of cytosine in water was greater than for adenine, the extent by which salt increased the hydration, was restricted. Sodium chloride might have enhanced the adsorption by intensifying the cytosine-cytosine association. Probably, the maximal degree of association was achieved. Increased structure breakup (from NaCl to NaNO₃, NaI) therefore promoted hydration of the pairs or oligomers so that, compared with NaCl-addition, the amounts of cytosine adsorbed were reduced.

Co-adsorption

The importance of base-base associations was also shown by co-adsorption phenomena. Only small amounts of uracil and thymine were adsorbed by montmorillonite, but the extent of adsorption was increased by an addition of adenine or cytosine (Table 2). The behavior was complex, and the results for one particular montmorillonite cannot be generalized. Table 2 gives two examples of the mutual influence of the bases.

Table 3. Two-step adsorption of bases.

Smectite	Step 1		Step 2			Adsorption
	Base A	Amounts adsorbed[1]	Base B	Amounts adsorbed[1]	Ratio B/A	of B on pure smectite[1]
C. Montmorillonite, Niederschönbuch	Adenine	32.2	Thymine or uracil	12.4	0.39	5.5
				37.1	1.15	6.2
	Cytosine	20.0	Thymine or uracil	0	0	5.5
				7.4	0.37	6.2
B. Montmorillonite, Wyoming, SWy-1	Adenine	34.9	Thymine or uracil	23.4	0.67	2.3
				9.0	0.26	2.7
	Cytosine	5.0	Thymine or uracil	0	0	2.3
				9.0	1.80	2.7
A. Hectorite, California, SHCa-1	Adenine	19.0	Thymine or uracil	0	0	2.2
				7.4	0.39	5.7
	Cytosine	1.6	Thymine or uracil	4.4	2.75	2.2
				7.4	4.63	5.7

[1] mmole/mole silicate.

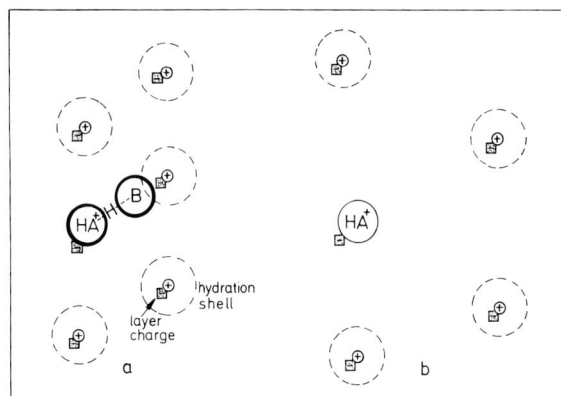

Figure 6. Model showing the influence of layer charge on the base-pairing on the surface. (a) Pairs of protonated and unprotonated base (HA⁺–B or HA⁺–A) form when the unprotonated base (B or A) can bridge between HA⁺ on an exchange site and the hydration shell around a neighboring interlayer cation; (b) only protonated base molecules are bound on exchange sites if the distances between possible binding sites are too long (layer charge too low, as shown in the model) or too short (layer charge too high).

Adenine enhanced the adsorption of thymine on both montmorillonites and enhanced the adsorption of uracil on montmorillonite B. The adsorption of adenine itself was reduced by both thymine and uracil, and the ratio of thymine (uracil) to adenine increased from about 0.1 to 0.5. A mutual influence of cytosine and thymine (uracil) on adsorption was only observed for montmorillonite B.

Two possible configurations of adenine-thymine pairs are slightly more stable than the pairs adenine-adenine (Pullman et al., 1966). Adenine-thymine pairs may thus have been preferentially adsorbed. Consequently the ratio thymine to adenine (and also uracil to adenine) increased, but a 1:1 ratio was never reached.

The reduced co-adsorption of cytosine-thymine is similar to the base-pairing in nucleic acids (complementary bases: adenine-thymine, noncomplementary: cytosine-thymine). The association between cytosine molecules was very strong and probably stronger than for cytosine-uracil and cytosine-thymine pairs. The adsorption of thymine and uracil was therefore less promoted by cytosine. Similarly, the preferential self-association of cytosine prevented co-crystallization with other bases (Pullman et al., 1966).

Two-step adsorption

The co-adsorption of thymine and uracil by adenine or cytosine was confirmed by the results of a two-step procedure. The smectites were first equilibrated with 1×10^{-3} M adenine or cytosine solution, washed with water, dried (65°C, $p \leq 10$ Pa), and then equilibrated with 1×10^{-3} M thymine or uracil solution. In separate runs, no or very small amounts of adenine or cytosine were desorbed during washing and equilibrating with

thymine or uracil solutions. The adsorption of thymine and uracil in the second step was augmented if adenine was pre-adsorbed in the first step (Table 3). The fact that pre-adsorption of adenine impeded thymine adsorption on hectorite clearly demonstrates that the extent of co-adsorption was sensitive to the layer charge and charge distribution.

Pre-adsorption of cytosine impeded the adsorption of thymine and increased only slightly the degree of uracil adsorption, probably due to the pronounced surface aggregation of cytosine.

SUMMARY

The adsorption of the nuclein bases adenine, cytosine, thymine, and uracil on smectites exhibit several unusual features. In aqueous solution the bases form strong hydrogen bonds with water molecules and are associated as pairs and oligomers. Adsorption is therefore affected by agents that change the water structure and the association of the bases. Several salts increase considerably the amount of bases adsorbed. The promoting effect of the salts is apparently related to their "breaking" action towards the water structure. Similarly, the extent of co-adsorption (increased adsorption of thymine and uracil in the presence of adenine and, to a lesser extent, cytosine) is dependent on the solution properties such as concentration, pH, and type and concentration of salts.

The level of adsorption appears to be a function of the layer charge. Hence, lateral interactions on the smectitic surface must play an important role. A speculative explanation is presented in Figure 6. Here, base-pairs are assumed to form only if one or more polar groups of one base take part in the hydration shell around an interlayer cation, whereas the other base in protonated form replaces an interlayer cation. The hydrogen bonding of one base to an organic cation already on an exchange site was documented by Doner and Mortland (1969). Thus, one nuclein molecule apparently bridges between a hydrated interlayer cation and the protonated base on an exhange site. Nuclein bases containing several polar groups are particularly suitable candidates for such interactions. Compatibility between the equivalent area and the area of base-pairs thus allows optimal aggregation of protonated and unprotonated species, probably as clusters. The clustering promotes the adsorption. This process appears to be similar to the phenomenon of surface condensation (Mingelgrin and Tsvetkov, 1985), but the clustering of the adsorbed species here is strongly controlled by the surface-charge density. Lower charge densities (equivalent area > base-pair area) favor the adsorption of non-associated bases, and the promoting effect of clustering is lost. At higher charge density, the formation of base-pairs is also rendered less probable because of the denser packing of the cations.

Due to the random variation of the charge density, only a part of the exchange sites possess the right geometrical conditions. Thus, in the pre-adsorption and two-step experiments, a 1:1 ratio of two complementary bases cannot be attained.

ACKNOWLEDGMENT

We are grateful to "Deutsche Forschungsgemeinschaft" for financial support.

REFERENCES

Bernal, J. D. (1951) *The Physical Basis of Life:* Routledge and Kegan Paul, London, 80 pp.

Cairns-Smith, A. G. (1966) The origin of life and the nature of the primitive gene: *J. Theoret. Biol.* **10**, 53–88.

Chan, S. I., Schweizer, M. P., Ts'o, P. O. P., and Helmkamp, G. K. (1964) A nuclear magnetic resonance study of the selfassociation of purine and methylpurine. *J. Amer. Chem. Soc.* **86**, 4182–4188.

Doner, H. E. and Mortland, M. M. (1969) Intermolecular interaction in montmorillonites: NH–CO systems: *Clays & Clay Minerals* **17**, 265–270.

El-Amamy, M. M. and Mill, Th. (1984) Hydrolysis kinetics of organic chemicals on montmorillonite and kaolinite surfaces as related to moisture content: *Clays & Clay Minerals* **32**, 67–73.

Farmer, V. C. and Mortland, M. M. (1966) An i. r. study of the coordination of pyridine and water to exchangeable cations in montmorillonite and saponite: *J. Chem. Soc. A,* 344–351.

Hofmann, U. (1961) Geheimnisse des Tons: *Ber. Dt. Keram. Ges.* **38**, 201–207.

Lagaly, G. (1981) Characterization of clays by organic compounds: *Clay Miner.* **16**, 1–21.

Lailach, G. E. and Brindley, G. W. (1969) Specific co-adsorption of purines and pyridines by montmorillonite: *Clays & Clay Minerals* **17**, 95–100.

Lailach, G. E., Thompson, R. D., and Brindley, G. W. (1968a) Adsorption of pyrimidines, purines, and nucleosides by Li-, Na-, Mg-, and Ca-montmorillonite: *Clays & Clay Minerals* **16**, 285–293, 295–301.

Lailach, G. E., Thompson, R. D., and Brindley, G. W. (1968b) Adsorption of pyrimidines, purines, and nucleosides by Co-, Ni-, Cu-, and Fe(III)-montmorillonite: *Clays & Clay Minerals* **16**, 295–301.

Luck, W. A. P. (1978) Zur Struktur des Wassers und wässeriger Systeme: *Progr. Colloid Interf. Sci.* **65**, 2–28.

Luck, W. A. P. (1984) Structure of water and aqueous systems: in *Synthetic Membrane Processes,* G. Belfort, ed., Academic Press, New York, 21–72.

Mingelgrin, U. and Tsvetkov, F. (1985) Surface condensation of organophosphate esters on smectites: *Clays & Clay Minerals* **33**, 62–70.

Mortland, M. M. (1966) Urea complexes with montmorillonite: an infrared absorption study: *Clay Miner.* **6**, 143–156.

Mortland, M. M. and Raman, K. V. (1968) Surface acidity of smectites in relation to hydration, exchangeable cation and structure: *Clays & Clay Minerals* **16**, 393–398.

Oparin, A. I. (1957) *The Origin of Life on the Earth:* Academic Press, New York, 495 pp.

Pullman, B., Claverie, P., and Caillet, J. (1966) Van der Waals-London interactions and the configuration of hydrogen-bonded purine and pyrimidine pairs: *Biochemistry* **55**, 904–912.

Stul, M. S. and van Leemput, L. (1982) Particle-size distribution, cation exchange capacity and charge density of deferrated montmorillonites: *Clay Miner.* **17**, 209–215.

Thompson, T. D. and Brindley, G. W. (1969) Adsorption of pyrimidines, purines, and nucleotides by Na-, Mg-, and Cu(III)-illite. *Amer. Mineral.* **54**, 858–868.

Touillaux, R., Salvador, P., Vandermeersche, C., and Fripiat, J. J. (1968) Study of water layers adsorbed on Na- and Ca-montmorillonite by the pulsed nuclear magnetic resonance technique: *Israel J. Chem.* **6**, 337–348.

Ts'o, P. O. P., Melvin, I. S., and Olson, A. C. (1963) Interaction and association of bases and nucleosides in aqueous solutions: *J. Amer. Chem. Soc.* **85**, 1289–1296.

van Olphen, H. and Fripiat, J. J. (1975) *Data Handbook for Clay Materials and Other Non-Metallic Minerals:* Pergamon Press, Oxford, 19–29 pp.

Vogt, K. and Köster, H. M. (1978) Zur Mineralogie, Kristallchemie und Geochemie einiger Montmorillonite aus Bentoniten: *Clay Miner.* **13**, 25–43.

Weiss, A. (1981) Replication and evolution in inorganic systems: *Angew. Chem. Internat. Ed.* **20**, 850–861.

Yariv, S. and Heller-Kallai, L. (1975) Comments on the paper: the adsorption of aromatic, heterocyclic and cyclic ammonium cations by montmorillonite: *Clay Miner.* **10**, 479–481.

Proceedings of the International Clay Conference, Denver, 1985, L. G. Schultz, H. van Olphen, and F. A. Mumpton, eds.,
The Clay Minerals Society, Bloomington, Indiana, 370–374 (1987).

SURFACTANT ADSORPTION AND RHEOLOGICAL BEHAVIOR OF SURFACE-MODIFIED SEPIOLITE

A. Alvarez,[1] J. Santarén,[1] R. Pérez-Castells,[1] B. Casal,[2]
E. Ruiz-Hitzky,[2] P. Levitz,[3] and J. J. Fripiat[3]

[1] Departamento de Investigación y Desarrollo, TOLSA, S.A., Madrid, Spain

[2] Instituto de Fisico-Química Mineral, C.S.I.C., Madrid, Spain

[3] Centre de Recherche sur les Solides à Organization Cristalline Imparfaite, C.N.R.S., Orléans, France

Abstract—Adsorption isotherms of a non-ionic surfactant (Triton X-100) and a quaternary cationic surfactant (Sanisol TPR) by aqueous dispersions of sepiolite gels have been measured at 25°C. The adsorption mechanisms of these two surfactants were different. Triton X-100 was adsorbed by hydrophyllic interaction, whereas the cationic surfactant was adsorbed in two stages. The first stage, to about 10 meq/100 g in the adsorption isotherm, involved cation exchange, as shown by a correlation between the amount of Mg extracted and the amount of adsorbed surfactant. The mechanism of adsorption in the second stage is not clear and was more complicated than that for smectite-quaternary systems.

The Brookfield viscosity and sediment volume of surface-modified sepiolite dispersions were found to depend on the nature and the amount of adsorbed surfactants and on the chemical nature of the solvent. Stable and highly viscous suspensions of Sanisol-modified sepiolite were obtained in organic solvents of low polarity (e.g., xylene), whereas Triton-modified sepiolite suspensions of high viscosity were obtained in polar organic solvents, such as triethanol amine.

Key Words—Adsorption, Cation exchange, Rheology, Sepiolite, Surfactant, Viscosity.

INTRODUCTION

By milling high-purity sepiolite, a commercial product has been developed that forms gels in water and in saline solutions (Alvarez et al., 1984a, 1984b, 1984c). Adsorption of surface-active agents yields "oleophillic" products, which form gels in organic solvents, even in those having low polarity. Surface-modified clay minerals that contain surface active agents are potentially of considerable industrial interest. They stabilize dispersions in organic media thereby preventing sedimentation of the suspended solid particles and provide suitable consistency and flow behavior to these media. For various applications, organo-layer clays have been studied extensively and applied commercially, but surfactant-treated chain-structure clay minerals such as sepiolite have been described less, and the available references are mainly in the patent literature (Sawyer, 1961; Matsumoto and Wada, 1979).

The present paper describes the adsorption of a cationic and a non-ionic surfactant on sepiolite and surveys the rheological and sedimentation behavior of dispersions of the resulting adsorption complexes in water and in organic solvents.

MATERIALS AND METHODS

Materials

The sepiolite was "rheological grade" (Pangel S, supplied by TOLSA, S.A.) and contained more than 95% sepiolite. It had a large N_2 BET surface area (340 m²/ g) and was <10 μm in size. Its cation-exchange capacity was about 10 ± 2 meq/100 g. The surface area available to water was determined to be 475 ± 25 m²/g by Fripiat et al. (1984).

The cationic surfactant, Sanisol TPR, was supplied by Sinor Kao. It is a benzyl dimethyl alkylammonium chloride with the following approximate chain-length distribution: C12, 1%; C14, 3%; C16, 30%; C18, 26%; and oleic C18, 40%. Its molecular or equivalent weight is 412. The non-ionic surfactant Triton X-100 was supplied by Rohm & Haas. It is an octylphenol polyoxythylene containing 10 ethoxy segments. Its molecular weight is 628.

Solvents of analytical grade supplied by Merck and by Fluka were used.

Methods

Adsorption isotherms were determined at 25°C for dispersions of 6% w/w of sepiolite in surfactant solutions at initial concentration in the range 5×10^{-4} to 5×10^{-2} N. The dispersions were gently stirred for 3 hr and then allowed to stand overnight. The modified clay was collected by vacuum filtration. The surfactant concentration was measured in the filtrate by UV spectrophotometry using the absorption at 258 or 277 μm (maximum absorption wave lengths of Sanisol TPR and Triton X-100, respectively) with a Varian-Cary 2300 spectrophotometer. For Sanisol, the Cl^- concentration in the filtrate was measured at each step of the

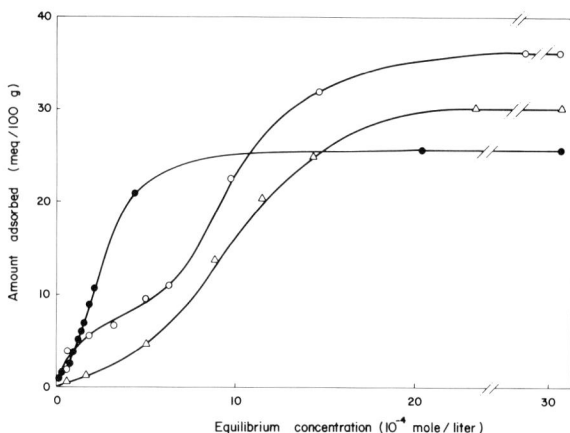

Figure 1. Adsorption isotherms of Triton X-100 (●) and Sanisol (○ total amount, △ SCl) in 6% (w/w) sepiolite-water dispersion at 25°C.

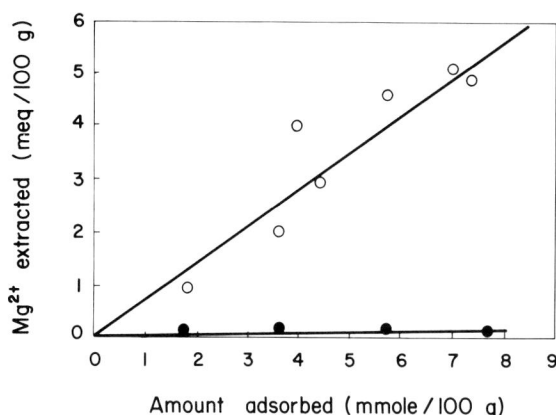

Figure 2. Amount of Mg^{2+} extracted vs. S^+ (meq) (○) and Triton X-100 (mmole) (●) adsorbed at 25°C.

adsorption isotherm with an ion selective electrode. The amount of Sanisol adsorbed in the cationic form (S^+) was determined from the difference between the total amount adsorbed ($S^+ + SCl$) and the amount adsorbed as salt (SCl).

The concentration of Mg^{2+} released by the sepiolite was determined in the filtrate by atomic absorption spectrophotometry.

A practical measurement of the rheological behavior of the 6% w/w suspension was obtained with a Brookfield RVT viscometer at a spindle speed of 5 rpm. In organic solvents a Helipath system at a spindle speed of 1 rpm was used. All the measurements were carried out at 20°C. Sediment volumes were obtained in 1% w/w suspensions by gravitation after 1 and 3 days, respectively.

RESULTS

Adsorption isotherms, sediment volumes, and flow behavior in aqueous solution

The adsorption isotherms for Sanisol and Triton X-100 in 6% w/w sepiolite aqueous dispersion are shown in Figure 1. The isotherms differ mainly in the range of low equilibrium concentrations where the nonionic surfactant isotherm is nearly linear. The cationic surfactant isotherm, however, has an inflection point near 10 meq/100 g. In this range Sanisol interaction with sepiolite involved cation-exchange, because, as shown in Figure 2, the amount of Mg^{2+} released increased linearly with the amount of Sanisol cation S^+ adsorbed. As expected, the adsorption of Triton X-100 was not accompanied by release of Mg^{2+}.

The sediment volumes of water slurries containing 1% sepiolite covered with increasing amounts of Triton X-100 or Sanisol are shown in Figures 3A and 3B, respectively. Using the cationic surfactant, the sediment volume reached a minimum at ~2 meq/100 g

and then increased to a value greater than that of untreated sepiolite until 10 meq/100 g of Sanisol was adsorbed. It then remained constant upon further Sanisol adsorption. The sediment volume of the Triton system decreased and then remained constant after adsorption of about 5 mmole/100 g.

The adsorption of surfactants strongly influenced the viscosity of 6% (w/w) water slurries of surface-modified sepiolite, as shown in Figure 4. As the surface coverage increased, the Brookfield viscosity at 5 rpm (about 15,000 cps for the pure sepiolite slurry) decreased sharply and tended towards a plateau at about 2000 cps for both types of molecules. This plateau was noted at about 5 mmole TX-100 and about 15 meq Sanisol adsorbed, respectively.

Flow behavior and sediment volumes of surfactant-treated sepiolite in organic solvents

Using surfactant-treated sepiolite, gels were prepared in organic solvents. In Figure 5 viscosity data are shown for systems containing a variety of organic solvents and both Sanisol- and Triton-modified sepiolite at maximum surface coverage. In all systems the viscosity increased exponentially with solid content.

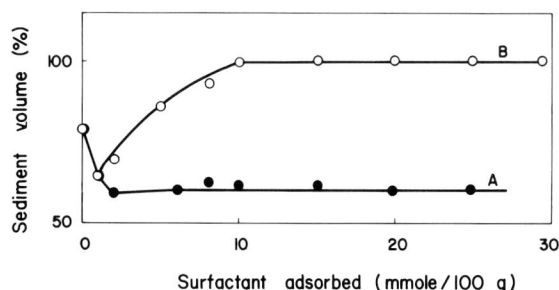

Figure 3. Sedimentation volume (%) of 1% (w/w) aqueous dispersions of sepiolite after adsorption of the indicated amount of surfactants, (A) Triton X-100 and (B) Sanisol.

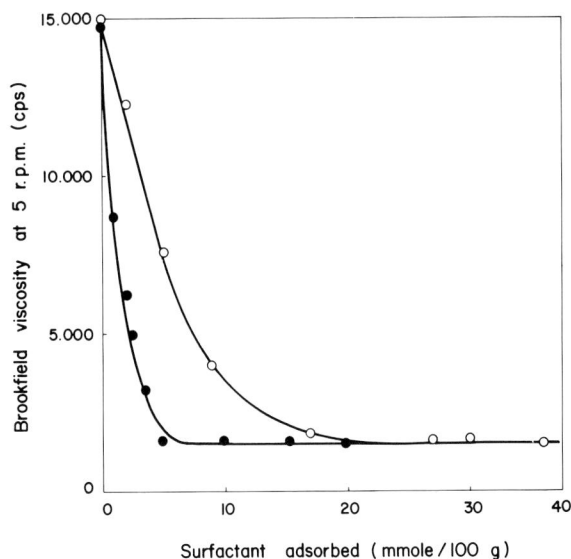

Figure 4. Brookfield viscosity (cps at 5 rpm, 20°C) of 6% (w/w) aqueous dispersions of sepiolite as a function of the amounts of surfactants adsorbed: (●), Triton and (○) Sanisol.

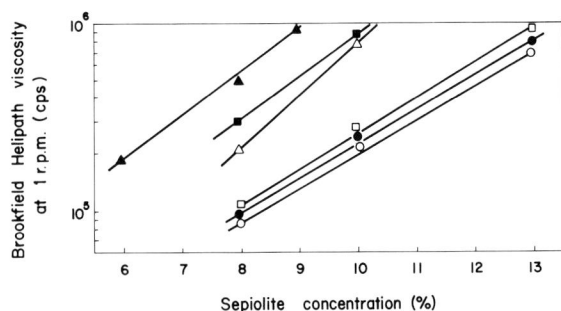

Figure 6. Brookfield-Helipath viscosity (cps) of (A) Triton-modified sepiolite in triethanol amine (1 rpm) and (B) Sanisol-modified sepiolite in xylene (1 rpm) at different water contents of the sepiolite.

Figure 6A shows that the viscosity of an 8% suspension of sepiolite containing different amounts of adsorbed Triton in triethanolamine and at clay moisture contents of 6% and 11% was in the range 300×10^3 to 500×10^3 centipoise. Therefore, neither the amount of Triton adsorbed in excess of that necessary to lower the viscosity in water to its minimum value (about 5 mmole/100 g) nor a clay moisture content as high as 11% significantly influenced the viscosity. High viscosities were also obtained for a 10% Sanisol-treated sepiolite dispersion in xylene (Figure 6B), but in this system the effect of moisture was appreciable. With increasing water content of the clay, the maximum viscosity shifted to systems with smaller amounts of adsorbed Sanisol.

The sediment volume of the Sanisol-xylene systems

increased slowly with surface coverage at low moisture content (2.7%), whereas it increased rapidly in systems containing 5.5% clay moisture. At low surface coverage, however, the sediment volume was less than that found at a clay moisture content of 2.7% (see Figure 7).

As a general note, the actual "surface coverage" in the organic solvent systems may have been less than that of the original treated materials due to a possible partial desorption in the presence of these solvents. The composition of the surface phase, however, was not determined.

DISCUSSION

Adsorption mechanisms

In sepiolite fibers two pore systems are distinguishable (Figure 8). The true zeolitic pores due to the structural characteristics are B-type pores, whereas A-type pores are due to defects in the arrangements of structural units. Although the surface areas of A- and B-type pores may be considered as internal surface, the surface of A-type pores actually contributes to the external surface area of the fibers.

From the viewpoint of the adsorption mechanism of surfactants, the available surfaces are indeed the external surfaces of the fibers as well as the surfaces of some of the A-type pores. This is evidenced by the variation of the N_2 BET surface area with the amount of adsorbed Triton X-100, as shown in Figure 9. The terminal SiOH groups were most probably the active centers for adsorbing the non-ionic surfactant. Assum-

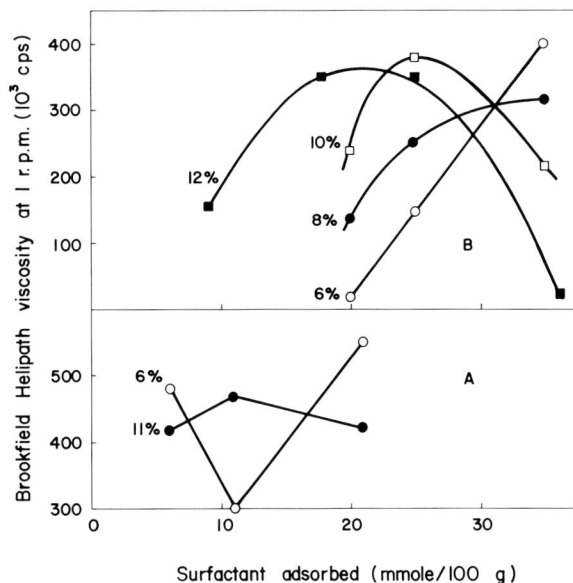

Figure 5. Brookfield-Helipath viscosity (cps at 1 rpm, 20°C) of Triton-modified sepiolite (10 mmole/100 g) in (△) triethanolamine, and Sanisol-modified sepiolite (28 meq/100 g) in (○) xylene, (●) benzene, (□) toluene, (■) di n-butyl phthalate, and (△) olive oil.

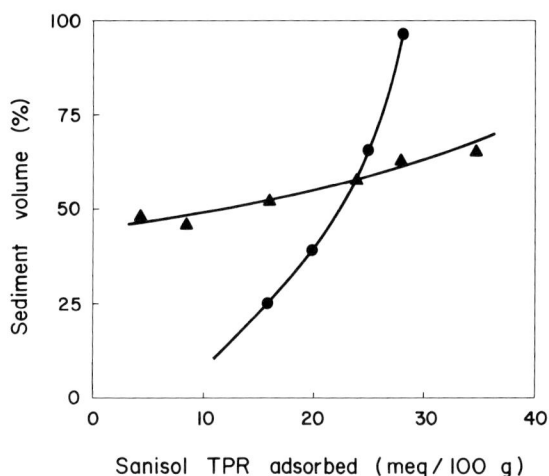

Figure 7. Sediment volume of 3% (w/w) sepiolite in xylene as a function of Sanisol-TPR adsorbed in presence of (▲) 2.7% water and (●) 5.5% water.

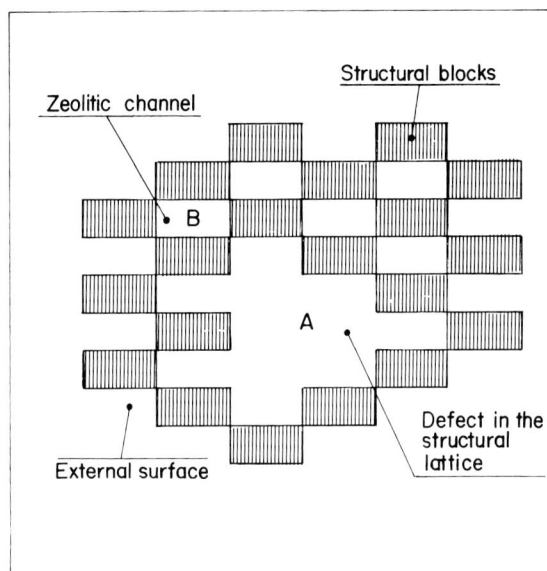

Figure 8. Schematic representation of the pore system of sepiolite fibers. Total nitrogen BET surface area is 340 m²/g. Total area available to water, as obtained from nuclear magnetic resonance measurement, is ≥450–500 m²/g (Fripiat et al., 1984).

ing, for example, an apparent molecular packing of 48 Å² per Triton X-100 molecule at the plateau, as obtained by Levitz (1985) for silica, the remaining external surface area of sepiolite on which micellar-like aggregates may be formed is about 73 m²/g for 25 × 10^{-3} mole adsorbed on 100 g of sepiolite, in good agreement with the data in Figure 9.

The adsorption of TX-100 on a non-microporous silica gel, close to the plateau in the corresponding isotherm, took place as surface micellar-like aggregates covering a mean surface area of about 10^4 Å² (Levitz, 1985). Assuming a similar process for sepiolite, good compatibility between the adsorbed layer and water was achieved in a large part of the adsorption isotherm near the plateau. This compatibility was probably the result of a part of the polyoxyethylene chains in the surface micelles being directed towards the solvent.

The adsorption mechanism of Sanisol on sepiolite appears to be more complicated than for smectites and quaternary ammonium compounds. In the latter systems, the first stage of cation-exchange adsorption was practically quantitive, and the clay surface became oleophillic because the hydrocarbon chains pointed towards the water phase. In a second stage, additional cations became associated with the previously adsorbed cations by van der Waals interaction of the hydrocarbon chains with the cationic groups pointing towards the water phase, and the clay-organic complex became hydrophillic again, with reversed charge.

In the Sanisol-sepiolite system, the first stage also involved ion exchange to 10 meq adsorbed per 100 g of sepiolite. At low amounts of Sanisol adsorbed (~2 meq/100 g), interaction between particles by hydrophobic bonds was possible, thereby decreasing the sediment volume (Figure 3B). Further adsorption of Sanisol did not eliminate the oleophillic properties of the

surface, because the complex was dispersible in hydrocarbons. In this adsorption step, the amount adsorbed as micellar-like aggregates increased (Figure 1).

Flow and sedimentation properties of the aqueous and organo-systems

The adsorption of the non-ionic Triton, as described above, should have resulted in particle repulsion in

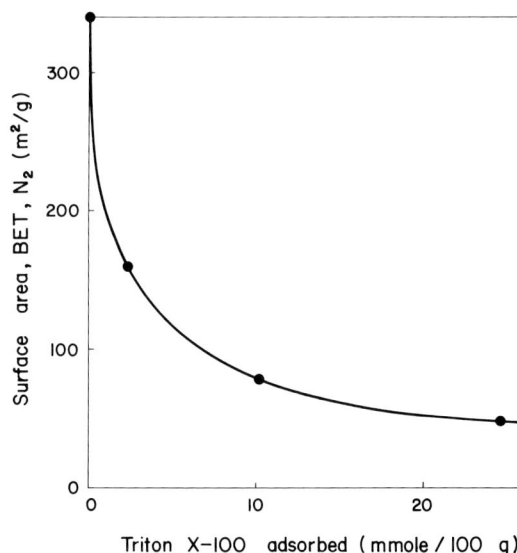

Figure 9. N₂ BET surface area of sepiolite with increasing coverage in Triton X-100.

aqueous systems, probably due primarily to steric hindrance effects ("entropy stabilization"). Such peptizing effect was indeed demonstrated by the lowering of the viscosity and the decreasing sediment volume (Figures 3A and 4). Because of the polar character of the Triton-covered surface, dispersions could only be prepared in polar organic solvents, not in hydrocarbons. The effect of some water in these systems was insignificant (Figure 6A).

In aqueous systems of Sanisol-treated sepiolite, exchange adsorption of the organic cation eliminated the diffuse electrical double layer, because the cations were strongly adsorbed at the surface. Therefore, the systems flocculated as shown by the increasing sediment volumes, despite the small dip in the curve at very low coverage (Figure 3B). Contrary to expectation, the viscosity decreased. Perhaps, the flocs were easily broken down in the Brookfield viscometer. The effect of water on the viscosity of Sanisol-coated sepiolite in xylene was possibly due to a modification in the polarity of the S^+Cl^- species. The good compatibility between the organic chain of Sanisol and the apolar solvent, however, permitted the formation of an intricate network of coated fibers and, thus, of a rigid gel.

More detailed rheological information, particularly on the yield stress and on thixotropic behavior, are needed to solve this discrepancy between sediment volume and flow behavior.

CONCLUSIONS

Sepiolite-surfactant complexes have several attractive properties of fundamental as well as applied interest. These properties are due to the fibrous habit of the complexes and to the heterogenous nature of their surfaces, which allow cation-exchange processes and hydrophillic interactions with polyethoxy chains. The experiments described herein show that the mechanism of surfactant adsorption and the colloid chemical consequences are more complicated than for analogous smectite systems because of the structural differences between the two types of clays. Further work is required to obtain a better understanding of the behavior of the sepiolite systems.

REFERENCES

Alvarez, A., Aragón, J. J., and Esteban, M. A. (1984a) Procedimiento de fabricación de un producto de sepiolita de grado reológico: *Spanish Patent* **534,838,** 11 pp.

Alvarez, A., Aragón, J. J., and Pérez-Castells, R. (1984b) Procedimiento de fabricación y purificación de un producto de sepiolita de grado reológico: *Spanish Patent* **534,891,** 12 pp.

Alvarez, A., Aragón, J. J., and Pérez-Castells, R. (1984c) Procedimiento de fabricación de alta concentración en sólidos de un producto de sepiolita-palygorskita de grado reológico: *Spanish Patent* **534,317,** 12 pp.

Fripiat, J., Letellier, M., and Levitz, P. (1984) Interaction of water with clay surfaces: *Phil. Trans. R. Soc. Lond. A* **311,** 287–299.

Levitz, P. (1985) Processus d'association des molécules amphiphites non ioniques: adsorption à l'interface solide hydrophileau et micellisation en phase aqueuse: Thèse d'état, Univ. Orléans, Orléans, France, 112 pp.

Matsumoto, K. and Wada, T. (1979) Sampi compound: *Japan Kokai* **54-75507,** 9 pp.

McAtee, J. L. (1959) Inorganic-organic cation exchange on montmorillonite: *Amer. Mineral.* **44,** 1230–1236.

Sawyer, E. W. (1961) Agent for gelling organic liquids and organic liquid gelled therewith. *U.S. Patent* **3,049,498,** 5 pp.

Proceedings of the International Clay Conference, Denver, 1985, L. G. Schultz, H. van Olphen, and F. A. Mumpton, eds.,
The Clay Minerals Society, Bloomington, Indiana, 375–381 (1987).

INTERACTION OF KAOLINITE AND MONTMORILLONITE WITH NEUTRAL POLYSACCHARIDES

C. Chenu,[1] C. H. Pons,[2] and M. Robert[1]

[1] Station de Science de Sol, I.N.R.A., 78000 Versailles, France

[2] Laboratoire de Cristallographie, Université d'Orléans, 45000 La Source, France

Abstract—The adsorption of polysaccharides by clays has been widely studied, but its consequences on clay fabric and clay aggregation have received little attention. Therefore, the interactions between St. Austell kaolinite and Wyoming montmorillonite and the microbial exopolysaccharides, dextran and scleroglucan, were investigated. Adsorption isotherms were determined by an anthron procedure, and the fabrics of the complexes were examined by X-ray powder diffraction (XRD), low-angle X-ray scattering (LAXS), and scanning electron microscopy (SEM). The wet-state organization of clay particles was preserved as much as possible. The aggregative effect of the polysaccharides was investigated using a wet sieving test.

The adsorption of scleroglucan was much higher than that of dextran. Three factors related to the structure of the polysaccharides are proposed to explain this difference, namely, the influence of comformation, apparent molecular weight, and solubility.

XRD experiments showed that both polysaccharides were adsorbed only on the external surfaces of the clay minerals. Dextran and scleroglucan changed the degree of parallel stacking in Ca-montmorillonite quasicrystals, but the particle arrangement of the starting clays observed by SEM was roughly preserved. The water-stable aggregation, however, increased with scleroglucan. These results on clay-polysaccharide complexes demonstrate the important role played by polysaccharide structure (e.g., tertiary and quaternary structure) in adsorption and clay aggregation.

Key Words—Adsorption, Aggregation, Fabric, Kaolinite, Montmorillonite, Polysaccharide.

INTRODUCTION

Polysaccharides play a major role in soil aggregation. Although several authors in an attempt to understand this action have studied the adsorption of polysaccharides on clays (e.g., Clapp *et al.*, 1968; Guckert, 1973; Guidi *et al.*, 1977; Olness and Clapp, 1975; Parfitt, 1972; Parfitt and Greenland, 1970), little is known about the changes in clay fabric and behavior induced by polysaccharides (Clapp and Emerson, 1972; Saini and MacLean, 1966). According to Martin (1971), Harris *et al.* (1966), and Tisdall and Oades (1979), the most effective polysaccharides are the microbial gums. These polymers possess high water-retention capacities and form viscous solutions and gels, properties which are related to their high molecular weights and to their tertiary and quaternary structures (Rees, 1977).

Microbial gums were therefore selected for a study of the interaction between polysaccharides and clays; namely, adsorption phenomena and the effect of polysaccharide adsorption on clay fabric and aggregation.

MATERIALS AND METHODS

St. Austell kaolinite and Wyoming montmorillonite, saturated with either Na or Ca, were used in this study. The <2-μm fraction was separated by repeated sedimentation and decantation. The suspension was washed three times with an equal volume of 1 M sodium or calcium chloride solution and then repeatedly with 10^{-3} M sodium or calcium chloride solutions until this salt concentration was reached in the clay suspension. The clay:water ratio was then 1:100.

The polysaccharides used were dextran, a bacterial secretion (produced by *Leuconostoc mesenteroides* B 512 F, from SIGMA) and scleroglucan, a fungal slime (from *Sclerotium glycanicum,* from CECA, France). Both are neutral glucose polymers having molecular weights of about 2×10^6 and 1.5×10^6, respectively. Dextran consists of 5% alpha(1–3) glucose side chains linked to an alpha(1–6) glucose backbone (Jeanes *et al.,* 1954). The geometry of the linkages and the random branching lead to a random coil conformation. No gels form (Powell, 1979). Scleroglucan is composed of a beta(1–3) glucose main chain having regularly attached beta(1–6) glucose side chains (Rinaudo and Vincedon, 1982) (Figure 1). In solution, scleroglucan occurs as aggregated triple helixes (Yanaki and Norisuye, 1983) and forms viscous solutions and gels (Bluhm *et al.,* 1982).

The adsorption experiments were carried out by mixing 10 ml of clay suspension with 10 ml of polysaccharide solutions of known concentration. The suspensions were shaken for 2 hr at 20°C and centrifuged at 12,000 *g* for 10 min. The polysaccharide concentration in the supernatant solution was determined using a modification of the anthron colorimetric method

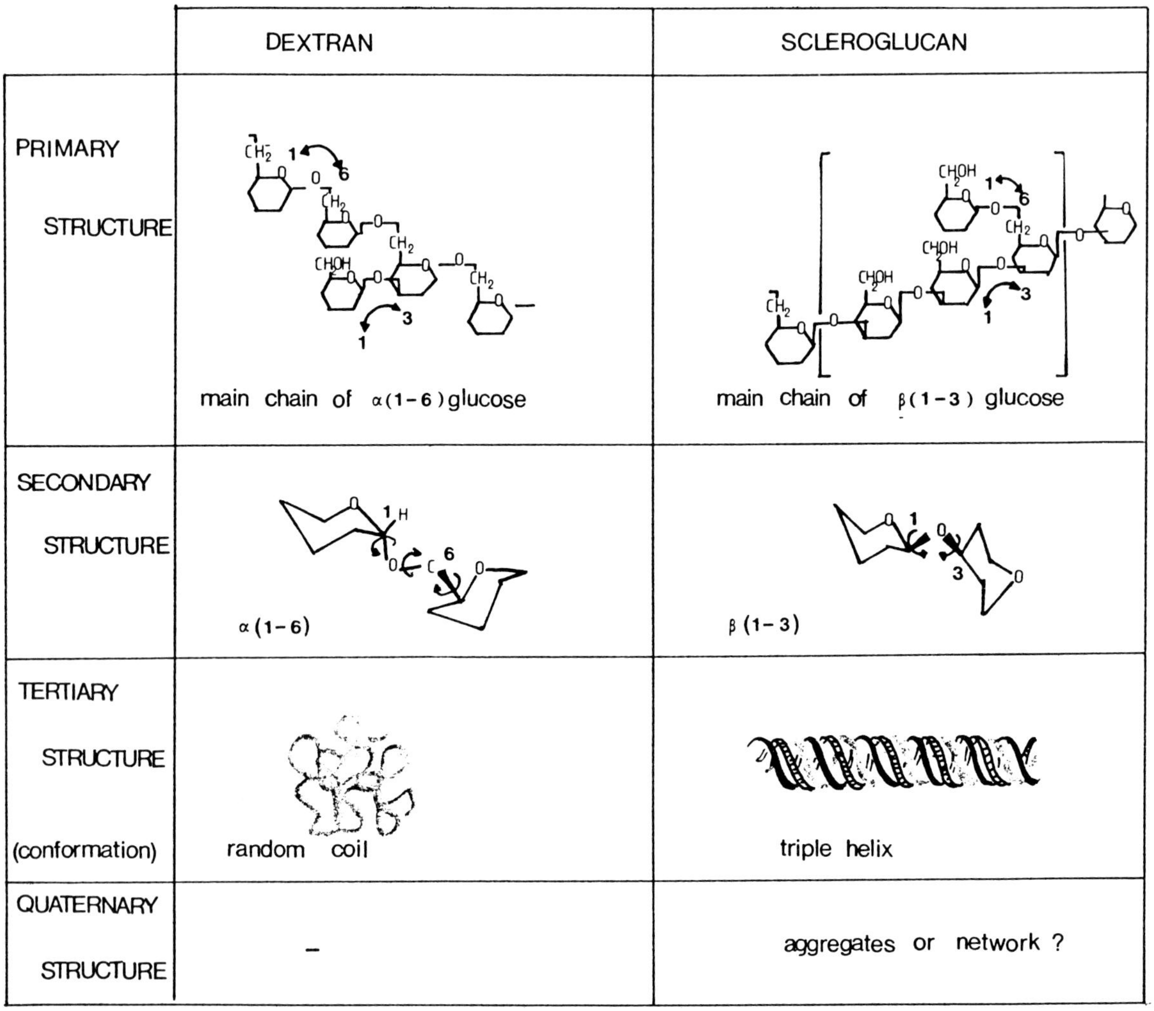

	DEXTRAN	SCLEROGLUCAN
PRIMARY STRUCTURE	main chain of α(1-6) glucose	main chain of β(1-3) glucose
SECONDARY STRUCTURE	α (1-6)	β (1-3)
TERTIARY STRUCTURE (conformation)	random coil	triple helix
QUATERNARY STRUCTURE	–	aggregates or network ?

Figure 1.   Molecular structure of dextran and scleroglucan.

of Morris (1948). One milliliter of the polysaccharide solution was mixed with 6 ml of the anthron solution (0.1% anthron in 85% sulfuric acid) and heated for 20 min at 80°C. After cooling the solution with ice, its optical density was measured at 620 nm. Because the optical density is proportional to the carbohydrate content, the polysaccharide concentration was determined with reference to standard glucose solutions.

The complexes were then submitted to different pressures (0.032 to 10 bar) to evacuate the water, using a modification of the Richards apparatus (Richards, 1947; Tessier and Berrier, 1979) and were characterized in the moist and the dry states by the following methods. By X-ray powder diffraction (XRD), the 001 spacings of the clay complexes were determined. Moist montmorillonites were examined by low-angle X-ray scattering (LAXS) (which depends on interlayer distances and the degree of parallel layer stacking within the clay aggregate) using a synchroton beam at L.U.R.E. (Orsay,

France) and the method of experimental diagram analysis proposed by Pons et al. (1982). Moist samples were also examined by scanning electron microscopy (SEM) using a cryoscan device after the samples were frozen in liquid Freon.

The state of aggregation was measured using a wet sieving test. First, dried samples were cut into 1-mm cubes; 500-mg portions of these artificial aggregates were then wetted with 10 ml of distilled water and sieved at 200 μm under water, as described by Hénin et al. (1958). The weight of the aggregates remaining on the sieve after this treatment was taken as the percentage of water-stable aggregates.

RESULTS

Adsorption

The adsorption of dextran on the clays was low (Figure 2), and the results were not precise because of the "difference" method used. Scleroglucan exhibited high-

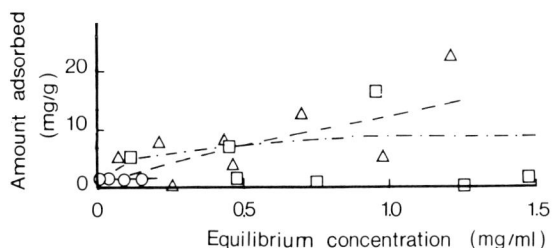

Figure 2. Adsorption isotherms of dextran on kaolinite and montmorillonite. O = Ca-kaolinite; △ = Ca-montmorillonite; □ = Na-montmorillonite.

affinity type isotherms, in which a steep slope is followed by a plateau (Figure 3). The maximum amount of scleroglucan adsorbed was 38 mg/g for Na-kaolinite, 41 mg/g for Ca-kaolinite, 149 mg/g for Ca-montmorillonite, and 385 mg/g for Na-montmorillonite. These results for dextran adsorption are markedly different from previously reported data. Parfitt and Greenland (1970), Clapp et al. (1968), and Guidi et al. (1977) reported adsorptions ranging from 67 to 457 mg/g for montmorillonite (Table 1).

The influence of different experimental conditions on the adsorption (Table 1) was measured by using different samples of dextran (from Pharmacia, Sweden) and by using various pretreatments of the clay; e.g., repeated washing with distilled water to investigate the hydrolysis effects (Shainberg, 1973) and freeze drying. The time dependence of the adsorption, and the influ-

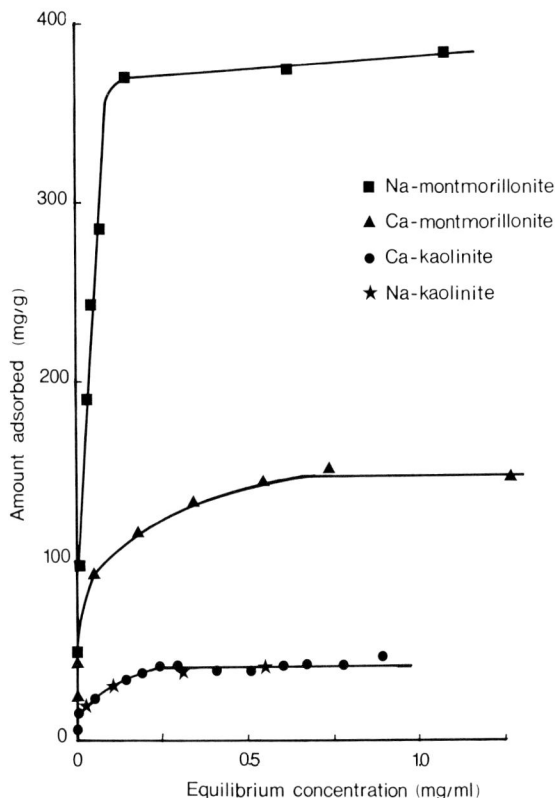

Figure 3. Adsorption isotherms of scleroglucan on kaolinite and montmorillonite.

Table 1. Literature data on dextran adsorption by Ca- and Na-montmorillonite and on experimental designs used in this investigation.

		Parfitt and Greenland (1970)	Clapp et al. (1968)	Olness and Clapp (1975)	Guidi et al. (1977)	Present paper
Results Amount of dextran adsorbed (mg/g)	Na-montmorillonite Ca-montmorillonite	457	250 110	443	96 67	25 20
Material						
Dextran	bacteria	Leuconostoc mesenteroïdes BF 512				
	mol. weight	2×10^6				
	origin	A. Jeanes collection			Pharmacia	Sigma
Clay mineral	nature	Wyoming montmorillonite				
	preparation	washed with distilled water				washed with 10^{-3} M solutions
	conservation	freeze dried				in suspension at 4°C
Experimental designs Mixing time (hr) Temperature (°C) Clay concentration (%) Presence of flocculating salts Centrifugation (rpm)		16 2 0.5 0	0.5 1 Na_2SO_4, 0.1 M 15,000	4 0 15,000	4 27 0.5 0	2 20 0.5 0 10,000

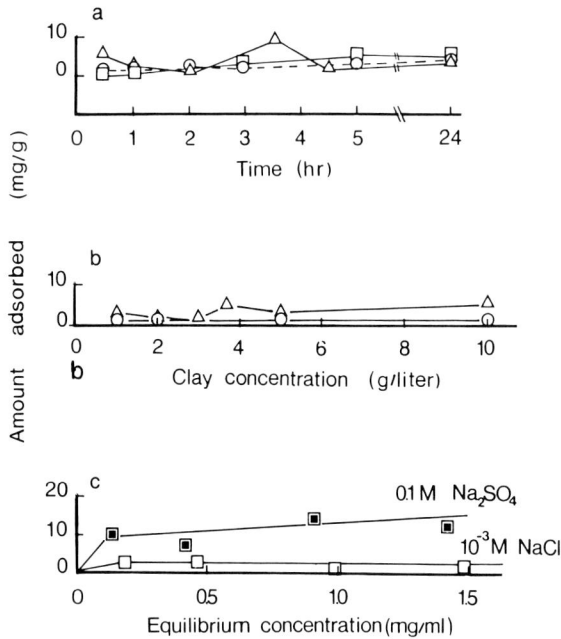

Figure 4. Influence of (a) time, (b) clay concentration, and (c) flocculating salts on the adsorption of dextran on kaolinite and montmorillonite. O = Ca-kaolinite; △ = Ca-montmorillonite; □ = Na-montmorillonite.

ence of clay : water ratio, flocculating salts (Figure 4), and the centrifugation conditions were also investigated. None of these treatments significantly affected the adsorption. Perez and Proust (1985) reported similar results; however, differences between the present results and those in the literature concerning the adsorption of dextran on clay minerals may be due to differences in the clay samples used and the methods used to prepare the dextran samples.

Fabric

XRD data showed that the 001 spacing was unchanged, suggesting that adsorption occurred only on external surfaces of the montmorillonite. LAXS indicated that the parallel stacking in Ca-montmorillonite amounted to 16 layers and that it remained constant during evacuation drying and rehydration (Figure 5). As demonstrated by Pons *et al.* (1982) and Ben Rhaïem *et al.* (1987), the particle thickness of untreated Ca-montmorillonite, however, increased during dehydration, but did not return to the initial thickness on rehydration (Figure 5). Therefore, the adsorption of polysaccharides on the external surface of primary particles hindered their parallel stacking during drying (Figure 6).

The SEMs (Figures 8a and 8b) show that dextran did not change the particle arrangement in the kaolinite or the montmorillonite. In the complexes of Ca-kaolinite and Ca-montmorillonite and scleroglucan (Figures 8c and 8f), the polymer formed strands 0.2–1 μm

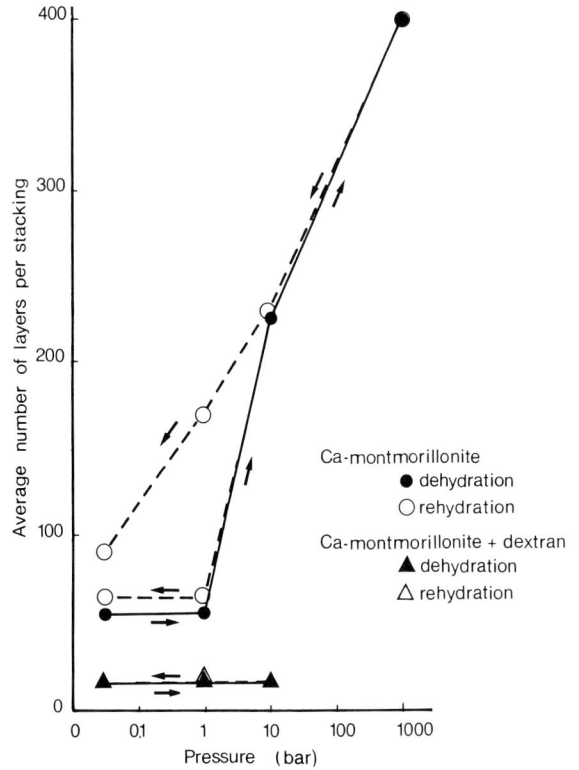

Figure 5. Effect of dextran on the thickness of Ca-montmorillonite stackings during drying and rehydration, as determined by low-angle X-ray scattering.

in length and 0.01 μm in thickness. Similar strands were observed by Koreeda and Harada (1974) in electron micrographs of various polysaccharides, including beta(1–3) glucans, but the precise morphology observed for both types of treated clays could not be distinguished either as the true morphology of the hy-

Figure 6. Effect of polysaccharides on the thickness of Ca-montmorillonite stackings.

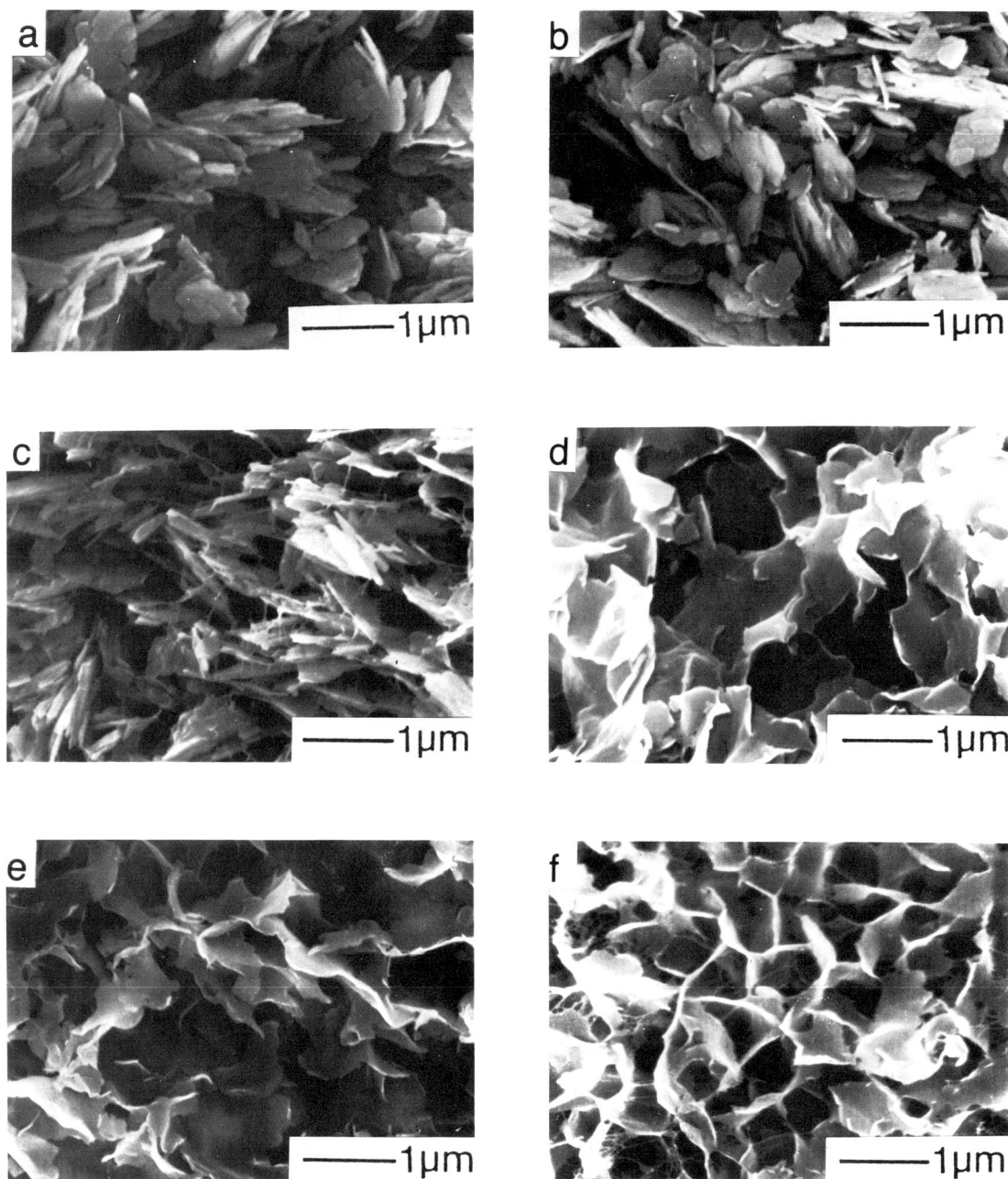

Figure 7. Scanning electron micrographs of clay-polysaccharide complexes. (a) Ca-kaolinite; (b) Ca-kaolinite + 1% dextran; (c) Ca-kaolinite + 5% scleroglucan; (d) Ca-montmorillonite; (e) Ca-montmorillonite + 2% dextran; (f) Ca-montmorillonite + 15% scleroglucan. (Moist state; pressure = 0.032 bar.)

drated polymer or as an artefact due to freezing and/or dehydration (Roth, 1977). The Ca-montmorillonite quasicrystals appeared to be smaller after the scleroglucan was added (Figures 7d and 7f), which is consistent with the results obtained by LAXS. In both clays, however, the characteristic arrangement of the particles was preserved; i.e., the card house structure

of kaolinite and the three-dimensional network of montmorillonite (Tessier and Pédro, 1982).

With scleroglucan, even at low concentrations, the percentage of water-stable aggregates increased for Ca-kaolinite, but the dextran was not very effective in this respect (Figure 8a). The aggregate stability of Ca-montmorillonite, which was much higher than that of Ca-

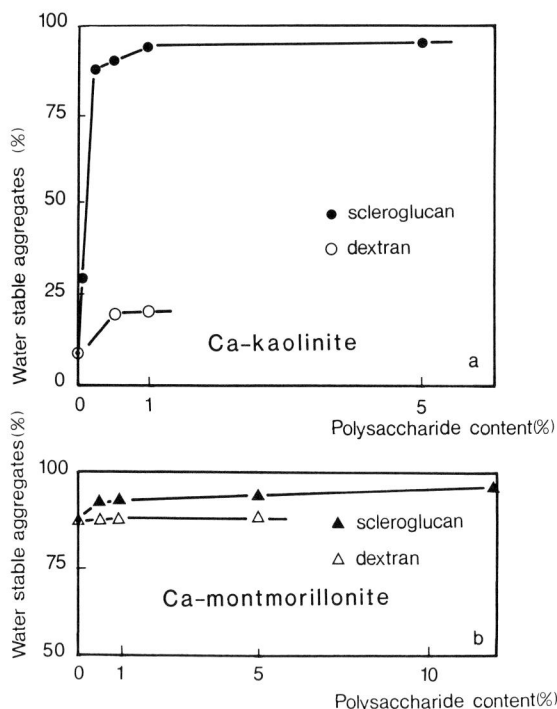

Figure 8.   Effect of polysaccharides on the water stability of Ca-kaolinite and Ca-montmorillonite.

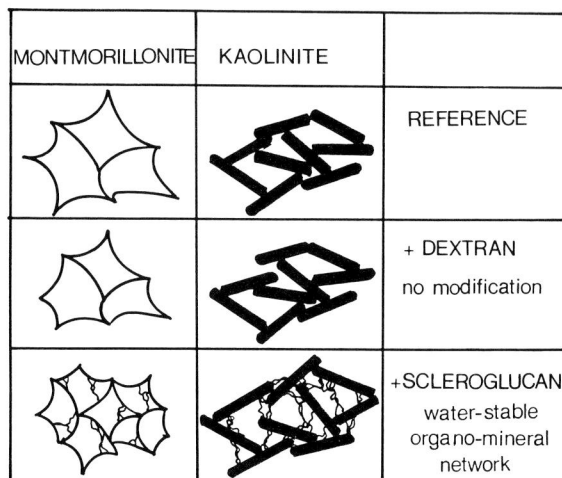

Figure 9.   Schematic representation of the fabric of Ca-kaolinite and Ca-montmorillonite in presence of polysaccharides.

kaolinite, was not affected by dextran and only slightly affected by scleroglucan (Figure 8b).

DISCUSSION

Both polysaccharides were adsorbed on the external surfaces of the clay minerals, but in very different amounts. The higher adsorption of scleroglucan vs. dextran can be explained as follows:

1. Adsorption of neutral glucose polymers on clay minerals involves van der Waals forces and hydrogen bonds (Theng, 1979). These are low-energy forces, but they can become very effective if numerous links are formed between the macromolecule and the solid surface (La Mer and Healy, 1963). Consequently, close contact of the polymer and the clay surface is essential, and the triple helix of scleroglucan gives rise to a greater degree of contact with the clay surface than does a random coil.
2. Aggregation of the scleroglucan triple helixes increases its apparent molecular mass and, thus, may enhance adsorption (La Mer and Healy, 1963; Silberberg, 1968).
3. Scleroglucan is less soluble than dextran, and it is well known that reduced solubility promotes polymer adsorption (Silberberg, 1968).

The relative importance of these three mechanisms is still difficult to assess, and conformation, apparent mass, and solubility are all related to the structure of the polysaccharides (for the relation between the structure and solubility of glucans, see Rinaudo and Vincedon, 1982). The amount of scleroglucan adsorbed by the clay minerals seems to have depended on the external surface available on the solid (12 m^2/g for Na- and Ca-kaolinite; 50 m^2/g for Ca-montmorillonite), but further investigations are needed.

Because dextran and scleroglucan were adsorbed in significantly different amounts, they did not have the same effect on the clay minerals. Their influence on clay fabric, as summarized in Figure 9, was complex. Dextran and scleroglucan did not affect the interlamellar space, but they modified the size of Ca-montmorillonite quasicrystals and on another scale did not significantly change the particle arrangement.

The most striking effect of these polysaccharides was the increased amount of water-stable aggregates of the clay minerals, particularly kaolinite, for scleroglucan-treated materials. This phenomenon indicates either a reduced wettability of the clay mineral due to the adsorbed polymer or the formation of polysaccharide bridges between the clay particles. In the electron microscope, these bridges were noted as strands. The macromolecular characteristics of scleroglucan (high molecular weight, linear structure, and a habit of aggregates or networks) may have promoted the formation of such bridges.

CONCLUSIONS

1. The adsorption of neutral polysaccharides modifies the fabric of kaolinite and montmorillonite and influences their physical behavior.
2. Recent investigations have established that most fungal slimes are neutral and resemble scleroglucan (Deslandes et al., 1980; Dubourdieu et al., 1981; Barreto-Bergter and Gorin, 1983). Thus, our results

confirm the major role of fungi in soil aggregation as reported by Tisdall and Oades (1979, 1982). The fungi appear to act by physical entanglement of soil particles (Tisdall and Oades, 1979, 1982), but, according to the present results, their secretions may also have a pronounced effect on soil structure.

3. Because the behavior of dextran and scleroglucan is different, it is difficult to understand the interactions of polysaccharides with soils if only one of these molecules is used as a model. Dextran, which has often been used in soil studies, does not possess the gel-like properties or the structure of most microbial gums. We believe that it is therefore not very representative of the microbial polysaccharides that are naturally present in soils.

4. The structure of polysaccharides and more particularly their tertiary and quaternary structures appear to be one of the most important factors in the interaction of clays and polysaccharides.

REFERENCES

Barreto-Bergter, E. and Gorin, P. A. J. (1983) Structural chemistry of polysaccharides from fungi and lichens: *Adv. Carbohydrate Biochem.* **41**, 67–103.

Ben Rhaïem, H., Pons, C. H., and Tessier, D. (1987) Factors affecting the microstructure of smectites. Role of cation and history of applied stresses: in *Proc. Int. Clay Conf., Denver, 1985,* L. G. Schultz, H. van Olphen, and F. A. Mumpton, eds., The Clay Minerals Society, Bloomington, Indiana, 292–297.

Bluhm, T. L., Deslandes, Y., Marchessault, R. H., Perez, S., and Rinaudo, M. (1982) Solid state and solution conformation of scleroglucan: *Carbohydrate Res.* **100**, 117–130.

Clapp, C. E., Olness, A. E., and Hofmann, D. J. (1968) Adsorption studies of a dextran on montmorillonite: in *Trans. 9th Int. Congr. Soil Sci., Adelaide, 1968 Vol. 1,* Angus and Robertson, Sydney, 627–637.

Clapp, C. E. and Emerson, W. W. (1972) Reactions between Ca-montmorillonite and polysaccharides: *Soil Sci.* **114**, 210–216.

Deslandes, Y., Marchessault, R. H., and Sarko, A. (1980) Triple helical structure of (1–3) beta-D glucan: *Macromolecules* **13**, 1466–1471.

Dubourdieu, D., Ribereau-Gayon, P., and Fournet, B. (1981) Structure of the extracellular beta-D glucan from *Botrytis cinerea: Carbohydrate Res.* **93**, 294–299.

Guckert, A. (1973) Contribution à l'étude des polysaccharides dans les sols et de leur rôle dans les mechanismes d'aggregation: Thèse Doct. d'Etat, Univ. Nancy, Nancy, France, 124 pp.

Guidi, G., Petruzzelli, G., and Giachetti, M. (1977) Molecular weight as an influencing factor on the adsorption of dextrans on sodium and calcium montmorillonite: *Zeit. Planzenernährung Bodenkunde* **141**, 367–377.

Harris, R. F., Chesters, G., and Allen, O. N. (1966) Dynamics of soil aggregation: *Adv. Agron.* **18**, 107–168.

Hénin, S., Monnier, G., and Combeau, A. (1958) Méthode pour l'étude de la stabilité structurale des sols: *Ann. Agron.* **9**, 73–92.

Jeanes, A., Haynes, W. C., Wilham, C. A., Rankin, J. C., Melvin, E. H., Austin, M. J., Cluskey, J. E., Fisher, B. E., Tsuchiya, H. M., and Rist, C. E. (1954) Characterization of dextrans from ninety strains of bacteria: *J. Amer. Chem. Soc.* **76**, 5041–5052.

Koreeda, A. and Harada, T. (1974) Study of the ultrastructure of gel forming (1–3) beta-D glucan (Curdlan-type polysaccharide) by electron microscopy: *Carbohydrate Res.* **33**, 396–399.

La Mer, V. K. and Healy, T. W. (1963) Adsorption flocculation reactions of macromolecules, at the liquid-solid interface: *Rev. Pure Appl. Chem.* **13**, 112–113.

Martin, J. P. (1971) Decomposition and binding action of polysaccharides in soil: *Soil Biol. Biochem.* **3**, 33–41.

Morris, D. C. (1948) Quantitative determination of carbohydrate with Dreywood's anthron reagent: *Science* **107**, 254–255.

Olness, A. and Clapp, C. E. (1975) Influence of polysaccharide structure on dextran adsorption by montmorillonite: *Soil Biol. Biochem.* **7**, 113–118.

Parfitt, R. L. (1972) Adsorption of charged sugars by montmorillonite: *Soil Sci.* **113**, 417–421.

Parfitt, R. L. and Greenland, D. J. (1970) Adsorption of polysaccharides by montmorillonite: *Soil Sci. Soc. Amer. Proc.* **34**, 862–866.

Perez, E. and Proust, J. E. (1985) Effect of a nonadsorbing polymer on colloid stability: force measurement between mica surfaces immersed in dextran solution: *J. Phys. Lett.* **46**, L79–L84.

Pons, C. H., Tessier, D., Ben Rhaïem, H., and Tchoubar, D. (1982) A comparison between X-ray studies and electron microscopy observations of smectite fabric: in *Proc. Int. Clay Conf., Bologna, Pavia, 1981,* H. van Olphen and F. Veniale, eds., Elsevier, Amsterdam, 173–179.

Powell, D. A. (1979) Structure, solution properties and biological interactions of some microbial extracellular polysaccharases: in *Microbial Polysaccharides and Polysaccharases,* R. C. W. Berkeley, G. W. Gooday, and D. C. Elwood, eds., Academic Press, London, 117–160.

Rees, D. A. (1977) *Polysaccharide Shapes:* Chapman and Hall, London, 80 pp.

Richards, L. A. (1947) A pressure membrane apparatus—construction and use: *Agric. Eng.* **28**, 451–454.

Rinaudo, M. and Vincedon, M. (1982) ^{13}C NMR structural investigation of scleroglucan: *Carbohydrates Polymers* **2**, 135–144.

Roth, Y. L. (1977) Physical structure of surface carbohydrates: in *Surface Carbohydrates of the Procaryotic Cell,* I. W. Sutherland, ed., Academic Press, London, 5–26.

Saini, G. R. and MacLean, A. A. (1966) Adsorption flocculation reactions of soil polysaccharides with kaolinite: *Soil Sci. Soc. Amer. Proc.* **30**, 697–699.

Shainberg, T. (1973) Rate and mechanism of Na-montmorillonite hydrolysis in suspensions: *Soil Sci. Soc. Amer. Proc.* **37**, 688–693.

Silberberg, A. (1968) Adsorption of flexible macromolecules. IV. Effect of solvent-solute interaction, solute concentration, and molecular weight: *J. Chem. Phys.* **48**, 2835–2851.

Tessier, D. and Berrier, J. (1979) Utilisation de la microscopie électronique à balayage dans l'étude des sols. Observation de sols humides soumis à différents pF: *Sci. Sol* **1**, 67–82.

Tessier, D. and Pédro, G. (1982) Electron microscopy of Na-smectite. Role of layer charge, salt concentration and suction parameters: in *Proc. Int. Clay Conf., Bologna, Pavia, 1981,* H. van Olphen and F. Veniale, eds., Elsevier, Amsterdam, 165–176.

Theng, B. K. W. (1979) *Formation and Properties of Clay-Polymer Complexes:* Elsevier, Amsterdam, 362 pp.

Tisdall, J. M. and Oades, J. M. (1979) Stabilization of soil aggregates by the root systems of ryegrass: *Aust. J. Soil Res.* **17**, 429–441.

Tisdall, J. M. and Oades, J. M. (1982) Organic matter and water-stable aggregates in soils: *J. Soil Sci.* **33**, 141–163.

Yanaki, T. and Norisuye, T. (1983) Triple helix and random coil of scleroglucan in dilute solution: *Polymer J.* **15**, 389–396.

INDUSTRIAL
AND
ENVIRONMENTAL
APPLICATIONS

Proceedings of the International Clay Conference, Denver, 1985, L. G. Schultz, H. van Olphen, and F. A. Mumpton, eds.,
The Clay Minerals Society, Bloomington, Indiana, 385–390 (1987).

CHARACTERIZATION SCHEME FOR CLAY MINERALS USED IN THE CERAMIC TILE INDUSTRY

ALIGI DE PRETIS,[1] DINO MINICHELLI,[2] AND FLAVIA RICCIARDIELLO[1]

[1] Istituto di Chimica Applicata e Industriale, Università di Trieste
via Valerio, 34127 Trieste, Italy

[2] Svilluppo e Ricerca per la Ceramica, Snc, via dei Burlo 1, 34100 Trieste, Italy

Abstract—Eighty samples of the general type of clay raw materials used in the ceramic tile industry of Sassuolo, Italy, for single-firing tile production were subjected to chemical, physical, and technological analyses to determine the minimum number of parameters needed to select suitable raw materials. Most of the samples came from the Westerwald basin in the Federal Republic of Germany, the Provins basin of France, and several areas of Sardinia. The unfired raw materials were milled to the same size (100% <500 μm) and characterized by X-ray powder diffraction, X-ray fluorescence, thermal, and organic carbon analysis. The samples were then pressed into tiles and fired in a modern industrial roll kiln at a temperature of 1200°C. Porosity measurements, reflected light microscopic determinations, and mechanical tests were performed on the sintered bodies. A flow sheet detailing the progression from sulfur analysis, to thermogravimetry, to organic carbon and chemical analysis, to X-ray powder diffraction, to bending strength determination, to water absorption and shrinkage measurements, to dilatometry, and finally to porosity determination and optical microscopy was developed. This approach established for each type of analysis a range of acceptable values, permitting the successive rejection of unsuitable samples by a step-wise process.

The application of this method permitted a reduction in the number of potentially suitable samples from 80 to 22, thereby reducing the number of laboratory and industrial tests needed for further evaluation. By this process, sample evaluation times were dramatically shortened.

Key Words—Ceramic tile, Geotechnical properties, Kaolin, Porosity, Thermal properties, Water absorption.

INTRODUCTION

The Italian ceramic floor and tile industry, which is largely concentrated in the Sassuolo district, is committed to reducing production costs by the adoption of energy-saving technology. This new technology has affected all stages of the manufacturing cycle, particularly the firing process. The proper choice of the ceramic batch composition has therefore become very important, and starting materials must be selected carefully according to their physicochemical and technological characteristics. Materials are currently selected on the basis of the following experimental determinations: (1) sulfur, organic carbon, and chemical analysis; (2) thermogravimetry; (3) X-ray powder diffraction; (4) bending strength; (5) water absorption and linear shrinkage; (6) dilatometry; (7) porosity; and (8) optical microscopy.

The number of tests is therefore large, and generally, a large number of materials from different producers must be evaluated. This work is time-consuming and requires well-equipped research laboratories, which are generally not available in the ceramic industries of the Sassuolo district. The present work was therefore undertaken with the objective of reducing sample-evaluation time by a succession of experimental tests that would permit a stepwise elimination of unsuitable materials.

EXPERIMENTAL

Materials

Eighty clay mineral samples were obtained from suppliers and subjected to the following tests. The prices of the materials and the size and uniformity of the mineral deposits (important data from an economic point of view) were not taken into consideration in this investigation.

Methods

The following tests were performed on all samples ground to <500 μm and dried at room temperature:

1. Quantitative chemical analyses were made of each sample by atomic absorption spectroscopy (AA), after an alkali melting of the sample.
2. X-ray powder diffraction (XRD) was carried out using a vertical goniometer (Philips) and Ni-filtered CuKα radiation, at a scanning rate of 1°2θ/min.
3. Thermogravimetric (TGA) and differential thermogravimetric (DTG) analyses were made at a heating rate of 10°/min.

385

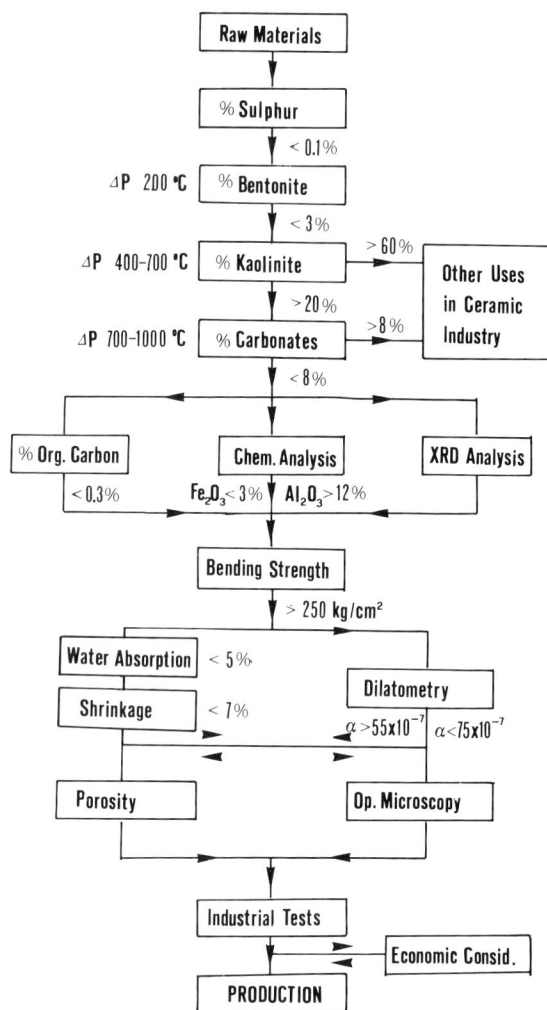

Figure 1.   Flow sheet of the testing scheme.

4. Linear shrinkage (L.S.) and water absorption (W.A.) were determined at different firing temperatures.
5. Bending strength was determined by means of a Netzsch three-point bending-strength device.

TESTING SEQUENCE

To minimize the amount of analytical work, a sequence of tests, together with limiting values for the acceptability of the materials, was developed. The sequence is shown as a flow sheet in Figure 1. The sequence of the tests and their limiting values are discussed below.

Initial testing

Sulfur content. The sulfur content of a sample is extremely critical (Aliprandi, 1975) and was therefore selected as the first parameter to be determined. Because of the high heating rate in the roller kiln, complete oxidation of the sulfur is difficult to achieve. Incomplete oxidation may lead to undesired formation of the well-known "black-heart" (Kingery, 1960). Our industrial experience has shown that the largest amount of sulfur permissible in these materials is 0.1%; all samples with higher sulfur content were therefore immediately rejected.

Bentonite content. The bentonite content of a sample is also critical, because the presence of bentonite causes the material to swell and break during drying and preheating (Emiliani and Emiliani, 1982). The presence of bentonite can be detected in X-ray powder diffraction diagrams; a peak at 6°–8°2θ is typical of this phase. Quantitative estimates of bentonite were carried out by a TGA method at temperatures <200°C (MacKenzie, 1972b). Those materials for which the weight loss at 200°C did not exceed 3% were accepted, because our industrial experience has shown that this is the maximum acceptable bentonite content.

Kaolinite content. The kaolinite content is of basic importance in tile production. To estimate the kaolinite percentage of a sample, the difference between the weight loss at 400°C and 700°C, on the TGA curve, was calculated and related to the transformation of kaolinite into metakaolinite (MacKenzie, 1972a). Very plastic materials (i.e., kaolinite contents of >60%) were considered too refractory for the working temperature in roller kilns and were therefore rejected. Also, materials with kaolinite contents of <20%, were rejected because they are unsuitable for tile shaping.

Carbonate content. Clays for single firing must contain a low percentage of carbonate (calcite + dolomite), because these minerals promote porosity in the fired materials. Calcite and dolomite were determined quantitatively from the difference between the weight loss at 700°C and that at 1000°C on the TGA curve. Materials with carbonate contents of >8% were rejected.

4. Total sulfur (SA) was determined to a precision of ±0.005%.
5. Organic carbon was determined by volumetric chemical analysis.

The following tests were carried out on the powders shaped with 4–7% admixed water into 50 × 100 mm tiles, under 250 kg/cm² pressure and subsequently fired in an electrical furnace at temperatures ranging from 1000° to 1250°C:

1. The pore-size distribution (PSD) was obtained using a ERBA 2000 mercury porosimeter.
2. Samples mounted in resin (polyester) and polished using the usual metallographic method were examined by optical microscopy.
3. Dilatometric analyses (TEA) were made in the temperature range 20° to 1000°C and the linear thermal expansion coefficient was calculated up to a temperature of 400°C.

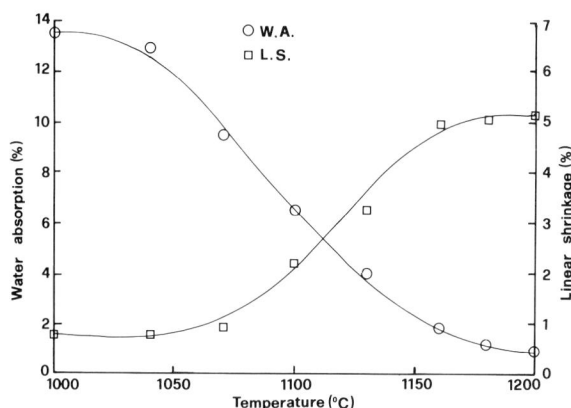

Figure 2. Typical curves of linear shrinkage (L.S.) and water absorption (W.A.) vs. temperature.

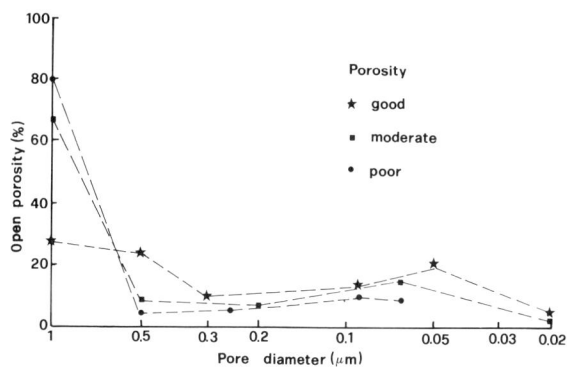

Figure 3. Typical porosimetric curves.

This first series of rapidly performed analyses allowed about 30% of the materials examined to be rejected.

Secondary testing

The remaining samples (~70%) were tested as described below.

Organic carbon content. The organic carbon content is not as critical as those components mentioned above, but it should not be high to avoid gas evolution in the preheating step. Materials with organic carbon contents of >0.3% were rejected.

Quantitative chemical analysis. Chemical analyses were made on all samples. Of particular interest are the percentages of Al_2O_3 (related to the kaolinite content), Na_2O and K_2O (connected with the melting point of the material), and, especially, Fe_2O_3 (related to the color of the fired product). All materials with Fe_2O_3 contents of ≥3% were rejected as unsuitable for white fired-tile production, as were those with Al_2O_3 contents of ≤12%, as not being sufficiently plastic.

Mineralogical analysis by XRD. Mineralogical analyses by XRD were previously discussed by Minichelli and Ori (1980) and Minichelli (1982). In general, to be suitable for tile production, a clay must be rich in quartz and kaolinite, contain some illite, and be as free as possible from carbonates and montmorillonites. The presence of feldspar, which lowers the melting point, is also desirable.

Following this second series of tests, an additional 20% of the materials were rejected.

Tertiary testing

The 41 samples remaining were pressed into tiles and fired at different temperatures. The fired tiles were examined as follows:

Linear shrinkage (L.S.) and water absorption (W.A.).

Linear shrinkage and water absorption determinations are of great practical importance. In fact, as the temperature gradually increases until equilibrium between liquid and solid phases is attained (just before melting) the % W.A. decreases, whereas the % L.S. increases. Examples of such curves are shown in Figure 2. The control of firing temperature in the industrial roller kiln is critical; during the production cycle, temperature fluctuations of ±20% are common. To avoid dimensional problems in the tiles, it is therefore desirable that, for the range of the maximum firing temperature, L.S. and W.A. curves be linear and parallel.

Bending strength (B.S.). The bending strength test employed was A.S.T.M. standard C 674-71. Materials with B.S. values ≥250 kg/cm² were considered suitable.

Dilatometric analysis. The linear thermal expansion coefficient α was calculated between 20° and 400°C. These data were of interest particularly for selecting the most suitable glazes to be applied. Generally, the value varied between 55×10^{-7} cm and 75×10^{-7} cm.

Visual check. A visual check was also necessary to observe the possible presence of faults, pin-holes, bubbles, warpings, etc.

After these tertiary tests on the fired material, 25% of these samples were rejected.

Final testing

Samples that had passed all the above tests and were thus considered suitable for further industrial testing, were subjected to the following two final examinations.

Porosimetric analysis. Using a 2000 ERBA mercury porosimeter, the total porosity and the pore-size distribution were determined. The pore-size distribution data are closely connected to bending strength values (Rice, 1977). As can be seen in Figure 3, tiles with a low total porosity (<3%) and a narrow pore distribution near 0.5 μm were graded as good; the tiles with a

Table 1.   Test results for all samples examined.[1]

Sample and source[2]	Sulfur (%)	Bent. (%)	Kaol. (%)	Carb. (%)	Org. C (%)	Fe_2O_3 (%)	Na_2O, K_2O (%)	Al_2O_3 (%)	Bending strength (kg/cm^2)	Water absorp. (%)	Porosity[3]	Disposition[4]
1 W	0.03–0.1	<2	40–60	2–5	>0.3	0.8–2	2–4	18–22	>300	1.5–2.5	Good	REJECTED
2 W	<0.01	<2	27–40	<2	<0.1	0.8–2	2–4	18–22	250–300	2.5–5	Poor	ACCEPTED
3 W	0.01–0.03	2–3	<20	2–5	0.1–0.3	<0.8	2–4	12–18	250–300	<1.5	Int.	ACCEPTED*
4 W	<0.01	<2	20–27	<2	0.1–0.3	<0.8	<2	<12				ACCEPTED
5 W	0.03–0.1	<2	20–27	<2	0.1–0.3	<0.8	<2	12–18	<250			REJECTED
6 W	<0.01	2–3	27–40	2–5	0.1–0.3	0.8–2	2–4	18–22	<250			REJECTED
7 W	<0.01	2–3	20–27	<2	<0.1							REJECTED
8 W	0.01–0.03	<2	<20									REJECTED*
9 W	<0.01	<2	27–40	<2	<0.1	>3						REJECTED
10 S	0.01–0.03	>3										REJECTED
11 S	0.01–0.03	<2	<20									REJECTED
12 S	>0.1											REJECTED
13 W	<0.01	<2	40–60	<2	<0.1	<0.8	2–4	18–22	<250			REJECTED
14 W	0.01–0.03	<2	40–60	<2	<0.1	<0.8	2–4	12–18	<250			REJECTED
15 W	<0.01	<2	20–27	<2	<0.1	<0.8	<2	18–22	250–300	1.5–2.5	Good	ACCEPTED
16 W	0.01–0.03	<2	20–27	2–5	<0.1	<0.8	2–4	18–22	250–300	2.5–5	Int.	ACCEPTED
17 W	<0.01	<2	27–40	2–5	0.1–0.3	<0.8	2–4	18–22	250–300	>5		REJECTED*
18 W	<0.01	<2	27–40	<2	<0.1	0.8–2	2–4	18–22	250–300	>5		REJECTED*
19 W	0.01–0.03	<2	27–40	<2	0.1–0.3	>3						REJECTED*
20 W	0.01–0.03	<2	27–40	<2	<0.1	>3						REJECTED*
21 W	<0.01	<2	<20									REJECTED
22 W	0.01–0.03	2–3	<20									REJECTED
23 S	0.03–0.1	>3										REJECTED
24 S	>0.1											REJECTED
25 S	>0.1											REJECTED*
26 Pd	0.01–0.03	<2	<20	>8								REJECTED
27 S	>0.1											REJECTED*
28 P	<0.01	<2	40–60	5–8	>0.3	>3						REJECTED*
29 W	<0.01	<2	27–40	2–5	0.1–0.3							REJECTED
30 W	0.01–0.03	<2	27–40	2–5	>0.3							REJECTED
31 W	0.01–0.03	<2	27–40	<2	<0.1	0.8–2	<2	12–18	<250			ACCEPTED
32 W	0.01–0.03	2–3	27–40	<2	0.1–0.3	<0.8	2–4	18–22	250–300	2.5–5	Int.	REJECTED
33 W	<0.01	<2	27–40	<2	0.1–0.3	0.8–2	2–4	18–22	<250			REJECTED
34 S	0.01–0.03	>3										REJECTED
35 S	>0.1											REJECTED*
36 W	0.03–0.1	2–3	>60	5–8	>0.3	0.8–2	2–4	>22	>300	1.5–2.5	Good	REJECTED
37 W	0.03–0.1	2–3	40–60	<2	0.1–0.3	0.8–2	2–4	>22	>300	2.5–5	Poor	ACCEPTED
38 W	<0.01	<2	40–60	<2	<0.2	0.8–2	2–4	18–22	250–300	>5		ACCEPTED*
39 W	<0.01	2–3	40–60	<2	<0.2							REJECTED*
40 W	<0.01	<2	27–40	<2	<0.2							REJECTED*
41 P	0.03–0.1	2–3	>60	<2	<0.2	<0.8	2–4	12–18	<250	1.5–2.5	Int.	REJECTED
42 W	<0.01	<2	27–40	<2	<0.2	<0.8	2–4	12–18	250–300			REJECTED
43 W	0.01–0.03	<2	27–40	<2								ACCEPTED
44 W	0.01–0.03	<2	<20									REJECTED

Table 1. Continued.

Sample and source[2]	Sulfur (%)	Bent. (%)	Kaol. (%)	Carb. (%)	Org. C (%)	Fe$_2$O$_3$ (%)	Na$_2$O, K$_2$O (%)	Al$_2$O$_3$ (%)	Bending strength (kg/cm²)	Water absorp. (%)	Porosity[3]	Disposition[4]
45 W	<0.01	<2	27–40	5–8	0.1–0.3	0.8–2	2–4	12–18	250–300	2.5–5	Int.	ACCEPTED
46 W	<0.01	<2	20–27	<2	0.1–0.3	<0.8	2–4	18–22	250–300	<1.5	Good	ACCEPTED
47 W	0.03–0.1	<2	27–40	2–5	>0.3							REJECTED
48 W	<0.01	<2	27–40	<2	<0.1	0.8–2	2–4	18–22	>300	2.5–5	Int.	ACCEPTED
49 W	<0.01	<2	27–40	<2	0.1–0.3	<0.8	<2	12–18	<250			REJECTED
50 W	<0.01	<2	27–40	2–5	0.1–0.3	<0.8	2–4	18–22	250–300	2.5–5	Good	ACCEPTED
51 W	<0.01	<2	27–40	<2	0.1–0.3	0.8–2	2–4	18–22	<250			REJECTED
52 W	<0.01	2–3	20–27	<2	0.1–0.3	0.8–2	2–4	12–18	250–300	2.5–5	Int.	ACCEPTED
53 W	0.01–0.03	<2	27–40	2–5	<0.1	0.8–2	2–4	>22	250–300	>5		REJECTED*
54 W	0.01–0.03	<2	27–40	2–5	0.1–0.3	0.8–2	2–4	18–22	<250			REJECTED*
55 P	0.01–0.03	2–3	>60									REJECTED*
56 P	0.01–0.03	2–3	27–40	5–8								REJECTED
57 W	<0.01	<2	20–27	<2	<0.1	<0.8	4–6	12–18	250–300	2.5–5	Int.	ACCEPTED
58 W	<0.01	<2	27–40	<2	<0.1	0.8–2	4–6	12–18	250–300	2.5–5	Good	ACCEPTED
59 W	0.01–0.03	<2	20–27	<2	<0.1	0.8–2	4–6	<12				REJECTED
60 W	<0.01	<2	20–27	2–5	<0.1	0.8–2	4–6	<12				REJECTED
61 W	<0.01	<2	20–27	<2	<0.1	<0.8	2–4	18–22	250–300	2.5–5	Poor	ACCEPTED*
62 W	<0.01	<2	20–27	<2	0.1–0.3	0.8–2	2–4	12–18	250–300	>5		REJECTED
63 S	>0.1											REJECTED
64 S	>0.1											REJECTED
65 S	0.03–0.1	2–3	40–60	2–5	>0.3	0.8–2	2–4	18–22				REJECTED
66 W	<0.01	<2	27–40	5–8	<0.1	0.8–2	2–4	18–22	<250			REJECTED
67 W	0.03–0.1	<2	27–40	<2	0.1–0.3	2–3	2–4	18–22	250–300	2.5–5	Int.	ACCEPTED
68 W	0.01–0.03	<2	20–27	<2	<0.1	0.8–2	2–4	18–22	250–300	1.5–2.5	Poor	ACCEPTED
69 W	<0.01	<2	20–27	<2	<0.1	0.8–2	2–4	18–22	250–300	2.5–5	Int.	ACCEPTED
70 S	0.03–0.1	<2	27–40	2–5	0.1–0.3	0.8–2	4–6	>22	>300	2.5–5	Good	ACCEPTED
71 S	>0.1											REJECTED
72 P	0.03–0.1	<2	27–40	2–5	0.1–0.3	0.8–2	<2	18–22	250–300	>5		REJECTED*
73 P	0.03–0.1	<2	27–40	<2	0.1–0.3	<0.8	<2	18–22	<250			REJECTED
74 P	0.03–0.1	<2	27–40	2–5	<0.1	<0.8	<2	12–18	<250			REJECTED
75 Pd	<0.01	<2	<20									REJECTED
76 S	>0.1											REJECTED
77 W	<0.01	<2	40–60	2–5	0.1–0.3	0.8–2	4–6	>22	250–300	2.5–5	Poor	ACCEPTED*
78 W	0.03–0.1	<2	40–60	2–5	>0.3							REJECTED
79 S	<0.01	<2	27–40	2–5	0.1–0.3	>3						REJECTED*
80 S	<0.01	<2	20–27	2–5	0.1–0.3	<0.8	2–4	>12				REJECTED

[1] Bent. = bentonite % = weight loss at <200°C; Kaol. = kaolinite % = weight loss at 400°–700°C; Carb. = carbonate % = weight loss at 700°–100°C;
[2] W = Westerwald, Germany; S = Sardinia, Italy; P = Provins, France; Pd = Piedmont, Italy.
[3] Good = low total porosity and particle sizes narrowly distributed around 0.5 µm; Poor = medium to high porosity and/or broad distribution of particle sizes; Int. = intermediate between Good and Poor.
[4] REJECTED* = suitable for other uses; ACCEPTED* = with reservations.

medium or high total porosity (3–5%) and/or narrowly distributed pores were graded as poor.

Optical microscopy. A microscopic observation of the fired tiles was used to confirm the uniformity of the texture and to control the degree of sintering. Such tests were carried out using an optical microscope on samples mounted in resin and lapped for metallographic analysis.

SUMMARY

The test data for all 80 samples are listed in Table 1. The table illustrates the elimination process. For example, sample 12 was immediately eliminated on the basis of a sulfur content >0.1%; sample 10 which passed the sulfur test was rejected on the basis of a high bentonite content (>3%).

After all tests had been completed, only 22 of the 80 samples were considered for industrial tests. The adoption of the proposed elimination system reduced the number of total experimental tests from 880 to 465.

The results obtained in the subsequent production stages confirmed the validity of this selection scheme.

REFERENCES

Aliprandi, A. (1975) *Principi di Ceramurgia:* E.C.I.G. Publ. Co., Genoa, Italy, p. 730.

Emiliani, T. and Emiliani, R. (1982) *Tecnologie dei Processi Ceramici:* Ceramurgia Publ. Co., Faenza, Italy, p. 20.

Kingery, W. D. (1960) *Introduction to Ceramics:* Wiley, New York, p. 504.

Mackenzie, R. C. (1972a) *Differential Thermal Analysis, Vol. 1:* Academic Press, London, p. 153.

Mackenzie, R. C. (1972b) *Differential Thermal Analysis, Vol. 2:* Academic Press, London, p. 507.

Minichelli, D. (1982) The quantitative phase analysis of clay minerals by X-ray diffraction: modern aspects of individual routine control: *Clay Miner.* **17,** 401–408.

Minichelli, D. and Ori, G. (1980) Uno studio sull'analisi razionale delle aregille mediante diffrazione ai raggi X: *Ceramica Inf.* **175,** 668–672.

Rice, R. W. (1977) The compressive strength in ceramics: in *Treatise on Materials Science and Technology,* R. K. MacKrone, ed., Academic Press, London, p. 200.

Proceedings of the International Clay Conference, Denver, 1985, L. G. Schultz, H. van Olphen, and F. A. Mumpton, eds.,
The Clay Minerals Society, Bloomington, Indiana, 391–395 (1987).

INFLUENCE OF SODIUM CHLORIDE ON
THERMAL REACTIONS OF HEAVY CLAYS DURING FIRING

B. Fabbri and C. Fiori

Istituto di Ricerche Tecnologiche per la Ceramica
Consiglio Nazionale delle Ricerche, 48018 Faenza, Italy

Abstract—The reactions occurring during the firing of a calcareous illitic-chloritic clay and of a kaolin clay in the presence of 2% NaCl were studied by gravimetric, thermal, chemical, and X-ray powder diffraction analysis. The NaCl reacted with both clays to form volatile hydrochloric acid. For the kaolin clay, this reaction led to the complete removal of chlorine at about 520°C. For the calcareous illitic-chloritic clay, however, part of the product HCl was volatilized and part promoted the decomposition of carbonate impurities and formed potassium chloride, which volatilized at a higher temperature. The addition of the NaCl resulted in a significant decrease in the amount of residual potassium and a progressive elimination of residual Cl in samples fired above 750°C.

The emission of alkali-element compounds from NaCl-containing calcareous clays used for heavy clay products causes degradation of the refractory lining of tunnel kilns in brickworks. To minimize this degradation, it is proposed to inject steam-saturated air into an appropriate zone of the kiln where the temperature is less than 500°C to enhance the formation and subsequent emission of HCl in this "low" temperature range, thereby reducing the amount of alkali-element compounds volatilized at higher temperatures.

Key Words—Chlorite, Hydrochloric acid, Illite, Kaolin, Potassium, Sodium chloride, Thermal treatment, Volatilization.

INTRODUCTION

Degradation of part of the internal refractory lining of tunnel kilns in brickworks is due to chemical attack by alkali-element compounds, mainly potassium compounds (Biffi and Fiori, 1983). These compounds form from the calcareous illitic-chloritic clay raw materials during the firing process; the emission consists mainly of volatile potassium compounds, but it is directly connected with the presence of NaCl in the raw materials (Fiori and Fabbri, 1983). In addition, alkali-element compounds from the clay and volatile sulfur-bearing compounds from the fuel react to form alkali sulfate deposits in other parts of the kiln lining and efflorescent salt deposits in the heavy clay products (Fiori *et al.*, 1984). The emission of the potassium compounds begins at about 750°C, at which temperature about half of the chlorine has already been released from the clay. The chlorine emissions, independent of the effect of alkali elements, is probably due to the dehydroxylation of the clay minerals (Fiori and Fabbri, 1983).

The aim of the present investigation was to determine the changes taking place during the firing of NaCl-containing clay raw materials and to suggest possible modifications of the firing process to prevent the emission of alkali-element compounds that cause the degradation of the kiln linings.

EXPERIMENTAL

An Italian clay used for heavy clay production, washed free of soluble salts, and a kaolin clay from Zettlitz, Czechoslovakia, were mixed separately with 2% analytical grade NaCl. The mixtures were prepared by dispersing the clay powders in acetone in an agate mortar until complete evaporation of the solvent. The chemical and mineralogical compositions of the starting clay materials are listed in Table 1. The chemical analyses were made by flame atomic absorption spectroscopy (Fabbri, 1979); the mineralogical constituents were determined by X-ray powder diffraction and quantified from the elemental analyses by means of a computerized procedure based on the least squares solution of a mathematical system (Krajewski *et al.*, 1985; Fabbri *et al.*, 1985).

Differential thermal (DTA) and thermal gravimetric analyses (TGA) were carried out on all samples, with or without added NaCl, to 1000°C with a Netzsch instrument. Several cylindrical specimens (25-mm diameter, 10-mm thickness) were prepared from the mixture of the illitic-chloritic clay + NaCl by dry pressing the powder at about 300 kg/cm² (about 30 MPa). Some of the cylinders were fired in an electric furnace at various temperatures, applying a heating rate of about 5°C/min, after which they were held for 3 hr at the maximum temperature. Other cylinders were fired at 700°C for different periods of time. After cooling in the furnace, the weight loss and K and Na contents, expressed as K_2O and Na_2O, and Cl remaining in the fired product were determined. The concentrations of K and Na were determined by flame emission spectroscopy (Fabbri, 1979); the Cl concentration was de-

Table 1. Chemical composition and mineralogical composition of washed illitic-chloritic calcareous clay and kaolin.[1]

Chemical composition (wt. %)			Mineralogical composition (wt. %)			
	Clay	Kaolin		Clay		
SiO_2	44.75	47.35	Quartz	21	Calcite	19
Al_2O_3	12.49	37.00	Albite	14	Dolomite	10
TiO_2	0.65	0.20	Illite +		Fe-oxides	4
Fe_2O_3	5.31	0.80	muscovite	23	Others	3
MgO	3.76	0.25	Chlorite	6		
CaO	14.01	0.65				
Na_2O	1.64	0.03		Kaolin		
K_2O	2.05	1.05				
L.O.I.	15.28	12.60	Kaolinite	84	Illite	11
Total	99.94	99.93	Others	5		

[1] Mineralogical composition was determined by a computerized rational procedure (Krajewski et al., 1985).

termined indirectly from the amount of silver (determined by atomic absorption spectroscopy) used to precipitate AgCl. For this analysis, samples were fused with NaOH in a nickel crucible (Fabbri and Donati, 1981).

X-ray powder diffraction (XRD) analyses of the original clays and the fired products were made using Ni-filtered CuKα radiation.

RESULTS

Thermal analyses

TGA curves obtained at a heating rate of 2°C/min for the calcareous illitic-chloritic clay, with and without added NaCl, are reported in Figure 1. The curve for the clay without added NaCl shows four stages of weight loss which are characteristic of this type of raw material: (1) Below about 250°C, 3% of the weight was lost due to the removal of moisture and adsorbed water from the clay minerals. (2) Between 250° and 400°C, the sample lost about 0.8% of its original weight, probably as a result of the combustion of a small amount of impurity organic material. (3) Between 400° and 675°C, the sample lost 4.8% of its original weight which can be attributed to the dehydroxylation of the clay minerals, particularly illite. (4) Between 675° and 800°C, a further weight loss of about 10% was likely due to the thermal decomposition of impurity calcite and dolomite. Above 800°C, no further weight loss was observed to the maximum temperature of the experiment (1050°C).

The TGA curve of the same sample to which 2% NaCl had been added (Figure 1, curve B) has roughly the same shape as that of the original sample, but a fifth weight loss is present at 800°–900°C. This weight loss, as much as 1.5%, is probably due to the emission of alkali-element compounds, mainly KCl, on the basis of chemical analyses described below. The TGA curves for the kaolin sample, with and without added NaCl, are shown in Figure 2. Both curves show only one weight loss at about 550°C, but they differ in that the kaolin + NaCl sample shows a higher weight loss than

that of the original kaolin sample, corresponding to the amount of NaCl added to the sample.

DTA curves of these samples, shown in Figures 3 and 4, confirm the TGA results. The DTA curve of the calcareous illitic-chloritic clay + NaCl mixture shows evidence of the low-temperature decomposition of carbonates. The DTA curve of the kaolin + NaCl mixture shows a broad endothermal peak at low temperatures, probably due to the release of volatile HCl, as discussed below.

Chemical analyses of fired products

The concentrations of K_2O, Na_2O, and Cl in the fired products were corrected to a no-weight-loss basis, by using the following equation:

$$C.V. = D.C.(100 - I.L./100),$$

where C.V. = corrected value, D.C. = determined concentration, and I.L. = ignition loss during firing. The values so obtained represent the amount of K_2O, Na_2O, and Cl (in grams) in the residue after firing 100 g of sample. These values, along with weight loss during firing (L.O.I.), are plotted against maximum firing temperature in Figure 5 and against firing duration at 700°C in Figure 6.

The data plotted in Figure 5 suggest that: (1) below 500°–550°C, the quantities of K_2O, Na_2O, and Cl did not vary and that the weight loss was due mainly to the dehydroxylation of illitic materials; (2) above 500°–550°C, but below about 750°C (where the weight loss was due chiefly to the decomposition of carbonate minerals), the quantities of K_2O and Na_2O did not vary appreciably, but the Cl content decreased rapidly; (3) above 750°C, the K_2O and Cl contents decreased significantly, whereas that of Na_2O decreased only slightly, clearly indicating that the weight loss in this range was due to the simultaneous emission of potassium and chlorine.

Figure 6 shows that the duration of heating at 700°C had little influence on the amount of K_2O and Na_2O in the fired product; only the amount of chlorine de-

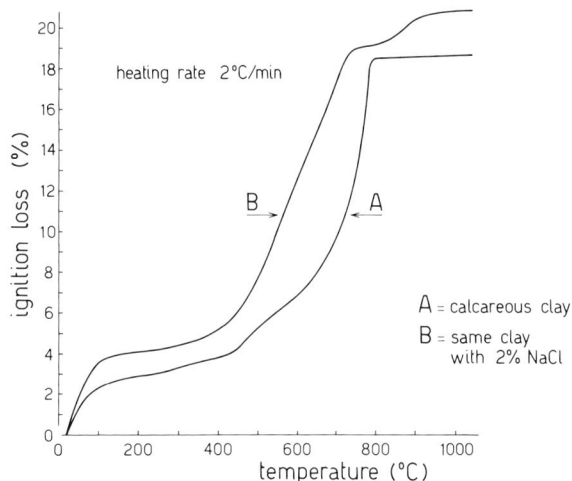

Figure 1. Thermal gravimetric analysis of the calcareous illitic-chloritic clay with and without 2% NaCl.

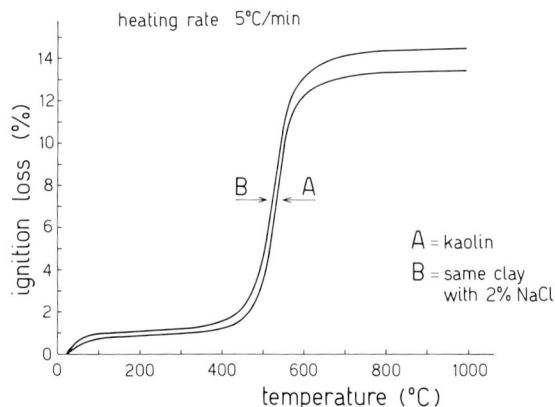

Figure 3. Differential thermal analysis of the calcareous illitic-chloritic clay with and without 2% NaCl.

creased significantly. Firing at 700°C for 24 hr, however, was still not sufficient to drive off all the chlorine.

X-ray powder diffraction analyses

XRD analyses are summarized in Table 2. In addition to the low-temperature decomposition of carbonate minerals and the formation of calcium silicates, such as gehlenite, diopside, and anorthite, the data show that NaCl is not stable in these mixtures at temperatures of more than about 500°C.

DISCUSSION

The stability of a small amount of NaCl in the clay mixtures is limited to temperatures below 500°C, even though the melting point of this compound is about 800°C. The disappearance of crystalline NaCl at about 500°C coincides with the beginning of the clay dehydroxylation process and with the beginning of the chlo-

rine loss from the mixtures. For the carbonate-free kaolin clay, the dehydroxylation process coincides with the complete removal of all chlorine added as NaCl, as is evident from the difference TGA curve of the samples with and without added NaCl (Figure 7). Figure 7 also shows that at a heating rate of 5°C/min, the maximum emission of Cl was at 520°C. Therefore, according to De Keyser and Degueldre (1955), Heller-Kallai (1978), Heller-Kallai and Frenkel (1979), and Yariv et al. (1982), the kaolinite and chlorides of alkali metals reacted on heating to form HCl, with the water

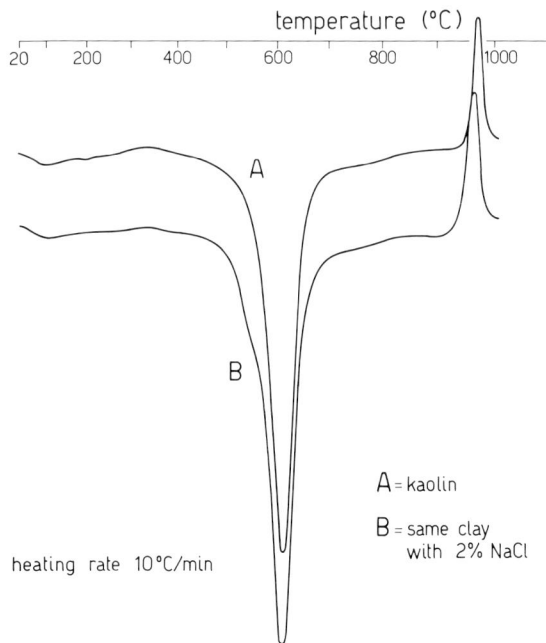

Figure 2. Thermal gravimetric analysis of kaolin from Zettlitz with and without 2% NaCl.

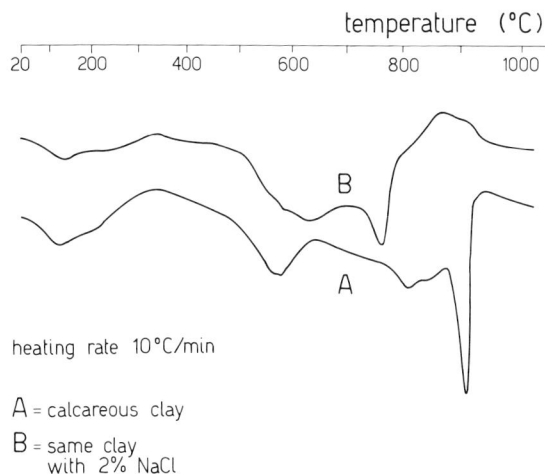

Figure 4. Differential thermal analysis of kaolin from Zettlitz with and without 2% NaCl.

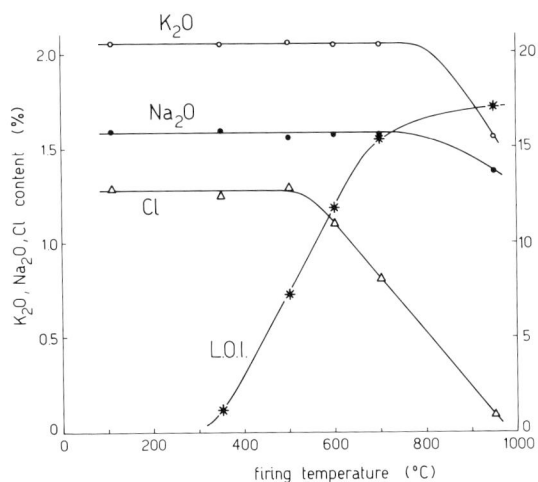

Figure 5.   Corrected concentrations of K_2O, Na_2O and Cl in fired calcareous clay with 2% NaCl vs. maximum firing temperature; ignition weight loss (L.O.I.) is also plotted.

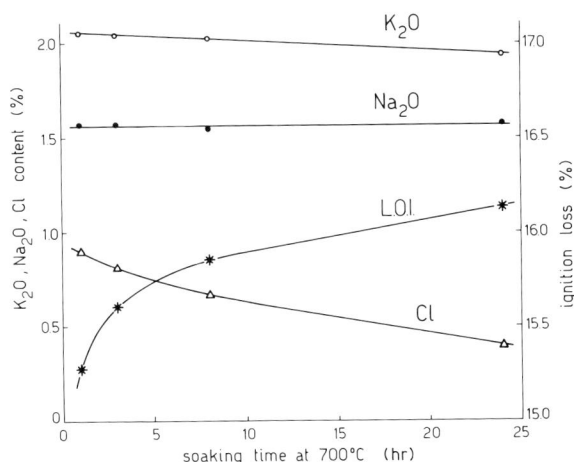

Figure 6.   Corrected concentrations of K_2O, Na_2O and Cl in the calcareous clay with 2% NaCl and fired at 700°C vs. the time during which specimens were kept at the maximum firing temperature; ignition weight loss (L.O.I.) is also plotted.

liberated by the dehydroxylation of the clay acting as a solvent, as represented by the following equation:

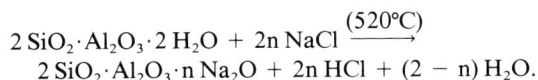

$$2\,SiO_2\cdot Al_2O_3\cdot 2\,H_2O + 2n\,NaCl \xrightarrow{(520°C)}$$
$$2\,SiO_2\cdot Al_2O_3\cdot n\,Na_2O + 2n\,HCl + (2-n)\,H_2O.$$

For the calcareous illitic-chloritic clay, an analogous reaction took place between the clay minerals and NaCl, even though these minerals released less water than the kaolin at 500°–600°C. For the calcareous illitic-chloritic clay, the removal of HCl at about 520°C was not complete because part of the product HCl reacted with the carbonates at a temperature considerably less than the thermal decomposition of these materials (Webb and Heystek, 1957; Garn, 1965; Webb and Kruger, 1970). To explain the subsequent emission of potas-

sium together with chlorine, potassium extracted from the illitic minerals must have reacted with the chlorine to produce KCl which volatilized near its melting point. Inasmuch as the chief mineral phases in the fired products were anorthite and diopside, the following type of reaction must have taken place:

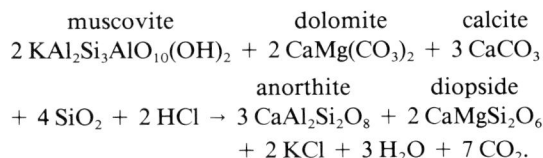

muscovite   dolomite   calcite
$2\,KAl_2Si_3AlO_{10}(OH)_2 + 2\,CaMg(CO_3)_2 + 3\,CaCO_3$

anorthite   diopside
$+ 4\,SiO_2 + 2\,HCl \rightarrow 3\,CaAl_2Si_2O_8 + 2\,CaMgSi_2O_6$
$+ 2\,KCl + 3\,H_2O + 7\,CO_2.$

The small amount of sodium lost at the highest temperature may have been due to the volatilization of a small amount of NaCl which had insufficient time to

Table 2.   Summary of X-ray powder diffraction analyses of products obtained by firing the calcareous illitic-chloritic clay with 2% NaCl at different temperatures.

	Firing temperature (°C)				
Mineral phase	350	500	600	700	950
Quartz	+++++	+++++	+++++	++++	++++
Calcite	+++	++	+		
Dolomite	++	+			
Albite	++	++	++	++	++
Illite + muscovite	++++	++++	+++	+	
Chlorite	++	+			
Halite	+				
Anorthite					+++
Gehlenite				+	++
Diopside				+	++++

+++++ = very abundant.
++++ = abundant.
+++ = present.
++ = rare.
+ = trace.

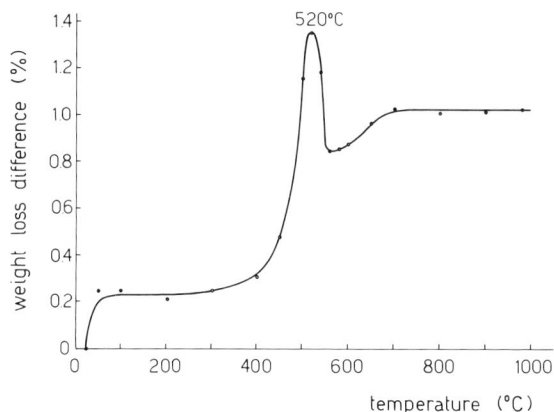

Figure 7. Weight loss difference between the two thermal gravimetric curves of kaolin samples with and without 2% NaCl, respectively.

react with the clay minerals and/or carbonates. New sodium chloride could have formed, however, from the sodium of the clay minerals in a manner similar to that postulated for the formation of KCl.

CONCLUSIONS

Further investigations are necessary, of course, for a more thorough understanding of the chemical reactions that took place during firing, but from a technological point of view, the experimental data presented here suggest a possible means of minimizing the emission of alkali-element compounds that damage kiln linings during the firing of bricks made of such NaCl-bearing calcareous clays. Such a process would involve reducing the amount of HCl available for reaction with the carbonates and illitic clays by maximizing the production and subsequent emission of HCl vapors at low temperature (500°C). Sufficient water would have to be available at these temperatures, however, for the formation of the acid, along with a high air flux around the bricks to remove the HCl produced. Insufflating steam-saturated air into the pre-firing zone

of the tunnel kiln might well inhibit the emission of alkali-element vapors at the higher firing temperatures.

REFERENCES

Biffi, G. and Fiori, C. (1983) Attacco alcalino ai refrattari in forni a tunnel per laterizi: *Ceramurgia* **13**, 3–16.

De Keyser, W. L. and Degueldre, L. (1955) The action of potassium chloride and sodium chloride on kaolin at high temperature: *Industrie Chim. Belge* **20**, 145–149.

Fabbri, B. (1979) Dosaggio dei costituenti chimici di materiali silicoalluminiferi: messa a punto di una metodologia in spettrofotometria di assorbimento atomico: *Ceramurgia* **9**, 57–64.

Fabbri, B. and Donati, F. (1981) Sample fusion at low temperature for the potentiometric determination of fluorine in silicate materials: *Analyst* **106**, 1338–1341.

Fabbri, B., Fiori, C., Krajewski, A., Valmori, R., and Tenaglia, A. (1985) Comparison between traditional mineralogical and computerized rational analysis of ceramic raw materials: in *Proceedings of "Science of Ceramics 13"* (in press).

Fiori, C. and Fabbri, B. (1983) Emissione di vapori alcalini durante la cottura di argille per laterizi contenenti NaCl: *Ceramurgia* **13**, 252–260.

Fiori, C., Biffi, G., and Fabbri, B. (1984) Cause dell'attacco alcalino ai refrattari dei forni a tunnel per laterizi: *La Ceramica* **37**, 13–20.

Garn, P. D. (1965) *Thermoanalytical Method of Investigation:* Academic Press, New York, 606 pp.

Heller-Kallai, L. (1978) Reactions of salts with kaolinite at elevated temperatures. I: *Clay Miner.* **13**, 221–235.

Heller-Kallai, L. and Frenkel, M. (1979) Reactions of salts with kaolinite at elevated temperatures—Part 2: in *Proc. Int. Clay Conf., Oxford, 1978,* M. M. Mortland and V. C. Farmer, eds., Elsevier, Amsterdam, 629–637.

Krajewski, A., Fabbri, B., Fiori, C., and Valmori, R. (1985) Analisi mineralogica di materie prime ceramiche con computer: *Ceramica Informazione* **20**, 391–396.

Webb, T. L. and Heystek, H. (1957) The carbonate minerals: in *The Differential Thermal Investigation of Clays,* R. C. Mackenzie, ed., Mineralogical Society, London, 329–363.

Webb, T. L. and Kruger, J. E. (1970) Carbonates: in *Differential Thermal Analysis, Vol. 1, Fundamental Aspects,* R. C. Mackenzie, ed., Academic Press, London, 303–341.

Yariv, S., Mendelovici, E., and Villalba, R. (1982) The study of the interaction between cesium chloride and kaolinite by thermal methods: in *Thermal Analysis, Proc. 7th Int. Conf. Thermal Analysis, Vol. 1,* B. Miller, ed., Wiley, Chichester, Great Britain, 533–540.

Proceedings of the International Clay Conference, Denver, 1985, L. G. Schultz, H. van Olphen, and F. A. Mumpton, eds.,
The Clay Minerals Society, Bloomington, Indiana, 396–399 (1987).

PROPERTIES OF TALCS AND STEATITES,
WITH EMPHASIS ON CERAMIC APPLICATIONS

Karl-Heinz Schüller[1] and Heinz Kromer[2]

[1] Fachhochschule Nürnberg, Fachbereich Werkstofftechnik
Nürnberg, Federal Republic of Germany

[2] Staatliches Forschungsinstitut für Angewandte Mineralogie
Technischen Universität München, Regensburg
Federal Republic of Germany

Abstract—Because talcs and steatites are both hydrated magnesium silicates, they cannot be distinguished chemically; however, talcs appear to differ from steatites by their coarser grain size. X-ray powder diffraction patterns of talcs also show more intense basal (00*l*) reflections than steatites. Coarse-grained talcs appear to be less suitable than steatites for the fabrication of ceramic bodies that are shaped by dry pressing because the coarse grains tend to orient during the forming process, causing higher shear stresses during firing. Steatites are more suitable than talcs in these applications because their finer grain sizes facilitate random rather than parallel particle orientation. Because steatite rocks may display different crystal size and technical properties, these parameters must be carefully considered when selecting suitable ceramic raw materials.

Key Words—Ceramic, Crystal size, Particle orientation, Steatite, Talc, X-ray powder diffraction.

INTRODUCTION

Talc ($Mg_3Si_4O_{10}(OH)_2$) occurs in nature in two textural varieties. In the present paper the term talc is used for the relatively coarse grained, platy form of the mineral, and the term steatite is used for the fine-grained more homogeneously textured form. Schüller (1971) pointed out that talcs differ from steatites in certain properties. For example, talcs have higher mean coefficients of thermal expansion than steatites, and their basal X-ray powder diffraction (XRD) reflections are more intense than those of steatites, regardless of the direction of scatter. Such XRD observations agree with those of von Gehlen (1965), who investigated the scatter of XRD reflections in various directions in order to measure particle orientation. von Gehlen also found the basal reflections of talcs to be more intense than those of steatites. Perdikatsis (1972) was not able to explain the difference between the properties of talcs and steatites merely by the differences in grain size and suggested that the differences were possibly related to structure, as suggested earlier by Schüller (1971). In fact, Perdikatsis (1972) deduced six theoretical arrangements of ions in the talc structure, two of which he identified in natural talcs. Moreover, he observed a lattice transformation in single crystals of talc held for several months without mechanical stress.

It is still not clear whether the differences between talcs and steatites are due to structural differences, because no methods are available to determine the crystal structure of such fine-grained materials as steatites. Whereas Schüller's (1971) study was limited to seven

samples, the present investigation made use of 50 samples from various sources worldwide. It also determined which properties of steatites affect their use as ceramic raw materials.

SAMPLES AND TESTS

A suite of 50 commercially available samples was investigated, from which eight representative types were selected for detailed examination. These eight samples are listed in Table 1 by place and mode of origin, mean crystal size, and chemical composition. The selected samples are fairly pure and approach the theoretical talc composition of 63.36% SiO_2, 31.89% MgO, and 4.75% H_2O. Few commercial talcs are found purer than those selected for this study.

X-ray powder diffraction (XRD) and differential thermal analyses (DTA) revealed few minerals other than talc in the samples. Polarized light microscopy proved to be a more sensitive technique for detecting traces of chlorite, calcite, dolomite, quartz, apatite, zoisite, goethite, and opaque minerals in various samples (Endlicher *et al.,* 1985). Optical microscopy was also useful for comparing crystal sizes of the samples, which ranged from extremely fine grained to extremely coarse grained, as shown in Figures 1–6.

For XRD patterns, particle orientation was kept to a minimum by applying as little pressure as possible during the preparation of the powder sample mounts. The 002 and 020 reflections at about 4.66 and 4.55 Å, respectively, were used to characterize the samples. Several sets of XRD patterns obtained from mounts

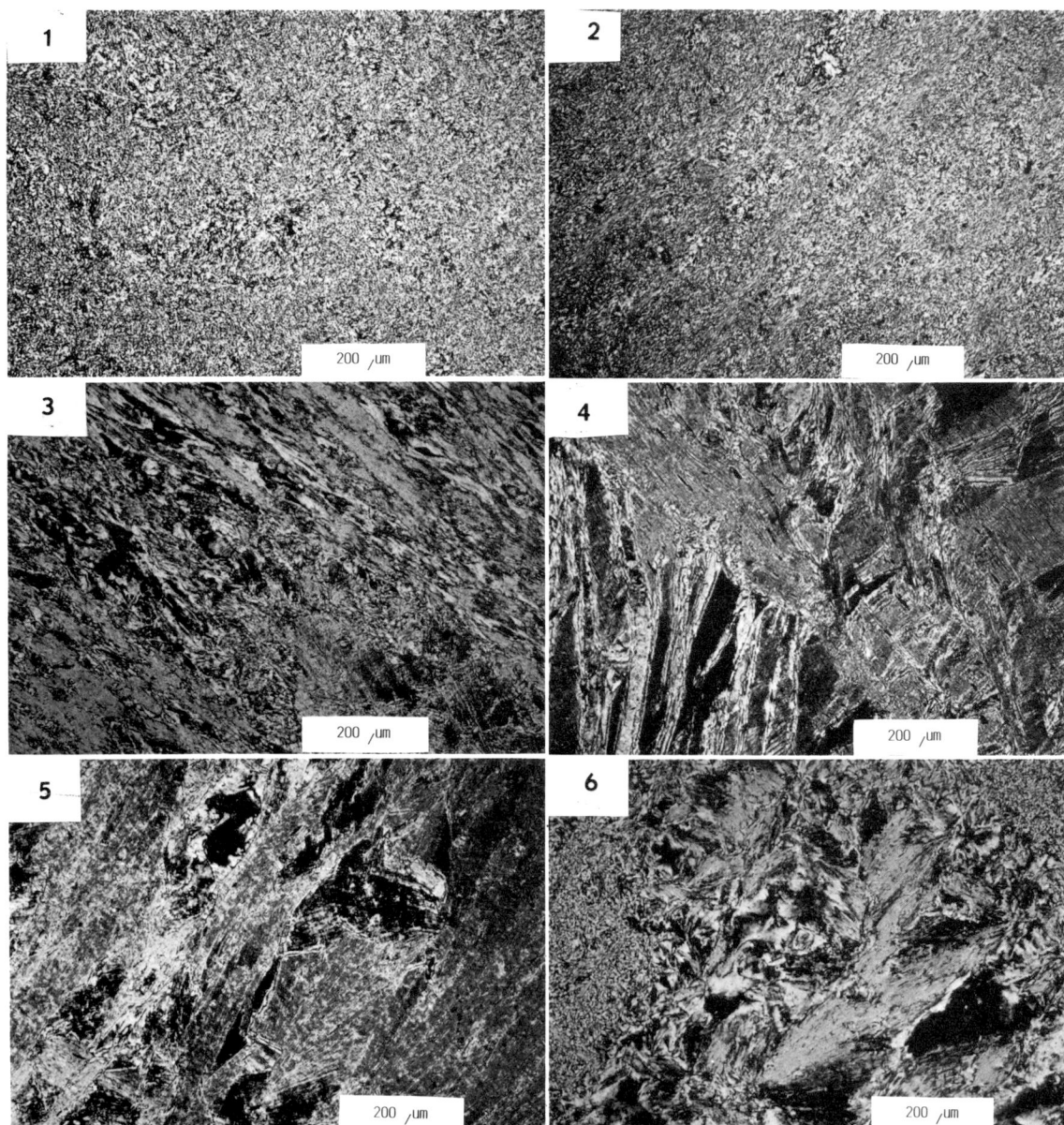

Figure 1. Photomicrograph of thin section of steatite sample 44.
Figure 2. Photomicrograph of thin section of steatite sample 5; note the distinct crystal orientation.
Figure 3. Photomicrograph of thin section of talc sample 9; note the distinct orientation of medium-sized crystals.
Figure 4. Photomicrograph of thin section of talc sample 11, consisting of extremely coarse-grained crystals.
Figure 5. Photomicrograph of thin section of talc sample 34, consisting of extremely coarse-grained crystals.
Figure 6. Photomicrograph of thin section of steatite sample 23; note the inhomogeneous particle size distribution.

prepared by different sample preparation methods confirmed the differences in XRD intensities of the samples. Table 2 shows 002/020 peak-height and peak-area ratios and the total area beneath both reflections, expressed in arbitrary units. The two reflections partly overlap, especially in the patterns of coarse-grained samples which have an intense 002 reflection. Therefore, the peak-area ratios were estimated from heights and mean widths only, and the value for the areas were used only to provide a rough estimate for several samples.

RESULTS

Table 2 shows that the larger the crystal size, the greater the intensity of the 002 reflection, in relation to the intensity of the 020 reflection. For example,

Table 1. Description of specimens discussed in this study.

Sample	Origin	Rock type	Parent rock	Mean crystal size (μm)	Chemical composition (wt. %)						
					SiO_2	Al_2O_3	Fe_2O_3	CaO	MgO	LOI[1]	H_2O[2]
44	Yellowstone, U.S.A.	Steatite	Carbonate	15	62.49	0.09	1.47	0.05	30.72	4.95	4.50
5	Bombay, India	Steatite	Carbonate	30	62.10	0.22	0.63	0.10	31.19	5.35	4.40
15	Perth, Australia	Steatite	Carbonate	150	62.55	0.30	0.99	0.05	30.99	4.80	4.30
9	Bombay, India	Talc	Carbonate	280	62.12	0.28	0.58	0.05	31.61	4.80	4.40
11	Bombay, India	Talc	Ultramafic	700	61.34	0.12	1.02	0.05	30.94	4.72	4.30
34	Pinerolo, Italy	Talc	Carbonate	1000	61.31	0.29	0.57	0.05	31.20	6.10	4.10
23	Australia	Steatite	Carbonate	8–300[3]	60.99	0.78	0.98	0.10	31.03	5.44	4.00
16	Zambia	Steatite	Carbonate	20–150[3]	61.90	0.26	0.64	0.05	31.44	4.87	4.80

[1] LOI = Loss on ignition.

[2] H_2O was determined separately from the total loss on ignition to demonstrate that other volatile components are present in small quantities in the sample.

[3] Maximum particle size in inhomogeneous representatives of steatite; see text.

samples 11 and 34 have by far the largest crystal sizes, and they also have the largest peak heights and area ratios; both are talcs (Table 1). Sample 9 is anomalous because although it is of medium grain size, compared with other samples, it nevertheless has an especially intense 002 peak, comparable with the peak intensities of much coarser grained samples. This anomaly is probably due to the distinct particle orientation in the raw sample, as shown in Figure 3, which probably carries over into the XRD powder mount. Samples 44, 5, 23, and 16 are fine grained and have small XRD intensity ratios and total peak areas; all are classified as steatites (Table 1). Sample 15, although classified as a steatite, has a particle size and XRD properties that are intermediate between those of talc and steatite.

The total combined areas of the 002 and 020 peaks also increase with mean particle size; talcs give the largest areas, steatites the smallest.

DISCUSSION

X-ray powder diffraction and particle size

This study shows that talcs, because of their relatively large particle size, are sensitive to crystal ori-

Table 2. Comparison of X-ray powder diffraction reflections 002 and 020.

Sample	Mean crystal size (μm)	Peak height ratio 002/020	Peak area ratio 002/020	Area of reflections 002 and 020, arbitrary units[1]	
				A_{002}	A_{020}
44	15	0.9	0.6	16	27
5	30	2.4	1.2	34	28
15	150	4.2	2.8	46	17
9	280	13	11	83	8
11	700	16	11	92	8
34	1000	16	11	92	8
23	8–300	2.5	1.8	34	19
16	20–150	1.6	0.9	22	24

[1] Rough estimate because of overlap of XRD lines.

entation during the preparation of XRD mounts. In contrast, the finer-grained steatites are much less prone to particle orientation. The intensities of the 002 and 020 reflections are apparently controlled by particle orientation. Nevertheless, the increase in total area of both of these reflections with increasing crystal size appears also to be an indication that the basal reflections of talcs are more intense than those of steatites, as was shown by von Gehlen (1965) and Schüller (1971).

Samples 23 and 16 have a wide range of crystal sizes, as can be seen in Figure 6 for sample 23; some crystals are as large as 300 μm (Tables 1 and 2). The XRD data (Table 2), however, show that these two samples are similar to steatites rather than to talcs, even though they contain a significant proportion of talc-like crystals.

STEATITES AS CERAMIC RAW MATERIALS

Coarse-grained talcs tend to orient strongly during ceramic forming processes, causing high shear stresses during firing. Therefore, relatively coarse grained talcs are not as well suited as fine-grained steatites for ceramic bodies that must be shaped by pressing operations. Distinct differences in particle orientation, however, can be found among steatites. Extremely fine grained steatites, such as sample 44 (Yellowstone), generally retain their random particle orientation during ceramic processing. The lower specimen in Figure 7

Figure 7. Pressed steatite bodies formed from raw materials of slightly different crystal size and, thus, different textures. Upper sample is slightly coarser grained.

Figure 8. Pressed steatite bodies fabricated with different pressure resulting in different particle orientation and uniformity of texture. Lower sample with slaty texture had the higher pressure.

has a uniform texture and was formed from an extremely fine grained steatite, resembling sample 44. Steatites consisting of only slightly coarser crystals tend to develop a distinct particle orientation and slaty cleavage during ordinary forming processes (upper specimen, Figure 7). This slaty textured product was made from a commercial steatite, which, because of its slightly coarser grain size, required special treatment to prepare satisfactory ceramic products.

Even the highest quality steatites are sensitive to fabrication conditions. Normally, very fine grained steatites (e.g., sample 44, Yellowstone, Figure 1) retain a homogeneous, random particle orientation, but on application of increasingly greater pressures, such materials tend to develop a slaty texture. Figure 8 shows pressed specimens from the same high-quality com-

mercial steatite (not discussed here). The products differ only by the amount of pressure applied during shaping and, thus, in their degree of particle orientation. The tendency towards a slaty texture as displayed by the lower specimen must be considered in selecting fabricating conditions, especially if large ceramic bodies and/or complicated shapes are to be formed by a dry-pressing technique. Under improper conditions stress can be non-uniformly distributed, and particles will orient in those areas subjected to the greatest amount of stress.

Fine-grained steatites having marked particle orientation tend to behave like coarse-grained talcs during ceramic forming processes. XRD patterns, however, can be used to predict particle orientation in natural steatites and thus to grade them according to the probable utility as ceramic raw materials.

ACKNOWLEDGMENT

R. B. Hall kindly helped revise the manuscript. We thank him warmly.

REFERENCES

Endlicher, G., Kromer, H., and Schüller, K.-H. (1985) Mikroskopische Untersuchungen zu Mineralbestand und Gefüge von Talken und Specksteinen: *Fortschr. Miner.* **63,** p. 57.

von Gehlen, K. (1965) Messung von Gefügeregelungen im Steatit: *Ber. Dt. keram. Ges.* **42,** p. 397.

Perdikatsis, B. (1972) Über die Talk- und Protoenstatit-Struktur und ihre strukturellen Beziehungen: *Ph.D. thesis,* University Erlangen, Nürnberg, 80 pp.

Schüller, K.-H. (1971) Untersuchungen an Talken und Specksteinen: *Sprechsaal Keramik, Glas, Email, Silikate* **104,** 421–426.

Proceedings of the International Clay Conference, Denver, 1985, L. G. Schultz, H. van Olphen, and F. A. Mumpton, eds.,
The Clay Minerals Society, Bloomington, Indiana, 400–404 (1987).

INDUSTRIAL APPLICATIONS OF SEPIOLITE FROM VALLACAS-VICÁLVARO, SPAIN: A REVIEW

EMILIO GALÁN

Departamento de Geología, Universidad de Sevilla, 41071 Sevilla, Spain

Abstract—The palygorskite-sepiolite group of clay minerals has a wide range of industrial applications. Sepiolite, in particular, has desirable sorptive, colloidal-rheological, and catalytic properties which can be modified and improved by appropriate thermal and acid treatments to extend its range of commercial uses. Sepiolite from the Vallecas deposit in Spain has been mined since the late 1600s, where it was used in the manufacture of pipes and later as a ceramic paste in the production of porcelain. Currently, the Vallecas sepiolite is used as a deodorizing agent for animal wastes, a decolorizing agent for mineral and vegetable oils, an adsorbent of undesirable polar compounds in cigarette filters, an adsorbent for toxins in the gastrointestinal tract, a catalytic support in petrochemical processes, a thixotropic agent in cosmetic preparations, a suspension agent in fertilizers and drilling muds, a filler and carrier in rubber and animal feeds, and an excipient in pharmaceutical products.

Key Words—Adsorption, Applications, Catalysis, Rheology, Sepiolite, Vallecas, Spain.

INTRODUCTION

Sepiolite, a relatively rare phyllosilicate, possesses some particularly desirable sorptive, colloidal-rheological, and catalytic properties that are of considerable commercial interest. Few exploitable deposits of this mineral have been discovered; the principal deposits of commercial significance are located (1) near Las Vegas, Nevada (Post, 1984); (2) in Amboseli, Kenya (Stoessell and Hay, 1978; Hay and Stoessell, 1984); (3) in Hunan Province, China (Zhang *et al.*, 1985); (4) Eskişehir, Turkey; and (5) Vallecas-Vicálvaro, Madrid, Spain (Galán and Castillo, 1984). The Eskişehir deposit has been mined for more than 2000 years (Kuzvart, 1984), but is now nearly exhausted. Other deposits, such as those in the Masai district, Tanzania; Grant County, New Mexico; and El Bur, Somalia; have been exploited intermittently, but, in general, have had relatively low outputs.

The largest deposit of sepiolite discovered to date is the Vallecas-Vicálvaro deposit near Madrid, Spain. This deposit contains more than 90% of the world's known reserves of sepiolite. Spain is currently the world's leading producer of this mineral (Galán, 1979). Production reached 200,000 tons in 1978 and more than 300,000 tons in 1983; 90% of the processed sepiolite was exported (Anonymous, 1983).

In the last 20 years, research on the industrial uses of sepiolite has increased. The following is a review of the most interesting industrial applications of sepiolite; in particular, the applications which have been developed for the Vallecas sepiolite.

THE VALLECAS SEPIOLITE

The Vallecas-Vicálvaro sepiolite deposit is located 10 km northeast of Madrid and occupies approximately 7 km². The deposit is of continental origin and was formed in the Tajo basin during the Miocene Period. It is part of a facies composed of fine detrital materials—the detrital subfacies of a transition facies—according to Galán and Castillo (1984). The origin of the sepiolite and other magnesium phyllosilicates (saponite, stevensite, and palygorskite) found in this basin have been and continue to be investigated extensively. Galán and Castillo (1984) attributed the formation of these minerals to chemical sedimentation in a silica- and magnesium-rich environment containing smaller quantities of Al at a pH of about 8–9. Precipitation of these authigenic minerals took place under semiarid or seasonally arid conditions during periods of tectonic calm in playa lakes and lacustrine environments.

The Vallecas sepiolite has been mined since the late 1600s. The softer and more compact variety of the ore was first used in the manufacture of pipes and cigarette filters and later as a building material in "rustic houses" (Vilanova, 1875). Between 1735 and 1808, ceramic paste for the famous "porcelains" of the Buen Retiro, Madrid, was prepared by mixing Vallecas sepiolite with clay from Capodimonte, near Naples, Italy (Prado, 1864). The full industrial exploitation of the Vallecas sepiolite began in 1945 following the first investigations of the mineralogical and technical properties of this mineral by Lacazette (1947) and Martin Vivaldi and Cano (1953).

The purity of the Vallecas sepiolite ranges from 65 to 95%. The most significant impurities are: Mg-smectites, illite, calcite, quartz, cristobalite, feldspars, and goethite. Minor amounts of chlorite, palygorskite, and dolomite are also present in some samples (Galán and Castillo, 1984). A chemical analysis of a nearly pure sample (~95% sepiolite) is listed in Table 1. Energy-

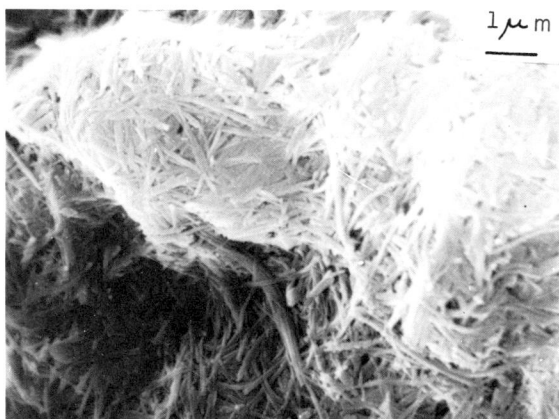

1 μm

Figure 1. Scanning electron micrograph of the Vallecas sepiolite.

Table 1. Chemical analysis of the Vallecas sepiolite.

	1	2
SiO_2	62.05	61.13
Al_2O_3	1.68	4.71
Fe_2O_3	0.46	1.36
TiO_2	0.11	0.15
MnO	0.00	0.03
CaO	0.51	1.16
MgO	23.80	20.41
Na_2O	0.30	0.34
K_2O	0.65	1.16
H_2O^+	10.90	9.94
Total	100.46	100.39

1. >95% sepiolite. 2. Commercial sepiolite (70–80%) for granular absorbents.

dispersive X-ray analysis of 10 different sepiolite particles in this sample showed no traces of Al (unpublished data from this laboratory); hence, the presence of Al in bulk samples can be used as an impurity index for the Vallecas ore.

The cation-exchange capacity (CEC) of this "pure" sample is 9.5 meq/100 g; the exchangeable cations are Mg and Ca. The N_2 surface area (BET method) is 284 m^2/g for samples degassed at room temperature, of which 139 m^2/g is the external surface area and 145 m^2/g is the surface area of the micropores. The mean micropore diameter is 15 Å, and the predominant radius of the mesopores is about 45 Å (unpublished work from this laboratory). Scanning electron microscopy shows that the sepiolite fibers are usually curved (Figure 1).

A chemical analysis of an absorbent-grade product, containing 70–80% sepiolite, is also listed in Table 1. The CEC of this material is 26 meq/100 g (TOLSA, personal communication, 1981), which is higher than that of the "pure" sample described above because of the presence of clay mineral impurities, especially Mg-smectite.

APPLICATIONS

The fabric, surface area, porosity, crystal morphology, structure, and composition of sepiolite are closely linked to the physicochemical properties which are the basis for the technological applications of this mineral. The structure of sepiolite (Brauner and Preisinger, 1956) is sensitive to thermal treatment, and different types of adsorbed water are lost successively as temperature increases. The structure collapses at 300°C, however, dehydroxylation occurs at about 800°C (Preisinger, 1963; Nagata et al., 1974; Serna et al., 1975; Prost, 1975). The mineral is also sensitive to acid treatment, which partially destroys the structure (Rodriguez Rei-

noso et al., 1981; Jimenez López et al., 1978). Both heat and acid treatment alter the surface characteristics and porosity of sepiolite, and some of the mineral's most useful properties (sorptive, colloidal, and catalytic) can be enhanced by these treatments.

Sorptive applications

The sorptive properties of absorbent granules containing about 75–80% sepiolite are shown in Table 2. These granules are used commercially as a deodorizing agent for animal litters; for controlling the concentration of ammonia in farm environments; as pesticide and herbicide carriers; as bleaching agents for mineral and vegetable oils, paraffins, fats, butter, and wine; and as an absorbent for spills on floors.

The "activation" of sepiolite for absorbent use normally involves heating it to 200°–300°C, at which temperature hygroscopic and zeolitic water is removed (Martin Vivaldi and Fenoll, 1970). The sorptive properties and mechanical strength of the granules are substantially improved by such treatment. At about 300°C, the sorptive capacity of sepiolite decreases because of the collapse of the pores, especially the micropores. The surface area also decreases to about 140 m^2/g at this temperature (Fernandez Alvarez, 1970, 1978), which corresponds to the formation of anhydrous sepiolite (Serna et al., 1975).

Sepiolite is an effective decolorizing agent (Huertas, 1969). It physically retains colored particles during filtration or percolation and adsorbs colored compounds and converts them catalytically into colorless substances. Sepiolite is substantially "activated" by extrusion under high pressure at low moisture content prior to heat treatment. The reasons for this "activation" are not well understood.

Applications of sepiolite in cigarette filters take advantage of its selective adsorptivity for nitriles, acetone, and other dangerous polar gaseous compounds over less polar compounds, such as aromatic hydrocarbons which enhance tobacco flavor. Sepiolite can also be used as an adsorbent for toxins, bacteria, and liquid in the gastrointestinal tract as can palygorskite and kaolinite (Martindale, 1982). Sepiolite suspensions effectively coat the walls of the stomach and intestines (Alvarez, 1984).

Catalytic applications

Clay catalysts are normally produced from smectites, halloysite, and kaolinite. Of these, smectite (bentonite) is by far the most widely employed, but recently, because of their high surface area, mechanical strength, and thermal stability, sepiolite granules are increasingly used as catalyst carriers. Many patents have been issued wherein sepiolite is described as a support for Co, Ni, Fe, Zn, Cu, Mo, W, Al, and Mg in different catalytic processes, including hydrogenation, desulfuration, denitrogenation, and demetallization; the production of butadiene from ethanol and hydrocarbons from methanol; and various hydrocracking processes (see Alvarez, 1984). Sepiolite is used in both the natural and thermally and acid-modified forms.

The catalytic activity of clay minerals is primarily a function of their surface activity. Silanol groups (Si–OH) present on the surface of sepiolite particles (Serratosa, 1979) have a certain degree of acidity and can act as catalysts or reaction sites. These groups occur at intervals of about 5 Å along the fiber axis of the mineral. Acid treatment of sepiolite results in the removal of adsorbed cations, an increase in surface area, and the alteration of pore-size distribution and crystallinity (Jimenez López et al., 1978). Heating lowers the surface area at temperatures >300°C, but increases the strength of the granules, which is an important property in fluid-bed cracking.

Rheological applications

The Vallecas sepiolite produces a stable suspension of high "viscosity" (1000–40,000 cps at 5 rpm in a Brookfield viscometer) at relatively low concentrations (2–10%) in water or in other liquids of high or medium porosity. These suspensions show non-Newtonian behavior which depends mainly on the concentration and pH of the sepiolite and on the type and concentration of the electrolyte (Alvarez, 1984).

Because sepiolite suspensions are thixotropic, sepiolite is a useful thickening agent in cosmetics (e.g., milks, masks), adhesives, fertilizer suspensions, and as a fluid carrier for pregerminated seeds. Sepiolite is also used in drilling muds. Its advantage over other clays, such as bentonite, is that sepiolite-based muds are less sensitive to salts, i.e., the desired rheological properties remain relatively constant despite high electrolyte concentrations over a wide range of pHs <8. At pH >9,

Table 2. Properties of absorbent sepiolite granules, Vallecas, Spain.

	6/30[1]	30/60[2]
Bulk density (g/liter)	560	615
Water absorption (%) (Ford test)	94	120
Oil absorption (%) (Ford test)	58	95
Shell index (kg/cm^2)[3]	3.9	—
Moisture (%)	12 ± 3	12 ± 3

[1] 6 × 30 mesh.
[2] 30 × 60 mesh.
[3] Mechanical strength index.

peptization causes a sharp decrease in viscosity, and the rheological behavior becomes Newtonian. Sepiolite has a "mud yield" of >150 bbl/ton in saturated salt water (Alvarez, 1984).

Filler applications

After surface modification with organic compounds, sepiolite can be used as a reinforcing filler in rubber. González Hernandez et al. (1978) compared the behavior of sepiolite as a filler in SBR (styrene, butadiene rubber), NR (natural rubber), CR (chloroprene), NBR (nitrile butadiene rubber), and EPDM (ethylene propylene diene monomer) with that of kaolin. In general, the mechanical properties and ageing characteristics of the sepiolite-filled rubbers were similar or slightly better than kaolin-filled products.

Sepiolite can also be used in pharmaceutical products as an excipient on the basis of its high, active surface, at which active products can be retained and released at suitable rates. The surface activity of other clays is commonly degraded by oxidative reactions catalyzed by iron at the clay surface or in the clay structure (Cornejo et al., 1980). Such degradation usually does not occur if sepiolite is used (Hermosin et al., 1981).

Sepiolite has been investigated for animal nutrition applications because of its sorptive, free-flowing, anticaking, and atoxic properties and its chemical inertness. The principal opportunities in this field appear to be in formulations for growth promotion, supplement carrying, feed binding, and production stimulation. In addition, sepiolite concentrations in animal feeds from 0.5 to 3% in cattle and poultry feed result in increased weight gains of about 7% for pigs and 10% for broiler chickens and rabbits (Alvarez and Perez Castell, 1982). The improvement in feed efficiency could be a consequence of an increase in digestibility of the proteins caused by a slower flow of nutrient material through the intestines due to the formation of a gel with the sepiolite. The sepiolite may also control the ammonium level, thereby preventing ammonia intoxication, and absorb toxins in the gastrointestinal tract.

CONCLUSIONS

Sepiolite is currently used in more than 100 different applications derived mainly from its sorptive, rheological, and catalytic properties. Most of these uses are similar to those of the other clays (bentonite, palygorskite) more traditionally employed, but in general the sorptive capacity of sepiolite is higher than those of the other clays. New commercial products are being investigated, especially those that make use of the organophilic material sepiolite in paints, plastics, greases, rubbers, plastisols, and cosmetics. The most important new areas are those involving catalysis, agriculture, and environmental protection.

According to Alvarez *et al.* (1985), modified sepiolite could be used to form a stable complex with type-I collagen, thereby allowing the monomeric collagen to be isolated for the production of biomaterials capable of substituting wholly or in part for organic supporter tissues. Sepiolite can be used, according to these same authors, as a specific carrier of methanogenic bacteria for the production of biogas from waste waters and, because of its rheological properties, for stabilizing coal-liquid mixtures. In addition, sepiolite may also find application as a noncarcinogenic substitute for asbestos.

For most of these applications, more fundamental investigations are needed to obtain a better understanding of the physicochemical properties that control the reaction mechanisms, especially surface properties and porosity characteristics. Many of the sorptive and catalytic properties of sepiolite are also not well understood. An interesting and nearly unexplored field of investigation is that of sepiolite–drug interaction, which may take place in sepiolite-containing pharmaceutical products. More research is also needed on the organophilic sepiolites—their preparation and the behavior and properties of their modified surface.

REFERENCES

Alvarez, A. (1984) Sepiolite: properties and use: in *Palygorskite-Sepiolite: Occurrence, Genesis, and Uses,* A. Singer and E. Galán, eds., Elsevier, Amsterdam, 253–287.

Alvarez, A. and Perez Castell, R. (1982) Sepiolite in the field of animal nutrition: in *Proc. 5th Int. Cong. Industrial Minerals, Madrid, 1982,* Metal. Bull. Ltd., London, 37–45.

Alvarez, A., Castillo, A., Perez Castell, R., Santaren, J., and Sastre, J. L. (1985) New trends in the use of sepiolite: *Prog. Abstracts, Int. Clay Conf., Denver, 1985,* p. 7.

Anonymous (1983) *Estadistica Minera de España:* Ministerio de Industria y Energia, Madrid, 250 pp.

Brauner, K. and Preisinger, A. (1956) Struktur und Entstehung des Sepioliths: *Tschermaks Min. Petr. Mitt.* **6,** 120–140.

Cornejo, J., Hermosin, M. C., White, J. L., Garnet, P. E., and Hem, S. L. (1980) Oxidative degradation of hydrocortisone in presence of attapulgite: *J. Pharm. Sci.* **69,** 945–948.

Fernandez Alvarez, T. (1970) Superficie específica y estructura de poro de la sepiolita calentada a diferentes temperaturas: *Proc. Reunion Hispano-Belga. Min. Arcillas,* C.S.I.C., Madrid, 202–209.

Fernandez Alvarez, T. (1978) Efecto de la deshidratación sobre las propiedades absorbentes de la palygorskita y sepiolita. *Clay Miner.* **13,** 325–335.

Galán, E. (1979) The fibrous clay minerals in Spain: in *8th Conf. Clay Mineral. Petrol, Teplice, 1979,* J. Konta, ed., Univ. Carolinae, Prague, 239–249.

Galán, E. and Castillo, A. (1984) Sepiolite-palygorskite in Spanish Tertiary basins: genetical patterns in continental environments: in *Palygorskite-Sepiolite: Occurrence, Genesis and Uses,* A. Singer and E. Galán, eds., Elsevier, Amsterdam, 87–124.

González Hernandez, L., Ibarra Rueda, L., and Royo Martinez, J. (1978) Sepiolita: nueva carga inorgánica de procedencia nacional para mezclas de caucho: *I Congreso de Química del Automovil,* Barcelona, 46–58.

Hay, R. L. and Stoessell, R. K. (1984) Sepiolite in the Amboseli basin of Kenya: a new interpretation: in *Palygorskite-Sepiolite: Occurrence, Genesis and Uses,* A. Singer and E. Galán, eds., Elsevier, Amsterdam, 125–136.

Hermosin, M. C., Cornejo, J., White, J. L., and Hem, S. L. (1981) Sepiolite—a potential excipient for drugs subject to oxidative degradation: *J. Pharm. Sci.* **70,** 189–192.

Huertas, F. (1969) Minerales fibrosos de la arcilla. Su genética en cuencas sedimentarias españolas y sus aplicaciones tecnológicas: Ph.D. thesis, Univ. Madrid, 284 pp.

Jimenez López, A., López González, J. D., Ramirez Saenz, A., Rodriguez Reinoso, F., Valenzuela Calahorro, C., and Zurita Herrero, L. (1978) Evolution of surface area in a sepiolite as a function of acid and heat treatments: *Clay Miner.* **13,** 375–386.

Kuzvart, M. (1984) *Industrial Minerals and Rocks:* Elsevier, Amsterdam, 454 pp.

Lacazette, F. (1947) *Estadística Minera y Metalúrgica de España:* Consejo de Mineria, Ministerio de Industria, Madrid, 49 pp.

Martin Vivaldi, J. L. and Cano, J. (1953) La sepiolita. I. Las sepiolitas españolas: *An. Edafol. Fisiol. Veg.* **12,** 827–855.

Martin Vivaldi, J. L. and Fenoll, P. (1970) Palygorskites and sepiolites (hormites): in *Differential Thermal Analysis,* R. C. Mackenzie, ed., Academic Press, London, 553–573.

Martindale, W. (1982) *The Extra Pharmacopoeia:* 28th ed., Pharm. Soc. Great Britain, Pharmaceutical Press, London, 2025 pp.

Nagata, M., Shimoda, S., and Sudo, T. (1974) On dehydration of bound water of sepiolite: *Clays & Clay Minerals* **22,** 285–293.

Post, J. L. (1978) Sepiolite deposits of the Las Vegas, Nevada area: *Clays & Clay Minerals* **26,** 58–64.

Prado, F. (1864) *Descripción Fisiográfica y Geológica de la Provincia de Madrid:* Junta General de Estadística, Imprenta Nacional, Madrid, 148 pp.

Preisinger, A. (1963) Sepiolite and related compounds: its stability and applications: in *Clays and Clay Minerals, Proc. 10th Natl. Conf., Austin, Texas, 1961,* Ada Swineford, ed., Pergamon Press, New York, 365–371.

Prost, S. (1975) Etude de l'hydratation des argiles: interactions eau-mineral, et mechanisme de la retention de l'eau: These, Université de Paris VI, A.O. 11487, 100 pp.

Rodriguez Reinoso, F., Ramirez Saenz, A., López González, J. D., Valenzuela Calahorro, C., and Zurita Herrera, L. (1981) Activation of a sepiolite with dilute solutions of HNO_3 and subsequent heat treatments. III. Development of porosity: *Clay Miner.* **16,** 315–323.

Serna, C., Ahlrichs, J. L., and Serratosa, J. M. (1975) Folding in sepiolite crystals: *Clays & Clay Minerals* **23,** 452–457.

Serratosa, J. M. (1979) Surface properties of fibrous clay minerals (palygorskite and sepiolite): in *Proc. Int. Clay, Conf., Oxford, 1978,* M. M. Mortland and V. C. Farmer, eds., Elsevier, Amsterdam, 99–109.

Stoessel, R. K. and Hay, R. L. (1978) The geochemical origin of sepiolite and kerolite at Amboseli, Kenya: *Contr. Miner. Petrol.* **65,** 255–267.

Vilanova, J. (1875) Anal. Soc. Esp. Hist. Nat., IV Acta 46 (cited in Calderón, S. (1910) *Los Minerales de España, Vol. 2,* Junta Ampl. Estud. Invest. Cien., Madrid, 405–409).

Zhang, R., Qiu, C., Peng, Ch., Dai, G., and Yang, Z. (1985) The characteristics of magnesium-rich clay in Liling area, Hunan province and a discussion on its genesis. *Bull. Yichang Inst. Geol. Mineral Resources CAGS* **9,** 1–13.

Proceedings of the International Clay Conference, Denver, 1985, L. G. Schultz, H. van Olphen, and F. A. Mumpton, eds.,
The Clay Minerals Society, Bloomington, Indiana, 405–409 (1987).

EFFECT OF AGITATION CONDITIONS ON THE FLOCCULATION OF KAOLIN WITH POLYACRYLAMIDES

A. W. Paterson[1] and R. O. Heckroodt

Department of Materials Engineering, University of Cape Town
Rondebosch, Republic of South Africa

Abstract—The mechanism of floc formation in kaolin-polyacrylamide systems and the subsequent floc disruption in the course of continued agitation were investigated by using two stirring systems, one based on a magnetic stirrer and the other on an impeller stirrer. For similar turbulent conditions in the impeller zone, the energy input of the magnetic stirrer was 16 watt/m^3 compared with 1 watt/m^3 for the impeller stirrer. This difference in energy input resulted in differences in floc formation and disruption which were manifested by differences in supernatant turbidities, levels of absorption of flocculants, and the size and shape of the flocs. Thus, energy input appeared to be more important in floc disruption than the degree of turbulence. In the kaolin-polyacrylamide systems studied, the adsorption of flocculant increased with agitation time and an upward inflection in the curve was noted at higher doses of flocculant. When the flocculant was totally adsorbed, floc disruption proceeded rapidly.

Because of the nonequilibrium nature of flocculant adsorption, kinetic factors related to agitation dominate the adsorption behavior. Thus, flocculation by polymer bridging cannot be considered independently of the agitation conditions imposed on the system, especially if the extent of polymer adsorption is being determined. For example, short periods of moderate agitation may produce a well-flocculated suspension, whereas a longer period of agitation (or higher energy input) may produce a suspension of relatively higher turbidity and higher flocculant adsorption. In comparative studies of flocculation, therefore, the conditions of agitation (which may be described readily for simple vessels and impellers) must be clearly specified.

Key Words—Adsorption, Agitation, Flocculation, Kaolin, Polyacrylamide, Turbidity.

INTRODUCTION

The importance of agitation conditions on the flocculation of suspended solids by polymeric bridging has been noted by a number of researchers (Tomi and Bagster, 1980; Reich and Vold, 1959; Parker *et al.,* 1972; Jankovics, 1965). These studies indicate that flocculant adsorption in these systems is a nonequilibrium process. Therefore, in the continuum of flocculation behavior, i.e, the initial formation of flocs and their subsequent disruption by continued agitation, the amount of flocculant adsorbed at any one time can give insight into the processes involved, particularly if it is correlated with the turbidity of the suspension. Unless these results, however, are themselves related to the factors controlling the agitation process, they are only of limited value. Unfortunately, in many studies which tried to give general expression to the effect of agitation on the flocculation processes, ill-defined stirring systems were used, making such correlations dubious (e.g., Linke and Booth, 1960; Healy, 1961; Reich and Vold, 1959; Thomas, 1964; McCarty and Olsen, 1959). In the present study two agitation systems were used in the flocculation of solids by a non-ionic polyacrylamide. The conditions of turbulence in each system were established, using well-known stirrer relationships.

EXPERIMENTAL CONDITIONS

Two standard laboratory stirrers were used: an impeller stirrer with four vertical blades, and a magnetic stirrer using a Teflon-coated stirrer bar. The stirred volume in the tanks may be divided into three zones: a bulk zone, an impeller zone, and an impeller-tip zone (Tomi and Bagster, 1980; Davies, 1972). Although the impeller-tip zone constitutes only about 5% of the total volume of agitated fluid, the energy dissipation factor is 50 times greater in that zone, as compared with a value of about 0.25 watts/m^3 for the bulk zone. The energy dissipation rate can be considered to be a measure of the turbulent conditions in the stirred vessel that consume the energy put into the system by the impeller. It is numerically equivalent to the power input per unit mass of fluid (P_m) and may be determined from the dimensionless power number (P_0) which is used by chemical engineers to define stirring systems independently of the size and fluid volume for design purposes (Rushton *et al.,* 1950a, 1950b). P_0 and P_m are defined as follows:

For baffled tanks,

$$P_0 = \frac{P}{N^3L^5},$$

---

[1] Present address: National Institute for Materials Research, Council for Scientific and Industrial Research, P.O. Box 395, Pretoria 0001, Republic of South Africa.

or for unbaffled tanks,

$$P_0 = \left(\frac{P}{N^3L^5\rho}\right)\left(\frac{1}{N^2L}\right)^{\frac{(a-\log Re_t)}{b}},$$

where N = the number of impeller revolutions in unit time, L = the impeller diameter, Re_t = the Reynolds number for stirred tanks (NL^2/ν), a and b = constants related to the impeller type, ρ = the fluid density, and ν = the kinematic viscosity.

The power input per unit mass of fluid, P_m, is equivalent to the energy dissipation rate discussed earlier, and can be calculated using:

$$P_m = P_0N^3L^5/\pi a_t^2 H,$$

where a_t = the tank radius and H = the fluid-air interface height, or where $\pi a_t^2 H$ is the fluid volume.

It is clear that the peripheral speed of the impeller tip is an important factor in determining the characteristics of a flocculation process. To have comparable peripheral speeds during this study and thus comparable turbulent conditions, different rotational speeds were used for the two systems.

The stirring conditions were analyzed using empirical stirrer relationships modified for unbaffled tanks and the diagrams and tables of Rushton et al. (1950a, 1950b). The characteristics of the two stirrer systems are given in Table 1 from Paterson (1982).

The Reynolds number of a system is a dimensionless number which gives an indication of the balance between inertial and viscous forces in a fluid. Different Reynolds numbers may be defined for different systems; for example, the tank Reynolds number is defined:

$$Re_t = NL^2/\nu,$$

where N = the impeller velocity in revolutions per second, L = the impeller diameter, and ν = the kinematic viscosity. The tank Reynolds numbers were the same for the two systems, which means that the turbulence in the bulk of the suspension was the same for the two systems. The different impeller geometries and rotational velocities, however, implied that the power input per unit mass of fluid (P_m) would be different for the two systems, particularly in the impeller and impeller-tip zones. This difference permitted the relative importance of turbulence and energy input on the progress of flocculation to be evaluated.

A sodium kaolin was prepared by ion exchange from a beneficiated primary kaolin (Heckroodt, 1979). The flocculant was a non-ionic polyacrylamide with a molecular mass of $10-20 \times 10^6$ amu. Distilled deionized water was used throughout the study. Suspensions were prepared by dispersing 1 g of kaolin in 90 ml of water with the impeller stirrer at 240 rpm for 15 min. Without changing the stirring speed, 10 ml of flocculant of an appropriate concentration was introduced by pipet

Table 1. Characteristics of the two stirring systems.

	Magnetic stirrer	Impeller stirrer
Rotational speed (rpm)	960	240
Impeller swept volume (mm³)	3400	24,500
Reynolds number, Re_t	10,000	10,000
Dimensionless power number, P_0	1.40	1.20
Power input per unit mass of fluid (P_m)		
Mean (watt/m³)	0.56	0.24
Impeller zone (watt/m³)	16	1
Maximum intensity (watt/m³)	329	20

to give a final flocculant concentration in the suspension of 0.5, 1, 2, or 4 mg/g of kaolin. In a series of runs the suspensions were stirred for periods ranging from 1 min to more than 1 hr. Another series of runs was made using the magnetic stirrer at 960 rpm from the moment the flocculant was added. In each run, immediately after the stirring was stopped, the suspension was equally divided into two identical 50-ml measuring cylinders. The various measurements were generally carried out on one of these parts of the suspension; the other part was retained for periodical checking of the reproducibility of the measurements.

After 2-min settling, a 20-ml aliquot of the supernatant liquid was removed at the 25-ml mark of one of the cylinders by a pipet having a 90° bend in its tip. The residual flocculant in this aliquot was determined by quantitative precipitation of polyacrylamide from solution by reaction with chlorine radicals. This technique was developed by I. R. Macefield of Allied Colloids, Bradford, United Kingdom (unpublished). In this method, 1 ml of an acidic stock solution (made up of 26.3 ml glacial acetic acid, 4.2 ml concentrated hydrochloric acid, and 73.7 ml deionized water) was added to the aliquot and mixed by shaking; after 15 s, 1 ml of a hypochlorite stock solution was added (sodium hypochlorite containing 1.3% available chlorine). The turbidity caused by the precipitated polymer was determined after 8 min. Calibration graphs were prepared to allow a correction to be made for the small amount of suspended solids that were sometimes present in the supernatant liquid. Any polymer associated with the suspended particles was not expected to contribute significantly to the overall turbidity measurement (Paterson, 1982). The appearance of the flocs was recorded photographically.

RESULTS

Figures 1 and 2 show the adsorption of flocculant by the kaolin for the two systems as a function of time and different flocculant concentrations. In general terms, the adsorption behavior was similar for the two systems, but important differences were noted. For the magnetic stirrer, the initial rate of adsorption was rel-

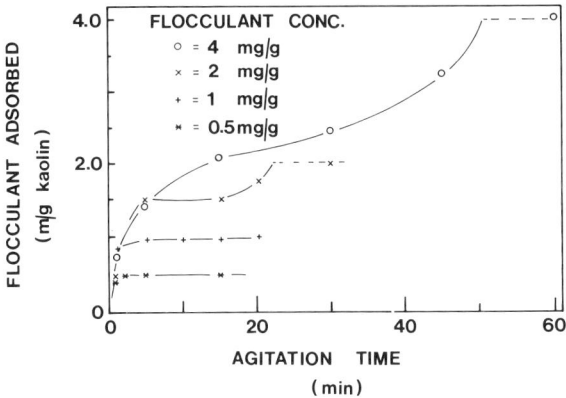

Figure 1. Effect of agitation period on the adsorption of polyacrylamide for the magnetic stirrer.

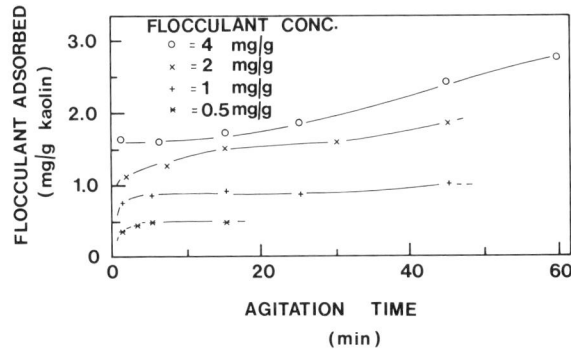

Figure 3. Effect of agitation period on the supernatant turbidity for the magnetic stirrer. JTU = Jackson turbidity units.

atively high, but decreased to a plateau region before full adsorption was achieved. At low concentrations (<1 mg/g) practically all the polyacrylamide adsorbed quickly, but as evidenced by the plateau region, small amounts of flocculant remained unadsorbed for quite a long time. For example, at initial concentrations of flocculant of 1 mg/g, the last trace of flocculant was adsorbed only after about 20 min of stirring, whereas an agitation time of >60 min was needed for complete adsorption at an initial concentration of 4 mg/g. For the impeller stirrer, the initial adsorption was generally more rapid than for the magnetic stirrer, but then the rate of adsorption abruptly decreased and complete adsorption took place only after much longer stirring times than if the magnetic stirrer had been used.

The turbidity of the supernatant liquid (determined after a settling time of 15 min) as a function of agitation time is shown in Figures 3 and 4. The behavior of the two systems is here also very similar: an initial decrease was followed by an increase in turbidity as the agitation time was increased. A minimum in turbidity was reached quickly if the initial flocculant concentration was low, but the actual turbidity was still high at this minimum point. Furthermore, as the initial flocculant

concentration was increased, the time to achieve minimum turbidity also increased. The clarity at the point of minimum turbidity improved rapidly at first, but then deteriorated slightly at initial flocculant concentrations >2 mg/g. The important difference between the two systems is that the point of minimum turbidity was reached more quickly and with higher clarity if the magnetic stirrer was used.

The manner by which the disruption of the flocs progressed with increasing agitation was different in the two systems. The degradation of the flocs was rapid and severe if the magnetic stirrer was used, as can be seen by comparing Figures 5 and 6, which show the effect of agitation time on the size of the flocs at an initial flocculant concentration of 2 mg/g. The disruption caused by the magnetic stirrer resulted in a progressive reduction in the size of the flocs; substantial disruption took place already at 15 min. For the im-

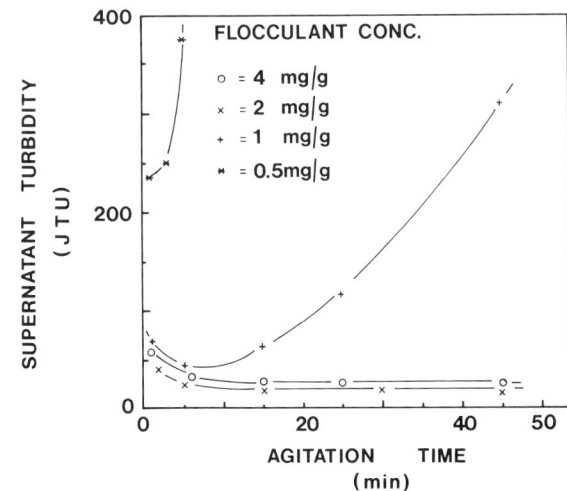

Figure 2. Effect of agitation period on the adsorption of polyacrylamide for the impeller stirrer.

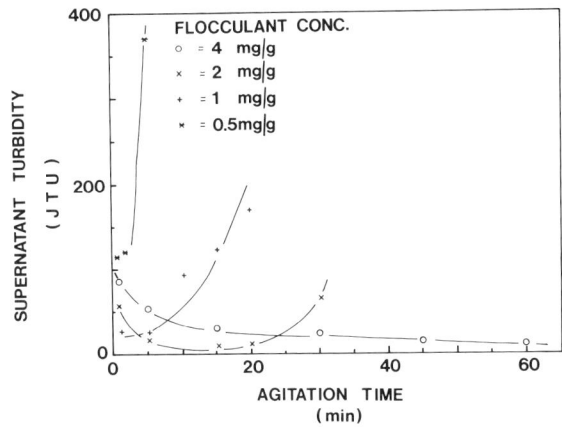

Figure 4. Effect of agitation period on the supernatant turbidity for the impeller stirrer. JTU = Jackson turbidity units.

peller stirrer, a slight but noticeable reduction in size was noted after about 5 min, but the disruption was obvious only after 60 min of agitation.

DISCUSSION

The processes by which flocs form and are subsequently disrupted depend on interactions between the suspended particles and the flocculant. These interactions are dominated by the agitation conditions. The rapid adsorption of flocculant and the associated floc formation, as well as the subsequent floc disruption, are not equilibrium processes; hence, the role of agitation is discussed accordingly.

In the present study, the sequence of events appeared to start with the adsorption of the polyacrylamide and the formation of links between the kaolin particles. The mechanism of bridging flocculation is well understood, and a number of reviews on this topic were presented by Ives (1978). In the systems studied, bridging critically determined the nature of the aggregated suspension, although some modification was expected from the effects of van der Waals and electrical double layer interactions as well as entropy stabilization. Furthermore, bridging flocculation apparently depended on a number of external variables that profoundly affected the bulk physical nature of the aggregation of particles.

In stirred systems in general, the initial rate of polymer adsorption and the concomitant formation of flocs are very dependent on the degree of turbulence, because the degree of turbulence determines the rate of collision and the rate at which free particles are entrapped by the existing flocs. The degree of adsorption is not very specific and is influenced by many factors. Adsorption is not a stoichiometric process, but for each suspension a relatively narrow range of polymer addition is required for efficient flocculation. If the amount of flocculant is insufficient, the adsorption is completed quickly, but some unflocculated free particles remain. In the presence of excess flocculant, adsorption continues to a level that depends on the degree of excess and the suspension is fully flocculated.

The degree of turbulence in the two stirring systems examined here was of the same moderate intensity, and the rate of adsorption of the polyacrylamide during the intial stages of flocculation reflects this similarity. For small polyacrylamide additions (<1 mg/g), practically complete adsorption was achieved within a few minutes, but the turbidity was still high, which indicates that flocculation was incomplete. For larger polyacrylamide additions, full adsorption was not achieved, but a limit was approached that was dependent on the total amount of polymer. Thus, nearly 100% adsorption was achieved by the addition of 1 mg of polyacrylamide per gram of suspension, but only about 75% was achieved for a 2-mg/g addition, and only 50% for a 4-mg/g addition of the polymer. The time necessary

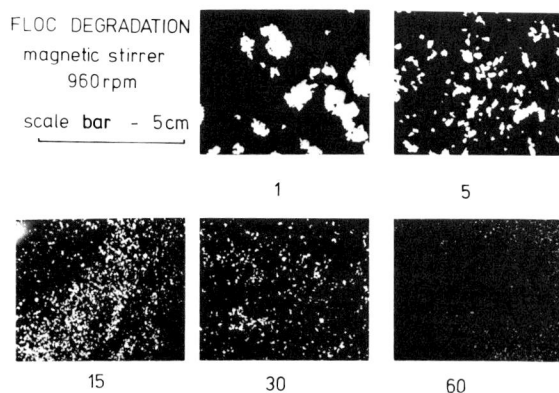

Figure 5. Photographs showing degradation of flocs with agitation (time given in minutes) for the magnetic stirrer.

to achieve this level of adsorption also increased with increasing excess of polymer.

Flocs subjected to continued agitation tended to break down by rupture or erosion. The rate of disruption was determined by the amount of energy available, i.e., the power input, but if excess polymer was available, the disruption was countered by the formation of new bridges, and, thus, ruptured flocs were repaired. The net effect was that the disruption was not initially apparent and became evident only when the excess polymer had been depleted. Thus, the polymer chains involved in the rupture of a floc tended to collapse and adsorb onto the surface of one of the fragments. As a consequence, they were not generally available to reform the bridges. The fragments, therefore, tended to become "stabilized."

The power input into the magnetic stirrer system was much higher than that into the impeller system; thus, the rate of disruption of flocs was much higher in the former system. Excess polymer was, therefore, consumed faster to repair the greater number of ruptures and the breakdown became apparent earlier, as

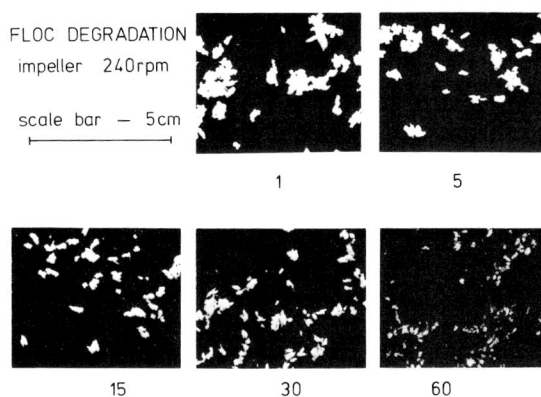

Figure 6. Photographs showing degradation of flocs with agitation (time given in minutes) for the impeller stirrer.

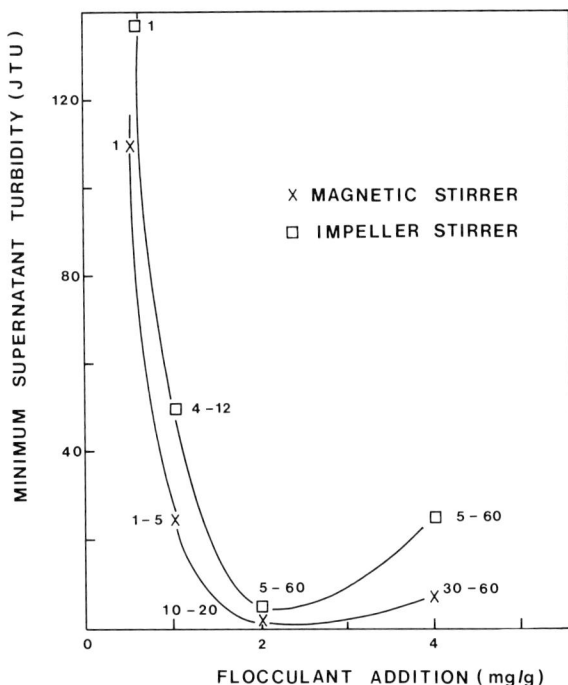

Figure 7. Minimum supernatant turbidity achieved with increasing polyacrylamide additions. Approximate starting and finishing times of the minima are indicated at each dosage. JTU = Jackson turbidity units.

suspension. The detailed response of each particular mineral species will depend on the type of cation on the mineral surface, the degree of hydrolysis, the anionic character of the flocculant, and the pH of the suspension. If a credible assessment of the impact of any one of these variables is to be made, particularly as they relate to flocculant adsorption, a detailed analysis and description of the agitation conditions are of prime importance. Similarly, for industrial applications, the point at which flocculants are introduced into the systems must be chosen with care to avoid breakdown and stabilization of the flocs by subsequent continued agitation.

REFERENCES

Davies, J. T. (1972) *Turbulence Phenomena:* Academic Press, New York and London, 121–173.

Healy, T. W. (1961) Flocculation-dispersion behaviour of quartz in the presence of polyacrylamide flocculant: *J. Coll. Sci.* **16**, 609–617.

Heckroodt, R. O. (1979) The properties of the Cape kaolins: *Ann. Geol. Surv. South Africa* **13**, 9–13.

Ives, K. J., ed. (1978) *The Scientific Basis of Flocculation:* NATO Advanced Studies Institute Series E, No. 27, Sijthoff and Noordhoff, Netherlands, 165–191.

Jankovics, L. (1965) Effect of agitation and molecular weight on polymer adsorption and deflocculation: *Appl. Polymer Sci.* **9**, 545–552.

Linke, W. F. and Booth, R. B. (1960) Physical chemical aspects of flocculation by polymers: *Trans. Amer. Inst. Min. Mech. Pet. Eng.* **217**, 364–371.

McCarty, M. F. and Olsen, R. S. (1959) Polyacrylamides for the mining industry: *Mining Eng.* **10**, 61–65.

Parker, D. S., Kaufmann, W. J., and Jenkins, D. (1972) Floc breakup in turbulent flocculation processes: *San. Eng. Div., Proc. Amer. Soc. Civ. Eng.* **98**, 79–99.

Paterson, A. W. (1982) The flocculation of kaolin with polyacrylamides: Ph.D. Thesis, University of Cape Town, Cape Town, Republic of South Africa, 52–54, 64–65.

Reich, I. and Vold, R. D. (1959) Flocculation-deflocculation in agitated suspensions: I. Carbon and ferric oxide in water: *J. Phys. Chem.* **63**, 1497–1501.

Rushton, J. H., Costlich, E. W., and Everett, H. J. (1950a) Power characteristics of mixing impellers, Part I: *Chem. Eng. Progress* **46**, 395–404.

Rushton, J. H., Costlich, E. W., and Everett, H. J. (1950b) Power characteristics of mixing impellers, Part II: *Chem. Eng. Progress* **46**, 467–476.

Thomas, D. G. (1964) Turbulent disruption of flocs in small particle size suspensions: *J. Amer. Inst. Chem. Eng.* **10**, 517–523.

Tomi, D. T. and Bagster, D. F. (1980) The behaviour of aggregates in stirred vessels: *Mat. Sci. Eng.* **12**, 3–19.

shown by the turbidity of the supernatant and by visual observation of the size of the flocs. The plateau region on the adsorption-time curves was thus shorter in the magnetic stirrer system.

The flocculation achieved by the two systems is compared in Figure 7, which shows the minimum turbidity that was achieved with different concentrations of polyacrylamide. As far as this aspect is concerned, the two systems were similar, which reflects the similarity of the turbulent conditions. The duration of the condition of minimum turbidity was, however, different and reflected the influence of energy input on the subsequent disruption of the flocs. Thus, the conditions of agitation, especially energy input, determined the condition of the floc suspension.

It is clear that flocculation by polymeric bridging is greatly influenced by the agitation conditions in the

Proceedings of the International Clay Conference, Denver, 1985, L. G. Schultz, H. van Olphen, and F. A. Mumpton, eds.,
The Clay Minerals Society, Bloomington, Indiana, 410–414 (1987).

SODIUM SULFATE-BEARING SOILS AND THE PROBLEM CAUSED IN PRESSURE FILTRATION DURING SALT EXTRACTION

Ali A. Hassani Pak

Mining Engineering Department, University of Tehran, Tehran, Iran

Abstract—Fifty-four samples of sodium sulfate-bearing soil excavated from mines in the Salt Desert (Dasht-e-Namak) of Iran were examined to investigate difficulties encountered during sodium sulfate extraction by the partial dissolution of the raw materials in water, using pressure filtration. From mineralogical and chemical composition and particle-size distribution of the samples, the chief causes of the extreme decrease in filtration rate were found to be: (1) water adsorption and enhanced dispersion of clay minerals, particularly smectites, in a Na_2SO_4-rich slurry, which caused the formation of colloidal particles and hence decreased the filter cake porosity and permeability; and (2) disintegration of aggregates of insoluble particles during agitation of the slurry, which caused a reduction of the average particle size and, hence, of the cake porosity.

The filtration rate was increased by the addition of lime to the slurry and by the reduction of particle disintegration through reducing the agitation time of the slurry from 30 to 3 min. This treatment caused only a small decrease in dissolution efficiency of sodium sulfate (from 91 to 85%).

Key Words—Clay dispersion, Filtration, Particle size, Porosity, Sodium sulfate.

INTRODUCTION

Many sodium sulfate deposits of the northeastern part of the Salt Desert (Dasht-e-Namak), Iran, are in the form of a thin, near-surface layer, 2–40 cm thick. The concentration of sodium sulfate in this enriched layer generally ranges from 20 to 25% (Hassani Pak, 1984). Sodium sulfate in this layer is present chiefly as mirabilite, $Na_2SO_4 \cdot 10H_2O$, and glauberite, $(Na_2,Ca)SO_4$, which coexist with gypsum, halite, and detrital minerals such as clay minerals, quartz, feldspars, and calcite. The enriched layer usually is covered by a soil layer as much as 20 cm thick. The soil layer contains about 1% sodium sulfate and the same detrital mineral constituents as the underlying enriched layer. The enriched layer overlies a relatively uniform sand-silt layer a few meters thick in which the concentration of sodium sulfate decreases from a maximum of a few percent at the top to near zero in the first half meter. During mechanical strip mining, the concentration of sodium sulfate in the mined material may be as low as 10–15% due to mixing of the overlying soil with the enriched layer.

Sodium sulfate can be extracted from this ore by various methods, the most efficient of which involves the following steps: (1) Sodium sulfate is dissolved in water by mixing the raw materials and water in attrition tanks. Agitation of the slurry accelerates the dissolution process and also increases the dissolution efficiency of sodium sulfate. (2) The soluble phases are separated from the insoluble phases by pressure filtration. (3) The filtrate is heated to evaporate water and to precipitate sodium sulfate. The problem with this method is that during the separation step, the permeability of the cake that is formed on the filter cloth is so low that even a thin layer causes a sharp decrease in filtration rate, and the process soon must be discontinued.

The object of the present study was to determine the factors that caused the extreme decrease in filtration rate and then to consider possible changes to obtain acceptable filter productivity.

SAMPLING AND ANALYTICAL METHODS

Fifty-four samples (each 2–4 kg) were taken initially from two types of mined sodium sulfate-bearing soils: (1) Soils were sampled which had been mined about five years before this study and stored in the open in piles 3–15 m thick. This type of soil was not homogeneous and was zoned from top to bottom, due to long-term action of atmospheric precipitation that redistributed the sodium sulfate, especially near the surface (Figure 1). The average concentration of sodium sulfate in this type of soil was about 7%. (2) Soils were sampled that were mined just before and during this study and which were relatively homogeneous and had a concentration of sodium sulfate of ~20%.

Each of the 15 samples in Figure 1 was composed of three or four samples, each of which consisted of nine 200–400-g samples which were collected by grid sampling or random channel sampling. Three additional samples were prepared as follows:

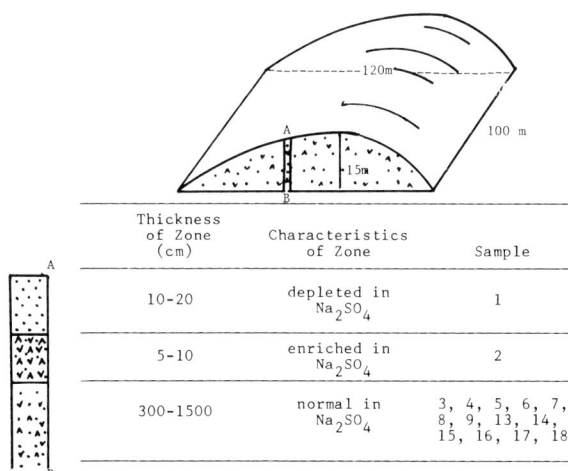

Figure 1. Sample-collecting scheme for mined mounds of type I soils.

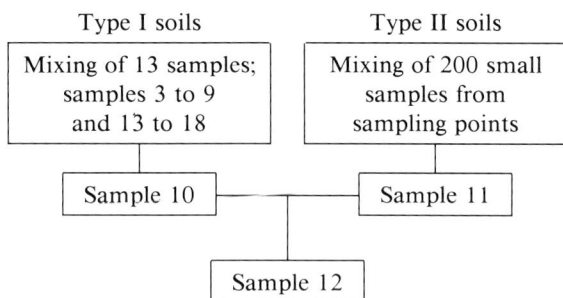

Thickness of Zone (cm)	Characteristics of Zone	Sample
10-20	depleted in Na$_2$SO$_4$	1
5-10	enriched in Na$_2$SO$_4$	2
300-1500	normal in Na$_2$SO$_4$	3, 4, 5, 6, 7, 8, 9, 13, 14, 15, 16, 17, 18

Type I soils	Type II soils
Mixing of 13 samples; samples 3 to 9 and 13 to 18	Mixing of 200 small samples from sampling points
Sample 10	Sample 11
	Sample 12

Quantitative X-ray diffraction (XRD) analysis was made to estimate the mineralogical composition of the soils, particularly the abundance and nature of expandable clay minerals. The relative abundances of soft minerals such as gypsum and hard minerals such as quartz in the soils was also of interest. The method described by Carver (1971) was used to determine the abundance of clay minerals. The percentage of quartz was determined by the method of Giachett and Tozzin (1983). A rough estimation was made of the feldspar content, mainly alkali feldspar, by XRD using an internal standard. A heavy liquid separation was used to estimate the amount of heavy minerals.

For each of the 18 samples, the water-soluble and cold-acid-soluble fraction were separately analyzed for SO$_4^{2-}$, Cl$^-$, Na$^+$, Ca^{2+}, Mg^{2+}, and K$^+$. The acid-soluble fraction was also analyzed for CO$_2$. The corresponding amounts of gypsum, halite, calcite, and the sum of sodium sulfate from mirabilite + glauberite were estimated from these data. The water-soluble fractions were prepared by adding 0.5 g of each pulverized sample to 250 ml of water and heating the mixture to 50°C with agitation for 30 min. A considerable fraction, if not all, of the gypsum in each sample may have dissolved; consequently the total soluble fraction shown

in Table 1 should not be regarded as consisting of only the common water-soluble phases. It should be noted that in both the water and cold-acid dissolutions, some of the CaSO$_4$ may have originated from glauberite. Exchangeable Ca, Mg, and Na were measured to determine the predominant exchangeable cation of these soils.

To determine the grain-size distribution of the insoluble phases of the soils and filter cake, samples were analyzed by the gravitational settling method, using ethanol as the fluid medium instead of water to avoid dissolution of gypsum, especially very fine grains which were produced during agitation of slurry in the attrition tanks.

RESULTS

Table 1 lists chemical analyses and calculated salt mineralogies of six more important samples investigated. From the chemical data, the hypothetical concentrations of sodium sulfate were calculated for each sample assuming that: (1) chlorine was present only as NaCl; (2) sodium was present both as NaCl and Na$_2$SO$_4$; (3) calcium was present in gypsum only, but during cold-acid dissolution it was present in both gypsum and calcite (XRD analysis indicated only trace, hence negligible, amounts of dolomite); and (4) potassium and magnesium were present in sulfate phases only. To determine the accuracy of the calculated sodium sulfate concentration, the relative error associated with each sample was also calculated and is shown in the same table.

Assuming that the dissolution efficiency of sodium sulfate in cold acid was 100%, the data in Table 1 suggest that the dissolution efficiency of sodium sulfate in water ranged from 71.1 to 96.4%, averaging 91%. The lowest and highest values of this range are from the same samples that showed the lowest and highest sodium sulfate concentrations, 1 and 19.3%, respectively. The abundance of calcium sulfate in the cold-acid-soluble fraction ranged from 17.1 to 25.1%, corresponding to 21.5 to 31.6% gypsum in the samples.

Quantitative X-ray powder diffraction analyses of six samples are shown in Table 2. The XRD analyses were carried out after extraction of soluble phases, and the values were adjusted to represent percentages in the total sample. These data indicate that the total percentage of clay minerals ranged from 17 to 28%, of which 7–20% is smectite. The total amount of quartz, feldspars, and heavy minerals ranged from 36 to 50%.

The exchangeable sodium, calcium, and magnesium ion contents of eight samples are shown in Table 3. These data indicate that 88–95% of the measured exchangeable cation is due to the sodium cation. This amount of exchangeable sodium is in agreement with the prolonged contact of the clay minerals with high concentrations of sodium-bearing solutions.

Table 1. Chemical analysis of water-soluble and cold-acid-soluble fractions of six samples and calculation of sodium sulfate concentration.

Sample	Cl¹ (%)	NaCl (%)	Na in NaCl¹ (%)	Na¹ (%)	Na in Na₂SO₄ (%)	Na₂SO₄ (%)	CO₂¹ (%)	CaCO₃ (%)	Ca in CaCO₃ (%)	Ca (%)	Ca in CaSO₄ (%)	CaSO₄ (%)	K¹ (%)
						Water soluble							
1	0.03	0.05	0.02	0.25	0.23	0.71	—	—	—	2.79	2.79	9.5	0.02
2	0.20	0.33	0.13	5.39	5.26	16.3	—	—	—	2.07	2.07	7.0	0.03
5	0.53	0.87	0.34	2.66	2.32	7.2	—	—	—	2.59	2.59	8.8	0.02
10	0.51	0.84	0.33	2.61	2.25	7.0	—	—	—	2.46	2.46	8.4	0.02
11	0.33	0.54	0.21	6.21	6.00	18.6	—	—	—	2.04	2.04	6.9	0.02
12	0.38	0.62	0.24	4.55	4.21	13.0	—	—	—	2.10	2.10	7.1	0.02
						Cold acid soluble							
1	0.03	0.05	0.02	0.34	0.32	0.99	3.18	7.21	2.90	8.47	5.57	18.9	0.02
2	0.20	0.33	0.13	5.61	5.48	17.00	2.59	5.88	2.35	7.38	5.03	17.1	0.03
5	0.53	0.87	0.34	2.83	2.49	7.70	2.34	5.13	2.13	9.52	7.39	25.1	0.02
10	0.51	0.84	0.33	2.88	2.55	7.90	2.16	4.90	1.96	9.26	7.30	24.8	0.02
11	0.33	0.54	0.21	6.44	6.23	19.30	0.98	2.22	0.89	5.91	5.02	17.1	0.02
12	0.38	0.62	0.24	4.83	4.59	14.20	1.53	3.47	1.39	7.56	6.17	20.4	0.02

Sample	K₂SO₄ (%)	Mg¹ (%)	MgSO₄ (%)	SO₄ in Na₂SO₄ (%)	SO₄ in CaSO₄ (%)	SO₄ in K₂SO₄ (%)	SO₄ in MgSO₄ (%)	Σ SO₄ used (%)	SO₄¹ (%)	ΔSO₄ (%)	ΔSO₄/(SO₄)¹ (%)	Total soluble (%)
						Water soluble						
1	0.04	0.05	0.23	0.54	6.69	0.024	0.17	7.42	7.10	0.32	4.5	10.5
2	0.06	0.07	0.32	11.00	4.97	0.036	0.25	16.25	16.59	0.34	2.0	23.9
5	0.04	0.12	0.55	4.87	6.21	0.024	0.42	11.52	11.60	0.08	1.0	17.2
10	0.04	0.06	0.27	4.79	5.90	0.024	0.21	10.92	10.99	0.02	0.2	16.3
11	0.04	0.09	0.41	12.60	4.90	0.024	0.32	17.84	18.13	0.29	1.6	26.3
12	0.04	0.05	0.23	8.84	5.04	0.024	0.17	14.70	14.54	0.16	1.0	20.8
						Cold acid soluble						
1	0.04	0.05	0.23	0.67	13.36	0.024	0.17	14.22	14.54	0.32	2.0	27.02
2	0.06	0.07	0.32	11.50	12.07	0.036	0.25	23.86	24.58	0.72	3.0	30.64
5	0.04	0.12	0.55	5.22	17.73	0.024	0.42	23.39	22.35	1.04	4.0	39.39
10	0.04	0.06	0.27	5.35	17.52	0.024	0.21	23.10	22.77	0.33	1.5	38.35
11	0.04	0.09	0.41	13.08	12.05	0.024	0.32	25.47	24.78	0.69	3.0	39.51
12	0.04	0.05	0.23	9.63	14.80	0.024	0.17	24.62	24.04	0.58	2.5	38.91

¹ Denotes measured quantity; other data were calculated.

The cumulative grain-size distribution curves of the samples are shown in Figure 2. The grain size ranged mainly from that of coarse silt to that of clay, but most of the material in each of the samples was of medium-silt size. From a comparison of mineralogical results and particle size distributions, the clay size fraction of each sample appears to contain only about half of the total amount of clay minerals present, i.e., about 50% of the clay minerals present in these samples occur as particles >2 μm in size. This would be an unusual result for recently formed clay minerals were it not that

Table 2. Results of quantitative X-ray powder diffraction analyses (in wt. %).

Sample	Quartz	Illite	Kaoli-nite	Smec-tite	Clay	Fl + HM¹	Total
1	32	8	5	10	23	18	73
5	27	6	3	8	17	17	61
9	22	10	5	7	22	18	62
10	21	8	5	13	26	15	62
12	21	11	4	11	25	15	62
17	22	2	6	20	28	16	66

¹ Feldspar + heavy minerals.

Table 3. Exchangeable sodium, calcium, and magnesium for selected samples (meq/100 g).

Sample	Total exchange cation	Exchangeable Ca + Mg	Exchangeable Na	Exchangeable Na Total exchangeable cation (%)
1	21.76	1.29	20.47	94
5	19.20	1.55	17.65	92
6	20.48	1.06	19.42	95
9	22.40	2.22	20.18	90
10	21.76	2.67	19.09	88
11	20.80	1.14	19.66	95
12	19.84	1.42	18.42	93
17	20.48	1.47	19.01	93

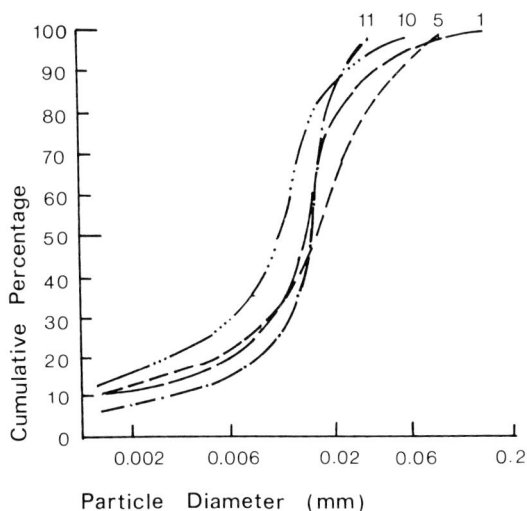

Figure 2. Cumulative size-distribution curves of some selected samples.

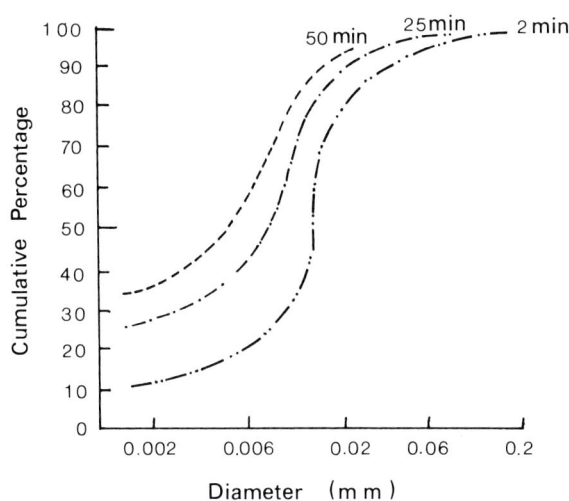

Figure 3. Effect of agitation time of the slurry on size of particles.

these formed in an evaporitic environment where it is not unusual for illite and kaolinite particles to form as particles >2 μm (C. E. Weaver, Georgia Institute of Technology, Atlanta, Georgia, personal communication, 1985).

IMPROVEMENT OF FILTRATION RATE

Lime treatment

Because it is well known that filtration of colloidal systems can be improved by treatment with strong flocculents, particularly divalent cations if negative colloids such as clays are involved, we studied the effect of lime in the filtration systems. Lime, as a <44-μm powder, first was mixed with water and then the mixture was used for the dissolution process and the making of slurry. As is seen in Table 4, the magnitude of V/Aθ (filtrate volume per unit time per unit filter area) increased with increased amount of lime added, but

because the pH of the slurry increased as well, some limit on the amount of lime added had to be applied, based on an acceptable corrosion rate for pipes and apparatus. For the plant related to these particular tests, the amount of lime added did not exceed 0.5% by weight.

Reduction of disintegration caused by agitation

The insoluble phases in the slurry consisted of minerals having a relatively large range of hardness. The Mohs hardness of the main constituents ranged from

Table 4. The result of lime added to the slurry on the filtration rate at constant temperature.

Test no.	Lime added (%)	V/Aθ[1] (cm/s)	pH
1	0.1	0.30	8.1
2	0.2	0.41	8.5
3	0.3	0.52	8.9
4	0.4	0.79	9.3
5	0.5	0.83	9.7
6	0.6	1.10	10.2
7	0.7	1.50	10.6
8	0.8	1.82	10.9
9	0.9	2.40	11.3
10	1.0	3.80	11.7

[1] Filtrate volume per unit time per unit filter area.

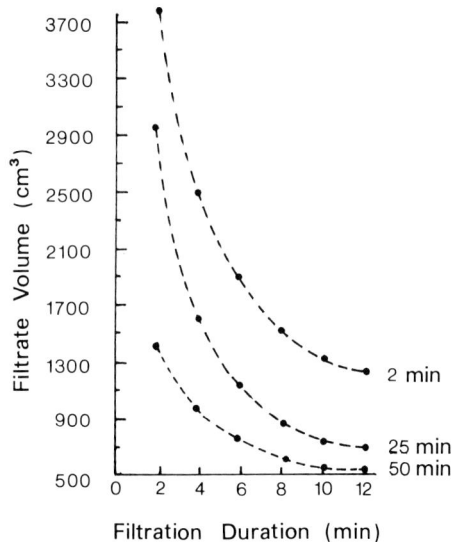

Figure 4. Effect of agitation time of the slurry on filtration rate.

2 (gypsum) to 7 (quartz). Agitation of the slurry containing 30–35% by weight of such insoluble phases for 30 min, as recommended by the designer of the apparatus for optimum dissolution of sodium sulfate, caused the soft particles to be abraded by the hard particles. Consequently, the particle-size range broadened and the mean particle size decreased as a function of agitation time. Thus, the porosity of the cake was reduced, and the average ratio of particle surface to particle volume increased, both tending to increase the specific cake resistance. In addition, more stirring of the slurry increased the dispersion of clay minerals, especially smectite, and enhanced the problem.

To evaluate the effect of agitation time on the particle-size distribution in the cake, particle-size analyses were made on three mud-cake samples deposited separately on the filter cloth after 2, 25, and 50 min of agitation of the same slurry (Figure 3). As is seen in Figure 3, agitation of the slurry significantly decreased the average particle size. Because specific cake resistance is sensitive to this variation, the agitation time was reduced. On the other hand, reduction of agitation time had an unfavorable effect on the efficiency of sodium sulfate dissolution. Results obtained from chemical analyses indicated that sodium sulfate dissolution efficiency rose sharply to an average of 85% during the first 3 min of agitation of the slurry and then increased gradually to an average of 91% after 30 min of agitation. Therefore, the agitation time was reduced from 30 to 3 min. The data on Figure 4 verify that increasing the agitation time of the slurry had a negative effect on the filtration rate.

SUMMARY

In pressure filtration of a slurry containing various minerals, the filtration rate can be affected markedly by the mineralogical composition of the crystalline phases. In the present investigation, enhanced dispersion of clay minerals, particularly fine-grained smectites, was found to decrease the filter cake permeability. To minimize this effect, preconditioning of the slurry by lime treatment was applied. This treatment enhanced the flocculation of the fine particles and the formation of larger agglomerates, which were more filterable.

The presence of soft and hard minerals in the slurry and the subsequent degradation of particles by agitation also increased the specific cake resistance and hence decreased the filtration rate. To minimize the degradation, agitation time of the slurry was reduced from 30 to 3 min. This reduction was at the expense of a decrease in sodium sulfate dissolution efficiency from 91 to 85%.

ACKNOWLEDGMENTS

The author is indebted to F. A. Mumpton, H. van Olphen, and H. W. Olsen for their critical reading of the manuscript and helpful editorial suggestions and many recommendations. The writer also thanks M. Wampler and C. Pollard for their valuable discussions and comments.

REFERENCES

Carver, E. (1971) *Procedures in Sedimentary Petrology:* Wiley, New York, 653 pp.

Giachett, L. and Tozzin, N. (1983) *Quantitative Powder X-ray Diffraction of some Ceramic Materials:* V. Vincenzini, ed., Elsevier, Amsterdam, 185–195.

Hassani Pak, A. A. (1984) Sodium sulfate exploration project in the northwest part of Salt Desert (Dasht-e-Namak), Iran: *Open File Rept.,* Sulfatic Company, Tehran, Iran, 84 pp.

Proceedings of the International Clay Conference, Denver, 1985, L. G. Schultz, H. van Olphen, and F. A. Mumpton, eds.,
The Clay Minerals Society, Bloomington, Indiana, 415–421 (1987).

ELEMENT DIFFUSION AND OXIDATION PHENOMENA ALONG PERMEABLE FRACTURES IN CLAY FORMATIONS AT MONTEROTONDO, ITALY

F. Antonioli,[1] M. D'Alessandro,[2] F. Mousty,[2] A. Saltelli,[2] and D. A. Stanners[2]

[1] Ente Nazionale Energie Alternative, Casaccia, Roma, Italy

[2] Commission of the European Communities, Joint Research Centre
Ispra Establishment, 21020 Ispra, Varese, Italy

Abstract—Clay formations are being considered in a number of European countries as possible geological host rocks for radioactive wastes. Risk analysis studies have pointed out that fracturing of the formation, followed by water intrusion, could lead to the release of radioactivity. A field study was conducted on samples taken around clay fractures to determine the diffusion profiles of several elements on both sides of the fracture surface and to date existing fractures through which water has percolated. Data collected on samples taken from the face of a clay quarry in central Italy, where oxidized fractures exhibit distinct signs of water percolation, show that: (1) analyses of Fe, Cs, Te, U, and Co are effective in discriminating between perturbed and unperturbed clay in which elements have been diffusing; and (2) the order of magnitude of the active lifetime of the percolation in quarry-face samples ranges between a few tens and a few hundreds of years. Cs, Te, U, and Fe concentration profiles were found to have distribution coefficients strongly dependent on the oxidation-reduction (redox) conditions. A stepwise profile was evident that corresponds to the redox front. The Co profile, however, was apparently not affected by the redox conditions of the medium inasmuch as it shows a more regular trend than those of the other elements.

Key Words—Clay, Diffusion, Fractures, Nuclear waste disposal, Oxidation.

INTRODUCTION

The geological formations chosen to host repositories of high-level radioactive waste must guarantee waste isolation from man's environment for extended periods of time. Of the many lithologic types examined, argillaceous rocks, if properly chosen, appear among the most suitable because of their high retention capability for radionuclides, their intrinsic impermeability, and their capability of plastic response to imposed stresses (Bonne *et al.*, 1985).

The high retention coefficient of clay is a well-known property that allows clay to be an effective barrier against radionuclide migration. Migration is also prevented by the lack of water movement; i.e., within compact and homogeneous clays a permeability as low as 10^{-9}–10^{-10} ms^{-1} allows ground-water displacement of only a few meters per hundred thousand years, assuming a typical hydraulic gradient of 0.1–0.01 m/m (Bonne *et al.*, 1979).

As long as the rock remains physically homogeneous and whole, isolation of waste is guaranteed, but fracturing can obviously increase permeability significantly (D'Alessandro and Gera, 1986). Although the possibility of permeable fractures forming in plastic clay is still an open question (Chiantore and Gera, 1986), faults and fractures that are locally permeable are common in clays, not only in outcrops or quarry faces, but also at depth (Chapman and Gera, 1985). In many deposits traces of oxidation along fissures in quarry faces seem

to indicate that the clay has been affected by ground-water circulation. This possibility raises an important question about the origin of the oxidation zones. Two different explanations are possible: (1) the oxidation zones were formed over a geological time period at the original depth of burial as a result of water percolating through the formation; or (2) oxidation was rapid and very close to the surface and was strictly due to the quarry being opened. In the latter process, the oxidation bands must postdate the beginning of mining.

The question of the origin of such oxidation zones was also addressed in previous studies (Antonioli and Lenzi, 1984; Saltelli and Antonioli, 1985), carried out on the Plio-Pleistocene blue clays of Monterotondo (north of Rome). Here, in an active quarry front, several fractures are characterized by striking oxidation bands along their borders that suggest the circulation of oxidizing solutions. About 100 m of a blue-grey clay sequence is exposed in this quarry. A pronounced sandy component is present at the top of the sequence. Minor erosion is apparent at the top of the formation only. The mineralogical composition of this clay is shown in Table 1. From the lower to the upper portion of the sequence, the preconsolidation pressure, P_c, and the plasticity index, IP, vary between 20 and 30 kg/cm^2 and 33 and 22%, respectively. These values do not vary continuously, but show a step in the middle of the formation that may reflect environmental difference

Table 1. Mineralogical composition of Monterotondo clay.

Mineral	Wt. %	Clay minerals	Wt. %
Clay minerals	50–60	Smectite	40–50
Quartz	~10	Illite	25–30
Calcite	20–25	Chlorite	~10
Dolomite	~5	Kaolinite	<10
Feldspar	~5	Interstratified	<10

between the Pliocene and Pleistocene sedimentation (Mortari, 1977). During the exploitation of the quarry, a set of vertical intersecting fractures, sandwiched between two red-yellow oxidation bands, were uncovered.

From the analytical results, an oxidation process appears to have been active along fissure borders, but it had no apparent effect on the mineralogical composition of the clay. The only recognizable effect was the change in trace element content (e.g., Co, S, Cs). As expected, the total Fe concentration was higher in the oxidized zone than in the reduced zones, and the Fe^{2+}/Fe^{3+} ratio increased from the fracture outwards into the blue clay. Furthermore, the oxidation bands showed a retention capability for Cs and Sr higher than that of the reduced material.

In these previous studies a preliminary estimate of the lifetime of the fracture, carried out on the basis of the Sr profile, gave a duration of a few hundred years; thus, the order of magnitude of the water circulation time through the fractured clays (assumed as the age of the permeable fracture) seemed to relate to a historical, rather than a geological time-span.

In the present study, some results from the above mentioned investigations were checked, and new samples of clays showing significant oxidation bands (from the same Monterotondo quarry) were studied. Chemical analyses, mainly of trace elements, were made, and a mathematical model of uranium migration through the fractured clay was developed to provide a general understanding of the diffusion phenomena in these clays.

EXPERIMENTAL

Description of samples and sampled zone

Samples A and B (Figure 1) were taken in a fracture from a level in the quarry corresponding to an original depth below ground level of about 20–30 m. The fracture was part of a network of oxidized discontinuities forming a mesh 2–5 m wide, clearly visible on the floor of the quarry. The oxidation halos bordering the fracture surfaces were 3–7 cm thick, and the color change from the brown-yellow oxidation bands to the grey bulk rock was generally discontinuous. The halos appeared as a sequence of stripes of different shades showing well-defined borders between each other (sample A). In other samples the brown-yellow bands changed

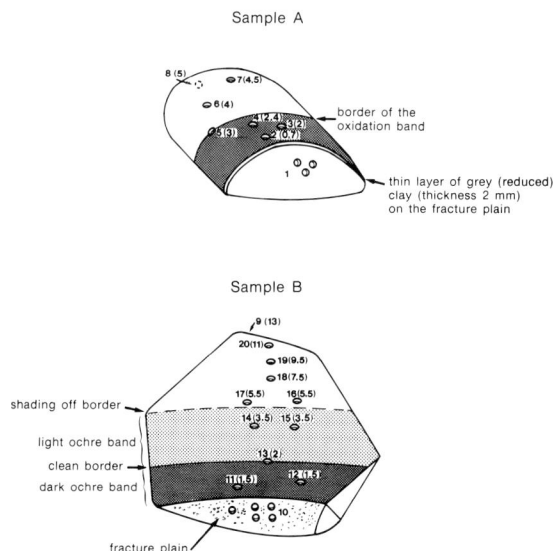

Figure 1. Collection schemes for clay samples A and B for neutron activation analysis. Sample numbers and the distance from the fracture plain (cm) are shown.

continuously into grey bands, passing from the oxidized to the reduced zone. In some areas the fractures appeared to be healed by a precipitate of hydrous calcium sulfate (sample B), possibly due to the oxidation of pyrite. The fractures were filled with a thin layer (~2 mm) of reduced clay, sandwiched between the two oxidation bands (samples A and B).

A larger sample C (Figure 2) was taken from a wall of the quarry at an original depth of 30–40 m, where the fractures showed a more regular, sub-parallel trend. In this sample the oxidation zone was a band of varying thickness, having the darkest portion in its thickest part. Three different traverses, I, II, and III were examined to evaluate effects of the apparently different degrees of severity of oxidation.

Chemical analysis

Depending on the elements to be determined and on their concentration, three different analytical techniques were used: electrothermal atomic absorption spectroscopy (ETAAS), inductively coupled plasma-emission spectroscopy (ICPES), and neutron activation analysis (NAA). Where possible, key elements, such as Co and Fe, were determined by two of the above-mentioned techniques.

Apparatus

The equipment used for ETAAS and ICPES consisted of a Perkin Elmer Atomic Spectrometer Model 5500, equipped with a graphite furnace (HGA-500), an automatic sampler model AS-40, a Model 023 recorder, a Data System 10 computer with a PR-100 printer, and a plasma-therm RF generator HF P2500F.

Figure 2. Sample C, showing an irregular oxidation band; I, II, and III are the traces of three different traverses. Sample width is about 19 cm.

For NAA, the solid samples were irradiated at the TRIGA reactor at Pavia University, and the gamma activity was measured using a hyperpure ORTEC Ge detector (10% efficiency) connected to a Silena-Cicero multichannel analyzer and coupled with a PDP 11/23 PLUS computer.

Standard solution and reagents

Standard solutions for calibration purposes were prepared by suitable dilution of BDH chemicals (standard solutions = 1000 mg/liter); other reagents were of analytical grade. All preparations utilized deionized water from a Millipore-Milli-Q water purification system.

Sample dissolution technique for ETAAS and ICPES

About 0.5 g of an accurately weighed portion of each clay sample was mixed in a platinum crucible containing 2.5 g of sodium tetraborate as the fusion medium. The mixture was heated at 1150°C in a muffle furnace for at least 1 hr until a clear melt was obtained. After cooling, the crucible was placed in a beaker containing 150 ml of 1.5 M nitric acid. The mixture was stirred and heated until dissolution was complete. Afterwards, the volume was adjusted to exactly 250 ml, and aliquots were taken for the determinations.

For ETAAS, the addition method (at least 4 points) was used, whereas for ICPES each analytical line was checked before analysis for possible interference from the matrix.

The specificity of the NAA procedure did not allow simultaneous determination of all the elements. For samples A and B, the NAA technique was used to measure the concentrations of the following elements: Fe, Ce, Cs, Yb, Sc, Th, Lu, Ba, Co, Te, and Eu. For sample C, NAA was used only for the determination of U. Destructive chemical analyses by ETAAS and ICPES were used for Co, Fe, Mn, Sr, V, and Zn. NAA results are expressed as Bequerels/gram (Bq/g), whereas results of the destructive analyses, ETAAS and

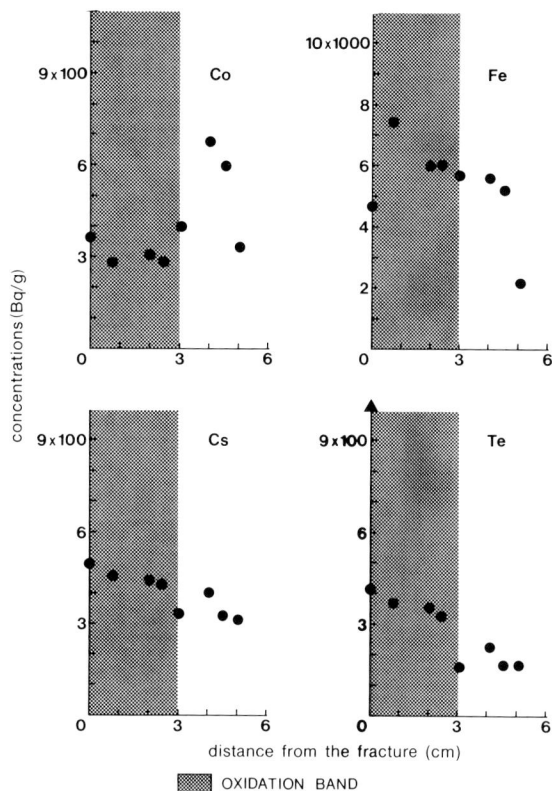

Figure 3. Profiles of Co, Fe, Cs, and Te for sample A.

ICPES, are given in ppm (except for Fe, which is expressed in percent).

RESULTS AND DISCUSSION

The sampling scheme is reported in Figures 1 and 2. Concentration profiles were obtained for Fe, Co, Mn, U, Cs, and Te (Figures 3–8). For all other elements concentration profiles could not be correlated with the leaching phenomenon, due either to a flat profile or to excessive scatter of points. The general trend of the profiles in Figures 3–8 shows an element enrichment in the oxidized zone, with Co (Figures 3–5) and Mn (Figure 7) being the only exceptions. Although data points for Co are highly scattered, its depletion in the oxidized zone is clear in all samples. The enrichment of other elements can be explained as a migration from the bulk of the clay to the oxidized zone which apparently has a higher retention capability (Antonioli and Lenzi, 1984).

As can be expected, Fe shows the most pronounced enrichment in the oxidized zone, due to the precipitation of colloidal ferric hydroxide. It is reasonable to attribute the increased exchange capacity and element retention in the oxidized zone to the colloidal ferric hydroxide, the concentration of which reaches 7% in this zone (Antonioli and Lenzi, 1984). The flat profiles

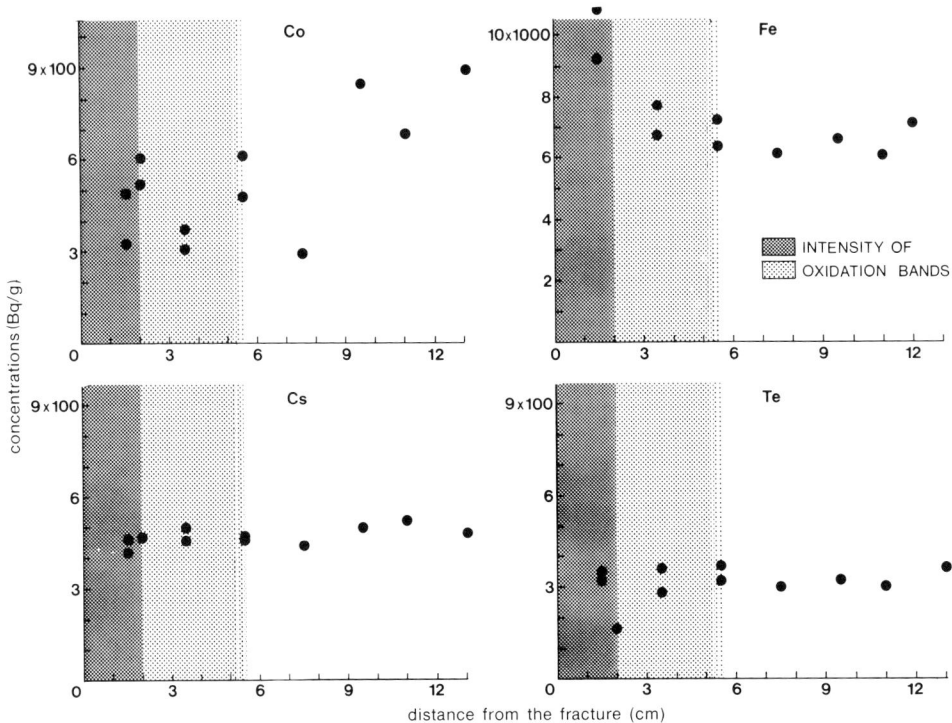

Figure 4.   Profiles of Co, Fe, Cs, and Te for sample B.

of Cs and Te in sample B (in contrast with those of sample A) can be explained if sample B is assumed to be older in terms of percolation time (thicker oxidation band). In these circumstances, element redistribution might have taken place. For the larger sample C, the differences in thickness of the oxidized zone are reflected by the shape of the different profiles, traverse I being the sampling direction across the halos in its thickest part and traverse III being the sampling direction across the halos in its thinnest part. The Mn depletion in the oxidized zone in all profiles can be explained by the presence of Mn in the form of the MnO_4^- anion, which is scarcely sensitive to the exchange capacity of the medium.

Co also appears to be less sensitive to the oxidation-reduction potential and to the exchange capacity of the oxidized zone. Uranium profiles were determined on sample C only (traverses I and II) due to the very specific irradiation needed for the activation analysis. The analytical results from these two traverses suggest a maximum concentration of U close to the border of the oxidized zone. An interpretation of this trend will be given below in the section describing the mathematical model of U migration.

Fracture age assessment

One of the objectives of the present analysis was to estimate the length of time during which the rock was affected by water percolation. Such information can be

obtained from the thickness of the oxidized zone, or more generally, for a given element, from the thickness affected by diffusion of that element (Saltelli and Antonioli, 1985). If transport obeys the simple Fick's law,

$$\partial C/\partial t = D \ \partial^2 C/\partial x^2,$$

where D is an effective diffusion coefficient and C, x, and t are concentration, space, and time, respectively. D, the diffusion depth, L, and the diffusion time, T, are roughly related as follows:

$$L \cong 2(DT)^{0.5}.$$

Table 2 shows values of T calculated from the above relationship, using best estimates from Lanza et al. (1985), and L, as measured in the experimental profiles. Despite the considerable scatter, the calculated time spans give an idea of the active lifetime of the permeable fracture, which seems to pertain to a historical rather than a geological time scale.

Migration model

Due to the considerable scatter in the data for all profiles, a curve-fitting exercise was not possible. The difficulty was increased by the heterogeneity of the clay medium, which necessitated the use of different retention factors for the reduced and the oxidized zones. Actually, the transition between these zones was continuous, so that the retention factor is better described in terms of a function of space and time. A complete

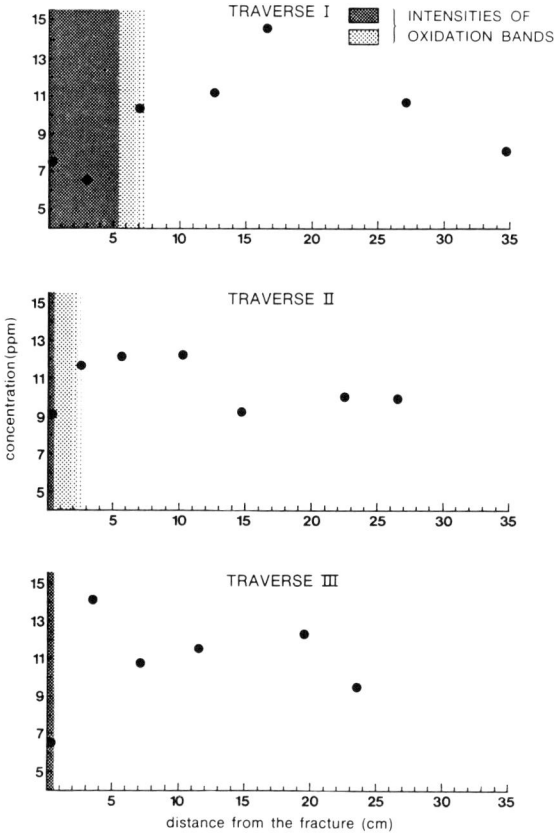

Figure 5. Co profiles for sample C.

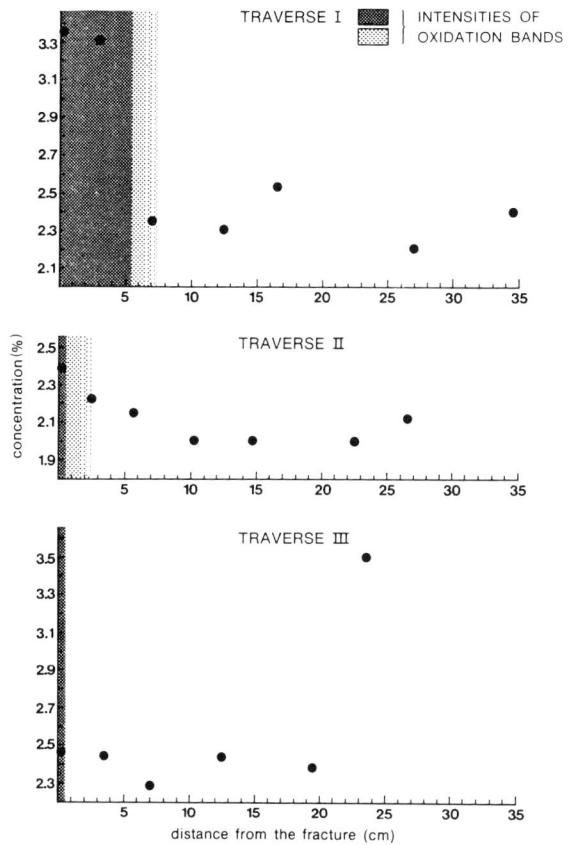

Figure 6. Fe profiles for sample C.

model should take into account the diffusion of oxygen inside the clay with accompanying chemical reactions and the diffusion of a specific metallic constituent outside the clay. In general, the two fluxes will be coupled, the activity of metal being a function of the oxygen activity and vice-versa. The chemical interaction between the metal and the substratum will be a function of the oxygen content. For example, most of the metals are likely to adsorb on the ferric hydroxide, the chief product of the oxidation reaction. Instead of giving a full description of the system, a qualitative fitting exercise was attempted using a simplified model that considered: (1) oxygen diffusion inside the clay, assuming fast reactions independent of the diffusion of the clay constituents; and (2) metal diffusion outside the clay, assuming fast chemical reactions with the substratum, the equilibrium constant being a function of oxygen concentration.

The first hypothesis is probably an oversimplification, inasmuch as the oxygen strongly reacts with Fe, which is itself diffusing. If oxygen uptake was a fast reaction, the transport can be modeled using an effective diffusion coefficient rather than a kinetic constant. The second hypothesis was introduced in the model from the relationship,

$$R(x) = R_1 + (R_2 - R_1)[C_{O_2}(x,t)]/C^0_{O_2}, \qquad (1)$$

where $R(x)$ = the metal retention coefficient at a distance x from the interface (nondimensional); R_1 = the limiting value for the oxidized zone; R_2 = the limiting value for the reduced zone; $C_{O_2}(x,t)$ = the oxygen concentration (M) at a distance x and time t; and $C^0_{O_2}$ = the oxygen concentration at the water-clay interface. The retention coefficient R can also be defined as:

$$R(x) = \bar{C}_M/C_M, \qquad (2)$$

where $\bar{C}_M$ and C_M are the metal concentrations in the substratum and in the solution, respectively, expressed in the same units (M).

Eq. (1) is coupled with the transport equations

$$\partial C_M/\partial t = [D_M/R(x)][\partial^2 C_M/\partial x^2] \qquad (3)$$

$$\partial C_{O_2}/\partial t = D_{O_2}(\partial^2 C_{O_2}/\partial x^2) \qquad (4)$$

and the boundary conditions

$$C_{O_2}(0, t) = C^0_{O_2}$$
$$C_M(0, t) = C^0_M \qquad \text{(interfacial concentrations)}$$
$$C_{O_2}(x, 0) = C^b_{O_2}$$
$$C_M(x, 0) = C^b_M \qquad \text{(bulk concentrations)}.$$

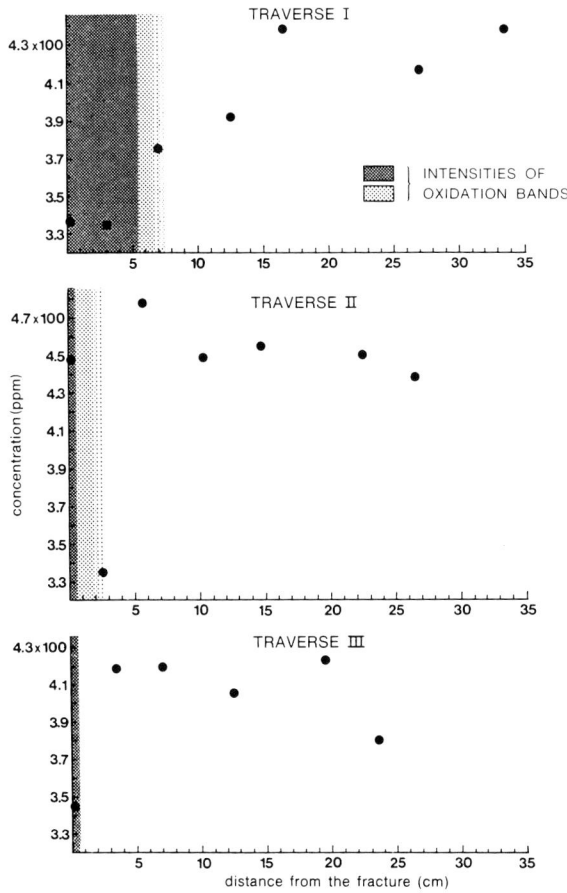

Figure 7. Mn profiles for sample C.

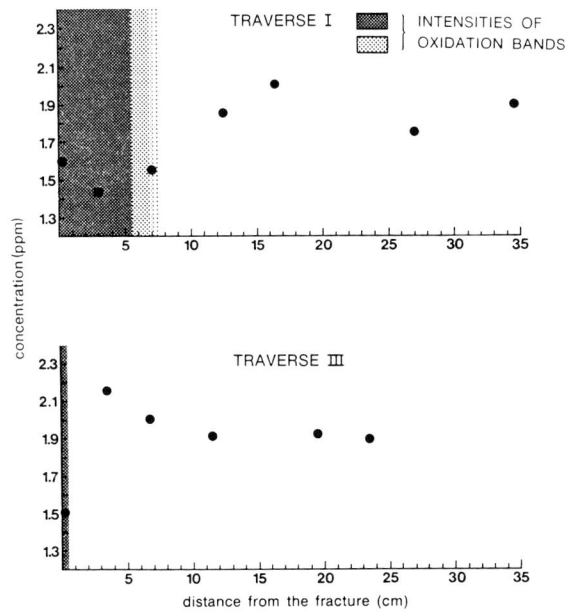

Figure 8. U profiles for sample C.

In the above formulae, D_{O_2} is an effective diffusion coefficient, incorporating a retention coefficient, whereas the effective diffusion of the metal is explicitly calculated as $D_M/R(x)$, where D_M accounts for the molecular diffusion and for the so-called tortuosity factor.

The uranium curves were chosen for the fitting exercise, because they exhibited a regular trend from a plateau inside the clay, to a concentration increase close to the border of the oxidized zone, and a decrease again at the fracture plane (Figure 8).

The fit for traverses I and II is plotted in Figure 9, where the full lines are the C_{O_2} and C_M profiles. The local maximum of C_M is due to the competing action

of the U leaching out of the clay and of the increased retention coefficient which follows the O_2 penetration.

The values of the several parameters employed in the fit are given in Table 3. Because several parameters were adjusted to obtain the fit, even the good agreement with the experimental profiles cannot be taken as a proof of the correctness of the model. The results are nevertheless encouraging and suggest that the main features of the physical system have been reproduced.

CONCLUSIONS

Despite the large variability in element migration data for the system under examination, the order of magnitude of the active lifetime of the fracture (i.e., duration of the migration process) can be reasonably well estimated. That the fracture system of Monterotondo is of a historical age was confirmed (Saltelli and Antonioli, 1985). Thus, the network of permeable discontinuities in the Monterotondo clay was found to be a local and shallow feature, probably due to the unloading resulting from the quarry exploitation, rather than a network of large fractures capable of enhancing

Table 2. Fracture age computed from migration profiles.

Element	Diffusion coefficient (cm²/sec) Lanza et al. (1985)	Fracture age (yr) Profile I	Profile II	Profile III
U	1.2×10^{-8}	39–203	—	0.2–97
Co	2–4.8×10^{-9}	570–1015	?	1–36
Fe	2×10^{-9}	194	16–99	—
Mn	2×10^{-9}	1015	36–99	1–36

Table 3. Parameter values used in the U-migration model.

$C^0_{O_2}$	=	0.781 E-4 M
$C^b_{O_2}$	=	0
C^b_M	=	0.017 M
C^0_M	=	0.0083 M
D_{O_2}	=	7.931 E-9 cm²/s
D_M	=	3.448 E-7 cm²/s
R_1	=	163 (ϕ)
R_2	=	100 (ϕ)

the medium. This system is a good example of "natural analog" for the problem of radionuclide migration through geologic media. Natural analogs typically give qualitative rather than quantitative information.

With reference to the effects of water circulation in fractured clay, the observations on the Monterotondo quarry should be extended by a survey of other clay formations. The final outcome should be the identification of an element (or set of elements) whose profile can be considered as a "marker" of water percolation through fractured clays, which will allow permeable fractures in the clays to be recognized if striking chromatic evidence of the percolation phenomenon is lacking.

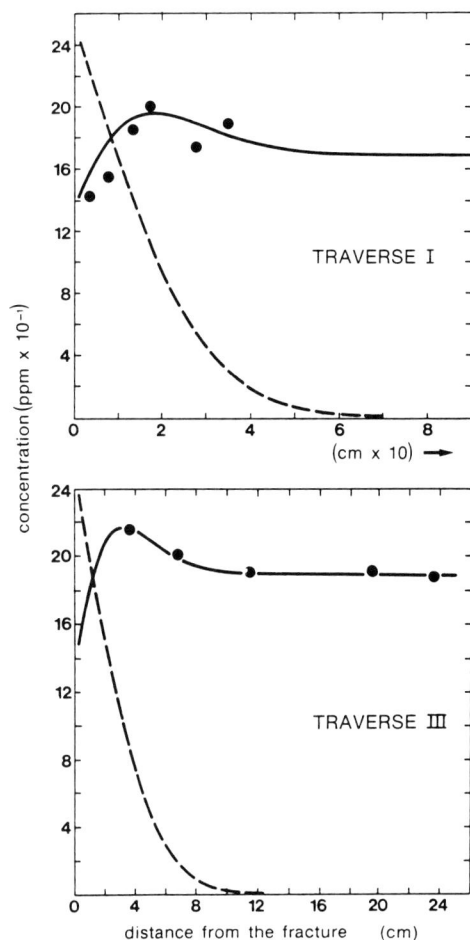

Figure 9. Results of mathematical model: —— = uranium profile; - - - - = oxygen profile; ● = experimental data.

the clay permeability at depth. At depth the clay (even if fractured) should maintain its impermeability.

The element profiles provide useful information on the migration of chemical elements in clay. In the study of the oxidation zones bordering fractures, one effectively deals with a long-term migration experiment which cannot possibly be performed on a laboratory time-scale.

Due to the extreme heterogeneity of the system clay/oxygen–clay/water, and the wide spread of the data, such a system gives qualitative information on the general transport mechanisms, rather than quantitative data on the various diffusion parameters. Studying the U profile in the fracture, for example, does not readily provide a set of transport parameters (K_D, D . . .), but rather points out the predominant transport mechanism and the effect of oxidation-reduction potential of

REFERENCES

Antonioli, F. and Lenzi, G. (1984) Aspetti geochimici relativi ai fenomeni di ossidazione presenti in alcune fratture in formazioni agrillose: ENEA, Casaccia, Roma **Rept. RT/PAS/84/8,** 49 pp.

Bonne, A., Black, J., Gera, F., Gonze, P., Tassoni, E., and Thimus, J. F. (1985) Characterisation and behaviour of argillaceous rocks: in *Proc. 2nd European Community Conf. on Radioactive Waste Management and Disposal, Luxembourg, 1985,* R. Simon, ed., Cambridge University Press, New York, 487–405.

Bonne, A., Heremans, R., Manfroy, P., and Dejonghe, P. (1979) Investigations entreprises pour préciser les caractéristiques du site argileux de Mol comme lieu de rejet souterrain pour les déchets radioactifs solidifiés: in *Proc. IAEA-NEA Int. Symp. Underground Disposal of Radioactive Waste, Otaniemi, 1979,* IAEA, Vienna, 41–58.

Chapman, N. and Gera, F. (1985) Disposal of radioactive wastes in Italian clays: mined repositories or deep boreholes?: in *Radioactive Waste Management and the Nuclear Fuel Cycle, Vol. 6,* Harwood Academic Publishers, New York, 51–78.

Chiantore, V. and Gera, F. (1986) Fracture permeability of clays: *in situ* observations: in *Radioactive Waste Management and the Nuclear Fuel Cycle 7,* Harwood Academic Publishers, New York, 1–25.

D'Alessandro, M. and Gera, F. (1986) Geological isolation of radioactive waste in clay formations: fractures and faults as possible pathways for radioactive migration: in *Radioactive Waste Management and the Nuclear Fuel Cycle, Vol. 7,* Harwood Academic Publishers, New York, 381–406.

Lanza, F., Parnisari, E., and Ronsecco, C. (1985) Study of the release of various elements from HLW glasses in contact with porous media: in *Proc. Workshop on the Source Term for Radionuclide Migration from HLW or Spent Nuclear Fuel, Albuquerque, 1984,* T. O. Hunter and A. B. Mueller, eds., *Sandia Natl. Lab. Rept.* **SAND 85-0380,** 151–163.

Mortari, R. (1977) Elementi per una nuova interpretazione della preconsolidazione delle argille: *Geol. Appl. Idrogeol.* **12,** 189–199.

Saltelli, A. and Antonioli, F. (1985) Radioactive waste disposal in clay formations: a systematic approach to the problem of fractures and faults permeability: in *Radioactive Waste Management and Nuclear Fuel Cycle, Vol. 6,* Harwood Academic Publishers, New York, 101–120.

Proceedings of the International Clay Conference, Denver, 1985, L. G. Schultz, H. van Olphen, and F. A. Mumpton, eds.,
The Clay Minerals Society, Bloomington, Indiana, 422–426 (1987).

EFFECTS OF ANION ADSORPTION ON MECHANICAL PROPERTIES OF CLAY-WATER SYSTEMS

RUNE WENDELBO AND IVAN TH. ROSENQVIST

Institute of Geology, University of Oslo, Postboks 1047, Blindern, Oslo 3, Norway

Abstract—The precipitation of sulfate as a result of acid rain appears to decrease the mechanical stability of clay soils. The sulfate anions are said to have a dispersive effect on clays through edge adsorption analogous to the action of phosphate. A comparative study was therefore made of the interaction of chloride, sulfate, and phosphate with montmorillonite and kaolinite. In addition, some measurements were performed with gibbsite, the surface of which may be considered to be a model for clay mineral edge surfaces, where Al–OH groups are exposed. The adsorption of chloride and sulfate was measured by the liquid scintillation technique, using the radiotracers ^{36}Cl and ^{35}S.

Measurements of critical coagulation concentrations, Bingham yield stresses, and relative sediment volumes show that the dispersive power of sulfate is intermediate between that of chloride and phosphate. Adsorption data show that little sulfate is adsorbed compared with phosphate. Therefore, the dispersive power of sulfate, especially at concentrations >0.1 N may be partly due to a secondary effect, i.e., divalent anions reduce the cation activity and thereby the flocculating power of cations more than do monovalent anions. The effect of H^+ that accompanies sulfate in acid rains has not yet been investigated; however, the present study suggests that if chloride in clays is exchanged by equivalent amounts of sulfate, the clay will become more dispersed.

Key Words—Acid rain, Adsorption, Chloride, Dispersion, Geotechnical properties, Kaolinite, Montmorillonite, Rheological properties, Sulfate.

INTRODUCTION

Vast amounts of sulfate infiltrate European and North American soils as a result of sulfate-polluted acid-rain precipitation. As a consequence, anion exchangers in soils should be expected to bind more sulfate. Clays are well known to adsorb anions onto the edge- or "broken bond" surfaces that have a positive net charge at pHs of <7 (van Olphen, 1977). Reported anion-adsorption capacities vary from negative values to >50 μeq/g, depending on pH, type of clay, and type of anion. Phosphate, and particularly polyphosphate, is strongly adsorbed and reverses the charge of the edge surfaces, resulting in increased repulsion between particles so that the clay forms stable suspensions at increased electrolyte levels. In contrast, chloride is thought to be inert towards clay minerals.

The adsorption behavior of sulfate on clays is less known than that of phosphate and chloride, but positive adsorption has been reported on kaolinite (Aylmore *et al.*, 1967; Rao and Sridharan, 1984) and on montmorillonite (de Haan, 1965). Bergseth (1961) found that the electrophoretic velocity of clay particles is significantly higher in sodium sulfate than in sodium chloride solution. The degree of dispersion induced by sulfate in artificial clay-water systems may indicate possible consequences for clay-rich deposits, some of which are known to be mechanically metastable—the so-called quick clays (Rosenqvist, 1984).

The aim of the present work was to investigate sul-

fate retention by clays and to correlate this retention with some mechanical properties of clay-water systems. Because only the effects of sulfate interaction with pure, citrate-washed clays was investigated, the results must be regarded as the basis for the evaluation of one single factor out of many in a complex natural system. For example, the lowering of the pH, which accompanies acid-rain infiltration, will counteract dispersion of clays; this simultaneous effect will be investigated later.

BACKGROUND

Measured adsorption of anions on clays is the sum of positive adsorption on edge surfaces and negative adsorption (NA) on basal surfaces of crystallites (Figure 1). This explains why negative anion-exchange capacities (AEC) have often been reported, e.g., by Wada and Okamura (1977) and Pissarides *et al.* (1968). NA can be computed if the specific basal surface area is known (Schofield, 1947). The distance of exclusion of co-ions, "d_{ex}," is inversely proportional to the square root of the counter-ion concentration. Furthermore, divalent co-ions are excluded 33% more than monovalent co-ions. de Haan (1965), Helmy (1963), and Helmy *et al.* (1980) developed the theory further for mixed ionic systems and systems with high clay concentration.

Positive adsorption is thus the difference between measured adsorption and calculated NA (Figure 1). It

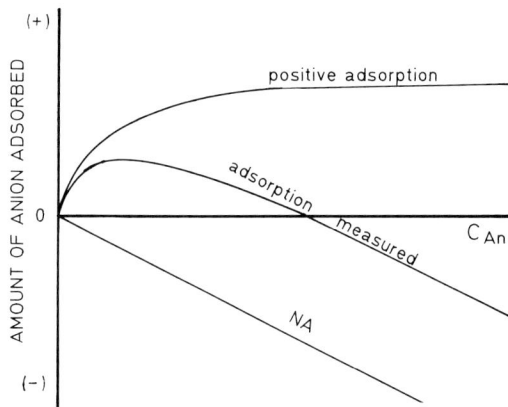

Figure 1. Schematic illustration of anion adsorption on clays as a function of concentration. Amount excluded (NA) is proportional to concentration at a constant level of total salt.

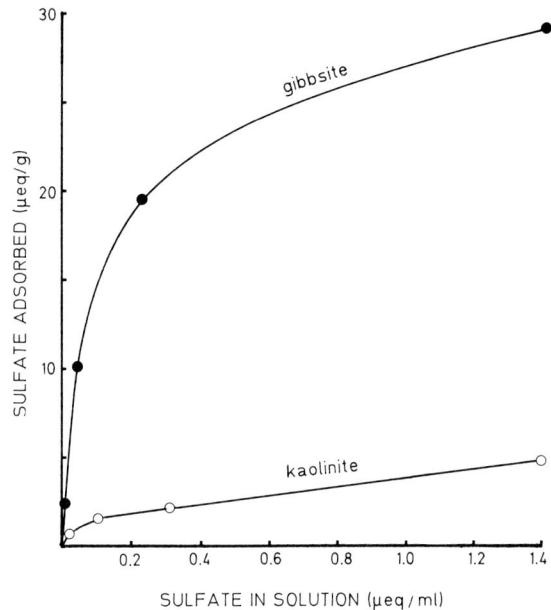

Figure 2. Adsorption isotherms for sulfate on kaolinite sample KGa-1 (pH ~ 5) (O) and gibbsite (pH ~ 7) (●) in 0.0014 N total electrolyte concentration. NaCl was used to keep Na-concentration constant.

is this positive adsorption that should be correlated with the results of mechanical tests on clay suspensions and water-rich sediments. Natural clay-soil samples have been avoided in this work inasmuch as NA cannot be properly calculated due to the lack of a method for the correct determination of wetted, anion excluding surface area in such samples.

MATERIALS AND METHODS

Materials

Standard clays from the Source Clays Repository of the The Clay Minerals Society were used for the analytical measurements, including: Wyoming bentonite (SWy-1) and Georgia kaolinite (KGa-1). Basic information about these clays was given by van Olphen and Fripiat (1979).

Most of the mechanical tests were carried out on a Norwegian montmorillonite (MN) having a N_2 BET surface area of 25 m²/g in the sodium form, and a Zettlitz La Standard kaolinite (KZ) from Germany having a N_2 BET surface area of 16 m²/g. To justify a direct comparison between analytical and mechanical test data, some of the mechanical tests were repeated with samples of SWy-1 and KGa-1. Although the test data on the two montmorillonites and the two kaolinites were different, they agreed qualitatively.

A synthetic gibbsite was prepared according to the method of Gastuche and Herbillon (1962). Its N_2 BET surface area was about 40 m²/g. Gibbsite was studied because its surface may be considered a model for the gibbsite-type layers in the clay minerals.

Pre-treatment of clay materials

The clay was cleaned using the citrate method of Mehra and Jackson (1960). The citrate treatment was necessary because traces of oxides will change the bulk AEC dramatically (Sumner and Reeve, 1966). The clay

was then equilibrated three times with 1 M NaCl or other salts in order to obtain homoionic clays. The suspension was then desalted by centrifugation and dialysis, as described by Schramm and Kwak (1980). Desalting was stopped at desired salt levels, which were determined conductometrically in the dialysate after equilibration. The >2-μm size fraction was removed by sedimentation, and the clay concentration in the suspension was determined gravimetrically after drying a sample at 110°C and correcting for salt content.

Adsorption measurements by tracer procedures

The distribution of chloride or sulfate between bulk solution and clay was measured by using anions tagged with [36]Cl and ^{35}S, respectively, and counting in a Beckman LS-8000 liquid scintillation counter. Clay-free equilibrium solution was separated from the clay in the following ways.

For montmorillonite at salt concentrations <0.1 N. Dialysis was performed with an Amicon DC2 dialyzer equipped with hollow-fiber cartridge H1P100-43 having a cut-off diameter of about 5 nm. To save time and to minimize hydrolysis and degradation of clay, the outer solution (dialysate) was prepared as closely to the equilibrium composition as possible. Still, for the experimental conditions chosen, it took as much as 4 hr of dialysis to ensure a difference in concentration <0.1%. Fortunately, samples containing 1% clay were counted without any loss of counting efficiency. Thus, after equilibration, samples from suspension and di-

Table 1. Positive adsorption of chloride and sulfate on clays and gibbsite before and after addition of 2 mg phosphate/g mineral.[1]

System	Amounts adsorbed (μeq/g)	
	Cl[-]	SO$_4$[2-]
Kaolinite KGa-1	0.3	3.0
Kaolinite KGa-1 + P	0.15	0.95
Montmorillonite SWy-1	1.2	5.9
Montmorillonite SWy-1 + P	0.36	2.2
Gibbsite	11	30
Gibbsite + P	5	12

[1] Salt concentration was 0.003 N in all batches.

alysate were compared directly, correcting for the clay volume only. Clay concentration and equilibrium salt concentration were the only variables that had to be determined to compute positive and negative adsorption.

For montmorillonite at salt concentrations >0.1 N and for kaolinite and gibbsite. Suspensions of montmorillonite at >0.1 N salt concentration and kaolinite and gibbsite tended to clog the prefilter and the hollow fibers. Separation was most easily achieved by sedimentation, and samples were taken from the supernatant liquid and compared with prepared standards. The exclusion volume in these batches was relatively small, so the error due to double layer overlap in the sediments was neglected. The sediments were too concentrated for liquid scintillation counting; hence careful addition of the tracer and control of the total volume were essential.

Mechanical tests

Critical coagulation concentrations (c.c.c.) were determined by means of flocculation series experiments. Bingham yield stress (σ_B) was measured using a Brookfield rotation viscosimeter model LVF. Relative volumes of sediment were measured in test tubes. Inasmuch as some of the experimental conditions for these tests were arbitrarily chosen, satisfactory reproduc-

Table 3. Bingham yield stress for different clay suspensions before and after addition of polyphosphate.

System[1]	Bingham yield stress (N/m^2)	
	NaCl[2]	Na$_2$SO$_4$[2]
19% Kaolinite KZ	40	17
19% Kaolinite KZ + PPh[3]	23	5
2.3% Montmorillonite MN	46	14
2.3% Montmorillonite MN + PPh[4]	3	0

[1] See text for source of clays.
[2] 0.2 N in montmorillonite suspension; 0.3 N in kaolinite suspension.
[3] 10 mg Na-polyphosphate (PPh)/g kaolinite.
[4] 40 mg Na-polyphosphate (PPh)/g montmorillonite.

ibility was expected for relative differences in single series of experiments only. By careful sample preparation, differences of >5% should be regarded as significant.

RESULTS AND DISCUSSION

Adsorption data from tracer experiments

The sulfate adsorption isotherm for kaolinite (Figure 2) agrees well with other published data, considering that kaolinites of different origins were used (e.g., Rao and Sridharan, 1984; Aylmore et al., 1967). Computed NA contributed less than 20% of total sulfate adsorption on kaolinite due to the high ratio of edge surface area to basal surface area. The net adsorption of Cl[-] was much smaller than that of SO$_4$[2-], and the measurements gave negative adsorption, although less than the computed NA. Addition of 2 mg polyphosphate/g kaolinite incompletely desorbed both sulfate and chloride (Table 1); hence, this amount of phosphate apparently did not block all edge adsorption sites. Even if complete desorption had been achieved, it would have been difficult to determine the NA experimentally, due to the small specific surface of kaolinite. Sumner and Reeve (1966) found that after deferrification and adjustment of pH to >9, desorption of Cl[-] was

Table 2. Critical coagulation concentrations for different clay-salt systems.

Mineral[1]	Critical coagulation concentration (meq/liter suspension)			
	NaCl	Na$_2$SO$_4$	Na$_2$HPO$_4$/NaH$_2$PO$_4$[2]	CaCl$_2$
Kaolinite KZ (pH ~ 6.0)	3	4	20	0.15
Kaolinite KZ (pH ~ 8.8)	90	100	150	—
Kaolinite KZ + PPh[3]	175	200	—	—
Montmorillonite MN	12	15	—	0.2
Montmorillonite MN + PPh[4]	190	255	—	—
Gibbsite (pH ~ 6.2)	100	3	0.5	100
Gibbsite (pH ~ 7.2)	3	0.7	0.08	—

[1] See text for source of clays.
[2] Na$_2$HPO$_4$ was used at pH >7; NaH$_2$PO$_4$ was used at pH <7.
[3] 20 mg Na-polyphosphate (PPh) added per gram kaolinite.
[4] 40 mg Na-polyphosphate (PPh) added per gram montmorillonite.

complete. Extrapolation of the data in Figure 2 to 0.003 N solution yields a much larger sulfate distribution coefficient than the data in Table 1, probably due to the two-week aging of the suspension between the citrate washing and the isotherm (Figure 2).

From Cl^--exclusion measurements with montmorillonite (SWy-1) specific surface areas of 650, 690, and 705 m^2/g were calculated, respectively, without phosphate present, and after the addition of 2 and 10 mg of NaH_2PO_4/g clay. These data are in good agreement with theory and other published data, e.g., Bolt and Warkentin (1958). The positive adsorption data in Table 1 were calculated assuming 705 m^2/g as the total anion-excluding surface area of the montmorillonite.

All minerals adsorbed more sulfate than chloride (Table 1). On addition of 2 mg of phosphate, sulfate desorption amounted to 60–68% for all minerals. Thus, the amount of phosphate adsorbed was apparently proportional to the amount of sulfate adsorbed. For chloride, the desorption was 50–72%, but these data were less precise due to the small differences measured.

In one experiment, only trace amounts of sulfate were added to batches of samples SWy-1 and KGa-1 in 10^{-3} N NaCl. About 10% of the sulfate was positively adsorbed on the montmorillonite, which was less than the calculated NA. In contrast, about 70% was positively adsorbed on the kaolinite, indicating the existence of sites for chemisorption of sulfate on the kaolinite, but not on the montmorillonite.

Measurements of critical coagulation concentrations

Critical coagulation concentrations were measured for several different anions with or without added Na-polyphosphate. Br^-, I^-, NO_3^-, and ClO_4^- gave essentially the same results as Cl^-. Differences due to size and polarizability were expected, but not detected, possibly due to insensitivity of the method. Results for chloride, sulfate, and phosphate are presented in Table 2. Both in the absence and presence of phosphate, c.c.c. values were higher for sulfate than for chloride indicating some deflocculating effect of the sulfate. This effect may in part have been caused by edge adsorption of sulfate and in part by reduction of the cation activity and, hence, the flocculating power of cations in the presence of divalent anions, such as sulfate. The validity of the Schulze-Hardy's rule is clear, i.e., the valency of the counter-ion determines the c.c.c., as long as chemisorption does not take place.

Measurements of Bingham yield stress

A Bingham yield stress (σ_B) is found in more viscous suspensions, and is thus a good measure of the cohesion of the clay, i.e., the degree of particle association in a gel structure. Results for 0.2 N salt-2.3% montmorillonite suspensions and 0.3 N salt-19% kaolinite suspensions are shown in Table 3. After addition of poly-

Table 4. Relative sediment volumes for different clays, with and without polyphosphate present.

System[1]	Relative sediment volume	
	NaCl[2]	Na$_2$SO$_4$[2]
Kaolinite KZ	1.00	0.77
Kaolinite KZ + PPh[3]	0.43	0.38
Montmorillonite MN	1.00	0.95
Montmorillonite MN + PPh[4]	0.79	0.76

[1] See text for source of clays.
[2] 0.25 N in montmorillonite suspension; 0.2 N in kaolinite suspension.
[3] 10 mg Na-polyphosphate (PPh)/g kaolinite.
[4] 40 mg Na-polyphosphate (PPh)/g montmorillonite.

phosphate (40 mg/g montmorillonite and 10 mg/g kaolinite), the suspensions were left 24 hr before measurement, because the exchange of sulfate was anticipated to be slower than the exchange of chloride.

The results show that sulfate counteracted gel formation, in agreement with the flocculation experiments. Face-to-face association (the formation of fewer and thicker plates) should have resulted in a reduced σ_B, but the sulfate probably did not affect the clay in this way. Again, both edge adsorption and reduced Na-activity in sulfate solution relative to chloride solution should have been taken into account, but more detailed work is required to determine these effects quantitatively.

Measurements of relative sediment volumes

The results presented in Table 4 show that all sulfate systems had more compact sediments than corresponding chloride systems, in agreement with the c.c.c. experiments. The interpretation of these results is the same as for the σ_B results: the same effect was measured, and the correlation was good.

CONCLUSIONS

Mechanical test results clearly show that if Cl^- in a clay suspension is exchanged by equivalent amounts of SO_4^{2-}, the clay will become more deflocculated. More sulfate than chloride was adsorbed by clays, apparently on the edge surfaces. Deflocculation in sulfate systems may be due to: (1) edge adsorption of sulfate ions and (2) reduction of the activity and, hence, the flocculating power of the cation in solution. The relative importance of these factors probably varies for different clays and different electrolyte levels.

Experiments with field materials and different types of "rain" must be performed before such precipitation can be held responsible for increased erosion and for landslides.

REFERENCES

Aylmore, L. A. G., Karim, M., and Quirk, J. P. (1967) Adsorption and desorption of sulfate ions by soil constituents: *Soil Sci.* **103**, 10–15.

Bergseth, H. (1961) Vergleich der Elektrophoretischen Wanderungsgeschwindigkeit für Teilchen eines quelfähigen und eines nicht quelfähigen Schichtsilikates: *Kolloid Z.* **179,** 67–69.

Bolt, G. H. and Warkentin, B. P. (1958) The negative adsorption of anions by clay suspensions: *Kolloid Z.* **156,** 41–46.

Gastuche, M. C. and Herbillon, A. (1962) Étude des gels d'alumine, cristallisation en milieu désionisé: *Bull. Soc. Chim. Fr.* **1962,** 1404–1412.

de Haan, F. A. M. (1965) The interaction of certain inorganic anions with clays and soils: *Thesis, University of Wageningen, Wageningen, The Netherlands,* 167 pp.

Helmy, A. K. (1963) Calculation of negative and positive adsorption in some clay electrolyte systems: *J. Soil Sci.* **14,** 217–224.

Helmy, A. K., Natale, I. M., and Mandolesi, M. E. (1980) Negative adsorption in clay-water systems with interacting double layers: *Clays & Clay Minerals* **28,** 262–266.

Mehra, O. P. and Jackson, M. L. (1960) Iron oxide removal from soils and clays by dithionite-citrate system with sodium bicarbonate: in *Clays and Clay Minerals, Proc. 7th Natl. Conf., Washington, D.C., 1958,* Ada Swineford, ed., Pergamon Press, New York, 317–327.

Pissarides, A., Steward, J. W. B., and Rennie, D. A. (1968) Influence of cation saturation on phosphorus adsorption by selected clay minerals: *Can. J. Soil. Sci.* **48,** 151–157.

Rao, S. M. and Sridharan, A. (1984) Mechanism of sulfate adsorption by kaolinite: *Clays & Clay Minerals* **32,** 414–418.

Rosenqvist, I. Th. (1984) The importance of pore water chemistry on mechanical properties of clay soils: *Phil. Trans. R. Soc. Lond.* **A311,** 369–373.

Schofield, R. K. (1947) Calculations of surface areas from measurements of negative adsorption: *Nature* **160,** 408–410.

Schramm, L. L. and Kwak, J. C. T. (1980) Application of ultrafiltration dialysis to the preparation of clay suspensions: *Clays & Clay Minerals* **28,** 67–69.

Sumner, M. E. and Reeve, N. G. (1966) The effect of iron oxide impurities on the positive and negative adsorption of chloride by kaolinites: *J. Soil. Sci.* **17,** 274–279.

van Olphen, H. (1977) *An Introduction to Clay Colloid Chemistry:* 2nd ed., Wiley, New York, 318 pp.

van Olphen, H. and Fripiat, J. J. (1979) *Data Handbook for Clay Materials and other Non-metallic Minerals:* Pergamon Press, Oxford, 346 pp.

Wada, K. and Okamura, Y. (1977) Measurements of exchange capacities and hydrolysis as means of characterizing cation and anion retention by soils: in *Proc. Int. Semin. Soil Environ. Fertil. Manage. Intensive Agric., Tokyo, 1977,* The Soc. Soil Manure, Tokyo, Japan, 811–815.

Proceedings of the International Clay Conference, Denver, 1985, L. G. Schultz, H. van Olphen, and F. A. Mumpton, eds.,
The Clay Minerals Society, Bloomington, Indiana, 427–435 (1987).

STABILIZATION OF SENSITIVE CLAYS (QUICK CLAYS) USING Al(OH)$_{2.5}$Cl$_{0.5}$

O. R. Bryhn, T. Løken

Norwegian Geotechnical Institute, P.O. Box 40, Taasen
0801 Oslo 8, Norway

M. G. Reed

Chevron Oil Field Research Company, P.O. Box 446
La Habra, California 90631

Abstract—A natural illitic, chloritic, silty clay having an undisturbed shear strength of 20 kPa was stabilized using Al(OH)$_{2.5}$Cl$_{0.5}$ (denoted OH-Al) alone or together with KCl or K$_2$SO$_4$. Results of field tests on three-year-old *in situ* stabilized clay columns and the surrounding undisturbed clay were compared with earlier field and laboratory tests. The clay columns were stabilized by remolding the soil by more or less complete mixing of the chemicals with an auger of 0.5-m diameter. After three years the shear strengths measured on 1-m-long sections of the columns had increased to 40 kPa. On 8-cm high, homogeneous, mixed samples from the same column, the shear strength was as much as 230 kPa. In the clay surrounding the columns, the undisturbed shear strength increased gradually from about 20 kPa in the quick clay to about 40 kPa within a few centimeters from the columns, and the remolded shear strength increased from 0.1 to 12 kPa. As a result, the sensitivity decreased from 200 to 8. The improvement in the geotechnical properties in the soil surrounding the columns was apparently caused by increased acidity and increased salt content in the pore water. The salt in the pore water originated by the diffusion of H$^+$, K$^+$, Ca^{2+}, Cl$^-$, and SO$_4^{2-}$ from columns and by subsequent ion-exchange reactions with the clays releasing Na$^+$, K$^+$, Mg^{2+}, and Ca^{2+}. Because of this ion diffusion, the surrounding clay completely changed its geotechnical behavior. The total stabilized cross section of each column increased from the original 0.2 m^2 to more than 4 m^2 after three years.

Key Words—Cation diffusion, Geotechnical properties, Polymeric hydroxy-aluminum, Quick clay, Shear strength, Stabilization.

INTRODUCTION

Soft, sensitive marine clays in Canada, Finland, Sweden, and Norway have always caused large geotechnical problems. Even 3–4-m-deep excavations or 2–3-m-high fillings can cause stability problems and ground failures. In natural slopes, small surface failures may result in large landslides similar to the Rissa landslide, where a 330,000-m^2 area containing seven farms and five houses slid, together with 5×10^6 m^3 of quick clay in less than an hour (Gregersen, 1981; Løken and Gregersen, 1981).

It is well known that the geotechnical properties of such sensitive clays are influenced by the chemical state of the clay (see e.g., Rosenqvist, 1955; Bjerrum, 1954; Løken, 1970; Torrance, 1975). Methods for stabilization of such clays by adding salts or lime, which increase the remolded (disturbed) shear strength, and also in some cases the undisturbed shear strength, have been used (Bjerrum *et al.,* 1967; Moum *et al.,* 1968; Eggestad and Sem, 1976; Broms and Boman, 1979; Holm, 1979; Mitchell, 1982). From investigations of the stiff, weathered top crust of these clays, the importance of the presence of iron and aluminum hy-

droxides was discovered by Moum (1968). He artificially weathered clay by storing two sets of natural quick clays in either an O$_2$ atmosphere or in a N$_2$ atmosphere. After three months, the mechanical properties and the chemical state of Fe were investigated. The N$_2$-stored samples showed very little change, whereas the O$_2$-stored samples showed increase in shear strength and an oxidation of Fe^{2+} to Fe^{3+}. These results were compared with laboratory tests in which several chemicals were mixed with quick clay (Løken, 1970). Separate additions of Fe(OH)$_3$ and Al(OH)$_3$ had less effect than that observed in the oxidation test, indicating that the composition of these hydroxides was different in the two tests.

To prevent dispersion and swelling of clay minerals in petroleum reservoirs, Reed (1972) used polymeric aluminum hydroxides. The influence of similar chemicals on the geotechnical properties of clays was investigated in the laboratory by Foster and Gazzard (1975). They found that especially small molecules of polymeric hydroxy-aluminum were efficient in stabilizing suspensions of kaolin. Recently, a number of papers by clay mineralogists and soil scientists have described the effects of polymeric aluminum hydroxides on the charge and cation-exchange capacity (CEC) of clay minerals. Most of these papers, however, deal with unspecified OH-Al at very low concentrations (see, e.g., Oades, 1984).

In 1980, a project was started at the Norwegian Geotechnical Institute on quick clay stabilization using polymeric aluminum hydroxides, including $Al(OH)_{2.5}Cl_{0.5}$ (denoted OH-Al). The project consisted of laboratory experiments as well as full-scale, *in situ* tests using OH-Al alone or together with KCl or K_2SO_4 (Table 1). The aim of this part of the investigation was to gain experience with the long-term effects of having columns of stabilized clay adjacent to undisturbed clay. The laboratory experiments (Table 1) included different chemical additions to six natural clayey soils. The rate of soil improvement was determined as the rate of increased shear strength. The results of the OH-Al-treated samples were also compared with those of the lime-treated samples. The results from the first part of the project (Bryhn et al., 1983) covered the laboratory mixtures and the *in situ* tests investigated after 8 and 40 weeks. The present paper reviews these tests and reports the results of continued investigations after nearly three years (147 weeks).

POLYMERIZATION OF OH-Al IN SOILS

According to Hsu and Bates (1964), the polymer formula of the cation part of $Al(OH)_{2.5}Cl_{0.5}$ is $[Al_{24}(OH)_{60}(H_2O)_{24}]^{12+}$; however, $[Al_{16}(OH)_{36}(H_2O)_{24}]^{12+}$ was postulated by Hem and Robertson (1967), $[Al_{13}O_4(OH)_{24}(H_2O)_{12}]^{7+}$ by Waters and Henty (1977), and $[Al_8(OH)_{20}(H_2O)_{10}]^{4+}$ by Zlateva et al. (1975). Hsu and Bates (1964) presumed that the ions consist of gibbsite-like sheets wherein rings of 6 Al atoms are coordinated with H_2O and OH.

Polymerization and depolymerization are controlled by the activity of ions in solution and the pH of the environment (Yariv and Cross, 1979; Hsu, 1977). Further, as stated by Yariv and Cross (1979): "The polymerization reaction rates are extremely slow in dilute solutions but become faster in more concentrated solutions. The reactions are catalyzed by various soluble salts and solid surfaces and are therefore fast in many geologic environments."

Dissolved OH-Al gives a solution having a pH of ~3.5, but it cannot form a gel at pH <5. When added in small amounts to sensitive clays, it acts like a large cation, making the clay plastic. Above about 15 g/100 g pore water the mixture turns grainy and becomes stiff within minutes or hours, but only if enough base is present to keep the pH above ~5.

Polymeric OH-Al reacts like an ideally shaped and charged cation in the exchange processes, releasing Ca^{2+}, Mg^{2+}, K^+, and Na^+. These ions are free to move into the surrounding quick clay, thereby causing a secondary improvement of the clay's geotechnical properties.

The addition of large quantities of salt together with OH-Al yields clay mixtures having low strength. Hsu (1977) described a possible inhibition of polymerization in the presence of high concentrations of Cl^-, which may explain the formation of low-strength mixtures.

The addition of SO_4^{2-} will most probably have a reaction which is similar to that of phosphate. According to Hsu (1968), phosphate ions replace water molecules at the edge of the polymer. At low concentrations of phosphate, the positive charge of the polymer is not neutralized and the solution is stable. At the isoelectric point, no electric repulsion or attraction occurs between the ions, and the solution will precipitate. At high phosphate concentration, the polymer will have a negative charge and break into smaller anionic units, giving rise to a stable solution.

MATERIALS AND METHODS

OH-Al alone or with additives were mixed with different natural soils and used *in situ* in the field in an area containing quick clay. The laboratory and *in situ* experiments and the chemicals used are summarized in Table 1. Homogeneous mixtures of chemicals and clays were tested mechanically in the laboratory by unconfined compression tests on 8-cm-high cylindrical casts. The procedure and the results of short-term tests were described in more detail by Bryhn et al. (1983).

Field experiments at Emmerstad

A quick clay area at Emmerstad, about 40 km south of Oslo, Norway, was chosen for the field experiments. This area consists of a small valley with steep, forested, bedrock slopes on both sides and a marine clay filling the floor. The area is used as farmland and is about 10 m above sea level. The upper 2–2.5 m of the clay deposit has been weathered into a top crust, which overlies the soft quick clay. Geotechnical data for the quick clay are summarized in Table 2.

Columns of stabilized clay were prepared in the field with the Linden-Alimak equipment (Broms and Boman, 1979) in August 1981. The columns were 50 cm in diameter and extended from about 1 to 5.5 m below the surface. Different chemicals in different chemical-clay ratios were used. The depth interval of 2.5–3.5 m of the columns themselves and the clay outside the columns were tested both mechanically and chemically. Altogether, 73 columns were placed in four groups, two of which were investigated after 8 weeks, one after 40 weeks, and one after 147 weeks (Figure 1).

In the first group (Figure 1), the columns were excavated in October 1981. Direct shear, unconfined compression, and bending tests were performed in the field on full diameter columns. In addition, unconfined compression tests on small pieces of the columns were made in the laboratory. In the second group, only unconfined compression tests were made on small pieces of the columns. In the third group, excavated in 1982

Table 1. Experimental series and chemicals used (Bryhn et al., 1983).

	OH-Al	OH-Al + KCl	OH-Al + K$_2$SO$_4$	OH-Al + CaCO$_3$	OH-Al + CMC	OH-Al + CaSO$_4$	OH-Al + misc.[1]	CaO	CaO + KCl	CaO + CaSO$_4$	CaO + misc.[2]	Geo-techn. tests	Geo-chem. tests
Preliminary laboratory experiments	X											X	
Preliminary diffusion experiments	X	X						X	X			X	X
Deep stabilization at Emmerstad:													
Section 1	X							X				X	
Sections 2, 3, and 4	X	X	X					X				X	X
Laboratory stabilization:													
Emmerstad	X	X	X	X	X	X	X	X	X	X	X	X	
Ellingsrud	X	X	X	X	X		X	X	X			X	
Hønefoss	X	X	X	X	X	X	X	X	X	X		X	
Torrekulla	X	X	X	X	X	X		X		X		X	X
St. Leon	X	X	X	X	X	X		X		X		X	
Drammen	X	X	X	X	X	X	X	X		X		X	

[1] OH-Al + NH$_4$Cl, NH$_4$NO$_3$, (NH$_4$)SO$_4$, FeSO$_4$, bentonite, iron powder.

[2] CaO + bentonite, silica dust.

(after 40 weeks), the shear strengths from unconfined compression tests were determined on small pieces of the columns. The clays adjacent to the columns were also tested both mechanically and chemically as described below.

In the fourth group, excavated in 1984 (after 147 weeks), full-scale unconfined compression tests were performed on 50-cm-diameter samples in addition to similar tests on small pieces. In addition, the adjacent clay was tested both mechanically and chemically as described below.

As mentioned above, the movement of cations into the undisturbed quick clay may have a dramatic effect upon the mechanical properties of the clay. To examine these effects more fully, undisturbed and remolded shear strength by fall cone test, liquid and plastic limit, and water content were measured in the adjacent clay. Ca^{2+}, Mg^{2+}, K$^+$, and Na$^+$ were measured by atomic absorption spectroscopy. Cl$^-$ was measured with a chloride selective electrode, and SO$_4^{2-}$ was measured colorimetrically on pore water. Resistivity and pH were measured directly in the wet clay.

SHORT-TERM TREATMENT RESULTS

Prior to the field experiment a pilot stabilization project, including a large number of laboratory tests on different soils and stabilizing chemicals, was performed (see Table 1). A brief summary of results covering the Emmerstad clay reported by Bryhn et al. (1983) follows, in addition to a description of the short-term treatment results of the field project.

The laboratory results of the Emmerstad clay stabilized with only OH-Al are shown in Figure 2. No significant change in the shear strength was measured by the unconfined compression tests during storage. For comparison, the results of treatments using OH-Al + K$_2$SO$_4$ (4:1) as a stabilizing agent are shown in

Figure 3. The increase in strength with time (within the shaded area) was significant for the greatest additions. Similar treatments using OH-Al + KCl (5:4) clearly resulted in an improvement in strength with time, from 1 to 8 weeks (see Figure 4).

The results from the field treatments are also included in these illustrations. For the columns stabilized with OH-Al (Figure 2), small pieces tested by unconfined compression tests showed a wide distribution of

Table 2. Description of Emmerstad clay.

Water content (w) (%)	43–49
Liquid limit (w$_L$) (%)	33
Plastic limit (w$_p$) (%)	19
Undisturbed shear strength by fallcone method (s$_u$) (kPa)	15–25
Remolded shear strength by fallcone method (s$_r$) (kPa)	0.1–1
Sensitivity (arbitrary units) (s$_t$)	20–250
Undrained shear strength by unconfined compression test (s$_u$) (kPa)	13–20
Strain at failure (ϵ_f) (%)	2
Salt content in pore water, equivalent NaCl (g/liter)	1.0
Inorganic carbon as CaCO$_3$ (wt. %)	2.5
Organic carbon (wt. %)	0.3
Grain size distribution: sand (%)	1–6
silt (%)	60–72
clay (%)	27–34
Bulk mineralogy	
17-Å, expanding mixed-layer minerals (%)	—
10-Å illite or muscovite (%)	25
7-Å chlorite (%)	10
10–14 Å, mixed-layer minerals (%)	traces
Quartz (%)	40
K-feldspar (%)	5–10
Plagioclase (%)	10–15
Amphibole (%)	0–5
Sum (%)	95–105

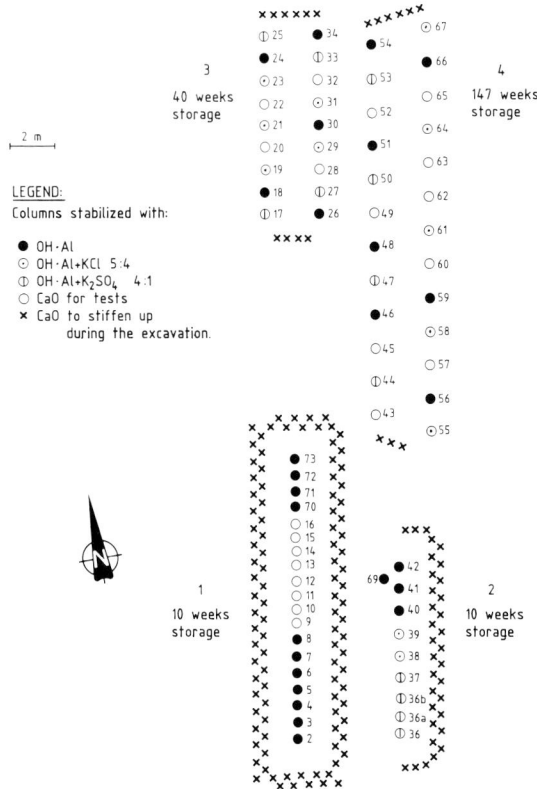

Figure 1.   Positions of the four groups of coloumns in test area installed early summer 1981.

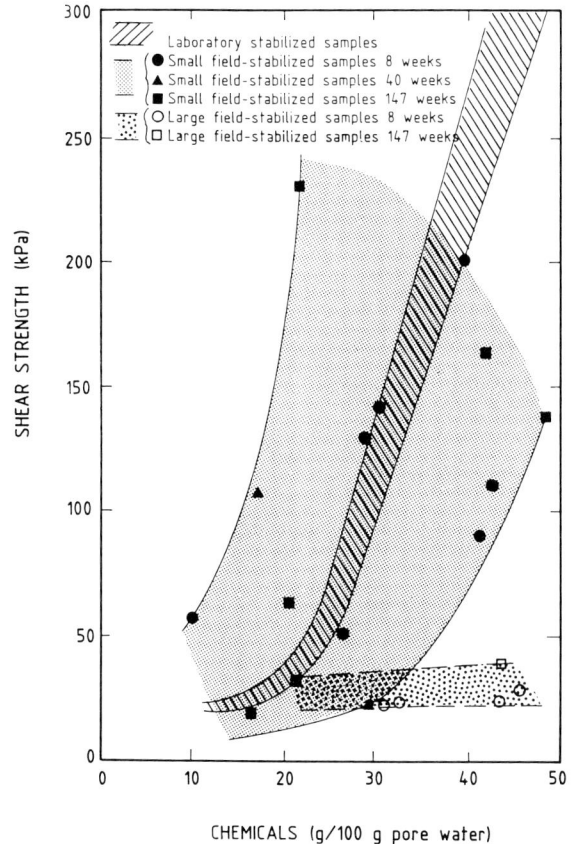

CHEMICALS (g/100 g pore water)

Figure 2.   Unconfined compression tests on clay stabilized with OH-Al. Shear strength as a function of added OH-Al. Field tests are based on a natural water content of 47%.

shear strengths. The storage time, however, seemed to have no influence. The results were similar for the columns stabilized with OH-Al + K_2SO_4 (Figure 3), but samples stabilized with OH-Al + KCl (Figure 4) were usually too weak to be tested.

Only four columns stabilized with OH-Al alone were tested in full scale (50-cm diameter), unconfined compression equipment. The results showed much lower shear strengths than those for comparable small specimen tests as shown in Figure 2.

THREE-YEAR STABILIZATION
TREATMENT RESULTS

The fourth group of 25 columns was excavated in 1984 after 147 weeks of treatment. Out of these, 7 columns were stabilized with OH-Al, 4 with OH-Al + K_2SO_4, 5 with OH-Al + KCl, and 9 with CaO. Full-scale compression tests were performed on 8 columns; small-scale compression tests were performed on samples from 16 columns. The effect of the diffusion into the adjacent clay was tested on 18 columns.

Results from the column treatments

Columns stabilized with only OH-Al were brittle and difficult to prepare, especially for full-scale tests. Only two columns had no visible cracks after they were re-

moved from the ditch. The shear strength results shown in Figure 2 illustrate that the large, full-scale, field-stabilized samples after 147 weeks were somewhat stronger (shear strength ~40 kPa) than similar samples after 8 weeks. On the other hand, the 147-week, small-scale, field-stabilized samples obtained a shear strength of ~150 kPa, somewhat higher than that of the previously tested field samples, but not as high as those of the best laboratory-mixed samples.

Of the columns stabilized with OH-Al + K_2SO_4 and OH-Al + KCl, only small-scale field samples were collected. The columns treated with OH-Al + K_2SO_4 showed some improvement after storage and a shear strength of ~130 kPa (Figure 3) was obtained, which is nearly the same as that for the OH-Al stabilized columns. The columns stabilized with OH-Al + KCl were in rather poor mechanical condition, and only two samples treated by the addition of large amounts of chemicals were tested, giving a shear strength of 65 kPa on very brittle samples (Figure 4).

Results from clay surrounding the columns

The strength properties of undisturbed clay outside the columns after 147 weeks are illustrated in Figure

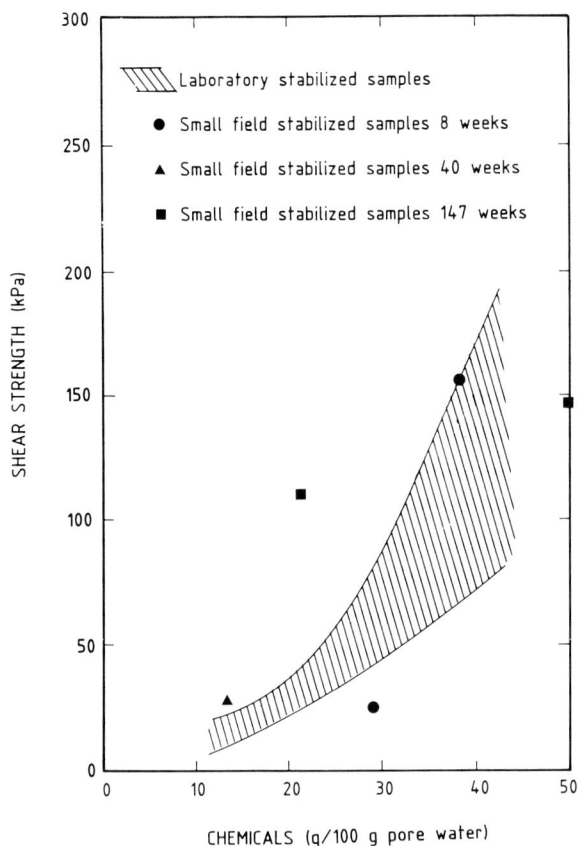

Figure 3. Unconfined compression tests on clay stabilized with OH-Al + K$_2$SO$_4$ (4:1). Shear strength as a function of added chemicals. Field tests are based on a natural water content of 47%.

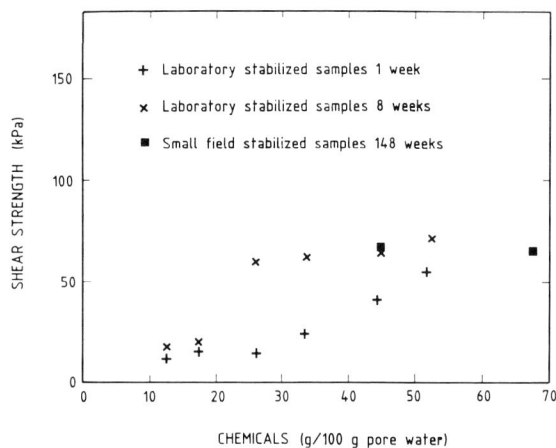

Figure 4. Unconfined compression tests on clay stabilized with OH-Al + KCl (5:4). Shear strength as a function of added chemicals. The field tests are based on a natural water content of 47%.

5 for the three different types of chemical additions. The effects of the chemical diffusion were determined by comparing properties of samples taken near the column with those taken at a distance of 110 cm where no further changes could be measured.

The geotechnical properties are shown in Figure 5. The variation in water content was large, however, a trend is noticeable. Outside the columns stabilized with OH-Al and with OH-Al + K$_2$SO$_4$, the water content ranged from 43 to 49%. Outside the OH-Al + KCl stabilized columns, however, the water content at a distance of 0–80 cm was significantly less, which can be explained by an osmotic movement of water into the column (see Figure 5a).

The undisturbed shear strength increased slightly as far as 75 cm from the treated column. The addition of OH-Al increased the shear strength the least, whereas the addition of OH-Al + K$_2$SO$_4$ increased the shear strength somewhat, and the addition of OH-Al + KCl showed the greatest strength increase outside the column. Here, the shear strength increased from about 23 kPa in the quick clay to 37 kPa at a few centimeters from the column (see Figure 5b).

Marked increases in remolded shear strength were noted along with a corresponding change in sensitivity. Outside the columns treated with OH-Al, the remolded shear strength of the original quick clay was 0.1 kPa. An increase in remolded shear strength was measured as far as 80 cm, and from 0 to 50 cm, the remolded shear strength was fairly constant at ~3 kPa. Outside the columns treated with OH-Al + K$_2$SO$_4$ and OH-Al + KCl, an even greater effect was noted. OH-Al + KCl treatment gave the most pronounced results, and an effect could be measured as far as 100 cm. The average remolded shear strength was 6 kPa in the zone a few centimeters from the column (see Figure 5c).

The sensitivity of 200 for the original quick clay was reduced to 7–12 in the region 0–50 cm from the column, the OH-Al + KCl treatment being the most effective (see Figure 5d). As expected, the plastic limit value of 20% changed little outside the column, but the liquid limit value changed from about 34–38% in the quick clay to as much as 53–58% within a few centimeters from the columns.

The geochemical parameters measured outside typical columns stabilized with OH-Al, OH-Al + KCl, and OH-Al + K$_2$SO$_4$ are shown in Figure 5. The concentration of Ca^{2+} and Mg^{2+} (Figures 5e and 5f), are of the order of 10 ppm in the original quick clay. Just outside the columns the concentrations of Ca^{2+} and Mg^{2+} are close to 2500 ppm and 500 ppm, respectively, for all three different chemical mixtures, showing that OH-Al treatment alone results in the release of as much Ca^{2+} and Mg^{2+} as the treatment with the OH-Al mixed with K$_2$SO$_4$ or KCl.

Determination of the K$^+$ and Na$^+$ contents vs. distance outside the column (Figures 5g and 5h) shows that the K$^+$ concentration increased from 30–50 ppm to 200 ppm, 300 ppm, and 1100 ppm outside columns stabilized with OH-Al, OH-Al + K$_2$SO$_4$, and OH-Al +

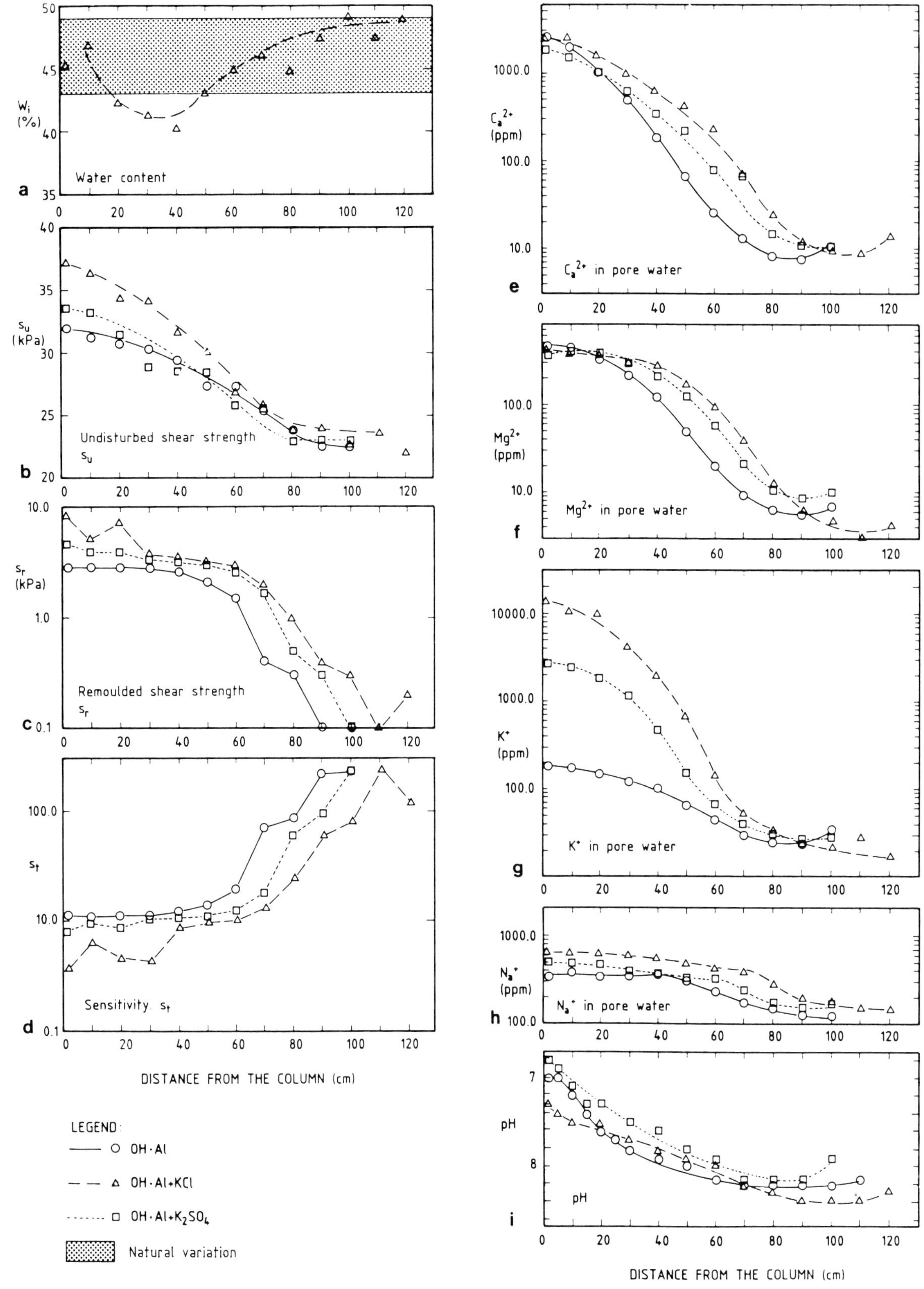

a — Water content

b — Undisturbed shear strength s_u

c — Remoulded shear strength s_r

d — Sensitivity s_t

e — Ca^{2+} in pore water

f — Mg^{2+} in pore water

g — K^+ in pore water

h — Na^+ in pore water

i — pH

DISTANCE FROM THE COLUMN (cm)

LEGEND:

— ○ OH·Al

— — △ OH·Al+KCl

···· □ OH·Al+K₂SO₄

Natural variation

KCl, respectively. The results were influenced by the addition of K$^+$ in the columns treated with OH-Al + K$_2$SO$_4$ and the OH-Al + KCl. The concentration of Na$^+$ increased from 100–150 ppm in the original quick clay to 350 ppm, 500 ppm, and 700 ppm, just outside the columns treated with OH-Al, OH-Al + K$_2$SO$_4$, and OH-Al + KCl, respectively.

The pH was similar for all the chemical additions, ranging from 6.7–7.0 adjacent to the columns, to 8 at a distance of 40–60 cm outside the columns, and to the background level of 8.2 in the quick clay (Figure 5i). The Cl$^-$ measurements unfortunately were not reliable due to a systematic analytical error. The results obtained in meq/liter are much higher than the sum of cations in meq/liter. Therefore only trends of Cl$^-$ diffusion can be seen.

For the columns treated with OH-Al + K$_2$SO$_4$, the maximum penetration of SO$_4{}^{2-}$ was 50–70 cm from the column surface. The concentration level was low indicating a high degree of fixation inside the columns.

DISCUSSION

The polymerization of OH-Al in a clay environment is complex. The mechanical property data on the clay mixtures showed that OH-Al additions of less than 15 g/100 g of pore water resulted in purely cation-exchange reactions. For additions of more than 15 g/100 g of pore water, a gel formed. Due to the relatively high alkaline-buffer capacity of the clay, the OH-Al molecules apparently dissolved in the pore water and polymerized rapidly at concentrations of about 23 g OH-Al/100 g of pore water or greater. Here, well-developed gels formed which encapsulated water and its dissolved components. The ions in the pore water thus were apparently trapped within the polymer OH-Al structure and their diffusion out of the treated columns was inhibited.

The use of a mixture of OH-Al + KCl as a stabilizing agent produced different effects. The high Cl$^-$ concentration apparently inhibited polymerization or possibly decreased the size of the polymer sheet and thus resulted in a softer gel. The somewhat higher pH found outside the OH-Al + KCl columns supports such a reaction. The polymer that formed in the chloride-rich solution gave a poorly stabilized clay within the column.

The maximum diffusion from the column treated with OH-Al + K$_2$SO$_4$ occurred at the same amount added as for the OH-Al treatment alone, and gave the same pH values outside the column. Within the columns treated with OH-Al + K$_2$SO$_4$, the shear strength data indicate a higher degree of polymerization than for the columns treated with OH-Al + KCl, but clearly lower than for those treated with OH-Al alone.

The reactions outside the columns reflected the chemical additions and reactions inside the column. The 2.5% CaCO$_3$ (shell fragments, foraminiferas, etc.) in the clay dissolved in the low pH pore solution produced by the addition of the OH-Al. Just outside the columns, the maximum concentrations of Ca^{2+} were 2650, 2500, and 2380 ppm for columns treated with OH-Al, OH-Al + KCl, and OH-Al + K$_2$SO$_4$, respectively.

The concentration of the released cations just outside the column treated with OH-Al are plotted in Figure 5. The effect of K$_2$SO$_4$ and KCl additions compared to OH-Al alone show that the concentrations of Mg^{2+} and Ca^{2+} were not changed, whereas that of Na$^+$ increased slightly.

From a soil stabilization point of view, the mechanical properties of the clay are most important. The slight increase in the undisturbed shear strength in the clay surrounding the columns is noteworthy, but the most important effect is the marked improvement of the remolded shear strength. This improvement means that if a treated clay is loaded to failure, it will no longer behave as a liquid. This state is reached if the remolded shear strength is >0.6 kPa or if the clay has a sensitivity of <30. The effect of stabilization can be measured as the distance from the column to where the sensitivity is 30; this distance is plotted against the amount of chemicals added in Figure 6.

For the columns treated with only OH-Al, the optimum stabilizing effect was estimated to be at the addition of about 30–35 g/100 g of pore water (175 kg/m^3 clay). Here, sensitivity values of <30 extended 70 cm into the clay (see Figure 6). If the amount of chemical was only 14 g/100 g of pore water (75 kg/m^3 clay) or as much as 50 g/100 g of pore water (275 kg/m^3 clay), the effect was limited to within 55 cm from the column.

Columns stabilized with OH-Al + K$_2$SO$_4$ behaved similarly, but the diffusion distance was greater than for pure OH-Al. For additions of OH-Al + K$_2$SO$_4$, the optimum effect was obtained at a distance of 75 cm for an addition of 38–45 g/100 g of pore water (200

←

Figure 5. Comparison of geotechnical and geochemical tests outside selected field-stabilized columns made of clay mixed with OH-Al, OH-Al + KCl (5:4) and OH-Al + K$_2$SO$_4$ (4:1) after 147 weeks, as a function of distance from the column. (a) = water content; (b) = undisturbed shear strength by fallcone test (s$_u$); (c) = remolded shear strength by fallcone test (s$_r$); (d) = sensitivity s$_t$ = s$_u$/s$_r$; (e) = Ca^{2+} concentration of the pore water; (f) = Mg^{2+} concentration of the pore water; (g) = K$^+$ concentration of the pore water; (h) = Na$^+$ concentration of the pore water; (i) = pH measured directly by an Orion combined electrode. Ca^{2+}, Mg^{2+}, K$^+$, and Na$^+$ measured by atomic absorption spectroscopy.

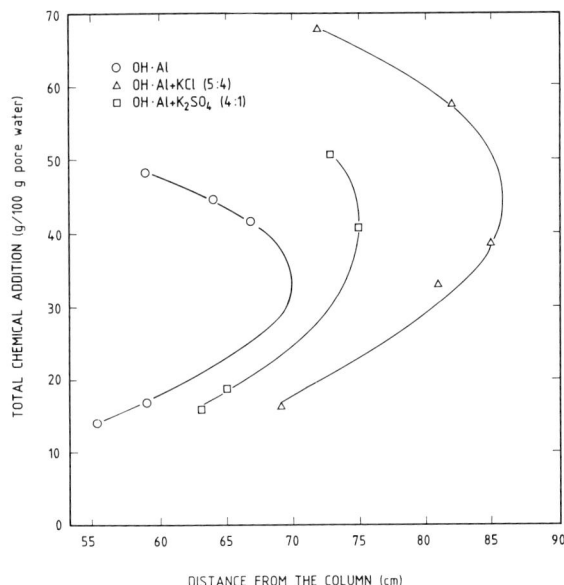

Figure 6. Distance from the column at which the sensitivity is less than 30, as a function of amount of added chemicals.

kg/m³ clay). The effect of OH-Al + KCl treatment reached even deeper into the clay. Here, the optimum was obtained 86 cm from the column by an addition of 40–50 g/100 g of pore water (225–250 kg/m³ clay). Similar to the OH-Al additions, treatments with OH-Al + K₂SO₄ and OH-Al + KCl gave reduced diffusion distances for low as well as high chemical dosages (Figure 6).

The somewhat surprising reduced distances observed when using higher quantities of chemicals was explained above as a retention by the OH-Al gel. Some of these captured ions will probably be released over an extended period of time, resulting in movement of the optimum effect towards higher chemical additions.

CONCLUSIONS

Stabilization of sensitive clays with polymeric hydroxy-aluminum seems very promising. The laboratory samples mixed with OH-Al alone reached a maximum shear strength of 500–600 kPa. Mixtures with K₂SO₄ and, especially, KCl, gave much lower shear strengths (as much as 200 kPa and 80 kPa, respectively).

Field-stabilized columns of clay using OH-Al showed an inhomogeneous pattern. The shear strengths of small homogeneous pieces varied considerably and were, under certain circumstances, even better than laboratory-stabilized samples. Full-scale tests on comparable columns showed low shear strengths due to inhomogeneities and cracking during excavation of the columns. Columns treated with OH-Al + K₂SO₄ and, especially, OH-Al + KCl were rather weak.

Stabilization of the clay adjacent to treated columns

appears promising. Diffusion of ions either added or resulting from ion exchange, produced increased remolded shear strengths of the clay. Consequently, the sensitivity was also reduced. From a stabilization point of view, sensitivities of <30 are acceptable. These values were measured as far as 70, 75, and 86 cm from the column treated with OH-Al, OH-Al + K₂SO₄, and OH-Al + KCl, respectively. The cross section of the final column plus the stabilized adjacent clay was increased from 0.2 to nearly 4 m². The optimum engineering choice of chemicals for stabilization, considering both the strength of the column and the effect in the surrounding clay, appears to be OH-Al for short-term stabilizations (e.g., temporary excavations) and OH-Al + K₂SO₄ for long-term improvement (e.g., permanent fillings).

REFERENCES

Bjerrum, L. (1954) Geotechnical properties of Norwegian marine clays: *Géotechnique* **4**, 49–69.

Bjerrum, L., Moum, J., and Eide, O. (1967) Application of electro-osmosis to a foundation problem in a Norwegian quick clay: *Géotechnique* **17**, 214–235.

Broms, B. B. and Boman, P. (1979) Lime columns—a new foundation method: *Amer. Soc. Civ. Eng. Proc.* **105**, No. GT 4, 539–556.

Bryhn, O., Løken, T., and Aas, G. (1983) Stabilization of sensitive clays with hydroxy-aluminium compared with unslaked lime: in *Proc. 8th European Conf. Soil Mech. Found. Eng., Helsinki, 1983, Vol. 2,* H. G. Rathmayer and K. H. O. Saari, eds., Balkema Publ., Rotterdam, 885–896.

Eggestad, Å. and Sem, H. (1976) Stability of excavations improved by salt diffusion from deep wells: in *Proc. 6th European Conf. Soil Mech. Found. Eng., Wien, 1976, Vol. 1.1,* Org. Comm. VI ICSMFE, Wien, 211–216.

Foster, R. H. and Gazzard, I. J. (1975) The influence of polymeric hydroxy-aluminium cations on the consolidation, shear strength and consistency limits of clay soils: *Géotechnique* **25**, 513–525.

Gregersen, O. (1981) The quick clay landslide in Rissa, Norway: in *Proc. 10th Int. Conf. Soil Mech. Found. Eng., Stockholm 1981, Vol. 3,* Balkema Publ., Rotterdam, 421–426.

Hem, J. D. and Robertson, C. E. (1967) Form and stability of aluminum hydroxide complexes in dilute solution: *U.S. Geol. Surv. Water-supply Pap.* 1827-A, A-1–A-53.

Holm, G. (1979) Lime column stabilization—experiences concerning strength and deformation properties: *Väg- och Vattenbyggaren* **25**, No. 7–8, 45–48.

Hsu, Pa Ho (1968) Interaction between aluminum and phosphate in aqueous solution: in *Advances in Chemistry Series* **73**, 115–127.

Hsu, Pa Ho (1977) Aluminum hydroxides and oxyhydroxides: *Minerals in Soil Environments,* J. Dixon and S. B. Weed, eds., Soil Sci. Soc. Amer., Madison, Wisconsin, 99–142.

Hsu, Pa Ho and Bates, T. F. (1964) Formation of X-ray amorphous and crystalline aluminum hydroxides: *Mineral. Mag.* **33**, 749–768.

Løken, T. (1970) Recent research at the Norwegian Geotechnical Institute concerning the influence of chemical additions on quick clay: *Geol. fören. Stockholm, Förhandl.* **92**, Pt. 2, 133–147.

Løken, T. and Gregersen, O. (1981) The Rissa landslide—

quick clay in Norway: Norwegian Geotechnical Institute, Oslo, Norway, 16 mm movie.

Mitchell, J. K. (1982) Soil improvement—state-of-the-art report: in *Proc. 10th Int. Conf. Soil Mech. Found. Eng., Stockholm, 1981, Vol. 4,* Balkema Publ., Rotterdam, 509–565.

Moum, J. (1968) Chemical environment compared with mechanical behaviour: in *Proc. Geotechn. Conf. Shear Strength Properties of Natural Soils and Rocks, Oslo, 1967, Vol. 2,* Norwegian Geotechnical Institute, Oslo, Norway, 125–126.

Moum, J., Sopp, O. I., and Løken, T. (1968) Stabilization of undisturbed quick clay by salt wells: *Väg- och Vattenbyggaren* **14**, No. 8, 23–29.

Oades, J. M. (1984) Interactions of polycations of aluminum and iron with clays: *Clays & Clay Minerals* **32**, 49–57.

Reed, M. G. (1972) Stabilization of formation clays with hydroxy-aluminium solutions: *J. Petrol. Tech.* **24**, 860–864.

Rosenqvist, I. T. (1955) Investigations in the clay-electrolyte-water system: *Nor. Geotech. Inst. Publ.* **9**, 125 pp.

Torrance, J. K. (1975) On the role of chemistry in the development and behavior of the sensitive marine clays of Canada and Scandinavia: *Can. Geotech. J.* **12**, 326–335.

Waters, D. N. and Henty, M. S. (1977) Raman spectra of aqueous solutions of hydrolyzed aluminium(III) salts: *J. Chem. Soc. Dalton Trans.* **3**, 243–245.

Yariv, S. and Cross, H. (1979) *Geochemistry of Colloid Systems for Earth Scientists:* Ch. 3, Springer, Berlin, 157–206.

Zlateva, I., Trendafelov, D., and Nikolov, G. S. (1975) Partially hydrolyzed aluminum chloride studied by spectral and DTA methods: *Bulg. Acad. Sci. Comm. Dept. Chem.* **8**, 433–442.

Proceedings of the International Clay Conference, Denver, 1985, L. G. Schultz, H. van Olphen, and F. A. Mumpton, eds.,
The Clay Minerals Society, Bloomington, Indiana, 436–440 (1987).

ENVIRONMENTAL GEOCHEMISTRY OF SELENIUM AND HEART DEATHS IN CHINA AND THE UNITED STATES

Marion L. Jackson

Department of Soil Science, College of Agricultural and Life Sciences
University of Wisconsin–Madison, Madison, Wisconsin 53706

Chang S. Li

Institute of Environmental Geochemistry, Academia Sinica
Beijing, P.O. Box 934, People's Republic of China

Ji Z. Zhang

Anhui Agricultural College, Hefei, People's Republic of China

Abstract—The geochemistry of selenium in soils on loess deposited as marine sediments involves the depletion by erosion of Se adsorbed on clays and iron oxides and its accumulation in lowlands. This depletion has affected the human heart-death rates (HDR) in Shaanxi Province, China, in western counties of the state of Wisconsin, and elsewhere. In Shaanxi province, a high age-adjusted, annual HDR of 140/100,000 people (due to cardiomyopathy or "Keshan disease") was found to be accompanied by blood-Se values of 20 ng/cm³ for people living in eroded, clay-poor loess hills. HDRs persisted (without Keshan disease) to middle ages, but were accompanied by somewhat higher blood-Se values. For people living in areas where some clays had been deposited on the Wei River terrace, HDRs were less (66–72) and were accompanied by blood-Se values of 133–79 ng/cm³ (n = 85), respectively. These values are at the low end of the blood-Se levels found in many other countries. From equations derived from these data, a blood-Se level of 274 ng/cm³ can be projected for the onset of such heart diseases, a value still less than half of that considered to be the toxic level of Se.

In western Wisconsin counties, containing loess derived in part from seleniferous (2–40 μg/g) Cretaceous shale from Minnesota, the population had lower annual HDRs (283–315/100,000 people) than in counties in which low-clay (sandy) soils predominate (HDR = 343–406). A soil extraction by 0.1 N NaOH gave Se values of 57–91 ng/g for soils in areas having the highest HDRs and 102–120 ng/g in areas having the lowest HDRs. Thus, topography, erosion of clay, bedrock source, and crop depletion of clay adsorbed Se appear to affect human blood levels of Se and heart-death rates.

Key Words—Blood selenium, Environmental health, Heart death, Keshan disease, Loess, Selenium, Soil.

INTRODUCTION

Environmental geochemistry of selenium in soil clays was first concerned with toxicity of Se in forage plants (Lakin and Byers, 1948), but attention has shifted to the deficiency of Se in livestock feed (Muth and Allaway, 1963; Hoekstra, 1975) and in food for human beings (Mertz, 1981; Xiu *et al.*, 1982). Great variation in heart death rates among counties, states, and provinces had been related to clay content, weathering, and erosion of land in China, Florida, Wisconsin, and elsewhere in the United States (Jackson *et al.*, 1985, 1986). Formerly inter-regional migrations of people and shipments of food were widely thought to preclude such geographic differences. Analyses show, however, that foods are extremely variable in amounts of some essential trace elements and occasionally are too high in some toxic elements (Mertz, 1981). Intensively farmed areas of the eastern and northwestern states show Se deficiency in livestock ("white muscle disease") (Muth

and Allaway, 1963) and high human ischemic heart-death rates where Se is low (Li and Jackson, 1985). A general acceptance of a relation of the geographical distribution of soil Se deficiency to human heart deaths in China (Levander, 1982) was followed by epidemiological evidence of a similar relationship in Finland (Salonen *et al.*, 1982). That Se functions in a vital reducing enzyme in all animals was established by Rotruck *et al.* (1973). The present paper points out the effects on human heart-death rates of soil erosion and intensive cropping of soils developed in loesses. The loesses of Shaanxi province in China and western Wisconsin counties were originally high in clay-adsorbed Se.

METHODS

An eroded loessial area of Shaanxi Province, north-central China, is noted for its high incidence of cardiomyopathy of livestock compared to the uneroded

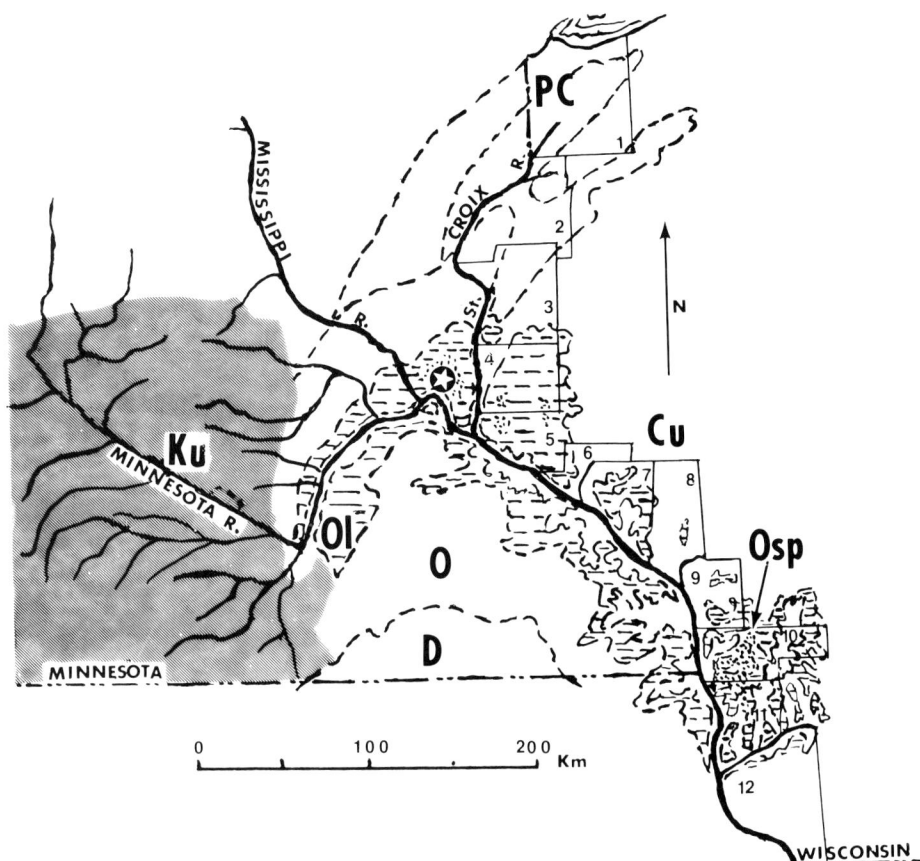

Figure 1. Cretaceous (Ku, Pierre Shale) provenance of seleniferous marine smectitic clay laid down in Wisconsinan loess along the Mississippi River in western Wisconsin. Ol = Prairie du Chien dolomite, which diverted the Minnesota River to St. Paul, Minnesota (star); Osp = St. Peter Sandstone cap rock in Vernon County; O = other Ordovician rocks; Cu = Upper Cambrian sandstone; PC = Precambrian rocks; D = Devonian rocks. The numbered counties are listed in Table 3.

loess plain. The eroded area was found to be coincident with high human heart-death rates (HDR). The environmental geochemistry appeared to be related to hill and valley topography and to erosion of clay and iron oxides, in which Se was adsorbed, from the hills into the Wei River valley. HDR and hair- and blood-Se levels were measured in the various topographic areas.

The Se in loess in southwestern Wisconsin was traced to its source in Cretaceous argillaceous shale, which contains 2 to 40 μg/g Se (Kubota, 1983). Dolomite bedrock (Stose, 1960) deflected the Minnesota River course northeastward to St. Paul, Minnesota (Figure 1) into the Mississippi River which supplied loess along western Wisconsin, southward (U.S. National Research Council, 1952). The counties north of this point are underlain by Cambrian and Precambrian rocks which yield much less clay (i.e., more sand) in the soils (Stose, 1960). The age-adjusted annual human HDRs (as well as cancer death rates; Jackson et al., 1985) for each county were obtained for 1979–1981 from the Wisconsin Division of Health, Madison, Wisconsin, as rates per 100,000 persons per year.

The soil-Se availability index was determined by extracting 5 g of soil in 50 ml of 0.1 M NaOH, with shaking for 1 hr, centrifuging, and filtering. The extract was analyzed for Se by complexing it with 2,3-diaminonaphthalene (Olson et al., 1975). The NaOH desorbs Se compounds adsorbed on clays and iron oxides.

RESULTS AND DISCUSSION

Shaanxi province

The environmental geochemical and geomorphological relationships to health are somewhat comparable in the study areas of Shaanxi province, China, and western Wisconsin (Figure 2). In Shaanxi province, erosion of loess hills combined with intensive land cultivation at Huanglong and Yanan has resulted in a smaller content of Se in food and, hence, blood-Se contents of as little as 21 ng/cm³ and HDRs of 101–140 (Table 1). In contrast, 48 ng/cm³ blood Se was found in people living in the uneroded loess area at Yulin along with a lower HDR of 79 (without Keshan

Table 1. Human cardiovascular death rates and Se contents of blood and hair in Shaanxi province, China.[1]

County	KD[2]	Cardio-vascular death rate[3] (deaths/yr)	Se content Blood (ng/cm³)	Se content Hair (ng/g)	Se content Cases (n)
Yanan	+	140.4	21	90 ± 22	385
Huanglong	+	101.2	22	91 ± 24	15
Shangluo	+	86.4	26	103 ± 35	26
Xianyang	−	83.5	33	124 ± 32	310
Yulin	−	79.4	48	169 ± 23	272
Hanzhong	−	77.0	69	233 ± 49	67
Weinan	−	72.5	79	264 ± 48	42
Xian	−	66.3	133	425 ± 44	43
Total[4]					1160

[1] From Wang (1982), Xiu *et al.* (1982), and unpublished; in Chinese.

[2] KD(+) = area affected by Keshan disease (cardiomyopathy in children aged 1–10 yr; KD(−) = area not affected by Keshan disease.

[3] Per 100,000 persons.

[4] $(r = -.64; p < .05)$; HDR = 113.4 − 0.134 (hair Se).

disease). The Se adsorbed on iron oxides is partly available to plants (Cary and Allaway, 1969) and, therefore, loessial clay films coated with iron oxides (Roth *et al.,* 1969) are probably moderately efficient suppliers of Se to the food chain. Erosional losses of clay colloids from the Shaanxi loess, in consequence, produces a depletion of soil Se and is associated with an increase in HDR (Figure 2).

Most of the people affected by Keshan disease (KD+) are rural populations whose traditional staple food is whole grain wheat or corn grown on the local soils, and they are dependent on the soil-food chain for their Se. Adequate vitamin E (a reducing compound which supplements the role of Se; Hoekstra, 1975) may be supplied by the oil of the whole grain used as the

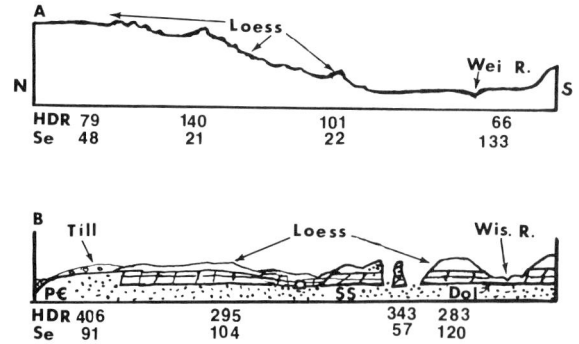

Figure 2. Geomorphology shown in cross sections of (A) Shaanxi (People's Republic of China) and (B) Wisconsin (United States) as related to HDR; Se in human blood (ng/cm³) is given for Shaanxi but blood Se is not available for the western Wisconsin area represeented, and therefore the soil-Se availability index (ng/g) is given (Table 3). SS = sandstone, Dol = dolomite; PC = Precambrian sandstone and Keewanawan conglomerate.

staple food in Shaanxi province. The need of Se supplementation at blood Se levels of 8–26 ng/cm³ in China to prevent elevated HDR during the first one-fourth of life-span is, however, accepted (Levander, 1982; Li and Jackson, 1985). Consequently, the question of Se deficiency and HDR in the mid-life period is of interest. Highly variable blood-Se levels (8–7800 ng/cm³) have been reported world-wide (Table 2) but have not been fully evaluated in terms of HDR. Salonen *et al.* (1982), however, examined a Finnish population ranging from 35 to 59 years of age, selected from 1100 persons to average in cohort pairs at blood-Se levels of 63 and 76 ng/cm³, respectively (Table 2). During the course of only seven years, the HDR was 22% greater for the 63 ng/cm³ blood-Se set as compared to the 76 ng/cm³ set. A high HDR is well known for New Zealand people having 56–100 ng/cm³ blood Se (Table 2).

Table 2. Range in human blood Se contents in various localities.

Locality	Blood Se contents[1] Mean (ng/cm³)	Blood Se contents[1] Range (ng/cm³)	Blood Se contents[1] Cases (n)	Reference
China, rural (KD+)[2]	17	8–26	274	Li and Jackson (1985)
(KD−)	47	32–83	136	Li and Jackson (1985)
urban	174	102–255	87	Li and Jackson (1985)
toxic	3000	700–7800	39	Li and Jackson (1985)
United States, urban	206	100–340	210	Li and Jackson (1985)
Finland	70	63–76	256	Salonen *et al.* (1982)
New Zealand, South Island	68	56–84	215	Griffiths and Thomson (1974)
North Island	69	50–100	24	Watkinson (1974)
Sweden	120	SD ± 20	6	Watkinson (1974)
Canada	180	100–350	250	Watkinson (1974)
Guatemala	230	140–330	13	Watkinson (1974)
United Kingdom	320	260–370	8	Watkinson (1974)

[1] (blood Se) = 0.335 (hair Se) − 9 (Li and Jackson, 1985).

[2] KD(+) indicates Keshan disease (cardiomyopathy) present in young children 1–10 years of age; KD(−) indicates Keshan disease absent.

Table 3. Age-adjusted heart death (HDR), 1979–1981, and soil-Se availability index of western-most counties of Wisconsin in order north to south.

County[1]	HDR[2] (deaths/yr)	Soil-Se availability index (ng/g)	Geomorphological setting
1. Douglas	406	91	Sandy soils from Precambrian rocks
2. Burnett	318	97	Sandy soils with thin silt
3. Polk	317	102	Thin silty soils over sand
4. St. Croix	294	106	Deep silty[3] soils over dolomite
5. Pierce	315	101	Silty[3] soils over dolomite
6. Pepin	313	103	Silty[3] soils over dolomite
7. Buffalo	295	104	Deep silty[3] soils over dolomite
8. Trempealeau	317	96	Silty[3] soils with sand owing to erosion
9. La Crosse	286	104	Deep silty[3] soils over dolomite
10. Vernon	343	57	Eroded silty soils on sandstone
11. Crawford	283	120	Deep silty[3] soils over dolomite
12. Grant	284	120	Deep silty[3] soils over dolomite

[1] Numbers refer to Figure 1.
[2] Per 100,000 persons.
[3] Silt from Cretaceous Pierre Shale, with depth increasing from 0.5 to >4 m from Pierce County south.

In Shaanxi province, the highest reported blood- and hair-Se levels (133 ng/cm^3 and 425 ± 44 ng/g, respectively) were found for people living in Xian on the Wei River terrace. These Se levels were associated with a HDR for this population of 66 (Table 1). The HDR of Shaanxi province correlates inversely with the Se content of hair (r = −.64; p < .05). If hair-Se and blood-Se levels are calculated for HDR = 0 from the regression equations (Tables 1 and 2), hair Se = 846 ± 90 ng/g and blood Se = 274 ± 30 ng/cm^3. These values are nearly double the levels in Xian (Table 1). They are well below the toxic level of blood Se (Table 2).

Wisconsin

The environmental, geochemical, and geomorphological relationships to HDR were compared for western Wisconsin with those for Shaanxi (Figure 2, B; Table 3). The heart-death rate in the United States is much higher than in China because of the higher incidence of other causes of death in China. The annual HDR averages 321 ± 33/100,000 people for all 72 Wisconsin counties (Jackson *et al.,* 1985).

Loess of western Wisconsin south of St. Paul, Minnesota, was derived in major part from the Cretaceous argillaceous marine deposit (Pierre Shale, Figure 1) which contains 2–40 μg/g Se (Kubota, 1983). The HDR of Crawford County is 283; this area is covered by 1–3 m of loess, for which the soil-Se index is 120 ng/g (Figure 2, B). In contrast, the HRD of Douglas county is 406, and the soil-Se index is 91 ng/g. The difference in Douglas county is attributable to the origin of the till from little-weathered Precambrian sandstone and Keewanawan conglomerate, rocks low in clay. Vernon County stands out (Table 3), having HDR of 343, even though it originally had a thick loess cover. This county has a caprock of St. Peter Sandstone (Figure 1) which made the land unusually susceptible to steep gullying

by erosion down into the Cambrian sandstone (Figure 2). The erosion, together with intensive agriculture on the resulting ridges and valleys of mixed sand and loess, has resulted in a soil-Se index of only 57 ng/g. People living on the loess deposits of Pierce to Grant counties along the Mississippi River and Polk and Burnett counties along the St. Croix River have low to intermediate HDRs (Table 3), illustrating the importance of Se-rich marine clay provenance of loess in the lowering of HDRs. Significantly, the Wisconsin cancer death rate (Jackson *et al.,* 1985) is also minimal in the Se-rich loess belt of Wisconsin (lowest in Pierce county, of 72 counties).

The soil-Se index was used to test further the effect of soil clay content and of crop removal of Se. In an unharvested forest of Forest County the soil-Se index was 228 ng/g, whereas an extensively cropped alfalfa field nearby tested only 91 ng/g. A cropped ferruginous clay soil in Brown County was found to contain 211 ng/g Se compared to 64 and 106 ng/g Se for sandy soils in Washara and Washburn counties, respectively.

PERSPECTIVE

Eaton and Konner (1985) showed that the human population has adopted nutritional changes faster than the human DNA has changed; depletion of clay-adsorbed Se (and Cu; McBride, 1985) appear to have caused nutritional changes that affect heart death rates. Similarly, Schneider and Reed (1985) called attention to limitations (50% by HDR) of human life expectancy (74 ± 4 years), far below the potential life span of 115 years. Although plants have adapted not to need Se (Mertz, 1981) during 420 million years since migrating to the land, animals (cattle, deer, man), during 250 million years, have not and thus suffer muscle diseases (HDR). The highest longevity areas in the United States (U.S. National Research Council, 1981) are areas of

Cretaceous rocks (high Se in marine clays) of the Great Plains where the blood Se levels are the highest (Levander, 1982). The lowest longevity area in the United States is the southeastern coastal plain where clay-poor, sandy soils predominate.

The option exists to raise human blood Se through supplementation (organically bound Se; Schrauzer and McGinness, 1979). Products are available "over-the-counter" with RDA trace element dosage guidelines (U.S. National Academy of Sciences, 1980). An earlier model (cf. Se), that iodine deficiency caused goiter, was discovered in 1825, yet a 160-year time lapse for its full correction is demonstrated by the presently existing 200 million goiter cases (Jackson and Lim, 1982). More blood Se data are urgently needed; however in the mean time, the method for a soil-Se index described here can provide some guidance on the relative status of Se in soils.

ACKNOWLEDGMENTS

This work was supported by the Academy of Sciences, People's Republic of China, and the College of Agricultural and Life Sciences, University of Wisconsin–Madison. Soil clay data were gathered with the assistance from the National Science Foundation grant EAR-8405422. Gratitude is expressed to Chrystie Jackson for her editorial and typing assistance.

REFERENCES

Cary, E. E. and Allaway, W. H. (1969) The stability of different forms of selenium applied to low-selenium soils: *Soil Sci. Soc. Amer. Proc.* 33, 571–574.

Eaton, S. B. and Konner, M. (1985) Paleolithic nutrition: a consideration of its nature and current implications: *N. Engl. J. Med.* 312, 283–289.

Griffiths, N. M. and Thomson, C. D. (1974) Selenium in whole blood of New Zealand residents: *N.Z. Med. J.* 80, 199–202.

Hoekstra, W. G. (1975) Biochemical function of selenium and its relation to vitamin E: *Fed. Proc.* 34, 2083–2089.

Jackson, M. L., Li, C. S., and Martin, D. F. (1986) Do land characteristics affect heart and gastrointestinal cancer death rates among Florida counties: *Fla. Sci.* 49, 82–97.

Jackson, M. L. and Lim, C. H. (1982) The role of clay minerals in environmental sciences: in *Proc. Int. Clay Conf., Bologna, Pavia, 1981,* H. van Olphen and F. Veniale, eds., Elsevier, Amsterdam, 641–653.

Jackson, M. L., Zhang, J. Z., and Li, C. S. (1985) Land characteristics associated with diverse human heart and digestive system cancer death rates among Wisconsin counties: *Trans. Wis. Acad. Sci. Arts Lett.* 73, 35–41.

Kubota, J. (1983) Soils and plants and the geochemical environment—4: in *Applied Environmental Geochemistry,* I. Thornton, ed., Academic Press, New York, 103–122.

Lakin, H. W. and Byers, H. G. (1948) Selenium occurrence in certain soils in the United States, with a discussion of related topics: *7th Rept., U.S. Dept. Agric. Tech. Bull.* 950, 1–36.

Levander, O. A. (1982) Selenium: biochemical actions, interactions, and some human health implications: in *Clinical, Biochemical and Nutritional Aspects of Trace Elements,* A. S. Prasad, ed., Alan R. Liss, New York, 345–368.

Li, C. S. and Jackson, M. L. (1985) Selenium in human blood affects heart death rates in China and USA: *Trace Substances in Environmental Health—XIX,* D. D. Hemphill, ed., University of Missouri, Columbia, Missouri, 264–276.

McBride, J. (1985) Copper: a missing link in coronary heart disease? *Agric. Res.* 33(9), 6–9.

Mertz, W. (1981) The essential trace elements: *Science* 213, 1332–1338.

Muth, O. H. and Allaway, W. H. (1963) The relationship of white muscle disease to the distribution of naturally occurring selenium: *J. Amer. Vet. Med. Assoc.* 142, 1379–1384.

Olson, O. E., Palmer, I. S., and Cary, E. E. (1975) Modification of the official fluorometric method for selenium in plants: *J. Assoc. Off. Agric. Chem.* 58, 117–121.

Roth, C. B., Jackson, M. L., and Syers, J. K. (1969) Deferration effect on structural ferrous-ferric iron ratio and CEC of vermiculites and soils: *Clays & Clay Minerals* 17, 253–264.

Rotruck, J. T., Pope, A. L., Ganther, H. E., Swanson, A. B., Hafeman, D. G., and Hoekstra, W. G. (1973) Selenium: biochemical role as a component of glutathione peroxidase: *Science* 179, 588–590.

Salonen, J. T., Alfthan, G., Pikkarainen, G. A. J., Huttunen, J. K., and Puska, H. P. (1982) Association between cardiovascular death and myocardial infarction and serum selenium in a matched-pair longitudinal study: *Lancet* 1982-II, 175–179.

Schneider, E. L. and Reed, Jr., J. D. (1985) Life extension: *N. Engl. J. Med.* 312, 1159–1168.

Schrauzer, G. N. and McGinness, J. E. (1979) Observations on human selenium supplementation: in *Trace Substances in Environmental Health—XIII,* D. D. Hemphill, ed., University of Missouri, Columbia, Missouri, 64–67.

Stose, G. W. (1960) *Geologic Map of the United States:* U.S. Geological Survey, Washington, D.C.

U.S. National Academy of Sciences (1980) *Recommended Dietary Allowances (RDA):* 9th ed., National Academy Press, Washington, D.C., 185 pp.

U.S. National Research Council (1981) *Aging and the Geochemical Environment:* National Academy Press, Washington, D.C., 141 pp.

U.S. National Research Council, Committee for the Study of Eolian Deposits (1952) *Map—Pleistocene Eolian Deposits of the United States, Alaska and Parts of Canada:* Geol. Soc. Amer., Washington, D.C., 337 pp.

Wang, M. Y. (1982) Geographical differentiation of selenium content in hair from children and youngsters in China: *Acta Nutrimenta Sinica* 4, 201–207 (in Chinese).

Watkinson, J. H. (1974) The selenium status of New Zealanders: *N.Z. Med. J.* 80, 202–205.

Xiu, G. L., Xiue, W. L., Zhang, P. Y., Feng, C. F., Huong, S. Y., and Liang, W. D. (1982) Selenium status and dietary selenium content of populations in the endemic or non-endemic areas of Keshan disease: *Acta Nutrimenta Sinica* 4, 183–190 (in Chinese).

AUTHOR INDEX

GEOGRAPHICAL INDEX

Proceedings of the International Clay Conference, Denver, 1985, L. G. Schultz, H. van Olphen, and F. A. Mumpton, eds.,
The Clay Minerals Society, Bloomington, Indiana, 443–456 (1986).

COMPREHENSIVE SUBJECT INDEX[1]

L. G. SCHULTZ AND F. A. MUMPTON

---

[1] All items are indexed to the first page of the article in
which they appear.

SBS
£ 50.50